Advances in
Cryogenic Engineering
Materials

VOLUME 36, PART A

An International Cryogenic Materials Conference Publication

Advances in
Cryogenic Engineering
Materials

VOLUME 36, PART A

Edited by
R. P. Reed and F. R. Fickett
National Institute of Standards and Technology
U.S. Department of Commerce
Boulder, Colorado

SPRINGER SCIENCE+BUSINESS MEDIA, LLC

The Library of Congress cataloged the first volume of this title as follows:

Advances in cryogenic engineering. v. 1–
  New York, Cryogenic Engineering Conference; distributed
  by Plenum Press, 1960–
    v. illus., diagrs. 26 cm.
  Vols. 1–   are reprints of the Proceedings of the Cryogenic Engineering
Conference, 1954–
    Editor: 1960–   K. D. Timmerhaus.

  1. Low temperature engineering—Congresses.   I. Timmerhaus, K. D., ed. II
Cryogenic Engineering Conference.
TP490.A3   660.29368   57-35598

Proceedings of the Eighth International Cryogenic Materials Conference
(ICMC) held July 24–28, 1989, at UCLA, in Los Angeles, California
ISBN 978-1-4613-9882-0   ISBN 978-1-4613-9880-6 (eBook)
DOI 10.1007/978-1-4613-9880-6

© 1990 Springer Science+Business Media New York
Originally published by Plenum Press, New York 1990
Softcover reprint of the hardcover 1st edition 1990

CONTENTS

PART A

REVIEW

GENERAL SUPERCONDUCTOR THEORY, MEASUREMENT, AND PROCESSING

LOW TEMPERATURE SUPERCONDUCTORS - NbTi

LOW TEMPERATURE SUPERCONDUCTORS - NEWER MATERIALS

HIGH TEMPERATURE SUPERCONDUCTORS - YBCO

HIGH TEMPERATURE SUPERCONDUCTORS - Bi-BASED

PURE METAL CRYOCONDUCTORS

MAGNETIC MATERIALS

PART B

REVIEWS

COMPOSITES: TEST METHODOLOGY

COMPOSITES: IRRADIATION EFFECTS

COMPOSITES - PROPERTIES OF POLYMER-BASED

COMPOSITES: METAL BASED

METALS: TEST ROUND ROBINS, METHODOLOGY AND THEORY

AUSTENITIC STEELS: FATIGUE AND CREEP

AUSTENITIC STEELS: STRUCTURE/PROPERTIES

FERRITIC STEELS

INDEXES

FOREWORD

The eighth International Cryogenic Materials Conference
(ICMC) was held on the campus of the University of California
at Los Angeles in collaboration with the Cryogenic Engineering
Conference (CEC) on July 24-28, 1989. The continuity of the
bond between these two major conferences in the field of cryogenics is indicative of the extreme interdependence of the
subject matter. The main purpose of the conference is sharing
of the latest advances in low temperature materials science
and technology. However, the many side benefits which accrue
when this many experts gather, such as identification of new
research areas, formation of new collaborations which often
cross the boundaries of both scientific discipline and politics,
and a chance for those new to the field to meet the old-timers,
may override the stated purpose.

This 1989 ICMC Conference was chaired by R. M. Scanlan
of the Lawrence Berkeley Laboratory. J. W. Morris, of U. C.
Berkeley served as Program Chairman with the assistance of seven
other Program Committee members. Special contributions to the
Conference were made by Committee members D. C. Larbalestier
of the U. of Wisconsin and D. O. Welch of Brookhaven National
Laboratory in the form of a Symposium on Critical Currents in
High Temperature Superconductors which preceded the formal
opening of the Conference.

We especially appreciate the contributions of the CEC Board
and its Conference Chairman, T.H.K. Frederking of UCLA, to the
organization of this joint conference. UCLA hosted the conference; the local arrangements and management, under the direction
of D. C. Hustvedt, were excellent.

The combined conferences are held biennially. The next
CEC/ICMC will be hosted by the University of Alabama at
Huntsville and held on June 9-13, 1991, on the UAH campus.
J. B. Hendricks of Alabama Cryogenic Engineering will be the
CEC Chairman and F. R. Fickett of the National Institute of
Standards and Technology will be the ICMC Chairman. The ICMC
Board sponsors conferences on special topics in others years.
In 1990 two topical conferences will be held: (1) High-
Temperature Superconductors - - Materials Aspects will be held
May 9-11, in Garmish-Partenkirchen, FRG, under the chairmanship
of H. C. Freyhardt, (Institut fur Metallphysik, U. of Goettingen,
Goettingen, FRG) and co-sponsored by ICMC and the Deutsche
Gesellschaft Metallkunde e. V. and (2) Nonmetallic Materials
and Composites at Low Temperatures, V. at Heidelberg, FRG, on
May 17-18, organized by G. Hartwig (KfK, Karlsruhe, FRG).

Participation in the conference continues to increase with about 780 registrants for the combined conference. The ICMC attracted more than 180 contributed and invited papers covering all aspects of cryogenic materials research. Plenary papers addressed two aspects of superconductivity applications, the first, by A. Das Gupta from DoE, discussed materials problems related to large-scale applications of the high temperature materials, and the second, by C. Taylor of LBL, described the array of modern low temperature superconductors and their applications in present-day devices. Low temperature super-conductors continued to dominate the contributed papers in superconductivity, representing about 60% of the total, with the remainder treating various aspects of the high temperature materials. In both areas, the subjects discussed ranged from the very practical to the truly exotic with the material taking forms ranging from wires and bulk to very thin film structures. International cooperation was evident in the full session devoted to the VAMAS studies. The topic of cryoconductors appeared this year with two sessions devoted to these normal conductor composites which take advantage of the very low resistivity of pure metals at low temperatures to provide a rugged, high current conductor for application in liquid hydrogen.

Special topical sessions were held on mechanical test standards at low temperatures for structural composites. Shear and compression testing and irradiation effects were emphasized. Contributions from the United States, Japan and Europe, hope-fully, will lead toward more uniformity in testing these complex materials.

This volume is, once again, the largest ever published by ICMC, a measure of the growing interest in the area of low temperature materials research and the importance of this research to numerous international programs. Our thanks goes to Mrs. Lynn Preston for her diligent work in the assembly of this volume.

F. R. Fickett
R. P. Reed

BEST PAPER AWARDS

Awards for the best papers of the 1987 ICMC proceedings, Advances in Cryogenic Engineering - Materials, volume 34, were presented at the 1989 conference. Selection is made by the awards committee from nominations of the editors in the categories of superconductors, structural materials, and student paper. It was a pleasure to present these three awards. We thank the authors for their exemplary contributions.

ANALYSIS OF AC-LOSS MEASUREMENTS ON SUPERCONDUCTORS USING COIL CONFIGURATIONS

A.J.M. Roovers, H. A. van den Brink and L.J.M. van de Klundert
University of Twente

MANGANESE-MODIFICATION OF GAMMA'-STRENGTHENED IRON-BASE SUPERALLOYS FOR CRYOGENIC APPLICATIONS

Keijiro Hiraga and Keisuke Ishikawa
National Research Institute for Metals Tsukuba Lab.

A MODEL FOR THE RESISTIVE CRITICAL CURRENT TRANSITION IN COMPOSITE SUPERCONDUCTORS

William H. Warnes
Applied Superconductivity Center
University of Wisconsin

1989 INTERNATIONAL CRYOGENIC MATERIALS

CONFERENCE BOARD

THE VAMAS INTERCOMPARISON IN THE AREA OF SUPERCONDUCTING AND

CRYOGENIC |STRUCTURAL MATERIALS

K. Tachikawa

Faculty of Engineering
Tokai University
Hiratsuka, Kanagawa 259-12, Japan

ABSTRACT

The international cooperation on the characterization of relevant
properties of superconducting materials and those of cryogenic structural
materials necessary for the construction of superconducting equipments are
successfully proceeding under the framework of VAMAS intercomparison.
The progress and present status of the VAMAS intercomparison are outlined
in this paper.   Fairly large coefficient of scatter in critical current
of Nb_3Sn wires was obtained depending on the wire manufacturing process
and the measurement conditions.   The strain in the specimen was
considered to be a major origin of the scatter in critical current.
Meanwhile, relatively good agreement was obtained in the round robin test
of AC loss measurement of Nb-Ti wires.   The results of tensile and
fracture toughness testing of SUS 316LN and YUS 170 steels at 4.2K also
show rather small scatter.   An intercomparison on the electrical strain
gauge calibration at cryogenic temperatures has been recently initiated.

INTRODUCTION

The Versailles Project on Advanced Materials and Standards, commonly
referred to as VAMAS aims at promoting the development of economic growth
and employment through more rapid application of new technologies.   In
response to the request of the VAMAS Steering Committee, an international
technical working party (TWP) on superconducting and cryogenic structural
materials was organized in 1986 (area No.6 of VAMAS cooperation) [1].   The
TWP meeting has been convened five times since then, 1st in April 1986
at KfK, FRG, 2nd in July 1987 at NBS, USA, 3rd in May 1988 at Sunshine
City, Japan, 4th in July 1988 at Univ. of Southampton, UK and 5th in July
1989 at UCLA, USA.   Two subgroups, i.e. subgroup I for superconducting
materials and subgroup II for cryogenic structural materials, were
organized in the TWP.   The performance of following intercomparison
(round robin test) programs has been agreed at the TWP.

1. Round robin test programs on superconducting materials:
1.1 Critical current , Ic, measurement in Nb_3Sn multifilamentary wires.
1.2 AC loss measurement in Nb-Ti multifilamentary wires.

Advances in Cryogenic Engineering (Materials), Vol. 36
Edited by R. P. Reed and F. R. Fickett
Plenum Press, New York, 1990

1

Table 1   List of Participant Labs in the Round Robin Tests on
Superconducting Materials

Ic measurment on Nb_3Sn Wires	AC Loss Measurment on Nb-Ti Wires
Europe	Europe
Atomnist. Ost. Univ. (Austria)	Atominst. Ost. Univ. (Austria)
Inst. Expml. Phys. Ost. Univ. (Austria)	Alsthom D. E. A. (France)
S. C. K. /C. E. N. (Belgium)	KfK (F. R. G.)
S. N. C. I., C. N. R. S. (France)	Siemens (F. R. G.)
KfK (F. R. G.)	C. I. S. E. (Italy)
Siemens (F. R. G.)	Univ. Twente (The Netherlands)
Vakuumschmelze (F. R. G.)	Clarendon Lab. (UK)
E. N. E. A., Centro di Frascati (Italy)	
Univ. Nijmegen (The Netherlands)	
Clarendon Lab (UK)	
Rutherford Appleton Lab (UK)	
USA	USA
Brookhaven National Lab.	Battelle
Francis Bitter National Mag. Lab.	Brookhaven National Lab
Lawrence Livermore National Lab	N. I. S. T.
N. I. S. T.	
Univ. Wisconsin	
Japan	Japan
Electrotechnical Lab.	Center Res. Inst. Elec. Power Ind.
Furukawa Electric Co.	Electrotechnical Lab.
J. A. E. R. I.	J. A. E. R. I.
Hitachi	Kyushu Univ.
Kobe steel.	Nihon Univ.
N. R. I. M.	N. R. I. M.
Osaka Univ.	Tohoku Univ.
Tohoku Univ.	Tokai Univ.
	Toshiba

2. Round robin test programs on cryogenic structural materials:
1.2 Tensile measurement at 4.2K on SUS 316LN and YUS 170 steels.
1.3 Fracture toughness measurement at 4.2K on the same steels.
1.4 Electrical strain gauge calibration at cryogenic temperatures.

ROUND ROBIN TEST ON THE CRITICAL CURRENT IN Nb_3Sn WIRES

Critical current measurements on multifilamentary Nb_3Sn wires were successfully completed with 24 participant labs from 9 different countries listed in Table 1. The distribution of test samples and accumulation of resulting Ic data in EC, Japan and USA were performed by the respective central labs; BCMN, NRIM, and NIST. The BCMN in Belgium served as the central lab in Europe, although the BCMN itself did not perform the Ic measurement.

Three samples wires were supplied, one from each of EC, Japan and USA; these samples are labelled as sample A, B and C in no specific order. The specification and respective heat treatment conditions of three wires

	Sample A	Sample B	Sample C
Fabrication process	Bronze	Bronze	Internal Sn
Wire diameter (mm)	0.8	1.0	0.68
Composite component	NbTa/CuSn	Nb/CuSnTi	Nb/Cu/Sn
Cu/non-Cu ratio	0.22	1.68	0.88
Bronze/Cores ratio	2.8	2.5	3.1
Filament diam. (μm)	3.6	4.5	2.7
No. filament	10,000	5,047	5,550
Heat treatment	700℃×96h	670℃×200h	700℃×48h

are indicated in Table 2.   Upper critical fields , Hc$_2$, for sample A and
sample B are enhanced by additions of Ta and Ti, respectively.   Each
laboratory was asked to prepare two specimens from each wire.   One of
these they heat treated themselves, whilst the other they sent for heat
treatment at a central laboratory (central heat treatment).   The round
robin test procedure of central heat-treated specimen is illustrated in
Fig. 1.

The round robin test results were intercompared to examine the effects
of test holder materials, specimen mounting methods, measurement details
etc. on the scatter in measured Ic.   To ensure a meaningful inter-

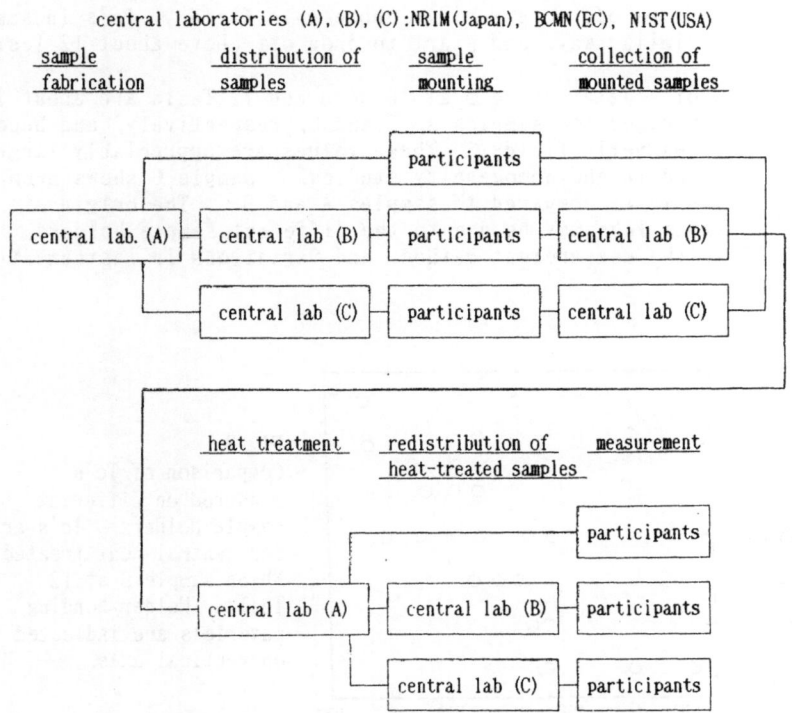

Fig. 1 Test process chart of round robin test for sample A of central reaction.

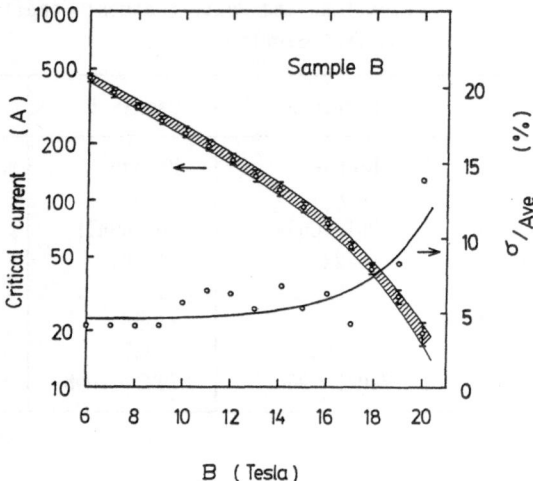

Fig. 2   Averaged Ic at 10 μV/m versus magnetic field for central heat-treated
Nb_3Sn sample B. The bars on the Ic curve indicate standard deviation
σ at respective field. σ/ave. is coefficient of scatter.

comparison, homogeneity in Ic values of sample A and B was inspected, and
found to be within ±ca. 1 %, good enough for the intercomparison.   The
homogeneity in Ic of sample C was not inspected.   Preliminary reports on
the present round robin test on Ic have been published in elsewhere [2,3].

Fig. 2 is an example of the test results that shows averaged Ic data
and coefficient of data scatter on sample B as a function of magnetic
field.   The semilogarithmic plots show that for samples A and B the logs
of the Ic decreases almost linearly with increasing field up to about 15
Tesla, and then drops off more rapidly above this field.   Ic's in sample
C decrease in a similar way, and start to drop off above about 12 Tesla.

Coefficients of scatter of Ic's at 10 μV/m and 12 Tesla are about 7, 5
and 19% of the averages for samples A, B and C, respectively, and become
larger at higher magnetic fields.   These values are appreciably larger
than those obtained in the homogeneity studies.   Sample C shows much
larger scatter in Ic as compared to samples A and B.   The origin of
scatter in Ic among labs may be due to the different sample holders,
sample mounting, the measurement method, and variations in instrument
calibration.

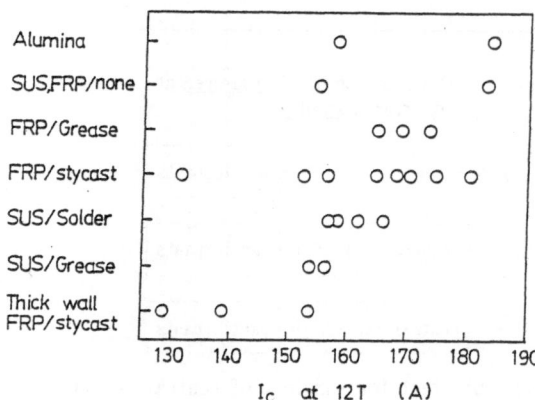

Fig. 3
Comparison of Ic's
measured on different
sample holders. Ic's are
for central heat-treated
Nb_3Sn sample B at 12
Tesla. Holder/bonding
materials are indicated
on vertical axis.

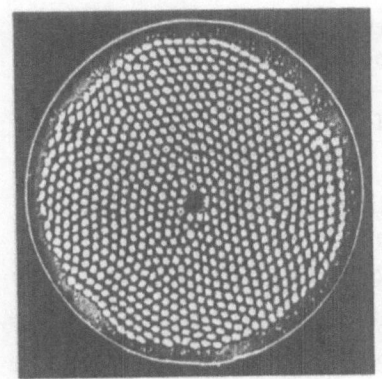
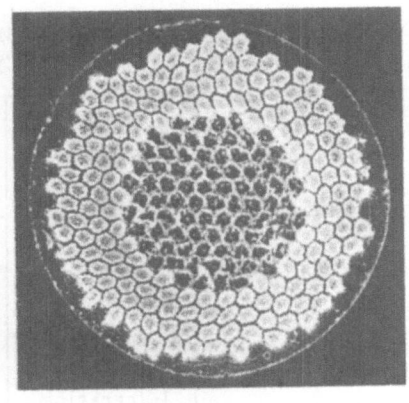

Fig. 4 Cross-sections of wires E and F. Left: wire E, outer diam. 0.35mm,
filament diam. 6.3 µm, No. of filaments 760. Right: wire F,
outer diam. 0.14mm, filament diam. 0.50 µm, No. of filaments 14280.

As an example, Fig. 3 demonstrates the effect of holder materials on
the scatter in Ic for sample B. The figure suggests that there seems to be
a systematic effect on Ic originated by holder materials. Goodrich and
Bray studied more quantitatively the effect of wall thickness of FRP
holder on Ic[4].   Strain effect measurements on all sample wires have been
also carried out in high magnetic fields.   The tensile strain sensitivity
of Ic is largest for sample C and smallest for sample B.   This may be
related partly to the Hc_2 of these samples; sample C has a relatively low
Hc_2 of ~ 19 Tesla, while it is ~ 24 Tesla for samples A and B.   The
imposition of compressive strain to sample C may cause a large depression
in Ic, whilst that to sample B may cause a smaller depression.

In Ic measurement, the strain in a specimen varies from lab to lab,
depending on the thermal contraction of the holder material and the
details of the fixing of the specimen to the holder.   The scatters in Ic
of round robin test results may principally be assigned to the strain
effect of samples.

ROUND ROBIN TEST ON THE AC LOSS IN Nb-Ti WIRES

Following the critical current measurement in Nb_3Sn wires, an intercomparison program for AC loss measurement on Nb-Ti multifilamentary wires
was started with 19 participant labs from 8 different countries listed in
Table 1.   Taking into account a large variety of measurement techniques
and the absence of a predominant practice, this program assumes that the
measurement be carried out by means of exising techniques and on-site
apparatus at each participant lab.

Four Nb-Ti sample wires, two from Japan and one from each of EC and
USA, have been supplied.   These wires are coded as wire D, E, F and G in
this paper.   The wires D and E are for accelerator magnet and pulsed
magnet use, respectively.   Relatively large AC losses are expected in
these wires.   Meanwhile, the wires F and G are for power frequency
application, and expected to have much smaller AC losses than those of
wires D and E. Fig. 4 shows the cross-sections of wires E and F. The
outer diameter and filament diameter of wire D are 0.74 mm and 4.6 µm
(10980 filaments), respectively, while those of wire G are 0.20 mm and
0.18 µm (242892 filaments), respectively.   The wires F and G were wound
into test coils, and then circulated for the round robin test.   The
length of wires in the coil F and coil G were 590 m and 250 m, respectively.

Table 3　Comparison of Hysteresis Loss Measurement for a
0→1→0 Tesla Cycle on Nb-Ti Test Wires

Laboratory	Measurement Method	Cycle	Hysteresis Loss (mJ/m)	
			Wire D	Wire E
a	Integration	1/2	5.0	0.47
b	Calorimetric	1/2	5.7	0.54
d	VSM	1	4.3	0.48
e	Integration	1/2	4.39	0.412
f	VSM	1/2	4.78	0.475
g	Integration			0.7*
h	VSM	1/2		0.46
i	Integration	1/2		0.54
k	Integration	1/2	4.88	0.508
m	Integration	1/2	5.89	0.531
n	Integration	1/2	5.6	0.60
o	VSM	1/2	4.5	0.45

* Bmax=0.42T, full cycle loss.　1/2: half cycle　[0-1]Tesla
1 : full cycle [-1→+1]Tesla

Electric, magnetic and calorimetric AC loss measurements were performed
on these wires.　This round robin test is still underway, but preliminary
results on samples D and E are summarized in Table 3.　The average
hysteresis losses are 4.9 and 0.50 mJ/m for sample D and E , respectively
with corresponding standard deviations 0.49 and 0.051 mJ/m, respectively.
The coefficient of scatter in the present AC loss measurement is ca. 10 %
which is rather small referring the variety of measuring details.

Meanwhile, an example of coil loss measurement for wire G is illustrated in Fig. 5 [5].　A calorimetric measurement was performed with
induced current of different frequency.　The loss is much smaller than
that of wire D and E as expected.　A tan δ method was also performed for
the loss measurement in coils F and G.　A summarized result for Japanese
participant labs will be published in this issue [6].　The round robin test

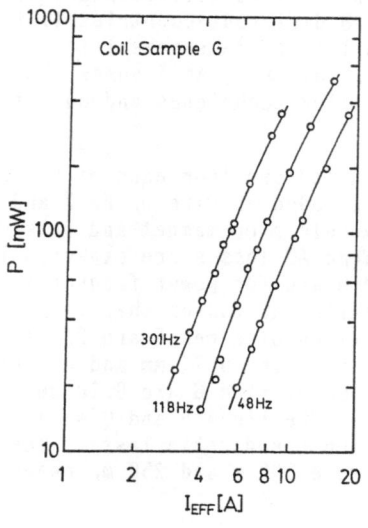

Fig. 5
Total power loss in Nb-Ti coil G as
a function of transport current of
three different frequencies.

Table 4 List of Participant Labs in the Round Robin Tests on
Cryogenic Structural Materials

Tensile Measurement	Fracture Toughness Measurement	Strain Gauge Calibration
Europe	Europe	Europe
Tech. Univ. Wien (Austria)	KfK (F.R.G.)	Atominst. Ost. Univ. (Austria)
KfK (F.R.G.)		Inst. Expml. Phys. Ost. Univ (Austria)
EMPA (Switzerland)		KfK (F.R.G.)
Rutherford Appleton Lab. (UK)		Ansaido (Italy)
		IMGC (Italy)
		INFN-LASA (Italy)
		EMPA (Switzerland)
		Rutherford Appleton Lab. (UK)
USA	USA	USA
Lawrence Livermore Natl. Lab.	N.I.S.T.	N.I.S.T.
Master. Res. Engr. Inc.		
N.I.S.T.		
Teledyne Engr. Service		
Japan	Japan	Japan
Hitachi	Hitachi	Hitachi
J.A.E.R.I.	Kobe Steel	Kobe Steel
Kawasaki Steel	N.R.I.M.	N.R.I.M.
Kobe Steel	Tohoku Univ.	Univ. Tokyo
Nippon Kokan	Toshiba	
Nippon Steel	Univ. Tokyo	
N.R.I.M.		
Tohoku Univ.		
Toshiba		
Univ. Tokyo		

on AC loss measurement will be completed in the near future, and a recommended test procedure will be proposed based on the discussions at the TWP
meeting.

ROUND ROBIN TESTS ON CRYOGENIC STRUCTURAL MATERIALS

Two austenitic steels, SUS 316LN and YUS 170 steels, were chosen as the
round robin test materials. These steels are commercially available and
have high yield strength and good fracture toughness favorable for
cryogenic structure use. The participant labs in the tests are listed in
Table 4. Tensile and fracture toughness test specimens were machined at
NRIM according to specifications designed by each participant. Each
participant applied a different measurement method; there were no strict
requirements on the measurement details, except that the strain rates were
recommended to be low. Instead, detailed measurement conditions were
reported along with the mechanical properties.

In general, tensile data reported are in a relatively good agreement
with each other. Statistical analysis showed small scatter for Young's
modulus, yield strength, and tensile strength of both steels. For these
properties the ratio of the standard deviation to mean value was between
0.01 and 0.06. This means the current test methods have been fairly

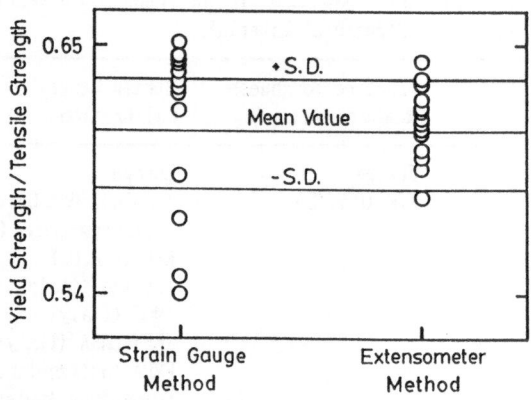

Fig. 6 Comparison of scatter in yield strength normalized by tensile strength between strain gauge and extensometer methods for SUS 316LN steel at 4.2 K.

improved for precise and reliable determination of liquid helium temperature properties. The scatter of elongation and reduction of area was also not too large, and was about 0.05 to 0.12 [7].

It seems that the scatter in tensile strength measurement arises from load cell calibration error. Strain gauges were found to give a larger scatter of data on the yield strength than did extensometers. Fig. 6 compares the scatter in yield strength data for the two methods. In the determination of elongation, we found that the influence of 'serration', indicative of localized deformation, was important.

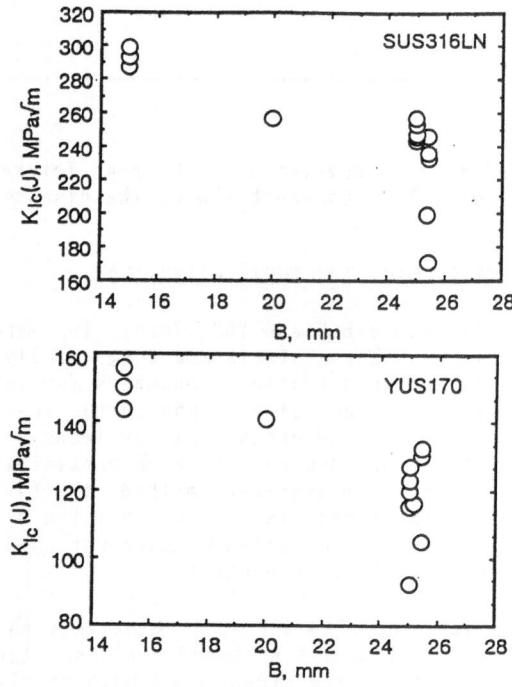

Fig. 7 Effect of specimen thickness on fracture toughness of SUS 316LN and YUS 170 steels at 4.2K.

The fracture toughness measurement results are being analyzed. The scatter in fracture toughness data reported seems to be attributed to the measurement variables including the machine type, the strain rate, the deformation control-mode, etc. Fig. 7 indicates the relationship between fracture toughness and sample thickness for both steels. The complete result of the fracture toughness round robin test was reported at the TWP meeting in July, 1989.

In addition, round robin test on the electrical strain gauge calibration at cryogenic temperatures has been recently initiated with 13 participant labs from 7 different countries as listed in Table 4. The technical format for the round robin test has been distributed to the participant labs, and the test hardwares will be circulated shortly.

CONCLUSIONS

The round robin test on critical current measurement in multifilamentary Nb_3Sn wires with 24 participant labs from 9 different countries was successfully completed. The coefficient of scatter of Ic depends both on the wire manufacturing process and the measuring techniques. Several parameters affecting Ic have been extracted. The strain in the sample was suggested to be a major origin for the scatter in Ic.

The round robin test on AC loss measurement in multifilamentary Nb-Ti wires was carried out using electric, magnetic and calorimetric methods. The results so far obtained show relatively small coefficient of scatter. The round robin test will be completed in the near future, and the results will be summarized.

In the area of cryogenic structual materials, the round robin test on tensile testing at 4.2 K was completed, and summarized reports was published. The test data showed good agreement among participant labs. Following this, the round robin test on fracture toughness at 4.2 K using the same steels has been completed. Effects of testing parameters on the results are being analyzed. More recently the round robin test on strain gauge calibration at cryogenic temperatures has been started.

Now, new round robin test programs are being considered for different measurements and materials. These may involve Ic measurements on new and large scale conductors, strain effect measurement on superconducting materials, tensile and fracture toughness measurement on Ti and Al alloys, etc. Complementary round robin tests on Ic and AC loss measurements with more strictly defined experimental conditions are also under consideration.

The application of supercondutivity should be developed under a concept of long term project whereby international cooperation would play an important role. The methods of performance characterization for superconducting and cryogenic structural materials would be effectively established through the international cooperation. The worldwide technical exchanges would be also enhanced through such cooperation.

ACKNOWLEDGMENTS

The author, chairman of the VAMAS TWP, would like to extend his sincere thanks to all the participants involved in the above international round robin test programs.

REFERENCES

1. H. Jones, _Cryogenics_ 26:488 (1986).
2. K. Tachikawa, K. Itoh, H. Wada, D. Gould, H. Jones, C. R. Walters, L. F. Goodrich, J. W. Ekin and S. L. Bray, _IEEE Trans. Magn._ 25:2368 (1989).
3. K. Tachikawa, _Cryogenics_ 29:710 (1989).
4. L. F. Goodrich and S. L. Bray, _Cryogenics_ 29:699 (1989).
5. A. J. M. Roovers and L. J. M. van de Klundert, private communication.
6. K. Itoh, H. Wada, T. Ando, E. Shimizu, D. Itoh, M. Iwakuma, K. Yamafuji, A. Nagata, K. Watanabe, Y. Kubota, T. Ogasawara, S. Akita, M. Umeda, Y. Kimura and K. Tachikawa, in this issue.
7. K. Nagai, T. Ogata, K. Ishikawa, K. Shibata and E. Fukushima, _Cryogenic Materials '88_, Ed. by R. P. Reed, Z. S. Xing and E. W. Collings, 2:893 (1988).

THE RESISTIVE STATE OF A RANDOM WEAK LINK NETWORK

M.G. Blamire and J.E. Evetts

Department of Materials Science and Metallurgy, Cambridge University
Pembroke Street, Cambridge CB2 3QZ, U.K.*and*
Interdisciplinary Research Centre in Superconductivity
Cambridge University, West Site, Madingley Road, Cambridge CB3 0HE, U.K.

While the equilibrium response of superconducting weak link networks has been studied for a number of years, the dynamic or resistive state has received little attention. This paper develops a model for the resistive state in such networks, and shows that at low voltages there exist phase-locked regions within the network, which are separated by phase-slip paths which generate the observed voltage. The characteristic dimension of these phase-locked regions Λ, is shown to determine the power-law or n-value response of the resistive state of such networks. The results of this model agree well with experimental results on sintered ceramic superconductors and fabricated tunnel junction arrays.

INTRODUCTION

The accepted picture of polycrystalline high temperature superconductors is of high quality superconductivity associated with the oxide grains or subgrannular regions, with only weakly superconducting connections between the grains or subgrains. The simplest of the conceptual pictures which has emerged is that of a composite constructed of perfectly superconducting material connected at each interface by a Josephson weak link. A phenomenological model based on this idea has produced reasonable agreement with some experimental data[1]. There are also a number of existing theoretical models of superconductivity in such networks based on analyses in terms of percolation theory which predict the behaviour of T_c and diamagnetism in superconductor / non-superconductor composite structures[2-5]. These do not however address the practical problem of critical current densities and the magnetic field suppression of these currents, or the dynamic behaviour of the system above its zero-voltage state.

This paper describes a theoretical method of treating the dynamic state of such arrays which successfully describes the main observed features of the resistive response of such networks, including non-linearities in voltage vs current curves which can be approximately fitted by a power law of the form $V=kI^n$ or "n-value", the magnetic field suppression of the critical current and the change of n-value with magnetic field.

The simplest limit of such a network is a one-dimensional chain of weak links. Such series arrays have been fabricated for many years and their resistive response is well-understood: Fig. 1 shows a typical V vs I characteristic from a 1-D SIS tunnel junction array. As the applied current reaches the critical current i_c of each of the junctions there is an increase in the voltage of the array corresponding to twice the gap voltage Δ. The junctions have a distribution of critical currents and so this results in a curvature of the characteristic. This portion of the curve can be analysed by conventional resistive transition techniques to give plots of the n value ($dLog(V)/dLog(I)$) against various parameters; and since the voltage increment due to each junction entering the resistive state is constant, plotting dV/dI gives directly the probability distribution of i_cs $f(i_c)$.

If the weak links all have the same i_c the transition will be sharp; however, for a normal $f(i_c)$ the n-value of the transition will depend on σ, the standard deviation of $f(i_c)$, and inversely on the mean critical current $<i_c>$. The application of a magnetic field to such a system results in the partial or complete suppression of a number of weak links within this distribution.

Fig. 2 shows a plot of n vs the electric field (E) for a sample of $YBa_2Cu_3O_7$; which shows the characteristic structure of decreasing n with increasing E exhibited by all weak link networks including series arrays of tunnel junctions. In a previous publication[7] we showed that the resistive behaviour of oxide superconductor samples tended to that of a series array as the sample compaction density was decreased. In order to directly compare different samples it is necessary to use reduced units n/n_0 and B/B_0. where n_0 is the n-value in zero magnetic field at a particular electric field criterion, and B_0 is the characteristic width of the decay of critical current with magnetic field; this is shown in Fig. 3. The agreement between the low density sample and the series array shows that provided the different field behaviour and $f(i_c)$ distributions are taken into account then these two apparently different types of network are comparable and that a simple measure of the behaviour can be drawn from plots such as this.

Advances in Cryogenic Engineering (Materials), Vol. 36
Edited by R. P. Reed and F. R. Fickett
Plenum Press, New York, 1990

11

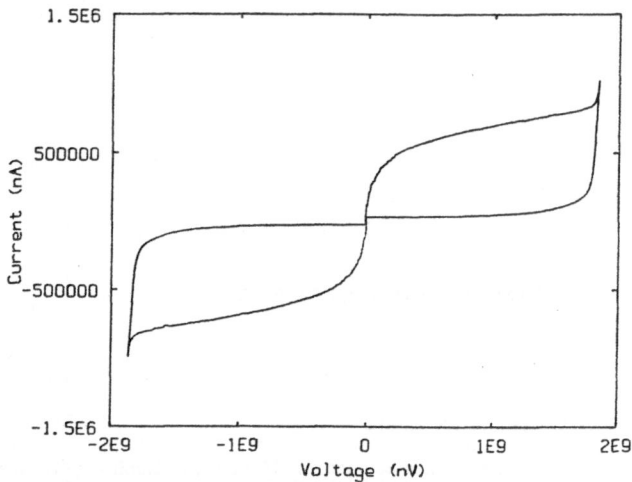

Fig. 1. Current vs voltage plot for a series array of 600 Nb/AlO$_x$/Nb tunnel junctions. The critical current portion of the curve shows a smooth curvature due to random variations in device size.

However, to further understand the behaviour of samples such as these, it is necessary to consider the underlying mechanisms of current transfer within multiply-connected weak link networks. The following sections describe a model for the resistive behaviour of such systems and will then show a quantitative comparison between the theoretical results and the experimental data.

FLUX VORTEX PINNING ARRAYS

The standard model of a weak link network consists of a rectangular network with identical microbridges or other weak links lying between each node. Plots of critical current against magnetic field have been obtained from microfabricated test structures which show a characteristic peak structure in the field dependence of the critical current corresponding to the presence of integral number of flux vortices within each cell[4,5]. J_c vs B peaks of this type are also known to occur in measurements on systems described as ordered pinning arrays[8,9]; indeed as the ratio of hole diameter to spacing increases then the simplest weak link network consisting of a regular array of holes in a superconducting film becomes identical to flux vortex pinning array structures. There is thus at least one limit in which the response of WLNs can be modelled on the basis of vortex pinning. This section will consider the behaviour of imperfect pinning arrays as a means of introducing the general WLN problem.

In such systems there is a well-defined zero-voltage critical current density at all magnetic fields; being that required to displace vortices (or defects in an otherwise commensurate vortex lattice) from pinning centres and maintain continuous vortex flow throughout the sample[10]. Vortex behaviour in these systems is translationally isotropic on a scale larger than the pin spacing and the behaviour can be summarised as follows: below the zero-voltage critical current J_{co} there is no continuous vortex motion, and above this value all vortices or lattice defects (which mediate the dynamic response of the vortex lattice when this is incommensurate with the pinning array) move with a common average velocity.

We will now consider the effect of superimposing a random potential on such a system, so that the pinning array now consists of potential barriers which are similarly shaped and uniformly spaced, but which vary in height in a random fashion. In such a 2-D array there is a time-averaged translational

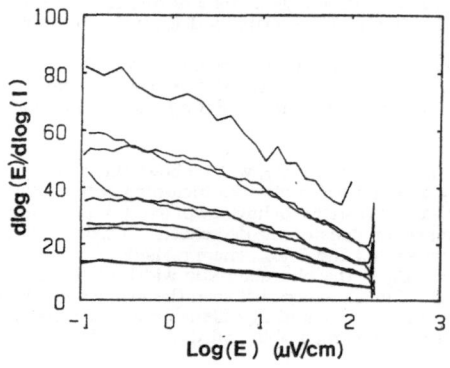

Fig. 2. Plot of n-value (dLog(V)/dLog(I)) vs E for 93% dense oxide superconductor sample at magnetic field values (from top) 1,2,3,4,6mT.

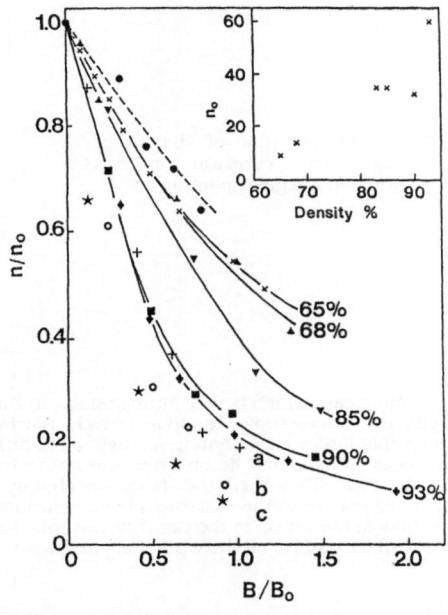

Fig. 3. Reduced n-value vs reduced magnetic field. Dashed line shows data from tunnel junction array shown in Fig. 1. Points a,b,c represent theoretical curves with parameters $C=1$, $\sigma=1$; $C=1$, $\sigma=0.5$; $C=2$, $\sigma=1$ respectively. Inset is the zero field n-value vs sample compaction density.

isotropy provided that vortex motion is continuous at all points within the sample, since all vortices will (in time in an infinite array) experience the same random pinning potential. If this condition of continuous motion is satisfied then the motion of a single vortex will represent the behaviour of the vortex lattice as a whole. Since vortices have zero inertia, the equation of motion for a vortex is simply $f=\eta v$, where η is the vortex flow viscosity, and f is the net driving force on the vortex which in this example may be expanded so that $f(x)=f_0(x)+g(x)$, where f_0 is the sum of the periodic pinning potential and the Lorentz driving force and $g(x)$ represents the net force due to the random pinning potential and the vortex lattice restoring force[11]. This equation of motion can be integrated (Appendix A) to give the average vortex velocity and hence the electric field

$$E = \frac{B.L}{\eta}\left[\int_0^a \frac{1}{f_0(x)}\sum_{n=0}^{L/a} 1+\left(\frac{g(na)}{f_0(x)}\right)^2 + \dots\ dx\right]^{-1} \tag{1}$$

This shows that as the random force term becomes small compared to periodic term (which includes the Lorentz driving force) then the electric field tends to

$$E= \frac{B.a}{\eta}\left[\int_0^a \frac{1}{f_0(x)}\ dx\right]^{-1} \tag{2}$$

which is the electric field obtained from the perfect periodic pinning array. Thus we can say that as the current is decreased from a high value, well above J_{co}, the electric field of the random pinning array initially decreases faster than the perfect periodic array, and that the magnitude of this field depression increases as the magnitude of the random potential variation increases.

So far we have presented a qualitative argument for the existence, for a pinning array in the resistive or dynamic state, of regions within which the electric field is zero (phase-locked regions) separated by channels or paths along which vortex motion occurs. Treatments of weak link networks (WLN) are usually based on consideration of the static energy Hamiltonian of an array of coupled loops containing two or more weak links. The derivation of the dynamic characteristics of such networks has proved intractable. The starting point here is to assert the formal equivalence of an irreversible superconductor in the mixed state, and a WLN. There is a continuum of physical systems ranging from a random pinned vortex lattice through an inhomogeneous type II system with zero pinning regions or even normal inclusions or voids to an interconnected system of microbridges and finally to a random interconnected array of SNS or SIS tunnel junctions. Since the problems of interaction and non-locality in a pinned elastic vortex lattice have been fully treated[12-14] without introducing the additional concept of frustration it is clear that a similar treatment for WLNs can lead to a transparent interpretation of the dynamic state. In what follows we present a model that yields the dimensions of the phase-locked regions in a random WLN, and thence predicts the dynamic response of the WLN.

WEAK LINK NETWORKS

In discussing this problem two key terms will be employed: (a) the global critical current distribution $f(i_c)$; this is the probability distribution of the critical currents in all the junctions in the material which in an isotropic material will be a scalar distribution but directional in textured or non-isotropic materials; (b) the macroscipic current flow; the direction of flow of the total current, averaged over the whole sample width.

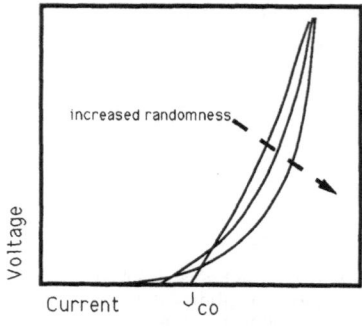

Fig. 4. Schematic plot of the effect of increasing the random potential superimposed on a regular pinning array.

Anisotropic weak link network

Before considering the general WLN there is a particular limiting case which is very illuminating. In this limit the WLN is to be imagined as a rectangular lattice with the macroscopic transport current flowing parallel to one of the principal axes. The lines which make up this lattice each contain a single weak link. The simplifying feature in this limit is to assume that the weak links which lie on arms *normal* to the macroscopic current flow are completely superconducting at all times. Thus the lattice is represented by a series array of sets of weak links; each set lies on a line or plane *normal* to the macroscopic current flow, and quantum coherence distributes the current over all weak links in the set up to their critical current. The critical current of each set (I_c) is thus simply the sum of the critical currents of all the weak links in the set.

$$I_{cm} = \sum_{0}^{K} i_{ck,m} \tag{3}$$

where K is the number of weak links *normal* to the net current flow direction. The distribution of critical currents in the sets $F(I_c)$ is gaussian by the central limit theorem, and consequently the standard deviation of $F(I_c)$ is given by $\sigma_{F(Ic)} = \sigma_{f(ic)}/\sqrt{K}$. The problem thus reduces to a series array of sets, each of which behaves as a single weak link and the distribution of whose critical currents is given by a standard statistical summation over the original global distribution. The effect of this is to produce a much sharper distribution of total critical currents in the series array with a consequent increase in n-value. However, unless $f(i_c)$ is a ∂ function, $\sigma_{F(Ic)}$ and hence n will still be finite and the V̇ vs I characteristic will still exhibit curvature and a transition from zero to finite voltage state which occurs at different currents in different parts of the array. This refutes the argument[5], implicit in some existing work, that a 2-D or 3-D coupled arrangement of weak links must inevitably reach a critical state simultaneously.

The effect of magnetic field is to alter the overall distribution of critical currents in the way discussed under the one-dimensional limit above. Since the weak links remain coupled normal to the macroscopic current flow by fully superconducting links, the same statistical treatment described above can be applied to this new $f(i_c)$ to give the distribution of critical currents in the sets of weak links which form the series array.

Isotropic weak link network

We now consider the effect of replacing the network described above with one in which weak links lie along all segments of the array. In this case it is those weak links, transverse or normal to the mean current flow, with the lowest critical current which will make the significant difference, in that they will limit the current transfer which gave rise to the complete summation of sets of weak link critical currents described in the section above. However, a proportion of the transverse weak links (even those with very low i_cs) will not be significant: if the local environment of a transverse weak link is sufficiently symmetrical, then there will be only minimal current transfer across it.

The important quantity determining the behaviour of this lattice is the mean spacing between weak links which affect transverse current transfer. This allows us to think in terms of a transverse current transfer decay length Λ. This characteristic length was introduced earlier, when it was defined as characterising the non-linearity of vortex motion paths and it was shown that the non-linearity gave rise to a partitioning of the network into regions of constant voltage. We now equate the current decay length with the mean size of these phase-locked regions normal to the local current flow, since the mean of an exponential probability decay of the form $p=e^{-x/\Lambda}/\Lambda$ is also Λ.

Current within a particular region can transfer laterally with a probability which decreases with a decay length Λ since the sample is partitioned into regions of constant voltage with this mean transverse width. Consequently, instead of the problem being viewed as a simple series array of sets of weak links summed across the whole width of the sample (as in 2.1 above), the flow structure can now be imagined as a system of multiple series arrays, each of width Λ. Since the summation width has been reduced from that of the whole sample to Λ, the variance of the sets of critical currents will increase. If the sample is homogeneous on a scale larger than Λ, then in the definition of Λ, we have already accounted for all current exchange, and so the current flow is completely described by an appropriate distribution of series arrays of the form described. Thus the dynamic response of the WLN is determined by a summation of range Λ over the global critical current distribution. Λ must itself derive directly from the global critical current distribution and the topology of the WLN (since these completely specify the problem). A first order calculation of Λ is presented below.

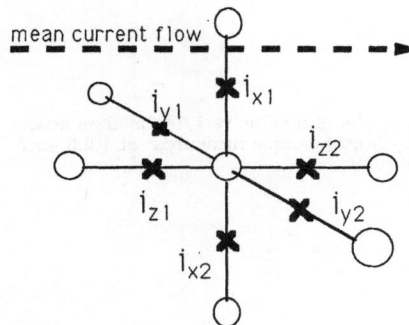

mean current flow

Fig. 5. A single node from a 3-D cubic lattice. The weak links which lie on each of the internode arms are represented by **X**s.

THE CALCULATION OF Λ FROM THE PROBABILITY DISTRIBUTION

This calculation of Λ is based on the probability of transverse current transfer from a single node exceeding the critical currents of the available weak links. As such it is the first order approximation to the complete solution. The treatment starts by again assuming a cubic lattice of weak links, with the net current applied parallel to one of the principal axes. Fig. 5 shows a node viewed from the direction of net current flow. In general the critical current i_{z1} of the junction leading to the node in the direction of mean current flow will be different to the critical current of the junction leading out of the node in the z direction (i_{z2}). In other words $|i_{z1}-i_{z2}| \neq 0$. This first order approximation is based on a number of assumptions:

(1) An equal applied current is taken to flow initially through all nodes before any redistribution; the calculation is therefore self-consistent in terms of the total current flow.

(2) If this applied current exceeds either of the critical currents i_{z1} and i_{z2} then in order to maintain the critical state in the region of this node a finite current (I_t) must be transferred to or from neighbouring nodes in the XY plane.

(3) In this first order approximation, we assume that the neighbouring nodes, and their associated junctions are capable of receiving the current without themselves entering a finite voltage state.

(4) The probability of the critical state being exceeded in the region of the original node is the probability of the critical currents of the four transverse junctions being less than I_t. The total transverse critical current is I_{xy} ($=i_{x1}+i_{x2}+...$); therefore we wish to determine $p_c(I_t>I_{xy})$.

Of these assumptions only 3 and 4 need careful consideration. (3) describes a first order approximation which can be extended to cover any degree of accuracy required; so that the next order would consider the probability of reverse current transfer from neighbouring nodes in all directions in the manner of a typical mean-field type calculation. The relationship between the local transfer probability calculated in (4) and Λ is less obvious, but within the first order approximation the only information describing Λ is $p_c(I_t>I_{xy})$, and so it isreasonable to derive one from the other, provided the latter is defined as before as the critical state decay length normal to the direction of local current flow. Using the standard relationship between a decay probability and decay length we can relate the two parameters by $1-p_c = e^{-r/\Lambda}/\Lambda$ where r is the internode spacing. To calculate the n-value we assume that the critical currents of junctions lying within Λ in any direction may be summed. Following the previous example, this will produce a $1/\sqrt{K}$ reduction in $\sigma(i_c)$; thus

$$\sigma_{F(I_c)} = \sigma_{f(i_c)}\sqrt{(r/\Lambda)^2} \qquad (4)$$

It has been shown elsewhere[6] that for a gaussian critical current distribution n is inversely proportional to standard deviation, so that

$$n \propto \left(\sigma_{f(i_c)}\ln(1-p_c)\right)^{-1} \cong 1/\sigma_{f(i_c)}p_c \qquad (5)$$

which implies that

$$n \propto 1/p_c \qquad (6)$$

for a given global probability distribution.

Thus as the probability of critical state decay at any node increases, the observed n-value will decrease. Since an increase in the transport current will lead to a increase in p_c, it can be seen that a fall in n with increasing current would be expected from this result.

MONTE-CARLO CALCULATION OF Λ

A monte-carlo solution of the inequalities describing p_c was performed. This calculation sampled the critical currents I_{z1-2} and I_{xy1-C} from a global critical current distribution, where C is the transverse connectivity (C=4 for a cubic lattice). The following criteria were then applied to each set of sampled values

(a) If $I \leq I_{z1}$ and $I \leq I_{z2}$ where I is the mean current flowing longitudinally through each node, then the result was false.

15

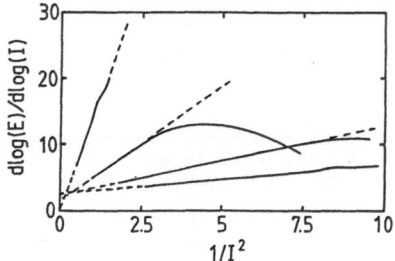

Fig. 6. plot of n value vs $1/I^2$ for 93% dense $YBa_2Cu_3O_7$ sample measured at 1,2,3,4mT (from top).

(b) If $I_{z1} \leq I < I_{z2}$ or $I_{z2} \leq I < I_{z1}$ then if

$$I - \min(i_{z1}, i_{z2}) > \sum_{k=1}^{C} i_{xy\,k} \qquad (7)$$

where min(a,b) represents the smaller of critical currents a or b, the result was true, otherwise false.

(c) If $I \geq I_{z1}$ and $I \geq I_{z2}$ then the result was true.

After completing a number of tests the number of true outcomes divided by the total number of trials was taken as the value of p_c. The n-value was then calculated from p_c using equation (6) above.

This calculation was performed for comparison with two different types of data. Firstly results obtained from sintered $YBa_2Cu_3O_7$ wires which clearly form a 3-D network, and secondly microfabricated 2-D SIS tunnel junction arrays. The experimental details are described in the next section.

EXPERIMENTAL RESULTS

Experimental results were obtained from measurements of sintered $YBa_2Cu_3O_7$ wires fabricated by a polymer precursor processing route. Both the preparation route and the experimental arrangement have been described more fully elsewhere[7]. The samples had a uniform circular cross-section of $\approx$ 0.8mm diameter. By varying the sintering times and temperatures the density of the wires could be varied over the range 65-93% of the theoretical bulk value. Measurements were performed in liquid nitrogen and the magnetic field was provided by a copper wound magnet external to the dewar.

The tunnel junction arrays were prepared from $Nb/AlO_x/Nb$ whole wafers by the SAWW processing route[15]. The mask set was designed with a single 46x148 2-D rectangular array, and two 400 junction 1-D series arrays on the same 0.5"x0.125" chip. Although the masks were designed to produce constant $5x5\mu m^2$ junction areas, the small size and consequent lithographic distortion was sufficient to ensure a random variation in area and critical current. These samples were measured at 4.2K.

Plots of voltage vs current were digitally recorded at various values of applied magnetic field. Log E vs Log J plots were constructed from this data and numerically differentiated to give the n-value. The Log E vs Log J plots invariably showed significant departures from linearity, and to demonstrate the variation of n-value with current and voltage, plots of dLogE/dLogJ vs logE and dLogE/dLogJ vs I^{-2} were constructed. Typical plots are shown in figures 2 and 6. Plots of the latter type invariably show a characteristic linearity which emerged from the theoretical model discussed above. The plots against Log E show an approximately linear decrease with increasing E for all values of applied magnetic field and for all values of sample density.

The behaviour of the critical current as a function of applied field was examined at different voltage criteria. All samples showed an extremely rapid decrease in critical current with applied field. The interpretation of this result in terms of a simple model of interconnected tunnel junctions has been discussed previously[1,7], and this variation is ascribed to a progressive increase in the mean junction area as the sample density is increased.

COMPARISON OF THEORETICAL AND EXPERIMENTAL RESULTS

In a previous publication we briefly described the application of the current transfer model to experimental results from high temperature superconductors and showed that there was good qualitative agreement. In this case we used an $f(i_c)$ which was gaussian in zero magnetic field with mean unity and variable standard deviation S. The weak links were taken to be SIS tunnel junctions with a Fraunhofer dependence of the critical current on the magnetic field, and for simplicity it was assumed that the critical current was proportional to the junction area. The plot of n/n_0 against B/B_0 (fig. 3) also shows these modelled results determined for different connectivities and $f(i_c)$ distributions. The curves have a very similar characteristic shape, and the negative gradient is increased by either increasing the connectivity C or reducing the standard deviation S. The values of connectivity used in the model calculations are lower than would be expected in a close-packed three dimensional sample, but this reflects the fact that a simple gaussian distribution of junction areas was used. In a real sample we would expect a distribution of critical currents which was much more heavily skewed towards zero[1], reflecting the fact that (particularly in low density samples) a majority of the weak links are poorly conducting. In this sense, the form of the critical current distribution and the connectivity can be used interchangeably to describe the overall behaviour, the use of a gaussian distribution and a low connectivity imply a procedure which pre-selects for high current weak links and treats low current paths as insulating.

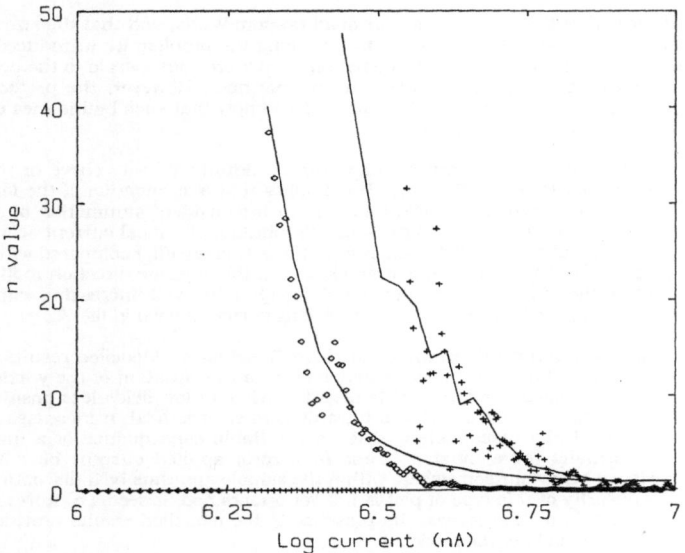

Fig. 7. Plot of n-value vs log of the transport current. The points represent data from a 2-D tunnel junction array at two different magnetic fields (0mT crosses and 1.3mT diamonds). Solid lines represent theoretical curves modelled using the 1-D critical current distributions at the respective fields.

The use of a probably unrealistic gaussian distribution with a rather simplistic magnetic field variation has meant that quantitative comparison with experiment was not possible. This was the main motivation for the investigation of the fabricated 2-D arrays. In this case the 1-D arrays could be measured and analysed to directly give $f(i_c)$. Since the 1-D and 2-D arrays were identical in construction apart from a meander etch mask which removed the transverse connections, this allows a calculation of the 2-D response using a known $f(i_c)$ at each value of magnetic field. A plot of n vs Log(I) obtained from the 2-D array at two different field values is shown in Fig. 7. Superimposed on this plot are the results of the modelling calculations, using the 1-D $f(i_c)$. The quantitative agreement is good, even allowing for the rather noisy original data, and it is imprtant to note that there are no free variables in the modelling technique. The extent of the agreement is discussed in the next section, together with an interpretation of the results.

THE INTERPRETATION OF THE RESULTS

In section 2 we showed that a periodic pinning array with random barrier height gave rise to a low electric field behaviour in which the sample was partitioned transversely and longitudinally into phase-

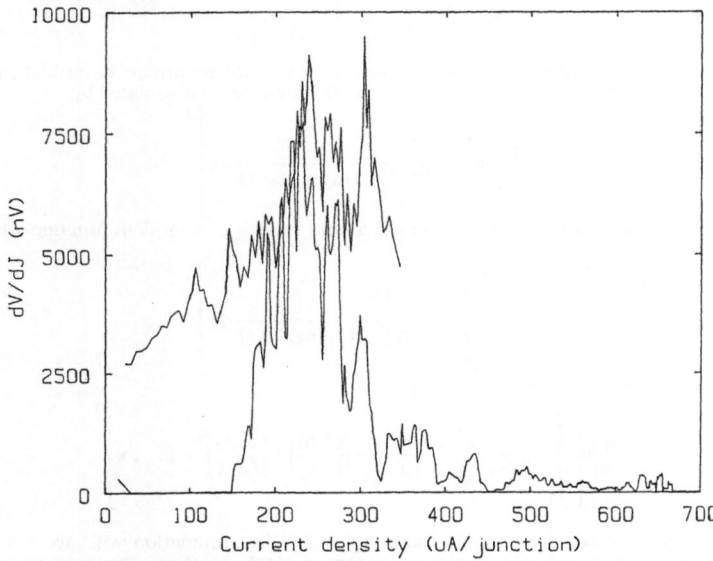

Fig. 8. Plot of critical current distributions from 1-D (upper curve) and 2-D (lower curve) tunnel junction arrays.

17

locked regions. We believe that this is the behaviour of all random WLNs, and that the size of these regions determines the dynamic response of the network. In modelling the problem we introduced the transverse current decay length Λ, which in the first order calculation simply provides a scale to the problem, but does not directly predict the existence of closed phase-slip boundaries. However, the prediction of a mean transverse distance between transverse phase-slip centres does imply that such boundaries exist, since such centres must be connected to form lines or planes.

The results for the 2-D array are of central importance. Firstly V vs I curve of the 2-D array is considerably sharper than that for the 1-D array. This implies that a summation of the form described in the model is in fact occurring. Secondly, a field and current independent summation of critical currents across the width of the 2-D array would lead to gaussian distribution of critical current sets as described in section 2.i. A plot of the critical current distribution from the 2-D array $F(I_c)$ compared with the 1-D $f(I_c)$ is shown in Fig. 8. $F(I_c)$ is neither gaussian, and is even skewed in the opposite direction to $f(I_c)$. This implies (as the model predicts) that the range of summation is not fixed but falls with increasing current. The same effect is reflected also in the rapid decrease in n with increasing current shown in fig. 7.

Fig. 6 shows the variation of n with current for the high T_c samples. Modelled results show strikingly similar behaviour, and the qualitative agreement extends even to the variation of the y axis intercept with applied magnetic field; at high fields the intercept is negative, while at low fields it is positive. This overall behaviour indicates that as the current is reduced towards zero electric field, n increases without bound. This infinite n-value is physically reasonable and is an inevitable consequence of a finite zero-voltage critical current, since formally there must at some (non-zero) applied current be a transition from superconducting to resistive behaviour somewhere within the sample and thus be a discontinuity in the V vs I curvature. While the linearity of this type of plot is not yet understood, it seems a universal feature of all data including the 2-D tunnel junction arrays. Its presence in the modelled results provides a measure of independent support for the validity of the model.

CONCLUSIONS

We have described and used a model for the resistive behaviour of a multi-dimensional WLN. The quantitative agreement between the model and data for 2-D tunnel junction networks is very good and is already of value in assessing the interconnectedness of different samples of sintered high temperature superconductor. The model should apply to all systems of strong superconductivity interconnected by weak links and should allow the investigation of more complex samples, such as anisotropically compacted material. The development of higher order approximations is being investigated.

ACKNOWLEDGEMENTS

This work was supported by the Science and Engineering Research Council, Trinity College Cambridge and the Royal Society. The high T_c samples were provided by ICI and the tunnel junction arrays were fabricated by D. Glowacka. We are grateful to P.L. Sampson and B.A.Glowacki for useful discussions and experimental data.

APPENDIX A

This appendix describes an approximate solution to the vortex equation of motion (section 2). Since $E=B \times v$ and so $<E>=B.<v>$ where $<v>$ is the mean vortex velocity normal to the net current flow, then

$$E = \frac{B.L}{\eta}\left[\int_0^L \frac{1}{f_0(x)+g(x)}dx\right]^{-1} \qquad (A1)$$

where L is the length of he integration path. Using a mean field treatment described in a previous publication[11] and splitting the integral into one over a sum this may be approximated by

$$E = \frac{B.L}{\eta}\left[\int_0^a \sum_{n=0}^{L/a} \frac{1}{f_0(na+x)+g(na+x)}dx\right]^{-1} \qquad (A2)$$

Now since f_0 is a periodic function with wavelength a, and since g is a random function with no phase information this may be rearranged to give

$$E = \frac{B.L}{\eta}\left[\int_0^a \frac{1}{f_0(x)}\sum_{n=0}^{L/a} \frac{1}{1+g(na)/f_0(x)}dx\right]^{-1} \qquad (A3)$$

This can then be expanded as a Taylor series to give

$$E = \frac{B.L}{\eta}\left[\int_0^a \frac{1}{f_0(x)}\sum_{n=0}^{L/a} 1 - \frac{g(na)}{f_0(x)} + \left(\frac{g(na)}{f_0(x)}\right)^2 - \left(\frac{g(na)}{f_0(x)}\right)^3 + ... dx\right]^{-1} \qquad (A4)$$

Since g is a random function, the periodic sampling under the summation will give a value for the summation which provides an unbiassed measure of the spread of this function. Furthermore, as the mean vortex velocity increases g will tend to equal weighting about zero and hence the odd terms of this summation to zero. The complete expression for the electric field is then as given in equation 2.

18

REFERENCES

1. R.L. Peterson and J.W. Ekin, Phys. Rev. B, **37**, 9848 (1988)
2. J.R. Clem and V.G. Kogan, Jpn. J. Appl. Phys. **Supp. 26-3**, 1161 (1987)
3. R.G. Steinmann and B. Pannetier, Europhys. Lett. **5**, 559 (1988)
4. Y.Y.Wang, B. Pannetier and R. Rammal, J. Phys. France, **49**, 2045 (1988)
5. C.J. Lobb, D.W. Abraham and M. Tinkham, Phys. Rev. B, **27**, 150 (1983-)
6. C.J.G. Plummer and J.E. Evetts, IEEE Trans. Magn. **MAG-23**, 1179 (1987)
7. J.E. Evetts, B.A. Glowacki, P.L. Sampson, M.G. Blamire, N. McN. Alford and M.A. Harmer, IEEE Trans. Magn. **MAG-25**, 2041 (1989)
8. A.T. Fiory, A.F. Hebard and S. Somekh, Appl. Phys. Lett. **32**, 73 (1978)
9. P. Martinoli, Phys. Rev. B, **17**, 1175 (1978)
10. M.G. Blamire, J. Low Temp. Phys. **68** ,335, (1987)
11. M.G. Blamire and J.E. Evetts, Phil. Mag. B. **51**, 421-438, 1985
12. E.H. Brandt, J. Low Temp. Phys. **28** ,263 & 291, (1977)
13. A.I. Larkin and Yu.N. Ovchinniov, Sov. Phys. JETP, 38, 854 (1979)
14. A.M. Campbell, Jpn. J. Appl. Phys. **Supp. 26-3**, 2053 (1987)
15. M.G. Blamire and J.E. Evetts, IEEE Trans. Magn. **MAG-25**, 1123 (1989)

BOSE-EINSTEIN GAS MODEL FOR T_C AND ENERGY GAP FOR MOST

SUPERCONDUCTORS, ESPECIALLY THE CERAMIC OXIDES

Mario Rabinowitz

Electric Power Research Institute
Palo Alto, CA 94303

ABSTRACT

A generalized Bose-Einstein gas model is sufficient to derive reasonable estimates of T_c, the energy gap, and the coherence length for all classes of superconductors such as the ceramic oxides, the metallics, heavy-fermion metals, metallic hydrogen, and neutron stars for 3-dimensional, quasi-2 and quasi-1 dimensional states. Analysis of the new high-temperature ceramic oxide superconductors determines upper limits of T_C using as input the number density of conduction electrons and the effective mass of the charge carriers. This calculation for the ceramic oxides yields 10K, 40K, and 300K in 3, Q2, and Q1 dimensions. Interpreting the ceramic oxide case as one in which the interchain interactions are equal to the intrachain interactions, leads to

$$T_c = (T_{c1} \ T_{c2})^{1/2} = (300K \times 40K)^{1/2} = 110K = 100K$$

as the approximate transition temperature for these materials when there is clearly a combined linear and planar structure. It is noteworthy that without specifying a coupling mechanism or coupling strength, this general model does well in calculating transition temperatures, and coupling strengths over nine orders of magnitude (1K to 10^9K and meV to MeV) from the heavy fermion metals to neutron stars.

INTRODUCTION

For an idealized non-interacting point-particle Bose-Einstein gas, the condensation temperature is considerably higher than the pairing temperature of electrons in a conductor. Thus there is no manifestation of superconductivity until the electrons pair, because there can be no condensation until bosons are formed. Therefore calculation of a pairing temperature would be equivalent to determination of the superconducting transition temperature, T_C. If the opposite were true, and the condensation temperature were less than or equal to the pairing temperature, it would suffice to calculate the condensation temperature to find the critical temperature, T_C. In this case since superconducting properties could not manifest themselves until the condensation temperature were reached, it would be difficult to ascertain that the electrons had paired.

This paper is presented in the spirit that the feature common to superconductivity in all of its manifestations may be the condensation temperature, rather than the pairing temperature. Thus it may be quite secondary or incidental that there are a variety of pairing mechanisms with a range of pairing strengths. In the midst of a panorama of coupling mechanisms and diversity of strengths,it may suffice to calculate the condensation temperature and in the process also obtain the pairing strength. It may be possible to do all this without reference to any particular coupling model.

Advances in Cryogenic Engineering (Materials), Vol. 36
Edited by R. P. Reed and F. R. Fickett
Plenum Press, New York, 1990

21

At this stage of this theoretical approach, no attempt is made to discriminate between superconductors and other conductors. Although it is generally believed that the original BCS theory discriminates well between superconductors and non-superconductors, it in fact does not for two reasons.

In the original BCS theory[1],

$$T_c = 1.14 \, (h\nu/k)exp[-\{N(E_F)V\}^{-1}] \quad where, \tag{1}$$

h is Planck's constant, k is the Boltzmann constant, ν is the average phonon frequency (much like the Einstein frequency), $N(E_F)$ is the density of Bloch states of one spin (i.e. ignoring spin) per unit energy for electrons at the Fermi energy of the normal metal, and V is the constant average matrix element (neglecting anisotropic effects) for scattering interactions of pairs making transitions in the region $-h\nu < \varepsilon < h\nu$. Because of the exponential dependence of T_c, the parameters must be known quite accurately, as there is a correspondingly larger uncertainty in the calculated value of T_c. So lack of sufficiently accurate input data is one reason why the BCS prediction is sometimes off. The second reason is that for some materials like the alkali metals where the parameters are well known, BCS theory predicts superconductivity at easily attainable temperatures, but none has been found at even the lowest measurable temperatures.

The first objective of this analysis is to derive an upper limit for T_c for different classes of superconducting materials in different dimensionalities. There can be many mechanisms which poison T_c to lower temperatures, but it is of some value to be able to put an upper limit on T_c for a given class of superconductors. The second object is to derive an approximate coupling strength i.e. the energy gap for these classes. As will be shown, the ability of this simple theory to predict general ranges of critical temperatures is particularly striking. In addition it gives reasonable predictions in areas such as the heavy fermion (heavy electron) metal superconductors and the ceramic oxide superconductors where neither the BCS theory nor other theories do very well.

CHARGED THREE-DIMENSIONAL QUANTUM GAS

In a Bose gas, particles may be expected to start a Bose-Einstein condensation into a ground state in momentum space when the quantum mechanical wavelength λ is much greater than the interparticle spacing. If this is to occur for real particles such as electrons at T_c, it is necessary to modify the usual condensation derivation which yields a condensation temperature $>> T_c$. Superconductivity can occur if the Fermi particles pair near the Fermi level. When this happens, the paired particles have integral spin and obey Bose-Einstein statistics rather than Fermi-Dirac statistics.

At temperature $T \leq T_c$, we have

$$\lambda/4 \geq (n_s)^{-1/3}, \quad where \tag{2}$$
$$n_s \sim (kT_c/E_F)n. \tag{3}$$

n is the number density of free charged particles, n_s is the number density of particle pairs whose condensation temperature is T_c, k is the Boltzmann constant, and E_F is the Fermi energy.

$$E_F \doteq (h^2/8m)n^{2/3}, \tag{4}$$

where m is the effective mass of the charge carrier.

For a particle pair gas, of momentum p, mass 2m, and incremental energy E near the Fermi level:

$$\lambda = h/p = h/[2(2m)E]^{1/2} = h/[4m(f/2)kT_c]^{1/2}, \tag{5}$$

where f is the number of degrees of freedom per particle.

Combining eq's. 2,3,4, and 5:

$$T_c \sim h^2 n^{2/3}/(8f)^3 mk = E_F/64f^3 k = (h^6 D_F^2)/[2^{13}\pi^2(fm)^3 k], \tag{6}$$

where D_F is the density of states at the Fermi surface.

In order to estimate T_c for some materials like the ceramic oxides, we should consider the possibility that the conduction paths for superconductivity may be two and/or even one dimensional. The oxide superconductors such as $Y_1 Ba_2 Cu_3 O_{7-y}$ are not strictly of lower dimensionality, as mechanisms such as Josephson tunneling through non-conducting regions tend to produce in equilibrium a Fermi energy corresponding to three dimensions E_F rather than to two- or one- dimensional Fermi energies E_{F2} or E_{F1}. Eq. 6 is also obtained for lower dimensions, as these systems are 3-dimensional with respect to E_F and thermal equilibrium, and only the degrees of freedom of the paired fermions are restricted to 2 and 1 dimensions. Anisotropy may produce a sufficiently higher effective mass in some directions as to practically reduce the dimensionality. These systems may be considered to be of quasi-lower dimensionality where only the degrees of freedom f_2 and f_1 of the paired fermions are restricted in their nontunneling conduction paths.

For the ceramic oxides eq.6 yields 10K, 40K, and 300K in 3, quasi-2, and quasi-1 dimensions. Interpreting the ceramic oxide case as one in which the interchain interactions are equal to the intrachain interactions, leads to

$$T_c = (T_{c1} \ T_{c2})^{1/2} = (300K \times 40K)^{1/2} = 110K \doteq 100K \tag{7}$$

as the approximate transition temperature for these materials when there is clearly a combined linear and planar structure.

MODEL INDEPENDENT ENERGY GAP

Most superconductors have an energy gap, 2Δ. Non-interacting point particles would not be expected to have an energy gap. However, for any real neutral or charged gas, there is an interaction potential energy which can result in an energy gap. The particles exhibit no flow viscosity when $kT < 2\Delta$, because this minimum energy must be supplied before they can leave the states they are in.

Without assuming any particular coupling mechanism or coupling strength, let us introduce a minimum excitation energy or binding energy Δ per particle which must be provided to break up a particle pair. The energy gap is designated 2Δ, since both particles must be taken out of the superconducting state. One cannot be left behind, as only particle pairs can exist in the superconducting state. In the BCS theory, the energy gap is

$$2\Delta_{BCS} = 3.52 kT_c . \tag{8}$$

Twice the number density of particle pairs is roughly the fraction $(\Delta/2)/E_F$ of conduction electrons that have a strong interaction near the Fermi level, where 2Δ is the energy gap. This fraction may be a function of the class of superconductors as the number of states expelled from the gap is related to the size of the gap and the density of normal states.

Thus, in place of eq. 3,
$$2n_s \sim (\Delta/2)n/E_F . \tag{9}$$
Eqs. 2,4,5 & 9 yield

$$2\Delta = 8(4fkT_c)^{3/2}E_F^{-1/2} = 128(2m)^{1/2}(fkT_c)^{3/2}h^{-1}n^{-1/3} . \tag{10}$$

As before, eq. 10 is obtained for quasi-2 &quasi-1 dimensions.[2,3]

Table 1. Comparison of Estimated T_C and 2Δ with Data for Wide Range of Matter

Material	n	Dim.	Est. T_C,K	T_C^*,K	2Δ,meV	$2\Delta_{BCS}$,meV
Heavy elec. metal	$10^{29}/m^3$	3	0.5-5	0.5-1.5	0.68	0.304
		Q2	2-20	-	2.9	-
		Q1	10-100	-	11	-
Mettalic Supercond.	$10^{29}/m^3$	3	50	23.2	10	7.04
		Q2	200	-	140	-
		Q1	1000	-	560	-
Ceramic Oxide	$10^{28}/m^3$	3	10	13	19	3.95
		Q2	40	35	44	10.6
		Q1	300	125	110	38.0
Li-Be-H	$5\times10^{29}/m^3$	3	200	-	160	60.8
		Q2	500	-	330	-
		Q1	4000	-	2600	-
Metallic H	$10^{30}/m^3$	3	300	120-260	180	79.0
		Q2	800	-	520	-
		Q1	7000	-	4800	-
Neutron Star	10^{43}- $10^{29}/m^3$	3	10^8-10^9	10^{10}-10^{11}	1.8×10^{10}	3.04×10^9
		Q2	10^8-10^9	-	10^{11}	-
		Q1	10^9-10^{10}	-	10^{11}	-

Table 1 shows that eqs. 6 & 10 do well in representing experimental and calculated results over a range of T_C and Δ of 9 orders of magnitude. T_C^* is the experimental or literature value [4-12] of critical temperature . Although only one significant figure may be warranted for Δ, two are listed for the purpose of comparing Δ and Δ_{BCS} .

For Table 1, eq.10 is considered independent of eq.6, and only experimental values of T_C should be used as input into eq.10. However, if eqs. 6 and 10 are combined, then

$$2\Delta = 8kT_C \tag{11}$$

in good agreement with the ceramic oxides. Using experimental values of T_C causes the ratio $2\Delta/kT_C$ to vary.

CONCLUSIONS

As shown by Table 1, this simple theory does well in predicting critical temperatures and energy gaps over a range of nine orders of magnitude. The agreement with experiment is good. The theory agrees well with the BCS predictions where BCS does well. In addition this theory makes reasonably good predictions where BCS does not, as in the case of the heavy fermion metals and the ceramic oxides. Equation 6 is indicative of the T_c that may be expected for a given class of materials in a given conduction dimensionality. The experimental and literature data are presented in support of the estimates calculated from this equation. Because of the low number density of electrons, the ceramic oxides are not likely to be three-dimensional superconductors except at the lower transition temperatures.

It is both interesting and noteworthy that this theory does so well, without specifying a coupling mechanism or strength, and with only three variables f, n, and m, which can be determined from experiment. The variation of m from that of a free electron mass may be able to account for small differences in T_c within a given class of materials, as would more detailed knowledge of the Fermi surface in general. Particularly important are mechanisms which act to impair T_c, i.e. poison the superconductivity.

The mass variation appears to work well for the heavy electron metals, where $T_c \propto 1/m$, as predicted by eq. 6. Crystallographic data can help determine f. The combination of different dimensions, such as Q1 and Q2 in the ceramic oxides, may be an important factor in determining T_c.

REFERENCES

1. J. Bardeen, L.N. Cooper, and J.R. Schrieffer, Phys.Rev. 108:1175 (1957).
2. M. Rabinowitz, Int'l. J. Theo. Phys. 28:137 (1989).
3. M. Rabinowitz, In Proc.: EPRI Workshop on High-Temp. Superconductivity, (1987).
4. Z. Fisk, D.W. Hess, C.J. Pethick, D. Pines, J.L. Smith, J.D. Thompson, and J.O. Willis, Science 239:33 (1988).
5. Z. Fisk, H. Rott, T.M. Rice, and J.L. Smith, Nature 320:124 (1986).
6. D.C. Johnston, H. Prakash, W.H. Zachariasen, and V. Viswanathan, Mater. Res. Bull. 8:777 (1973).
7. A.W. Sleight, J.L. Gillson, and P.E. Bierstedt, Solid State Commun. 17:27 (1975).
8. Z.Z. Sheng and A.M. Hermann, Nature 322:55 (1988).
9. A.W. Overhauser, Phys.Rev. B35:411 (1987).
10. J. Jaffe and N.W. Ashcroft, Phys. Rev.B27:5852 (1983).
11. V. L. Ginzburg, Com. Astrophys. & Space Phys. 1:81 (1969).
12. V. L. Ginzburg, and D.A. Kirzhnits, Sov. Phys. JETP 20:1346 (1965).

FLUX CREEP IN MULTIFILAMENTARY CONDUCTORS OF NbTi and Nb_3Sn*

A. K. Ghosh, Youwen Xu and M. Suenaga

Brookhaven National Laboratory
Upton, New York 11973

ABSTRACT

Long term decay of the sextupole field has been observed in accelerator dipole magnets, and this effect has been linked to "flux creep" in the superconductor. To study this problem, the decay of the magnetization of multifilamentary conductors of NbTi and Nb_3Sn have been measured as a function of time. Measurements show that as a function of $\ln t$ this decay cannot always be characterized by a single decay rate. Long time decay rates are sometimes approximately half that which is observed at short times < 1000 secs. Creep rates are found to: (1) scale with filament diameter, (2) change slowly with field, (3) change rapidly as H is backed off from the critical state and (4) is fairly insensitive to temperature for $0.3T_c < T < 0.8T_c$. Results of this investigation are compared to sextupole field decay observed in magnets.

INTRODUCTION

In recent years, the problem of flux creep in type II superconductors has attracted attention following the observation of large time dependent magnetization decay in high T_c material[1]. However, this decay, which had been observed by early researchers[2,3] of type II superconductors did not generate much interest in traditional low T_c superconductors like NbTi, since the effect was small and well understood within the context of the critical state model[4] of critical currents and the Anderson[5] theory of flux creep.

During the commissioning of the ring of superconducting magnets of the TEVATRON Collider at Fermilab, Finley et al.[6] reported that the chromaticity of the beam changed slowly over a long period of time during particle injection at 150 GeV. While the high energy particles are being injected into the ring, the magnets are at a fixed dipole field of ~ 0.66T. However, the chromaticity changes are related to the sextupole harmonic field b_2. At injection field, most of the b_2 arises from the persistent current magnetization of the superconductor, which in this case is NbTi. It was subsequently established by measurements on model TEVATRON dipole magnets[7], that b_2 indeed decays with a logarithmic time dependence at a rate which is consistent with the observed $\ln t$ chromaticity changes seen in

---

* Work done under the auspices of the U. S. Department of Energy

Advances in Cryogenic Engineering (Materials), Vol. 36
Edited by R. P. Reed and F. R. Fickett
Plenum Press, New York, 1990

27

the Collider operation.[8]   There are now other published reports that record similar behavior in Superconducting Super Collider SSC model dipole magnets[9] and Hadron-Electron Ring Accelerator HERA magnets[10]. Kuchnir[11] et al. have recently reported on flux creep measurements on the TEVATRON dipole cable. Their data, which extend to fields ~ 0.3T, show that the ln$t$ creep rate as a fraction of the initial magnetization is fairly small, ~ 0.2-0.3%.

In this paper, we report on magnetization measurements as a function of time for NbTi and Nb_3Sn multifilamentary wires in transverse applied fields.

THEORETICAL BACKGROUND

The Critical State Model

The magnetization of multifilamentary NbTi and Nb_3Sn is well understood within the framework of the critical state model.[4] In this model, once the critical state is established, the irreversible or hysteretic part of the magnetization is related to the critical current density of the super-conductor, J_c. If $M^+(M^-)$ denotes the increasing (decreasing) applied field branch of the magnetization loop, then the irreversible part of the magnetization, M_i, at a fixed temperature is given by

$$2M_i(H) = (M^+ - M^-) \tag{1}$$

where for a cylindrical wire sample in transverse field

$$M_i = (2/3\pi) \lambda J_c(H)d \tag{2}$$

λ is the volume fraction of superconducting filaments of diameter d in the wire. J_c is a function of both temperature and field.

$$2M_e = (M^+ + M^-) \tag{3}$$

is the equilibrium magnetization at a given H which is assumed to be time independent and reversible with H.

Flux Creep

The long time persistent current decay in type II superconductors that was observed by Kim et al.[2] was explained by Anderson's[5] theory of flux creep due to thermally activated motion of flux lines which are normally pinned at defect sites in the material. The theoretical details of this mechanism was described by Beasely et al.[3] They verified their analyses by careful measurements of magnetization decay of solid cylindrical rods of PbTl alloys using a superconducting quantum interference (SQUID) device to record the flux changes. They showed that the observed ln$t$ dependence of the flux change is given in terms of the effective pinning energy U(∇B) which depends on the B field gradient in the cylinder. In a recent paper, Welch et al.[9] have reexamined the theoretical treatment of Beasley et al.[3] and have expressed the creep rate as

$$dM/d(\ln t) = - (kT/U_o^*) M_i (1 \pm \delta) \tag{4}$$

where $U_o^* = - |B| (\partial U/\partial |\nabla B|)$ is an apparent pinning potential. The relationship of U_o^* to the true barrier height, Up, depends on the nature of U-versus-driving force relation. For non-linear dependence of U on $|\nabla B|$, U_o^* as determined from Eq. (4) can be much less than U_p. δ is a parameter which takes into account small differences in creep rate for increasing and decreasing fields and is usually $\ll$ 1.

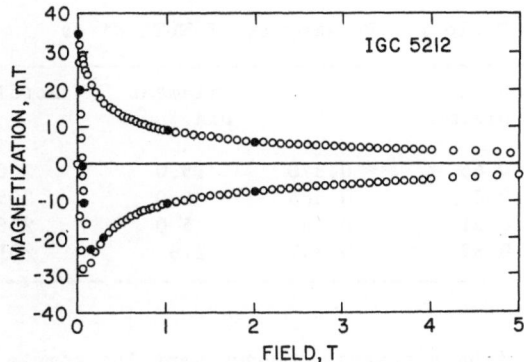

Figure 1. Magnetization of a 15 μm filament diameter NbTi wire sample in transverse field. The field cycle followed is 0→5T→0→0.3T. The filled circles indicate the fields at which decay measurements are made.

EXPERIMENTAL PROCEDURES

Magnetization measurements were carried out in a commercial SQUID magnetometer (Quantum Design), with the applied field normal to the wire sample. Scan length, the distance over which the sample travels through a set of detection coils is set such that the variation in H at this setting is < 0.05%. After a temperature is set, field is incremented gradually to avoid overshoot in H. At each field setting the superconducting magnet is switched into the persistent mode prior to measurement. Flux creep is measured during the hysteresis cycle by stopping at several magnet field values and measuring M over time lengths ~ 12000 secs. The first measurement is made ~45 secs after the field is locked in the persistent mode. In the text, magnetization and field are given in tesla(T) which is equivalent to $\mu_o M$ and $\mu_o H$, where $\mu_o = 4\pi \times 10^{-7}$ H/m is the permeability of vacuum.

RESULTS

In this section are presented the decay rate measurements made on a series of SSC prototype NbTi multifilamentary wires and a Nb_3Sn internal-tin wire. Since the Tc of Nb_3Sn is higher than NbTi, the Nb_3Sn sample was measured at several temperatures ranging from 3K to 12K. Preliminary results of the temperature dependence of the creep rate are given.

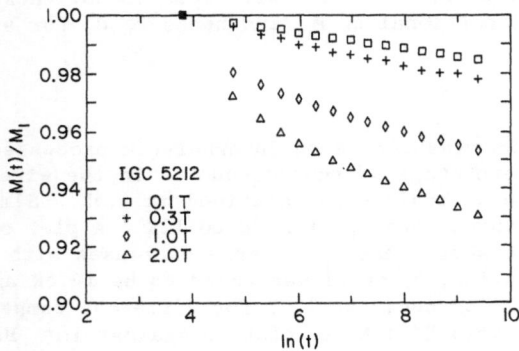

Figure 2. Magnetization decay at several fields for IGC5212 sample at 4.5K. M_1 is the first measurement for the time sequence. Time is measured in seconds.

Table 1. Parameters of NbTi Wires

	Wire Dia.mm	λ	Filament Dia. μm	$Jc(5T)$ A/mm^2
IGC5212	0.65	0.370	15.0	2400
SC12006	0.81	0.385	6.0	2650
OST1738	0.81	0.385	5.0	2560
SCN2120	0.65	0.357	2.6	2010

NbTi

Fig. 1 shows the 4.5K hysteresis measurement for sample IGC5212 which
has 15 μm filaments with an interfilament spacing of > 4 μm. At various
fields indicated by filled circles in Fig. 1, decay measurements were made.
An example of the long time decay as seen in the magnetization measurements
for the initial field sweep of the sample is shown in Fig. 2, where we have
plotted the ratio $M(t)/M_1$ against $\ln t$. M_1 is the first measurement taken
in the time sequence. We note that the decay is not always linear in $\ln t$.
However, for long times > 1000 secs, the creep rate is fairly constant.

The creep rate observed for IGC 5212 is summarized in Fig. 3. We note
the following: (1) when the filaments are fully penetrated and the critical
state has been established, i.e., for H > 0.2T the creep rate is a slowly
varying function of field. Since $(dM/d\ln t)$ is proportional to (M_1/U_o^*),
then from the known field dependence of M_1 we infer that U_o^* and hence Up is
decreasing slowly with field. (2) After a cycle to high field and back to
H=0, a subsequent increase in H~0.02T drives the creep rate to almost zero.
A further increase in H changes the creep rate from negative to positive,
as the magnetization currents are fully reversed and the critical state is
re-established in the opposite direction. In the critical state, the
differences in the rates for field increasing and decreasing is < 10%,
thereby showing that $\delta \ll 1$.

To examine the behavior of creep rate as the filament diameter is
changed, we measured the magnetization decay of four wires whose parameters
are listed in Table 1. The wires were all processed so as to produce a
high transport J_c at 5T. The comparison is made at 1T and 2T where the
filaments are in the critical state and known to behave independently.
SCN2120 has 0.5% Mn doped copper in the interfilament region to suppress
proximity coupling of the filaments.[13] Table 2 lists the average creep
rate measured for the various wires. From this table we note that although
M_1 changes by almost an order of magnitude, the normalized creep rate R =
$(1/M_1)$ $(dM/d\ln t)$ varies only by ~ 30-40%. This establishes the fact that
the creep rate is proportional to M_1, and hence to d, for similarly
processed wires.

Nb$_3$Sn

The sample of Nb$_3$Sn measured is an internal-tin processed wire with 3.5
μm filaments. The percentage of superconductor in the wire is ~ 13%. In
Fig. 4 is shown the hysteresis curve obtained at 4.5K. Similar curves were
obtained for temperatures ranging from 3K to 12K. A plot of M_1 at 1T and
2T versus temperature shows that M_1 linearly decreases with temperature for
T > 4K. By extrapolation, T_c at 1T was found to be 14.5K and at 2T it was
13.2K. The magnetization decay at 1.0T for different temperatures is shown
in Fig. 5 where the ratio $M(t)/M_1$ is plotted against $\ln t$ Here too we
observe that for a long time the decay is constant in $\ln t$. Although the
ratio of $M(t)/M_1$ decreases with T, it is found that the long time decay
rate is independent of T for 3K < T < 12K. Details of these measurements
is being published elsewhere.

Table 2. Flux Creep Data at 1.0T and 2.0T

	1.0T			2.0T		
Sample	M_i (mT)	dM/d(ln$t$) (mT)	R	M_i (mT)	dM/d(ln$t$) (mT)	R
IGC5212	9.95	0.059	0.0059	6.56	0.052	0.0079
SC12006	4.63	0.034	0.0073	2.87	0.026	0.0091
OST1738	4.46	0.035	0.0079	2.77	0.028	0.0101
SCN2120	1.65	0.0153	0.0093	0.96	0.0094	0.0098

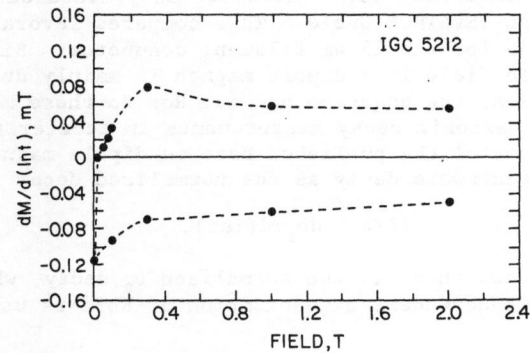

Figure 3. The creep rate dM/d(ln$t$) for IGC5212 plotted as a function of applied field. Note the rapid change in rate as the field is cycled from 0 to 0.3T.

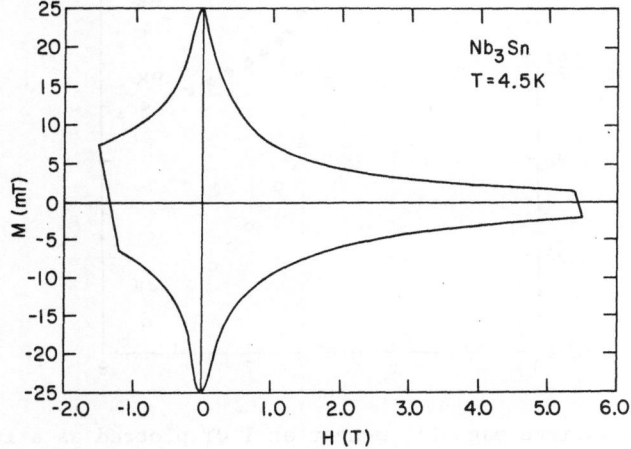

Figure 4. Magnetization hysteresis for a Nb_3Sn sample at 4.5K.

Table 3. b_2 decay rates in dipole magnets

Magnet	Fil. Dia. of Conductor (μm)	Field (T)	Cycle to Field	$\frac{1}{b_2}\frac{db_2}{d(\ln t)}$
(0.8 m Model)[7]		0.32	$0 \to 0.32$	0.0065
TEVATRON	9	0.65	$0 \to 0.65$	0.014
		0.65	$0\to4\to0.4\to0.66$	0.036
HERA Dipole[10]	14	0.17	not known	0.04
SSC Model[9] Dipole @ LBL	6	0.33	$0 \to6.6\to0.05\to0.33$	0.06

DISCUSSION

The creep rate observed for NbTi is small and not very different from those reported by Kuchnir et al.[11] At 0.15T they measured an R ~ 0.002 for the 9 μm filament TEVATRON Cable. This compares favorably with low field values ~ 0.0035 for the 15 μm filament conductor. Since at low fields, the sextupole field in a dipole magnet is mainly due to the filament magnetization, the question now is: How do these measurements compare with the b_2 harmonic decay measurements in accelerator magnets? In Table 3 is listed some of the published data on dipole magnets, where we have expressed the sextupole decay as the normalized decay rate

$$(1/b_2)\ db_2/d(\ln t).$$

From this table we note that (1) the normalized b_2 decay, which is assumed to be due to the time dependent magnetization of NbTi is usually quite

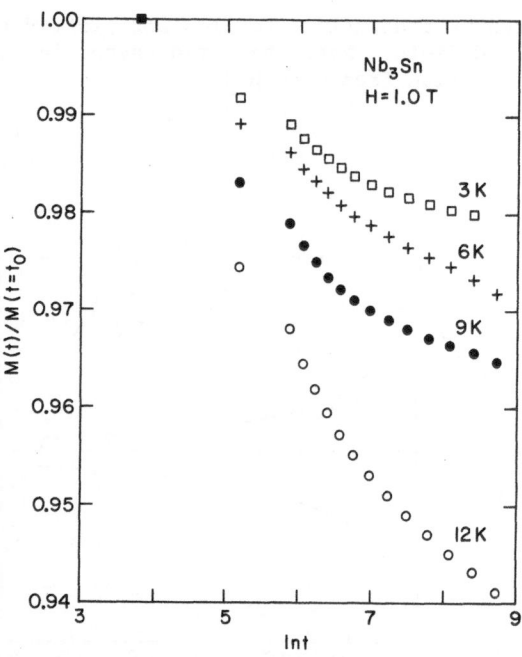

Figure 5.   Normalized magnetic moment at 1.0T plotted as a function of ln$t$ for various temperatures as indicated in the figure.

large compared to the short sample data presented here and (2) from the TEVATRON magnet data, the b_2 decay at the same field of 0.65T is seen to depend on the cycle followed to reach that field. Clearly, to correlate short sample measurements of $dM/d\ln t$ to b_2 changes, it is necessary to compute the effect of measured magnetization decay rates for the magnet cross section including the effect of magnet history at each point in the cross section.

CONCLUSION

Creep rate measurements on NbTi and Nb$_3$Sn have shown that the rate $(dM/d\ln t)$: (1) scales with filament diameter (2) is a slowly varying function of H in the critical state (3) rapidly reduces as the field is backed away from the critical state and (4) is fairly insensitive to temperature for $0.3T_c < T < 0.8T_c$. These conclusions are essentially similar to those reported by Beasley et al.[3] Correlation of short sample data to magnet observations have yet to be established.

ACKNOWLEDGEMENTS

The authors would like to thank D. O. Welch for his many discussions on the problem of flux creep.

REFERENCES

1. Y. Yeshuran and A. P. Malozemoff, _Phys. Rev. Letts._, 60: 2202 (1988).
2. Y. B. Kim, C. F. Hempstead and A. R. Strand, _Phys. Rev. Letts_. 9:306 (1962)
3. M. R. Beasley, R. Labusch and W. W. Webb, _Phys. Rev._, 181:682 (1969).
4. C. P. Bean, _Phys. Rev. Letts_. 8:250 (1962).
5. P. W. Anderson, _Phys. Rev. Letts_. 9:309 (1962).
6. D. A. Finley, D. A. Edwards, R. W. Hanft, R. Johnson, A. D. McInturff, and J. Strait, Proceeding of the 1987 IEEE Particle Accel. Conf., Washington, D.C., 151 (1987).
7. R. W. Hanft, B. C. Brown, D. A. Herrup, M. J. Lamm, A. D. McInturff and M. J. Syphers, _IEEE Trans. on Magnetics_ 25: 1647 (1989).
8. D. A. Herrup, M. J. Syphers, D. E. Johnson, R. P. Johnson, A. V. Tollestrup, R. W. Hanft, B. C. Brown, M. J. Lamm, M. Kuchnir and A.D. McInturff, _IEEE Trans. Magnetics_ 25: 1643 (1989).
9. W. S. Gilbert, R. F. Althaus, P. J. Barale, R. W. Benjegerdes, M. A. Green, M. I. Green and R. M. Scanlan, _IEEE Trans. on Magnetics_ 25: 1459 (1989).
10. K. H. Mess and P. Schmüser, DESY report HERA 89-01 (1989).
11. M. Kuchnir and A. V. Tollestrup, _IEEE Trans. on Magnetics_ 25: 1839 (1989).
12. Youwen Xu, M. Suenaga, A. R. Moodenbaugh and D. O. Welch, submitted to _Phys. Rev_. (1989).
13. A. K. Ghosh and W. B. Sampson, _IEEE Trans. on Magnetics_ 24: 1145 (1988).

THEORY OF FLUX PENETRATION EFFECTS BELOW H_{c1} IN

MULTIFILAMENTARY SUPERCONDUCTORS

A. J. Markworth*, E. W. Collings*, J. K. McCoy*, and D. Stroud**

*Metals and Ceramics Department   **Department of Physics
Battelle Memorial Institute   The Ohio State University
Columbus, Ohio 43201   Columbus, Ohio 43210

ABSTRACT

Results are presented of magnetic-field calculations for multifilamentary (MF) superconductors, taking into account the penetration of magnetic flux into individual filaments. The filaments are assumed to be circular in cross section, infinitely long, and arranged parallel to one another on a hexagonal lattice. The externally applied magnetic field is taken to be directed perpendicular to the filament length. The problem is simplified by treating the medium as being effectively uniform in a plane perpendicular to the filament length, and a self-consistent formalism is used to calculate the interfilamentary magnetic field. Flux penetration into the filaments is described using the London theory. Magnetic fields outside and inside the filaments are calculated as functions of filament radius, the London penetration depth, the area fraction covered by filament cross sections, and the shape of the filament bundle.

INTRODUCTION

The magnetic field in an MF superconducting body is affected by several factors, such as the geometry of the filament array, the filament size and interfilamentary separation distance, the magnitude of the London penetration depth, the cross-sectional shape of the filament bundle, and the strength and direction of the externally applied field. In an earlier work,[1] the interfilamentary magnetic field was calculated for the case of a system of infinitely long, parallel filaments, having equal circular cross section, and arranged on a hexagonal lattice. The externally applied magnetic field was taken to be perpendicular to the filament-length direction, and the filaments were assumed to totally exclude the magnetic flux. Effects of filament-bundle shape were included by considering a bundle that was elliptical in cross section.

The assumption of total flux exclusion from the filaments was equivalent to assuming that the magnetic field around each filament did not exceed the value of the lower critical field, H_{c1}, and in addition, that the filament diameter was very large compared to the London penetration depth. However, in cases of practical interest, the filament diameter may be quite small (less than one μm), in which case flux penetration may indeed be significant. In the present work, the same problem as before is examined except that the flux-exclusion assumption is relaxed, the London theory being used to describe the field inside a given

Advances in Cryogenic Engineering (Materials), Vol. 36
Edited by R. P. Reed and F. R. Fickett
Plenum Press, New York, 1990

35

filament. Effects on magnetic properties of varying the geometric factors that define the filament bundle are considered, particularly as related to penetration of magnetic flux into a filament.

CALCULATION OF THE FIELD AROUND A SINGLE FILAMENT IN AN INFINITE ARRAY

A portion of an essentially infinite, hexagonal array of filaments is shown in cross section in Fig. 1, the cross-sectional plane being taken as the xy plane. The host matrix is assumed to be a normal metal having unit permeability. Following our previous approach, as well as that used in other applications,[2] we regard this complex, two-phase system as being effectively uniform in the xy plane and characterized by constant permeability, μ_0, in this plane. An effectively uniform (constant) magnetic field, H_0, arising from external sources, is taken to exist within this region and to be directed parallel to the x axis.

With this picture in mind, we focus our attention on a single filament, of radius a, and examine the details of the magnetic field in its vicinity. (Note that all filaments are equivalent, since we consider the system to be infinite in extent.) As before,[1] we construct a hexagonal Wigner-Seitz cell around the filament, as shown in Fig. 1, and for computational simplicity, replace this cell with a circle of radius b having the same area. The region outside this circle will be treated as the effectively uniform region discussed above, and the magnetic field inside the circle will be calculated on this basis. In this manner, we assess, at least to an approximation, details of the magnetic-field distribution around a given filament while still regarding the overall medium as effectively uniform.

The pertinent regions of space are illustrated further in Fig. 2, as are the polar coordinates, r and θ, used to define position. Region 1 is regarded as effectively uniform; Region 2 contains normal-metal host matrix; and Region 3 consists of the superconducting filament itself.

The magnetic field in Region i ($i = 1$ or 2) can be conveniently described in terms of a magnetic scalar potential, ϕ_i, which is given by

$$\phi_i = \left(\frac{\alpha_i}{r} + \beta_i r\right)\cos\theta \tag{1}$$

where α_i and β_i are constants to be determined by imposition of boundary conditions. The pertinent H field is given by $H_i = -\nabla\phi_i$. In addition, the B fields are given by $B_i = \mu_i H_i$ with $\mu_1 = \mu_0$ and $\mu_2 = 1$.

Clearly, we must set $\beta_1 = -H_0$. Also, self-consistency requires that we set $\alpha_1 = 0$ in keeping with our assumption that Region 1 is effectively uniform. We require that the tangential component of the H field and the normal component of the B field be continuous at $r = b$, which is the boundary between Regions 1 and 2.

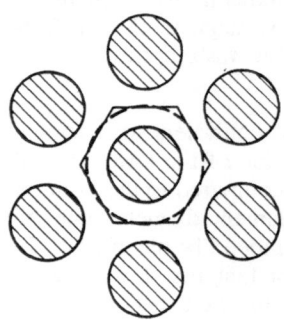

Fig. 1.

Portion of an infinite array of filaments arranged on a hexagonal lattice. The filaments are shown in cross section as shaded circles, with a hexagonal Wigner-Seitz cell constructed around one filament. The dashed circle is concentric with the enclosed filament and has the same area as the Wigner-Seitz cell.

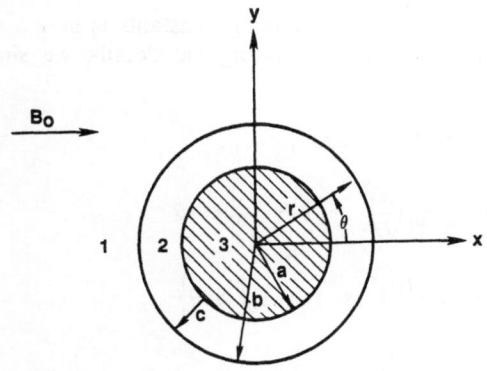

Fig. 2.   Illustration of the division of space into three distinct regions for magnetic-field calculations. Cylindrical polar coordinates used in the calculations are also shown, as is the effectively uniform field, B_0, assumed to exist in Region 1.

Magnetic-field calculations for Region 3 are more difficult because of its superconducting nature. Assuming that the field existing in Region 2 at the filament surface is nowhere larger than H_{c1}, we use the London equation, expressed as follows, for Region 3:

$$\mathbf{\nabla} \times (\mathbf{\nabla} \times \mathbf{B}_3) + \beta^2 \mathbf{B}_3 = 0 \tag{2}$$

where β (which has no relation to the β_i above) is the reciprocal of the London penetration depth. The boundary conditions pertinent to the filament surface ($r = a$) are that the normal and tangential components of the $\mathbf{B}$ field are continuous across this interface.

The field $\mathbf{B}_3$ can be calculated as follows: First, since $\mathbf{\nabla} \cdot \mathbf{B}_3 = 0$, it follows that Eq. 2 can be expressed as

$$\mathbf{\nabla}^2 \mathbf{B}_3 - \beta^2 \mathbf{B}_3 = 0 \quad . \tag{3}$$

We now assume that $\mathbf{B}_3$ can be expressed in the following general form:

$$\mathbf{B}_3 = \mathbf{e}_r u(r) \cos\theta + \mathbf{e}_\theta v(r) \sin\theta \tag{4}$$

where $\mathbf{e}_r$ and $\mathbf{e}_\theta$ are the usual polar-coordinate unit vectors and $u(r)$ and $v(r)$ are functions to be determined. One can substitute Eq. 4 into Eq. 3 and show that

$$\mathbf{B}_3 = \frac{C I_1(\beta r)}{r} \, (\mathbf{e}_r \cos\theta - \mathbf{e}_\theta \xi(\beta r) \sin\theta) \tag{5}$$

with

$$\xi(x) \equiv x \frac{I_0(x)}{I_1(x)} - 1 \tag{6}$$

where C is a constant and I_n is, in general, a modified Bessel function of the first kind of order n.

Determination of values for the remaining constants is now a straightforward matter of applying the boundary conditions. Omitting the details, we simply state the results, namely:

$$\mathbf{B}_1 = B_0(\mathbf{e}_r\cos\theta - \mathbf{e}_\theta\sin\theta) = B_0\mathbf{e}_x \ , \tag{7}$$

$$\mathbf{B}_2 = \frac{B_0}{2\mu_0}\left\{\mathbf{e}_r\left[-(1 - \mu_0)\left(\frac{b}{r}\right)^2 + 1 + \mu_0\right]\cos\theta\right.$$

$$\left. - \mathbf{e}_\theta\left[(1 - \mu_0)\left(\frac{b}{r}\right)^2 + 1 + \mu_0\right]\sin\theta\right\} \ , \tag{8}$$

$$\mu_0 = \frac{1 + \lambda + (1 - \lambda)\xi(\beta a)}{(1 + \lambda)\xi(\beta a) + 1 - \lambda} \ , \tag{9}$$

$$\lambda \equiv (a/b)^2 \ , \tag{10}$$

$$C = \frac{2B_0}{1 + \lambda + (1 - \lambda)\xi(\beta a)} \cdot \frac{a}{I_1(\beta a)} \ . \tag{11}$$

The quantity B_0 is equal to $\mu_0 H_0$. Moreover, λ, defined in Eq. 10, is clearly the area fraction occupied by filament cross sections in the xy plane. In addition, $\mathbf{e}_x$ is a unit vector in the x direction.

It should be emphasized that B_0 is *not* the externally applied field, B_a. The latter is the uniform field taken to exist in space, parallel to the xy plane, at distances far from the filament bundle. The relationship between these two fields can be determined by considering the filament array to be finite in extent, and characterized in the xy plane by the constant, effective permeability, μ_0, as given by Eq. 9. This should be valid as long as the characteristic dimensions of the bundle are large compared to the filament diameter. As an example, we consider a bundle that is elliptical in cross section, with semi-axes d and e parallel to the x and y axes, respectively. The medium outside the bundle is assumed to have unit permeability. Following a procedure similar to that used in Ref. 1, it can easily be shown that

$$B_0 = \frac{1 + \lambda + (1 - \lambda)\xi(\beta a)}{1 + \xi(\beta a) + \lambda\left(\frac{\rho - 1}{\rho + 1}\right)[\xi(\beta a) - 1]}B_a \tag{12}$$

where $\rho \equiv d/e$, and where the fields B_0 and B_a are both parallel to the x axis. Clearly, B_0 is a function of λ, βa, the bundle-shape parameter, ρ, and the magnitude, B_a, of the applied field.

Finally, we calculate one additional quantity of interest: the total magnetic flux, Φ, passing through unit length of one filament. This can be obtained from the expression

$$\Phi = \int_{-\pi/2}^{\pi/2}|\mathbf{B}_{3r}(a,\theta)|a \ d\theta \tag{13}$$

where $\mathbf{B}_{3r}$ is the r-component of $\mathbf{B}_3$. From Eqs. 5, 11, 12, and 13, we obtain

$$\Phi = \frac{4B_a a}{1 + \xi(\beta a) + \lambda\left(\frac{\rho - 1}{\rho + 1}\right)[\xi(\beta a) - 1]} \ . \tag{14}$$

One interesting feature of Eq. 14 is that Φ is independent of λ for a circular filament bundle (for which $\rho = 1$).

DISCUSSION

The effective permeability, μ_0, is seen in Eq. 9 to be a function of both λ and βa. One interesting limiting case is that for which $\beta a \to \infty$, which corresponds to complete flux exclusion. For this case, one can show that $\xi(\beta a) \to \beta a$, and μ_0 becomes dependent only upon λ, as seen from Eq. 9. Moreover, the value that μ_0 approaches is consistent with that obtained previously (Eq. 6 of Ref. 1). At the other extreme is the limiting case for which $\beta a \to 0$, which corresponds to complete flux penetration. For this case, $\xi(\beta a) \to 1$, and we find, as expected, that $\mu_0 \to 1$.

The variation of μ_0 with λ, as predicted by Eq. 9, is illustrated in Fig. 3 for selected values of βa. We make several observations: First, the tendency for μ_0 to decrease with increasing λ can be seen, although this tendency becomes less pronounced as βa becomes smaller. Second, effects of flux penetration become quite important as βa decreases (for a fixed value of λ), with μ_0 increasing toward unity as βa decreases toward zero. Finally, it should be noted that our model actually does not apply all the way to $\lambda = 1$; for example, nearest-neighbor circles arranged on a hexagonal array would touch at a value of $\pi/\sqrt{12}$ ($\simeq 0.907$) for the fractional area covered.

Flux-penetration effects are illustrated graphically in Fig. 4, in which lines of magnetic flux around a filament are plotted for selected examples. The uniform field seen to exist in Region 1 is a consequence of our "uniform-medium" approximation and is therefore not a physical representation of the actual field.

In Fig. 4, the dependence of flux penetration upon λ and β is demonstrated, the value of b being the same for all the cases illustrated. For example, the tendency for flux exclusion from the superconductor to occur is quite small for the examples for which $\beta = 1.5/b$. However, for the larger value of β (i.e., smaller penetration distance), flux exclusion is much more pronounced and its dependence upon λ can be seen as well. It should be noted, however, that the flux density in Region 1 is itself related in a complex manner to that which is incident upon the filament bundle, as seen from Eq. 12.

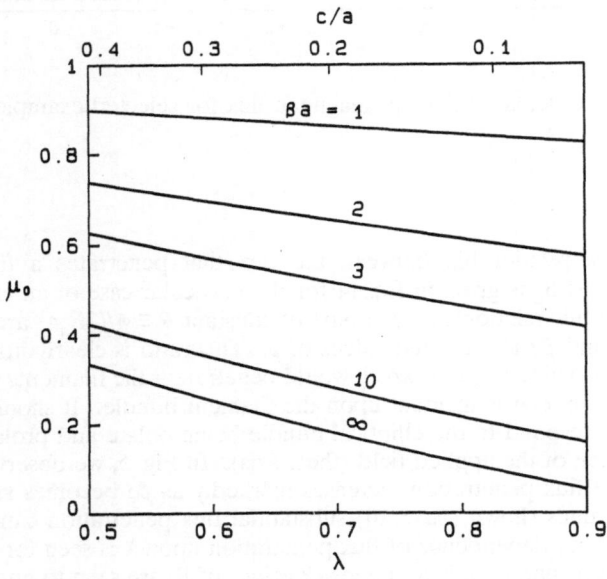

Fig. 3. Variation of μ_0 with λ for selected values of βa.

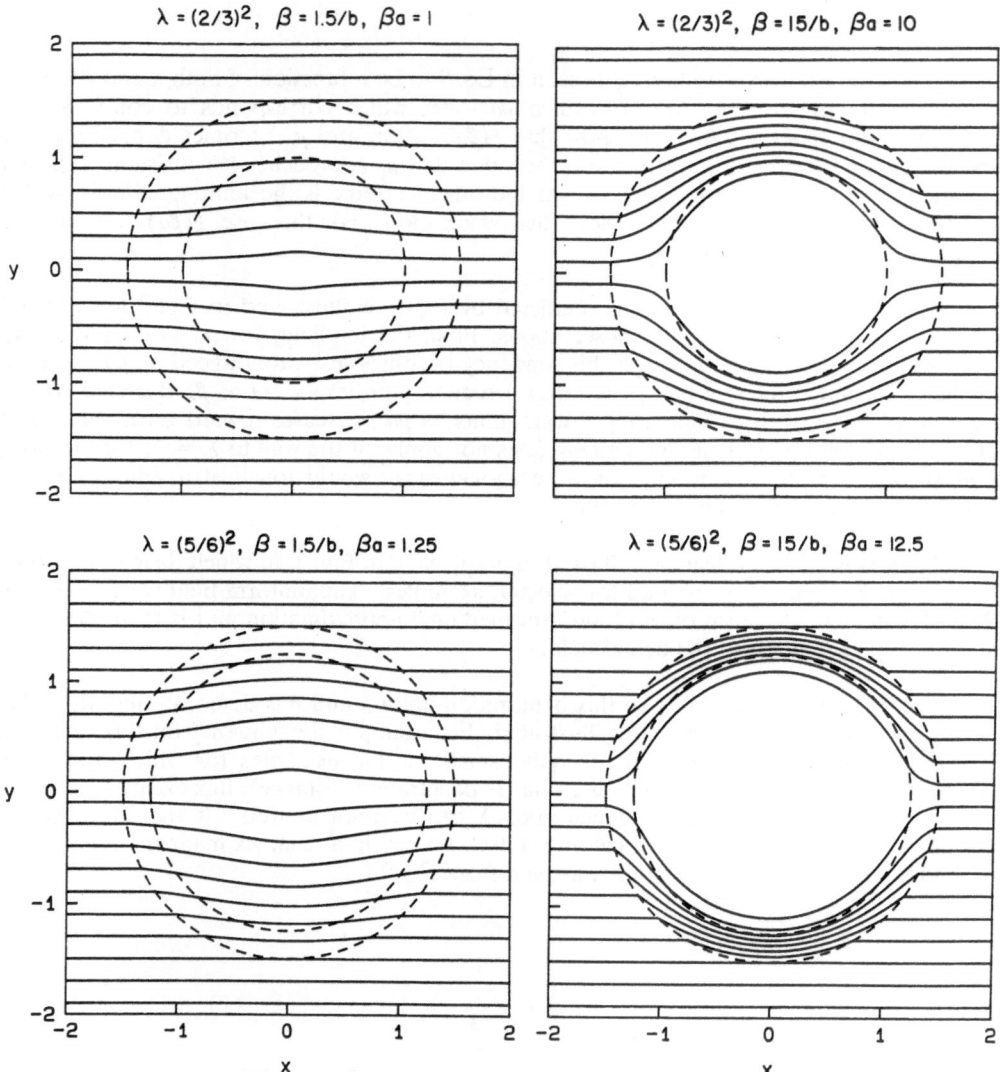

Fig. 4.   Calculated lines of magnetic flux for selected examples.

The *quantitative* relationship between the flux that penetrates a filament and the parameters βa, λ, ρ and B_a is given in Eq. 14 for the particular case of an elliptical filament bundle. To illustrate this relationship, contours of constant $R \equiv \Phi/(2B_a a)$ are plotted in Fig. 5 as a function of λ and βa for selected values of ρ. This ratio is clearly just the actual flux penetrating a filament divided by that which would penetrate if the filaments were completely "transparent" to the flux that is incident upon the filament bundle. It should be noted that $\rho < 1$ and $\rho > 1$ correspond to the elliptical bundle being oblate and prolate, respectively, relative to the direction of the applied field (the x axis). In Fig. 5, we observe the following: The relative extent of flux penetration increases markedly as βa becomes smaller, although even for the larger values shown ($\beta a \simeq 10$) substantial flux penetration can still occur. As expected from Eq. 14, no dependence of flux penetration upon λ is seen for $\rho = 1$. Finally, the contours themselves, particularly at the lower values of R, are seen to undergo significant variation as the value of ρ is changed.

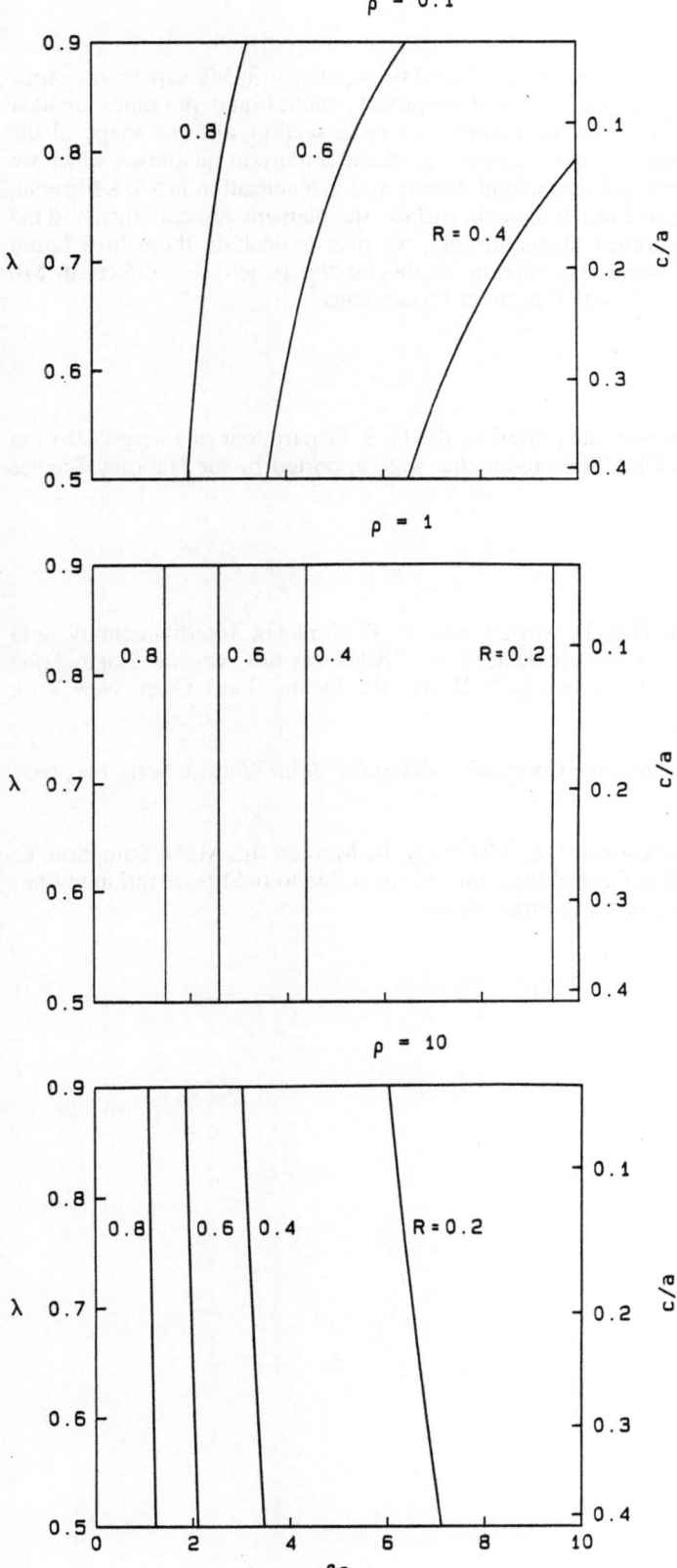

Fig. 5.

Contours of constant
$R = \Phi/(2B_a a)$ plotted
as a function of λ and
βa for selected values
of ρ.

CONCLUSION

We have shown that flux penetration can indeed be significant in MF superconductors, the factors of importance being the product βa of reciprocal penetration depth times filament radius, the area fraction λ covered by the filaments in cross section, and the shape of the filament bundle, exemplified here by the parameter ρ. Some important quantities, which we have not considered in this work, are the current density and magnetization inside a filament, the field enhancement in Region 2 at the filament surface, the filament susceptibility, and the magnetic-energy density and critical magnetic field; we plan to include these in a future study. Finally, we note that some experimental studies of flux-penetration effects in MF superconductors are described elsewhere in these Proceedings.[3]

ACKNOWLEDGEMENTS

The Battelle researchers were supported by the U. S. Department of Energy, Division of High-Energy Physics. The Ohio State researcher was supported by the National Science Foundation.

REFERENCES

1. A. J. Markworth, P. M. Hui, D. Stroud, and E. W. Collings, Interfilamentary field enhancement below H_{c1} in MF conductors, *in*: "Advances in Cryogenic Engineering Materials," Vol. 34, A. F. Clark and R. P. Reed, eds., Plenum Publ. Corp., New York (1988), p. 523.

2. R. M. Christensen, "Mechanics of Composite Materials," John Wiley & Sons, Inc., New York (1979).

3. E. W. Collings, A. J. Markworth, J. K. McCoy, K. R. Marken, Jr., M. D. Sumption, E. Gregory, and T. S. Kreilick, Critical field enhancement due to field penetration in fine-filament superconductors, *in* these Proceedings.

INTEGRITY TESTS FOR HIGH-T_c AND CONVENTIONAL

CRITICAL-CURRENT MEASUREMENT SYSTEMS[*]

L. F. Goodrich and S. L. Bray

Electromagnetic Technology Division
National Institute of Standards and Technology
Boulder, Colorado 80303

ABSTRACT

Critical-current measurement systems must be extremely sensitive to the small differential voltage that is present across the test specimen as it changes from the zero resistance state to the flux-flow resistance state. Consequently, these measurement systems are also sensitive to interfering voltages. Such voltages can be caused by ground loops and by common mode voltages. Specific methods for testing the sensitivity of critical-current measurement systems and for detecting the presence of interfering voltages are discussed. These include a simple procedure that simulates the zero resistance state and the use of an electronic circuit that simulates the flux-flow resistance state.

INTRODUCTION

The determination of a superconductor's critical current (I_c) requires the measurement of extremely low voltages,[1] on the order of 1 μV. Consequently, the I_c measurement system must be quite sensitive to the resistive voltage that appears as the test specimen changes from the superconducting to the normal state, and be insensitive to other sources of voltage that might otherwise corrupt the measurement. The I_c measurement system is susceptible to sources of interfering voltage that might be negligible for many other measurements. Ground loop and common mode voltages[2] are prime examples of these sources of interference.

Because of the nebulous character of these voltages, it is often difficult to predict, based simply on the design of the measurement system, whether or not the system is prone to these problems. However, some practical test methods that are useful for checking the sensitivity and accuracy of the measurement system and for detecting the presence of interfering voltages have been developed. The test methods do not directly indicate the sources of problems; they simply indicate their presence. Consequently, the solution of these measurement problems depends on knowledge of their likely sources. As a diagnostic tool, these tests are best used for evaluating the success of specific changes that are intended to alleviate these problems.

Advances in Cryogenic Engineering (Materials), Vol. 36
Edited by R. P. Reed and F. R. Fickett
Plenum Press, New York, 1990

43

There are two general test methods, the zero resistance test and the finite resistance test. The zero resistance test is used to detect the presence of interfering voltages, and the finite resistance test is used to evaluate the sensitivity and accuracy of the measurement system. Both tests can be conducted at either room or cryogenic temperature. The room temperature tests are usually more convenient, but less definitive because they do not depict the actual measurement conditions as precisely as the cryogenic tests.

Finally, the development of an I_c measurement system is complicated by the expensive and volatile nature of the liquid cryogen. This is particularly true when the cryogen is helium and the measurement system is computer controlled. During the first I_c measurements for a new or modified measurement system, large quantities of liquid helium are often expended during the inevitable debugging process. To address this problem, a simple electronic circuit that simulates the voltage-current (V-I) characteristic of a superconductor has been developed and tested. The simulator is an effective tool for developing the measurement system to a high level, before expending liquid helium. Also, for complex I_c measurement systems, the simulator has proven useful as part of a pre-operation check. In this way, problems can be detected and corrected prior to transferring liquid cryogen from the storage Dewar to the measurement system cryostat.

TEST METHODS

Finite Resistance Test

The finite resistance test is simply a four-wire resistance measurement where the superconductor specimen is replaced in the measurement system by an appropriate copper conductor. The idea is to measure a specimen that has a known resistance to assess the accuracy and sensitivity of the measurement system. In order to closely approximate the actual measurement conditions, it is important for the copper specimen's resistance, over the length that is spanned by the voltage taps, to be similar to that of the superconductor at its critical current. This allows testing of the measurement system at an appropriate voltage and current.

Zero Resistance Test

Interfering voltages are often difficult to detect because they are not always easily distinguished from actual specimen voltages. For example, the interfering voltages can have the character of a current-transfer voltage[3] or even a flux-flow voltage. This is particularly true of the high-transition-temperature (high-T_c) superconductors because their V-I characteristics are not as well understood as those of the conventional, or low-T_c, materials. The zero resistance test effectively simulates an ideal superconductor where the V-I characteristic is $V(I) = 0$.

Ideally, the test is carried out with the measurement system configured exactly as it would be for an I_c measurement, with one exception: the voltage tap leads are not both connected to the superconductor specimen. Rather, one of the leads is connected to the specimen and the second lead is connected to the first lead close to, but not in direct contact with, the specimen. This forms a null voltage tap pair (see Fig. 1). There are situations where the null voltage tap pair should have enhanced inductance to simulate the inductance of the differential voltage tap pair.[4] In this configuration the measured voltage should be equal to zero regardless of the current carried by the specimen. Any voltage that is detected, as the current is cycled, is an interfering voltage. The important point is that all of the electrical current paths that are present for an I_c measurement are preserved in this configuration and the common mode voltage applied to the input terminals of the voltmeter is also the same. The only difference is that the differential voltage has been eliminated to allow easy recognition of any interfering voltages.

If the current is increased from zero to some maximum value and then decreased back to zero with a linear current ramp, equal and opposite inductive voltages will be generated for increasing and decreasing current. These voltages are easily recognized because they are of constant magnitude. In the case of a continuous data acquisition system, where the specimen voltage is recorded while the current is being continuously increased, the inductive voltages can be reduced by tightly twisting the voltage leads, but they cannot be eliminated. However, the inductive voltage is just a dc offset and, thus, it does not affect the determination of the I_c. For discrete measurement systems, where the voltage data are acquired with the current held constant at selected set points, inductive voltages are not a factor. Like inductive voltages, thermal electric voltages are always present and, if they are held constant during the data acquisition cycle, do not affect the I_c measurement.

The interfering voltages that present a problem for I_c measurements are those that change with changing current. These are the ground loop and common mode voltages. The zero resistance test is effective for detecting the presence of these voltages.

A more convenient, but less definitive, zero resistance test can be made at room temperature using a copper conductor, as in the finite resistance test. Once again, the voltage tap leads must be connected together and then connected to a single tap to eliminate the differential voltage. It is still important to retain as much of the actual I_c measurement system as possible. For example, all instruments that are used during I_c measurements (chart recorders, computers, and so on) should be connected for the zero resistance test. These peripheral devices often have ground connections and are sometimes the source of interfering voltages. For some measurement systems, the test fixture and specimen are not electrically isolated from ground. When this is the case, the room temperature test should be made with the copper conductor connected to ground through an appropriate resistance.

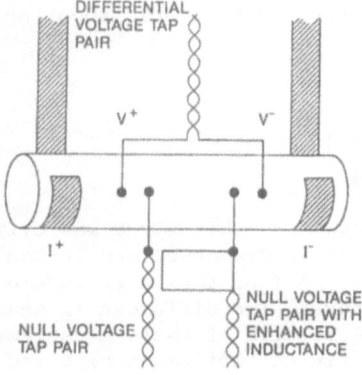

Fig. 1.   Illustration of sample and lead configuration for zero and finite resistance tests.

Superconductor Simulator

A simple electronic circuit was designed to simulate the intrinsic V-I characteristic of a superconductor. This circuit was used to characterize the response of a nanovoltmeter when subjected to a highly asymmetric periodic voltage that results from passing a dc-biased ac current through a superconductor.[4] A more general application of this circuit has been to aid in the development and testing of critical-current data acquisition systems.

Another important parameter in the determination of I_c is the measurement of the n value. The parameter n is defined by the approximate intrinsic voltage-current (V-I) relationship,

$$V = V_0(I/I_0)^n,$$

where I_0 is a reference I_c at a voltage criterion V_0, V is the sample voltage, I is the sample current, and n reflects the shape of the curve with typical values from 20 to 60. A higher number means a sharper transition. In the measurement of the V-I characteristic, the sensitivity of the voltmeter is the key factor. Voltage accuracy is less significant in the determination of the I_c for a sample with a high n value. For example, with n = 30, a voltage error of 10% translates into a 0.3% current error.

The details of the circuit design are given in Ref. 4. The input to the circuit comes from a shunt resistor connected in series with the current supply. The output current from the simulator passes through two shunt resistors, a "high output" and "low output." Typically, the nanovoltmeter being tested is connected to the low output resistor, which generates a signal in the microvolt range, and a recording instrument is connected to the high output resistor, which generates a signal that is 10^4 times as large as the low output signal. Another channel of the recording instrument is connected to the analog output of the nanovoltmeter. These two channels are then compared and the measurement system may be thus characterized under conditions similar to an I_c measurement. The simulated values of the I_c and n can be adjusted.

The simulator does not reproduce all of the possible elements of an actual superconductor's V-I characteristic. Current-transfer voltages and the complex voltage patterns associated with flux dynamics, for example, are not produced by the circuit. However, the simulator does produce the predominant element of a superconductor's V-I characteristic, its abrupt increase in voltage with increasing current. It also produces a very low and well defined output voltage. These two capabilities make the simulator very useful for the development and trouble shooting of I_c measurement systems.

EXAMPLES OF TEST RESULTS

Measurement System Details

The copper specimen used for this study was cylindrical and measured 14.6 cm in length and 8.9 cm in diameter, and it had a voltage tap separation of about 0.1 cm. All of the tests made using this specimen were conducted at room temperature. Two different I_c measurement systems were used for these tests. The details of these measurement systems are given elsewhere.[5] For both the finite resistance test and the zero resistance test, the current was steadily increased from zero to a maximum level and then steadily reduced to zero while the voltage was recorded with a digital processing oscilloscope.

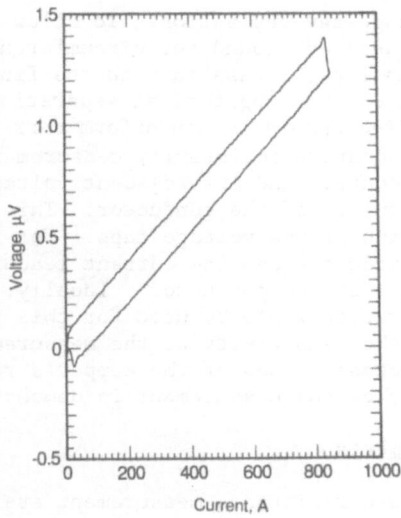

Fig. 2. Finite resistance test, voltage
versus current for a copper sample.

Finite Resistance Measurements

Figure 2 shows the results of a finite resistance test. The hysteresis
in the data is due to inductive voltage. The lower portion of the
hysteresis loop is for increasing current and the upper portion is for
decreasing current. The measured resistance is approximately 1.4 nΩ. Based
on the diameter of the copper conductor (8.9 cm) and the voltage tap
separation (0.1 cm), this implies a copper resistivity of 0.9 $\mu\Omega$-cm.

The actual resistivity of the copper is probably closer to 1.7 $\mu\Omega$-cm.
The discrepancy is probably due, primarily, to the lack of precision in the

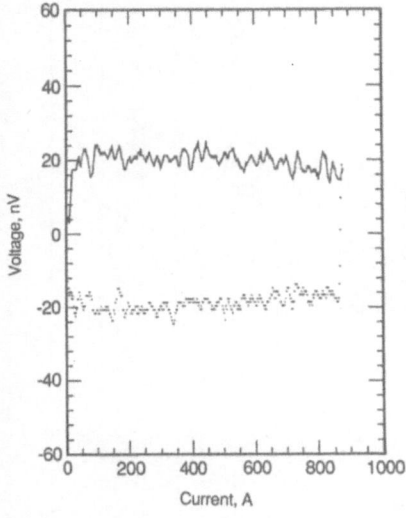

Fig. 3. Successful zero resistance test,
voltage versus current for a copper
sample.

measurement of the voltage tap separation. To allow space for soldering, the voltage taps are staggered around the circumference of the conductor. Given the relatively small tap separation and the finite size of the taps, an accurate measurement of the longitudinal separation is difficult. Another source of the discrepancy is nonuniform current distribution within the conductor. For an accurate resistivity measurement, the separation between either current contact and its adjacent voltage tap should be at least five times the diameter of the conductor. This allows uniform current distribution in the region of the voltage taps. For the conductor used in these tests, the separation between the current leads and voltage tap leads is less than the diameter of the conductor. Ideally, a longer conductor with a greater tap separation would be used for this test. Nonetheless, this test demonstrates the sensitivity of the measurement and, to the extent that the measured and actual values of the copper's resistivity are in the same range, the accuracy of the measurement is demonstrated.

Zero Resistance Measurements

The results of a zero resistance measurement are shown in Fig. 3. This test was made using the same measurement system that was used for the finite resistance test. The voltage scale for this plot is nanovolts. Again the hysteresis is caused by inductive voltage. The continuous curve (upper portion of the loop) is for increasing current and the discrete data are for decreasing current. The important point is that, if the inductive voltage is subtracted from the data, the measured dc voltage is essentially equal to zero. This measurement system has a voltage noise of about ±5 nV and a voltage measurement uncertainty of about ±2 nV ±2% of the signal.

In contrast, Fig. 4 shows the results of a zero resistance test for another measurement system. In this case, the voltage scale is in microvolts and, even if the inductive voltage is subtracted from the data, the measured voltage is not equal to zero. This is an example of an interfering voltage. In an actual I_c measurement, the interfering voltage would be detected along with any differential voltage, thus altering the I_c

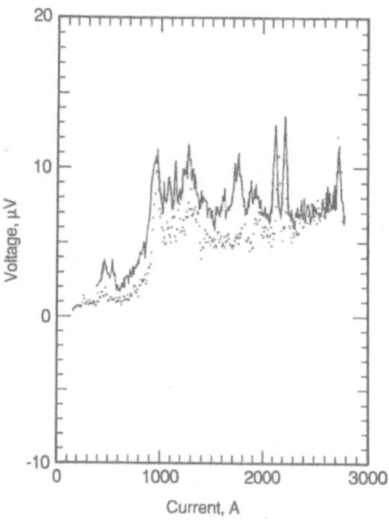

Fig. 4. Unsuccessful zero resistance test, voltage versus current for a copper sample.

measurement. In fact, the abrupt increase in voltage that occurs at approximately 900 A might be mistaken for a superconducting transition. The features of the increasing current (continuous curve) are reproduced for the decreasing current (discrete points).

DISCUSSION

Other examples of integrity tests can be found in Ref. 5, where various combinations of voltmeters, power supplies, and load grounding conditions are given. These combinations can change the results significantly; a voltmeter that works well with one current supply may not work with another. In general, if the load can be grounded near the test sample, the level of interfering voltages can be reduced. The resistance of the voltage tap leads can also be a factor; the higher the lead resistance is, the larger the interfering voltage.

The required cross sectional area of the copper test specimen used in the finite resistance test depends on the I_c of the superconductor, the selected I_c criterion, and the test temperature (room or cryogenic). For high-current systems the required size of the copper specimen can become impractical. However, low-current systems may require only a copper test specimen that is comparable in cross sectional area to that of the superconductor. The length of the copper specimen is also important to ensure an accurate measurement of its resistivity. It should be long enough, in comparison with its cross sectional area, to ensure uniform current distribution in the area of the voltage taps.

CONCLUSIONS

A set of simple procedures that will test the integrity of measurement systems used for critical-current determinations on high-T_c and conventional superconductors has been developed. These tests include a finite resistance, a zero resistance, and a superconductor voltage-current simulator. In the measurement of the critical current, voltage sensitivity is a key factor. The zero resistance test is the most effective test to detect the presence of interfering voltages such as ground loop or common-mode voltages and will determines the voltage sensitivity limit of a measurement system.

ACKNOWLEDGMENTS

The authors extend their thanks to W. P. Dube' for the idea of a zero resistance test, and to R. L. Spomer for drafting.

This work was supported by the Department of Energy, Office of Fusion Energy and Division of High Energy Physics.

REFERENCES

1. L. F. Goodrich and F. R. Fickett, Critical current measurements: a compendium of experimental results Cryogenics 22:225 (1982).
2. Ralph Morrison, "Grounding and Shielding Techniques in Instrumentation," John Wiley & Sons, New York (1977).
3. J. W. Ekin, Current transfer in multifilamentary superconductors. I. theory J. Appl. Phys. 49:3406 (1978).
4. L. F. Goodrich and S. L. Bray, Current ripple effect on superconductive d.c. critical current measurements Cryogenics 28:737 (1988).

5. L. F. Goodrich, S. L. Bray, W. P. Dube', E. S. Pittman, and A. F. Clark, "Development of Standards for Superconductors, Interim Report January-December 1985," NBSIR 87-3066, National Bureau of Standards, Boulder, Colorado (1987).

CONVERSION OF A 11 MN EXTRUSION PRESS FOR HYDROSTATIC EXTRUSION OF

SUPERCONDUCTING MATERIALS

T.S. Kreilick
Supercon Inc., Shrewsbury, MA

R.J. Fiorentino, E.G. Smith, Jr., W.W. Sunderland
Battelle Columbus Division, Columbus, OH

ABSTRACT

The purpose of this program was to demonstrate the feasibility of modifying an existing mid-size conventional extrusion press to permit, with a simple tooling change, hydrostatic extrusion of superconducting materials. Hydrostatic extrusion is considered by many the ideal method for low-temperature processing of superconductors in order to (a) maximize current densities and (b) successfully coextrude widely dissimilar materials that are typically within multifilament superconductor billets. Successful conversion of a mid-size press could then ultimately lead to a similar modification of a much larger existing production extrusion press. This would provide the U.S. with a production capability at a much lower cost than would otherwise be possible. Battelle Columbus Division (BCD), as a sub-contractor to Supercon Inc., undertook the task of converting a 11 MN press located at Battelle's Pacific Northwest Division (BPND). The converted press was designed to accommodate billets up to 94 mm in diameter x 559 mm long. Also, the converted tooling was designed to operate at extrusion pressures up to 1517 MPa. Feasibility of the converted press was demonstrated by successfully extruding Cu/NbTi, Al/NbTi, and Cu/Nb/Sn billets under conditions comparable to those used for hydrostatic extrusion of the same materials in a 39.1 MN production hydrostatic extrusion press. In addition, enhanced critical current densities in Cu/NbTi composites are shown for hydrostatically extruded materials.

INTRODUCTION

This research was initiated for several reasons:

(1) There was a growing recognition of the need for hydrostatic extrusion as the ideal method for low-temperature processing of superconductors in order to (a) maximize critical current densities, (b) successfully coextrude widely dissimilar materials such as aluminum-stabilized superconductors, and (c) to reduce intermetallic compound formation.[1]

(2) Although the hydrostatic extrusion process was actually being used in production in Japan[2-4] and Europe[5] on up to 39.1 MN production hydrostatic extrusion presses, there were only small laboratory presses in the United States.

The lack of comparable production hydrostatic extrusion facilities in the U.S. was due to two main reasons:

(1) The size of the superconductor market, especially prior to the proposed Superconducting SuperCollider (SSC) project, was never considered large enough to justify

Advances in Cryogenic Engineering (Materials), Vol. 36
Edited by R. P. Reed and F. R. Fickett
Plenum Press, New York, 1990

the purchase of a costly, new hydrostatic-extrusion-press facility by any U.S. company, particularly those companies engaged in toll-extrusion work of superconductors.

(2) A new hydrostatic extrusion press, as designed in the 1970's, was made to be used for hydrostatic extrusion exclusively. Since the hydrostatic extrusion process is neither required nor cost-effective for extrusion of many commodity materials and products, it does not make economic sense to invest in a costly new press that cannot be utilized sufficiently to obtain a reasonable return-on-investment.

It appeared that there was only one viable alternative to this technical/economic dilemma, i.e., conversion of an appropriate existing conventional press to permit both hydrostatic and conventional extrusion by a simple and quick change in tooling. Such a tooling change would involve mainly the container and ram, both of which are now changed routinely for conventional extrusion of billets of various sizes. This increased extrusion capability would expand the use of an existing press, thus making it far easier to justify the investment compared to a purchase of a new press to be used exclusively for hydrostatic extrusion.

A careful analysis of the situation indicated the following important reasons for conversion of a mid-size press before undertaking the conversion of a large production press:

(1) It was considered essential that the technical/economic feasibility of conversion be first demonstrated on a mid-size press before a private company and/or the U.S. Government could be convinced to fund the conversion of a large production press.

(2) The opportunity to test the conversion tooling on a mid-size press may lead to important design improvements that could be incorporated later in the design for the conversion of a production press.

The expertise for conversion of existing presses for hydrostatic extrusion resided at Battelle Columbus Division (BCD). This expertise was based on BCD's extensive experience, being at the forefront of hydrostatic extrusion technology in the U.S. since its early stages of development.[6-9]

PRESS CONVERSION

The specific mid-size press selected for conversion was a 11.1 MN horizontal Loewy extrusion press (see Figure 1) owned by the Department of Energy and located at Battelle's Pacific Northwest Division (BPND). Conversion of this conventional press for hydrostatic extrusion required the replacement of three major components:

- container
- stem or ram
- die.

The new container was designed to withstand fluid pressures up to 1517 MPa (220 ksi). It consists of four press-fitted rings with the required amount of interference to permit operation at the target maximum pressure. The bore surface of the liner was ground and honed to the high finish required for sealing the hydrostatic media at high pressures. The outer ring was designed so that the overall container could be inserted and keyed in place inside the standard container housing of the press.

The replacement stem was also made to withstand the target maximum pressure of 1517 MPa. The front end of the stem was designed with a stem cap to hold the seals for dynamic sealing of the hydrostatic fluid during the extrusion stroke. The seals consisted of a standard O-ring/miter ring combination commonly used in hydrostatic extrusion.

The basic extrusion die was designed to fit inside the liner bore of the container. The seal between the liner bore and the die can be achieved by several methods, including the standard O-ring/miter ring approach. However, Battelle chose its own proprietary method of sealing by means of a press-fit between the die and liner surface.

Figure 1 View of 11.1 MN conventional extrusion press proposed for conversion to permit hydrostatic extrusion.

Figure 2 Die holder, container and ram fabricated for the press conversion.

The container bore is sized to accommodate billets up to 94 mm (3.7 in) in diameter and 559 mm (22 in) long. This represents a billet length-to-diameter (l/d) ratio of almost 6:1, an amount much greater than that considered typical for conventional extrusion.

Figure 2 shows the die holder, container and ram fabricated for the press conversion. Figure 3 is a view of the opposite end face of the container and the four press-fitted rings are clearly evident. Figure 4 shows one method of inserting a billet into the container. In this case, the billet is inserted at the "die-side" or "muzzle-end" of the container. Once the billet is positioned between the container and die, the container is shifted over the billet and sealed tightly against the die. Fluid can then be injected into the liner bore through the collar on the "stem-side" of the container. If required, the billet can also be loaded on the "breech-end" or "stem-side" of the container.

HYDROSTATIC EXTRUSION TRIALS ON THE CONVERTED PRESS

Four composite superconductor billets were fabricated by Supercon for extrusion on the converted horizontal press. Three billets contained NbTi filaments in either copper or aluminum matrices. Filament hardness differed in the two copper-matrix billets (BPND-1 and BPND-2) because of preprocessing differences. The fourth billet was a Cu-Nb powder metallurgy (P/M) mixture with Sn reservoirs, all within a copper matrix. This discussion will concern only the copper matrix NbTi composites.

Figure 3 View of the "die-side" of the container and its four press-fitted rings.

Figure 4 View of billet loaded between the die and container.

Copper clad NbTi monofilament, incorporating a Nb diffusion barrier, was drawn to a diameter of 3.3 mm. One half of the material was annealed in vacuum for 1.5 hours at 800 °C. The other half of the material was heat treated for 40 hours at 375 °C in an argon atmosphere. All of the material was then drawn and shaped into a hexagonal cross-section 1.16 mm (0.0455 in) flat-to-flat. The "hexed" monofilament was straightened and cut to equal lengths for restacking as multifilamentary composites. The designed values of filament spacing to filament diameter (S/D) and copper to superconductor ratio are 0.15 and 1.54:1, respectively, for these composites. At a final wire diameter of 0.81 mm (0.0318 in) the filaments are 8.6 μm in diameter.

The diameter of the CDA alloy 101 copper cans used in the multifilament assembly were 94 mm (3.7 in). The total length of an assembled billet was 560 mm (22 in). A total of 3450 filaments were assembled in each of two billets. Billet #BPND-1 incorporates monofilaments that were annealed (800 °C) prior to assembly. Billet #BPND-2 incorporates monofilaments that were heat treated (375 °C) prior to assembly. Both billets were hot isostatically pressed (593 °C, 103 MPa, 2 hours) prior to extrusion. The second billet (#BPND-2) was heat treated once more (40 hours at 375 °C) prior to hydrostatic extrusion in the converted Battelle-Northwest press.

Both billets were pre-heated at 200 °C and extruded at a ratio of 18.5:1 yielding a product 21.8 mm in diameter. After loading the billet and fluid, the stem was advanced into the container until the fluid surrounded the billet and began to pressurize. The stem was then advanced in an automatic mode at a pre-set speed (6.1 mm/sec.), further pressurizing the fluid and extruding the billet. The extrusion stroke was continued for a given ram displacement where the stem was stopped and withdrawn from the container. The container was next moved off the die, exposing the unextruded butt. The butt was pulled away from the die and a rotary saw used to separate it from the extruded product. A continuous record of ram pressure and ram speed as a function of ram displacement was made for each trial. Differences in pressure levels recorded for billets BPND-1 and BPND-2, although not great, reflect the relative hardness of the NbTi filaments present. Breakthrough pressures for billets BPND-1 and BPND-2 were 1103 and 1241 MPa, respectively. These values are comfortably within the 1517 MPa capacity of the converted tooling.

Samples of the extruded Cu/NbTi stock from both billets were examined metallographically to verify that satisfactory coextrusion of the NbTi filaments and copper matrix took place. A cross-section macrograph of the stock from billet BPND-1 is shown in Figure 5. The filaments in this stock, as well as the BPND-2 product, were uniform in size, indicating they had fully coextruded with the copper matrix.

COMPARISON OF HYDROSTATICALLY AND CONVENTIONALLY
EXTRUDED Cu/NbTi MATERIALS

The strand diameter required for the inner cable of an Superconducting SuperCollider (SSC) dipole magnet is 0.81 mm. If one starts with a 305 mm diameter billet, the total strain to which the material is subjected is approximately 12. If the first precipitation heat treatment is delayed until a strain of 6, and the strain after the last heat treatment is 4, then there there is room for only three heat treatments.[10,11] More recently, Lee has shown that the pre-strain can be reduced to 4 if high homogeneity Nb 46.5 wt% Ti alloy is used with a 420 °C heat treatment.[12] Reduction of the pre-strain to lower values could form Widmanstätten α-Ti and/or ω phase precipitates which could lead to ductility problems.

An available strain of 12 assumes that relatively hot extrusion has the same strain effect as cold drawing. If cold work is not retained throughout conventional extrusion than the total strain available is significantly reduced from the calculations referenced above. It is the belief of the authors that some, but, not all of the cold work present at the time of the extrusion is retained. Experiments designed to help quantify the retained cold work are described below.

Lowering the temperature of conventional hot extrusion has long been identified as a possible means of increasing the retained cold work, but, lower temperatures must be compensated for in higher available press tonnages or smaller diameter billets. Insufficient bonding of the various components is one drawback to this approach. Supercon Inc. has introduced a hot isostatic pressing (HIP) step to enable bonding prior to extrusion at lower temperatures. In this way, one is able to retain the cold working gained during the extrusion process. If hydrostatic extrusion can be employed in both the monofilament extrusion and the multifilament extrusion the available strain space can then be increased to 18 or more.

It was with this knowledge base that two hydrostatically extruded Cu/NbTi composites were compared with two identical composites extruded utilizing conventional hot extrusion techniques.[13] The purpose for these billets was to determine the advantages of hydrostatic extrusion on the largest available press. This experiment was carried out concurrently with the mid-sized press conversion described above and timed to identify any potential problem areas prior to our trials on the converted press.

The diameter of each multifilament billet was 168 mm (6.61 in) with a nominal copper to superconductor ratio of 1.8:1. All four composites contain 516 NbTi filaments, each incorporating a Nb diffusion barrier and clad with copper designed to yield an average S/D ratio of 0.15:1.

The material for these billets was processed in a manner similar to that processed for billets BPND-1 and BPND-2, described above. One billet from each series contains filaments that received an anneal of the monofilament at an intermediate diameter prior to further drawing and shaping to a hexagonal cross-section. Annealing parameters were 800 °C for 1.5 hours. The second billet of each series contains filaments that received an α-Ti precipitation heat treatment (40 hrs at 375 °C) rather than an anneal at the same intermediate size followed by a second heat treatment (40 hrs at 375 °C) as a hexed monofilament. All four billets were hot isostatically pressed (593 °C, 103 MPa, 4 hrs) prior to extrusion to ensure bonding of all billet components.

The billets are identified as follows:

Billet Number	Method of Extrusion	Condition of NbTi	# of Heat Treatments
SCN-1	conventional	annealed	3 or 4
SCN-2	conventional	heat treated	5 or 6
SCN-3	hydrostatic	annealed	3 or 4
SCN-4	hydrostatic	heat treated	5 or 6

Conventional extrusion of billets SCN-1 and SCN-2 took place after a pre-heat at 593 °C. The extrusion ratio was 25.5:1 yielding a product 33.3 mm in diameter. Hydrostatically extruded billets SCN-3 and SCN-4 were processed to the same diameter following a pre-heat at 200 °C. Figure 6 shows a cross-section of the hydrostatically extruded billet processed with an anneal of the monofilament (Billet #SCN-3). This cross-section is representative of all four billets.

Figure 7 is a flow-chart of the thermomechanical processing sequence for the material that received an anneal (800 °C) of the monofilament prior to assembly as a multifilament (i.e. billets SCN-1 and SCN-3). Figure 8 is a flow-chart outlining the processing sequence for billets SCN-2 and SCN-4, those materials that received a heat treatment (375 °C) prior to assembly as a multifilament composite. For all four billets, when the total number of heat treatments is specified as "four" or "six" this is an indication that the material received a precipitation heat treatment at the "as extruded" diameter of 33.3 mm. All of the intermediate precipitation heat treatments were 40 hours at 375 °C and were spaced equally by a nominal strain of 1.0.

Figure 9 shows the critical current density (Jc) at 5T and 4.2 K versus the strain after the last intermediate precipitation heat treatment for the conventionally extruded materials. Figure 10 shows the same information for the hydrostatically extruded materials. "HTA" refers to "heat treatments after" the extrusion and "HTB" refers to "heat treatments before" the extrusion.

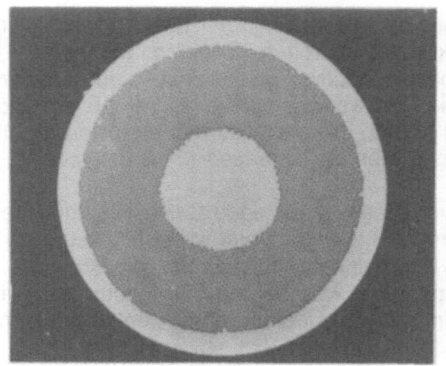

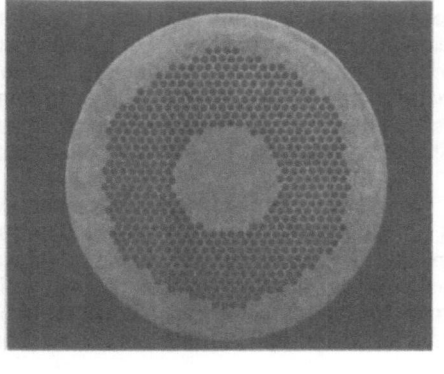

Figure 5 Cross-section of hydrostatically extruded Cu/NbTi multifilamentary composite (BPND-1).

Figure 6 Cross-section of hydrostatically extruded Cu/NbTi multifilamentary composite (SCN-3).

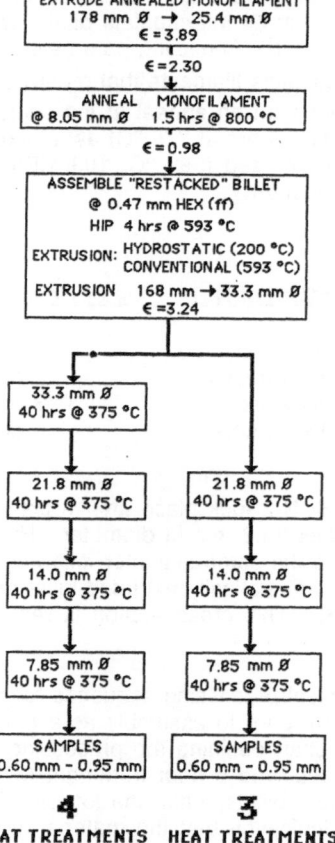

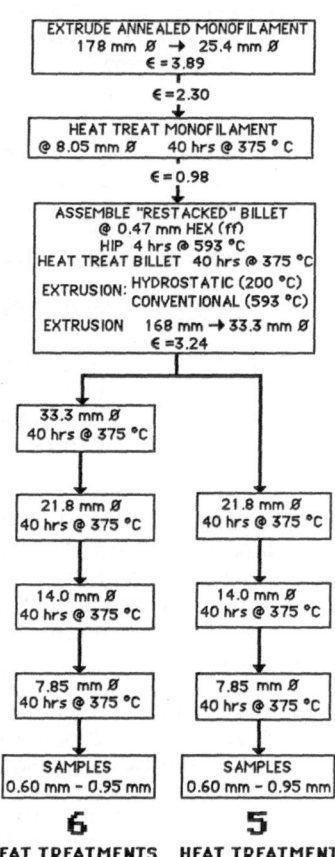

Figure 7 Flowchart of the thermo-mechanical processing for billets SCN-1 and SCN-3, both annealed at 800 °C prior to assembly as multi-filament composites.

Figure 8 Flowchart of the thermo-mechanical processing for billets SCN-2 and SCN-4, both heat treated at 375 °C prior to assembly as multifilament composites.

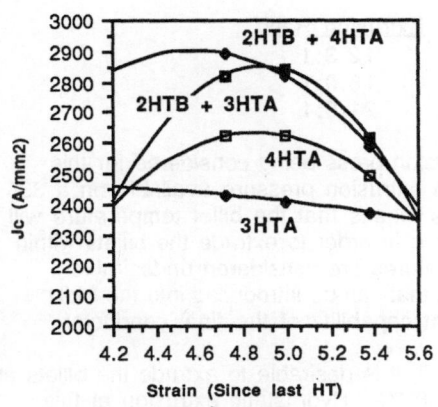

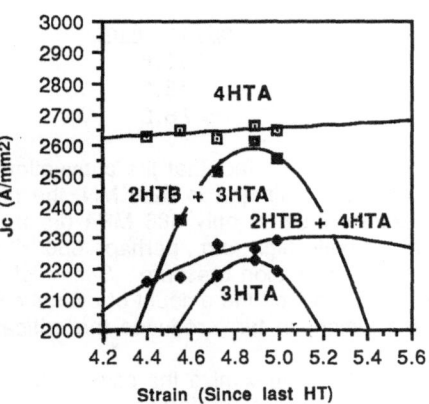

Figure 9 Critical current density (5T, 4.2 K) versus strain for billets SCN-1 and SCN-2 (conventional extrusion).

Figure 10 Critical current density (5T, 4.2 K) versus strain for billets SCN-3 and SCN-4 (hydrostatic extrusion).

The following observations can be drawn from this comparison of hydrostatic and conventionally extruded Cu/NbTi materials:

1) Hydrostatically processed material is superior to conventionally processed material in all comparisons, except in the case of four heat treatments after extrusion (4HTA).

2) Heat treatments carried out before extrusion are much more effective with hydrostatically processed material than with conventionally processed material. They show a behavior similar to applying all heat treatments after extrusion in finer filament work.[14] Also important to note, is the observation that conventionally extruded material <u>does</u> retain some degree of cold work throughout conventional extrusion.

3) Hydrostatically processed material, which is presently limited to 170 mm diameter, can be made to exhibit good properties with an acceptable number of heat treatments after extrusion.

4) Hydrostatic extrusion shows potential for higher Jc's in large diameter wires, particularly if the new heat treatments are applied and if larger presses could be converted.

TECHNICAL/ECONOMIC ASSESSMENT OF CONVERSION OF A PRODUCTION EXTRUSION PRESS

It is clear from the foregoing that the hydrostatic extrusion process offers important advantages for processing of Cu/NbTi superconducting materials. The conversion of an existing, larger, production extrusion press to permit hydrostatic extrusion would be important to the full range of metallic superconductors, and perhaps even to the high T_C ceramic superconductors sometime in the future. However, because the largest pending U.S. superconductor requirement is for copper-stabilized NbTi for the SSC, the main emphasis in this discussion will be on this material.

A current government effort in this regard is directed at the conventional extrusion of billets nominally 356 mm (14 in) diameter x 762 mm (30 in) long (excluding nose and tail portions). The plan is to extrude these billets down to rod approximately 76.2 to 101.6 mm (3 to 4 in) in diameter, the actual size depending on the availability of suitable draw benches for each of the U.S. superconductor manufacturers engaged in this effort. The nominal extrusion ratios calculated for three potential rod product sizes are:

Rod Size, mm	Extrusion Ratio
101.6	12.3:1
88.9	16.0:1
76.2	21.8:1

In view of the fact that the conventional extrusion press being considered for this extrusion work is limited to 62.3 MN, the maximum extrusion pressure available on a 356 mm diameter billet is only 586 MPa (85 ksi). This means that the billet temperature will have to be quite high, e.g., perhaps 650 °C or higher, in order to extrude the billets within the available extrusion pressure. Such high temperatures are considered undesirable because they reduce the amount of overall cold work that can be introduced into the NbTi filaments which, in turn, reduces the critical current capability of the final conductor.

In order to maximize the cold work in the NbTi, it is desirable to extrude the billets at minimal temperatures, preferably in the range of 200 °C. Hydrostatic extrusion at this temperature requires, for extrusion ratios of about 18:1 to 22:1, extrusion pressures in the order of 1103-1241 MPa (160-180 ksi), depending on the level of cold work in the NbTi at this point.

Therefore, if an existing extrusion press were to be converted, it would be prudent to design the extrusion tooling to operate at extrusion pressures as high as 1379 MPa (200 ksi) to allow for contingencies and other superconductor materials. With this in mind, the potential billet and product sizes that would be possible with hydrostatic extrusion on existing presses were calculated for three relatively high, but practical, extrusion pressure levels, i.e. 1103, 1241, and 1379 MPa. This information is given in Table 1 for existing presses that range from 34.3 to 137.9 MN (3850 to 15,500 tons) in capacity. The billet length is based on an estimate of the maximum filament length that could fit into the existing known length of container for each press size. This assumption ignores the possibility that still longer containers may be possible in some of the existing presses when converted. It should also be pointed out that still larger diameter billets could be extruded in all the presses cited, at extrusion pressures lower than 1103 MPa (160 ksi).

It can be seen from Table 1 that the billet weights are influenced by both the billet diameter and length that are possible with the various extrusion presses. In general, the higher the billet weight as well as the billet length-to-diameter ratio, the greater the product yield. Thus, the larger extrusion presses would be favored because the cost per pound of extrusion should be lower due to both the greater billet weights possible as well as the attendant greater product yield.

However, recommendation of a specific large press cannot be made until all important factors are properly considered. Among such factors are:

(1) company interest in the conversion option;

(2) conversion cost and the availability of a governmental subsidy;

(3) availability of the specific press for superconductor billet extrusion on a toll basis; and

(4) potential business for the converted press for both superconductor and non-superconductor billet extrusion.

It seems reasonable to expect that the willingness of a company to consider conversion would depend greatly on the amount of governmental subsidy, if any, and the projected increased usage of its press due to a hydrostatic extrusion capability. A factor that should be kept in mind is that the largest production hydrostatic extrusion presses available in the free world today are limited to 39.1 MN capacity. Thus, conversion of a substantially larger press, e.g. 62.3 MN (7000 tons) or more could, perhaps, lead to a significant increase in business volume from many new sources, both domestic and foreign, simply because of the uniqueness of such a press worldwide.

Table 1. Potential billet and product sizes possible with hydrostatic extrusion on existing extrusion presses at various assumed extrusion pressure levels

Existing Press Size, MN	Assumed Extrusion Pressures, MPa	Potential Billet Size			Product Size at 20:1 Extrusion Ratio	
		Diameter, mm	Est. Filament, Length(a), cm	Superconductor Weight(b), kg	Diameter, mm	Length, m
137.9	1103	399	152	1842	88.9	30.5
	1241	376	152	1637	83.8	30.5
	1379	358	152	1488	81.3	30.5
124.5	1103	378	152	1660	83.8	30.5
	1241	358	152	1642	81.3	30.5
	1379	345	152	1334	76.2	30.5
106.8	1103	351	127	1188	78.7	25.3
	1241	330	127	1052	73.7	25.3
	1379	315	127	957	71.1	25.3
77.4	1103	300	114	785	66.0	22.9
	1241	279	114	689	63.5	22.9
	1379	267	114	621	58.4	22.9
62.3	1103	269	89	485	61.0	17.7
	1241	254	89	435	55.9	17.7
	1379	239	89	386	53.3	17.7
48.9	1103	239	76	327	53.3	15.2
	1241	224	76	290	50.8	15.2
	1379	213	76	263	48.3	15.2
34.3	1103	198	66	195	43.2	10.1
	1241	188	66	177	40.6	10.1
	1379	179	66	159	38.1	10.1

(a) Estimate of maximum filament length possible with existing container length for each press size, excluding billet nose and tail.

(b) Estimate is based on an assumed copper-to-superconductor ratio of 1.5:1 and excludes billet nose and tail portions.

CONCLUSIONS

(1) The feasibility of converting a 11.1 MN conventional press to permit hydrostatic extrusion by a simple tooling change was successfully demonstrated.

(2) Feasibility was demonstrated by successful extrusion of Cu/NbTi, Al/NbTi, and Cu/Nb/Sn superconductor billets under conditions comparable to those used for extrusion of the same materials in a 39.1 MN production hydrostatic extrusion press.

(3) Successful conversion of this mid-scale press indicates that there should be very little or no risk involved in conversion of a much larger existing production press for hydrostatic extrusion.

ACKNOWLEDGEMENTS

The authors would like to acknowledge the support of their colleagues, past and present, at Supercon Inc. and Battelle Columbus Division. In addition, the important contributions of personnel at Battelle's Pacific Northwest Division, namely, C. Lavender and C. Bigelow are gratefully acknowledged. The authors would also like to acknowledge the financial assistance of the U.S. Department of Energy, Small Business Innovative Research program contract numbers DE-ACO2-84ER80184 and DE-ACO1-85ER80321.

REFERENCES

1. R.M. Scanlan, J. Royet, R. Hannaford, "Evaluation of Various Fabrication Techniques for Fabrication of Fine Filament NbTi Superconductors", IEEE Trans. MAG-23, 2, 1719 (Mar. 1987).

2. S. Sakai, G. Iwaki, Y. Sawada, H. Moriai, Y. Ishigami, "Recent Developments of the Cu/Nb-Ti Superconducting Cables for SSC in Hitachi Cable, Ltd.", Paper III-F-39, to be published in Proc. IISSC, New Orleans, LA, 8-10 Feb. 1989.

3. M. Ikeda, "Development of SSC Cable in Furukawa", Paper III-D-5, to be published in Proc. IISSC, New Orleans, LA, 8-10 Feb. 1989.

4. T. Fukutsuka, Y. Monju, Kobe Steel Co., Private Communication.

5. R.K. Maix, D. Salathe, S.L. Wipf, M. Garber, "Manufacture and Testing of 465 km Superconducting Cable for the HERA Dipole Magnets", IEEE Trans. MAG-25, 2, 1656 (Mar. 1989).

6. R.J. Fiorentino, B.D. Richardson, G.E. Meyer, A.M. Sabroff, F.W. Boulger, "Development of the Manufacturing Capabilities of the Hydrostatic Extrusion Process", Technical Report AFML-TR-67-327, Vol. 1 (Oct. 1967) Contract No. AF33(615)-1390.

7. R.J. Fiorentino, G.E. Meyer, T.G. Byrer, "The Thick-Film Hydrostatic Extrusion Process", Paper MF71-103, Society of Manufacturing Engineers (April 1971).

8. G.E. Meyer, R.J. Fiorentino, F.J. Jelinek, E.W. Collings, "Hydrostatic Extrusion of Nb_3Sn and NbTi Multifilamentary Superconducting Wire", Proc. of Inter. Conf. on Manufacture of Superconducting Materials, ed. R.W. Meyerhoff (Nov. 1976).

9. E.G. Smith, Jr., R.J. Fiorentino, E.W. Collings, F.J. Jelinek, "Recent Advances in Hydrostatic Extrusion of Multifilament Nb_3Sn and NbTi Superconductors", IEEE Trans. MAG-15, 1, 91 (Jan 1979).

10. M.I. Buckett, D.C. Larbalestier, "Precipitation at Low Strains in Nb 46.5 wt% Ti", IEEE Trans. MAG-23, 2, 1638 (Mar. 1987).

11. H. Kanithi, "Expectations and Limitations of J_c in Practical NbTi Conductors", Adv. Cryo. Engr., Vol. 34, eds. A.F. Clark, R.P. Reed, Plenum Press, 951 (1988).

12. P. Lee, "Adventures in Heat Treatment", presented to the 9th NbTi Workshop, Asilomar, CA (Jan. 1989).

13. E. Gregory, T.S. Kreilick, J. Wong, "Innovations in the Design of Multifilamentary NbTi Superconducting Composites for the SuperCollider and Other Applications", Paper III-D-9, to be published in Proc. IISSC, New Orleans, LA, 8-10 Feb. 1989.

14. T.S. Kreilick, E. Gregory, J. Wong, "Fine Filamentary NbTi Superconducting Wires", Adv. Cryo. Engr., Vol. 32, eds. A.F. Clark, R.P. Reed, Plenum Press, 739 (1986).

STRAIN EFFECTS IN VAMAS ROUND ROBIN TEST WIRES

K. Katagiri, K. Saito, M. Ohgami, T. Okada, A.Nagata+, K. Noto++,
K. Watanabe+++, K. Itoh*, H. Wada* K. Tachikawa**, J.W. Ekin†,
and C.R. Walters††,

ISIR, Osaka University, Ibaraki, Osaka 567, Japan

+Fac. Mining, Akita University, Akita 010, Japan

++Fac. Eng., Iwate University, Morioka, Iwate 020, Japan

+++IMR, Tohoku University, Sendai, Miyagi 980, Japan

*Nat. Res. Inst. Metals, Tsukuba, Ibaraki 305, Japan

**Fac. Eng., Tokai University, Hiratsuka, Kanagawa 259-12, Japan

†Nat. Inst. Stand. Tech., Boulder, CO 80303, USA

††Rutherford Appleton Lab., Didcot, OX11 0QX, UK

ABSTRACT

The strain characteristics of critical current, I_c, in three kinds of
VAMAS round robin test wires were evaluated. The multifilamentary Nb_3Sn
samples measured are: bronze route Ta added and internally stabilized wire
A, Ti added and externally stabilized wire B and internal tin diffusion
processed wire C, respectively. The strain for I_c peak ranged 0.20-0.30 %
and the reversible strain limit 0.8-1.1 %. The results obtained at 15 T in
three institutes, NIST, Rutherford Lab. and Osaka/Tohoku Univ., are
compared. Fairly good agreement was obtained. The strain sensitivity was
higher in the order of C, A and B. This can be mainly attributed to the
effect of addition of the third element. The correlation between the
strain sensitivity and the scatter of critical currents measured in the
round robin test participant laboratories is briefly discussed.

INTRODUCTION

Interlaboratory comparison of the critical current, I_c, measurements
using common multifilamentary Nb_3Sn sample superconducting wires (round
robin test) have been successfully carried out as an activity of VAMAS
(Versailles Project on Advanced Materials and Standards).[1-3] The summary
report suggested that there exists a strong correlation between the scatter
of I_c values and strain effects associated with the measurements. This
paper describes interlaboratory comparisons of the strain effects of VAMAS
wires measured in three laboratories and discusses how strain effects in the
conductor influence I_c measurements.

Advances in Cryogenic Engineering (Materials), Vol. 36
Edited by R. P. Reed and F. R. Fickett
Plenum Press, New York, 1990

Table 1. Specication of samples

	Wire A	Wire B	Wire C
Fabrication Method	Bronze	Bronze	Internal Sn
Wire Diameter (mm)	0.8	1.0	0.68
Structure	NbTa/CuSn	Nb/CuSnTi	Nb/Cu/Sn
Cu/non-Cu	0.22	1.68	0.88
Bronze/Cores	2.8	2.5	3.1
Filament Diam.(μm)	3.6	4.5	2.7
No. Filaments	10,000	5,047	5,550
Heat Treatment	973 K	943 K	973 K
	96h	200h	48h

EXPERIMENT

Samples

Three multifilamentary Nb3Sn sample wires were tested. The first and the second, A and B, are bronze route wires with Nb3Sn filaments containing the third additional elements of Ta or Ti, respectively. Wire A is stabilized by copper internally and B externally. The third, C, is an internal Sn processed wire with no third element addition. Brief sample specifications are given in Table 1. Further details have already been previously reported.[1]

Apparatus and Measurement

Three groups of laboratories participated in the strain effects evaluation tests using apparatus of their own. They are Inst. Sci. Ind. Res., Osaka Univ./Mater. Res. Inst., Tohoku Univ., National Inst. Standard Technology and Rutherford Appleton Lab.; these institutes are labeled disorderly as Laboratory I, II and III in this paper. Laboratories I and II used a short straight sample testing apparatus.[4,5] Axial tensile strain was applied in the magnet bore perpendicularly to the magnetic field; the I_c was measured using custom low-force clip-on gauge. The sample was cooled in a force-free state. Laboratory III used a long coil shape sample soldered on the outer periphery of a spring sample holder. Twisting the spring results in tensile or compressive strains in the sample.[6]

Only the data at magnetic fields of 15 T were provided by laboratory I, whereas more detailed results measured as a function of the magnetic field were presented by laboratories II and III. The results are compared based on the data at this field using an I_c criterion of 1 μV/cm. The data at 15 T of laboratory II are obtained through simple interpolation using points at 14 and 16 T.

RESULTS AND DISCUSSION

Strain Effect

As an example of results, the strain dependence of I_c in wire C is shown in Fig. 1. The I_c increases with strain to the peak, I_{cm}, at ε_m where compressive pre-strain in Nb3Sn induced by the other constituents with different thermal contraction coefficients on cooling is released. It decreases on further straining. The irreversible strain limit ε_{irrev} is determined as a strain beyond which the I_c value on unloading does not fall

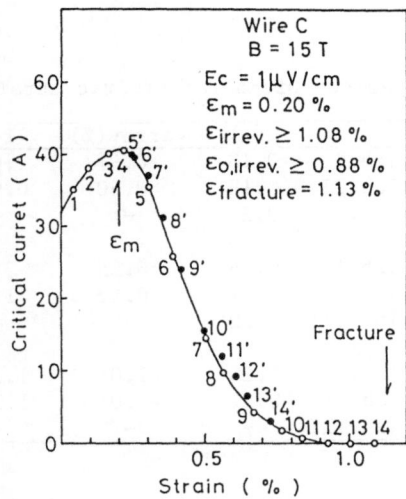

Fig. 1. Ic vs. strain characteristic for wire C (laboratory I).
(Open circles are obtained on loading, solid unloading)

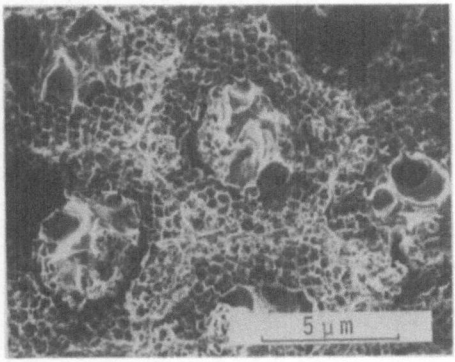

Fig. 2. Fracture surface of wire C. (4.2 K, 15 T).

on the curve obtained by previous loading. Figure 1 indicates $\varepsilon_m = 0.2$ %
and I_c is reversible until the wire is nearly strained to fracture, ε_f.
The highly reversible characteristics of this wire is ascribed to the wire
construction: The fractograph of this wire at 4.2 K, 15 T indicated that the
size of voids observed in the distributed Sn core is small, which does not
induce a significant strain concentration (Fig.2). This is in contrast with
the large voids observed in internal tin diffusion processed wires with
larger core size, in which the ε_{irrev} is small and the strain sensitivity of
I_c is high.[7,8]

Comparison of Measurements Among Laboratories

Values of ε_m, ε_{irrev}, and fracture strain ε_f obtained in each
laboratory are compared in Table 2. The prestrain value ε_m agrees within
0.06 % for laboratories I and II. The ε_m values differ somewhat from the

Table 2. Comparison of characteristic strain values

		$\varepsilon_m(\%)$	$\varepsilon_{irrev}(\%)$	$\varepsilon_f(\%)$
Wire A	lab.I	0.18	>0.97	1.02
	lab.II	0.18	>0.80	0.91
	lab.III	0.25	–	–
Wire B	lab.I	0.25	0.85	>3.5
	lab.II	0.31	0.82	>1.70
	lab.III	0.20	–	–
Wire C	lab.I	0.20	>1.08	1.13
	lab.II	0.21	>1.05	1.34
	lab.III	–	–	–

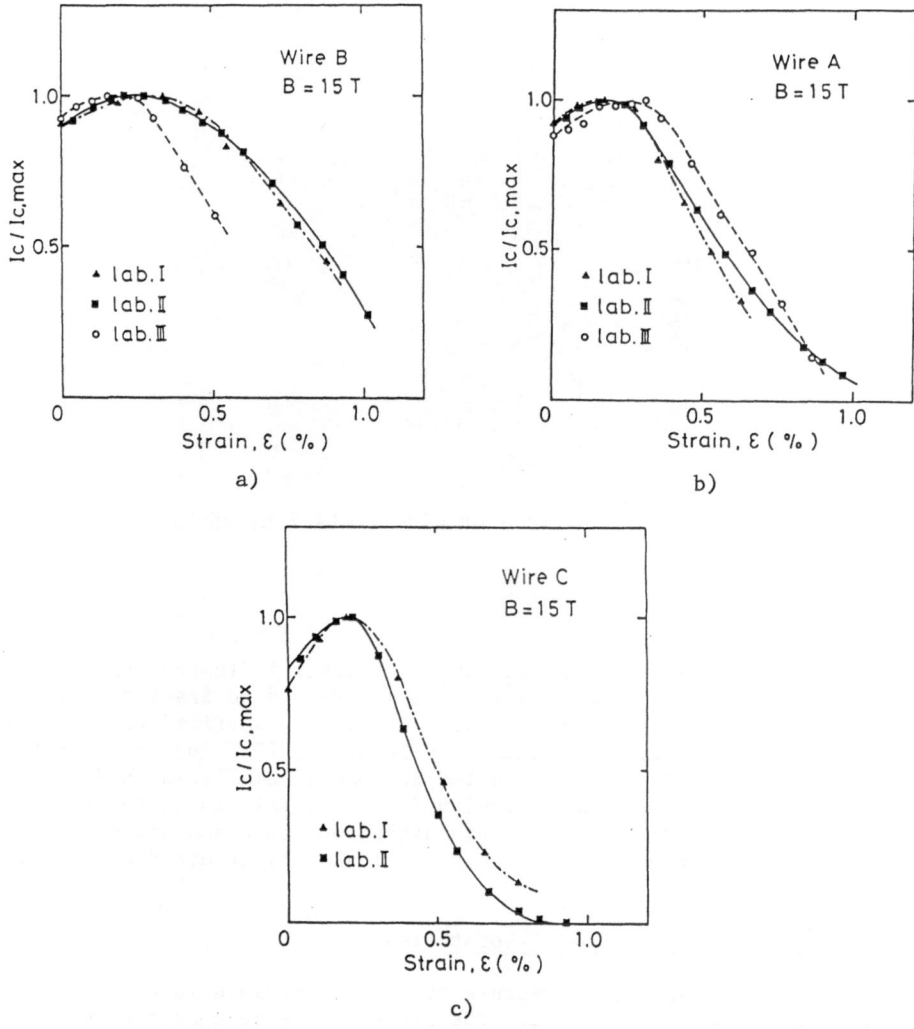

a)

b)

c)

Fig. 3. Strain dependence of I_c/I_{cm}.

results obtained by others for laboratory III. The irreversible strain limit values, or its lower limit values, agreed well, within 0.05 %, in wire B and are somewhat scattered in wires A and C. The reason for large scatter of ε_{irrev} is that the reversible strain limit coincides with the wire fracture strain at which I_c can be no longer determined. The fracture is controlled by the weakest defect in wires A and C, whereas the reversible strain limit in ductile wire B is determined by micro-fracture in the filaments prior to the final fracture.

The strain dependence of I_c for the three wires obtained by the three laboratory groups are shown in Fig. 3. The I_c is normalized to that at peak strain, I_{cm}. The agreement of ε_m, as mentioned above, as well as the strain sensitivity between laboratories I and II is quite good. To make a meaningful comparison with the data from laboratory III, in which the wire is soldered to a strip of copper firmly laid over the surface of the titanium spring, it appears necessary to fit the data to the prestrain values determined by laboratories I and II, which are obtained through force-free cooling.

The ε_m is largest in wire B in which both the bronze-to-core ratio, R, and the copper-to-non-copper ratio, S, are high. The smallest ε_m in wire A is presumed to be the consequences of low R and S and the position of the stabilizer.[9]

Strain Sensitivities and Scatter of I_c Measurements

Figure 4 shows comparison of tensile strain sensitivities for three wires obtained in laboratory I. The sensitivity is highest in the wire C and tends to decrease lower in the order of A and B. These are mainly controlled by their upper critical field, B_{c2}; 19 T, 24.5 T, and 26 T, respectively. The change in B_{c2} is attributed to the species and the amount of additional elements to Nb_3Sn; no addition, Ta, and Ti respectively. The strain sensitivity of I_c is known to be higher in the conductor with a higher R ratio and higher ε_m.[10] The ratio S is also supposed to contribute to it in the same direction. It seems worthy to note that I_c in the wires stabilized internally are more sensitive as compared with that in externally stabilized wires.[9] However, it appears that these effects are masked by the vast difference in B_{c2}.

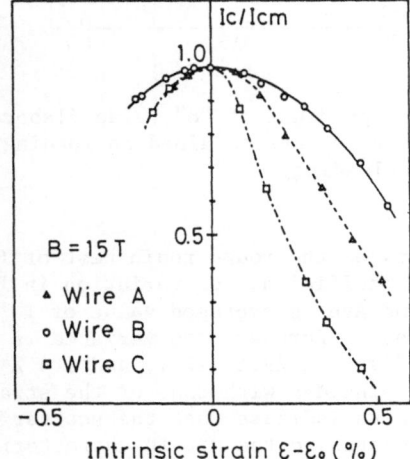

Fig. 4. Strain sensitivity of Ic (laboratory I).

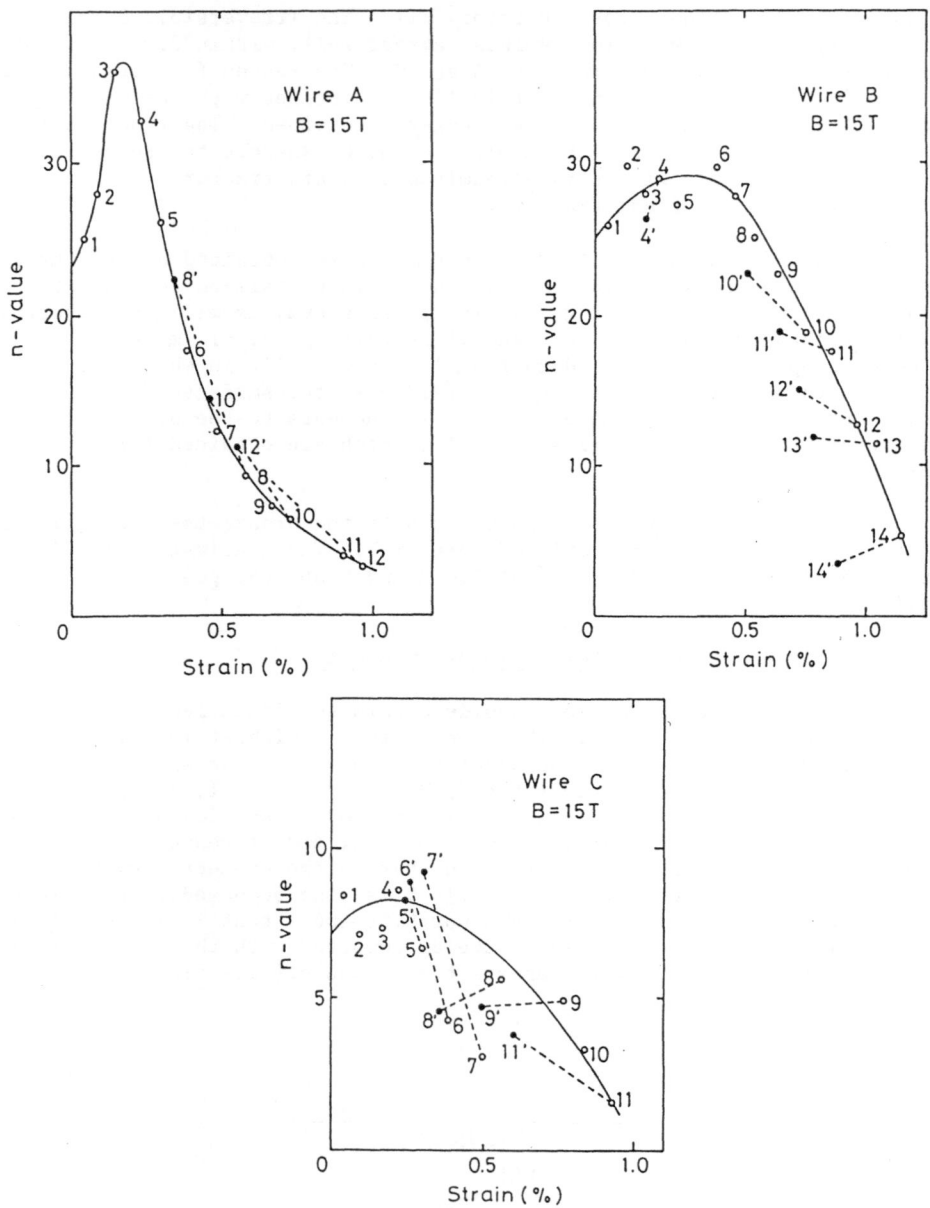

Fig. 5. Strain dependence of "n" value (laboratory I).
(Open circles are obtained on loading,
solid unloading)

According to a summary of the round robin test of I_C determined by the
criterion of 10 μV/m, the coefficient of variation in I_C data, s/Ave where
s is standard deviation and Ave is averaged value of I_C's, increases as the
magnetic field is increased. Further, the variance is larger in the order
of samples C, A, and B; s/Ave are 29.4, 12.1, and 5.4 % at 15 T,
respectively. The order coincides with that of the strain sensitivity of I_C
mentioned above. These facts indicate that the scatter in the I_C
measurements is closely related to the strain characteristics of the wire;
the intrinsic sensitivity of the conductor associated with the prestrain
during handling and/or cooling down. The systematic differences among

Table 3. Comparison of Icm

		Icm, A
	lab.I	115
Wire A	lab.II	132
	lab.III	114
	lab.I	116
Wire B	lab.II	124
	lab.III	107
	lab.I	40
Wire C	lab.II	41
	lab.III	–

laboratories in the round robin test have also been mainly attributed to the strain effect associated with the sample holder (Fig. 5 in the reference 1).

So long as the value of the prestrain experienced before the I_c measurement is small, the importance of the magnitude of ϵ_m is to be emphasized. Because the extent of change in I_c is directly associated with the strain sensitivity around the zero applied strain, the value of ϵ_m, the deviation from the peak, has significant meaning not only for the I_c value itself but also the scatter of I_c measurement.

It must be noted, however, that the values of I_{cm}, the I_c free from strain are also different among the laboratories (Table 3). This indicates that factors other than strain still can exist in the measurements of each laboratory; variation of magnetic field, parameters specific to the apparatus (excluding strain effect), recording equipments, and variation of sample itself. The parameters excepting the last one can partly be responsible for the systematic differences among laboratories. Difference in heat treatment condition also leads to variation of B_{c2}, I_{cm} and, therefore, the strain effect. More significant variation is induced in the samples with shorter heat treatment time in the order of wire C, A, and B. The situation for wire C is particularly bad because there is another problem with inadequate confinement of internal tin in short sample during heat treatment. The filament diameter (increasing in the order of wire C, A and B) may in part be responsible for a variation in the longitudinal homogeneity of the wires. The results of a homogeneity check for wires has been reported, except wire C. The scatter in wires A and B is rather low .

Strain Sensitivity of "n" Value

In the course of I_c round robin tests, n values specifying the relationship between the voltage V and the transport current I in a empirical expression $V \propto I^n$ has been also evaluated. A larger n value corresponds to a sharper transition in the specimen and is an index of good homogeneity.[11] The dependence of I_c on criterion voltage is higher for conductors of smaller n values.[12] Changes in n values determined within the voltage range of 0.5 to 5 μV/cm versus strain for three wires are shown in Fig. 5. Although the scatter of n is rather large, especially in wire C, n changes with strain as in the case of I_c. Because n reflects the homogeneity of the conductor, the irreversibility in n value is presumed to be caused by damage in the filaments. The change in n is larger in the wires C and A. The value of n near the peak I_{cm} is larger in wire A, but at strains other than the peak region it is rather small as compared to wire B.

The effect of the difference in position of stabilizer on the n value is not clear at present. The n values in wire C are erratic and small. A dull increase in voltage, i.e. smaller n value, will result in scatter of I_c if I_c criteria is of high accuracy, 1 μV/cm or smaller, or when the voltage base line in measurement has some fluctuation.

CONCLUSIONS

1. Interlaboratory comparisons of strain effects in three VAMAS sample superconductors showed fairly good agreement in prestrain ε_m, irreversible strain ε_{irrev}, and strain sensitivity of I_c.

2. Scatter of I_c measurement in the round robin test samples is mainly ascribed to strain effects; strain sensitivity and magnitude of prestrain.

ACKNOWLEDGMENTS

The authors are grateful to the members of HFLSM, Tohoku University for giving convenience to use 16.5 T-SM. This work is partly supported by Grant in Aid for Scientific Research No. 63050006, Ministry of Education, Science and Culture, Japan.

REFERENCES

1. K. Tachikawa, K. Itoh, H. Wada, D. Gould, H. Jones, C.R. Walters, L.F. Goodrich, J.W. Ekin and S.L. Bray, IEEE Trans. Magn., 25:2368 (1989).
2. K. Tachikawa, "Proc. 6th Japan-US Workshop for High Field Superconductors", Boulder, 1989, in press.
3. J.W. Ekin, ibid.
4. J.W. Ekin, Cryogenics, 20:611 (1980).
5. K. Katagiri, M. Fukumoto, K. Saito, M. Ohgami, T. Okada, A. Nagata, K. Noto, and K. Watanabe, Presented at the ICMC, Los ANGELES, CA, July 24-28, 1989.
6. C.R. Walters, I.M. Davidson, and G.E. Tuck, Cryogenics, 26:406 (1986).
7. M. Umeda, H. Yamazaki, M.Watanabe, Y. Kimura, Cryog. Eng., 22:110 (1987) (in Japanese).
8. K. Katagiri, K. Saito, M. Ohgami, T. Okada, A. Nagata, K. Noto, and K. Watanabe, in: "New Developments in Applied Superconductivity", p. 401 Y. Murakami ed., World Sci. Publ. (1989) SINGAPOLE.
9. K. Katagiri, M. Ohgami, T. Okada, T. Fukutsuka, K. Matsumoto, M. Hamada, K. Noto, K. Watanabe, and A. Nagata, Presented at the ICMC, Los ANGELES, CA, July 24-28, 1989.
10. T. Luhmann, M. Suenaga, D. O. Welch and K. kaiho, IEEE Trans. Magn. 15:699 (1979).
11. W.H. Warnes, D.C. Larbalestier, in: "Proc. Int. Symp. Flux Pinning Electromagn. Prop. Superconds.", T. Matsushita, K. Yamafuji and F. Irie, eds., P. 156, Matsukuma Press, Fukuoka (1985).
12. L.F. Goodrich and F.R. Fickett, Cryogenics, 225 (1982).

AN APPARATUS FOR EVALUATING STRAIN EFFECT OF CRITICAL CURRENT

IN SUPERCONDUCTING WIRES IN MAGNETIC FIELDS UP TO 16.5 T

K. Katagiri, M. Fukumoto, K. Saito, M. Ohgami, T. Okada,
A. Nagata*, K. Noto**, and K. Watanabe***,

ISIR, Osaka University, Ibaraki, Osaka 567, Japan
*Fac. Mining, Akita University, Akita 010, Japan
**Fac. Eng., Iwate University, Morioka, Iwate 020, Japan
***IMR, Tohoku University, Sendai, Miyagi 980, Japan

ABSTRACT

An apparatus to evaluate the strain effect of critical current, I_c, in practical superconducting wires was fabricated. The apparatus was installed in the 16.5 T superconducting magnet of High Field Laboratory for Superconducting Materials, Tohoku University. Total length of a specimen is 44 mm. Each end of it is soldered to a copper grip which also serves as a current terminal. The load is applied by moving the lever arm. The strain is measured by use of a extensometer set between the movable grip and the fixed one. The accuracy is 0.05 % strain. The voltage taps are soldered at a distance of 10 mm. The capacities in load and current of the apparatus are 500 N and 200 A, respectively. A good operation is confirmed by comparing results with those in the conventional apparatus. Reasonable agreement with the data obtained at NIST in VAMAS international round robin test samples gave a proof for reliability on the apparatus. Using this apparatus, strain characteristics of a wire fabricated by liquid infiltration method is evaluated.

INTRODUCTION

The influence of tensile stress upon the properties of superconducting materials, particularly Nb_3Sn, has been the subject of research. Changes in the critical current, I_c, above all, in the superconducting wires associated with mechanical behavior is important from the practical view point of magnet design and operation. The effect of strain is known to become more significant as the magnetic field is increased close to the upper critical field B_{c2}. Hitherto, there have been constructed many types of apparatus which facilitate the measurement of I_c in the course of tensile test at 4.2 K under high field above 12 T. Specking et al[1] constructed an apparatus which works in the gap of split pair magnet of the field up to 14 T in a cryostat. The tensile housing, "sword", is essentially the same as one in the conventional low magnetic field test consisted of a fixed copper block at the bottom and a movable one at the top which is pulled by driving device out side of the cryostat. The advantages of this apparatus are that the

Advances in Cryogenic Engineering (Materials), Vol. 36
Edited by R. P. Reed and F. R. Fickett
Plenum Press, New York, 1990

69

accuracy of strain measurement is high because the long straight sample is available and the instability caused by heat generation is minimized because the sample can be soldered in the region of low field.

Aiming at the measurements at higher field, Rupp[2] used an apparatus which work in the 2.5 cm bore of Bitter magnet capable of producing 23 T. Axial tensile strain transverse to the magnetic field is applied to the wire specimen by moving a wedge in the direction of bore axis separating two copper blocks on which each end of sample is soldered. Prior to the apparatus of Rupp, Ekin[3] succeeded in measurement of I_c change with strain including reversibility at 23 T using a compact apparatus, in which a movable copper block was pushed by lever arm. Strain in the wire is measured by a custom low-force extensometer attached to the copper blocks. Because the gauge length is short, the accurate strain measurement is the key technique. Walters[4] fabricated a sophisticated apparatus for extreme strain sensitivity, using thick spiral spring sample holder. Twisting the spring, long coil sample attached on the out side of the turn is strained.

In this paper, a brief description is given of another apparatus fabricated for measurement of strain effect of short straight samples at high field. Results of the operation check and an evaluation on a conductor under development are also presented.

APPARATUS

Superconducting Magnet

A superconducting magnet consisting of 30 double pancake wound with a surface diffusion processed Nb_3Sn tape conductor in the High Field Laboratory for Superconducting Materials, Tohoku University is used. The magnet generates high field up to 16.5 T in a 50 mm diameter bore. Detailed description has been given previously .[5]

Sample holder

A schematic illustration of the sample holder which can apply tensile load to the specimen is shown in Fig. 1. The 310 stable austenite stainless steel was used for structural component. The sample was gripped at either end by soldering on to two copper blocks of the length of 12 mm. The copper blocks serve as current terminals at the same time. The right block is fixed to the support structure through FRP insulator. The left block was mounted to the lower end of the lever arm. It can be moved by lifting the pull rod connected to a cam link which press the upper end of the lever arm.

Strain in the specimen was measured as a displacement of the copper block, using a custom 2-strain gauge extensometer (clip on gauge). The knife blades shaped at the ends of phosphor bronze arms of the extensometer were attached to the V-shaped grooves of both the copper block and a FRP calibration plate to be mentioned below. The extensometer applies only 0.5 N of spring tension to the sample. Linearity of the extensometer is excellent and no hysteresis was measured on calibrating experiment at room temperature. The feature of this holder is that the calibration of strain is made at 4.2 K and in the magnetic field on each I_c measurement. By pulling up the calibration rod, we can turn the calibration plate around the knife edge shaped in the right (fixed) copper block by certain angle corresponding to a calibration gap. This gives pre-determined amount of displacement of 0.185 mm in the extensometer arms (the left copper block is supported by the specimen). Releasing the calibration rod, the calibration plate is pressed again to the left side of the gap by the spring. Strain was determined to within 0.05 % strain. Load was applied to the pull rod

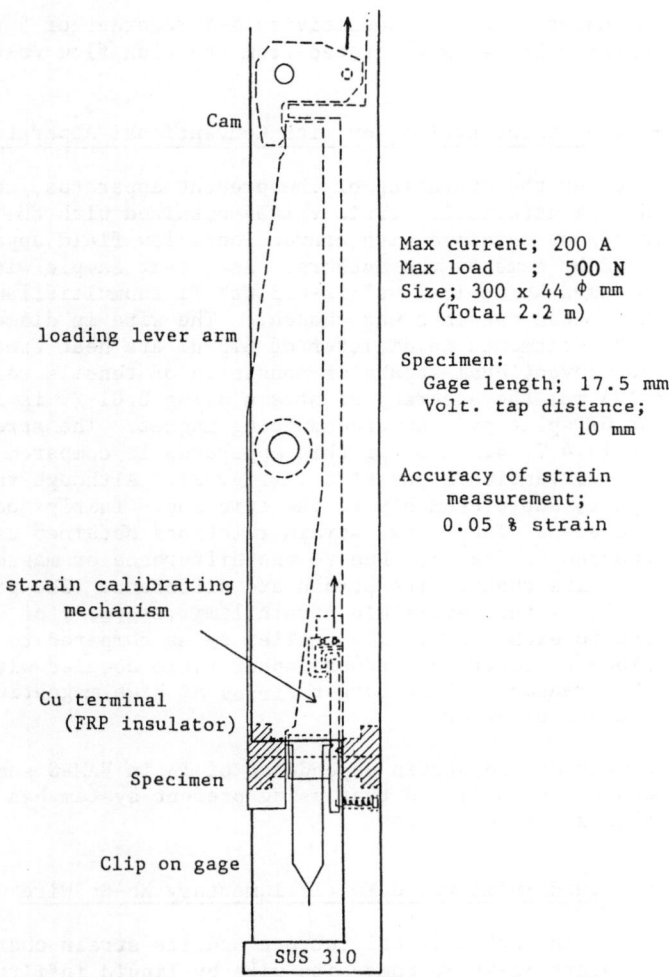

Max current; 200 A
Max load   ;   500 N
Size; 300 x 44 Φ mm
   (Total 2.2 m)

Specimen:
   Gage length; 17.5 mm
   Volt. tap distance;
                10 mm

Accuracy of strain
   measurement;
      0.05 % strain

Cam

loading lever arm

strain calibrating
mechanism

Cu terminal
(FRP insulator)

Specimen

Clip on gage

SUS 310

Fig. 1.   Schematic illustration of sample holder.

by a screw driven device via a load cell of 1 kN capacity out side of the cryostat, and was measured to within 1 %.   Stress-strain curves were plotted on an X-Y recorder.

After the sample was set, whole apparatus is purged by He gas in order to avoid freezing of air at movable portions in the apparatus.   The sample holder is slowly lowered in the cryostat (it took approximately 1 h), sliding through vacuum tight movable seal at the flange attached to the cryostat.   In order to ensure strain free cooling down, attention was paid to secure free rotation of lever arm during cooling; hollow pin (shrink fast) with a clearance of 0.1 mm was used.

Current up to 150 A was supplied to the specimen from a DC power source with a sweeper.   The Lorenz force was supported by a stainless steel block covered with FRP sheet exactly aligned in the height level of copper block. At the end of the measurement, the sample fractured at random points along the 17.5 mm length between the two copper blocks, indicating no stress concentration at the copper-solder grips.   Voltage taps separated by 10 mm were spot-soldered to the sample near the middle of the gauge length.   The Ic criterion used was 1μV/cm.   The critical current could be easily

71

determined on the chart of a high sensitivity X-Y recorder of 5 μV/cm range. The current transfer voltage is separated from the flux flow voltage, if necessary.

Comparison of Strain Effect Evaluation with Conventional Apparatus

In order to check the operation of the present apparatus, the stress strain curve and characteristic strain values obtained with the apparatus are compared with those obtained with conventional low field apparatus[6] which has been used by some of the authors. As a test sample with moderate I_c, a bronze processed Nb/Cu-13.2 wt% Sn-0.3 wt% Ti submultifilamentary wire prepared for VAMAS Japan specimen was chosen.[7] The wire in diameter of 0.4 mm consisted of 721 filaments in diameter of 6.5 μm are heat treated at 943 K for 200 h. The conventional apparatus consisted of tensile test rig with gauge length of 165 mm, the accuracy of strain being 0.01 %, is inserted within the gap of 6T split pair superconducting magnet. The stress-strain curve obtained at 14.4 T, 4.2 K using this apparatus is compared with that obtained using the conventional one at 6 T (Fig. 2). Although the stress is much the less higher, due presumably to the friction, fairly good coincidence can be seen. The I_c vs. strain relations obtained using the two apparatus are compared in Fig. 3. Due to the difference of magnetic field applied, the I_c and its change with strain are different. The prestrain, ε_m, of 0.2 % as well as the reversible strain limit, ε_{irrev} , of 0.55-0.60 %, however, are close to each other. The smaller ε_m as compared to the VAMAS Jpn wire is ascribed to lower copper/non copper ratio coupled with coarser filament size. The reason for the larger ε_{irrev} at high magnetic field has already been discussed elsewhere.

A good agreement of the strain dependence of I_c in VAMAS samples at a field of 15 T measured at NIST and that using present system has shown the reliability of this apparatus.[9]

Strain Effect in Liquid Infiltrated Multifilamentary Nb-Sn Wire

The superiority in both critical current and its strain characteristics in the multifilamentary Nb-Nb$_3$Sn composite wire by liquid infiltration method as compared with those in bronze processed has been reported.[10] Using the apparatus mentioned above, the strain characteristics in a wire fabricated in our group by liquid infiltration method using laboratory scale powder metallurgy processing technique[11] was measured at 15.5 T. The 13-15 wt% Sn composite, 10 mm in diameter was drawn down to 0.53 mm in diameter

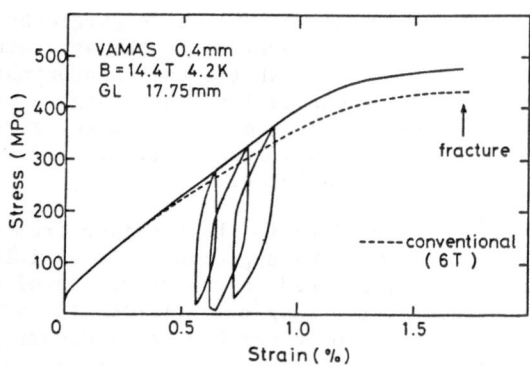

Fig. 2. Stress-strain curves of submultifilamentary wire for VAMAS Japan sample.

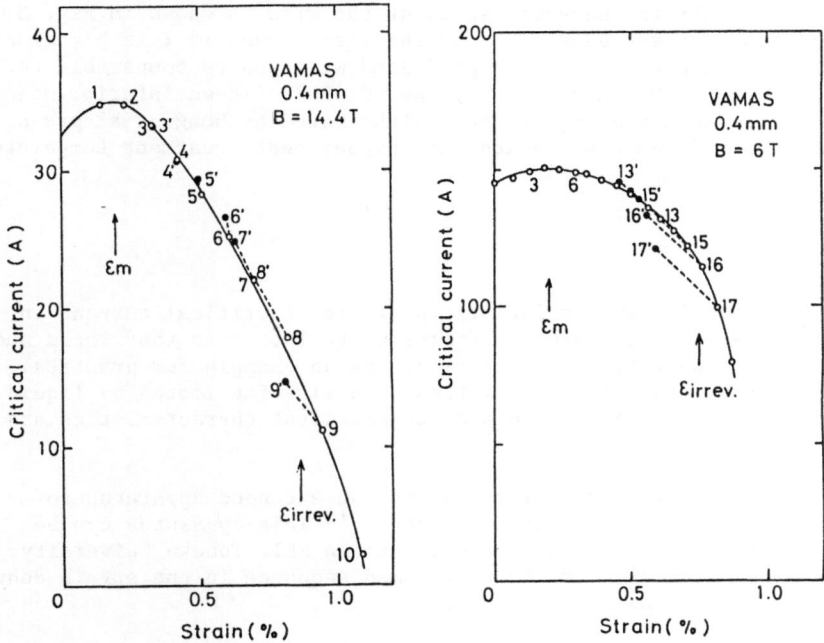

Fig. 3. Strain dependence of I_c.
a) measured using the present apparatus
b) measured using conventional apparatus

(Open circles are obtined on loading,
solid unloading)

together with Nb barrier and Cu stabilizer. Figure 4 shows a optical micrograph of the cross-section of the wire. Due to high oxygen content in starting Nb powder (900 ppm), reduction of the cross-sectional area was small (360). The heat treatment condition adopted by Hong et al[10] could not result in good I_c value. It was found that the two stage heat treatment of 1223 K x 10 min + 973 K x 5 day was effective for formation of superconducting A-15 phase. The T_c of 18.0 K with a 50 mK transition width and the B_{c2} of 23.5 T were obtained.

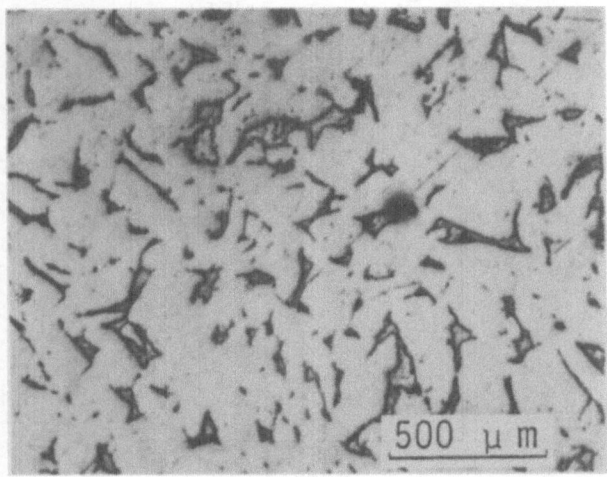

Fig. 4. Optical micrograph of the Nb-Sn composite as-infiltrated.

The I_C vs. strain characteristics of the wire is shown in Fig. 5. The non Cu critical current density J_C of the wire around 15 T is higher as compared with conventional bronze processed wire and is comparable to that of Hong. The ε_m is 0.1 % and the ε_{irrev} 0.65 %. These inferior strain characteristics as compared with those fabricated by Hong et al are ascribed to coarser Nb$_3$Sn filaments coupled with higher heat treatment temperature with longer duration.

SUMMARY

An apparatus for evaluating strain effect of critical current in superconducting wires in magnetic fields up to 16.5 T is fabricated and the fairly excellent operation is confirmed. As an example for practical application, the strain characteristics of a wire fabricated by liquid infiltration method is evaluated and its excellent characteristics are elucidated.

Using eventually identical mechanics, an extended apparatus for measurement up to 23 T is being constructed.[12] This apparatus can be operated remote from the HM-2 hybrid magnet in MRI, Tohoku University. The I_C was measured according to the programmed sequence in the strain-control or the load-control mode.

ACKNOWLEDGMENTS

The authors are grateful to the members of HFLSM, Tohoku University for giving convenience to use 16.5 T-SM. This work is partly supported by Grant in Aid for Scientific Research No. 63050006, Ministry of Education, Science and Culture, Japan.

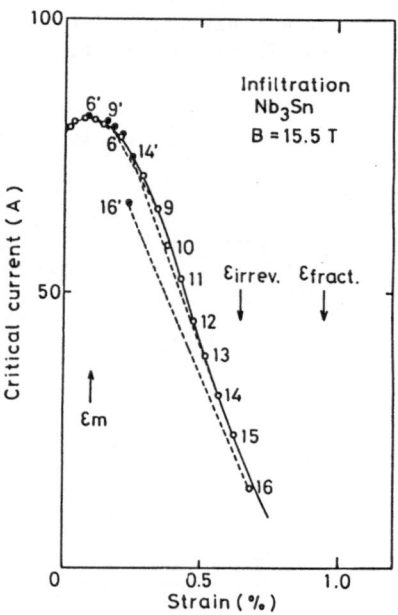

Fig. 5. Strain dependence of liquid infiltrated Nb$_3$Sn wire.

REFERENCES

1. W. Specking, and R. Flueckiger, J. de Physique, 45:C1-79 (1984).
2. G. Rupp, IEEE Trans. MAG-17:1099 (1981).
3. J.W. Ekin, Cryogenics, 20;611 (1980).
4. C.R. Walters, I.M. Davidson and G.E. Tuck, Cryogenics, 26:406 (1986).
5. K. Watanabe, K. Noto, T. Sasaki and Y. Muto, Sci. Rep. RITU, A-33:319 (1986)
6. M. Fukumoto, T. Okada, K. Yasohama, K. Yasukochi, Adv. Cryog. Eng., 30:867(1984).
7. K. Tachikawa, K. Itoh, H. Wada, D. Gould, H. Jones, C.R. Walters, L.F. Goodrich, J.W. Ekin and S.L. Bray, IEEE Trans. Magn., 25:2368 (1989).
8. K. Katagiri, K. Saito, T. Okada, A. Nagata, K. Noto, Adv. Cryog. Eng. 34:531 (1988).
9. J. Ekin, Proc. 6th Japan-US Workshop for High Field Superconductors, Boulder, 1989, in press.
10. M. Hong, G.W. Hull Jr., J.T. Holthuis, W.V. Hassenzahl and J.W. Ekin, IEEE Trans., MAG-19:912 (1983).
11. K. Watanabe et al, to be published in Sci. Rep. RITU.
12. K. Kamata et al, to be published in Proc. 11th Int. Conf. Magn. Tech., Tsukuba (1989).

THE CHARACTERIZATION OF Nb3Sn SUPERCONDUCTORS
FOR USE IN MAGNETS OF 19 T AND GREATER

L.T. Summers, M.J. Strum, and J.R. Miller

Lawrence Livermore National Laboratory
P.O. Box 5511, L-643
Livermore CA, 94550

ABSTRACT

Increased resolution of NMR spectrometry will require the use of very high field Nb3Sn superconducting magnets. Here we report the results of our investigation into mechanical and temperature effects on internal-Sn superconductors similar to those proposed for use in a 900 MHz, 21 T NMR magnet system. Thermal precompression was found to be about 0.225%, and the irreversible strain was about 0.8%. Fatigue degradation was not observed at cyclic intrinsic strains below 0.575%. Additions of reinforcing steel in cable conductors was found to reduce the critical current by as much as 50% compared to similar, unreinforced cables. Reduction of the testing temperature to 2.3 K did not increase the critical current in steel-reinforced cables to a level significantly above that of unreinforced samples.

INTRODUCTION

Nuclear Magnetic Resonance (NMR) spectrometry is useful for the study of chemical reactions, particularly those of interest to medical and pharmaceutical research. The operating limits of these devices are a function of magnetic field, higher fields providing opportunities for increased resolution. Carnegie-Mellon University has proposed development of a 900 MHz, 21 T NMR facility with a warm bore of about 7.6 cm diameter. In support of this effort we evaluated several internal-tin superconductors and superconducting cables similar those proposed for use in this 21 T magnet. These wires are also proposed for use in a 19 T prototype coil to be tested in LLNL's High Field Test Facility (HFTF).

Due to the high current density required for 21 T operation, the proposed magnets will be constructed using cabled superconductors potted with epoxy

Advances in Cryogenic Engineering (Materials), Vol. 36
Edited by R. P. Reed and F. R. Fickett
Plenum Press, New York, 1990

77

and bath-cooled at 1.8 K. The high Lorentz forces generated at these fields will put substantial mechanical loads on the superconducting wires. Since the magnet will be periodically de-energized for routine maintenance the effects of high strain cyclic loading were of interest. It is anticipated that the load cycles will be less than 100 over the useful lifetime of the NMR system.

We have also studied the effects of applied strain on critical current degradation, especially precompression effects caused by the inclusion of steel reinforcement in cables. It is well known that the strain sensitivity of the critical current is a strong function of field and as the operating field approaches H_{c2}, considerable degradation of J_c occurs at small applied strains. Operation of magnets at 21 T, close to H_{c2}, will require magnet designers to critically assess the effects of operational strain and precompression and account for possible degradation.

EXPERIMENTAL PROCEDURE

Two types of internal-Sn Ti-alloyed Nb_3Sn test wires were supplied by IGC. Although we anticipate that actual magnets will probably use Ta-alloyed Nb_3Sn, due to its higher upper critical field, our initial studies have focused on this readily available material. The first type of wire tested was a 7 subelement design drawn to 0.42 and 0.92 mm diameter. The second wire, used for testing cables, was also a 7 subelement at 0.615 mm diameter The specifications for both wires are shown in Table 1.

The critical current as a function of applied strain and fatigue loading was measured using a system consisting of a screw-driven pull rod, digitally-

Table 1. Specifications of wires used in this study

Seven subelement wire	
Diameters	0.42 and 0.92 mm
Nb Diffusion barrier	4% by volume
Stabilizer	50% by volume
Remaining non-copper fraction	50% by volume
	27% Nb (1.25% Ti)
	19.2% Sn
	53.7% Cu matrix
Cabled wires	
Diameter	0.615 mm
Nb Diffusion barrier	4.3% by volume
Stabilizer	62.5% by volume
Remaining non-copper fraction	33.2% by volume
	20.5% Nb (1.20% Ti)
	17.2% Sn
	62.3% Cu matrix

Table 2. Test cable specifications

Cable configurations:

	Packing factor	Reinforcement
1. 6 around 1	84%	none
2. 6 around 1	90%	none
3. 6 around 1	84%	central steel wire 0.615 mm diameter
4. 5 around 1	84%	central steel wire 0.430 mm diameter

Cable dimensions
1.   1.774 mm diameter
2.   1.714 mm diameter
3.   1.774 mm diameter
4.   1.517 mm diameter

controlled servo-motor and microcomputer. During testing, the pull rod and sample pass through a 12 T radial access superconducting magnet. The magnet is equipped with holmium pole pieces which raise the effective field at the specimen to 15 T. The details of this apparatus have been described previously.[1, 2]

Samples for strain and fatigue studies were given an internal-Sn heat treatment, found by testing, to give near optimum critical current at fields of about 14 T. This heat treatment consists of 4 steps: 200°C for 24 hours + 340°C for 48 hours + 660°C for 72 hours + 725°C for 8-12 hours depending on wire size.

Four round cables were supplied by IGC in several configurations two of which contained a steel reinforcing wire at the center of the cable pattern. The cable specifications are shown in Table 2.

The cables were mounted on 25 mm long, 50 mm diameter stainless steel spools with machined vee-grooves. Cable ends were sealed with a welding torch to prevent tin leakage. The samples were given a three step heat treatment, 200°C for 48 hours + 325°C for 24 hours + 700°C for 90 hours, in flowing argon followed by furnace cooling. This heat treatment was selected for inter-laboratory comparison of results and is not considered to be optimal.

The samples were tested on the reaction spools in a 14 T superconducting magnet oriented so that the applied field was normal to the cables. Voltage taps were placed 975 mm apart. Limited testing at temperatures below 4.2 K was achieved by pumping on a vacuum tight anticryostat which surrounded the test fixture. Temperature was monitored by carbon glass resistors placed in the helium bath near the specimens. The critical current for all samples was determined using a resistance criteria of .1 x 10^{-14} Ω•m. The critical currents reported are for the non-copper wire fraction.

RESULTS

J_c as a function of strain is shown in Figure 1. The initial critical current of the 0.42 and 0.92 mm wires were 529 and 486 A mm^{-2} respectively. For reference, a J_c versus strain curve calculated using Ekin's expression[3] is shown. The prestrain resulting from thermal contraction was approximately 0.225% for both wires. Critical currents (J_{cm}) at zero intrinsic strain (ε_m) were 610 and 570 A mm^{-2} for 0.42 and 0.92 mm diameter wires respectively. No irreversible strain degradation (ε_{irrev}) was observed below 0.8% intrinsic strain.

The results of stress-controlled fatigue tests are shown in Table 3. The results are reported as the ratio $J_c(\varepsilon)/J_{cm}$, where $J_c(\varepsilon)$ is the critical current at the applied field and strain and J_{cm} is the critical current at field and zero intrinsic strain. Constant $J_c(\varepsilon)/J_{cm}$ indicates no cyclic degradation. Decreasing $J_c(\varepsilon)/J_{cm}$ indicates that degradation has occurred during cyclic loading. No fatigue-dependent degradation was observed until the applied strain reached 0.8% (0.575% intrinsic).

The results of 4.2 K critical current measurements of cabled conductors are shown in Figure 2. The data for each type of cable is the average of two different samples. Good agreement was found between similar samples. Cable packing factor has no apparent effect on critical current. The cables containing steel reinforcement have significantly lower critical currents than the unreinforced cables.

Limited studies of temperature effects have been completed. The results of testing reinforced 6 x 1 specimens are shown in Figure 3. Each data point is the average of tests on two separate specimens. Comparison of Figures 2 and 3

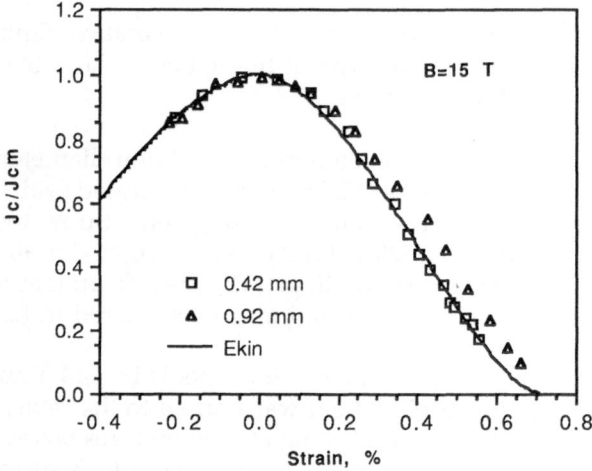

Figure 1. Normalized critical current as a function of intrinsic strain for Ti-alloyed Nb$_3$Sn (IGC). J_{cm} = 610 and 570 A mm^{-2} respectively for 0.42 and 0.92 mm diameter wires.samples. At 14 T, the reinforced 6 x 1 cables have critical currents that are approximately 50% lower than the unreinforced 6 x 1 samples.

Table 3. Fatigue effect on the critical current of internal-Sn $Nb_3Sn(Ti)$ at 15T

Stress MPa	Strain %	Cycles	$J_c(\epsilon)/J_{cm}$
188	0.4	1	0.90
		5	0.91
		25	0.91
		125	0.92
		1050	0.93
		1051	0.92
210	0.5	1	0.86
		2	0.84
		50	0.85
		51	0.86
228	0.6	1	0.78
		2	0.77
		50	0.77
		100	0.75
		101	0.76
245	0.7	1	0.71
		2	0.72
		50	0.71
		51	0.70
283	0.8	1	0.56
		2	0.53
		7	0.46

shows that reducing the operating temperature of the reinforced samples to 2.3 K yields critical currents that are comparable to the unreinforced samples at 4.2 K.

DISCUSSION

A thermal precompression of 0.225% is similar to that seen in other wires and is encouraging. The intrinsic strain to ϵ_{irrev} is comfortably high for magnet designs that emphasize high fields and accompanying high stresses. As anticipated, there was reasonable agreement with strain-induced J_c degradation predicted using Ekin's formalism. The slight difference seen between the two wire diameters may result from inaccuracy in measuring applied strains during testing.

Optimal magnet design would encourage the acceptance of high operating loads in order to minimize the magnet space needed for structural elements such as the magnet case or reinforcing epoxy insulation. For steel

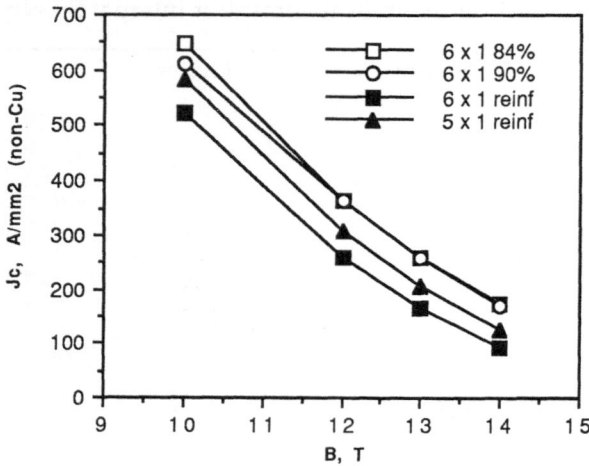

Figure 2. Critical current vs. field for cabled superconductors at 4.2K.

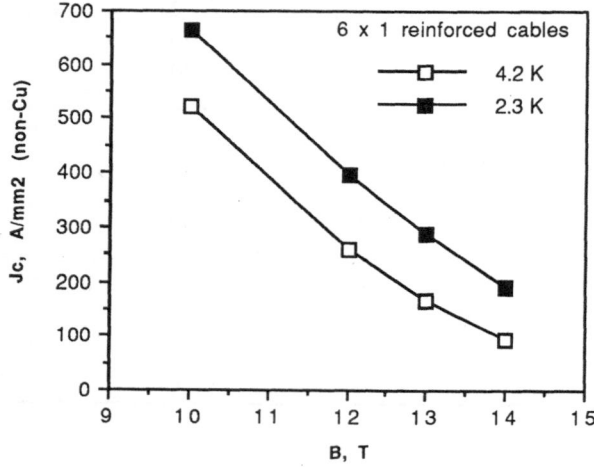

Figure 3. Effect of temperature on the critical current of reinforced 6 x 1 cables.

structures, prudent design would limit strains to about 0.3%, which is equivalent to about 1/2 the yield strength of high-strength cryogenic steels such as 316 LN or JBK-75. Additionally, strains above 0.25% may be considered impractical for organic insulators, such as G-10 or other epoxies. Designs which approach these strain limits have the advantage of relieving a large fraction of the thermal precompression. For the wires studied here, for which the precompression is about 0.225%, these strain limits would leave the conductor near zero intrinsic strain. For operation at extremely high fields where strain degradation is excessive, this represents an ideal magnet design scenario.

The performance of these wires under conditions of fatigue will not impact design of the 19 T prototype or 21 T NMR magnets where the number of load cycles is small. For the limited number of cycles tested no degradation was observed until applied strains of 0.8% were reached. This corresponds to 0.575%

intrinsic strain which provides a comfortable margin between ε_{irrev} and the optimal operating strain ε_m.

Comparison of the critical currents of reinforced and unreinforced cables points out one possible pitfall in this approach to stress management. The unreinforced 6 x 1 cables clearly have the highest critical currents while the 5 x 1 and 6 x 1 reinforced cables follow in order. It is interesting to note that the 5 x 1 and 6 x 1 reinforced cables are 9 and 14% steel by volume. Hence, the degradation seems to scale with volume fraction of steel in the conductor. At this point the evidence is not irrefutable, however, the indication is clear that incorporation of steel reinforcement may result in increased thermal precompression hence lower J_c's. The magnitude of this result is surprising since the cables were soldered to stainless steel drums which, by themselves, should introduce a high precompression on cooling. This indicates that co-winding steel and superconductor leads to very effective coupling and transmission of strain. If such precompression were present in an actual magnet only a small portion could be relieved by operating strains. Higher operating strains, giving near-zero intrinsic conductor strain, would not be possible since structural material operating limits would be exceeded. Quantitative estimates of the degree of precompression in the cables is complicated by soldering to the steel test fixtures.

In actual magnets, the cable space is filled with epoxy which has a low 4 K modulus ($\approx$ 5 GPa). To what extent reinforcing core precompression occurs in that case, then, is uncertain. Further testing under appropriate conditions is advisable. In addition, testing of unsoldered, unfilled cables would be interesting to determine the magnitude of direct mechanical interaction that exists between the reinforcement and superconductor. This may have consequence to cable-in-conduit conductor (CICC) design. A previous study of the effects of different core elements[4] proved inconclusive.

Data at temperatures below 4 K are still being collected at this time, and the limited information presented in Figure 3 is still inconclusive. However, it is interesting to note that a 1.9 K temperature reduction in the reinforced samples yields critical currents that are only slightly higher than the unreinforced samples at 4.2 K. Clearly, unreinforced cables have a significant advantage.

CONCLUSIONS

Thermal precompression in the 7 subelement Ti-alloyed internal-Sn wires tested was 0.225%. Irreversible strain degradation was not observed until the applied strains exceeded 0.8% intrinsic strain. Cyclic fatigue at applied intrinsic strains of less than 0.8% did not cause degradation of critical current while cyclic intrinsic strains greater than .575% resulted in immediate fatigue degradation of J_c.

Small changes in the conductor packing fraction showed no measurable effect on the J_c of small cables . Inclusion of steel reinforcement, on the other hand, was found to reduce J_c by as much as 50% at 14 T, presumably due to thermal precompression. This effect scaled with the volume fraction of steel included in the cable.

Decreasing the testing temperature was found to affect large increases in J_c, however cables containing 14% steel reinforcement (6 x 1) only slightly exceeded unreinforced cable critical currents even at 2.3 K. Steel reinforcement may potentially be unacceptable for high field magnets due to precompression effects.

ACKNOWLEDGEMENTS

The authors would like to thank Intermagnetics General Corporation for providing the test specimens. We wish to acknowledge the excellent assistance of J.E. Bowman, A. Duenas, C.E. Karlsen, R.C. Jenkins, and K.R. Tapscott of LLNL with testing, instrumentation, and data acquisition. We also wish to thank T.S.E. Summers for assistance with preparation of this manuscript. This work was performed under the auspices of the U.S. Department of Energy by the Lawrence Livermore National Laboratory under Contract W-7405-Eng-48.

REFERENCES

1. R. W. Hoard, S.C. Mance, R.L. Leber, E.N.C. Dalder, M.R. Chaplin, K. Blair, D.H. Nelson and D.A. Van Dyke, "Field Enhancement of a 12.5 T Magnet Using Holmium Poles," IEEE Trans. Mag., Vol. MAG-21, 448, 1985.

2. M.J. Strum, L.T. Summers, and J.R. Miller, "Ductility Enhancement in Unreacted Internal-Sn Nb_3Sn Through Low-Temperature Anneals," IEEE Trans. Mag., Vol. 25, 2208, 1989.

3. J.W. Ekin, "Strain Scaling Law for Flux Pinning in Practical Superconductors. Part 1: Basic Relationship and Application to Nb_3Sn Conductors," Cryogenics, V. 20, 611, 1980.

4. J.R. Miller and L.T. Summers, "The Effects of Various Conductor Components on Nb_3Sn Filament Prestrain in Cable-in-Conduit Conductors", Adv. Cryo. Eng., V. 34, 553, 1988.

STRAIN EFFECT IN AN INTERNALLY STABILIZED MULTIFILAMENTARY

(Nb,Ti)₃Sn SUPERCONDUCTING WIRE

K. Katagiri, M. Ohgami, T. Okada, T. Fukutsuka*, K. Matsumoto*,
M. Hamada*, K. Noto**, K. Watanabe***, A. Nagata****

ISIR, Osaka University, Ibaraki, Osaka 567, Japan
*Kobe Steel Ltd., Kobe, Hyogo 651, Japan
**Fac. Eng., Iwate University, Morioka 020, Japan
***IMR, Tohoku University, Sendai, Miyagi 980, Japan
****Fac. Mining, Akita University, Akita 010, Japan

ABSTRACT

Strain effect in a bronze processed multifilamentary (Nb,Ti)₃Sn
superconducting wire with Cu stabilizer at the center of the wire cross-
section was studied at a magnetic field of 15 T. The heat treatment
condition altered the strain characteristics, such as the stress-strain
curve and the I_c vs. strain curve of the conductor. The results are
compared with those of externally stabilized one. The difference in strain
characteristics due to the change in the location of stabilizer is briefly
discussed from the strain state of the (Nb,Ti)₃Sn filament in the composite
wire after cooling down to cryogenic temperature and during tensile
deformation.

INTRODUCTION

In order to fabricate a superconducting wire for the magnet to be
operated in persistent current mode at high fields, an internal copper
stabilized wire has been developed.[1] The wire is constituted with bronze
matrix and Nb filaments in the outer layer, stabilizing copper at the center
of wire separated with Nb barrier. Low resistivity joint of Nb filaments is
capable through etching bronze away by nitric acid. Although the critical
current density, J_c, of the internally stabilized wire had been improved by
modifying effective bronze ratio, 2.6×10^4 at 15 T for example, J_c was still
slightly lower compared to that in the externally stabilized wire of the
equivalent constitution. The strain state of the conductor was suspected to
be one of the reason for the low J_c. In this paper, the strain
characteristics of J_c in the internally stabilized wire is described.

EXPERIMENTAL

Two kinds of (Nb,Ti)₃Sn multifilamentary wire were manufactured by a
bronze method. One was stabilized with copper internally (wire I) and the
other externally (wire E). The cross-section as well as the specification

Advances in Cryogenic Engineering (Materials), Vol. 36
Edited by R. P. Reed and F. R. Fickett
Plenum Press, New York, 1990

85

Table. 1. Specification of wires

	Internally stabilized wire	Externally stabilized wire
Dia. of wire(mm)	0.7	0.85
Dia. of fil.(μm)	3.4	3.4
Number of fil.	7224	10285
Bronze/Nb core	3.3	2.8
Cu/non Cu	0.33	0.33

of the wires are shown in Fig. 1 and Table 1, respectively. The details of the wire fabrication method have already been described previously.[1]

The strain characteristics of the wires were measured at 15 T using an apparatus combined with 16.5 T superconducting magnet in HFLSM, Tohoku University.[2] Unfortunately, due to the magnet condition, data on wire E was obtained at the magnetic field of 14.4 T. Each end of specimen was soldered to a copper grip of 12 mm length and tensile load was applied perpendicularly to the magnetic field. The axial strain in the specimen was measured using an extensometer with the accuracy of 0.05 % strain. The gauge length of the specimen is 17.5 mm and the distance of the voltage taps 10 mm. The criterion of the critical current, I_c, was 1 μV/cm.

RESULTS AND DISCUSSION

Change of I_c with Strain

The stress vs. strain relation in the course of measurement of strain dependence of I_c at 15 T, 4.2 K changed with the heat treatment time. Longer heat treatment resulted in a steep slope of the curve excepting initial stage as well as short fracture strain. This is somewhat different from the result in the externally stabilized wires in which the longer heat treatment resulted in the gentle slope from the initial to final stage of the curve. This is to be interpreted in terms of competitive effects of growth of $(Nb,Ti)_3Sn$ layer for the longer heat treatment with simultaneous lowering of flow stress of bronze by depletion of Sn content[3] coupled with the specific behavior of stabilizing copper to be mentioned later. An SEM observation on the fracture surface revealed the marked growth of compound layer with the time of heat treatment (Fig. 2).

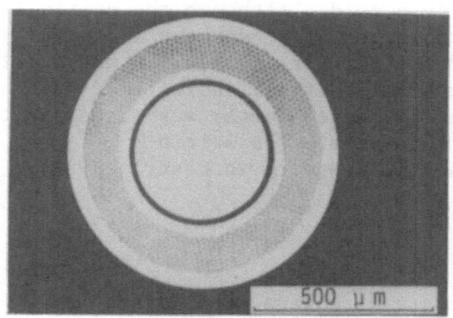

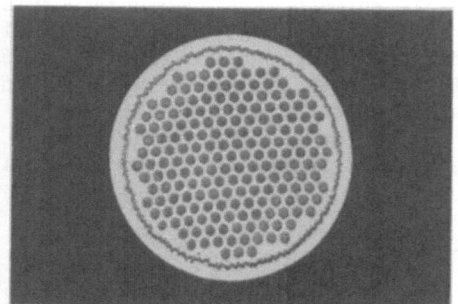

a) Internally stabilized   b) Externally stabilized

Fig. 1. Cross section of the wires.

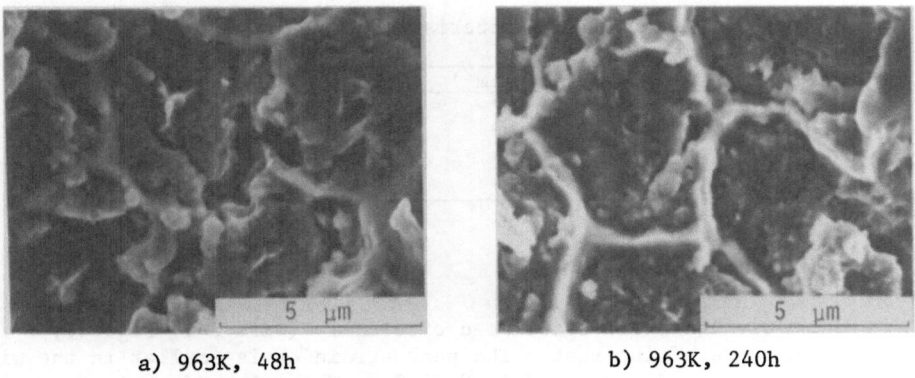

a) 963K, 48h   b) 963K, 240h

Fig. 2. (Nb,Ti) Sn layers.

Figure 3 shows the strain dependence of I_c in the internally stabilized
wires of different heat treatment time. When the heat treatment time at 963
K was changed from 48 h (wire I-s) to 240 h (wire I-l), the peak value of
I_c, I_{cm}, increases from 70 to 89 A. This is mainly due to the increase of
Nb_3Sn layer thickness. The strain value for I_c , ε_m, decreases and the
reversible strain limit ε_{irrev} as well as its intrinsic value $\varepsilon_{0,irrev}$ =
$\varepsilon_{irrev} - \varepsilon_m$ decrease. The former is a consequence of the increase of Nb_3Sn
layer thickness and decrease of Sn concentration in the matrix bronze.
The decrease in $\varepsilon_{0,irrev}$ indicates the strain limit for the damage of
$(Nb,Ti)_3Sn$ filaments is decreased due to thicker compound layer.[4] These are
consistent with the results obtained in the externally stabilized multifilamentary superconducting wires.[3-5] Compared with the strain dependence

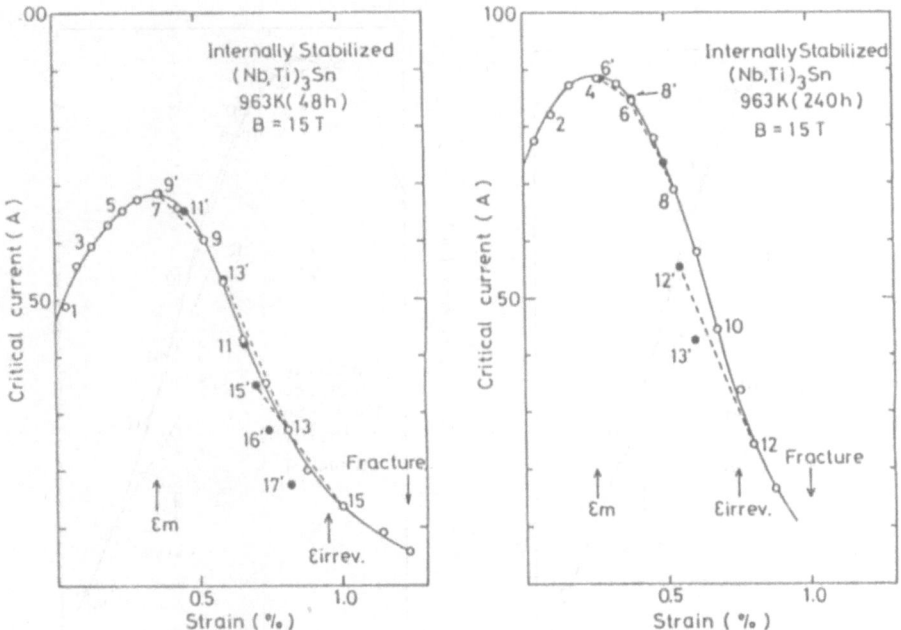

Fig. 3. Strain dependence of I_c in internally stabilized wires.
(Open circles are obtained on loading, solid unloading)

Table. 2. Strain characteristic values of wires

Wire	H.T.time	I_{cm}(A)	J_{cm}(A/cm^2)	ε_m(%)	$\varepsilon_{0,irrev}$(%)	ε_f(%)	n(max)
Int. Stab.	240h(I-1)	89	3.1×10^4	0.25	0.50	0.92	30
963K	48h(I-s)	70	2.4×10^4	0.35	0.6	1.12	20
Ext. Stab.	200h(E-1)	110	2.6×10^4	0.35	0.52	2.04	20
963K	50h(E-s)	105	2.5×10^4	0.32	0.47	1.86	20

non-Cu

of I_c in the equivalent wire stabilized externally (wire E-1, Fig. 4),
following points are of interest. The peak strain ε_m is smaller in the wire
I-1 (0.25 %) as compared to wire E-1 (0.35 %). The value of ε_m is known to
decrease with 1) reaction of (Nb,Ti)$_3$Sn, as described above, and with 2)
increase of spacing of filaments.[6] Because the bronze ratio of wire I-1 is
slightly larger as compared to wire E-1, the factor 1) leads to larger ε_m
and the factor 2) leads to smaller ε_m. Therefore, the difference in the ε_m
is not solely ascribed to the arrangement of stabilizer. The values
relevant to the strain characteristics\are summarized in Table 2.

Strain Sensitivity of I_c

Figure 5 shows the relation of I_c normalized to I_{cm} vs. intrinsic
strain in the wires I-s, I-1 and E-1. The strain sensitivity is defined as
a slope of curves. In compressive strain region ($\varepsilon_0<0$), the sensitivity in
wires I-s and I-1 are larger than E-1. In order to comprehend the role of
stabilizing copper on the strain dependence of I_c, a qualitative analysis
with rough approximation is tried here. The wire is supposed to be divided
into two parts; one is the region of superconductor (Nb$_3$Sn filaments
embedded in the bronze) which behaves elastically, and the stabilizer which

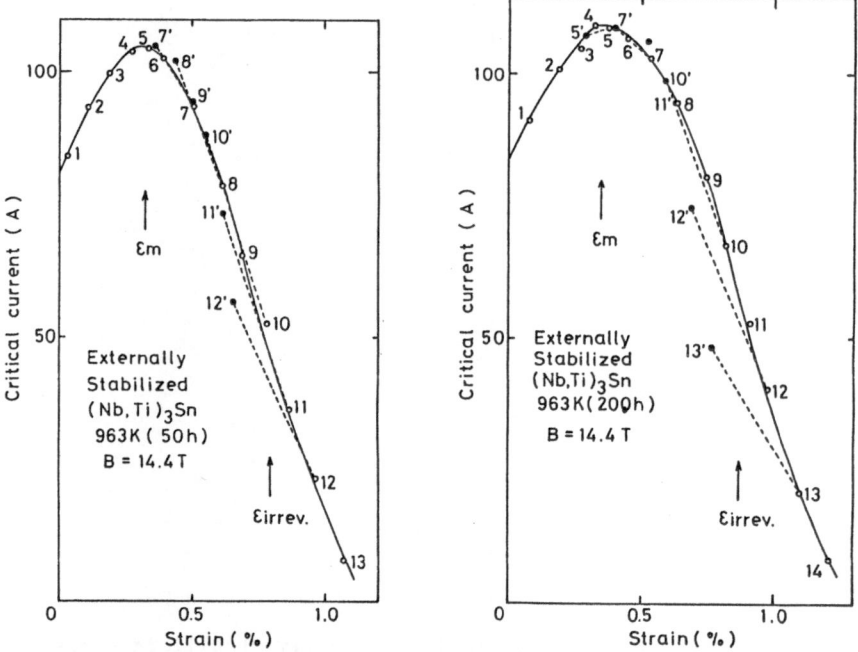

Fig. 4. Strain dependence of I_c in externally stabilized wires.

88

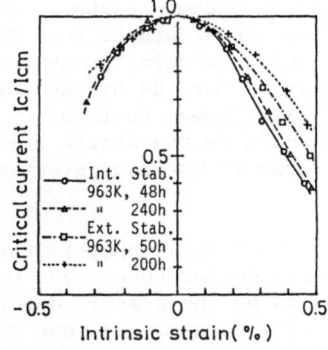

Fig. 5. Strain sensitivity of I_c.

behaves plastically. The coefficient of thermal contraction in the superconductor is known to be lower as compared to the others. The supposed shapes of components independent of each other and the expected strain state in the Nb_3Sn in wires before-, after cooldown, and elongated beyond ε_m the peak strain are schematically shown in Fig. 6. In the case of externally stabilized wire, the strain state as cooled down is much or less homogeneous in the three principal strain components, i.e., compressive axial, radial and tangential strain. If the wire is subjected to a strain above the yield point, stabilizer deform plastically. As the poisson's ratio is 0.3 for elastic deformation (superconductor) and 0.5 for plastic deformation (copper), larger reduction of diameter in copper will results in further compressive radial and tangential strain to the superconductor. On the other hand, when the internally stabilized wire is cooled down, the radial strain component is positive. This will result in higher effective (geometric averaged) strain for certain axial strain as compared to the case in the externally stabilized wire. Thus the strain sensitivity of the wire

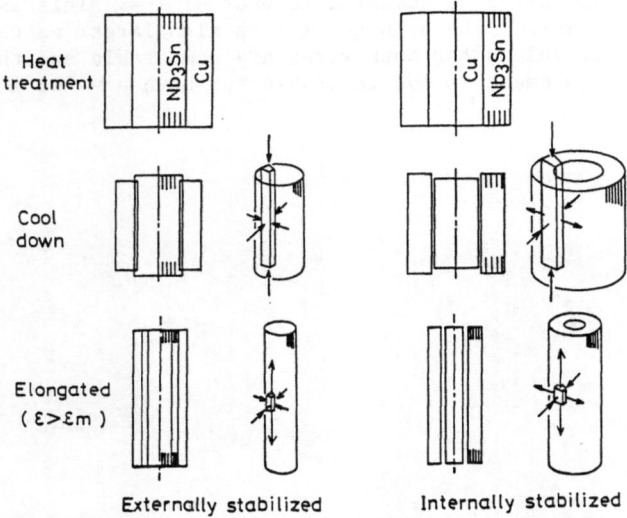

Fig. 6. Schematic illustration of deformation and strain state in the superconductors.

in the compressive strain region becomes higher. This situation is similar to that occurring in the filaments of bronze core Nb tube multifilamentary conductor.[7] The detailed computer calculation by Scanlan et al showed effective strains are much higher for the bronze core sample.[7] According to their result, however, they also showed that the change in the effective strain with axial strain is small in the bronze core sample as compared to the that of bronze matrix. This is not consistent with the present experimental results.

It must be pointed out that the copper in the internally stabilized wire is confined in the superconductor tube. The negative hydrostatic stress on cooling down and also by the plastic deformation can not be relaxed in spite of its low flow stress. So long as the separation between the superconductor and the stabilizer does not occur, this would not disappear. (Figure 7 shows the fracture surface of the wire I-1 indicating that the separation have taken place at fracture with significant necking.) In the case of externally stabilized wire, however, stress can partly be relaxed by the plastic deformation of copper.

It has been empirically shown that the stress dependence of I_c in multifilamentary wires is generally described in the equation[8]

$$I_c(\varepsilon) = I_{cm}(1-a|\varepsilon_0|)^u$$

where u=1.7, a=900 for $\varepsilon_0<0$ and a=1250 $\varepsilon_0>0$. The reason for this asymmetry across $\varepsilon_0=0$, to the authors knowledge so far, is not explained. One plausible reason would be the difference in the strain state in radial and tangential components in the compression range (from as cooled down to $\varepsilon_0=0$ axial strain) and in the tension range (from $\varepsilon_0=0$ to tensile strain). For the detailed discussion, a quantitative analysis using finite element method is required.

Strain Sensitivity of n Value

The n value specifying the relationship between the voltage V and the transport current I in empirical expression of $V \propto I^n$ has been calculated in these wires. Figure 8 shows strain dependence of n in three wires determined in the voltage range of 0.5 to 5μV/cm. Although the scatter of n is rather large, n changes with strain in the same manner as I_c. The n value in wire I-1 is larger as compared to that of I-s. This is because of higher I_c in the former. The n in the I-1 is also larger as compared to that in E-1. The I_c values for both wires are comparable and the magnetic field in the I_c measurement is not favorable for high n value in the former

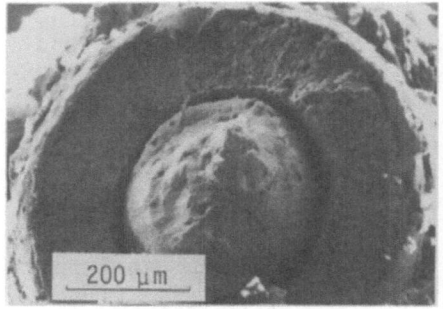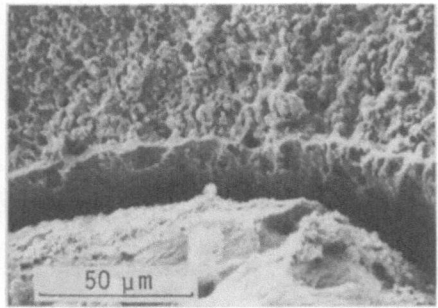

Fig. 7. Fracture surface of internally stabilized wire (4.2 K, 15 T).

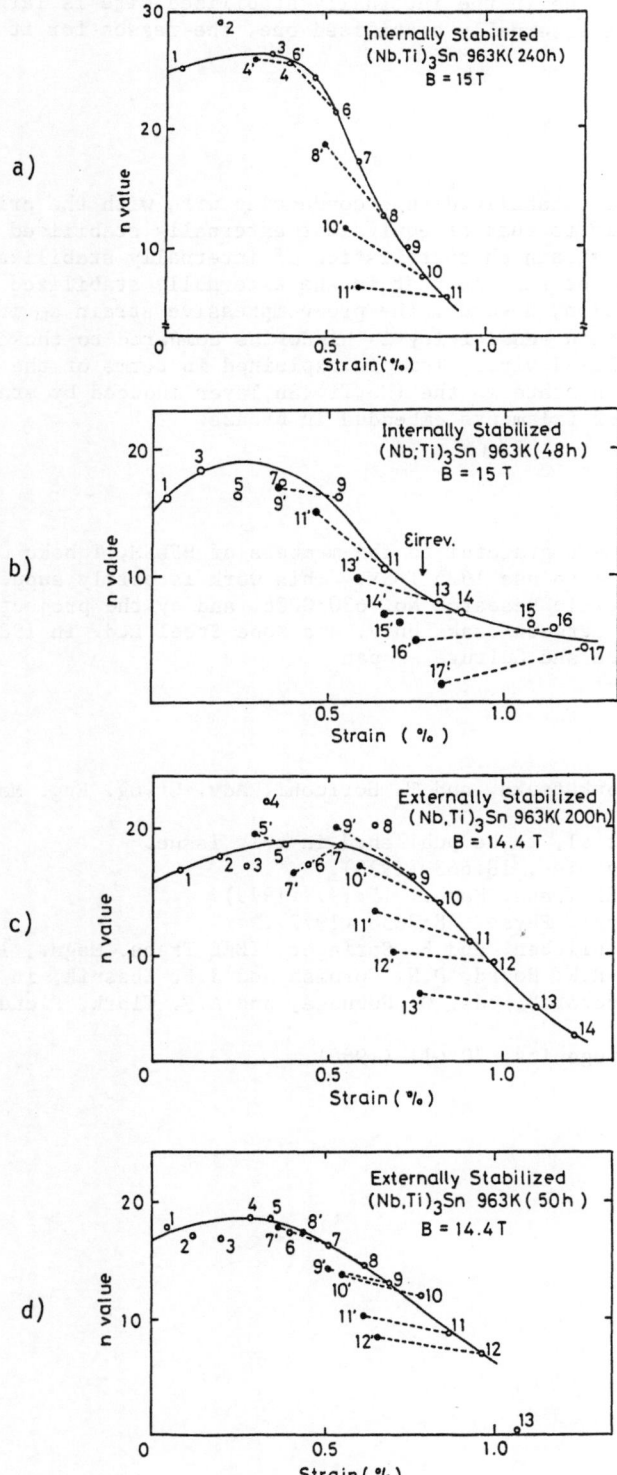

Fig. 8. Strain dependence of n-values.

(15 T vs. 14.4 T). Taking these facts into consideration, this result suggests that n value in the internally stabilized wire is larger than that in the equivalent externally stabilized one, the reason for it is not known at present.

SUMMARY

An internally stabilized superconducting wire with the critical current density comparable to that of equivalent externally stabilized one has been fabricated. The strain characteristics of internally stabilized $(Nb,Ti)_3Sn$ wires are almost the same as that in the externally stabilized one. In detailed examination, however, the pre-compressive strain ε_m in the wire is lower and the strain sensitivity is higher as compared to those in the externally stabilized wire. This is explained in terms of the change in the three axial strain state in the $(Nb,Ti)_3Sn$ layer induced by stabilizer confined inside of filaments embedded in bronze.

ACKNOWLEDGMENT

The authors are grateful to the members of HFLSM, Tohoku University for giving convenience to use 16.5 T-SM. This work is partly supported by Grant in Aid for Scientific Research No. 63050006, and by the project of cooperative work between Osaka Univ. and Kobe Steel Ltd. in 1989, Ministry of Education, Science and Culture, Japan.

REFERENCES

1. Y. Monju, T. Fukutsuka, and T. Horiuchi, Adv. Cryog. Eng. Maters., 32:977 (1986).
2. K. Katagiri et al, to be published in this issue.
3. G. Rupp, Cryogenics, 18:663 (1978).
4. J.W. Ekin, IEEE Trans. Magn., 15:197 (1979).
5. G. Rupp, J. Appl. Phys., 48:3858 (1977).
6. G. Rupp, K. Wohlleben, and E. Springer, IEEE Trans. Magn., 17:1622 (1981).
7. R.M. Scanlan, R.W. Hoard, D.N. Cornish and J.P. Zbasnik, in "Filamentary A15 Superconductors", eds. M. Suenaga, and A.F. Clark, Plenum, NY (1980) P.221.
8. J.W. Ekin, Cryogenics, 20:611 (1980).

EFFECTS OF TRANSVERSE STRESS ON THE CURRENT CARRYING CAPACITY

OF MULTIFILAMENTARY WIRES

H. Boschman and L.J.M. van de Klundert

University of Twente
P.O. Box 217, 7500 AE Enschede
The Netherlands

ABSTRACT

The influence of transverse compressive stress on the current carrying capacity of multifilamentary Nb_3Sn wires has been investigated on three short samples with a copper, bronze and mixed matrix. A 10 % current reduction has been observed at stresses ranging from 50 to 100 MPa for the three wires. In the test arrangement both applied force and deformation of the wires have been determined during the experiments. It has been found, that the critical current degradation can be described more unambiguously as a function of the deformation of the wire than as a function of applied load. This is caused by hysteresis effects in the stress-strain relation of the matrix material. There are indications that the length over which a compressive stress is applied is important for the observed current reduction. It has been found that degradation effects are more severe when small compression lengths are used.

INTRODUCTION

For the practical use of multifilamentary superconducting wires and cables, it is a necessity to have knowledge about their performance under mechanical loads. In magnets, windings are stressed in both axial and transverse direction due to Lorentz forces, cooling and prestressing. As a consequence the current carrying capacity of the superconductors can be affected, thus leading to a disappointing magnet performance. Therefore, measurements have been performed in which the critical current is determined while an external transverse compressive force is applied to superconducting wires. For round wires, this leads to a complicated stress pattern inside the wire. Nevertheless, in this paper applied transverse force is written as a stress (i.e. force divided by compressed length and diameter of the wire), in order to facilitate the comparison of the results of the different wires.

In the case of multifilamentary NbTi wires, effects caused by stress, are generally quite acceptable. According to Ekin[1], current degradation starts immediately when a wire is stressed axially. However, at fields below 7 T the critical current is not diminished below 95 % of its initial value, until the tensile strain of the wire is over 1 %, which corresponds with a stress of approximately 450 MPa. A transverse, compressive stress acting upon a NbTi wire does not result in a major current reduction either. It has

Advances in Cryogenic Engineering (Materials), Vol. 36
Edited by R. P. Reed and F. R. Fickett
Plenum Press, New York, 1990

93

been found[2], that a round wire with diameter 0.7 mm, which was transversely pressed between two parallel plates with a stress of 285 MPa, showed a decrease of only approximately 1.8 % of the maximum current at zero force (applied field: 6 T). Upon unloading, the current recovered almost completely to its initial value, although the cross section of the wire remained deformed permanently.

Degradation effects are much larger in the case of the brittle superconducting material Nb_3Sn. The effects of axial stress have been studied extensively during the past 20 years[1]. The current reduction pattern of multifilamentary Nb_3Sn wires is largely affected by the amount of compressive pre-stress on the filaments. When a tensile load is applied to a wire, first the filaments are freed from the pre-stress, which leads to an increase of the critical current. Typically, the maximum I_c is reached at a 0.1 to 0.3 % overall strain for practical Nb_3Sn conductors. After this point the filaments start to be under a regime of tensile stress and a rapid current reduction is the result (1 % intrinsic strain of the filaments means approximately 50 % I_c reduction at 7 T). Irreversible damage occurs to wires at intrinsic strains ranging from 0.4 to 0.7 %. Current degradation is reported to be much larger, when stress is applied in transverse direction. Ekin[3] examined a round multifilamentary Nb_3Sn wire with a 0.7 mm diameter. At 8 T a 10 % reduction of the critical current was found at a uniformly applied transverse stress of about 80 MPa. At the maximum applied stress of 180 MPa the critical current was only half of the original value in the unloaded case, but almost complete recovery was reported after unloading. Specking et al.[4,5] found similar results on preflattened wires, although it is not quite clear whether current degradation starts immediately and to what extent current recovery is present.

From the above-mentioned results it can be concluded that, especially for Nb_3Sn, mechanical loads can affect the current carrying capacity of superconducting wires to a large extent. However, in the case of transverse stress very little research has been done until now and it is not yet clear to what extent this current limiting effect obstructs practical application of Nb_3Sn wires. An aggravation of the situation can be expected when large local transverse loads occur, as can be the case at cross-over points in cabled conductors. Further research on this topic may well impose limitations with respect to size, materials and shape on the wires.

EXPERIMENTAL SET-UP

In order to study the influence of transverse compressive force on the current carrying behaviour of small samples, a test arrangement has been constructed which fits into the 7.6 cm bore of a 7 T dipole magnet (figure 1). Samples to be investigated are reacted in a U-formed shape and subsequently layed on a stainless steel plate. Current terminals are soldered to the ends of the sample and voltage taps are attached in the straight middle-section of the sample, which is approximately 7 cm. The direction of the magnetic field is perpendicular to the test section of the sample, in such a way that the Lorentz force, resulting from field and sample current, presses the wire onto the plate. The advantage of this design is that there is no need for epoxies or adhesives to fix the sample to its place during experiments. Usage of these materials would lead to an experiment in which applied stress is divided between wire and epoxy in an unknown way.

Part of the middle section is placed under a pressure block of stainless steel which acts as a lever. Transversal force can be applied onto the sample by pulling at a wire-rope outside the cryostat. Due to the leverage the applied force is multiplied with a factor ranging from 3 to 4,

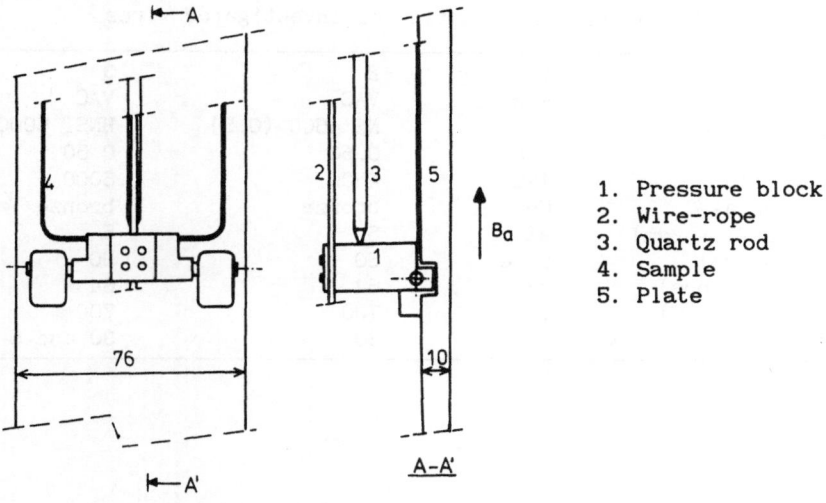

1. Pressure block
2. Wire-rope
3. Quartz rod
4. Sample
5. Plate

Fig. 1. Front and side view of test arrangement. Sizes in mm.

which depends on the position of the sample. The maximum externally applied force is approximately 500 N. Because several shaped levers can be used, the length over which the compressive force is exerted onto the sample is between 5 and 40 mm.

Another feature of this arrangement is a displacement indicator. The deformation of the wire under compression, which is in the order of μm's, is transformed into a rotation of the lever over a small angle. As a consequence, the vertical displacement of the pressure block is a measure for the wire deformation. A quartz rod which rests upon a fixed point on the upper side of the block, transfers the displacement to outside the cryostat, where it moves the gauge of the indicator. The latter device is connected by means of a quartz cylinder with the sample holder. It can move freely with respect to the cryostat's lid. In this way, effects of thermal shrinkage are minimized and it turns out that relative displacements of the lever with respect to the sample holder down to 1 μm can be detected. However, the absolute accuracy of this device is limited, because it is hard to determine the zero position of the undeformed round wire. Especially irregularities on the sample's surface allow large displacement under small pressure.

EXPERIMENTAL PROCEDURE

For all samples, the voltage-current characteristics have been registered at 4.2 K in several load cycles. In such a cycle, the critical current is first determined without applying any external force onto the sample. Next, voltage-current characteristics are measured while the force is gradually increased. Finally, the load is released in a couple of steps, in order to examine the recovery of the wire. This procedure is repeated several times, ever applying a larger maximum load.

While increasing the current, a small deformation of the sample (up to approximately 5 μm) takes place due to the Lorentz force which amounts about 1 to 3 kN/m. In the presentation of the experimental results, only the impressions of the samples at I_c are mentioned.

Table 1. Characterization of investigated wires

Wire nr.	1	2	3
Manufacturer	ECN	VAC	VAC
Manufacturer's code		NS 4500 (0.5)	HNST 6000 (0.6)
Diameter (mm)	0.59	0.50	0.60
Number of fil.	192	4500	6000
Matrix material	Cu	bronze	bronze/Ta/Cu
Diameter of fil. (μm)	22	5	5
Twist pitch (mm)	none	50	50
Reaction time (hrs)	48	64	64
Reaction temp. ($^{\circ}$C)	675	700	700
Pressed wire length (mm)	38	30	30 and 5

RESULTS

Table 1 shows the characteristics of the 3 wires that have been investigated. The voltage taps have been soldered to the samples outside the compressed area, except for wire 1. The length over which the voltage has been registered is approximately 35 mm. As a criterion for the critical current a 1 μV voltage drop has been adopted, independent of the compressed wire length. The accuracy of the critical current is $\pm$ 1 A. Note that the wires of Vacuumschmelze are industrially produced using the bronze route, while ECN follows its own process which leads to relatively large filaments containing a powder core.[6]

The measurements on both wire 1 and 2 were not completely successful: it was not possible to determine the critical current through the test section at zero load because of early quenching. For wire 2 this was due to the high resistivity of the matrix. This type of wire always has problems with respect to good conducting joints for high currents, especially when the joints and test section are in the same magnetic field. However, upon applying only a small compressive force, I_c could already be measured, indicating that the original quench current was only a little less than I_c.

Experimental results of wire 1 are shown in figure 2. It can be seen that I_c is reduced by 10 % when a stress of 45 MPa is applied to the wire (under the assumption that the initial critical current was 580 A). However, there is no clear relation between applied force and current degradation.

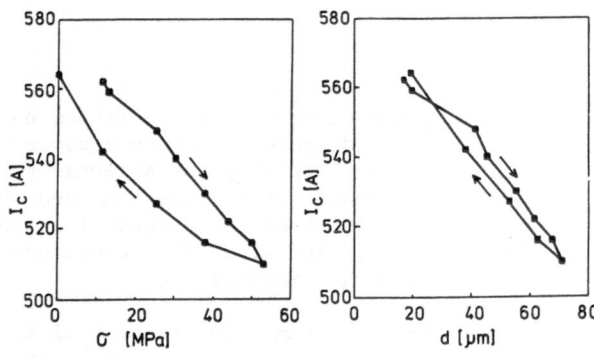

Fig. 2. Critical current as a function of applied stress and deformation for wire 1. Applied field: 6.5 T.

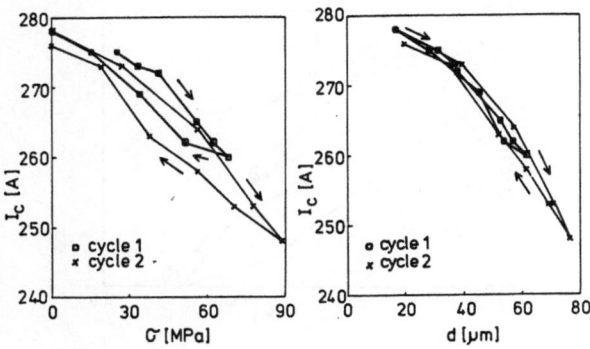

Fig. 3. Critical current as a function of applied stress and deformation
for wire 2. Applied field: 7.0 T.

When the load is decreased, the current turns out to be significantly
smaller than with increasing loads. However, in case the current is scaled
against the deformation of the wire, the relation between both variables is
much more unambiguous. This phenomenon is due to hysteresis effects in the
stress-strain relation of copper. The results show that current degradation
is influenced more directly by wire deformation than by applied force.

For wire 2 (figure 3) the same hysteresis effect is observed. Note,
however, that in the relation between current and stress a clear envelop can
be distinguisghed, for the increasing load line of cycle 2 approaches the
one of the first cycle. Therefore, expressing current degradation in terms
of applied stress makes sense when it refers to this envelop. For this wire,
the degradation amounts 10 % for approximately 80 MPa, when the initial I_c
is extrapollated to 280 A.

For the stabilized Vacuumschmelze wire (3) it was possible to measure
the critical current at zero load. The same behaviour with respect to the
hysteresis is seen in figure 4. At a compressive stress of 75 MPa a 5 %
current reduction has been measured. At these loads the current does not
recover completely to its original value, but a few amperes are lost
permanently. However, this does not necessarily mean that filaments are
damaged. In the relation between I_c and the deformation it can be seen that
the final point is still almost on the line, especially for cycle 1. On the
completely unloaded wire a small deformation is left, which leads to a
somewhat lower critical current.

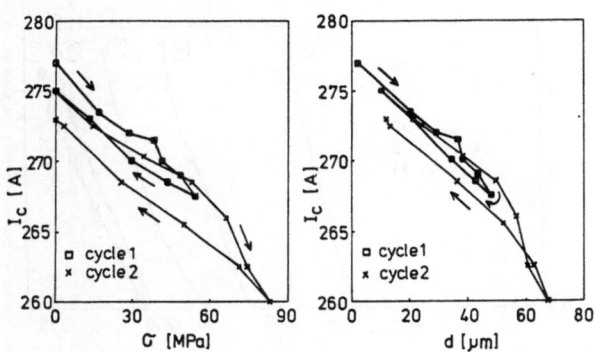

Fig. 4. Critical current as a function of applied stress and deformation
for wire 3, with compression length of 30 mm. Applied field: 7.0 T.

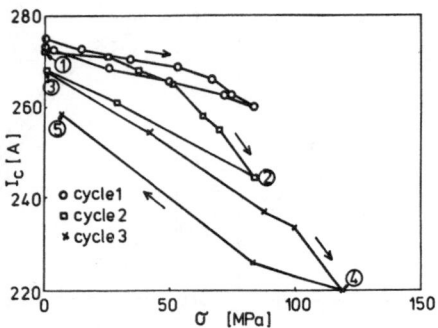

Fig. 5. Critical current as a function of applied stress for wire 3. Applied
field: 7.0 T. Cycle 1: compression length of 30 mm, cycle 2 and 3:
compression length of 5 mm. Numbers refer to fig. 6.

A further reduction of the critical current was achieved using another
pressure block which compresses the sample over a length of only 5 mm. Note
that the measurement concerns the same sample that had already been
impressed over a length of 30 mm. Because the critical current of a wire is
determined by the weakest spot in a wire, approximately the same dependence
of I_c on applied stress was expected. Figure 5 shows three consecutively
measured cycles. In the first cycle a 30 mm compression length was used and
small degradation effects were found. However, in case of the small pressure
block, I_c was reduced much more severely. This suggests, that the current
reduction depends on the length over which transverse stress is applied.

Another phenomenon which has been observed with the compression length
of 5 mm, concerns the shape of the voltage-current characteristics. In
figure 6 a few curves are scaled logarithmically in order to check the
relation[7]: $U \propto I^n$. Usually, a large n-value, which means a sharp transition
from the superconducting to normal state, is associated with a qualitatively
good wire. It can be concluded, that when the load on the wire is increased,
I_c is reduced, but the transition gets sharper. Although the fit to the
n-power law is not perfect, n-values have been determined between 1 and 5 μV
and show an increase after unloading. This latter result is in contrast with
intuition.

During the measurements with the 30 mm pressure length, the ECN-wire
had a constant high n-value of approximately 70. For the wires of
Vacuumschmelze only n-values below 30 were observed, except for the

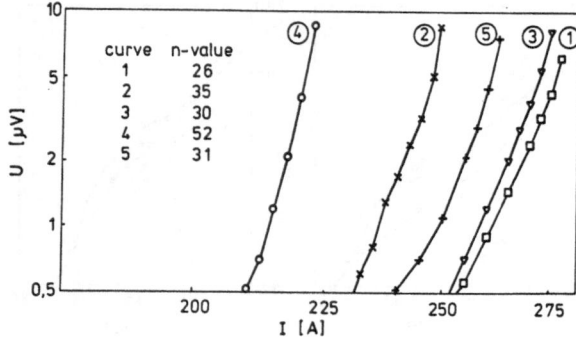

Fig. 6. Logarithmically scaled U-I curves of wire 3 with compression length
of 5 mm. The numbers of the curves refer to the points in fig. 5.

experiments with the short compression length. Also much higher currents were obtained by wire 1, but its sensitivity to transverse loads was a little larger. However, from the results, it is not yet clear to what extent the sensitivity of wires to transverse stress depends on composition of the wire, amount and sort of matrix material, twist pitch etc. Therefore more wires have to be investigated to deal with these factors.

FUTURE DEVELOPMENTS

Measurements of transverse stress effects on longer samples are expected to have a few advantages over ones on short samples, such as described above. First, due to longer sample lengths a higher voltage is obtained which facilitates a more detailed study of voltage-current characteristic (e.g. n-value). Moreover, measured voltages are less influenced by effects of current (re)distribution between the filaments, which can occur at the soldered joints and at the edges of the compression zone. Considering the influence of filament twist, samples should at least be a couple of twist pitches long.

If the homogeneity of the magnetic field is limited to a small plane, a wire can be wound spirally to obtain a long sample. It can be pressed between two parallel plates. On basis of this geometry, another press-arrangement has been constructed. Samples in this apparatus are not supported against the Lorentz forces, and need to be precompressed for fixation. The allowable amount of precompression without damaging the wire is indicated by experiments on short samples.

CONCLUSIONS

A considerable current degradation was found in multifilamentary Nb_3Sn wires under transverse stress. A 10 % current reduction has been found for stresses ranging from 40 to 100 MPa for three different wires. The results are in reasonable agreement with earlier results reported in literature. The history of loading is important for the relation between current and applied stress, because hysteresis effects in stress-strain relations. It turns out that deformation describes the current degradation more unambiguously.

A few phenomena observed are not completely understood. An increase in the n-value has been measured after compression of a wire. There are indications, that current reduction is dependent on the wire length over which stress is applied. Perhaps this is due to stress redistribution on the edge of the pressure block. Calculations on this issue should be made with respect to the stress distribution inside the wire. Preliminary results are to be published.[8]

ACKNOWLEDGEMENTS

The authors would like to thank H. ten Kate, P. Fornerod and H. de Jonge, who did part of the measurements for his scientific training, for their collaboration.

REFERENCES

1. J.W. Ekin, Mechanical properties and strain effects in superconductors, in: "Superconductor materials science: metallurgy, fabrication and application", S. Foner and B.B. Schwartz, eds., Plenum Press, New York, (1981).

2. H. Boschman, P.P.E. Fornerod and L.J.M. van de Klundert, The influence of transverse, compressive stress on the critical current of multifilamentary Nb₃Sn and NbTi wires, MAG-25, pp. 1976-1979, (1989).

3. J.W. Ekin, Effect of transverse compressive stress on the critical current and upper critical field of Nb₃Sn, J. Appl. Phys., vol. 62, pp. 4829-4834, (1987).

4. W. Specking, W. Goldacker and R. Flükiger, Effect of transverse compression on I_c of Nb₃Sn multifilamentary wire, Adv. Cryog. Eng., vol. 34, pp. 569-575, (1989).

5. W. Specking, F. Weiss and R. Flükiger, Effect of transverse compressive stress on I_c up to 20 T for binary and Ta alloyed Nb₃Sn wires, Proceedings 12th Symposium on Fusion Engineering, pp. 365-368, (1987).

6 E.M. Hornsveld, J.D. Elen, C.A.M. van Beijnen and P. Hoogendam, Development of ECN-type Niobium Tin wire towards smaller filament size, Adv. Cryog. Eng., vol. 34, pp. 493-498, (1989).

7. H. Boschman, H.H.J. ten Kate and L.J.M. van de Klundert, Critical current transition study on multifilamentary NbTi superconductors having a Cu, a CuNi or a mixed matrix, MAG-24, pp. 1141-1144, (1988).

8. H. Boschman, P. Fornerod, H.H.J. ten Kate and L.J.M. van de Klundert, Degradation of the critical current of multifilamentary Nb₃Sn wires under transverse mechanical load, to be presented at MT-11, (1989).

FURTHER STUDIES OF TRANSVERSE STRESS EFFECTS IN

CABLE-IN-CONDUIT CONDUCTORS

L.T. Summers and J.R. Miller

Lawrence Livermore National Laboratory
P.O. Box 5511, L-643
Livermore, CA 94550

ABSTRACT

The effect of transverse stress on critical current has been examined for
Cable-In-Conduit Conductors (CICC's) containing three active superconducting
composite strands in cables containing a total of 21 strands. In measurements
of this type reported previously, only soft copper was used for the inactive
strands, allowing the possibility that peaking of stresses at strand cross-over
points were avoided by deformation of the copper strands during CICC
fabrication and testing. In the present experiments, the degree of critical current
degradation was measured as a function of applied load for various void
fractions for cable patterns using stainless steel wires as the inactive strands.
The reduction of critical current, expressed as a function of load divided by the
projected area of the core of the superconducting composite strand, was found to
be similar to that observed in cables containing copper inactive strands. All the
CICC's tested show a higher sensitivity to transverse stress as compared to
single wires. At compressive loads of 50 MPa or less, the region of interest to
magnet designers, the critical current is, at worse, 79% of the critical current in
unloaded samples.   The sensitivity to transverse load is a function of CICC void
fraction, lower void fractions having less susceptibility to degradation. The
results of this investigation indicate that the performance of large magnets
employing CICC designs need not be seriously degraded due to transmitted or
self-induced Lorentz loads.

INTRODUCTION

The effect of axial strain on Nb_3Sn superconductors has been widely
studied and documented (see for example Ref. 1 and 2). This previous work has
led to the development of strain scaling laws for the prediction of H_{c2} and I_c as a
function of residual or applied strain and the prediction of multifilamentary

Advances in Cryogenic Engineering (Materials), Vol. 36
Edited by R. P. Reed and F. R. Fickett
Plenum Press, New York, 1990

101

conductor stress states based on the volume fractions and thermal-mechanical properties of the conductor constituents. [3,4,5]

Recently the effect of transverse stress on single wires of Nb₃Sn has been studied.[6,7,8] Investigators have found that degradation of I_c with stress apparently occurs much faster under conditions of transverse stress than equivalent conditions of axial stress. Significantly degraded performance is observed at compressive stress levels as low as 50 MPa in multifilamentary superconductors.[6]

This has raised serious question about the suitability of CICC's in large applications where transverse stresses, either transmitted from adjacent conductors or generated within a single CICC, could potentially degrade performance below acceptable levels. Recent work shows that transmitted loads are not a concern in CICC's.[9] The conductor conduit is extremely stiff with respect to the internal cable and bears a major fraction of loads transmitted from adjacent conductors. The average stress transmitted to the cable is only 3% of the average applied load and the peak stress is about 7%.

Although the effect of transverse stress is reasonable well understood for monolithic conductors and single wires, analysis of the effects of self generated loads in a CICC is complicated. This complication arises from the variation in the size of the load "footprint". Conductors lying on the inner face of the conduit in the direction of the J X B forces have a contact footprint on the conduit side that may be conveniently described by some aspect of their geometry, such projected cross sectional area, projected area of superconducting core, etc. The opposite face of these conductors in the direction of the interior of the CICC, as well as other conductors within the cable space, have load footprints that are described by the contact points between wires in the cable. The size of these contact points are a function of the cable size, cable twist pitch, and the amount of compaction during CICC fabrication.

Previously we reported the results of an investigation to experimentally measure transverse stress effects in CICC's of various void fractions. [10] Critical current degradation under transverse loads was found to be significantly worse than in single wires, however at low compressive stress ($\sigma < -50$ MPa) performance was not sufficiently degraded to preclude the use of CICC's in large applications such as magnets for fusion energy. The CICC's used in this investigation were fabricated using 3^3 cables of which 3 strands were superconductor and the remaining 24 strands were inactive copper. This configuration was chosen to reduce the effect of self generated lorentz loads, allowing control of transverse loads by external means under the experimenter direct control.

However, these CICC samples had several deficiencies. First, the position active superconducting strands with the conduit and cable bundle is random due to cable transposition. At certain locations the superconductor may lie along the inner wall of the conduit, while at other locations it may be found near the center of the cable. The geometry of wires near the center of the cable is also variable as the number of nearest neighbor wires is random. Thus the number number and type of load contacts variy from specimen to specimen and along the length of individual specimens. This was suspected of causing the large scatter in previously reported data. A second disadvantage of the earlier specimen design is the use of copper strands as the inactive elements. It was suggested that at high applied loads the soft copper could deform and increase the size of the load footprint thus reducing the effect of transverse compression.

In this investigation we tested CICC specimens fabricated using cables with a $(6 \times 1)^3$ cable pattern wherein the central strand is superconductor and the outer six strands are stainless steel. The hard stainless steel will not easily deform under applied load and will tend to minimize the size of the load footprint. Secondly the 6×1 cable first element, with the superconductor in the center, will mitigate geometry effects as the superconductor will always have 6 wire nearest neighbors and no contact with the inner face of the conduit.

EXPERIMENTAL PROCEDURE

We elected to evaluate transverse stress effects using sub-sized CICC conductors manufactured using 21 strand cables in a $(6 \times 1)^3$ cable pattern. Although we wanted to examine the effects of the internal, self generated load in a CICC, we also wanted to have external control of the transverse load, To accomplish that, we elected to test CICC's with a weakened jacket wall to allow direct transfer of an external load. Additionally, the cables inside were fabricated with only three active superconducting strands (the remaining   ere stainless steel) so that the self-generated transverse load was minimal ar .. ...e total transverse load on the strands was dominated by the externally applied load. The superconductor employed was a 0.9 mm diameter, modified jelly roll, binary Nb_3Sn with a non-copper volume fraction of 0.65.

The 27 strand cables were inserted into 304 stainless steel tubes and processed with a combination of swagging and Turk's head rolling to produce CICC's of square cross section. The rolling and swagging operations were terminated at reduction levels which produced conductors with helium void fractions of 0.40 and 0.30. Current contacts were attached to the specimens by swagging ETP copper fittings to the wire bundles protruding from the ends of the tubes. The CICC specifications are shown in Table 1.

The CICC were then given a reaction heat treatment at 700° C for 100 hours. After removal from the furnace opposite faces of the CICC were slotted with using an end mill. The slots were centered on the CICC face, were approximately 150 mm long, and sufficiently deep so as to leave only a thin foil

Table 1. Specifications of the CICC's used in this investigation

Void Fraction	Cable Pattern	External Dimension (Flat to Flat)	Wall Thickness
%		m m	m m
40	$(6 \times 1)^3$	6.33	0.71
30	$(6 \times 1)^3$	6.06	0.76

of the conduit at the bottom of the slot. The purpose of the slot was to remove supporting structural material so that the conduit was free to collapse on application of applied load. Virtually all of the applied transverse force is transmitted to the cable. This configuration and loading sequence efficiently mimics the internal J X B forces experienced by a much larger CICC.

The CICC's were tested in a 12 T split pair solenoid superconducting magnet equipped with a transverse load cage . The load cage is constructed of 304 stainless steel and consists of a movable ram that is actuated by a pressurized diaphragm. The diaphragm and cage assembly is immersed in LHe and is usable up to approximately 13.5 MPa, the solidification pressure of He at 4.2 K.

The loading forces are transmitted through the specimen and reacted against a fixed anvil attached to a tension tube and located at the opposite end of the load cage . The test specimen enters the assembly through a radial access port in the magnet and passes through a slot in the transverse load cage. The anvil and ram apply the load to a 38 mm length of the CICC. A diagram of the apparatus is shown in Figure 1.

The amount of force is measured indirectly using two temperature and field calibrated strain gages attached to the tension tube of the load cage and located 180° apart. Published values for the 4 K modulus of 304 stainless steel

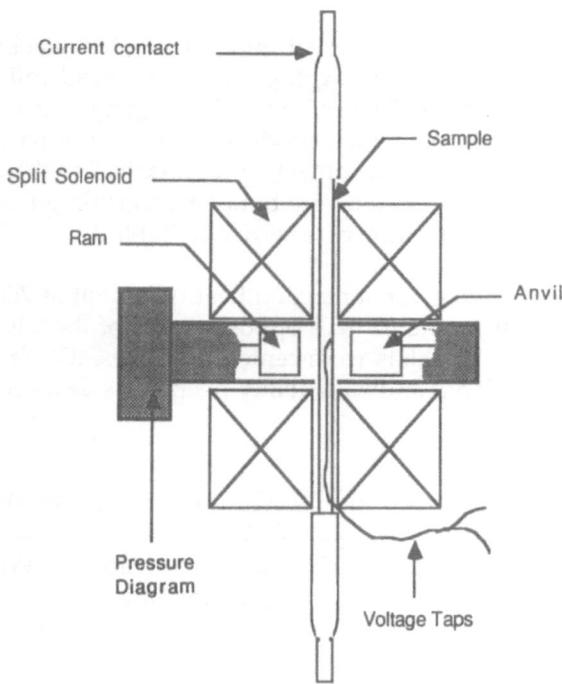

Figure 1. Sample arrangement within the test magnet and transverse load cage. (not to scale)

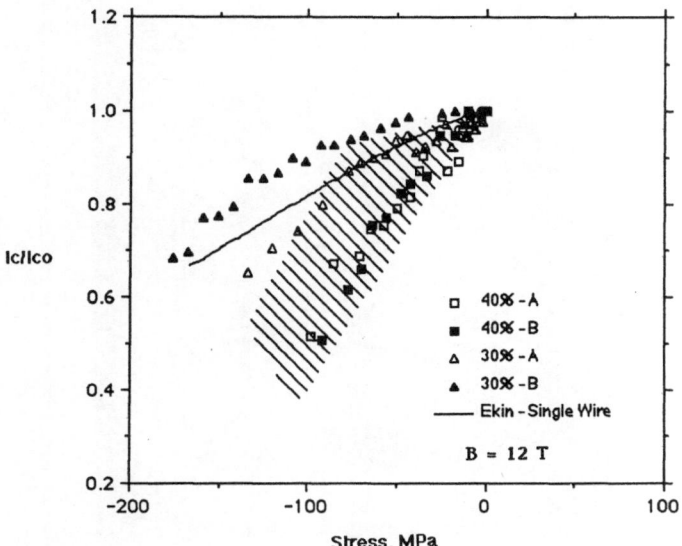

Figure 2. Normalized critical currents in $(6 \times 1)^3$ cables as a function of transverse stress and the void fractions shown. Data points A and B at each void fraction are for multiple specimens. The shaded area bounds a region in which data for 27 strand 3^3 cables using Cu inactive elements lies. Ekin's data for single strand wire is shown for reference.

were used for purposes of calculation. The stress in the cable is calculate using the projected area of the non-Cu cores.

Critical current was measured by voltage taps attached to the specimen conduit in the loaded section. The distance between voltage taps was 20 mm and I_c was determined using a voltage criteria of 1 μV cm^{-1}.

RESULTS

The data obtained from tests of 40% and 30% void fraction CICC's is shown in Figure 2. For reference a plot of Ekin's data for single round wires shown. His 10 T data was converted to 12 T using a procedure described previously.[10] The shaded region bounds the results of previous tests of 27 strand CICC's containing Cu inactive strands.

A cross section of a region in a 40% void specimen that was not loaded in compression is shown in Figure 3a. The three active strands of superconductor can be seen at the center of each 6 x 1 first element. Note the severe deformation of the superconductor that occurred during fabrication. A cross section of the same specimen in a region actively loaded during testing is shown in Figure 3b. Note the slits in the conduit wall that allow almost all the applied load to be transmitted to the cable. Deformation of the superconductor that occurs during testing is not discernable because of the deformation occurrnign during fabrication.

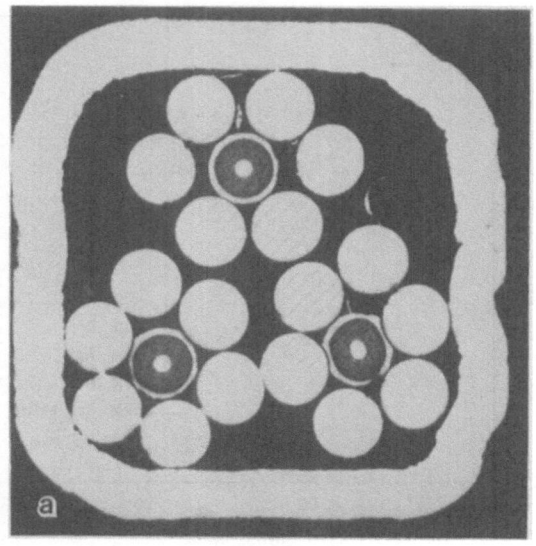

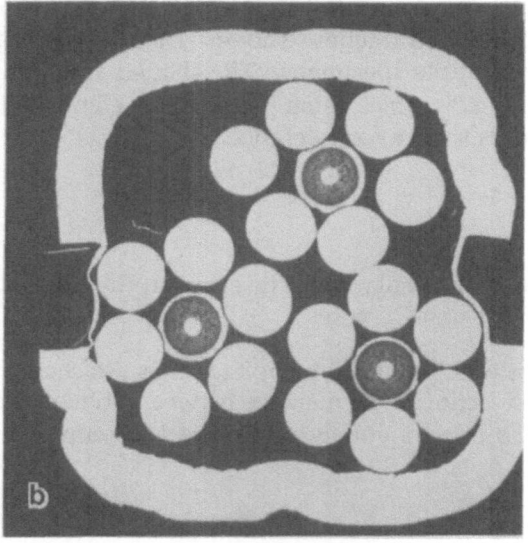

Figure 3. Cross section of a 40% void CICC specimen. A region of the specimen that was not loaded in compression is shown in 3a. An area of the same specimen that was loaded in transverse compression is shown in 3b.

DISCUSSION

While there is good agreement between the data for the two 40% void specimens, there is considerable scatter between the two 30% specimens. As with previous tests, the stress state of the Nb_3Sn wires may vary as a function of position within the CICC. This is particularly noticeable in the 30% void specimen B which has less sensitivity to transverse stress than single wires. Obviously one or more of the superconducting wires in this specimen is shielded from the applied load. Attempts to control the scatter by using 6 x 1 cable patterns have not been wholly successful.

Regardless of the scatter in data the effect of void fraction is clearly evident. The 40% void fraction specimens receive less compaction than the 30% void specimens during fabrication. Therefore, the cross over points between wires in the higher void samples do not deform as much as in the low void specimens. The load foot print is smaller in high void samples and the higher contact loading results in increased susceptibility to applied transverse loads.

Use of stainless steel wires in the cable may not have had the desired effect on transverse load sensitivity. One would assume that the inactive steel wires cannot be as easily deformed as the dead soft copper used in earlier studies. Ideally this would limit deformation at wire cross over points at high loads and result in more severe critical current degradation. As seen in Figure 2, however, all the data falls in or above the trend band for the cables made using copper wires.

A possible explanation for this is the high deformation that takes place during CICC compaction. The steel, although annealed, is harder to deform than the superconductor. When the CICC is swagged and rolled square allot of the deformation is taken up by the superconductor alone. Therefore the cross over points between the steel and Nb_3Sn are highly deformed prior to application of the transverse load. Point contact is not as severe and the samples show less susceptibility to transverse stress. Further testing using 25% and 35% void fraction specimens is being pursued and those results may confirm this hypothesis.

The impact of transverse stress effects on CICC's used in large applications is limited. As an example a 36 kA, 11 T superconductor proposed for ITER Toroidal Field (TF) coils will produce only a 30 MPa lorentz force due to self loading. The resulting degradation of critical current in this large conductor would be less than 20% according to this and previous data. In addition, a magnet designer also has the option to change the conductor height in the direction of the lorentz force thus reducing self loading.

CONCLUSIONS

Transverse stress effects scale with CICC void fraction. This is likely due to increasing point contact with increasing void fraction.

There is significant scatter in the data of some specimens tested. This indicates that geometric effects have not been fully mitigated by use of $(6 \times 1)^3$ cable patterns.

Use of inactive steel strands result in less sensitivity to transverse stress. this may result from high deformation at steel/Nb_3Sn cross over points during fabrication. This deformation masks the effect of the reduction of deformation anticipated during application of an external load.

ACKNOWLEDGEMENTS

The authors would like to thank J.T. Holthuis, J.K. Wu, and J.A. Jacobsen of the Lawrence Berkeley National Laboratory for their assistance with sample manufacture. We also wish to acknowledge the excellent assistance of J.E. Bowman, A. Duenas, C.E. Karlsen, R.C. Jenkins, and K.R. Tapscott of LLNL with testing, instrumentation, and data acquisition. This work was performed under the auspices of the U.S. Department of Energy by the Lawrence Livermore National Laboratory under Contract W-7405-Eng-48.

REFERENCES

1. J.W. Ekin, "Effect of Stress on the Critical Current of Nb_3Sn Multifilamentary Wire," Appl. Phys. Lett., 29, pp. 216, 1976.

2. C.C. Koch and D.S. Easton, "A Review of Mechanical behavior and Stress Effects in Hard Superconductors," Cryogenics, V. 17, pp. 391, 1977.

3. J.W. Ekin, Cryogenics, "Strain Scaling Law for Flux Pinning in Practical Superconductors. Part 1: Basic Relationship and Application to Nb_3Sn," 20 (11), pp. 611, Nov. 1980.

4. G. Rupp, "Parameters Affecting Prestrain and B_{c2} in Multifilamentary Nb_3Sn Conductors," Adv. Cryo. Eng., V. 26, pp. 522, 1979.

5. D. S. Easton, D.M. Kroeger, W. Specking, and C.C. Koch, " A Prediction of Stress State in Nb_3Sn Superconducting Composites," J. Appl. Phys., V. 51 (5), pp. 2748, May 1980.

6. J.W. Ekin, "Transverse Stress Effect on Multifilamentary Nb_3Sn Superconductor," Adv. Cryo. Eng., V. 34, pp. 547, 1987.

7. W. Goldacker and R. Flükiger, "Calculation of Stress Tensors in Nb3Sn Multifilamentary Wires," Adv. Cryo. Eng. V. 34, pp. 561, 1987.

8. W. Specking, W. Goldacker, and R. Flükiger, "Effect of Transverse Compression on I_c of Nb_3Sn Multifilamentary Wire," Adv. Cryo. Eng., V. 34, pp. 569, 1987.

9. C.R. Gibson and J.R. Miller,"Structural Characteristics of Proposed ITER TF Coil Conductor," IEEE Trans Mag., V. 25, 1725, 1989.

10. L.T. Summers and J.R. Miller, "The Effect of Transverse Stress on the Critical Current of Nb_3Sn Cable-in-Conduit Superconductors," IEEE Trans. Mag., V. 25, 1835, 1989.

JACKET MATERIAL EVALUATION FOR NET'S

WIND-AND-REACT SUPERCONDUCTOR

Walter J. Muster, Jakob Kübler, Christa Hochhaus

Swiss Federal Laboratories for Testing
Materials and Research (EMPA)
Dübendorf, Switzerland

ABSTRACT

The inner poloidal field coils of the Next European Torus (NET) are planned to be built with a Nb_3Sn cable-in-conduit conductor. To meet the severe degradation in the superconductivity of Nb_3Sn due to strain effects the coils will be produced with the wind-and-react technique. High service loads together with the need for good ageability and weldability limit the number of possible jacket material candidates.

Eight Fe and Ni base alloys with promising strength and toughness properties were tested at EMPA within a screening program for their low temperature mechanical behaviour after a heat treatment of 50 h/700°C (base material and flash butt welded specimens). Stainless steels tend to embrittle by intergranular precipitations under these circumstances, whereas the ductility and formability of precipitation strengthening Ni base alloys may cause problems.

INTRODUCTION

The proposed conductor for NET's central solenoid coils, designed for operation with 40 kA at a peak field as high as 12,5 T, is given in Fig. 1. The NET Team has specified from the working stresses and in consideration of the special fabrication procedure of the coils for an ideal jacket material the following requirements:
a) high yield strength (>1000 MPa) in base and weld at 4 K, tensile elongation >10%
b) good cold working properties for jacket forming and coil winding
c) excellent weldability
d) good fracture toughness (>130 MPa $\sqrt{m}$) and fatigue strength
e) suitable coefficient of expansion over a temperature range 1000 K - 4 K
f) compatibility wih the reaction schedule of Nb_3Sn, meaning that the properties of a) and d) can be guaranteed after a heat treatment of 50 h/700°C.

Advances in Cryogenic Engineering (Materials), Vol. 36
Edited by R. P. Reed and F. R. Fickett
Plenum Press, New York, 1990

109

Previous works showed that usual stainless steels such as AISI 304, 304 N, 316 LN, can be degraded detrimentally by such a heat treatment; on the other hand an addition of small amounts of V or Nb tends to preserve a tolerable toughness after ageing [1,2]. A new developped Ni base alloy ("Incoloy 9XA") has been reported to fulfill the needs for a material with the mentioned specifications in an optimal way[3].

To get more fundamental results to answer the question, which commercial alloys could meet the crucial mechanical requirements (points a, d, f) best and what metallurgical parameters influence the ageability most, a screening test program was set up, considering 5 CrNi(MoMn)N stainless steels, 2 Ni base and 1 Fe base alloys (all these three precipitation strengthening). Low temperature tensile and Charpy V-notch tests of base and flash butt welded material should give an idea of the change of the mechanical properties resulting from a treatment of 50 h/700°C; fractography and metallography would reveal the phenomena of possible embrittling and explain its microstructural reasons. The main interest was thereby focused on the base material properties of stainless steels, welding was considered only in secondary priority.

EXPERIMENTAL

Table 1 gives the composition of the 8 investigated alloys (all data by the materials suppliers).

Herein the steel S1 (316 LN + Nb) corresponds to the material described by Shimada/Tone[2], N1 to "Incoloy 9XA" (commercial designation "Incoloy 908"), N2 to "Inconel 728", whereas P1 is of the type described by Hiraga et al.[4]. S1, S2 and S3 correspond furthermore to the steels JK1, JKA1 and JN1 mentioned in a paper of Nakajima et al.[5], whereas the ageing behaviour and RT properties of S5 were already described by Heimann[6].

The content of impurities is in all alloys very low, namely below 0,005% for sulphur and below 0,026% for phosphorus; all alloys were produced with modern metallurgical procedures (ESR, vacuum treating etc.).

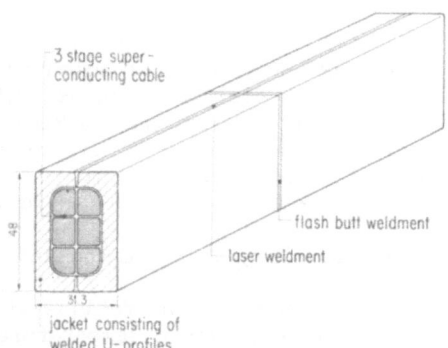

Fig. 1. Conductor assembly designed within a feasibility study for NET. The superconducting cable is situated in a jacket of a high performance structural material (stainless steel or Ni base alloy), that will be welded using drawn U-profiles.

Table 1. Chemical compositions of the investigated alloys (S...stainless steels, N... Ni base alloys, P... precipitation hardening Fe base alloy)

	C	N	Cr	Ni	Mo	Mn	Nb	V	Ti	Al	Fe
S1	0,010	0,17	17,2	12,3	2,0	1,2	0,05	-	-	-	rest
S2	0,023	0,268	25,0	14,0	0,68	0,49	-	0,30	-	-	rest
S3	0,026	0,34	24,2	14,7	-	4,2	-	-	-	-	rest
S4	0,019	0,17	17,9	13,1	2,6	1,9	-	-	-	-	rest
S5	0,016	0,33	22,3	16,5	3,2	5,9	0,18	-	-	-	rest
N1	<0,005	-	3,9	49,5	-	-	1,5	-	1,5	0,97	rest
N2	0,05	-	19,0	53,0	3.1	-	5,2	-	0,9	0,6	rest
P1	0,003	-	13,9	26,7	2,0	7,2	-	-	2,2	0,13	rest
											(wt-%)

For the tensile tests at 4 K a special screw driven testing machine with regulated crosshead displacement and a capacity of 200 kN was used, equipped with a He cryostat and controlled with a IBM AT 03 PC. The strain measurement was done by two capacitance gauges, its rate equaled in the domain of interest $1,6 \cdot 10^{-4} \text{s}^{-1}$. The Charpy V-notch tests were done at 77 K only, because the temperature preservation cannot be guaranteed strictly at 4 K during the specimen handling and the testing procedure itsself (deformation heat). As experience shows, the toughness data generally do not change very much though between 4 and 77 K; degradation by embrittling phases may eventually not be revealed at RT,but at 77 K the effects are undoubtedly recognizable, a fact that is clearly confirmed by Nohara et al.[1] and Shimada et al.[7].

MECHANICAL TESTS

The results of the tensile and the Charpy V-notch are summarized in table 2 resp. graphically in Fig. 2.

Table 2. Summary of the cryogenic mechanical properties of the 8 alloys in the aged (50 h/700°C) state. Averaged values from several measurements in the base (b) resp. flash butt welded (w) material.

	Rp0,2 (MPa), 4 K		Elongation (%), 4 K		Charpy impact value (J), 77 K	
	b	w	b	w	b	w
S1	994	1003	52	27	213	160
S2	1414	1458	7,6	2,7	19	9
S3	1501	1521	2,3	1,3	11	8
S4	1109	1120	48	27	161	63
S5	1714	1611	4,9	2,5	16	11
N1	1152	1076	31	18	57	31
N2	1403	1390	12	14	13	15
P	950	840	30	10,3	64	32

Apparently the toughness of all stainless steels is reduced by the ageing procedure in comparison with the state of delivery (annealed and partially with a low degree of cold work). In the other alloys (N1, N2, P), the heat treatment 700°C/50 h was used directly for precipitation strengthening and toughness in this state is somewhere in between the different stainless steels.

Considering that NET's specified fracture toughness of 130 MPa $\sqrt{m}$ corresponds to a Charpy energy of at least in the order of 50 to 100 J and that the needed elongation at 4 K was not or only hardly reached by S2, S3, S5 and N2 there are finally 3 alloys that can be designated as real jacket material candidates: S1, S4 and N1, whereas the yield strength of P1 drops under the limit, especially after the welding procedure.

DISCUSSION OF THE MECHANICAL RESULTS

A closer look to the chemical compositions in table 1 shows that the N content of the stainless steels lies between 0,17 and 0,34%, whereas C (content between 0,010 and 0,026%) seems to be negligible in comparison. This indeed holds for the yield strength, where in the corresponding Fig. 3 the influencee on the N content is visible, thus also confirming the linear dependence in a formula for the mean values

$$R_{p0,2} = 300 + 4000 \text{ N}$$

($R_{p0,2}$ in MPa, N in wt-%)

The main conclusion is, that with the exception of S3 the heat treatment affects the yield strength only in negligible extent.

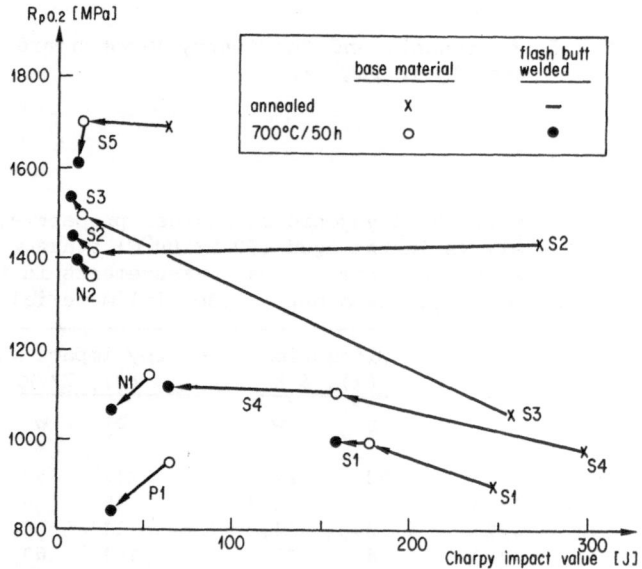

Fig. 2. Yield strength at 4 K vs. Charpy impact value at 77 K in a graph, demonstrating with arrows a drastic degradation of toughness starting from the annealed (only stainless steels) to the aged state (first base then flash butt welded material)

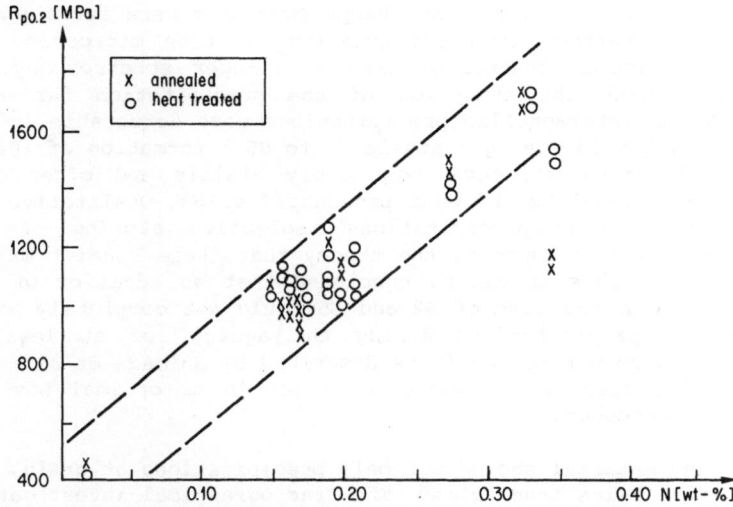

Fig. 3. Yield strength at 4 K vs N content. Data from
stainless steels S1 to S5, from literature[1,2]
and from unpublished additional investigations
by EMPA; comparison between annealed and aged
(50 to 75 h/700°C) state.

The situation is quite different for the Charpy toughness; the
severe degradation is already visible in Fig. 2. A specific discussion
is possible with the graphs of Fig. 4 presenting selected alloys with a
limited span in the N content (0,15 to 0,20%), but a variation of the C
content by a factor 3.4. It is obvious that the drop is mostly influenced by the carbon content, but can be reduced at least partially
by additional alloying of carbide forming elements such as Nb and V, a
fact that was already described by Nohara et al.[1] and Shimada/Tone[2].

PHYSICAL METALLURGY

All 8 alloys of table 1 were examined carefully by metallographi-
cal means (including electron microprobe) in all states (annealed, aged

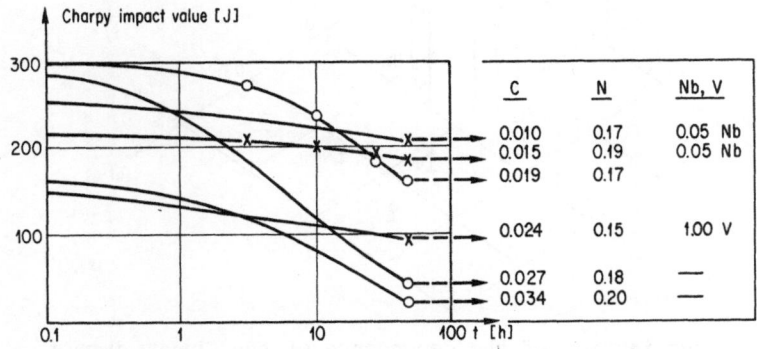

Fig. 4. Drop of the Charpy impact value, measured
at 77 K, in dependence of the time of an
ageing process at 700°C. On the right side
the corresponding the content of C, N and
additional alloyed Nb or V is given in wt-%.

and welded), whereas the tensile and Charpy specimens were investigated at their fracture surface with the scanning electron microscope and partially (intergranular fractures) also with Auger spectroscopy. It was found, that with the exception of the precipitation hardening alloys P and N2 no intermetallic precipitations were detectable in the light microscope, but in the aged steels S2 to S5 a formation of inter-granular precipitations was revealed, hardly visible and often only localized in the case of S2 and most pronounced at S3. Qualitative in-vestigations of isolated precipitations (selective etching of the austenitic steel matrix) supports the theory that these consist mainly of $M_{23}C_6$ carbides[8]. Thus it can be concluded that an addition of the elements Nb and V in the case of S2 and S5 could not completely avoid its intergranular precipitations during the ageing, or at least a stabilizing procedure before, as it is described by Shimada et al.[7] for example in similar steels, was not carried out in an optimal way and didn't render C innocuous.

Cold worked material showed not only precipitations at grain, but also at incoherent twins boundaries[8]. The fractographical investigation revealed that the steels S1 and S4 were the only ones that didn't break intergranularly. Thus the above mentioned poor toughness properties of the steels S2, S3 and S5 can also be correlated with the mode of fracture. Fig. 5 finally gives a three-dimensional graphical view, confirming that the weight of the C content is about ten times higher in its influence on the Charpy impact value than the N content.

The physical metallurgy of the precipitation hardening alloys will not be treated here any longer, for fundamental microstructural infor-mation on the alloy "Incoloy 9XA/908" the publication of Morra[9] is re-commended.

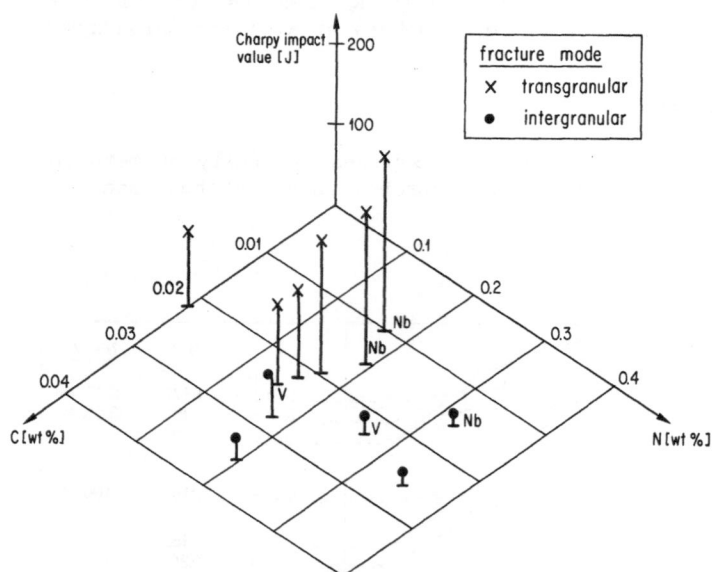

Fig. 5. Visualization of the dependence of the Charpy impact value, measured on aged (50 h/700°C) stainless steels at 77 K, on the content of the elements C and N. Some Some steels are stabilized with Nb or V. The high energies in this selected series of literature and EMPA data go along with transgranular fracture.

CONCLUSIONS

The low temperature toughness of CrNi(MoMn)N stainless steels aged for 50 h/700°C ist strongly influenced by C content. Similar to the sensitization for the intergranular corrosion of austenitic steels after such a heat treatment the failure mechanism is correlated with the carbide precipitation at the grain boundaries. On the other hand the low temperature yield strength of these alloys is governed by the nitrogen content confirming the corresponding linear dependence in the annealed state as well as in the aged. Flash butt welding reduces especially the Charpy impact values, but not in dramatic manner. The influence of inclusions and their spacings, as described by Simon and Reed[10], was not investigated within this study.

To fulfill the requirements of NET the N content should be adjusted at about 0,17% and that of C reduced below at least 0,020%; small amounts of Nb or V together with an optimized stabilizing procedure and an appropriate degree of cold work improving so the distribution characteristic of precipitations may lead to better results in the ageing behaviour.

The Ni base alloy "Incoloy 908" is to a certain extent limited in its ductility, but has a considerable potential because of its suitable expansion coefficient compared with the Nb_3Sn superconductors.

Detailed investigations planned for the near future on these two types of alloys will allow further conclusions.

ACKNOWLEDGEMENT

The investigations were funded mainly by Asea Brown Boveri Switzerland within a R & D contract with the NET Team in Munich/Germany. The authors are also grateful to the different companies supplying these high performance alloys and many data to their behaviour.

REFERENCES

1. K. Nohara, Advances in Cryogenic Engineering - Materials 28, Plenum Press, New York (1982), 117
2. M. Shimada and S. Tone, Advances in Cryogenic Engineering - Materials 34, Plenum Press, New York /1988), 131
3. M. M. Morra, Advances in Cryogenic Engineering - Materials 34, Plenum Press, New York (1988), 157
4. K. Hiraga, Advances in Cryogenic Engineering - Materials 32, Plenum Press, New York (1988), 111
5. H. Nakajima, Advances in Cryogenic Engineering - Materials 34, Plenum Press, New York (1988), 173
6. W. Heimann, Thesis RWTH Aachen/Germany (1973)
7. M. Shimada, Cryogenic Engineering 23 (1988), 183
8. B. Weiss and R. Stickler, Metallurgical Transactions 3 (1972), 851
9. M. M. Morra, Master Thesis, MIT (1989)
10. N. J. Simon and R.P. Reed, Advances in Cryogenic Engineering - Materials 34, Plenum Press, New York /1988), 165

THERMAL CONTRACTION OF FIBERGLASS-EPOXY SAMPLE

HOLDERS USED FOR Nb₃Sn CRITICAL-CURRENT MEASUREMENTS*

L. F. Goodrich, S. L. Bray, and T. C. Stauffer

Electromagnetic Technology Division
National Institute of Standards and Technology
Boulder, Colorado 80303

ABSTRACT

It is typical for Nb_3Sn-Cu superconductor specimens to be wound into coils on tubular specimen holders for critical-current measurements. If the thermal contraction of the holder is different than that of the specimen, axial strain may be applied to the specimen upon cooling from room to liquid-helium temperature. This strain can affect the measured critical current. The thermal contraction was measured for three different Nb_3Sn-Cu superconductors. Also, the thermal contraction was measured for several different specimen holders, all of which were made from fiberglass-epoxy composites. The specimen holder measurements were made using an electrical-resistance strain-gage technique, and they were confirmed by direct mechanical measurements. The tubes varied in diameter, wall thickness, and fabrication technique. Some of the tubes were made directly from tube stock, and others were machined from plate stock. The results of these measurements show that the thermal contraction of tube stock is strongly dependent on the ratio of its wall thickness to its radius, while the contraction of tubes machined from plate stock is relatively independent of these dimensions. Critical-current measurements of Nb_3Sn-Cu specimens mounted on these various holders show that the presence of differential thermal contraction between the specimen and its holder can significantly affect the measured critical current.

INTRODUCTION

Critical-current (I_c) measurements of Nb_3Sn-Cu superconductors can be affected by the relative thermal contraction of the material on which the test specimen is mounted. This measurement variable became apparent in the recent VAMAS (Versailles Project on Advanced Materials and Standards) interlaboratory comparative measurements (round robin) of the critical current of Nb_3Sn.[1] Differential thermal contraction between the specimen and its holder in cooling from room to liquid-helium temperature can cause either a tensile or compressive strain of the specimen, either of which can affect the measured I_c.[2,3] For coil specimens that are mounted on the surface of cylindrical holders, the strain is predominantly along the axis of the specimen. The amount of strain depends on the magnitude of the

---

*Contribution of NIST, not subject to copyright.

Advances in Cryogenic Engineering (Materials), Vol. 36
Edited by R. P. Reed and F. R. Fickett
Plenum Press, New York, 1990

117

differential contraction, the relative strengths of the specimen and holder, and the mechanical coupling between the specimen and its holder.

The literature contains considerable data on the compressive pre-strain of Nb_3Sn filaments caused by differential thermal contraction between the filaments and the matrix material. However, very little data on the overall thermal contraction of Nb_3Sn-Cu wires are presently available. The thermal contraction of a Nb_3Sn-Cu cable is given in Ref. 4; however, this conductor has a tungsten core that reduces its thermal contraction. Consequently, the thermal contraction of three different Nb_3Sn wires was measured as a part of this study.

Fiberglass-epoxy composites are commonly used sample holder materials. These materials are anisotropic in three mutually perpendicular directions. The three directions are associated with characteristics of the fiberglass fabric, and they are designated as the warp, fill, and normal directions. The fabric orientation for plate stock is shown in Fig. 1. The normal direction is perpendicular to the fabric planes, while the warp and fill directions are parallel to the fabric planes. The density of the fabric is not the same in both directions of the weave. The warp and fill directions are determined by the fabric's thread count. The number of threads per unit length of fabric is lower in the warp direction than in the fill direction. This structural anisotropy causes a three-dimensional variation in thermal contraction. The contraction in the fill direction is slightly greater than in the warp direction, but the contraction in the normal direction is considerably larger.[4,5]

Figure 1 also shows the typical fabric orientation for a rolled tube. The radial thermal contraction of the tube results from a competition between the larger contraction in the normal direction and the smaller contraction in the warp direction. Consequently, it depends on the ratio of the tube's wall thickness to its outside radius (wall-to-radius ratio). For thin-walled tubes the radial thermal contraction approaches that of a plate in the warp direction, and for thick walled tubes it approaches that of a plate in the normal direction.

A holder whose thermal contraction is relatively independent of its wall-to-radius ratio can be made by machining a cylindrical tube from thick fiberglass-epoxy plate stock with the axis of the tube perpendicular to the surface of the plate (plate tube). For this orientation (Fig. 1) the radial contraction is based on the contraction in the warp and fill directions, which are both similar to that of a Nb_3Sn-Cu wire. In addition to thermal contraction measurements of both rolled tubes and plate tubes, I_c measurements were made using both types of specimen holder to confirm the relationship between thermal contraction and the I_c measurement.

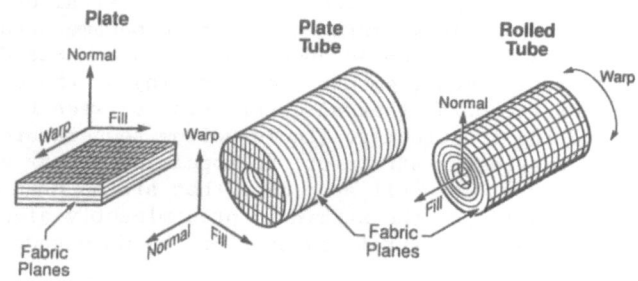

Fig. 1. Geometries of fiberglass-epoxy composites.

EXPERIMENTAL DETAILS

Fiberglass-Epoxy Composites

Two different types of fiberglass-epoxy composite were used in this study, NEMA (National Electrical Manufacturers' Association) G-10 and G-11. All of the rolled tubes were G-10 and the plate tubes were G-11. This was not a matter of choice but, rather, one of material availability. The thermal contraction of the G-11 is slightly less than that of G-10. Neither of these materials was cryogenic-radiation (CR) grade, which is designated as G-10CR or G-11CR. The manufacturing specifications for the CR grade materials are more stringent and their performance at cryogenic temperatures is more predictable.[5]

Thermal Contraction

All of the thermal contraction measurements were conducted between room temperature and liquid-nitrogen (LN_2) temperature. We assume that, if a material's contraction is well matched to a Nb_3Sn-Cu wire from room to LN_2 temperature, then the additional differential thermal contraction that occurs between LN_2 and liquid-helium temperature will be insignificant.

The Nb_3Sn-Cu thermal contraction measurements were made using quartz reference tubes and reacted Nb_3Sn-Cu specimens that measure approximately 35 cm in length. The specimen is placed in a quartz tube and attached to one of the tube's ends. At room temperature, the quartz tubes are approximately 1 mm longer than the specimens. This differential length is measured with a micrometer, the tube and specimen are cooled to LN_2 temperature, and the differential length is remeasured. The thermal contraction of the specimen is deduced from the measured change in the differential length and the known contraction of the quartz tube. The uncertainty of these measurements is estimated to be ±5%.

Two different methods were used to measure the thermal contraction of the fiberglass tubes. The first was a mechanical method, where the tube was submerged in liquid nitrogen, allowed to reach thermal equilibrium, removed from the nitrogen, and the diameter was then quickly measured with a precision micrometer. To address the thermally transient nature of these measurements, the tube's diameter was measured as a function of time while it was warming toward room temperature. These measurements allowed an extrapolation of the data to liquid-nitrogen temperature. In the case of the plate tubes, two orthogonal measurements were made at each time, one in the warp direction and one in the fill direction. The uncertainty of the mechanical measurements of thermal contraction was estimated to be ±10% of the measured value (for example, 0.20% ±0.02%).

The second type of thermal contraction measurement used electrical-resistance strain gages.[6] For this measurement, two well matched 350 Ω strain gages were used in a half-bridge configuration. One of the gages was bonded to the test specimen and the other was bonded to a reference material, a quartz tube, whose thermal contraction was known. All of the strain-gage measurements reported here were taken with the strain gages mounted on the circumference of the test specimen since this represents the contraction which is relevant for the Nb_3Sn-Cu coil. Both the test specimen and reference material were cooled to liquid nitrogen temperature and the resulting output from the strain bridge was measured.

Two factors will cause a change in the resistance of the strain gages when cooled from room to LN_2 temperature. First, the resistivity of the strain gages' grid alloy will change with temperature. Since this change in resistance will be nearly equal for matched strain gages, there will be little effect on the output of the strain bridge. The second source of change in resistance is caused by thermally induced strain. This strain results from differential thermal contraction between the strain gage and the test specimen. If the thermal contraction of the test specimen material is different than that of the reference material, the strain bridge will be unbalanced and its output will be indicative of the difference in thermal contraction between these two materials. Since the thermal contraction of the reference material is known, the thermal contraction of the test material can be deduced. The uncertainty of the strain-gage measurements of thermal contraction was estimated to be ±5% of the measured value (for example, 0.20% ±0.01%).

Critical Current

For the I_c measurements, a Nb_3Sn-Cu wire with a diameter of 0.68 mm was used. This conductor was made by an internal-tin diffusion process, and it has 37 sub-bundles of 150 Nb filaments. A single Ta diffusion barrier separates the filament region from the outer Cu layer. The specimen was wound onto a stainless steel tube and then vacuum heat treated at 700°C for 48 h. The stainless steel tube has a helical groove machined into its surface to retain the specimen and define its geometry. Following heat treatment, the specimen was removed from the stainless steel reaction holder and transferred to the fiberglass measurement holder. The outside diameter of the fiberglass holder is 3.12 cm and its surface is not grooved. A thin continuous layer of filled epoxy adhesive was painted over the surface of the specimen and holder. The typical specimen length was approximately 80 cm. Three pairs of adjacent voltage taps were placed along the center of the specimen. Each pair had a separation of about 10 cm, and there was a 1 to 2 cm gap between adjacent pairs. An electric field criterion of 10 μV/m was used for determining the critical current. The uncertainty of the I_c measurements may be as large as ±5% because of the strong systematic effects of the specimen holders.

RESULTS

Thermal Contraction Measurements

The results of the thermal contraction measurements are shown in Fig. 2 where the thermal contraction of the tubes' diameters (Δd/d) is plotted as a function of the tubes' wall-to-radius ratios. The thermal contraction of three different superconductors (the VAMAS I_c round robin conductors[1]) was also measured. The dashed horizontal line labeled "Nb_3Sn-Cu" indicates the measured thermal contraction (0.28%) of the conductor whose I_c was measured in this study. The measured thermal contraction of the other two conductors is 0.28% and 0.26%. The two types of data symbols that are labeled "Mechanical" and "Strain Gage" show the thermal contraction of the G-10 rolled tube specimens as measured by the two different techniques. With the exception of the "Mechanical" data, all of the data shown in this figure were obtained using the strain-gage technique.

The data show a strong dependence of the thermal contraction on the wall-to-radius ratio. The approximate thermal contraction of G-10CR plate stock in cooling from room to liquid-nitrogen temperature is 0.224%, 0.264%, and 0.687% in the warp, fill, and normal directions respectively, and it is 0.202%, 0.227%, and 0.585% for G-11CR.[4,5] As expected from the structural geometry of the rolled tubes, their thermal contraction fall between that of the warp and normal directions of plate stock. The measured thermal contraction of the G-11 plate tubes is shown for three wall-to-radius ratios and in two structural directions, warp and fill. In both directions, the

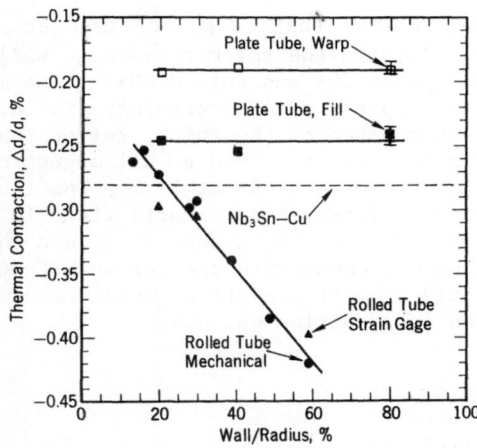

Fig. 2. Plot of room to liquid-nitrogen temperature thermal
contraction of G-10 rolled tubes and G-11 plate tubes
as a function of the tubes' wall-to-radius ratio.

thermal contraction is relatively independent of the wall-to-radius ratio
and it is comparable to that of plate stock. The error bars associated with
the high wall-to-radius ratio data points represent the range of values that
were measured on three different specimens with three repeat determinations
for each specimen.

Additional measurements were made using strain gages that were mounted
on the circumference of the plate tube halfway between the warp and fill
axes. The thermal contraction at these positions was within the
experimental uncertainty of the value measured in the warp direction. This
indicates that the effective circumferential thermal contraction of the
plate tube is closer to the value measured for the warp direction.
Mechanical thermal contraction measurements were also made on the plate
tubes and were within the experimental uncertainty of the strain-gage
measurements. Additional thermal contraction measurements were made on
tubes with different outer diameters to test the scaling of thermal
contraction with the wall-to-radius ratio. All of the data of Fig. 2 were
taken on tubes having outer diameters of about 3.18 cm and various inner
diameters. However, two rolled tubes with outer diameters of 11.4 and
12.4 cm, and wall-to-radius ratios of 14 and 8%, respectively, were also
measured. The thermal contraction of these two tubes was within the
experimental uncertainty of the rolled tube data of Fig. 2. Finally, an
11.4 cm outer diameter (94% wall-to-radius ratio) plate tube was measured
and was within the experimental uncertainty of the plate tube data of
Fig. 2.

Critical-Current Measurements

The results of the I_c measurements are shown in Fig. 3, where the I_c is
plotted as a function of applied magnetic field for several different
conductor specimens and specimen holders. There are three variables, in
addition to the applied magnetic field, for these I_c measurements: the
location where the specimen was reacted, "central" or "self"; the specimen
holder's wall-to-radius ratio, "thick" or "thin"; and the type of tube used
for the specimen holder, "rolled" or "plate". The self reacted samples were
reacted at NIST, whereas the central reacted samples were reacted at another
site and shipped to NIST for I_c measurements. The thick and thin
designations do not indicate specific wall-to-radius ratios; instead, they
indicate two general categories of tube geometry. All of the thin tubes

have wall-to-radius ratios that are no greater than 13%
(self 10% and central 13%) and the thick tubes have wall-to-radius ratios
that are at least 60% (plate 80% and rolled 60%). The measured I_c's are
nearly the same, within experimental uncertainty, for all of the specimens
except the one that was mounted on the thick, rolled tube. The measured I_c
is significantly lower for this specimen at all magnetic fields. The
difference is about 14% at 6 T and 40% at 12 T. The fact that the I_c
degradation increases with increasing magnetic field is consistent with a
strain effect. The magnitude of the change in I_c and the measured thermal
contraction are in good agreement with the strain effect.[3] At 12 T, a 33%
reduction was calculated from strain-effect measurements[1] that were made at
14 T, and the measured I_c reduction was 40%.

DISCUSSION

    A practical reality of round robin measurements is that the dimensions
of the specimen holders vary between different laboratories. For consistent
measurements, the specimen holders should be designed so that the I_c
measurement is insensitive to this variable. The thermal contraction of a
tubular specimen holder that is made from an anisotropic material can vary
with its geometry. This presents the potential for variations in the strain
state of the specimen and, thus, variations in the measured I_c. An apparent
solution to this problem is to use an isotropic material for the specimen
holders and to use a bonding technique that rigidly couples the specimen to
its holder. This will ensure that the strain transmitted to the specimen,
due to thermal contraction, is independent of the holder's geometry and,
thus, equivalent from laboratory to laboratory. This approach addresses the
issue of measurement consistency, but it does not address accuracy.

    The I_c measurement should, arguably, be made with a minimum of
externally applied strain on the superconductor. This requires a strong
bond between the specimen and holder to avoid specimen strain under the
influence of the Lorentz force, and it requires that the thermal contraction
of the sample holder be well matched to that of the superconductor. Also,

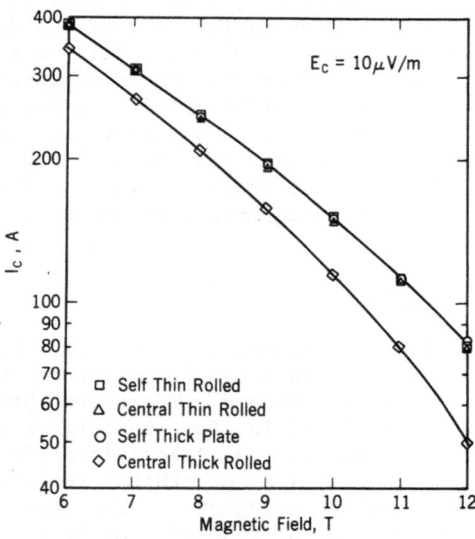

Fig. 3. A semilogarithmic plot of I_c at an electric field
        criterion of 10 μV/cm as a function of magnetic
        field.

the holder should, ideally, be made from an electrically insulating material to prevent current sharing with the test specimen. Unfortunately, an isotropic and insulating material with a thermal contraction similar to that of Nb3Sn-Cu is not readily available. Fiberglass-epoxy plate tubes are a practical alternative to the ideal isotropic specimen holder. The thermal contraction of a plate tube is slightly anisotropic; however, it is relatively independent of the tube's dimensions. Furthermore, the thermal contraction in the radial direction (the pertinent direction for a coil-type specimen) is similar to that of Nb3Sn-Cu. Based on the thermal contraction of G-10CR and G-11CR plate,[4,5] the thermal contraction of G-10 plate tubes may be slightly closer to that of a Nb3Sn-Cu wire than the G-11 plate tubes that were measured here. It is possible that the difference in the circumferential thermal contraction between the warp and fill directions will result in a spatial variation of the strain state of the Nb3Sn-Cu coil sample, but the I_c data indicate that this effect is not significant.

The plate tubes present some practical disadvantages. First of all, machining a specimen holder from plate stock is considerably more difficult than from tube stock. Also, the length of a plate tube is limited by the thickness of the available plate stock. This, in turn, limits the length of the superconductor specimen for a given coil diameter and pitch. Furthermore, short specimen holders are often incompatible with existing I_c test fixtures. Satisfactory specimen holders of greater length could perhaps be constructed by bonding a series of short plate tubes together. This technique might require an alignment of the warp and fill fibers between individual tube sections because of the anisotropic radial thermal contraction of plate tubes.

CONCLUSIONS

For I_c measurements, fiberglass-epoxy composites are suitable specimen holder materials. However, the design of the specimen holder should take into account the anisotropic nature of the material and the resulting variability in thermal contraction. These characteristics of the material can result in large variations in the measured I_c for specimens mounted on holders of different designs. A cylinder or tube can be machined from a thick fiberglass-epoxy plate with the axis of the cylinder perpendicular to the surface of the plate. This type of tube (plate tube) has a thermal contraction that is relatively independent of its dimensions and that is similar to that of a Nb3Sn-Cu specimen. Alternatively, the specimen holder can be made from tube stock and machined to a wall-to-radius ratio that results in a thermal contraction that closely matches that of the Nb3Sn-Cu specimen.

ACKNOWLEDGEMENTS

The authors acknowledge the contribution of R. L. Spomer, S. Bird, and T. Harris who made some strain-gage measurements on the G-10 rolled tubes as part of a Colorado School of Mines class project. The authors extend their thanks to M. Thoner (Vacuumschelze) for a discussion of the plate tube geometry, to R. M. Folsom for sample preparation and assistance with the measurements, to J. D. McColskey for instructions on mounting strain gages, and to R. Gerrans for drafting.

This work was supported by the Department of Energy, Office of Fusion Energy and Division of High Energy Physics.

An effort was made to avoid the identification of commercial products by the manufacturer's name or label, but in some cases these products might be indirectly identified by their particular properties. In no instance does this identification imply endorsement by the National Institute of Standards and Technology, nor does it imply that the particular products are necessarily the best available for that purpose.

REFERENCES

1. K. Tachikawa, K. Itoh, H. Wada, D. Gould, H. Jones, C. R. Walters, L. F. Goodrich, J. W. Ekin, and S. L. Bray, VAMAS intercomparison of critical current measurement in Nb_3Sn wires IEEE Trans. Magn. 25-2:2368 (1989).
2. G. Fujii, J. W. Ekin, R. Radebaugh, and A. F. Clark, Effect of thermal contraction of sample holder material on critical current Adv. Cryog. Eng. 26:589 (1980).
3. J. W. Ekin, Strain scaling law for flux pinning in practical superconductors. Part 1: basic relationship and application to Nb_3Sn conductors Cryogenics 20:611 (1980).
4. A. F. Clark, G. Fujji, and M. A. Ranney, The thermal expansion of several materials for superconducting magnets IEEE Trans. Mag. MAG-17:2316 (1981).
5. M. B. Kasen, G. R. MacDonald, D. H. Beekman, Jr., and R. E. Schramm, Mechanical, electrical, and thermal characterization of G-10CR and G-11CR glass-cloth/epoxy laminates between room temperature and 4 K Adv. Cryog. Eng. 26:235 (1980).
6. Measurements Group, Inc., Tech Note-513, Measurement of thermal expansion coefficient using strain gages (1986).

DEVELOPMENT OF Nb$_3$Sn SUPERCONDUCTING WIRE USING AN IN-SITU PROCESSED

LARGE INGOT

Y. Ikeno, M. Sugimoto, K. Goto, and O. Kohno

Fujikura Limited
1-5-1, Kiba, Koto-Ku, Tokyo 135, Japan

INTRODUCTION

The in-situ process for Cu-Nb has been developed on a laboratory scale, and recently several groups have produced multifilamentary wires with overall critical current density comparable to commercial continuous-fiber multifilamentary materials.[1] The mechanical properties of the in-situ wires are considerably better than those of the conventional multifilamentary superconducting wires.[2] For practical applications, it is necessary to demonstrate that the in-situ process can be adapted to industrial scale. Generally, small-diameter in-situ rods were chosen because good quality control of the Cu-Nb alloys fabrication could be achieved on the basis of prior experience with small chill casting.[3] To demonstrate the feasibility of scale-up Cu-Nb ingots were made by smelting and casting by the calucia process. In this paper we report our results on the development of this in-situ Nb$_3$Sn wires.

EXPERIMENT

The in-situ process begins with production of the Cu-Nb casting in which the Nb is present as randomly arrayed dendrites. Because of the nature of the solidification process, a size distribution of aligned Nb filaments produced by the wire drawing step.[4] Cu-Nb alloys were prepared by CaO smelting process which was made by vacuum induction smelting molten metal by pouring into CaO crucible at about 1750°C and then solidified the refine metal by pouring into CaO mold or Cu mold. Ingot (about 200 mm in length and 20 ∿ 150 mm in diameter) with 20 ∿ 40 wt% Nb were produced. The transverse section of castings into Cu mold are shown in Fig. 1, it is a typical example of Cu-30Nb ingots.

Ingots were swaged and drawn to 0.2 mm diameter, through suitable heat treatment. The wire was then Sn plated, heated to produce diffusion of the Sn, and reacted with different treatment scadules (time and temperature). The critical current was measured in a transverse magnetic field at 4.2K and defined as the current corresponding to a voltage of 1 μV/cm. The overall critical current density, Jc, was obtained by dividing the critical current by the total cross-section area of the specimen.

Advances in Cryogenic Engineering (Materials), Vol. 36
Edited by R. P. Reed and F. R. Fickett
Plenum Press, New York, 1990

125

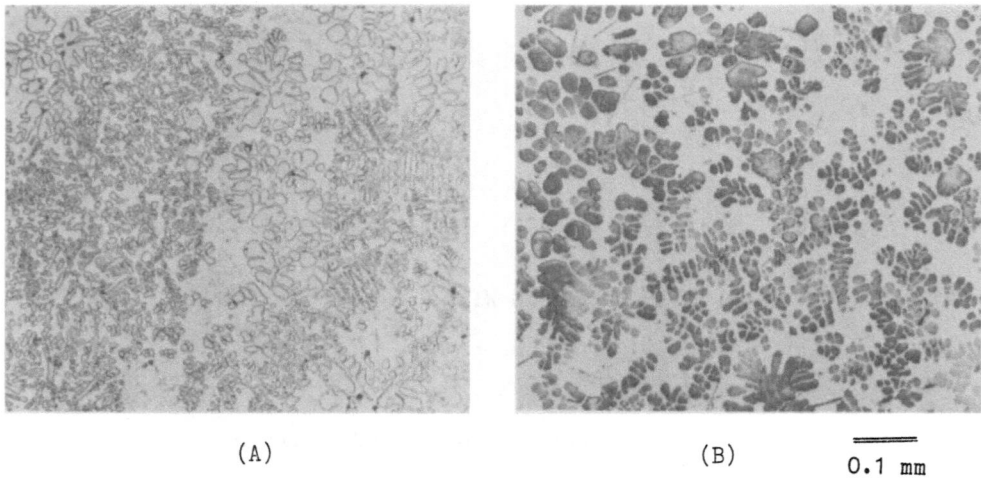

(A)   (B)   0.1 mm

Fig. 1 Transverse section showing Nb dendrite solidified into Cu mold.
(A) Cu-30Nb,30 mm ingot diameter. (B) 150 mm diameter.

The effect of bending strain and the effect of twist on Jc had been investigated. Furthermore, to improve the high field critical current density we have examined the effect of additives.

RESULTS AND DISCUSSION

Specimen

The as-solidified microstructure by CaO process are shown in Fig. 2. The Nb dendrite shows almost the same size regardless of the ingot size.The size was evaluated by deep etching the Cu matrix away and examining them in an SEM. What is obvious on comparing the Fig. 1 and 2, the average dendrite size of the casting into CaO mold increased by approximately four times that of Cu mold. Excessive oxygen and carbon pickup of only 186 ∿ 668 parts per million (ppm) and furthermore a carbon

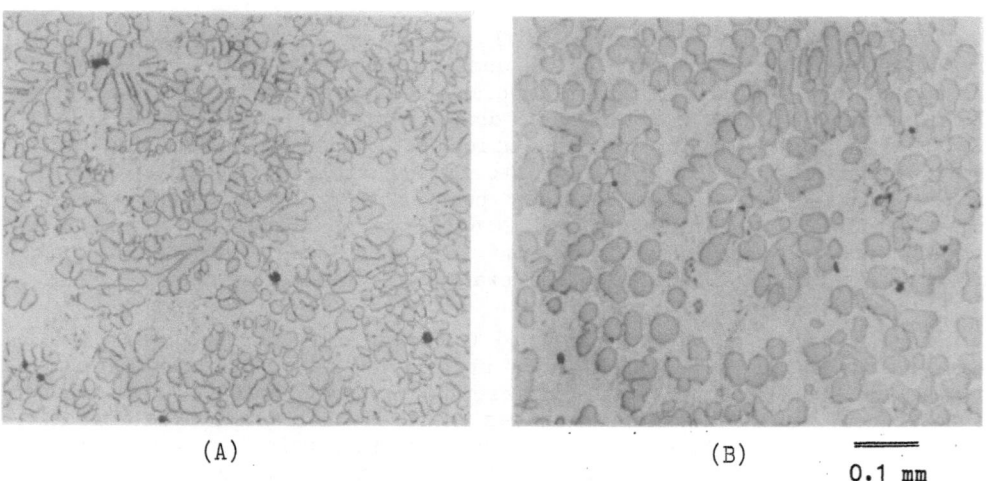

(A)   (B)   0.1 mm

Fig. 2 Transverse section showing Nb dendrite solidified into CaO mold.
(A) Cu-30Nb, 30 mm ingot diameter. (B) 150 mm diameter.

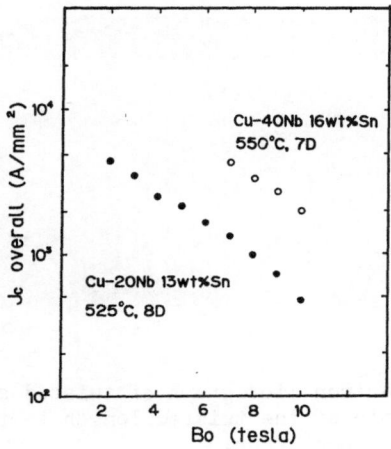

Fig. 3 Comparison of critical current on wires made from Cu-20Nb and Cu-40Nb wt% ingot.

pickup of 4~48 ppm. This level of oxygen and carbon contamination did not have any significant deleterious effect upon either the drawability of the Cu-Nb ingots or the resulting Jc.

Superconducting properties

The critical current for Nb concentration 20 and 40 wt% as a function of a transverse magnetic field, Bo, is shown in Fig. 3. A relatively higher Jc, 2 to 2.4 x 10^3 A/mm^2 at 10 Tesla, was obtained for the Cu-40Nb alloy. The influence of ingot size to Jc was not observed. Also the value of the Jc is very susceptible to the arrangement of Nb filament before diffusion heat-treatment. In this experiment, the arrangement of Nb filaments was altered by combining heat-treatment. Fig. 4 shows Jc values for wires which were changed by the degree of cold-working. It can be seen that the superconducting properties were improved by the optimizing degree of cold-working.

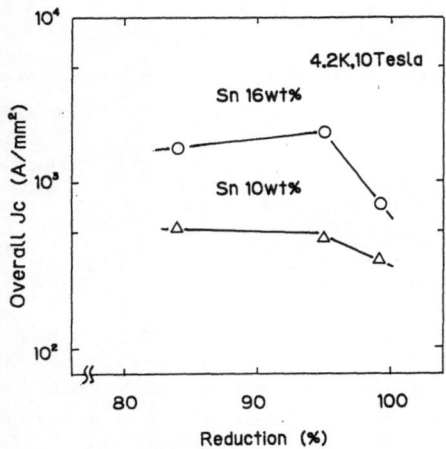

Fig. 4 Influence of cold working on the Jc at 10 Tesla for Cu-40Nb-16Sn and Cu-40Nb-10Sn wires submitted to final heat treatment at 800°C for 1 day.

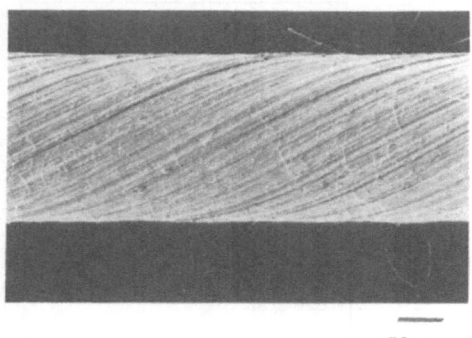

50 µm

Fig. 5 Scanning electron micrograph of twisted sample with Cu etched
out. The ratio of the twisted length to the wire diameter of
the order is about four.

The strain effect of the in-situ wire was measured with twisted
and untwisted wires. A ratio of the twisted length (lp) to the wire
diameter (dw) of the order from 30 to 4 was achieved. The final diameter
of all specimens was 0.2 mm. The twisted filaments were clearly observed
in photograph (Fig. 5). The effect of bending on Jc for twisted and
untwisted Cu-Nb wires are shown in Fig. 6, critical current reduced to
the Ic of the straight wire (Ico) versus bending strain are presented.
Untwisted wires were reacted at 525°C for 16 days and twisted wires were
reacted at 600°C for 6 days. The degradation of Jc for Cu-40Nb untwisted
wires had not observed until 1.2 % bending strain, but the results for
short-twisted wire at 10 Tesla showed considerable degradation in critical
current.

The effect of additives

There have been few studies for the improvement in Jc at high field
for the in-situ wires. The investigation has been carried out on the
influence of additives in Cu-40Nb ally to improve Jc at high field.
Fig. 7 shows the Jc for Cu-40Nb wires as a function of the applied
magnetic field, Bo. Above 15 Tesla, small improvements in Jc values of

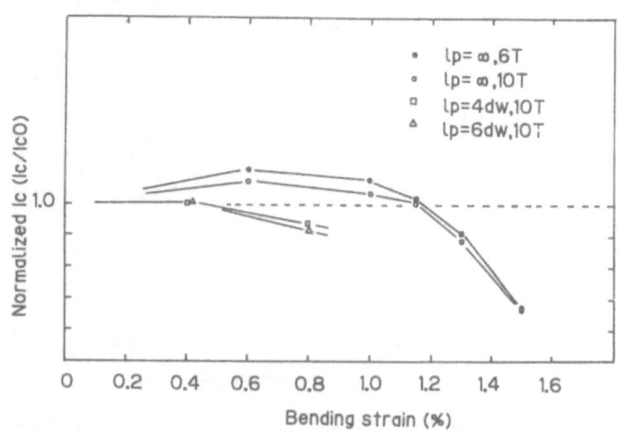

Fig. 6 Comparison of bending- strain degradation for twisted in-situ
Cu-Nb wires.

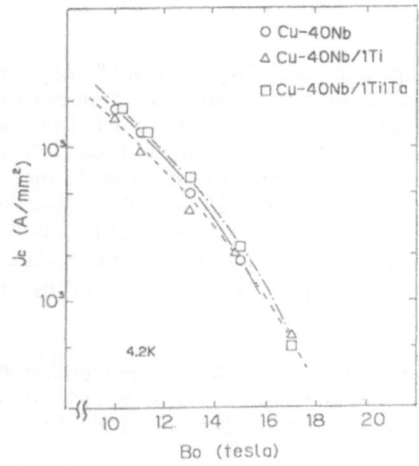

Fig.7 Overall critical current density, Jc, for in-situ Nb$_3$Sn wires
    as a function of the applied magnetic field, Bo.

Cu-40Nb wires are obtained by addition of 1wt%Ti and 1wt%Ti + 1wt% Ta.
These high field data are linear on a Jc$^{1/2}$B$^{1/4}$ versus B plot and
extrapolate to zero Jc at 18.9 Tesla.

Large wire diameter

    Adding Sn to the Cu-Nb wire by plating becomes more difficult as
the wire size increases because the diffusion period becomes quite long
and thicker Sn layer required tend to "ball-up" when the tin layer is
melted. We are currently working on a process to produce a cryostabilized
wire. Fig. 8A shows the transverse section of cryostabilized wire having
a 0.28 mm diameter. Fig. 8B shows the cross section of in-situ Nb$_3$Sn
wire the 1.1 mm diameter made by the Internally Tin Plating Process.[6]
Jc of the in-situ Nb$_3$Sn core of 0.28 mm wire was 2290 A/mm^2 (heat
treated at 550°C for 10 days) and Jc of 1.1 mm wire was 1675 A/mm^2 (600°C
for 7 days) at 10 Tesla, respectively.

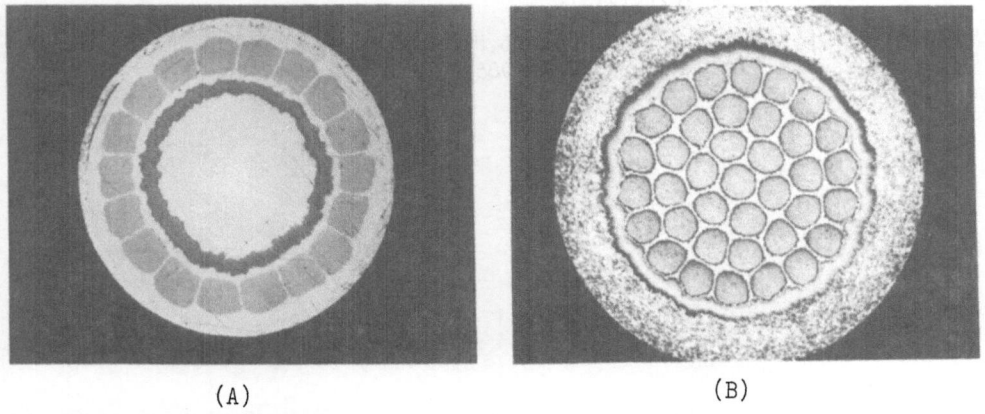

(A)   (B)

Fig. 8 Cross section of Cu stabilized conductor.(A) 0.28 mm in outer
    diameter. (B) 1.1 mm in outer diameter.

CONCLUSIONS

The "CaO smelting and casting process" appear to provide a cost-saving and reliable method for preparing dendrite Cu-Nb alloys suitable for the large scale production, 150 mm diameter large ingots have been studied and good characteristics have been found. Critical current density, Jc, for Cu-40Nb wires at 10 Tesla after reaction at 550°C for 7 days were 2400 A/mm^2. The degradation in Jc for Cu-40Nb untwisted wire had not observed until 1.2% bending strain, but Jc for bending strain for Cu-40Nb was affected significantly by short twist length (lp=4dw,6dw). Instead of a single core for small diameter wire, a multicored composite in larger diameter have been developed.

We have developed the in-situ Nb_3Sn conductor for power application, such as superconducting generator. More detailed studies of ac losses in the twisted in-situ wires are being carried out.

ACKNOWLEDGMENTS

The measurements of Jc at high magnetic field were performed at High Field Laboratory for Superconducting Materials RIISOM, of Tohoku University by use of 23 Tesla hybrid magnet HM-2. The authors are grateful to profesor Y.Muto, Dr.Watanabe and the staffs of RIISOM.

This work was performed as a part of " R&D on Superconducting Technology for Electric Power Apparatuses " under the Moonlight Project of Agency of Industrial Science and Technology, MITI, being consigned by New Energy and Industrial Technology Development Organization (NEDO).

REFERENCES

1. S.Foner,S.Pourrahimi,C.L.H.Thime,J.Otsubo,H.Zhang,T.P.Orlando,A.Zieba, A.Zaleski,S.Sekine,E.J.McNiff,Jr.,B.B.Schwarts,W.K.McDonald,R.Roberge and H.LeHuy,Adv. in Cryogenic Engineering Materials,vol 30,805(1983).

2. M.Fukumoto,T.Okada,K.Yasohama and K.Yasukouchi,Adv.in Cryogenic Engineering Materials,vol 30,867(1983).

3. J.J.Sue,J.D.Verhoeven,E.D.Gibson,J.E.Ostensen and D.K.Finnemore,Acta Metal.,vol 29,1791(1981).

4. C.C.Tsuei,Science 180,57(1973).

5. K.Agatsuma,K.Kaiho,K.Komuro,Y.Ikeno,N.Sadakata,M.Sugimoto and O.Kohno, IEEE Trans. on Magn. Mag-21,1040(1985).

MANUFACTURE AND EVALUATION OF TIN CORE

MODIFIED JELLY ROLL CABLES FOR THE US-DPC COIL

D. B. Smathers   M. M. Steeves, M. Takayasu,
M. B. Siddall   and M. O. Hoenig

Teledyne Wah Chang Albany   Massachusetts Institute
P. O. Box 460   of Technology
Albany, OR 97321   Cambridge, MA 02139

ABSTRACT

The US-Demonstration Poloidal Coil (US-DPC) is being built by the Massachusetts Institute of Technology as a first step test in developing an advanced cable in conduit conductor for the central solenoid (ohmic heating coil) of a Tokamak style Fusion Reactor. Three full-scale prototype pancake modules are being manufactured using Tin Core Modified Jelly Roll Nb_3Sn superconductor wire and will be operated in early 1990 at the Japan Atomic Energy Research Institute (JAERI). Teledyne Wah Chang Albany produced the wire (0.78 mm strand), supervised chrome plating and cabling operations and delivered three 225 strand, 168 meter long cables in 1988. Short sample tests show the wire to be uniform in properties from lot to lot. Wire extracted from the cables has also been tested; though the wire is significantly deformed in the cabling process at crossover points, the performance in these areas is not degraded by more than 5% relative to undeformed wire. We report on the chrome plating, cabling, short sample critical current and hysteresis loss data of strand and critical current data for cable strand.

INTRODUCTION

Teledyne Wah Chang Albany (TWCA) produced three full-scale prototype 225 chrome plated strand cables for the US-DPC double pancake modules being constructed by Massachusetts Institute of Technology (M.I.T.).[1,2] Our goal was to produce wire with uniform, consistent critical current and hysteresis loss. The tin core Modified Jelly Roll wire design has been previously discussed,[3,4] and the specification is listed below (Table I, Figure 1).

As a supplier, several challenges had to be met beyond producing more than 600 kg of uniformly good bare strand. The wire required a two micron chrome layer requiring that a chrome vendor for wire be developed. The chromed wire was abrasive and introduced new problems in cabling. We also had to respond to concerns regarding the strand mechanical properties.

TABLE I US-DPC STRAND SPECIFICATION

Wire Diameter	0.78 mm
Stabilizing Copper	54%
Guaranteed Critical Current	157 Amps 10T, 10 μV/m
Expected Critical Current	178 Amps 10T, 10 μV/m
Hysteresis Loss	
$\pm$3T Cycle	210 kJ/m^3 (Wire)
$\pm$7T Cycle	308 kJ/m^3 (Wire)
Twist Pitch	2.0 twists per inch
Chrome Plating	1 - 2 micron thickness

EXPERIENCE

Strand

Roughly three quarters of the wire used in the three cables has complete Ic and hysteresis loss data. In Figure 2, the calculated variation of $\pm$3 T hysteresis loss is plotted as a function of Jc at 10 Tesla, 10^{-13} Ωm. Our database was matched at 10^{-13} Ωm criterion. the prototypes described in reference 3 followed the prediction fairly well. All the data (33 sets) for the wire used in the cables fall within the small region marked production. The Quality Assurance/Control program employed gives us confidence all wire lots will perform as required.

The 10 μV/cm Ic (10T) values ranged from 164-178 amps and the $\pm$3 T hysteresis from 161-233 kJ/m^3.

One hundred percent of all wire ends shipped to the chrome plater were metallographically examined and a non-copper fraction determination made. These same sections were also examined for diffusion barrier integrity (a minimum of three sections mounted per wire end). Samples were provided from material adjacent to these wire ends for critical current and hysteresis loss testing both by TWCA and by M.I.T.

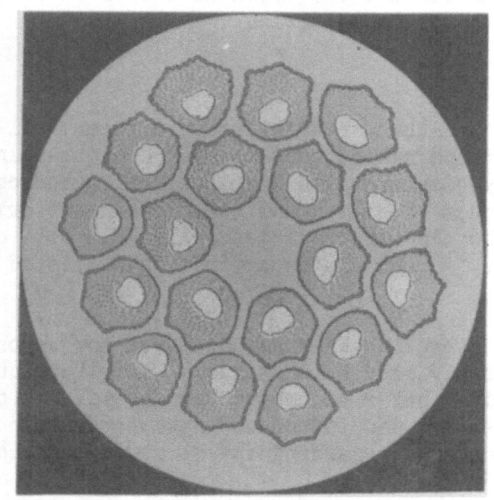

Figure 1.   US-DPC Conductor Strand - 0.78 mm diameter, 54% copper, two twists per inch.

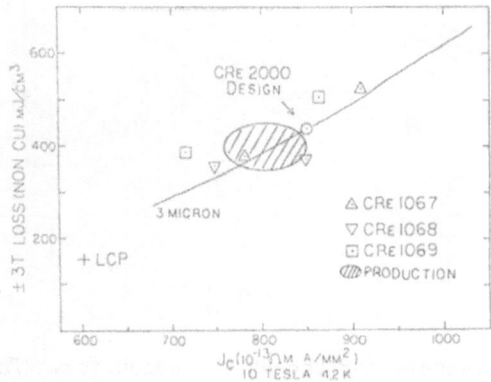

Figure 2. Non-copper hysteresis loss as a function of 10 Tesla critical current for the US-DPC conductor design, prototypes and production. Jc (10 μV/m) 75-100 A/mm² less than Jc(10⁻¹³ Ωm) for n between 25-30.

The chrome plated wire was divided into 975 meter lengths and layer wound on cabling spools at TWCA. Each chromed wire end was sectioned for layer thickness and adhesion.

Chrome

The availability of 1-2 micron chrome plating on wire in lengths exceeding about 1000 meters was non-existent in the U.S. In collaboration with M.I.T., Electrocoatings Inc. (Berkeley, California) established a chrome plating line with length limitations controlled only by the spooler. Many 5000 meter lengths were plated. We experienced a few minor start up problems but this did not impact the cables. We did observe that the chrome has a uniform smooth layer roughly one micron thick with a more irregular layer extending to 2 microns (Figure 3). All of the wire was covered with the one micron layer and most had the rougher texture extending to 2 microns. This rough outer morphology gives the plated wire a dull appearance and an abrasive quality. The layer is capable of withstanding 100% shear strain without flaking off the wire--though cracking is significant. This facility is no longer available due to a lack of commercial interest.

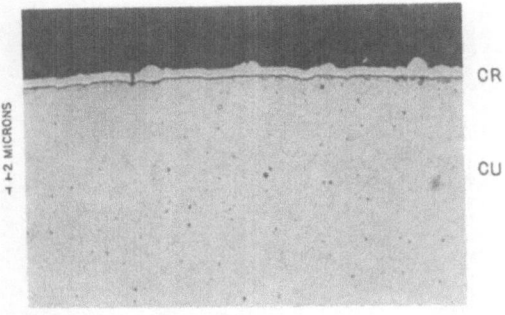

Figure 3. Chrome layer on the US-DPC conductor.

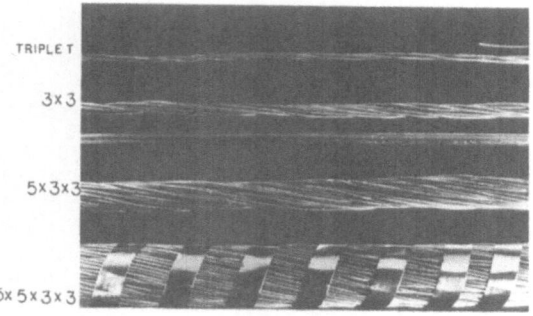

Figure 4.   Cabling elements for the US-DPC conductor. The 3.2 mm heater
was placed in the center of the cable in the final cabling
operation.

Cabling

An initial test cable using real superconductor strand (unplated)
indicated a tendency towards strand breakage. The cable pitches were
lengthened and split dies eliminated and the breakage was relieved. A
second test cable verified this and provided some cabled wire for short
sample tests.

The chrome plated wire provided extra problems in cabling. The
abrasive nature of the wire affected the compaction dies and increased the
pulling force required to compact the cable. We went to cabling with 45
units (975 m) to yield each 225 strand cable (3 x 3 x 5 x 5), Figure 4.
Maximum yield would produce a cable of about 192 meters. Our experience
was one strand break in each cable at the third stage caused by pinching
during heavy compaction. The excess length provided the opportunity to
yield in excess of the required minimum length (154 m) in each case. An
example of a pinched strand is shown in Figure 5. The wire is scissored
rather than tensilely elongated.

Cabled Strand Critical Currents

Examination of the cabled wire reveals no breaks but the wire is
significantly distorted in the process of compaction. Periodically, the
wire may get sandwiched and crimped leaving a severely deformed cross-
section. We disassembled a section of the second bare test cable and
extracted five wires at random, one from each of the 3 x 3 sub cables in

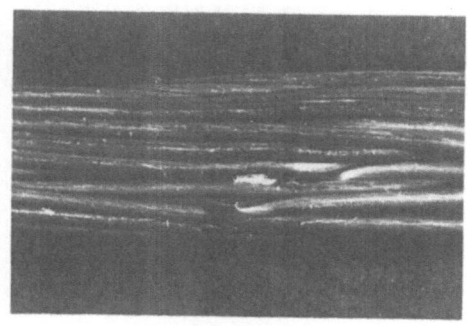

Figure 5.   Example of scissored wire strand from 5x3x3 level cable.

SAMPLE Ic (Amps 10 μV/m)

Field	Uncabled		Cabled					Cabled		
	Control A	Control B	I	II	III	IV	V	CRIMP A	CRIMP B	CRIMP C
2T	817	843	851	919	887	815	863	---	776	845
4T	497	511	519	558	538	495	523	511	487	505
6T	332	345	347	371	360	331	347	340	326	327
8T	238	240	242	261	244	231	238	235	225	227
10T	164	168	169	178	170	159	162	162	155	159
12T	110	111	108	117	109	106	104	104	103	102
n(10T)	22	28	26	23	24	23	23	23	18	22

the 45 strand third stage cable. None of these samples showed any severe deformations. We then searched the cable to provide a sample with severe deformation. The Jc results are listed in Table II along with representative uncabled strand. The undeformed cable strand average of 167 Amps (10 T, 10 μV/m) is consistent with the uncabled value of 166 Amps. The crimped samples were three strands from a single triplet. The average of 159 Amps is only 5% lower than the rest of the cable. Figure 6 shows how severely distorted the sample strands were.

Strand Mechanical Properties

Unreacted. During the production of wire and cable, a concern was voiced regarding the mechanical strength and elongation of as-drawn (heavily cold worked) tin core type composite wire.[5] We investigated the elongation, yield (0.2%) and ultimate tensile strength (UTS) of the wire

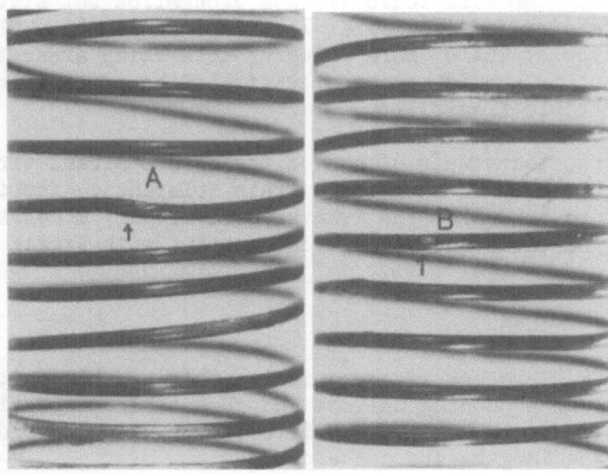

Figure 6.   Critical current coil sample with CRIMP A and B removed from unplated test cable. Note the jog in the wire at the crimp.

TABLE III TENSILE PROPERTIES - UNREACTED STRAND

Condition	Elongation	Yield (0.2%)	UTS	Vicker's Hardness 0.1 kg load Stabilizer Hardness
Bare .78 mm As Drawn	2-3%	413-551 MPa	565 MPa	99
Bare .78 mm Annealed	13%	386 MPa	393 MPa	75
Chrome .78 mm As Plated	1-2%	345-483 MPa	545 MPa	--
Bare/Cabled .78 mm No Anneal	1-3%	365 MPa	545 MPa	--

at various process steps. We observed not only a significant increase in elongation with minimal annealing (3 hrs/210°C) but also an associated drop in both yield and UTS. This latter effect was of great concern to us considering the failure mode we had observed in cabling, namely pinching. The aspected nature of the MJR filaments reduces the observed degradation of any wire flattening.

The elongation was measured by gauging the separation of notches cut into the wire surface. Wires with low elongations failed at the notches and may be artificially low.

Both plating and cabling soften the wire slightly. The plating may have a minimal annealing effect while the cabling may actually work soften the wire.

There was not enough time or excess material to manufacture a test cable of annealed wire. We argue that the increased strength and hardness of the wire will discourage crimping and be strong enough to survive cabling. Annealing the wire would improve elongation but also promote elongation by yielding and flowing at lower stress levels. The copper hardness would drop by 25% or more allowing more crimp areas to be produced. While this should not greatly affect the critical current it would reduce the cable flexibility.

Reacted. A reacted sample of one of the US-DPC strand wires (uncabled) was tested by J. Ekin[6] for strain tolerance. The intrinsic strain was found to be 0.38%, the irreversible strain, 0.84% and strain to failure 1.35%. The variation of critical current with strain was fairly flat.

CONCLUSIONS

The tin core MJR process is economical and is capable, with modest QA/QC procedures, of producing uniform and repeatable properties. The design can be tailored to meet both critical current and/or loss specifications, and lengths are not a problem.

Technology exists, within the U.S., capable of plating lengths exceeding 5 km with a 1-2 micron layer of chrome if demand is sufficient. The chrome is very adherent.

The chromed tin core wire can be cabled into fully transposed geometries, and though crimping may occur, there is minimal degradation on strand properties. We recommend, when prototype cables are manufactured, that sufficient overage be utilized to allow for difficulties.

We verified that the as drawn tin core wire has low elongations but that the elongations would be sufficient for reliable cabling. Both plating and cabling lead to slight softening. Though the elongation is low (2-3%), we feel that the hardness and strength seem more important for cabling to resist crimping and provide less strand/strand friction. This was verified by the successful cabling experience.

REFERENCES

1) M. M. Steeves, M. O. Hoenig, J. V. Minervini, C. R. Gibson, M. M. Morra, J. L. Martin, R. G. Ballinger, S. Autler, T. Ichihara, R. Randall, M. Takayasu and J. R. Hale, "The US-DPC, A Poloidal Coil Test Insert for the Japanese Demonstration Poloidal Coil Test Facility," IEEE Trans Mag 24, No. 2, 1307 (1988).

2) M. M. Steeves, M. O. Hoenig, M. Takayasu, R. Randall, J. E. Tracey, J. R. Hale, M. Mm. Morra, I. Hwang and P. Marti, "Progress in the Manufacture of the US-DPC Test Coil," IEEE Trans Mag 25 No. 2, 1738 (1989).

3) D. B. Smathers, P. M. O'Larey, M. M. Steeves and M. O. Hoenig, "Production of Tin Core Modified Jelly Roll Cable for the MIT Multipurpose Coil," IEEE Trans Mag 24 No. 2, 1131 (1988).

4) D. B. Smathers, "The Modified Jelly Roll Process for Manufactueing Multifilament Niobium-Tin Composite Superconductors," in International Symposium Tantalum and Niboium - Proceedings (Tantalum-Niobium International Study Center, Brussels, 1989), p. 707.

5) M. Strum, L. Summer and J. Miller, "Ductility Enhancement in Unreacted Internal-Sn Nb_3Sn Through Low Temperature Anneals," IEEE Trans Mag 25 No. 2, 2208 (1989).

6) J. Ekin, National Institute of Standards and Technology, Boulder, CO, Private Communication.

BRONZE-ROUTE Nb₃Sn SUPERCONDUCTING WIRES WITH IMPROVED J_C AND REDUCED BRIDGING

D. W. Capone II and K. DeMoranville

Supercon, Inc.
830 Boston Turnpike
Shrewsbury, MA 01545

INTRODUCTION

High-field applications which expose the superconductor to large time-varying magnetic fields, such as high performance generators and accelerator magnets, can benefit from a Nb₃Sn conductor, with good high-field J_C's, having fully decoupled filaments. By adapting billet assembly techniques developed for the production of fine filament NbTi conductors to the fabrication of bronze-route Nb₃Sn, a multifilament strand having unbridged filaments can be produced. The critical factors preventing filament bridging are the s/d (filament spacing/filament diameter) and the array quality, both of which can be precisely determined using these assembly techniques.

This paper summarizes the results of a study designed to optimize some of the processing issues necessary to produce high-J_C's in the 7-9 Tesla range for the advanced generator program. We have investigated the effect that changes in s/d have on filament sausaging, J_C, and fabricability. In more recent work, we have looked at the advantages to be gained by eliminating filament sausaging entirely. Finally, reducing the filament diameter to values below 2 μm are shown to greatly improve the critical current densities in these materials, independent of filament sausaging. J_C's at 4.2K in applied magnetic fields of 9T have exceeded 1500 A/mm² in recent samples.

FABRICATION

Nb₃Sn conductors are generally sought for their high current performance in moderate magnetic fields. Also, an additional margin of stability, afforded by the higher T_C (compared to NbTi conductors), is of benefit in many applications. For applications where the conductor is exposed to time varying magnetic fields, modification of the matrix materials to limit eddy current losses is needed. In addition, coupling losses must be reduced by limiting, or avoiding, filament coupling in the strand. One of the most common sources of coupling in Nb₃Sn materials is the bridging of filaments which can occur during reaction heat treatments. Filament bridging can arise mainly from a spacing that is inadequate to accommodate the 37% volume expansion which occurs during the reaction heat treatments. However, surface irregularities in the unreacted filaments can also produce bridging upon heat treatment.

Several issues factor into producing an optimum J_C while trying to avoid filament bridging. The optimum spacing is dictated by the trade off between maximizing the local bronze/Nb ratio (to reduce the amount of Sn diffusing over long distances), and minimizing the spacing (to minimize filament sausaging during the extrusion). In addition, an absolute minimum s/d is imposed by the 37% volume expansion which occurs

Advances in Cryogenic Engineering (Materials), Vol. 36
Edited by R. P. Reed and F. R. Fickett
Plenum Press, New York, 1990

139

Table 1. Billet details for materials processed in this work

Billet	Can dimensions o.d. (cm)	i.d. (cm)	Barrier	Bronze/Nb ratio Local	Overall	Total	Cu s/d(round)	# of filaments	Restack flat-flat(cm)
AF3	5.08	4.45	.051	0.44	2.5		0.20	1,112	0.0762
AF2	5.08	4.45	.051	0.55	2.5	.321	0.25	1,201	0.0762
AF6	5.08	4.45	.051	0.69	2.5		0.30	1,303	0.0762
AF1	5.08	4.45	.051	0.82	2.5	.327	0.35	1,406	0.0762
AF5	5.08	4.45	.051	0.96	2.5	.458	0.40	1,512	0.0762
AFH1	6.60	5.72	.051	0.78	2.5	.250	0.33	1,008	0.1143
AFH4	6.60	5.72	.051	0.78	2.5	.250	0.33	1,008	0.1143

during the reaction heat treatment. This is calculated to occur at an s/d less than 0.22. The range of the s/d's used in this study was chosen so as to cover these extremes. Five 5.08 cm o.d. billets were assembled with filament spacing/filament diameter (s/d) between 0.2 and 0.4. These billets had similar overall geometries and were constructed so as to maintain a fixed bronze/niobium ratio of 2.5:1 for all five billets. Also two, 6.60 cm diameter billets (#AFH1 and #AFH4) were assembled using a s/d ratio of 0.330 while maintaining a 2.5:1 bronze to Nb ratio. Billet # AFH1 was used for a comparsion of conventional extrusion (used for the five 5.08cm billets) and hydrostatic extrusion (used for billet AFH1). Billet #AFH4 will be used as a prototype for scaling up this conductor design to useful wire diameters. Table 1 details the billet compositions and assembly geometries.

Figure 1 is a cross section of billet AFH4 showing the basic geometry used for the billets in this work. Notice the significant filament sausaging which has occurred. This will be discussed further, below. All seven billets were assembled using single stacking of hexagonal subelements. This assembly technique has been the standard practice of NbTi conductors for many years and was adapted for the bronze route conductors in this study. In choosing this geometry, one is forced to provide additional bronze in other parts of the cross section to arrive at the 2.5:1 ratio desired. In this billet design, the bronze is located in the core of the conductor and in an anulus outside the filament array. Placing the bronze at remote locations can cause problems during reaction heat treatments, due to relatively long diffusion distances.

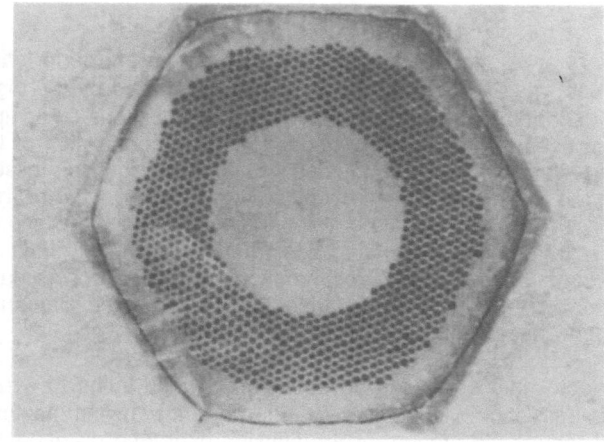

Figure 1 - Basic Geometry of Multifilament Billet (AFH4)

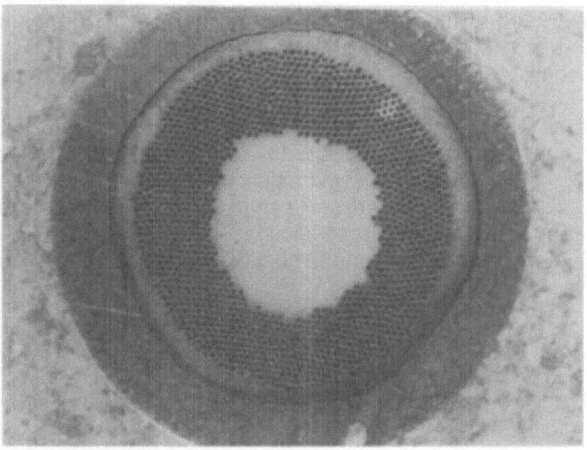

Figure 2a -Optical micrograph showing the cross section of billet AF6 at extruded size of 1.27cm. (Conventional Extrusion)

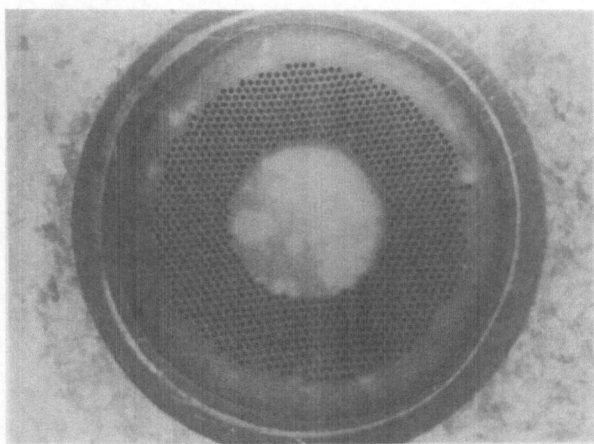

Figure 2b - Optical micrograph showing the cross section of billet AFH1 at extruded size of 1.905 cm. (Hydrostatic Extrusion)

PROCESSING

The general processing steps used in the fabrication of the Nb_3Sn materials for this study are described in this section. As mentioned above, the multifilament billets were assembled using a single stack approach. For billets AF1-AF6 hexagonal subelements with a flat to flat diameter of 0.076 cm were used. In billets AFH1 and AFH4 a subelement with a 0.114 cm flat to flat dimension were used. Prior to assembly, all of the subelements were degreased and chemically etched to ensure good metallurgical bonding during the subsequent processing steps. During handling, subelements were stored under dry nitrogen gas (from liquid nitrogen boil-off) until use. The billets were also stored under these conditions during assembly. After assembly, the stacked filaments were inserted into a copper extrusion can with a Ta barrier placed between all copper and non-copper areas. A copper nose and tail were then electron beam welded into place. The billets were hot isostatically presses (HIP) to enhance bonding of all the subelements. Billets AF1-AF6 were extruded conventionally at a temperature of 650°C. Two of the billets (AF1 and AF2) were extruded to 0.953 cm (28:1 reduction). The remaining three billets (AF3,

AF5 and AF6) were extruded at the same temperature to 1.27 cm in diameter (16:1 reduction). As mentioned in the previous section, all of these billets suffered from filament sausaging to some extent.

In order to reduce the amount of filament sausaging experienced during extrusion, billet #AFH1 was hydrostatically extruded at 400°C from 6.60 cm o.d. to a fnal size of 1.905 cm o.d. (7.1:1 reduction). The bulk of this extrusion was unsuccessful but approximately 1/4 of the billet was successfully extruded and samples from this were drawn to final wire size. In Figure 2a and b we show cross-sections of extrudate #AF6 and #AFH1 to show the substantial improvement in filament quality afforded by lower temperature, hydrostatic extrusion.

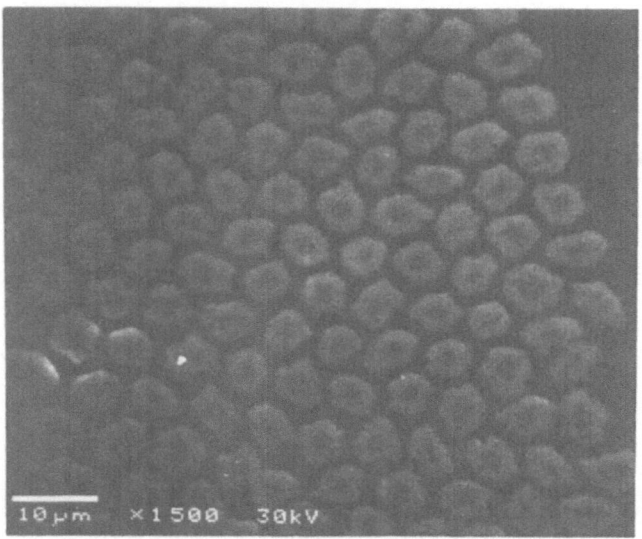

Figure 3a - SEM micrograph showing reacted filaments in a 0.012cm diameter wire after a 144 hour reaction heat treatment at 600°C.

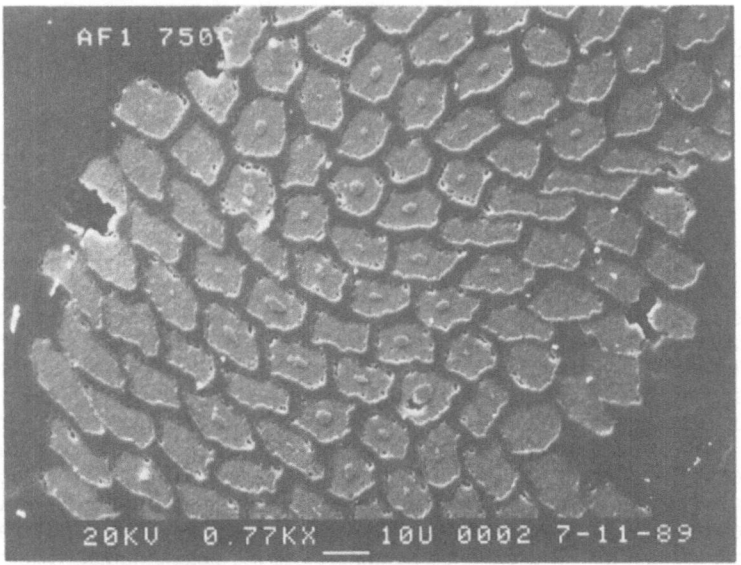

Figure 3b - SEM micrograph showing reacted filaments in a 0.036cm diameter wire after a 10 hour reaction heat treatment at 750°C.

All samples were reduced to wire size using conventional wire drawing techniques. At larger sizes (above ~0.64 cm) recovery anneals were performed after approximately 18-25% areal reduction. Below this size the recovery anneals could be spaced every 30-50% areal reduction. All recovery anneals were performed in a dry nitrogen atmosphere for one hour at 500°C. These limitations on reduction between recovery anneals were necessary to avoid radial fractures in the bronze matrix which can occur if excessive strain is imparted to the matrix during reductions.

PROPERTIES

To examine the superconducting properties of these materials, a series of reaction heat treatments have been performed on wires, from four of the seven billets processed in this study. Temperatures between 600°C and 750°C were used for times as short as 10 hours to as long as 310 hours. It must be noted that we are developing heat treatments which optimize the superconducting properties in the 7-9 Tesla range. Therefore, the conclusions reached do not necessarily apply to Nb_3Sn for high-field applications.

Two general features have been observed which are common to all the materials examined in this study: Firstly, in samples with approximately the same degree of filament reaction, lower reaction temperatures yield superior non-copper critical current densities. This is presumably due to smaller grain sizes in the reacted layer for lower reaction temperatures. Secondly, the lower reaction temperatures promote significantly more uniform layer growth throughout the filament array than with higher reaction temperatures. This can be seen in Figure 3a and b which compare the layer growth in two samples, one reacted at 600°C, the other at 750°C. In the 750°C case, the filaments close to the bronze reservoirs (inside and outside) have almost fully reacted to form Nb_3Sn while the interior filaments remain only partially reacted. When the lower reaction temperature of 600°C is used, uniform layer growth is achieved throughout the filament array. This, we believe, is due to the slower reaction rate at 600°C which allows sufficient time for Sn diffusion to occur from the bronze reservoirs to the filament array. Of course, the lower reaction temperatures will reduce the high-field performance of this material since the lower reaction temperature does not promote a well ordered A-15 compound. However, the high field properties of these materials might be improved by a short, higher temperature (750-800°C) heat treatment to order the A-15 phase [1].

The superconducting properties of these materials have been excellent, particularly for the samples reacted at lower temperatures. The initial heat treatment studies were performed on billets AF2, AF1, and AF5 having s/d's of 0.25, 0.35, and 0.40 respectively. It was immediately obvious that, of the material processed using conventional extrusion, billet AF1 was yielding better superconducting properties than the other two billets tested under all heat treatment conditions examined. We therefore began to concentrate the remainder of the testing program on billet AF1.

In Figure 4 we show the critical current density vs. reaction heat treatment time for samples from AF1, at 4.2K for magnetic fields of 5, 7, and 9 Tesla. The critical current density is determined using a 1 μV/cm electric field criterion and is calculated using the non-copper area as determined by a weigh-pickle-weigh technique. In this technique the sample is weighed on a microbalance, then the copper is chemically removed using a solution of HNO_3 and water, and weighed again. This process yields the fraction of copper to good accuracy, if performed correctly. The diameter of the Nb filaments before the reaction heat treatment is given in the figure along with the magnetic field and reaction heat treatment temperature.

Although there is some scatter in the data (which is unavoidable due to the small wire diameters examined) it is clear that at 9T J_c's in the range of 1300 A/mm^2 are achieved for these materials if the filament size is reduced to 1.6 μm or less. This is similar to results reported earlier by Furakawa [2].

As mentioned above, at an s/d = 0.35 a considerable amount of sausaging occurred during conventional extrusion at 650°C. We believed that this situation would be greatly improved if lower extrusion temperatures were used since the increased mechanical strength of the bronze matrix would provide sufficient support to prevent, or at least reduce, filament sausaging. As previously described, billet AFH1 was hydrostaticly extruded and was partially successful. The filaments in this billet displayed significantly reduced filament sausaging. In Figure 5 we show the critical current density vs. reaction heat treatment time for these wires with filament diameters indicated. A comparison with the data from billet AF1 shows some improvement in critical current density for samples with the same filament diameter. Unfortunately, smaller filament diameters were not produced due to the larger hex size used in the multifilamentary billet assembly, and the prohibitively small wire diameters needed to reach these filament sizes in the final conductor.

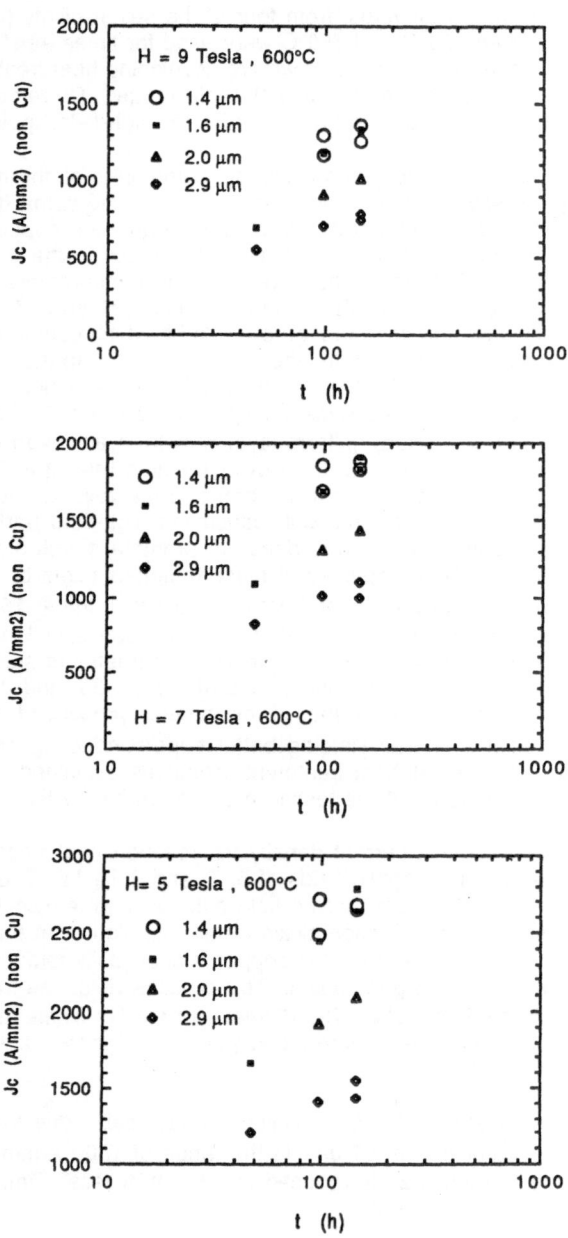

Figure 4 - Critical current density (non-copper) measured at 4.2K in the indicated magnetic fields vs. reaction time at 600° C for wires from billet AF1.

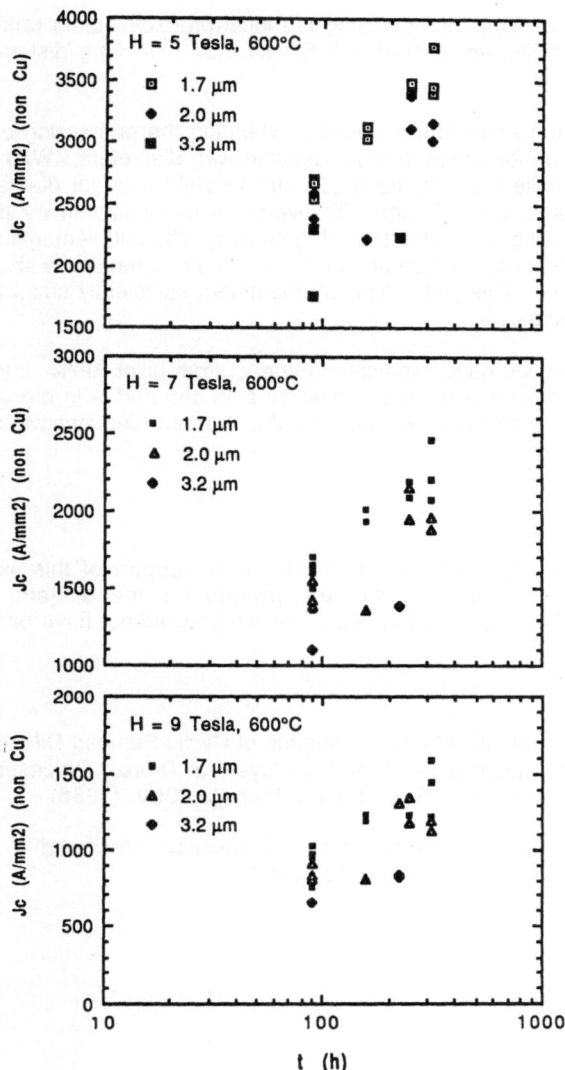

Figure 5 - Critical current density (non-copper) measured at 4.2K in the indicated magnetic fields vs. reaction time at 600° C for wires from billet AFH1.

DISCUSSION

We have addressed several major points during this work which are of interest to the continued development of bronze-route Nb_3Sn conductors in the U.S. Firstly, we have demonstrated that the single-stacked approach developed for NbTi conductors can be adapted to the fabrication of bronze-route Nb_3Sn conductors. This can allow precise control over geometrical factors in the billet, such as spacing and array uniformity. Secondly, we have examined the effects of s/d on the properties of these wires and find an optimum value, for 650°C extrusion temperatures, of about 0.35. Thirdly, we find that a reduced reaction heat treatment temperature yields substantially increased J_C's in bronze-route conductors at intermediate field ranges. This improvement, we believe, results from the reduced grain sizes and more uniform reaction layers produced by the lower temperatures. Finally, we have shown that reduced extrusion temperatures can yield further improvements by eliminating filament sausaging. This can also allow increased s/d's to be used, thus increasing the local bronze/niobium ratio, which decreases non-uniformities associated with Sn-diffusion over long distances within the strand.

Our work for the immediate future involves extending the processing of these materials so as to yield similar performance in more useable wire diameters. With the single stack approach, very large billets are required (30 cm) to yield filament diameters below 2 µm at reasonable wire diameters (~.7 mm). To overcome this limitation we have begun processing materials using a double-stacked approach. The subelements are fabricated using single-stacked methods. These are then restacked at a moderate size so as to be re-extruded a second time. This yields filament diameters sufficiently small to produce these results in real wire sizes.

At the time of writing we have assembled material from billet AFH4 into a second extrusion billet. This billet has been successfully extruded and is in process towards a final wire size near 0.76 mm o.d. At this size the filament diameter will be approximately 1.6 µm.

ACKNOWLEDGEMENTS

The authors are grateful to the Air Force for their support of this work under contract #F33615-86-C-2687. In addition, we are grateful for the diligent efforts of Dennis Corcoran and Robert Flanagan without which this work would not have been possible.

REFERENCES

1.   B.A. Glowacki; and J.E. Evetts, " Influence of Cyclic Pumped Diffusion on the Marphology and Microstructure of A-15 layers in Bronze Processed Multifilomentary Wire", IEEE Trans. Mag-25,2200 (1988)

2.   Y. Tanaka, "Practical Superconductor in Furakawa", presented at ICMC-88, Paper CA-2, Shenyang, China, June 7-10, 1988

AN INTERNAL TIN CONDUCTOR WITH Nb 1 WT % Ti FILAMENTS

E. Gregory, G. M. Ozeryansky, R.M. Schaedler, H.C. Kanithi
and B.A. Zeitlin

IGC Advanced Superconductors Inc
1875 Thomaston Ave
Waterbury, CT , 06704

D.W. Hazelton and W.D. Markiewicz

Intermagnetics General Corporation
Guilderland, New York,12084

ABSTRACT

A production program is underway at IGC Advanced Superconductors, Inc. to produce 1200 kg of internal tin wire for the high field test facility (HFTF) upgrade which is to be constructed by the Lawrence Livermore National Laboratory (LLNL).The details of the conductor design and the manufacturing process are presented. The niobium filaments are doped with 1% Titanium in order to optimize the high field current density. Representative samples from the initial production wire were reacted and their superconducting properties measured at magnetic fields from 3 to 20 Tesla. The details of the reaction heat treatment and critical current density J_c will be presented.

INTRODUCTION

Nb_3Sn, made primarily by the bronze process, has been used for many years to make small laboratory magnets [1], particularly Nuclear Magnetic Resonance (N.M.R.) magnets. Only in a few instances, however, has Nb_3Sn been used for large high field magnets. The reasons for this are partially technical and partially economic.

Technical scale-up problems were encountered with the bronze process but were overcome, to a large extent, in the manufacture of the conductor for the Westinghouse Large Coil. [2] The current density in this conductor was low relative to that produced more recently in Nb_3Sn made by other methods and the field of operation of the magnet was also low (8T).

IGC reported high current densities in fine filamentary bronze processed Nb_3Sn in 1979 [3], but sidelined the process because of the high cost of manufacture. More recently, precautions have been made to avoid prereaction, one of the drawbacks of the bronze process and relatively high J_c's have been achieved, particularly in the low field region [4,5,6]. These improvements have been achieved by better conductor design, addition of Ti or Ta, reduction in filament diameter and reduction in grain size.

These property increases have, however, done little, if anything, to lower the inherent high cost of the bronze process.

The internal tin process was developed by IGC [7] primarily to overcome this drawback.

Advances in Cryogenic Engineering (Materials), Vol. 36
Edited by R. P. Reed and F. R. Fickett
Plenum Press, New York, 1990

147

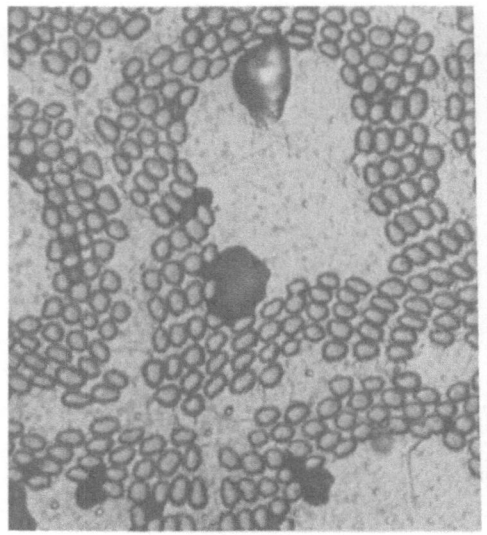
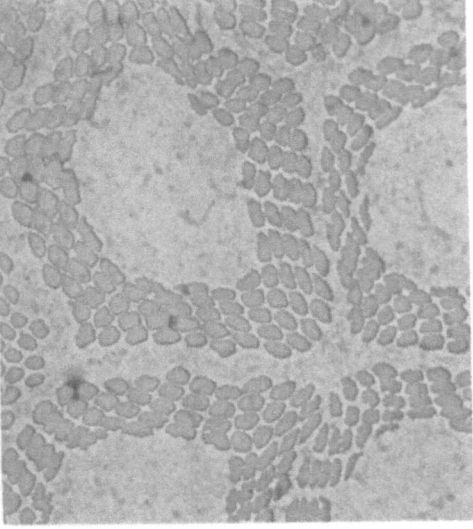

Figure 1a. Cross section of reacted internal tin material made with an LAR of 1.8.

Figure 1b Cross section of early reacted internal tin material with an LAR of 1

The principal advantage of Nb3Sn is to be found at fields above 12 T where NbTi and NbTiTa, even cooled to 1.8 K, no longer carry significant currents. The internal tin process was successfully developed for full scale applications using Nb filaments [7], although in the highest current density material filament bridging occurred [8]. For fields above 12 T, however, the addition of a small percentage of Titanium to the material is required to develop improved current carrying capability. Once the Ti was added to the Nb filaments, fabrication problems, which had not been evident in the earlier work, began to develop. The most obvious of these was poor piece length.

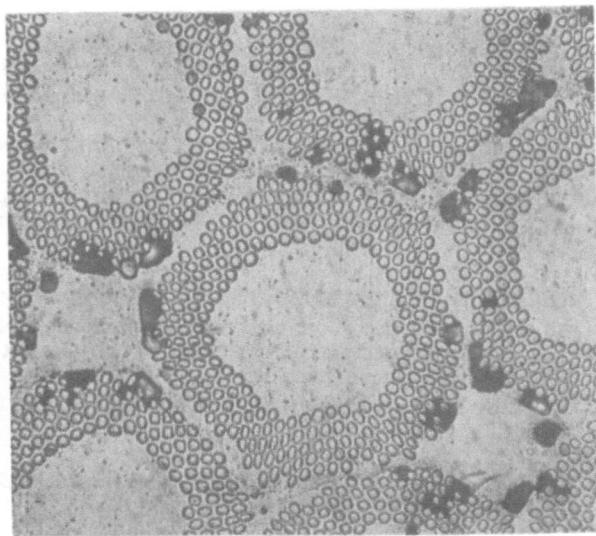

Figure 2. Cross section of reacted LLNL-HFTF Upgrade Nb3Sn showing uniform filament array and unbridged filaments.

In this paper, our work to reduce the bridging problem and to improve piece length, while maintaining high current density, is reported. Also mentioned is recent work on barrier quality, designed to improve product reliability. Our progress towards the production of 1200 kg of internal tin strand containing 1 wt % Ti in the filaments for the high field test facility (HFTF) upgrade at the Lawrence Livermore National Laboratory is also discussed.

THE BRIDGING PROBLEM

The bridging of filaments occurs when Nb, (or alloyed Nb) filaments are converted to Nb_3Sn as there is a volume expansion. This is a basic phenomenon, which is common to all methods of producing Nb_3Sn and may ultimately prove to be the factor limiting J_c for ac applications. This bridging is particularly prevalent where the filaments are closely spaced as in the high J_c materials; it is also exaggerated if the array of filaments is uneven.

Significantly reduced bridging was recently reported [9]. This was accomplished primarily by increasing the local area ratio (LAR) of Cu: Superconductor from 1 to 1.8. This simply increased the distance between the filaments but did little to improve the filament array. The inner ring of filaments in each subelement was still distorted and tended to bridge to the next ring, in places (Figure 1a). This was, however, a marked improvement over material made earlier (Figure 1b). The lowering of the overall J_c, resulting from an increase in LAR from 1 to 1.8, was quite significant.

Although the HFTF upgrade does not require particularly low losses, it is always desirable to have as uniform and as closely spaced a filament array as possible without a significant amount of bridging. To accomplish this we have returned to an LAR of 1 but, by changing the restack design, we have been able to increase the J_c without the occurrence of bridging. (Figure 2)

While magnetization data has not yet been obtained on this LLNL material, from a comparison microstructures after reaction (Figure 2 compared with Figures 1a and b), we are confident that the amount of bridging is very low.

THE PIECE LENGTH PROBLEM

In the early work with Nb filaments [7,10], scale up to full production was accomplished and the piece lengths obtained were adequate for economical production as well as for the customer's needs (up to 27 km.). Recently, however, as we have attempted to introduce material containing 1 wt. % Ti in the Nb into commercial operation, we have experienced wire breakage and piece length problems to the point where we have had difficulty in supplying material to customers in the specified lengths and on a timely basis. Many of our efforts in recent months have been directed towards decreasing wire breakage. We have been able to show dramatic piece length improvements by applying changes in design and fabrication procedures. We have recently drawn 40 kg., the first short, but full scale, restack assembly, to finished size in one piece.

THE PROBLEM OF BARRIER QUALITY

The recent procedures used for fabrication of the stabilizer have been to perform tubular extrusions on billets consisting of cryogenic grade copper with a thin diffusion barrier of niobium. In some cases where losses are important, it is desirable to have multiple layers, sometimes with tantalum, or vanadium on the inside and niobium on the outside. These stabilizers are placed near the inner diameter of the tube and separated by layers of copper.

This extrusion procedure for forming the stabilizer was introduced to ensure that as many surfaces as possible were adequately bonded prior to the final restack drawing. It was felt that this procedure was technically superior to a nested tube assembly, which is an alternative construction method, with cost and schedule advantages. The extruded stabilizers, however, proved to have problems of their own and a form of "sausaging" of at least some of the barrier layers was encountered. Design changes and alteration of the extrusion conditions, have now solved the problem for niobium and and work on different materials will be undertaken shortly.

Figure 3 shows a longitudinal cross section of the Nb barrier in some earlier material and Figure 4 a similar section showing the Nb barrier in the full scale extrusions of the stabilizer for the HFTF upgrade material. This improvement in barrier quality will certainly increase the reliability of the product and, while the barrier is not thought of as the prime cause of piece length problems, the improved material will certainly be less likely to fail either mechanically or electrically because of barrier discontinuities.

PRESENT STATUS OF THE LLNL - HFTF UPGRADE MATERIAL

This is an order for 1200 Kg of strand which will later be cabled and sheathed with a stainless tube to make a cable-in-conduit conductor (CICC).

The strand will have the following specifications:

Wire diameter	0.813 mm
Copper stabilizer	42.8 %
Niobium barrier	8.7 %
Filament diameter	3.6 μm
Local Area ratio Cu : Nb	1:1
# of Subelements	19
Total # of filaments	4902

In view of the problems encountered in recent projects, the decision was made to process one batch of material to final size ahead of the bulk of the order, incorporating all of the above described developments. The present status of the various components of the order is outlined below.

Subelements

Six 200 mm. O.D. billets, the total amount required for the project, have been fabricated to produce the required subelements. Figure 5a shows a portion of the cross section of one such extrusion showing a highly uniform filament array. One of these extrusions has had the tin inserted in it and has been drawn to ~7.6 mm. diameter for restacking into one of the full sized extruded stabilizer tubes. The remaining five extruded subelements are ready for tin insertion.

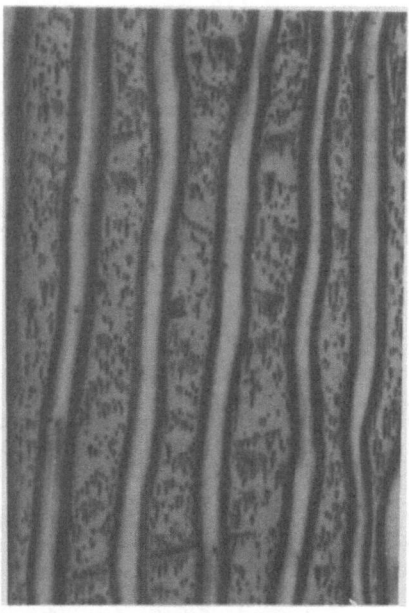

Figure 3. Longitudinal cross section of the layered Nb barrier in some earlier internal tin stabilizer.

150

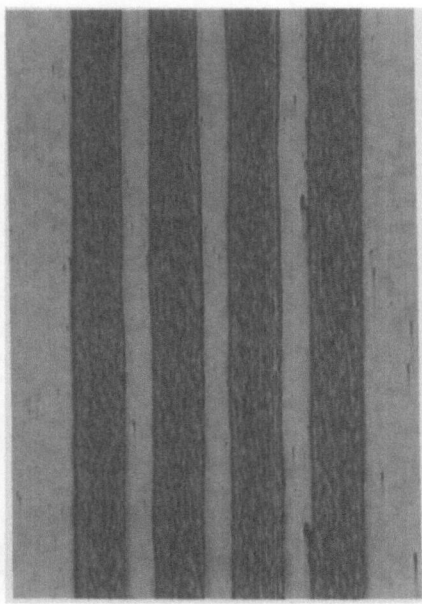

Figure 4. Longitudinal cross section of
the layered Nb barrier in the recent
LLNL HFTF upgrade stabilizer.

Stabilizer Tubes

Two 200 mm. O.D. stabilizer billets have been completed successfully and their
longitudinal cross sections show much improved barriers (Figure 4). Assemblies of the
remaining ten billets have been made, welded and will be extruded in the near future.

Restack Assemblies.

Two full sized restack assemblies have been made and the first one drawn down to
finished wire size.without breakage. Figure 5 b. shows a cross section of this material

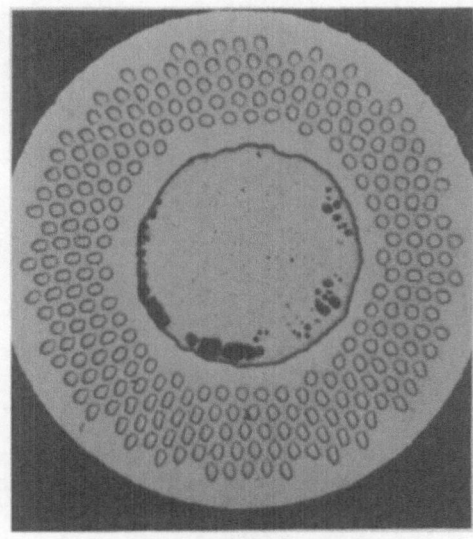

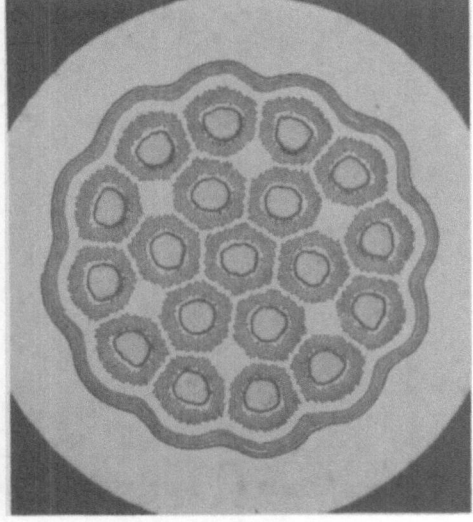

Figure 5a. Cross section of the subelement
used in the LLNL -HFTF

Figure 5 b. Cross section of first restack
material at 1.6 mm. diameter.

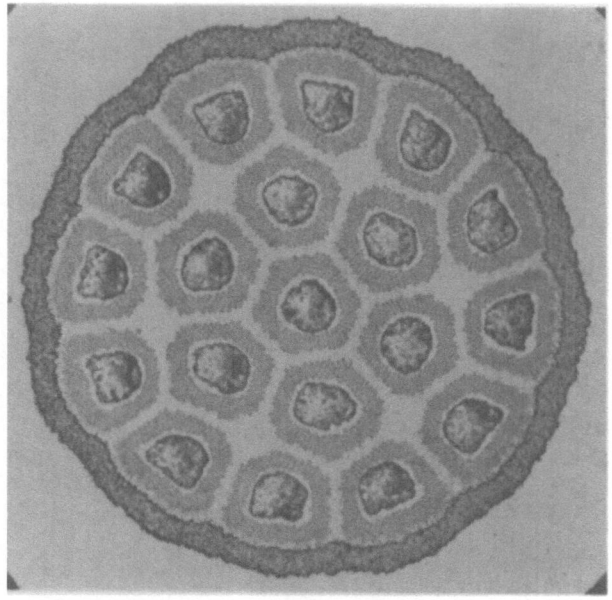

Figure 6. Cross section of trial sample before reaction.

unreacted at 1.6 mm. diameter. The filaments show very little distortion or bridging and the multiple layer barrier is of high quality. The remaining twenty two restack assemblies will be made in the same way.

Trial Experiments

Several small lengths of the subelement material, fabricated as described as above, were drawn to 2.4 mm. diameter, restacked in a nested copper and tantalum tube assembly with an O.D. of 19 mm. This was drawn down to 0.254 mm. diameter without any breakage using the newly developed designs and fabricating procedures. This could not be done using the previous procedures without considerable breakage.

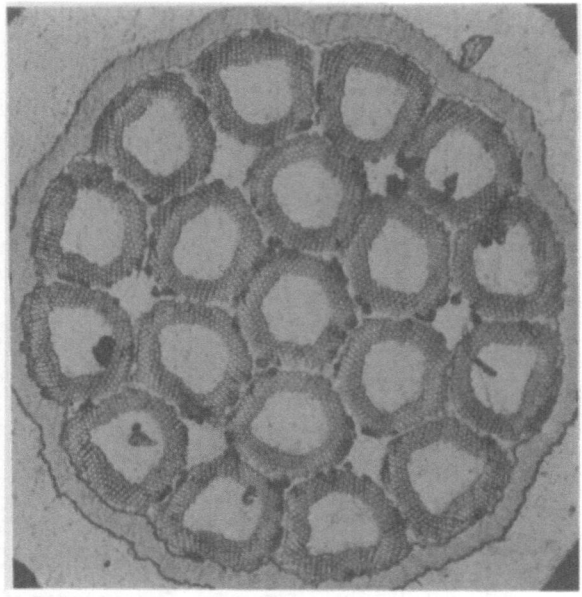

Figure 7. Cross section of trial material after reaction.

Table 1

Field (T)	4.2 K		3.0 K		2.0K	
	I_c (A)	J_c (A/mm^2)	I_c (A)	J_{c*} (A/mm^2)	I_c (A)	J_c (A/mm^2)
16	110.0	350				
17	73.4	234	113.6	362	134.0	427
18	57.4	183	73.6	235	104.8	334
19	36.0	115	54.0	173	69.4	221
20	28.5	90.8	37.7	120	50.7	167

Table 2

Field (T)	Sensitivity 0.1 µv/cm		Sensitivity 10^{-12} Ω · cm	
	I_c (A)	J_c (A/mm^2)	I_c (A)	J_c (A/mm^2)
16	110.0	350	101	322
17	73.4	234	67	213
18	57.4	183	51	163
19	36.0	115	29	92
20	28.5	90.8	22	70

Table 3

Field (T)	I_c (A)	J_c (A/mm^2)	n
3	1031	3286	29
4	801	2553	35
5	635	2024	34
6	525	1673	35
7	434	1383	30
8	362	1154	32

A portion of this material was held at 0.836 mm. diameter and heat treated according to the following procedure, which is not necessarily an optimized one for either high for low field applications.

24 hrs. @ 200 °C
48 hrs. @ 325 °C
20 hrs. @ 700 °C
4 hrs. @ 740 °C

Figure 6, shows a cross section of the material before reaction and Figure 7, a similar section after reaction. These photographs show a thick continuous barrier of tantalum and minimal distortion of the subelements. The filaments themselves are shown in Figure 2 and, although some are slightly "sausaged", they show very little lateral distortion and bridging.

The high field J_c properties of this trial material were measured from 16 to 20T at MIT's Francis Bitter National Magnet Laboratory (FBNML) using a Brookhaven National Laboratory (BNL) high field sample holder. The low field J_c properties were measured at BNL. The high field data at three different test temperatures is shown in Table 1, measured at a sensitivity of 0.1 µv/cm., a common sensitivity for reporting Nb_3Sn data for

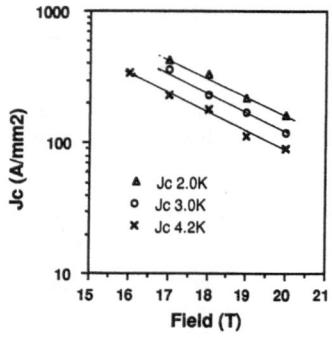

Figure 8. High field J_c at different temperatures.
Sensitivity 0.1 µv/cm.

fusion magnet applications. The wire diameter is 0.836 mm. and 57.2 % of the cross section is non-copper.

Table 2 shows the high field data at 4.2 K corrected to a sensitivity of $10^{-12} \Omega \cdot$ cm. to make it compatible with the low field data, compared with the Table 1 data.

Table 3 shows the low field data at 4.2 K and a sensitivity of $10^{-12} \Omega \cdot$ cm. together with "n" values. All J_c 's are reported as in the entire non copper area.

Figure 8 shows the plots of J_c at a sensitivity of 0.1 µv/cm. at 4.2 K, 3.0 K and 2.0 K in the field range 16 to 20 T. Figure 9 shows the plots of J_c at a sensitivity of $10^{-12} \Omega \cdot$ cm. at 4.2K from 3 to 20 T .

On the basis of these initial tests, the J_c at 16T (350 A/mm^2) meets the LLNL specification of 345 A/mm^2, at a sensitivity of 0.1 µv/cm. although heat treatment optimization has not yet been attempted. There is no conductor loss specification but we expect in the near future to obtain magnetization data from BNL which will confirm that minimal bridging and coupling is present.

SUMMARY

Changes to the internal tin process have been made recently to improve the filament array, reduce the subelement distortion, improve piece length and obtain higher J_c's, without filament bridging. The progress made to date on the LLNL - HFTF upgrade material is reported and preliminary data on trial samples show J_c in the field range 3 - 20 T . The LLNL J_c specification has been met.

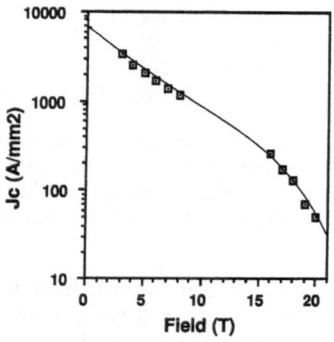

Figure 9. J_c data a 4.2 K from 3 to 20T
Sensitivity $10^{-12} \Omega \cdot$ cm.

ACKNOWLEDGEMENTS

Thanks are due to our colleagues at IGC Advanced Superconductors and Intermagnetics General Guilderland and our consultant, Dr. B. Avitzur, of Lehigh University, for their efforts in support of this work. We are also deeply indebted to M. Suenaga, W.B. Sampson, A.K. Ghosh, M. Garber and their colleagues at BNL for their help in testing and evaluating the data. The facilities of the the FBNML of MIT were indispensable for the work. The patience, tolerance and encouragement of J. Miller and L. Summers of LLNL is much appreciated.

REFERENCES

1. P. McDonald and W. Proctor, "The realization of the full potential of filamentary Nb_3Sn in practical magnets". Proc. ICEC 8, Genova, IPC Science and Technology Press Ltd., pp. 509-513, (1980).

2. P.A. Sanger, E. Adam, E. Gregory, W. Marancik, E. Mayer, G. Rothschild and M. Young, "Developments in Nb_3Sn forced flow conductors for large magnets", Trans. IEEE MAG-15, vol 1, pp. 789-791, (1979).

3. M.S. Walker, J.M. Cutro, B.A. Zeitlin, G.M. Ozeryansky, R.E. Schwall, C.E. Oberly, J.C. Ho and J.A. Woollan, "Properties and performance of fine filament bronze-process Nb_3Sn conductors", Trans. IEEE MAG-15, vol 1, pp. 80-82, (1979).

4. H. Tanaka "Predominance of bronze-processed superconductors". in Proc. ICFA Workshop, P. Dahl, ed. Rep. BNL 52006, Brookhaven National Laboratory, pp. 137-140, (1986).

5. H. Krauth, "Development of bronze route Nb_3Sn superconductors for accelerator magnets", in Proc. ICFA Workshop, P. Dahl, ed. Rep. BNL 52006, Brookhaven National Laboratory, pp. 147-149 (1986).

6. D.W. Capone II, K. DeMoranville and E. Gregory, " Bronze wrapped Nb_3Sn superconducting wires with improved J_c and fully decoupled filaments" ,Paper # FX 1, CEC/ICMC, Los Angeles, CA, (1989).

7. R.E. Schwall, G.M. Ozeryansky, D.W. Hazleton, S.F. Cogan and R.M. Rose, " Properties and performance of high current density Sn-core process MF Nb_3Sn", Trans. IEEE MAG-19, pp. 1135-1138, (1983).

8. R.B. Goldfarb and J.W. Ekin, "Hysteresis losses in fine filament internal-tin superconductors", Cryogenics, vol. 26, pp. 478-481, (1986).

9. H.C. Kanithi, L.R. Motowidlo, G. M. Ozeryansky, D.W. Hazleton and B.A. Zeitlin, "Low loss and high current Nb_3Sn conductors made by the internal tin method", Trans. IEEE, vol 25, pp 2204-2207, (1989).

10. B.A. Zeitlin, G.M. Ozeryansky and K. Hemachalam, "An overview of the IGC internal tin Nb_3Sn conductor", Trans. IEEE MAG-21, pp. 293-296, (1985)

DEVELOPMENT OF NIOBIUM-TIN CONDUCTORS AT ECN

E.M. Hornsveld and J.D. Elen

Netherlands Energy Research Foundation ECN
Petten, The Netherlands

INTRODUCTION

The development of niobium-tin conductors at ECN is based on a
superconducting wire, made according to the internal tin method, utilizing
$NbSn_2$ powder compound.

WIRE

In the internal tin method of ECN, the Nb_3Sn layers are formed inside the
filaments by a reaction of $NbSn_2$ powder with the surrounding niobium at $700^{\circ}C$.
The results of a series of experimental wires with 36 to 1332 filaments
were presented earlier.[1]

The niobium-tin wire is developed towards a smaller filament diameter for
two purposes: to reduce the remanent magnetization and to diminish the
strain sensitivity.

The first point is of importance for LHC in order to limit the multipole
field errors in the dipole magnets.[2]

Strain degradation is a limiting factor in manufacturing magnets by the
react and wind technique. This technique is used in the development of the
9 T and 12 T coils for the SULTAN test facility.[3]

We are presently engaged in a small pilot production of niobium-tin wire
with 192 filaments. The copper fraction is 55%. One of the benefits of the
internal tin method is the existence of a pure copper matrix between the
filaments.

The production is based on rod and wire drawing at a billet size presently
limited to 10 kg. The final diameter of the wire is either 0.9 mm to be
used in a model LHC cable[2] or 1.0 mm for the 9 T SULTAN coil.

From each wire samples are taken for measurements of the short sample
critical current. The statistics of the non-copper critical current density
at 11 T of 105 samples is shown in figure 1. The average value is 2120

Advances in Cryogenic Engineering (Materials), Vol. 36
Edited by R. P. Reed and F. R. Fickett
Plenum Press, New York, 1990

157

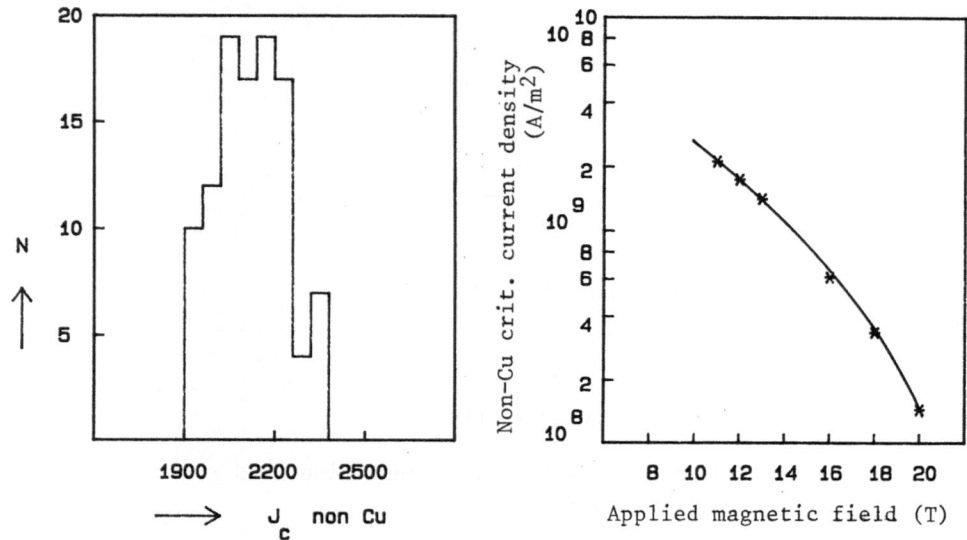

Fig. 1. Distribution of short sample non-copper critical current densities of 105 wire samples at 11 T, 4.2 K. The average value of Jc non-Cu is 2120 ± 116 A/mm².

Fig. 2. Non-copper critical current density vs. applied magnetic field of 192 filament Nb₃Sn wire.

A/mm² with a standard deviation of 5.5%. The plot of critical current density (non-copper) versus magnetic field is given in figure 2.

At a wire diameter of 0.9 mm the tubular Nb₃Sn filaments in the wire have an outer diameter of 33 µm, and an inner diameter of 23 µm. From the point of magnetization this is equivalent to solid filaments of 29 µm at full field penetration.[4]

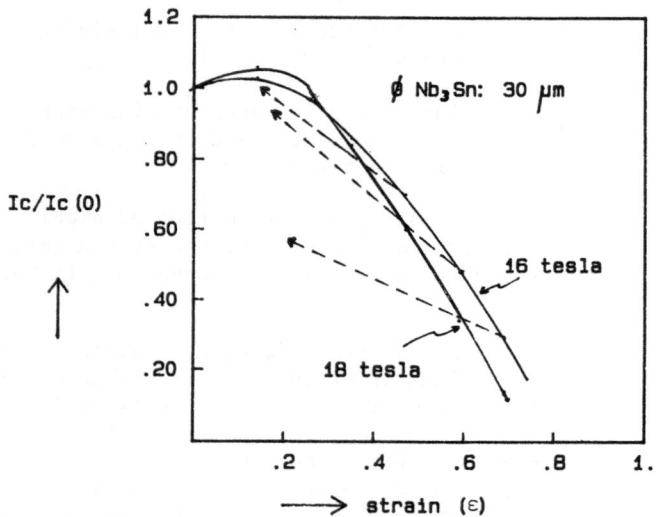

Fig. 3. Relative critical current values vs. strain at 16 and 18 T of a 192 filament Nb₃Sn wire. The value of ε irreversible is 0.6%.

An important feature in reducing the filament size in the niobium-tin wires is the decrease of the strain sensitivity. The irreversible strain value of 36 filament wire with filaments having an outer diameter of the Nb_3Sn layer of 60 um was approximately 0.2%. The irreversible strain value for the present 192 filament wire having Nb_3Sn layers with an outer diameter of 30 µm is measured to be 0.6%. In figure 3 the strain sensitivity of 192 filament wire measured at 16 and 18 tesla is shown.

CABLES

An experimental cable has been manufactured to be applied in a model dipole. The cable is made in a keystoned shape of 1.45/1.77 x 16.40 mm^2 using 36 strands of 192 filament niobium-tin wire with a diameter of 0.90 mm and twist pitch of 30 mm.

In cooperation with the Technical University of Twente the critical current versus applied magnetic field has been measured on a short sample of the cable in a laboratory 13 T magnet with a bore of 60 mm. using a flux transformer. The sample was reacted in a hairpin geometry.

The current in the cable is detected with a Hall element. A quench of the cable leads to a sudden drop of the Hall signal. The value of this current is taken as the critical current. The results of the measurements are shown in figure 4.

At 11 tesla a critical current of 19.3 kA has been measured. Based on the value of the critical current in a single wire of 2120 A/mm^2 non-copper, the calculated maximum current for the cable is 21.7 kA. As the cabling operation leads to a heavy compaction of the cable at the smaller side where two layers of 0.9 mm wire are compacted to 1.45 mm, a degradation of the critical current is expected. At 11 T the degradation is 11%. At 12 tesla the values for the measured critical current of the cable and the degradation due to compaction are 15.6 kA and 13%, while at 13 tesla these values are 11.8 kA and 19%.

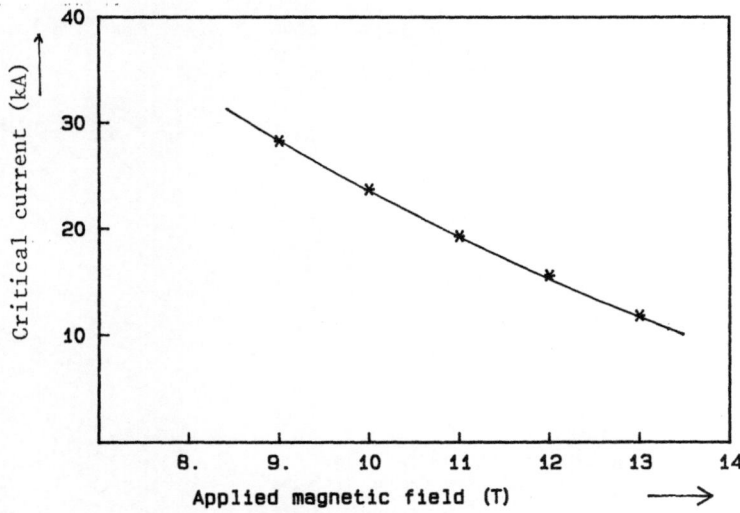

Fig. 4. Critical current vs. applied magnetic field of a 36 strand cable for an experimental dipole.

The Rutherford-type cable for the 9 tesla coil of SULTAN is made in a rectangular form with dimensions 18.1 x 1.84 mm^2, and consists of 36 strands of 192 filament wires at a diameter of 1.0 mm with a twist pitch of 30 mm. The compaction at the edges of the cable is less than in the case of a keystoned cable.

The operating current of the conductor will be 12 kA and the design critical current is 18 kA (9 T, 4.5 K).[3]
With respect to the short sample value of the wire (1130 A at 9 T, as extrapolated from the values at 11, 12 and 13 T) there is a very large margin to compensate for degradation due to the compaction of the cable.

ACKNOWLEDGEMENTS

The authors are very grateful to J.W. Ekin from NBS, Boulder for the measurements on strain sensitivity as well as to H.H.J. ten Kate from the Technical University of Twente for the critical current measurements of the cable with a flux transformer.

REFERENCES

1. E.M. Hornsveld, J.D. Elen, C.A.M. van Beijnen and P. Hoogendam, Adv. in Cryogenic Engineering Materials, Plenum Press, 34: 493 (1987)

2. R. Perin, IEEE Trans. Magn. 24: 734 (1988)

3. A. della Corte et al., Fusion Technology 1988, Proc. 15th SOFT, Elsevier Science Publ. 1989, p. 1476

4. J.A. Eikelboom, R.A. Hartmann and L.J.M. van de Klundert, IEEE Trans. Magn. 25: 1968 (1989)

EXPERIMENTAL RESEARCH Nb₃Sn-BASED

CORRUGATED SUPERCONDUCTING TUBE

P.I. Dolgosheev, V.A. Mitrochin, G.G. Svalov,
V.E. Sytnikov, N.A. Yakhtinskiy

All Union Scientific Research Institute of
Cable Industry, Moscow, USSR

G. Ziemek, P. Rhoner

Kabelmetal Elektro, Hannover, FRG

INTRODUCTION

For the last few years VNIIKP (Moscow, USSR) and Kabel-metal Electro (Hannover, FRG) have been jointly working on creation of flexible corrugated superconductors on the basis of the intermetallic compound Nb₃Sn /1,2/.

VNIIKP has created a cryogenic-physical facilities for testing lengths of flexible cable conductors for current values up to 100kA and has worked out the corresponding test methods. The first test results for samples of corrugated superconductors, created earlier, are presented in this report.

MANUFACTURE OF A CORRUGATED Nb₃Sn-BASED
SUPERCONDUCTOR AND ANALYSIS OF ITS MICROSTRUCTURE

The flexible corrugated conductor has been manufactured according to the following production process: the tape made from the alloy, containing niobium +3,0(%, atomic) of titanium, was coated with copper, using the mechanic-galvanic method; this composite was rolled to get a sample 60 m long and 0,6 mm thick. A corrugated tube with the outside diameter and length, being 33 mm and 30 m, correspondingly, was manufactured from alloy (Nb+3% Ti) + Cu using "Univema". The outside surface of the tube is galvanically covered with a layer of pure tin 20-25 mm thick. The process is described in detail in the work /2/.

A special stainless steel chamber was designed for heat treatment, inside which a corrugated conductor 30 m long was placed. The conductor was heated by direct current. The whole system was evacuated to the residual pressure 10^{-5} torr. A special cord made of aluminium oxide and silicon was winded on the conductor to insulate the latter from the metal walls of the chamber. Heat treatment conditions had to provide in the first stage degassing of the special cord and formation of a uniform bronze layer, and in the second stage, the formation of niobium-tin intermetallic compound. The following annealing

Advances in Cryogenic Engineering (Materials), Vol. 36
Edited by R. P. Reed and F. R. Fickett
Plenum Press, New York, 1990

161

conditions were worked out on the basis of the mentioned-above pre-conditions:

1. Slow heating by stepwise increase of temperature from 20°C to 250°C in the course of 24 hours.

2. Temperature increase up to 450°C in the course of 24 hours. Selection of short samples and choice of temperature of final annealing providing formation of an intermetallic compound.

3. Heating to temperature selected on the basis of test results for short samples and further quick cooling by filling the chamber with liquid or gaseous phase argon.

Short samples 33 mm long, chosen after fulfilment of the second stage were annealed at a temperature of 700°C for the period of 48 or 72 hours. Their electric-physical characteristics were measured after the process of annealing and the structure and chemical composition of different layers were studied (Table 1).

As a result of thermal diffusion treatment superconducting layers 7-13 mm thick were formed on all samples. A characteristic microstructure after annealing is presented in Fig. 1. X-ray analysis of Cu-K$_\alpha$ radiation using DRON-2 demonstrated that the superconducting layer consists mainly of Nb$_3$Sn and a small quantity of Nb$_3$Sn$_5$ phase. Lattice constant of Nb$_3$Sn is 5,287°A. Distribution of elements in different layers of the conductor after heat treatment has been studied using the electron microscope JSM-T330 with "Link" analysing system.

Data, presented in Table 1, show that after heat treatment both of a short sample (103) and of a long one (106) good correlation of the layers composition is observed. It should be mentioned that the superconducting phase of Nb$_3$Sn contains such alloying elements as copper and titanium. The results of the chemical analysis of tin concentration in Nb$_3$Sn and the lattice constant show that the forming compound Nb$_3$Sn is on the upper border of homogeneity region. The critical temperature of transition into a superconducting state con-

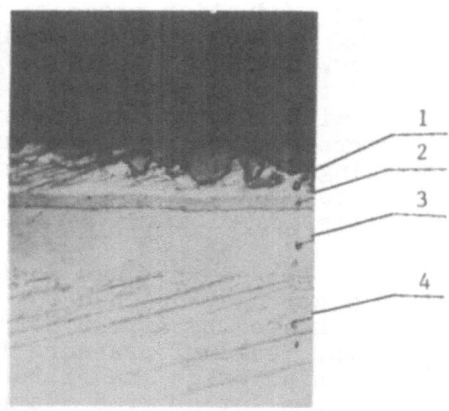

Fig.1. The microstructure of the sample 106 after thermal diffusion treatment (x 220). 1 - bronze layer Cu + Sn; 2 - superconducting layer Nb$_3$Sn; 3 - niobium-titanium layer; 4 - stabilizing copper.

Table 1. Chemical composition of corrugated current-carrying conductor (%, atomic)

Sample No	Heat treatment conditions	Bronze layer				Superconducting layer				Niobium-titanium layer			
		Cu	Sn	Nb	Ti	Cu	Sn	Nb	Ti	Cu	Sn	Nb	Ti
101	To=450°C t = 24	73,4	25,9	0,5	0,2	1,5	0,4	97,7	3,3	1,0	0,3	95,6	3,0
103	To=700°C t = 48	83,5	15,2	0,1	1,25	3,6	23,8	71,4	1,1	1,6	3,2	92,5	2,6
106	To=750°C t = 30	85,4	13,2	0,4	1,0	3,6	26,0	69,8	0,4	1,3	0,5	95,4	2,8

the sample cutted from a long tube

firms this opinion, because the measured value Tc = 16,7 K corresponds to tin concentration in the Nb_3Sn compound, which is 23,5 % at. /3/.

NON-CONTACT METHOD OF CRITICAL CURRENT MEASUREMENT

Non-contact methods are based on measurements of a magnetic field, generated by currents, induced in closed superconducting circuits with further calculation of linear density and critical current for established structural parameters of the sample and orientation of the external magnetic field induction vector.

Usually in investigations of the current critical density in superconducting cylinders the axis of the latter is parallel to the external magnetic field induction vector and the field of circular currents is measured as the difference of fields inside and outside the cylinder.

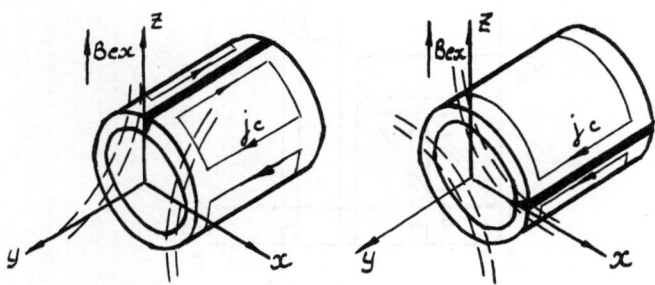

Fig.2. Arrangment of samples in the experiments on non-contact measurements of the current critical density. (a) The weld is in the plane ZOY; (b) The weld is in the plane XOY.

This method is not suitable for the considered case because of quick attenuation of circular currents on the non-superconducting weld. For this reason in our experiments the cylinder axis was oriented perpendicularly to the magnetic induction vector. In this case the screen current circuits are saddle-shaped. Two positions of the weld in relation to the external field vector are possible (Fig.2). If the weld is positioned in the plane XOY, two circuits of screen currents are induced by analogy with tests of a continuous superconducting cylinder. If the weld is positioned in the plane ZOY it divides one of the circuits into two smaller area circuits.

Field induction of screen currents in the centre of the sample is calculated on the basis of Biot-Savart's law for the following assumptions and conditions:

- the calculation is carried out for the case when the cross-section is filled with current, i.e. for the moment corresponding to transition of the sample into normally conducting stage;

- the metallographic analysis of the structure established that the Nb3Sn grains have equal axis in the plane of the underlayer, so it is assumed that $j_{cy}=j_{c\varphi}$;

- current critical density $j_c = const(\varphi, B)$.

The last assumption simplifies the calculations but introduces the error into the quantitative estimations, especially for small fields. But this is allowed because the main aim of non-contact measurements is comparison of small samples annealed in different conditions in order to choose the heat treatment conditions for long samples.

For the sample the length of which is equal to its diameter, equal to 33 mm, below are presented solutions of equations, connecting the induction of screen current field with linear density of the critical current for the case when the weld is in the XOY plane:

$$\Delta B_{z1} = 1,26.10^{-6} j_c; \qquad (1)$$

for the case when the weld is in the ZOY plane:

$$\Delta B_{z2} = 1,23.10^{-6} j_c, \qquad (2)$$

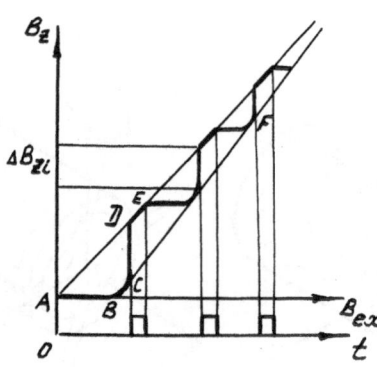

Fig.3. The experimental diagram for non-contact measurements of the current critical density. t - time of the heater switching on; B_{ex} - induction of the external magnetic field; B_2 - induction of the magnetic field in the centre of the sample.

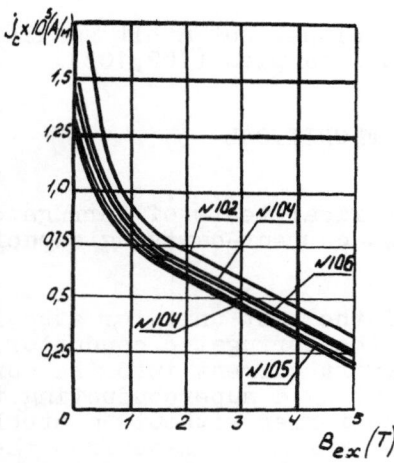

Fig.4. Dependence of linear critical current density on
the induction of the external magnetic field based
on the results of non-contact measurements.

where: ΔB_z - Z - induction component of screen current
field (T);
j_c - linear density of critical current (A/m).

The calculated coefficients were checked in a model experiment, using a cylinder with a gap, manufactured from a niobium-titanium tape with a set value of critical current density.

Fig.3 illustrates the sequence of experiments. The field
of the external solenoid changes monotonically with time. Initially ideal screening is observed (section AB), then the
field begins to penetrate into the sample from the end faces
(section BC) due to the small length of the sample. At the
point C the transition of the sample into normal state takes
place (section CD), then the heater is switched on in order
to exclude "diamagnetic" currents. During the work of the
heater screening is absent (DE). The heater is switched off
and the cycle is repeated.

Value ΔB_{zi}, substituted into the equation (1) or (2),
allows to calculate in a corresponding external magnetic
field. Plotting the envelope (CF) one can determine the critical current density for any induction value of the external
field.

Fig.4 shows the results of measurement of the critical
current density for cylindrical samples in a transverse magnetic field, calculated for the case presented in Fig.2b. The
tests were carried, chosen on the second stage of the experiment. Sample 102 was annealed in a vacuum furnace at a temperature of 700°C for 72 hours and sample 103 - for 48 hours.
After analysing test results it was decided to carry out heat
treatment of a sample 30 m long at a temperature of 750°C for
30 hours with the aim of decreasing the annealing time. After
annealing the tube was cut into three equal lengths, and a
sample was taken from each length (104,105,106) for testing
in order to control homogeneity of characteristics along the
full length. Fig.4 shows that the critical currents of three

last samples are very close, but a bit smaller, than the critical currents of model samples (102,103).

FOUR-CONTACT METHOD OF CRITICAL CURRENT MEASUREMENT

For further electrical tests of corrugated conductor, possessing the highest current-carrying capacity was chosen (sample 106).

Block diagram of the test unit for studying the current-carrying capacity of the corrugated conductor is presented in Fig.5. Input of transport current into the corrugated conductor (1) was carried out by a superconducting transformer (2). A superconducting transformer without a steel core was used, the primary winding of which was made from the wire $\emptyset$ 0,36 mm on the basis of NbTi with the number of turns W = 4510 and inductance value L = 0,353 H. The secondary winding of the transformer was a coaxial pair of conductors, shortcircuited between them on the ends. The inner conductor of the coaxial pair was a length of corrugated conductor 600 mm long. The external conductor (3) enveloped the superconducting transformer, placed on the corrugated conductor.

To decrease the inductive and resistive load components of the transformer, 15 tapes 7,2 mm wide, based on Nb_3Sn, were laid as the external conductor on the circumference with 18 mm radius parallel to the corrugated conductor axis; the length of P_3P_5 and P_7P_8 solded connections was 150 mm.

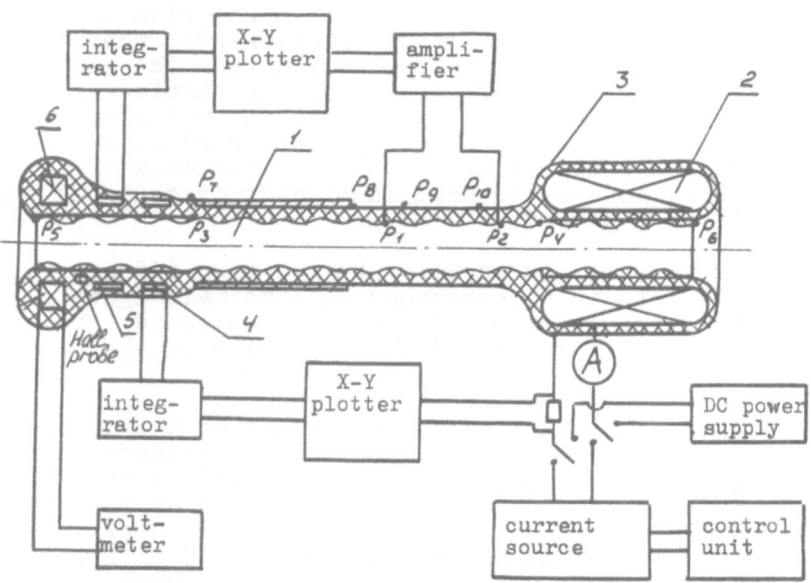

Fig.5. Block-diagram of the facilities for studying the current-carrying capacity of the conductor by four-contact method. 1 - superconducting corrugated conductor; 2 - superconducting transformer; 3 - superconducting tape; 4,5,6 - measuring coils; P_1-P_{10} - potential probes.

Current values in the corrugated conductor were measured with toroidal coils (Rogowski coils), preliminary calibrated, and were checked using the Hall converter. Volt-ampere characteristics of the corrugated conductor seals, and external conductor were studied using potential probes and Rogowski coils.

Studies of current characteristics of the corrugated conductor for impulse input of the current and for 50 Hz frequency gave the following results.

For impulse input of the current with the speed 50 A/c in the primary winding the current decay constant in the secondary transformer winding reached $\sim$120c, certifying high quality of the connections.

Critical current values for criteria 0,1; 1 and 5 μV/cm andquench reached 21,0 and 23,5; 31,7 and 33,0 kA, correspondingly. Resistance values of the connections P_3P_5 and P_4P_6 equel, correspondingly, 5,26.10^{-10} and 5,53.$10^{-10}\Omega$ and practically did not change when transport current value varied from 10 to 25 kA.

Magnetic field inductance values of the surface fo the corrugated conductor corresponding to the critical current for 1 μV/cm and the quench current reached 0,324 and 0,455 T, accordingly.

During current tests at a frequency of 50 Hz Measurement of electromagnetic energy losses was carried out by calorimetric method according to the quantity of evaporated helium using specially designed and manufactured gas-holders. Losses in the corrugated conductor were determined by subtraction of preliminary measured losses in the primary winding of the transformer and in Nb_3Sn -tapes of the external conductor from total losses. Losses in connections were also taken into account.

Energy losses in a corrugated conductor were large and reached, for example, 4300 μW/cm^2 at a linear current density of 2.560 A/cm.

High level of losses in the corrugated superconductor is explained by presence of bronze and residual excessive tin on the surface of Nb_3Sn-layer and by excessive roughness of the conductor surface (see Fig.1) and can be reduced by polishing of its surface.

The peak value of the current at a frequency of 50 Hz reached 15,3 kA, which corresponds to the rate of change of the magnetic field induction on the surface of the conductor, equal to 66,3 T/s. The corrugated conductor withstood prolonged current transmission of 14,3 kA at a frequency of 50 Hz without avalanche-type heating. The obtained data prove high stabilization level of the corrugated conductor.

SUMMARY

The results of experimental investigations of the corrugated superdonductor on the basis of Nb_3Sn certify good perspective of using the described technology in production of flexible superconducting conductors with a central channel for helium pumping in reference to cryogenic electrotechnical d.c. devices. Linear critical current density was equal to 2580 A/cm, and the critical current density was $\sim 2,6.10^6$ A/cm^2 in the field with $\sim 0,32$ T inductance at a temperature of 4,2 K. It is evident that the current-carrying capacity of the corrugated conductor and the critical characteristics of Nb_3Sn-layer can be greatly improved by increasing the conductor diameter and choice of optimum chemical composition of the initial substrate and conditions of thermal diffusion treatment during indirect heating. It should be mentioned that final conclusions concerning the developed technology can be made only after carrying out a wide range of investigations of influence of mechanical loads and deformations on the current-carrying capacity of the conductor. With reference to the alternating current the corrugated superconductor and its production process need further study in order to get a superconducting coating with a low level of electromagnetic energy losses.

REFERENCES

1. Ziemek G., Rhoner P (FRG), Meshchanov G.I., Peshkov I.B., Svalov G.G. Development of flexible superconducting cables for power electrical transmission lines. Superconductivity in equipment. The reports of the II all-Union Conference on utilization of Superconductivity. v.1. Superconducting machines and devices. Magnetic systems.Л.ЛНИВЦ , 1984, p.116-119.

2. G.I.Meshchanov, I.B.Peshkov, G.G.Svalov, P.Rohner and G.Ziemek, The manugacture of corrugated copper tubes with a Nb_3Sn layer. IEEE Trans. on Magn., 1985,v.MAG -21, N 2,p.324-327.

3. T.R.Sinlayson. Nb_3Sn-special-purpose electrical conductors. Metals Forum, 1985, 8, 1, p.3-13.

AC LOSS MEASUREMENTS OF TWO MULTIFILAMENTARY NbTi COMPOSITE STRANDS

E. W. Collings*, K. R. Marken, Jr.*, M. D. Sumption*†,
R. B. Goldfarb**, and R. J. Loughran**

* Battelle Memorial Institute, Columbus, OH 43201
† Permanent address: Ohio University, Athens, OH 45701
** National Institute of Standards and Technology, Boulder, CO 80303

ABSTRACT

As part of an interlaboratory comparative testing program conducted in support of the Versailles Agreement on Advanced Materials and Standards (VAMAS), transverse-field DC hysteresis loss measurements were made at liquid-helium temperatures at fields of up to 3 T (30 kG) on two samples of multifilamentary NbTi composite. The strands differed widely in filament number, were comparable in filament diameter, and one of them was provided with a Cu-Ni barrier between the filaments. The results have been analyzed, and magnetically deduced critical current density values obtained (for comparison with directly measured transport data) using various standard techniques. Based on these studies, a figure-of-merit for AC loss is recommended. The Cu-matrix strand, with its interfilamentary spacing of less than 1 μm, exhibited pronounced proximity-effect-induced coupling losses; this was not observed in the mixed-matrix strand which possessed not only a Cu-Ni barrier but also an interfilamentary spacing of typically 4 μm.

INTRODUCTION

In support of the Versailles Agreement on Advanced Materials and Standards (VAMAS), DC hysteresis loss measurements were made at liquid-helium temperatures in transverse magnetic fields of up to 3 T (30 kG) on samples prepared from two types of multifilamentary NbTi composite superconductors. The strands, designated herein as Sample E (henceforth SL-E) and Sample D (henceforth SL-D), had been manufactured in Japan and in the U.S.A., respectively. SL-E is a "mixed-matrix" strand in which the filaments are surrounded first by Cu and then by a thin (15 μm) eddy-current barrier of Cu-Ni alloy. SL-D consists of closely spaced NbTi filaments separated only by Cu. Photomicrographs of the strands are presented in Figs. 1 and 2. Noticeable in the magnified cross sections are the irregular shapes of the filaments, particularly in SL-E. For this reason, accurate filament cross-sectional areas (for the purpose of J_c calculation from the results of I_c measurement) could only be obtained by an etching-and-weighing procedure. Some specifications of the strands are listed in Table 1.

HYSTERESIS-LOSS MEASUREMENT

Measurements were made at Battelle, Columbus Division (BCD) and the National Institute of Standards and Technology (NIST) using vibrating-sample magnetometry. At both places the instruments were calibrated against pure Ni standards. At BCD,

Advances in Cryogenic Engineering (Materials), Vol. 36
Edited by R. P. Reed and F. R. Fickett
Plenum Press, New York, 1990

169

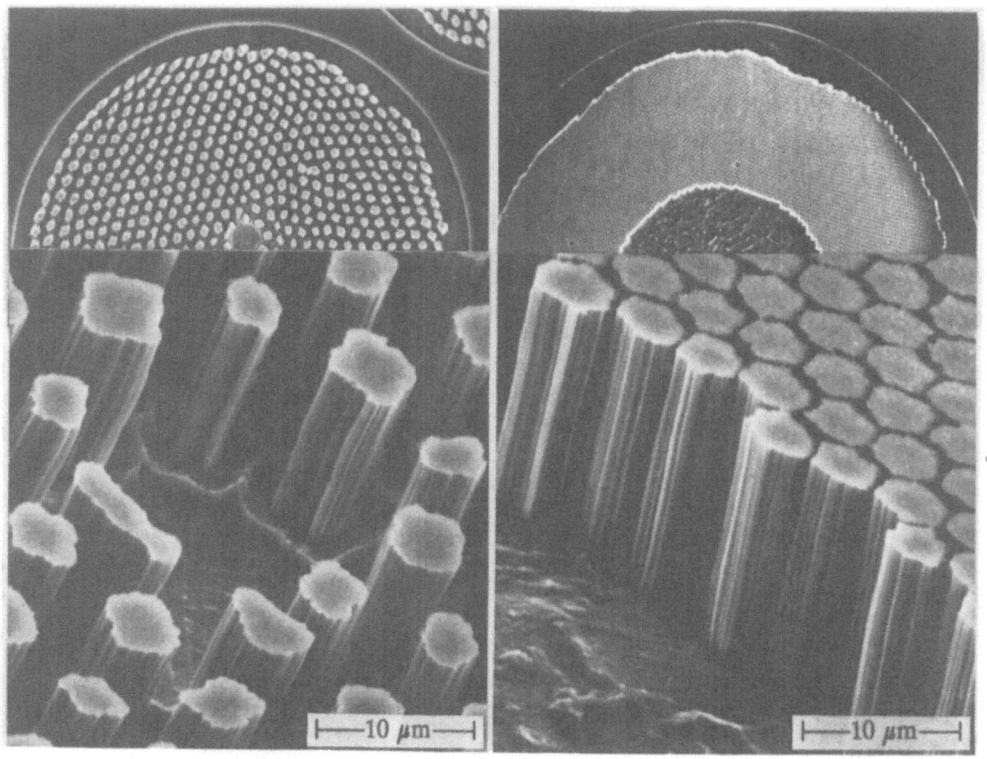

Fig. 1. Scanning electron micrograph of Sample E (SL-E).

Fig. 2. Scanning electron micrograph of Sample D (SL-D).

magnetization was measured in the slowly swept field (amplitudes 0.1 to 1.6 T; sweep rates 1.7 to 28 mT/s) of an iron-core electromagnet; data were recorded automatically at a field resolution of about 1/200th of the field-sweep amplitude. At NIST, point-by-point measurements were taken in stepped fields (up to 3 T) provided by a superconducting solenoid. The typical BCD sample consisted of an epoxy-potted 3-mm-diameter bundle of 6-mm-long pieces of strand. The typical NIST sample was formed by winding a strand or group of strands along the thread of a 5-mm-diameter nylon screw.

Table 1. Specifications of Strands under Investigation

Sample Code	Sample E (SL-E)	Sample D (SL-D)
Type	Mixed matrix	Copper matrix
Configuration	NbTi/Cu/CuNi	NbTi/Cu
Volume Ratio	21.5/44.1/34.4	42.0/58.0
Twist pitch, mm	6	13
Strand diameter, D, mm	0.35	0.742
Fil. diameter*, w, μm	5.79	4.62
Number of filaments	760	10,980
I_c at 3 tesla, A	50	675 **
J_c*** at 3 tesla, 10^5A/cm^2	2.50	3.67

* Measured by the etching-and-weighing technique on 304-cm (SL-E) and 149-cm (SL-D) lengths of strand. Measured density of bulk Nb-46.5Ti = 6.097 g/cm^3.

** Straight-sample measurement at NIST (manufacturer's supplied value, 653 A).

*** Based on I_c and the above-measured NbTi cross-sectional area.

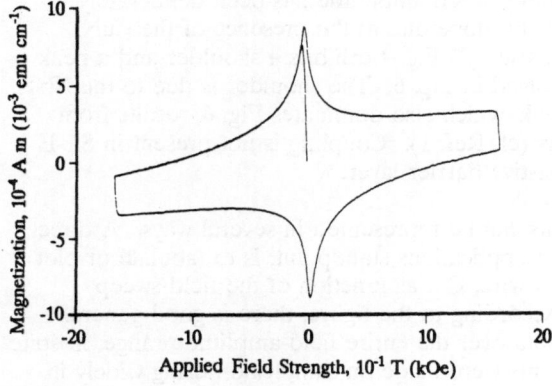

Fig. 3. Magnetization per unit length
of strand at 4.2 K for SL-E
-- typical BCD data.

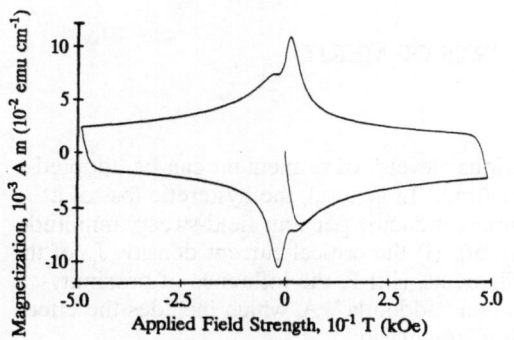

Fig. 4. Magnetization per unit length
of strand at 4.2 K for SL-D
-- typical BCD data.

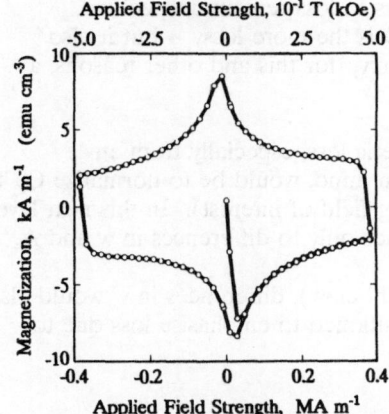

Fig. 5. Magnetization per unit volume
of strand at 4 K for SL-E
-- typical NIST data.

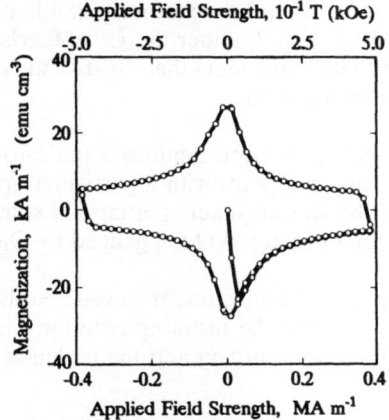

Fig. 6. Magnetization per unit volume
of strand at 4 K for SL-D
-- typical NIST data.

RESULTS AND DISCUSSION

Some typical hysteresis loops are presented in Figs. 3 to 6. Two features are noteworthy: (1) Fig. 3 (in which a large field-sweep amplitude has been deliberately selected) exhibits a pronounced paramagnetic slope due to the presence of the CuNi barrier material of which 34.4 vol.% is present; (2) Fig. 4 exhibits a shoulder and a peak near the origin, structure which is not resolved in Fig. 6. The shoulder is due to the NbTi filament magnetization while the sharp peak, which also dominates Fig. 6, results from coupling currents flowing in the Cu matrix (cf. Ref. 1). Coupling is not present in SL-E with its larger filament separation and resistive barrier layer.

The results of AC-loss measurements can be represented in several ways. A direct approach, and one which is useful from an applications standpoint, is to tabulate or plot the energy loss per cycle per unit length of wire, Q_ℓ, as function of the field-sweep amplitude, H_m. This is done in Fig. 7. According to the figure, there is good general agreement between the BCD and NIST data over the entire field-amplitude range, in spite of the fact that the two sets of measurements were made on samples differing widely in configuration. Furthermore, when measurements were made in both laboratories on the BCD samples, the results differed by less than 7% *.

AC-LOSS REPRESENTATIONS AND FIGURES OF MERIT

Representations of AC Loss

Depending on the purpose in mind, various "levels" of refinement can be adopted in reporting and comparing of AC-loss information. In general, the hysteretic loss of a multifilamentary strand per unit volume of superconductor per unit field-sweep amplitude may be regarded as a function, $Q(J_c, w, P, A)$, of: (i) the critical current density, J_c, of the superconductor, (ii) the diameter, w, of the filaments, (iii) P, the influence of proximity-effect coupling between the filaments, and (iv) an "addenda", A, which includes the effect of imponderables such as filament-cross-section irregularity.

Level-0. The simplest representational level, referred to here as Level-0, is Q_ℓ (Fig. 7) which derives directly from hysteresis-loss measurements on a known length, ℓ, of strand. At this level the various possible contributions to loss remain unspecified. According to Fig. 7, under the Level-0 criterion SL-D is the more lossy -- but it also possesses more filaments than SL-E and is able to carry, for this and other reasons, a larger critical current.

Level-1. A more significant indicator of hysteretic loss, especially from an engineering standpoint with a particular application in mind, would be to normalize Q_ℓ to ⁀ I_c of the strand (taken, perhaps, at some operating field of interest). In this next level of refinement hysteretic loss, gauged by Q_ℓ/I_c, responds only to differences in w and A.

Level-2. Finally, under Level-2 (to be discussed below), differences in w would also be absorbed, and the resulting criterion would be positioned to emphasise loss due to proximity-effect coupling and the addenda.

---

* At H_m = 0.97 T, the per-cycle hysteresis loss in SL-E as measured at NIST was 6.7% lower than the BCD-measured value; also at H_m = 0.97 T, the NIST-reported loss in SL-D was 1.4% higher than the BDC value.

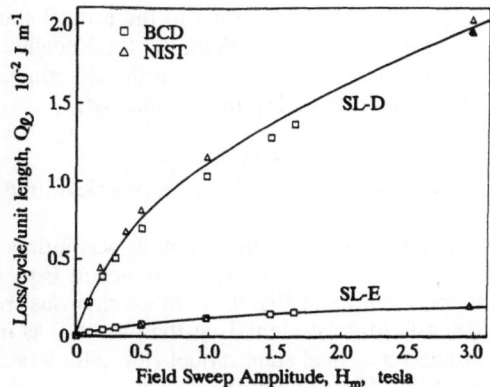

Fig. 7. The Level-0 criterion (loss per cycle per unit length of strand) applied to Sample D (SL-D) and Sample E (SL-E).

Figures of Merit for Hysteretic Loss

The Q_v/J_c (Level-1) Approach. Following the well-known critical-state model[2] as applied by Carr and co-authors (whose work is summarized in Ref. 3) to a multifilamentary composite strand in a transverse applied field, the hysteresis loss per unit volume of the superconducting component of the strand, Q_v, is given in SI units by:

$$Q_v = (8/3\pi) \, w \, J_c \, \mu_0 \, H_m$$
$$\text{(or } Q_v = (0.8/3\pi) \, w \, J_c \, H_m, \text{ in practical units)} \tag{1}$$

$$\text{i.e.,} \quad R_Q = Q_v/J_c = (8/3\pi) \, w \, \mu_0 \, H_m$$
$$\text{(or } R_Q = Q_v/J_c = (0.8/3\pi) \, w \, H_m, \text{ in practical units)} \tag{2}$$

It follows that a plot of R_Q versus H_m should, under the Bean approximation, be linear with slope proportional to w, the filament diameter. Departure of R_Q from its expected value, possibly as a result of proximity-effect coupling between the filaments, could then be expressed in terms of some effective filament diameter[4], w_{eff}. Within this framework, the following figures-of-merit (FOM) might be selected: (i) at a given H_m, the ratio w_{eff}/w; (ii) at a given H_m and H, the quotient $R_Q = Q_v/J_c(H)$. The use of $J_c(H)$ recognises that in practice J_c is not independent of field, and that for the purpose of a criterion may have to be measured at some field different from H_m. Finally, we note that in obtaining an experimental value of R_Q, it is convenient to replace Q_v/J_c by its identical equivalent, Q_ρ/I_c, the latter being a ratio of directly measured quantities, Fig. 8.

For a pair of "ideal" multifilamentary strands, say A and B, the quotient $R_{Q,A}/R_{Q,B}$ at a given H_m should be simply w_A/w_B, the filament-diameter ratio. In the case of the present strands, direct measurements have shown that $w_E/w_D = 1.25$ (see Table 1 and its footnote *). At $H_m = 3$ tesla, this may be compared to the magnetically obtained value of $R_{Q,E}/R_{Q,D} = 1.35$. The interfilamentary coupling present in SL-D (see Fig. 4) should cause $R_{Q,E}/R_{Q,D}$ to be less than the filament-diameter ratio. That it is not, suggests that factors not yet taken into consideration are masking the ability of the R_Q criterion to properly represent the presence of coupling. One such factor could well be a difference between the field-dependences of J_c for the two strands.

173

The $Q_v/\Delta M_v$ (Level-2) Approach. According to the critical-state model for a cylinder of diameter w in a transverse applied field, the total height, $\Delta M_v(H)$, of the NbTi-volume-normalized hysteresis loop (measured between the shielding and trapping branches) at some field H is related to $J_c(H)$, in SI units, by[5]:

$$\Delta M_v(H) = (4/3\pi) J_c w$$
$$(\text{or } \Delta M_v(H) = (0.4/3\pi) J_c w, \text{ in practical units}) \tag{3}$$

Bean's model was of course based on the premise that J_c was independent of field; nevertheless, even Bean[2] and many others to follow, employed Eqn. (3) to determine the field-dependence of J_c, as for example in Fig. 9. A more rigorous treatment of the critical state would be to introduce a field-dependent J_c at the outset -- as in the work of Ohmer and Heinrich[6], who constructed a critical state model (for cylinders in the field-parallel orientation) based on a modified Kim[7] equation.

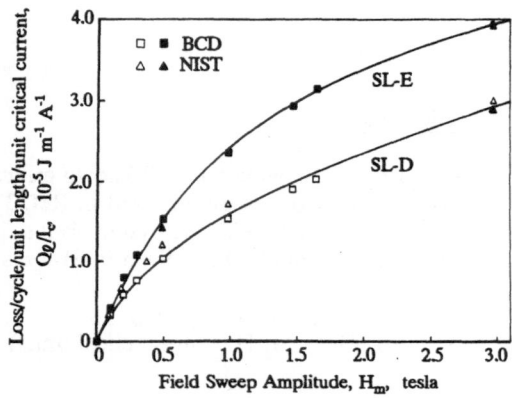

Fig. 8. The Level-1 criterion (loss per cycle per unit length per unit critical current at reference field) applied to SL-D and SL-E.

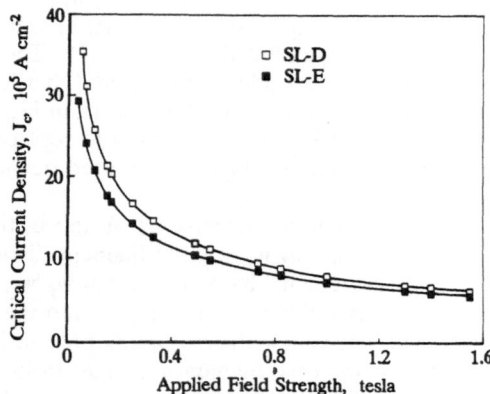

Fig. 9. Magnetization-determined J_c versus applied field strength for SL-E and SL-D.

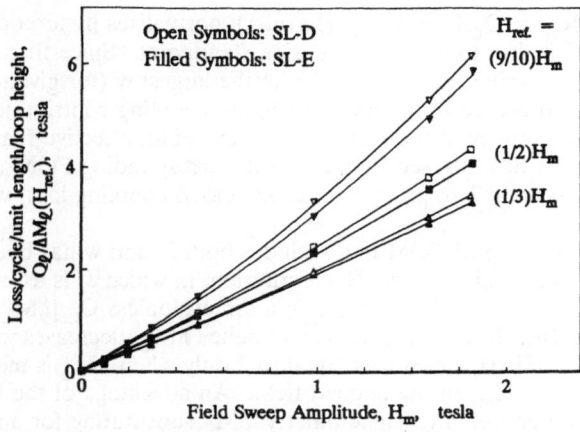

Fig. 10. The Level-2 criterion as function of M(H)-loop amplitude, H_m.

The proportionality between $\Delta M_v(H)$ and $J_c(H)$ provides an opportunity to convert R_Q into a quotient that can be derived solely from the hysteresis loop, without recourse to a separate J_c measurement. From Eqns. (1) and (3), and referring to either unit-volume or unit-length quantities, respectively:

$$R_M = Q_v/\Delta M_v(H) = Q_\varrho/\Delta M_\varrho(H) = 2\,\mu_0\,H_m$$
$$\text{(or } R_M = Q_v/\Delta M_v(H) = Q_\varrho/\Delta M_\varrho(H) = 2H_m, \text{ in practical units)} \qquad (4)$$

Under a strict Bean criterion (J_c = const.), ΔM is independent of H and R_M is linear with slope 2.0. Fig. 10 indicates the extent to which this is obeyed in practice. In formulating R_M as a new FOM it will be necessary to normalize Q_ϱ to a loop height measured at some arbitrarily selected reference field, H_{ref}. In Fig. 10, R_M is plotted versus H_m for three values of H_{ref}: $(1/3)H_m$, $(1/2)H_m$, and $(9/10)H_m$, respectively. An important advantage of R_M as an FOM lies in its independence of both J_c (as in the Level-0 criterion) and w (Level-1). Thus, R_M is expected to respond directly to the influence of proximity-effect coupling. Indeed, in Fig. 10, the curves for SL-D (which is coupled) all lie above their SL-E counterparts.

SUMMARY AND CONCLUSION

The hysteretic loss of two multifilamentary strands has been measured by vibrating-sample magnetometry. The results emphasize the importance of acquiring high-resolution data at low fields where J_c, and hence, the hysteresis-loop-height and the loss per unit field increment is greatest. Furthermore, if interfilamentary proximity-effect is present, high-resolution data are useful for resolving the coupling peak and its nearby NbTi shoulder.

Several representations of AC-loss criteria have been suggested. The simplest is Q_v or Q_ϱ (Level-0), which leaves unspecified all possible contributions to loss and which derives directly from hysteresis measurements on known amounts of superconductor. The next level of refinement takes current-carrying capacity into account, resulting in a criterion that responds to filament diameter and what might be termed "high-order" hysteretic losses such as those due to proximity-effect coupling, and perhaps filament-shape distortions and other "addenda". The final level of refinement yields a criterion that includes both I_c and w, and which emphasizes proximity-effect coupling and the addenda.

The criterion $R_Q = Q_\varrho/I_c = Q_v/J_c$ (Level-1), normalizes hysteretic loss to current-carrying capacity and is thus useful from a design standpoint. Since filament diameter, w, is not absorbed into the criterion, the strand with the largest w (for given I_c) will have the largest R_Q. Next, the presence of an interfilamentary-coupling contribution to AC loss could be identified directly by introducing the concept of an effective filament diameter[4], $w_{eff} = R_Q/(8/3\pi)H_m$ (in SI units, see Eqn. (2)); the corresponding FOM (a quantity intended to increase beyond 1 in proportion to increased coupling loss) would be w_{eff}/w.

Finally, a Level-2 form of FOM that includes both J_c and w has been suggested; it is $Q_\varrho/\Delta M_\varrho$ (see Eqn. (4)). Under a strict Bean criterion, in which J_c is assumed independent of H, ΔM_ϱ is constant and equal to $\Delta M_\varrho(H_m)$; a dimensionless $Q_\varrho/[\Delta M_\varrho(H_m).H_m]$ is then equal to 2.0 (see Fig. 10). But in practice J_c and hence ΔM_ϱ, decreases with H. Consequently $Q_\varrho/[\Delta M_\varrho(H).H_m]$ becomes equal to 2 only when ΔM_ϱ is measured at some intermediate value ($H < H_m$) of the applied field. An advantage of the Q/ΔM quotient is that it can be obtained entirely by magnetometry, ΔM_ϱ substituting for an auxiliary J_c determination. Another advantage of this approach lies in the fact (indicated in the first paragraph of this discussion) that the largest contribution to incremental hysteretic loss occurs at low fields, a region in which J_c may be out of the range of the current-transport measuring equipment.

In conclusion it is important to recognize, as a comparison of Figs. 8 and 10 indicates, the more inclusive the AC-loss criterion or FOM, the smaller is its variation from sample to sample. In general, when defining an FOM, normalization should be carried only far enough to achieve some specific objective, be it engineering or scientific in nature.

ACKNOWLEDGEMENTS

At BCD, the epoxy-potted magnetization samples were prepared by R. D. Smith. At NIST, L. F. Goodrich measured transport critical currents on SL-D, and R. L. Spomer assisted with acquiring the magnetization data. The research was sponsored by the U.S. Department of Energy, Division of High-Energy Physics.

REFERENCES

1. E. W. Collings, K. R. Marken Jr., M. D. Sumption, E. Gregory, and T. S. Kreilick, "Magnetic studies of proximity-effect coupling in a very closely spaced fine-filament NbTi/CuMn composite superconductor", paper in this conference.

2. C. P. Bean, "Magnetization of high-field superconductors", Rev. Mod. Phys. 36, 31-39 (1964).

3. E. W. Collings, Applied Superconductivity, Metallurgy, and Physics of Titanium Alloys, Vol. 1, Plenum Press, New York, 1988, p. 353.

4. S. S. Shen, "Magnetic properties of multifilamentary Nb_3Sn composites", in Filamentary A15 Superconductors, ed. by M. Suenaga and A. F. Clark, Plenum Press, New York, 1980, pp. 309-320.

5. W. J. Carr, Jr., and G. R. Wagner, "Hysteresis in a fine filament NbTi composite", Adv. Cryo. Eng. 30, 923-930 (1984).

6. M. C. Ohmer and J. P. Heinrich, "Magnetization of hysteretic superconductors for complete field penetration and critical state model with $J_c(H) = \alpha/H$", J. Appl. Phys. 44, 1804-1809 (1973).

7. Y. B. Kim, C. F. Hempstead, and A. R. Strnad, "Magnetization and critical super-currents", Phys. Rev. 129, 528-535 (1963).

AC LOSS MEASUREMENTS ON NbTi SUPERCONDUCTING WIRES FOR THE VAMAS ROUND ROBIN TEST

S. Zannella, P. Gislon*, V. Ottoboni,
A. M. Ricca, G. Ripamonti

CISE Spa, Segrate, Italy; *ENEA, Frascati, Italy

INTRODUCTION

A very important parameter of superconducting wires for ac applications is the energy loss dissipated when they are subjected to time-varying magnetic fields[1]. These losses add to the refrigeration load and in many cases may be the main heat load. Several efforts have been carried out in recent years to obtain superconducting composites suitable for ac applications[2,3,4]. Their use requires accurate measurements of the ac losses and the establishment of reliable standard techniques for their determination. One of the most effective ways to attain this goal is the use of round-robin tests in which samples from the same lot are measured by several laboratories and the results are correlated. Under VAMAS activity, a cooperative programme has been started for the measurement, with existing techniques and apparatuses, of ac losses on four reference NbTi multifilamentary wires.

The measurements have been carried out according to settled guidelines. The experimental results obtained at CISE on samples exposed to variable magnetic fields or supplied with 50 Hz transport currents are presented and discussed.

EXPERIMENTS

Samples

The characteristics of the NbTi multifilamentary wires tested are reported in Table 1. The Japanese A (sample E) and the American (sample D) wires are designed for pulsed magnets while the Japanese B (sample F) and the European (sample G) wires, having submicrons filaments and Cu-CuNi mixed matrix, are specially devoted to ac use at 50/60 Hz. We have measured their ac losses at 4.2 K during variable external magnetic fields or when supplied with 50 Hz transport current in the presence of a transverse dc magnetic field.

Advances in Cryogenic Engineering (Materials), Vol. 36
Edited by R. P. Reed and F. R. Fickett
Plenum Press, New York, 1990

Table 1. Characteristics of tested samples and coil geometries

sample characteristics	E (JPN A)	F (JPN B)	D (US)	G (EC)
wire diameter (mm)	0.35	0.14	0.742	0.2
filament diameter (μm)	6.3	0.5	4.6	0.175
number of filaments	760	14,280	10,980	242,892
NbTi/Cu/CuNi	21.5/44.1/34.4	1/1/3.5	1/1.38/0	0.02/0.79/0.19
twist pitch (mm)	6	1.9	13	0.8
critical current (A)	56 (3 T)	10 (1 T)	675 (3 T)	42 (1 T)
sample coil dimensions inner diameter (mm)	36.9	44.6	30.92	44.6
outer diameter (mm)	39.7	44.88	32.4	45
length (mm)	70	13	55	13
number of turns	721	14	55	14
number of layers	4	1	1	1

Experimental Apparatuses for ac Loss Measurements

Fig. 1 shows the experimental apparatus used to perform ac loss measurements in external fields varying in time with triangular waveform having a maximum sweep rate of 0.3 T/s. The technique used is based on the electrical method developed by Fietz[5] in which the losses are evaluated by the area of the magnetization cycles. Magnetization is a parameter that permits evaluation of ac losses, effective filament diameter and critical current density, thus providing a wide characterization of superconducting wires. The wire, wound on a cylindrical support, (see Table 1), is inserted between two concentric pick-up coils in the bore of a 8 T superconducting magnet with the wire axis perpendicular to the external field. Sample magnetization M(t) is measured as the integrated differential signal induced in the pick-up coils. The external field H(t) is measured by the voltage across a shunt resistor. These two signals are amplified and sent to a digital oscilloscope where the magnetization loop is displayed. The losses are evaluated by a computer interfaced to the digital oscilloscope. With this technique we have measured the losses of D and E samples. Fig.2 shows the experimental apparatus used for the measurements of ac losses due to a 50 Hz transport current supplied to the superconducting wire. The samples are wound in the form of a single layer, non-inductive coil (see Table 1). The still present inductive component of the terminal voltage across the sample is compensated by the output of a mutual inductance coupled to a magnet current lead which, by integration, also provides a signal proportional to the transport current. The compensated voltage and the current signals are amplified and then displayed on a digital oscilloscope. The losses may be evaluated either by an analogic microwattmeter or by digital sampling and multiplication of the instantaneous values of current and voltage. The self-field losses have been measured on F and G samples at different bias magnetic fields.

RESULTS

The hysteresis losses have been evaluated from magnetization loops at very low dB/dt values for different

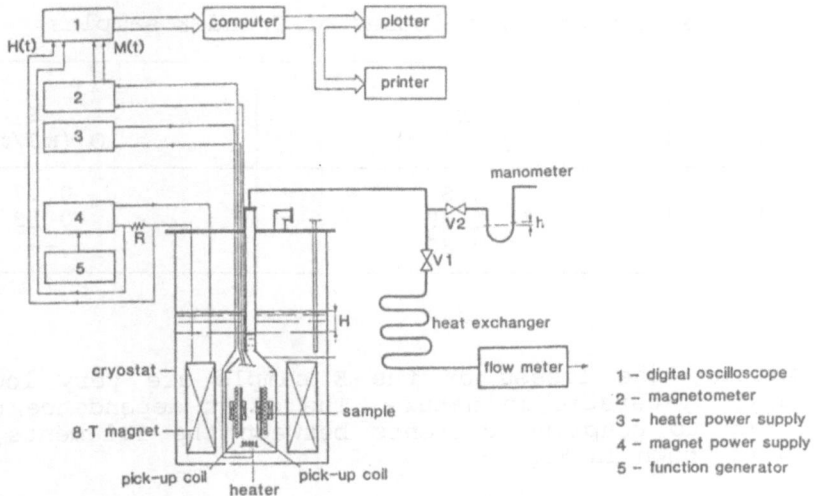

Fig. 1. Experimental apparatus for ac loss measurements of superconducting wires exposed to variable magnetic fields.

peak-to-peak ΔB values of the magnetic field. The results obtained on D and E samples are reported in Table 2. The results are in good agreement with the Bean formula[6] of hysteresis losses per cycle and unit length:

$$Q = (4/3\pi) \, \Delta B J_c \, d A_s \qquad (J/m)$$

where J_c is the critical current density, d the filament diameter and A_s the superconducting cross section of the wire. From this formula effective diameters of 4.6 μm and 5.7 μm are evaluated for samples D and E respectively.

Measurements of the hysteresis loss for 0-1-0 T cycle had been recommended for intercomparison, and our results seem to be in reasonable agreement with those reported from the other laboratories. Owing to the high transverse resistivity of the

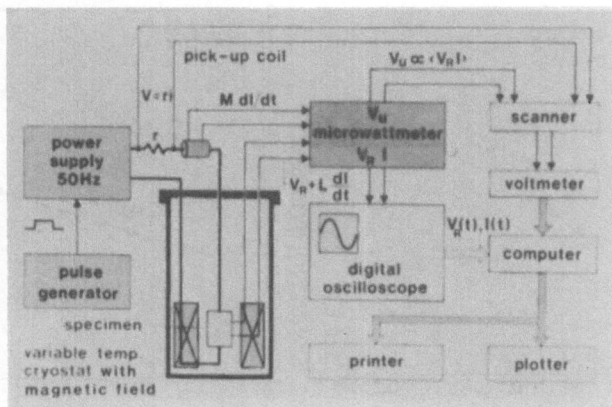

Fig. 2. Experimental apparatus for ac loss measurements of superconducting wires supplied with 50 Hz transport current and exposed to a transverse magnetic field.

Table 2. Hysteresis losses of D and E samples

$B_{min}-B_{max}$ (T)	D Q (mJ/m)	E Q (mJ/m)
0 - 1	4.8	0.47
2.5 - 3.5	1.39	0.13
4.5 - 5.5	0.88	--

mixed matrix, the losses of the E sample are very low and essentially hysteretic in nature. The dB/dt dependence of ac losses, due to coupling currents between the filaments, for sample D is shown in Fig. 3.

The results of self-field 50 Hz loss measurements on F and G samples as function of transport current at different static fields are shown in Fig. 4. For submicron filaments, energy losses different from the predicted ones should occur due to proximity effects[2]. The measured power losses per unit length are characterized by a power law $P \approx I^n$ with $2.6 < n < 3.3$, depending on the bias magnetic field; the slope of the loss curve steepens on approaching the critical current.

CONCLUSIONS

The increasing demand for superconducting wires for ac applications and pulsed magnets requires reliable standard measurement methods to evaluate their ac losses. Within the VAMAS round robin test we have measured ac losses of NbTi multifilamentary wires exposed to variable magnetic fields or subjected to 50 Hz transport currents. Our experimental results are presented and discussed. All the characterizations have been performed by electrical methods; calorimetric ac loss measurements will be also involved in the intercomparison.

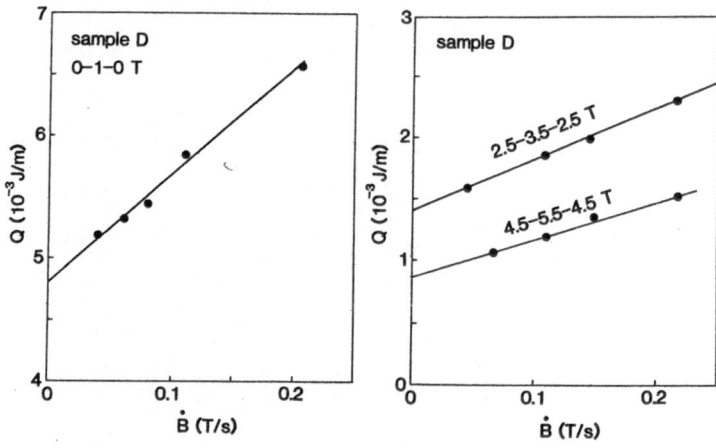

Fig. 3. dB/dT dependence of ac losses in D sample for different magnetic cycles.

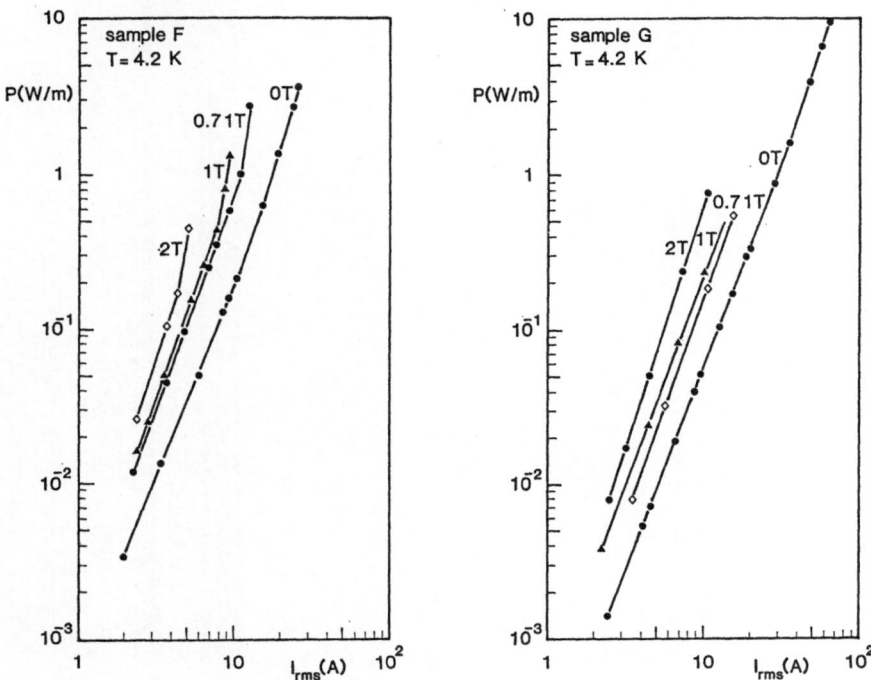

Fig. 4. 50 Hz self-field losses of F and G samples.

REFERENCES

1. I. Hlasnik,"Prospects of Multifilamentary Superconductors ac 50 Hz Applications", Journ.de Phys. 45(1):459 (1984)
2. J.R. Cave,"Electromagnetic Properties of Ultra-fine Filamentary Superconductors", Cryogenics 29:304 (1989)
3. A. Fevrier,"Latest News about Superconducting ac Machines", IEEE Trans. on Magn. 24 (2): 787 (1988)
4. N. Maki, K. Yamaguchi, M. Takahashi, R. Shiobara, "Development of Superconducting ac Generator", IEEE Trans. on Magn. 24 (2): 792 (1988)
5. W.A. Fietz,"Electronic Integration Technique for Measuring Magnetization of Hysteretic Superconducting Materials", Rev. Sci. Instrum. 36 (11): 1621 (1965)
6. C.P. Bean,"Magnetization of Hard Superconductors", Phys. Rev. Lett. 8 (6): 250 (1962).

AC LOSSES IN A PROTOTYPE NbTi CABLE FOR THE LHC DIPOLE MAGNETS

A.J.M. Roovers, A.J. van Pelt and L.J.M. van de Klundert

University of Twente, Dept. of Applied Physics
P.O. Box 217, NL-7500 AE Enschede
The Netherlands

ABSTRACT

In this paper loss measurements on a prototype NbTi cable for the Large Hadron Collider dipole magnets are presented. The cable is meant to be used for the outer layer of the magnets. It is a trapezium-shaped (key-stoned) Rutherford cable. The experimental results were obtained in a Twente test facility. Magnetization loss measurements as well as transport current loss measurements were performed. The results are compared with classical loss models. From the magnetization experiments the coupling time constant of the cable at different magnetic fields can be obtained. It is also possible to estimate the critical current density from these measurements. The critical current and the effective transverse resistivity are important for calculations concerning the loss during ramping the magnetic field and the field disturbance due to shielding currents in the filaments and to the coupling currents.

INTRODUCTION

At the University of Twente a test facility has been developed which can be used to determine the characteristics of high-current super-conducting wires and cables[1,2,3]. In this facility the losses in a prototype NbTi cable for the LHC dipole magnets have been measured. The Large Hadron Collider is an accelerator, which is planned to be built in the existing LEP tunnel at CERN[4,5]. The LEP tunnel, 27 km long, is meant to investigate collisions between electrons and protons. A proton-proton or proton-antiproton collider has become of interest for CERN. Building a hadron collider on top of the LEP, so using the existing infrastructure, is a cost-efficient option. In order to obtain the required energies of 8 to 9 TeV in the existing tunnel, 10 T dipole magnets need to be installed. The limited available space forces the use of "two-in-one" dipole magnets. Two apertures are included in the magnet system. At the moment two options are examined. The first is the Nb_3Sn option in which the magnets operate at about 4.5 K. A second option is one using the conventional NbTi technique. In this case however, the operating temperature should be approximately 2 K in order to be able to generate the required magnetic field. A high overall current density is needed because the maximum size of the magnets is very limited. For the same reason the dipole coils consist of two layers each carrying different current densities. The inner layer is made of a

Advances in Cryogenic Engineering (Materials), Vol. 36
Edited by R. P. Reed and F. R. Fickett
Plenum Press, New York, 1990

183

Table 1 Specifications of the LHC cable

Type of conductor	Rutherford cable
Outer dimensions	1.67/1.30 × 17 mm^2
Number of strands	40
Cabling length	190 mm
Strand diameter	0.84 mm
Number of filaments per strand	912
Filament diameter	14 μm
Superconductor	NbTi
Matrix material	Cu
Cu : NbTi	0.64 : 0.36
Insulation	none

26-strand Rutherford cable, having the dimensions 2.06/2.50×17.0 mm^2. The conductor for the outer layer (1.30/1.67×17.0 mm^2) consists of 40 strands. The conductor should be able the carry 17.5 kA at 8.5 T and 2 K. The cable tested is the one that is to be applied in the outer coil of the dipole magnets.

Magnetization losses have been measured in a frequency range of 0.1 to 50 Hz at magnetic field amplitudes up to 0.1 T. Transport current losses have been determined in the case of alternating transport currents. The magnetic background field was varied in the range of 0 to 5 T. The experimental results will be presented in this paper.

SPECIFICATIONS OF THE LHC CABLE

The LHC cable examined is a 40-strands Rutherford with a large aspect ratio (width/thickness). The cable is not exactly rectangular. It is more or less trapezoidal shaped. This particular shape is mainly used to obtain a high average critical current density in the magnet, without removing a part of the stabilization material. The characteristics of the cable are summarized in Tab. 1.

A cross-section of the cable is shown in Fig. 1. Solder has been attached to the outside of the cable.

TEST CONDITIONS

The magnetic background field points in the test set-up in radial direction. It means that the field is perpendicular to the wide side of the

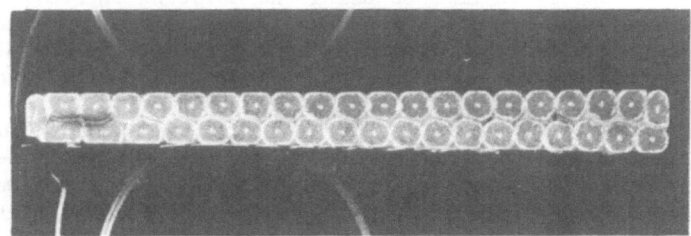

Fig. 1 A cross-section of the Rutherford cable for the model dipole magnets of the LHC accelerator at CERN.

conductor. The alternating magnetic field is parallel to the wide side. As long as the background field is large compared to the alternating field however, the direction of a transverse background field does not affect the loss behaviour. The transport current is induced by means of a transformer circuit. The losses were measured using the electric method, i.e., with pick-up coils for the magnetization losses and voltage taps for transport current losses. For more details about the test set-up is referred to Ref. 3.

EXPERIMENTAL RESULTS

The measurements have been performed at liquid helium temperature, 4.2 K. The actual temperature at which the conductor will be operated is 2 K. In spite of this difference in temperatures it is possible to relate the experiments at 4.2 K to the behaviour of the conductor at 2 K. The electrical resistivities are identical at these two temperatures. A different temperature only affects the critical current density. It was found however, that for large fields the critical current density versus field dependence at 2 K corresponds to the one at 4.2 K if the magnetic field is shifted about 3 T [6].

We were not able to determine experimentally the critical current due to premature quenching. This may be caused by wire motion: a part of the cable has not been impregnated. Furthermore the impregnated part of the cable in the test volume is not firmly fixed in the sample holder. Combining this with the large superconductor to matrix ratio makes instabilities due to movements likely to occur. A second explanation might be the current distribution among the different strands. If the cable is not fully transposed one strand might carry a larger current than the other strands. As a consequence this strand will reach its critical current before the others. If one strand quenches the others will follow if they cannot take over the current with a limited power dissipation. Because the strands are partially soldered together, so providing a low electrical resistance this between them, this is not the most likely explanation. A third explanation is the so-called self-field instability. A thermal disturbance invokes a current redistribution which causes a thermal run away. If we make an estimation of the quench current using the equations given by Veringa[7] for the self-field stability criterion for a round wire we find a quench current to critical current ratio of approximately 0.4. This agrees rather well with the quench currents we found (14 kA at 5 T).

Magnetization losses

The magnetization losses of the LHC cable have been measured at background fields of 2 and 5 T. They were determined by subjecting the conductor to a sinusoidal magnetic field parallel to the wide side of the cable. The loss was determined using the electrical technique. The results of the magnetization measurements at a background field of 5 T are depicted in Fig. 2. The losses per cycle per unit length of conductor are plotted as a function of the frequency for various amplitudes of the magnetic field.

The magnetization losses are dominated by the coupling loss. The time constants can be derived from the measurements   at 2 and 5 T. They were found to be 98 ms and 0.13 s respectively. This decrease of the coupling current time constant can be partially explained by the increase of the effective transverse resistivity due to magnetoresistivity. The magnetoresistivity of copper is rather well described by:

$$\rho(B) = \rho(0) \cdot (1 + 0.5 \; |B|)$$ (1)

Raising the magnetic field from 2 to 5 T results in an increase of the resistivity of copper of approximately 75 %. The increase of the effective transverse resistivity is approximately 35 %. This means that the effective transverse resistivity is determined partially by the copper matrix and partially by other barriers, like the contact resistance between the strands and the resistance of the solder. Considering the increase of the copper resistivity and the effective transverse resistivity, an almost equal contribution of the copper and other barriers is expected.

Campbell[8] gives an expression for the coupling loss in a monolithic superconductor:

$$Q_c = \frac{\pi \, \hat{B}^{a2}}{\mu_0} \left[\frac{\omega\tau}{1 + \omega^2\tau^2} \right] a \, b,$$ (2)

where a is the small side and b the wide side of the conductor. The typical time constant of the wire is given by:

$$\tau = \mu_0 \, \sigma_\perp \, L_p^2 \left[\frac{a}{4b} \right]^2 , \quad a \ll b .$$ (3)

In this expression L_p denotes the twist pitch and $\sigma_\perp$ the effective transverse conductivity.

In order to account for the hysteresis losses in the superconducting filaments, the losses given by Pang et al.[9], corrected for the shielding due to the coupling currents, can be used:

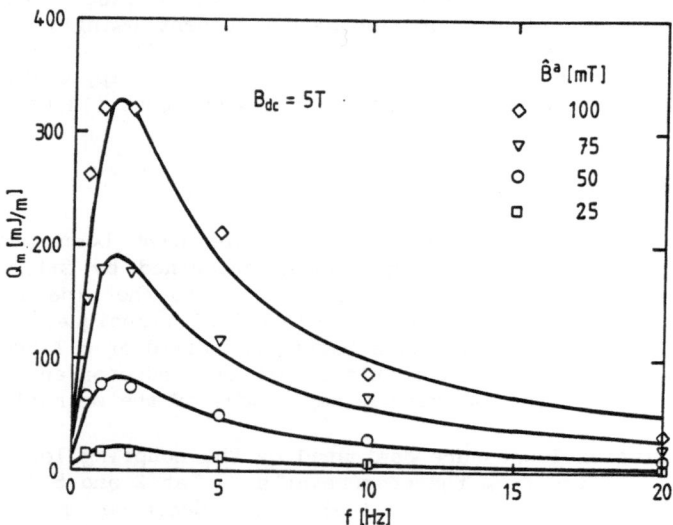

Fig. 2 *The magnetization losses per cycle per unit length of conductor in the LHC cable as a function of the frequency of the applied magnetic field. The magnetic background field equals 5 tesla. The solid lines represent the calculated losses per cycle according to Eq. 2 and 4.*

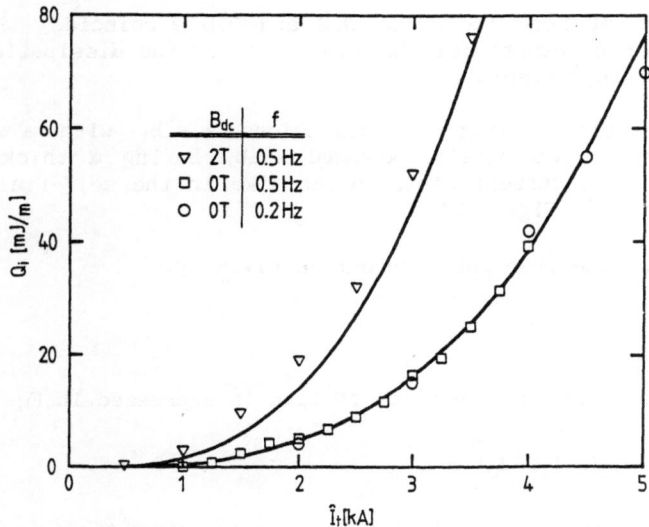

*Fig. 3 The transport current losses per cycle per unit length of conductor
in the LHC cable as a function of the amplitude of the transport
current. The losses were obtained using bipolar and unipolar
currents at zero and 2 T background field and frequencies of 0.2 and
0.5 Hz. The solid lines represent the self-field loss calculations
according to a slab model.*

$$Q_h = \frac{4\ N_{fil}\ \pi\ R_f^2}{3\ \mu_0}\ \hat{B}^a\ B_p\left[2 - \frac{B_p}{\hat{B}^a}\right],\tag{4}$$

where B_p, the penetration field, is given by

$$B_p = \frac{2\ \mu_0\ J_c\ R_f}{\pi}.\tag{5}$$

In Eqs. 4 and 5 N_{fil} represents the number of filaments and R_f its
diameter.

Substituting the fitted time constant in Eq. 2 yields a theoretical
curve which describes the coupling loss in the LHC cable very well.
Combining this with the existence of resistive barriers, it is most likely
to attribute the loss to inter-strand coupling currents. The effective
transverse conductivities at 2 and 5 T, calculated from Eq. 3, are
$4.4 \cdot 10^9\ (\Omega m)^{-1}$ and $6.0 \cdot 10^9\ (\Omega m)^{-1}$ respectively.

Transport current losses

The transport current losses have been determined for slowly changing
currents without applying an alternating magnetic field. The transport
current was changed sinusoidally in time. Figure 3 shows the transport
current loss for a bipolar current (between $-\hat{I}_t$ and $+\hat{I}_t$) having a frequency
of 0.2 and 0.5 Hz respectively, at zero background field. The transport
current loss has also been obtained at 0.5 Hz at a magnetic background
field of 2 T. In this case an unipolar current has been used: the current
varies between 0 and $2\ \hat{I}_t$.

The losses at zero field for 0.2 and 0.5 Hz coincide, which implies that under these circumstances the major part of the dissipation occurs in the superconducting material.

The LHC cable, having the dimensions a × b, with a ≪ b, can be approximated by an infinitely extended slab, having a thickness a (see Fig. 4 a). Assume a current distribution like in the self-field case of a solid conductor (see Fig. 4 b).

The amplitude of the transport current is given by:

$$\hat{I}^a = J_c \ (D-d_0) \ b \ . \tag{6}$$

The transport current as a function of time is expressed in Eq. 7:

$$I^a(t) = j_c \ (2d(t)-D-d_0) \ b \ . \tag{7}$$

The self-field loss can be calculated using the Poynting vector. In order to calculate the Poynting vector the electric and magnetic field at the surface of the conductor have to be known. The magnetic field can be calculated to be:

$$B_x(D,t) = \mu_0 \ \frac{I^a(t)}{b} \ . \tag{8}$$

Using Maxwell's equation $\partial_y \ E_z = -\partial_t \ B_x$ and $E_z(d(t)) = 0$ yields for the electric field at the surface:

$$E_z(D,t) = \mu_0 \ \frac{\partial I^a}{\partial t} \ \frac{[D-d(t)]}{b} \ . \tag{9}$$

Eliminating the moving boundary defined by d(t) expresses the electric field in terms of the transport current and its derivative:

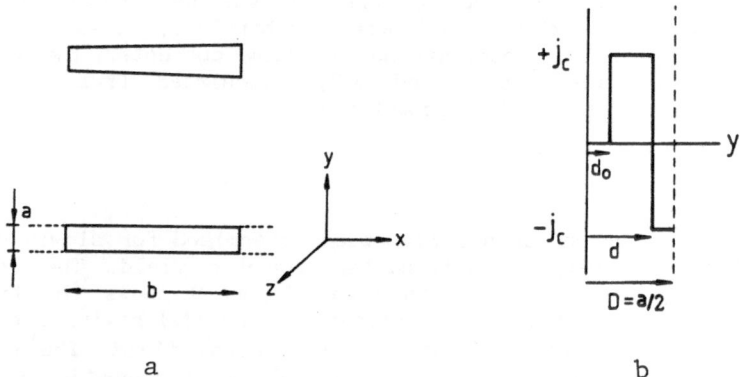

a   b

Fig. 4 A slab approximation of the LHC cable (a). The current distribution in a solid superconductor under self-field conditions (b).

$$E_z(D,t) = \mu_0 \frac{\partial I^a}{\partial t} \frac{D}{b} \left[\frac{\hat{I}^a}{2\,I_c} - \frac{I^a(t)}{2\,I_c} \right]. \tag{10}$$

Integrating the Poynting vector over the surface yields the self-field loss per unit length of conductor:

$$Q_{sf} = \frac{2}{3}\,\mu_0\,\frac{a}{b}\,I_c^2 \left(\frac{\hat{I}}{I_c} \right)^3 \tag{11}$$

Since we were not able to determine the critical current of the conductor experimentally, an estimation has to be made. M. A. Green[10] has collected various data of critical current density measurements in order to study the possibility of calculating the J_c, H_c, T_c surface, using reduced-state parameters. From these data we obtained critical current densities at 2 tesla $J_c(2T) = 4.5 \cdot 10^9$ Am^{-2} and at 0.2 tesla $J_c(0.2\,T) = 1.3 \cdot 10^{10}$ Am^{-2}. These critical current densities correspond to critical currents of $4.1 \cdot 10^4$ A and $1.2 \cdot 10^5$ A at 2 and 0.2 T respectively. The calculated losses according to Eq. 11, using the estimated critical currents, are depicted in Fig. 3 (solid lines).

DISCUSSION

The magnetization losses in the LHC cable are well described by the coupling loss formula for a rectangular conductor and the hysteresis loss formula for superconducting filaments. The effective transverse resistivity is expected to consist of two contributions. The first is the resistivity of the matrix material (copper), and the second is a magnetic field independent barrier, which may be due to the solder or the contact resistance between the individual strands.

The transport current losses have been measured in the case of an AC transport current at a background field of 0 and 2 tesla. The results have been compared with a simple self-field slab model. The theoretical and experimental results fit satisfactorily. The lack of difference between the the measurements at 0.2 and 0.5 Hz shows that the losses mainly occur in the superconducting material.

CONCLUSIONS

The magnetization losses due to a transverse alternating magnetic field were measured for different magnetic background fields. The field dependence of the effective transverse resistivity is less than would be expected in the case where the resistivity is purely determined by the copper matrix. The coupling losses are well described by the formula for rectangular multifilamentary composites. It is therefore likely to ascribe the coupling loss to the coupling of the individual strands.

The AC transport current loss, as measured in the LHC cable, is well described by a self-field slab model. It explains the frequency independence of the loss per cycle in the low-frequency range as well as the increase of the losses for a larger magnetic background field.

189

REFERENCES

1. A.J.M. Roovers, W. Uijttewaal, H.H.J. ten Kate, B. ten Haken and L.J.M. van de Klundert, A loss measurement system in a test facility for high-current superconducting cables and wires, IEEE Trans. on Magn., 24:1174-1177 (1988).

2. A.J.M. Roovers, W. Uijttewaal and L.J.M. van de Klundert, An analysis of loss measurement systems for high current superconductors, in: "Proceedings ICEC-12," Butterworth, Guildford (1988).

3. A.J.M. Roovers, An experimental study of AC losses in superconducting wires and cables, Thesis University of Twente, Enschede, The Netherlands (1989).

4. R. Perin, Progress on the superconducting magnets for the Large Hadron Collider, IEEE Trans. on Magn., 24:734-740 (1988).

5. D. Leroy, R. Perin, G. de Rijk and W. Thomi, Design of a high-field twin aperture superconducting dipole model, IEEE Trans. on Magn., 24:1373-1376 (1988).

6. D. Hagedorn, D. Leroy and R. Perin, Towards the development of high field superconducting magnets for a hadron collider in the LEP tunnel, in: "Proceedings MT-9", SIN, Villigen (1985).

7. H.J. Veringa, Intrinsic stability of technical superconductors, Thesis University of Twente, Enschede, The Netherlands (1981)

8. A.M. Campbell, A general treatment of losses in multi- filamentary superconductors, Cryogenics, 22:3-16 (1982).

9. C.Y. Pang, A.M. Campbell and P.G. McLaren, Losses in Nb/Ti multifilamentary composite when exposed to transverse alternating and rotating fields, IEEE Trans. on Magn., 17:134-137 (1981).

10. M.A. Green, Generation of the J_c, H_c, T_c surface for commercial superconductors using reduced-state parameters, Lawrence Berkeley Laboratory, University of California, doc. nr. SSC-N-502, LBL-24875 (1988).

AC LOSS MEASUREMENTS ON SUPERCONDUCTORS USING
A (LOW-INDUCTIVE) COIL GEOMETRY

A.J.M. Roovers, H.A. van der Vegt, K.W. Siekman, T. Wells,
L.J.M. van de Klundert, and J.L. Sabrié[*]

University of Twente, Dept. of Applied Physics
P.O. Box 217, NL-7500 AE Enschede, The Netherlands
[*]Alsthom, Belfort, France

ABSTRACT

In this paper loss measurements on various superconductors are presented. The measurements were performed on samples in a coil configuration. Most of the results will be used in the VAMAS Round Robin Test, in the course of which the experimental data of several laboratories throughout the world are compared. The characteristic loss parameters of the conductors are extracted from the experimental results, when possible, using classical loss theories.

Except from the loss measurements, also data concerning the critical currents of various conductors have been obtained. In order to investigate the effect of the filament diameter on the current carrying capacity of multifilamentary superconductors the results of critical current measurements on conductors, others than used in the VAMAS project, are also presented. Comparing these data indicates that the transport current at lower fields consists for fine filament conductors of bulk currents and surface currents.

INTRODUCTION

As part of the VAMAS Round Robin Test, concerning the AC behaviour of superconductors, the losses in coils made of these conductors are measured in several laboratories in the U.S.A., Japan and Europe. In this paper the loss measurements performed on the samples D and F (See Tab. 1) are presented.

The critical currents of the VAMAS samples E,F and G have been obtained in the magnetic field range of zero field to 6 T.

For reasons of comparison loss measurements and critical current measurements on other conductors were also performed.

SPECIFICATIONS OF THE CONDUCTORS

The critical currents and the losses of a number of superconductors have been measured. The specifications of the wires considered in this

Advances in Cryogenic Engineering (Materials), Vol. 36
Edited by R. P. Reed and F. R. Fickett
Plenum Press, New York, 1990

191

Table 1. The specifications of the conductors considered in this paper. A few of the conductors (D-G) are used in the VAMAS Round Robin Test.

Name	D_{wire}[mm]	D_{fil}[μm]	N_{fil}	η	L_p[mm]	matrix
Sample D	0.742	4.6	10,980	0.42	13	Cu
Sample E	0.35	6.3	760	0.215	6	Cu/CuNi
Sample F	0.14	0.50	14,280	0.182	1.9	Cu/CuNi
Sample G	0.20	0.175	242,892	0.186	0.8	Cu/CuNi
Sample X	1.00	0.875	242,892	0.186	4	Cu/CuNi
Sample Y	0.32	0.0388	9,393,931	0.138	?	Cu
Sample Z	0.12	0.58	14,496	0.33	0.8	Cu/CuNi

paper are summarized in Tab. 1. The conductors listed in the first block of Tab. 1 (D-G) are used in the VAMAS Round Robin Test. The other conductors are used for comparison purposes. All wires are NbTi multifilamentary conductors.

CRITICAL CURRENT MEASUREMENTS

The critical currents of the VAMAS conductors, E, F and G have been measured. The conductors were wound on a bobbin which was placed in the bore of a magnet. The voltage drop along the wire was measured over a length of 0.25 m. The critical currents were determined using the 1 μV/cm-criterion and the 10^{-14} Ωm-criterion. The results, in terms of the critical current density, are shown in Fig. 1.

The n-values of the conductors appeared to be independent, within the accuracy of the measurements, of the applied magnetic field. The n-values of the E, F and G sample were 30, 13 and 10 respectively. One should note the low n-value for the fine filament conductors. Figure 1 shows that the critical current densities of the wires having fine filaments increase rapidly when the magnetic field is lowered in the 0-2 T range. In this

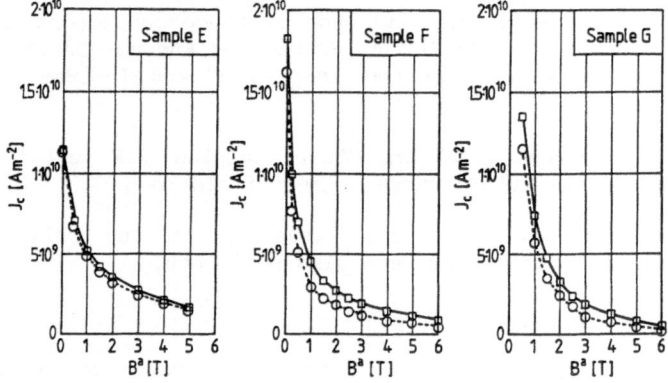

Fig. 1 The critical current densities of the samples E, F and G as a function of the applied magnetic field. The solid lines indicate the 1 μV/cm-criterion; the dashed lines the 10^{-14} Ωm-criterion.

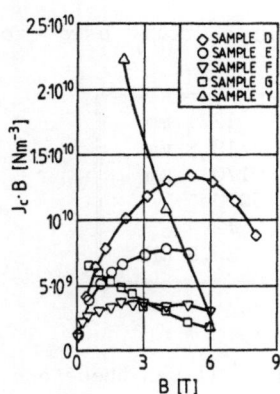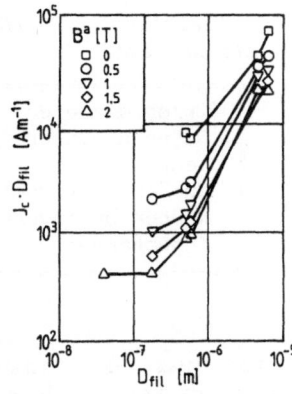

Fig. 2 a $J_c \cdot B$ as a function of B for several conductors. b $J_c \cdot D_{fil}$ as a function of D_{fil} at different fields. The filament diameter was not varied, but the data has been obtained from different conductors of various manufacturers.

region these conductors do not obey Kim's law. In the case of bulk currents the maximum volume pinning force is calculated by multiplying the critical current density by the magnetic field. For thick filament conductors this pinning force usually has its maximum in the range of 4 to 5 T. Plotting $J_c \cdot B$ as a function of B shows for fine filaments a maximum at much lower fields (See Fig. 2a).

The excessively large critical current densities at lower field for fine filament conductors may be ascribed to surface currents. The ratio $N_{fil}\pi D_{fil}/(N_{fil}\pi/4D^2_{fil})$ is much larger for these kinds of conductors than for those having thick filaments. If we assume the existence of both a bulk current and a surface current the critical current can be written as:

$$I_c = N_{fil} \left[J_{bulk,c} \frac{\pi}{4} D^2_{fil} + J_{surf,c} \pi D_{fil} \right] \qquad (1)$$

In Fig. 2b the critical current density multiplied by the filament diameter is plotted versus the filament diameter for most conductors listed in Tab. 1. The critical current data of sample D were taken from Goodrich[1]. For small magnetic fields the values of $J_c \cdot D_{fil}$ tend to become a constant for small diameters while for large diameters they are proportional to the filament diameter. This agrees well with the idea of coexisting bulk currents, which are determined by the manufacturing process, and surface currents, which are a characteristic for the material.

LOSS MEASUREMENTS

Sample G

The losses of sample G were measured in the case where an AC current was fed through a coil made of this conductor. In order to be able to extend the frequency range to higher frequencies the sample coil was part of a resonant circuit. For this purpose a capacitor was placed in series with the coil. The coil itself was part of a transformer. The data of the sample coil are given in Tab. 2.

193

Table 2 *The characteristics of the sample coil used for the loss measurements on sample G*

Inner diameter	12.1 mm
Outer diameter	19.2 mm
Length	170 mm
Number of turns	5076
Number of layers	12
Inductance	29.8 mH

The conductor was not insulated. It was therefore spaced by a non-conductive wire (diameter 0.14 mm) while it was wound on a bobbin. The individual layers were insulated by means of Kapton foil of 25 μm. The coil was wet-wound with a filled epoxy, STYCAST 2850 FT.

The losses were measured calorimetrically in the case of a sinusoidal changing transport current (frequency 48, 118 and 301 Hz). There was no magnetic background field. The results are shown in Fig. 3.

The dissipation appeared to be proportional to $I^{2.6}$. The losses per cycle for different frequencies coincide. The losses in the stabilizing material are, therefore, of no significance.

Sample X

Sample X is the same kind of conductor as sample F. It just did not pass the final drawing stages. Therefore the only differences are therefore the dimensions of the wire and its filaments. The losses of this conductor have been obtained using a sample coil as specified in Tab. 3. The coil consists of four layers of which two are wound in opposite directions to the others. The coil was designed in such a way that coupling with the DC magnet was minimized. The coil was part of the secondary circuit of a transformer. The current through the sample coil was induced by means of a toroidal transformer. A transformer circuit provides a possibility to induce a larger current through the sample than the maximum current of the power supply. In our set-up a current amplification of 20 was achieved. The current was measured by means of a Rogowski coil.

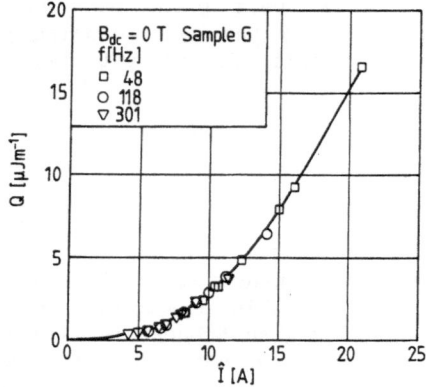

Fig. 3 *The losses measured on the coil made of sample G as a function of the transport current for different frequencies.*

Table 3 The characteristics of the bifilarly wound sample coil used for the loss measurements on sample X

Inner diameter	40.1 mm
Outer diameter	50.5 mm
Length	20 mm
Number of turns	+28/-22
Number of layers	4
Inductance	10 μH

Table 4 The characteristics of the bifilarly wound sample coil used for the loss measurements on sample D

Inner diameter	40.1 mm
Outer diameter	48.0 mm
Length	20 mm
Number of turns	+32/-26
Number of layers	4
Inductance	13 μH

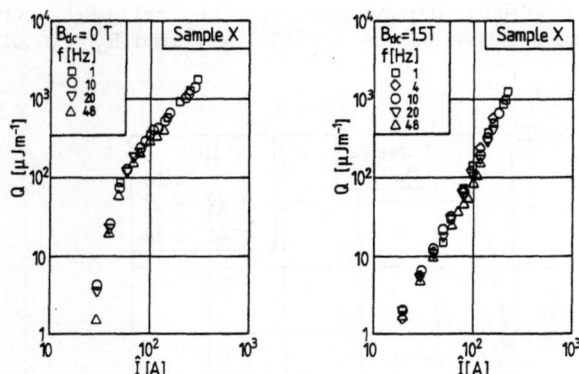

Fig. 4 The AC losses per cycle per unit length of conductor of sample X. The left figure shows the results at zero field and the right figure the results at a background field of 1.5 T.

The sample coil was placed in the bore of a DC magnet. Measurements were performed at zero field and at a background field of 1.5 T. The electrical as well as the calorimetric method was applied.

The experimental results are show in Fig. 4. The losses are presented as the average loss per cycle per unit length of conductor. The losses per cycle were found to be independent of the frequency.

Sample D

The losses of sample D were measured in the same set-up as used for sample X. The characteristics of the sample coil are listed in Tab. 4. The inner two layers were also wound in a direction opposite to the outer two layers.

The losses were measured at zero field and at a background field of 1.5 T. The results of the measurements at 1.5 T are shown in Fig. 5.

The losses show a behaviour which is typical for the case where the coupling losses are dominant.

DISCUSSION

The losses measured on sample G appeared to be proportional to $\hat{I}^{2.6}$. It suggests that the dissipation is due to the so-called dynamic resistance in the case where the wire is partly saturated. An extreme form of this loss mechanism is the self-field loss. A numerical program[2] which takes into account the current distribution and calculates the transport current loss and the magnetization loss predicts that for this geometry (combination of the local magnetic field and the transport current) the magnetization losses should be dominant for currents below 15 A. For large currents (where the transport current losses are larger than the magnetization losses) the program describes the loss behaviour well. For small currents the program calculates a dissipation proportional to the amplitude of the current (magnetization losses), while the experimental data show in this range the same current dependence as for large currents. In this program a field dependence of the critical current density according to Kim was assumed ($J_{c0} = 1.9 \cdot 10^{10}$ Am^{-2} and $B_0 = 0.25$ T)

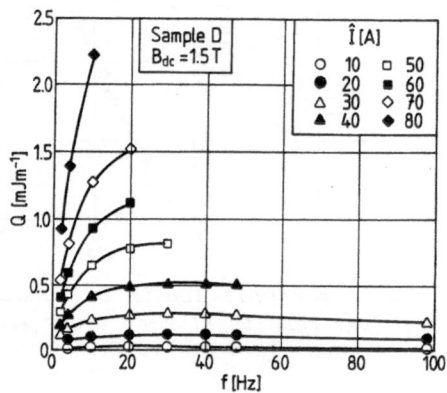

Fig. 5 The AC losses per cycle per unit length of conductor of sample D as a function of the frequency for several values of the transport current. The figure shows the results obtained at a background field of 1.5 T.

The measurements on sample X show a similar behaviour. The calculated losses at zero field, obtained with a numerical program in which a $J_c(B)$-function according to Kim with $J_{co} = 1.6 \cdot 10^{11}$ Am^{-2} and $B_0 = 0.3$ T was fitted, agree well with the experimental losses for larger currents ($I > 70$ A). For smaller current the experimental losses are significantly lower. The critical current density required to fit the experimental data however, is quite large.

At a background field of 1.5 T the losses are well predicted by the numerical code using a constant critical current density of $3 \cdot 10^9$ Am^{-2}.

The coupling current time constant of a conductor can be obtained using two methods. The first one is to find the frequency at which the loss per cycle has its maximum value. The second one is calculate the time constant from the slope of the loss per cycle versus the frequency dependence in the low frequency limit. In the latter case we need the average value of the squared magnetic field amplitude $<\hat{B}^{a2}>$. This quantity can be calculated from the field distribution function of the coil[3]. Knowing this value and so the coupling current time constant, the effective transverse conductivity is expressed as:

$$ \sigma_\perp = \frac{\partial Q}{\partial f} \left[\frac{\pi}{2} <\hat{B}^{a2}> L_p^2 R_w^2 \right]^{-1} , \tag{2} $$

where L_p denotes the twist pitch and R_w the radius of the conductor. Using Eq. 2 leads for sample D to an effective transverse conductivity at 1.5 T of $1.7 \cdot 10^9$ $(\Omega m)^{-1}$. Deducting the effective transverse conductivity from the frequency f_0 where the maximum of the loss per cycle occurs leads to:

$$ \sigma_\perp = \frac{1}{\pi f_0 \mu_0} \left[\frac{2\pi}{L_p} \right]^2 . \tag{3} $$

This maximum was found to be at a frequency of 30 ($\pm$3) Hz (at 1.5 T). This frequency yields an effective transverse conductivity of $2.0 \cdot 10^9$ $(\Omega m)^{-1}$ at a background field of 1.5 T. The experiments at zero field showed that the frequency f_0, at which the maximum of the loss per cycle occurs, is 20 ($\pm$2) Hz. From this frequency we calculated an effective transverse conductivity of $3.0 \cdot 10^9$ $(\Omega m)^{-1}$.

Schmidt[4] reports an RRR-value for sample D of 179 in absence of a magnetic field. This corresponds to a conductivity of the copper at 4.2 K of $1.1 \cdot 10^{10}$ $(\Omega m)^{-1}$. If we assume that the filaments act as insulators, the effective transverse conductivity would be $4.5 \cdot 10^9$ $(\Omega m)^{-1}$.

From the zero frequency limit, the hysteresis losses in the superconducting filaments can be obtained. These losses can be used to estimate the critical current density. For induced magnetic fields, much larger than the penetration field the critical current density can be extracted from the loss measurements according to:

$$ J_c = \frac{3 \lim_{f \to 0} Q}{2 D_w^2 \eta D_{fil}^2 <\hat{B}^a>} \tag{4} $$

The critical current density at 1.5 T according to Eq. 4 was estimated to be $5.7 \cdot 10^9$ Am^{-2}. Extrapolating the losses to zero frequency also yields the penetration field. From the penetration field we obtained the same value for the critical current density. This agrees well with the interpolated value of $5.3 \cdot 10^9$ Am^{-2} calculated from critical current measurements at 1 and 2 T reported by Goodrich[1].

CONCLUSIONS

From the critical currents measurements we can conclude that the n-values of fine filament conductors are substantially lower the the ones of thick filament conductors (which is usually in the order of 50). As a consequence the critical current density is less well defined, as it depends strongly on the criterion used.

The dependence of the critical current density on the applied magnetic field of fine filament conductors does not obey Kim's law[7]. The values of the critical current density are for these conductors at low fields much higher than for thick filament conductors. This is probably caused by surface currents which are more evident for fine filaments conductors, due to their large filament surface to volume ratio.

The loss measurements on fine filament conductors showed that the magnetization losses are not in accordance with the Bean model[8]. The magnetization behaviour of fine filament conductors is still not fully understood. The development of new models, which may include surface currents, is therefore desirable. This anomalous behaviour has already been noticed by other authors[5,6]. In the region (larger currents) where the transport current losses determine the loss behaviour the theoretical and experimental results agree well. This also holds for the coupling losses.

ACKNOWLEDGEMENTS

The authors would like to thank Alsthom, France, for kindly providing us with several samples.

REFERENCES

1. L.F. Goodrich, National Institute of Standards and Technology, Boulder, Colorado, U.S.A., private communications.
2. A.J.M. Roovers and L.J.M. van de Klundert, Current distribution and AC losses in twisted multifilamentary AC superconductors, IEEE Trans. on Magn., 25:2127-2130 (1989).
3. A.J.M. Roovers, H.A. van den Brink and L.J.M. van de Klundert, Analysis of AC loss measurements on superconductors using coil configurations, Adv. in Cryog. Eng., 34:909-916 (1988).
4. C. Schmidt, presented at this conference, paper BP-25.
5. F. Sumiyoshi et al., Anomalous magnetic behaviour due to reversible fluxoid motion in superconducting multifilamentary wires with very fine filaments, Jap. J. Appl. Phys., 25:L148-L150 (1986).
6. L. Cesnak et al., Losses in transformer-like coils wound from a very fine filament Nb-Ti superconductor, Cryogenics, 28:386-393 (1988).
7. Y.B. Kim, C.F. Hempstead and A.R. Strnad, Magnetization and critial supercurrents, Phys. Rev., 129:528 (1963).
8. C.P. Bean, Magnetization of hard superconductors, Phys. Rev. Lett., 8:250-253 (1962).

VAMAS INTERCOMPARISON OF AC LOSS MEASUREMENT: JAPANESE RESULTS

K. Itoh[1], H. Wada[1], T. Ando[2], E. Yoneda[3], D. Ito[3],
M. Iwakuma[4], K. Yamafuji[4], A. Nagata[5], K. Watanabe[6],
Y. Kubota[7], T. Ogasawara[7], S. Akita[8], M. Umeda[9],
Y. Kimura[9], and K. Tachikawa[10]

[1]National Res. Inst. Metals, Tsukuba, Ibaraki, Japan
[2]Japan Atomic Energy Res. Inst., Naka, Ibaraki, Japan
[3]Toshiba R & D Center, Kawasaki, Kanagawa, Japan
[4]Kyushu Univ., Fukuoka, Japan
[5]Akita Univ, Akita, Japan
[6]Inst. Materials Res. Tohoku Univ., Sendai, Japan
[7]Nihon Univ., Tokyo, Japan
[8]Central Res. Inst. Electric Power Industry, Tokyo, Japan
[9]Electrotechnical Lab., Tsukuba, Ibaraki, Japan
[10]Tokai Univ., Hiratsuka, Kanagawa, Japan

ABSTRACT

The first worldwide round robin test on ac losses was carried out, where 10 Japanese labs participated. 4 test samples, D, E, F and G, were prepared, and ac losses were measured as a function of either applied field amplitude or transport current. A variety of measurement methods were adopted, and an interim intercomparison of the results was made in terms of the hysteresis loss for 0 to 1 T field cycle as well as the coupling time constant. The standard deviation in hysteresis loss among labs including those of US and European labs so far reported was about 10 % for both high loss samples D and E. Time constants obtained among labs were in a relatively poor agreement with each other.

INTRODUCTION

The ac loss, or the electric power loss caused by a changing magnetic field, is an important parameter to such applications of superconductivity as ac machines, fusion generators and particle accelerators. Particularly, recent development in the fabrication of ultra-fine filament superconductors encourages ac power frequency applications. However, a large variety of measurement methods and variables on ac losses exist, and it is quite difficult to evaluate the results measured at different places.

The VAMAS (Versailles Agreement on Advanced Materials and Standards) technical working group in the field of superconducting and cryogenic structural materials has recently started a program on the intercomparison of ac loss measurement preceded by a similar one on the critical current measure-

Advances in Cryogenic Engineering (Materials), Vol. 36
Edited by R. P. Reed and F. R. Fickett
Plenum Press, New York, 1990

199

Table 1: Specifications of round robin test samples

Sample	D	E	F	G
Outer diam.	0.742 mm	0.35 mm	0.14 mm	0.2 mm
Fil. diam.	4.6 μ m	6.3 μ m	0.5 μ m	0.175 μ m
No. filaments	10,980	760	14,280	242,892
NbTi/Cu/CuNi	1/1.38/-	21.5/44.1/34.4	1/1/3.5	18.6/2/79.4
Twist pitch	13 mm	6 mm	1.9 mm	0.8 mm
Insulation	none	polyester	polyester	none

ment[1]. The purpose of the present program is to work out proposals which will eventually be useful for the standardization of the ac loss measurement. In this paper, the results so far obtained at Japanese labs are presented. The final results will be presented elsewhere together with US and European results.

TEST GUIDELINES

The ac loss in multifilamentary wires is usually understood in terms of hysteresis loss, coupling loss and eddy current loss. Hysteresis and eddy current losses are due to the properties of superconducting and stabilizing materials, respectively, while the coupling loss is a loss caused by the current flowing between superconducting filaments when a changing field is applied. In the present round robin test, there were few restrictions on the measurement methods and variables to allow as many labs as possible to participate in the test. In order to compare the results among labs, however, the measurement of the hysteresis loss for 0 to 1 T field cycle and the coupling time constant at around 1 T were recommended.

SAMPLES AND MEASUREMENT METHODS

Samples

4 multifilamentary NbTi conductors were chosen as test samples, one supplied from each of the US and Europe, and two from Japan. In this paper these samples are arbitrarily labelled as samples D, E, F, and G. Similarly, participant labs are labelled as lab g - lab o. Specifications of these samples are shown in Table 1.

Sample D, originally developed for accelerator magnet use, contains more than ten thousands of NbTi filaments embedded in a copper matrix, and has a high critical current density, J_o, 2500 A/mm^2 at 5 T. Sample E

Table 2: Specifications of coil samples

Sample coil	F1	F2	G1	G2	G3
Length of coil(mm)	160	160	170	170	170
Inner diam.(mm)	15.95	15.95	12.1	12.1	12.1
Outer diam.(mm)	21.65	21.70	19.2	19.2	19.9
N. of layers	10	10	12	12	12
N. of turns	9,940	10,001	5,076	5,015	4,817
Total length of wire(m)	~ 587	~ 591	~ 250	~ 247	~ 241
Inductance(mH)	--	--	29.8	29.5	--

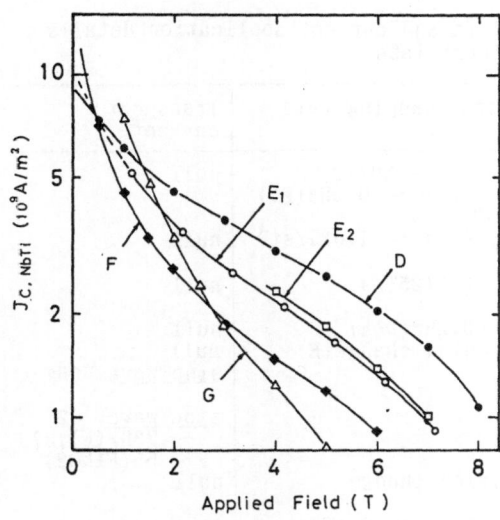

Fig. 1: J_c versus applied field curves for samples D, E, F and G. Curves E_1 and E_2 for sample E are obtained at labs h and l, respectively. Rest of curves are reported from US and European labs.

contains the largest filament diameter, $6.3\,\mu$m, among 4 samples, which usually results in relatively large hysteresis loss. However, since each filament in this sample is separated from others by the highly resistive Cu-Ni alloy sheath, the electric coupling between filaments should be suppressed. Samples F and G, both developed for ac power frequency use, have filaments of sub-μm in diameter, each separated by the Cu-Ni sheath. In sample F, a considerable amount of copper divided into pieces by Cu-Ni is placed in the center of the conductor, while little copper is incorporated in sample G. Both samples are twisted with a very short pitch.

An about 10 meter of sample D and an about 100 meter of sample E without windings were distributed to all of the participant labs. These lengths were suitable for electromagnetic measurements but not enough for calorimetric measurements. In order to enable both the calorimetric and electromagnetic measurements, 5 small solenoid coils with relatively large volumes were produced from samples F and G, and circulated among several participant labs. The specifications of these coil-shaped samples, labelled as F1, F2, G1, G2 and G3, are given in Table 2. All of these coil samples except G3 have almost identical dimensions and structures. Sample G3 was designed as the self-inductance come minimum for which electromagnetic measurements are more convenient. The current terminals were attached to these samples, enabling the current injection to the coil winding. In case of samples F1 and F2, a 0.2mm thick GFRP foil and a doubled 0.025mm thick Kraft paper were alternately inserted between winding layers. These two samples, impregnated with epoxy resin, had narrow cooling channels engraved around the GFRP coil bobbin. Samples G1, G2 and G3 were wet-wound with Al_2O_3 filled epoxy, 'STYCAST 2850FT' on a bobbin made of 'Celleron'. Their windings were spaced with a non-conductive wire of 0.14mm in diameter, since these conductors had no insulation on themselves. The winding layers were then insulated with a 0.025 mm thick 'Kapton' foil. Samples F1 and G2 were circulated from lab to lab, and ac loss measurements were carried out at each lab.

Measurement Methods

The test instruments for ac loss measurements used at the Japanese labs may be classified in four categories, that is, vibrating sample magnetometer (VSM), flux meter, tanδ meter and calorimeter. The VSM is

Table 3: Measurement methods, and field and current application details adopted at Japanese participant labs

Lab	Sample	Measurement method	Wave form and changing rate of field	Transport current
g	D,E,F1 and G2	flux meter	sine wave, 0.1 ∼ 1Hz(D), 0.2,1Hz(E), 0.1 ∼ 0.3Hz(F1) 0.1 ∼ 0.4Hz(G2)	null
		"	trapezoidal,0.04 ∼ 1.03T/s(D	null
h	E	VSM	triangular, 0.0125T/s	null
i	D&E	flux meter	triangular, 0.2Hz(D&E)	null
		"	triangular, slow change(E)	null
	F1&G2	liq He level	null	sine wave,50Hz
j	E,F1 and G2	tan δ meter	null	sine wave, 32 ∼ 73Hz(E),37 ∼ 55(F1&G2)
	F	VSM	triangular,slow change	null
k	D,E,F1 and G2	flux meter	triangular, 0.02 ∼ 0.4T/s	null
m	D&E	flux meter	sine wave, 0.05 Hz	null
		"	triangular, 0.023T/s	null
		"	sine + exponential decay, rise time:40ms, decay:110ms	null
	F1&G2	calorimeter	null	sine wave, 50Hz
n	D,E	flux meter	triangular, 0.016T/s(D&E)	null
			instant change,1.23 → 1 T(D)	null
o	D,E	VSM	triangular, 0.003T/s	null

an induction instrument and capable to measure static magnetic properties as a function of field, temperature and time. A bundle of short-cut wires were prepared as a specimen. The flux meter is also an induction instrument and capable to measure dynamic magnetic properties as a function of frequency or sweep rate dB/dt, and changing field amplitude. The tan δ meter is a method to measure coil loss by picking up the voltage across both ends of the sample coil. Detail of the method appears in Ref 2. The calorimetric measurement was carried out at labs i and m by observing the shift of liquid helium level in a dewar. In both of the tan δ and the calorimetric measurements the transport current was fed to the sample. The measurement methods and variables adopted are summarized in Table 3. The measurements under dc background fields were not made at Japanese labs.

Critical Current Measurements

Homogeneity studies on sample E were performed at labs h and l by measuring J_c's at 4.2 K and for various magnetic fields. At lab h 7 specimens of sample E were prepared, each 20 cm long and taken from every 200 m of the whole sample conductor length. The standard deviations of I_c and n value at field range of 2-7 T were within 1% and 10% of their averages, respectively, indicating the excellent homogeneity of sample E. Similar results were obtained at lab l, where J_c's of 12 specimens were measured at 4 to 10.5 T. The averaged J_c vs magnetic field curves are shown in Fig. 1, together with the curves for samples D, F and G which were reported from US and European labs; no homogeneity studies have been performed on these samples.

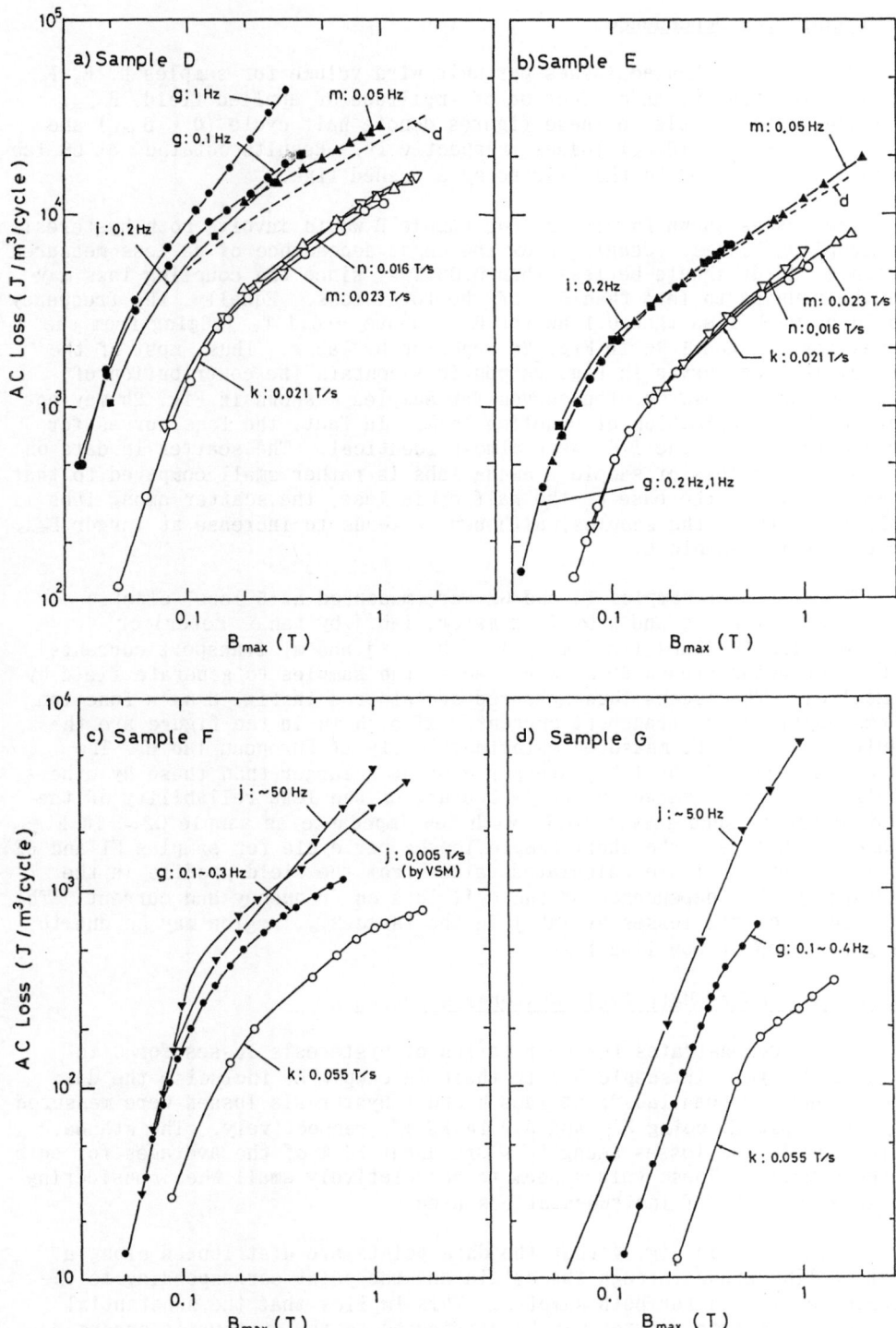

Fig. 2: ac losses versus field amplitude curves for; a)sample D, b)sample E, c)sample F and d)sample G.

RESULTS AND DISCUSSION

Field Amplitude Dependence

Figs. 2a-2d show ac losses per unit wire volume for samples D, E, F and G, respectively, as a function of amplitude of applied field, B_{max}. Open and solid symbols in these figures denote half cycle ($0 - B_{max}$) and full cycle ($-B_{max}, +B_{max}$) losses, respectively. Results obtained at US lab d are also presented in the figures by a dashed line.

The losses shown in Fig. 2a for sample D would involve both hysteresis and coupling losses. Judging from the dB/dt dependence of ac loss measured at lab g, dB/dt should be less than 0.03 T/s, since the coupling loss may then be reduced to less than 5 % of the total loss. Equally, the frequency should be much less than 0.1 Hz for B_{max} above ~ 0.1 T, judging from the curves for 0.1 and 1 Hz in Fig. 2a reported by lab g. Thus, most of the full cycle loss curves in Fig. 2a possibly contain the contribution of coupling loss. However, the curves for samples E shown in Fig. 2b may not contain the contribution of coupling loss. In fact, the loss curves for frequencies 0.2 Hz and 1 Hz were almost identical. The scatter in data on the full cycle loss of sample E among labs is rather small compared to that of sample D. In the case of the half cycle loss, the scatter among labs is small for both of the samples, although it tends to increase at larger B_{max}, especially for sample E.

The losses of samples F1 and G2 were measured at 5 labs electro-magnetically (labs g and k by flux meter, lab j by $\tan \delta$ meter) or calorimetrically (labs i and m). At labs i, j and m, transport currents with frequencies around 50 Hz were fed to the samples to generate field by themselves. The losses thus measured are plotted in Fig. 3 as a function of rms amplitude of transport current. Also shown in the figure are the results on sample G2 measured calorimetrically at European lab c. The losses on sample G2 by lab j are a factor of 2 larger than those by others in all the current range, probably because of the less reliability of the $\tan \delta$ meter on such sample coils with low impedance as sample G2. In Figs. 2c and 2d shown are the short sample losses per cycle for samples F1 and G2. The losses by lab j are calculated values from the field profile in the windings and the dependence of the coil-loss on frequency and current. The upper shift of the losses by lab j in the large B_{max} region may be due to the generation of coupling loss.

Intercomparison of Half Cycle Hysteresis Losses

Fig. 4 demonstrates the correlation of hysteresis losses for 0 to $+B_{max}(=1$ T) cycle in sample D with those in sample E, including the data from US and European labs[3]; at labs h and i hysteresis losses were measured only on sample E, being 5.6 and 5.0 in kJ/m^3, respectively. The standard deviations in the losses among labs are about 10 % of the averages for both of the samples. These values seem to be relatively small when considering the large variety of instrumentations used.

It is found in Fig. 4 that the data points are distributed along a straight line drawn through the origin and the point corresponding to the averages of losses for both samples. This implies that the substantial portion of the data scatter may be attributed to the systematic errors due to the variety of instrumentations and evaluation methods among labs. In fact, the standard deviation in the loss ratios of sample D to those of sample E is about 5 % which is a half of the standard deviation for each sample.

The hysteresis loss can be described in terms of the critical state model and calculated as a function of J_c, filament diameter, B_{max} and back ground dc field; this may not be true for ultra-fine filament wires where interfilament proximity effect becomes significant. J_c changes rapidly at fields below $B_{max}(= 1\ T)$. Its change is extremely large at fields close to zero field, leading to a drastic change in magnetization. The hysteresis loss is given by integrating the magnetization with respect to field. Thus, the precise determination of field strength around zero field is critical to a reliable hysteresis loss measurement.

In the case of the flux meter, a superconductor such as Pb, PbSn or Nb is used as standard sample on magnetization. An error may be introduced in determining the slope of magnetization vs field curve for the standard sample showing perfect diamagnetism. Another origin of errors may arise from the demagnetization factor which strongly depends on the shape of the sample and is thus difficult to be precisely estimated.

As described in the paragraph on the field dependence of ac loss, the losses measured may contain a certain amount of coupling loss. This, in addition to the uncertainty of sample temperature, could be important to the scatter in data. Detailed discussion on the scatter in the VAMAS ac loss data will be presented in the near future.

Coupling Time Constants

A direct measurement of the coupling time constant in sample D was carried out at lab n by plotting a decay profile of voltage generated in sensing coil to an instant change of applied field. The coupling time constants in sample D were also estimated indirectly from the coupling loss

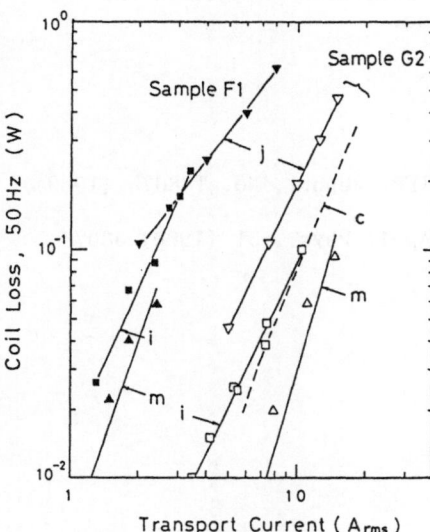

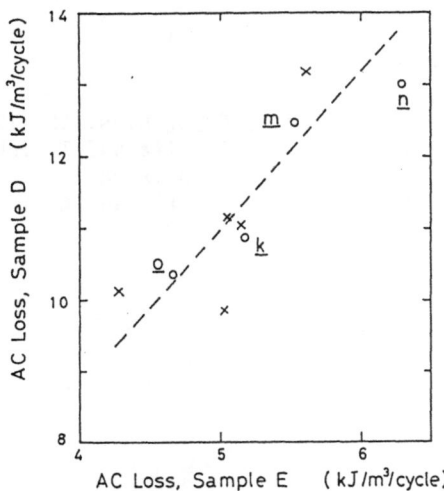

Fig. 3: Coil loss versus ac transport current curves for samples F1 and G2. Losses for 50 Hz estimated at lab j are derived from the frequency dependence of the loss.

Fig. 4: Correlation of hysteresis losses for 0 to $+B_{max}$ (=1 T) cycle in sample D with those of sample E; Japanese labs (o), US and European labs (x).

measurement data at labs g, k and m. The time constants obtained are 8.8 13.3 and 17 msec at labs g, k and n, respectively, much more scattered in comparison to the hysteresis losses measured. The origins of scatter are not clear at present and should be discussed in relation to the measurement and evaluation details including the demagnetization effect[4]. The time constant in sample E was measured only at lab m using a pulsed magnet method, which was $19\,\mu$ sec.

CONCLUSION

The round robin test on ac loss measurement was performed on 4 NbTi test samples, where 10 Japanese labs participated. The interim inter-comparison of the results were made on the hysteresis loss for 0 to 1 T field cycle and the coupling time constant. The standard deviation in the hysteresis losses among labs including US and European labs so far reported was about 10 % for both samples D and E. The standard deviation in the loss ratio of sample D to sample E was a half of that for each of samples, suggesting systematic errors in field determination and instrument calibration, etc, may be responsible for the scatter in data. The time constants obtained among labs showed a larger scatter than the hysteresis losses.

Analysis and comparison of the ac loss data measured as a function of field amplitude and transport current by means of a variety of measurement methods including tan δ meter, calorimeter, VSM and flux meter methods should be beneficial for establishing standard ac loss measurement methods. Detailed discussion will be made after all the participants have completed their measurements.

ACKNOWLEDGEMENT

The authors wish to thank all those who have been engaged in the present round robin test for their cooperation.

REFERENCE

1. K. Tachikawa, Cryogenics, 29 (1989) 710
2. H. Kasahara, S. Akita and T. Ishikawa, CRIEPI Report, No. T88075 (1989)
3. K. Tachikawa, this issue
4. F. Sumiyoshi, F. Irie and K. Yoshida, J. Appl. Phys., 51 (1980) 3807

ANOMALOUS LOW HYSTERESIS LOSSES IN NbTi SUPERCONDUCTORS
WITH VERY FINE FILAMENTS

C. Schmidt

Kernforschungszentrum Karlsruhe
Institut für Technische Physik
D-7500 Karlsruhe, FRG

ABSTRACT

Hysteresis losses of multifilamentary NbTi/Cu/CuNi mixed matrix conductors with very fine filaments were measured in the frame of the VAMAS
intercomparison. A calorimetric technique was developed for the purpose
allowing short sample measurements in the microwatt range. This method
allows an independent variation of the magnetic background field and the
alternating field amplitude. For low amplitudes, the losses were found
to be well below the values expected from the Bean model. A similar
behaviour, found recently in magnetization measurements, was explained
by reversible flux line motion. The present data are reasonably well
described by an existing theory. The paper includes also the results of
two samples having thicker filaments.

INTRODUCTION

In the frame of the VAMAS intercomparison of ac loss measurements
the hysteresis losses of NbTi multifilamentary conductors were measured
for different samples. While hysteresis losses of the samples having
larger diameter ($\gg 1$ μm) are described with a reasonable accuracy by the
Bean model of flux penetration, two of the samples with filament
diameters below 1 μm showed a distinct anomaly. For decreasing
alternating field amplitudes, the losses become much lower than
predicted by the Bean model. The deviation can be more than an order of
magnitude.

Reduced hysteresis losses were recently found in magnetization
measurements of ultrafine filament conductors[1,2] and in calorimetric
measurements on current carrying test coils.[2] The anomaly was explained
by "reversible flux motion" and "flux line - filament size effect" in
these materials. The purpose of the present paper is to give additional
experimental data measured calorimetrically on short samples, where the
alternating field amplitude and the background field can be varied as
independent parameters, and where no transport current flows in the
sample.

Advances in Cryogenic Engineering (Materials), Vol. 36
Edited by R. P. Reed and F. R. Fickett
Plenum Press, New York, 1990

207

Table 1. Superconducting samples

SAMPLE		D	E	F	G
Wire diameter	(mm)	0.74	0.35	0.14	0.2
No. of filaments		10 980	760	14 280	242 892
Filament diameter	(µm)	4.6	6.3	0.5	0.175
Twist pitch	(mm)	13	6	1.9	0.8
NbTi:Cu:CuNi ratio		1 : 3 .8 : −	1 : 2.1 : 1.6	1 : 1 : 3.5	1 : 0.11 : 4.3
I_c (B = 1 T)	(A)	1101	108	12.6	44.1
j_c (1 T) in NbTi	(GA/m²)	6.1	5.2	4.5	7.5
Rest resistivity ratio		179	15.6	16.3	15
Coupling loss constant at Bo = 1 T	(mWs²/T²)	22.5	0.025 ± 30%	-	-

The only quantitative theory known to the author which explains the reduced hysteresis losses was proposed by Takacs and Campbell.[3] A calculation according to this theory was performed and compared to the experimental results. The theory seems to describe the results, at least in a certain range of parameters, quite reasonably.

EXPERIMENTAL

The losses of some of the samples to be measured are several orders of magnitude below the resolution of the most sensitive experiment using the standard helium boil-off method.[4] We therefore developed a technique allowing a calorimetric measurement with a resolution of about 10^{-8} W. The technique is similar to a thermal conductivity measurement. It was first used for the measurement of large superconducting cables in the high loss range.[5] It ist however also suitable for very sensitive measurements, if the design of the experimental arrangement is appropriate.

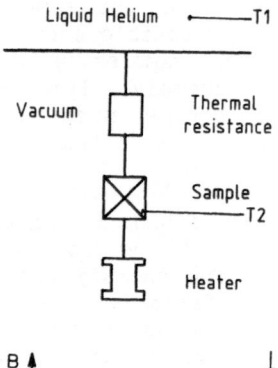

Fig. 1. Experimental arrangement, schematic, and definition of ac-field parameters. T1 and T2 are thermometers.

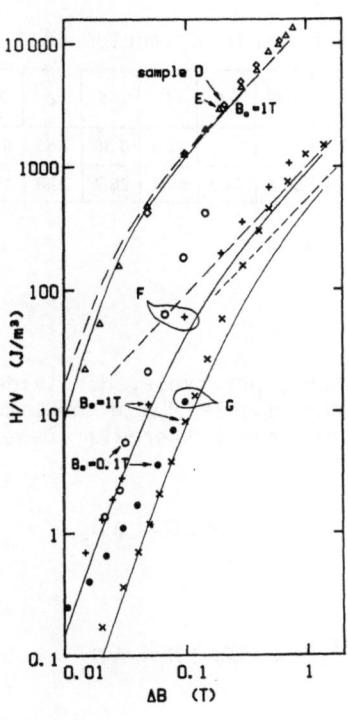

Fig. 2. Hysteresis losses per cycle and per unit NbTi volume as a function of the alternating field amplitude. Dashed lines according to the Bean model, solid lines according to the theory of Ref. 3, for the $B_O = 1$ T data.

A few meters of the superconducting wire are wound on a winding form. This sample is placed in a vacuum vessel and coupled to the liquid helium bath via a thermal resistance, see Fig. 1. The alternating field as well as the background field perpendicular to the wire axis is supplied by a superconducting coil. The figure gives also the definition of the ac-field parameters. The ac field leads to a temperature increase of the sample, which is a direct measure of the losses. The thermal resistance is here a copper wire of 0.1 mm dia. and the sample temperature is measured with an Allen Bradly carbon resistor. The temperature sensor must not be calibrated, and the value of the thermal resistance must not be know exactly. Calibration of the ac loss power is done with an ohmic heater connected to the sample. Further experimental details will be published elsewhere.[6]

Samples and data evaluation

Tab. 1 gives the characteristic parameters of the samples used in the VAMAS intercomparison. Sample D is a NbTi/Cu conductor, the other three samples are mixed matrix NbTi/Cu/CuNi conductors. Samples F and G are for 50 Hz application and have filament diameters below 1 μm.

The main loss component is, for samples E to G, hysteretic. Only sample D has large coupling losses, as it is expected for a pure Cu-matrix conductor. Sample E has only a small coupling loss contribution which was difficult to extract from the total losses. Samples F and G did'nt show a measurable contribution of coupling losses. The separation of coupling and hysteresis losses is described in [5]. The coupling loss power, Q, is, for not too high frequencies, proportional to $(f \cdot \Delta B)^2$. The coupling loss constants given in Table 1 are the values $Q/(f \cdot \Delta B)^2$ for 1 m of sample length.

Table 2. Critical currents of samples F and G. Criterion is 1 μV/cm

Sample	B(T)	0.2	0.25	0.3	0.4	0.5	0.6	0.8	1	1.5	2	2.5
F*	I_c(A)		27.8			19.6			12.6	9.35	7.53	6.28
G	I_c(A)	128		109	92.5	79.5	69.5	54.3	44.1	28.2	20.8	15.4

* from Ref. 8

RESULTS

Fig. 2 shows the hysteresis loss energy per cycle, H, divided by the NbTi volume, V, as a function of the double field amplitude, ΔB. According to the flux penetration model, the hysteresis losses can be expressed by [7]

$$H/V = \frac{\Delta B^2}{2 \mu_o} (\frac{4}{3} \beta - \frac{2}{3} \beta^2) \qquad for\ \Delta B < \Delta B_p$$

$$ \tag{1}$$

$$H/V = \frac{\Delta B^2}{2 \mu_o} (\frac{4}{3} \beta^{-1} - \frac{2}{3} \beta^{-2}) \qquad for\ \Delta B > \Delta B_p$$

with $\beta = \Delta B/\Delta B_p$.

$\Delta B_p = (4\ \mu_o/\pi)r \cdot j_c$ is here the penetration field, r the filament radius and j_c the critical current density in the superconductor. For $\Delta B \ll \Delta B_p$, the losses tend to a ΔB^3 dependence, whereas for $\Delta B \gg \Delta B_p$, a linear dependence on ΔB is approached.

The dashed lines in Fig. 2 are calculated for the $B_0 = 1$ T data of samples E to G. The curve for sample D, which is not much different from that of sample E, was omitted for clarity. For sample E, the penetration field is 0.02 T and the calculated curve describes reasonably well the data.

For the samples with very fine filaments, F and G, the penetration field is 1.8 and 1.0 mT, respectively. The condition $\Delta B \gg \Delta B_p$ is fulfilled in the whole range of measurement, and a linear dependence on ΔB is expected. As Fig. 2 shows, the experimental results are not at all described by Eq. (1). Only at high field amplitudes, H (ΔB) is more or less linear.

The flux penetration theory seems to be valid for these samples only at large ΔB values. The fact that the experimental values are here above the calculated curves, is not surprising. H is proportional to the filament radius and to the critical current density. The calculation was done with the nominal filament radius, calculated for ideally round filaments, whereas the effective filament radius is certainly higher. For j_c, the values measured according to the 1 μV/cm criterion were used (Table 2). This definition is however artificial and only of practical use. Hysteresis losses are determined rather by the local critical current density, which may be appreciably higher. The factor $r \cdot j_c$ could be used as a fit parameter to adjust the calculation. In this case an effective $r \cdot j_c$ higher by a factor of 1.5 and 2 would be required for sample F and G, respectively.

210

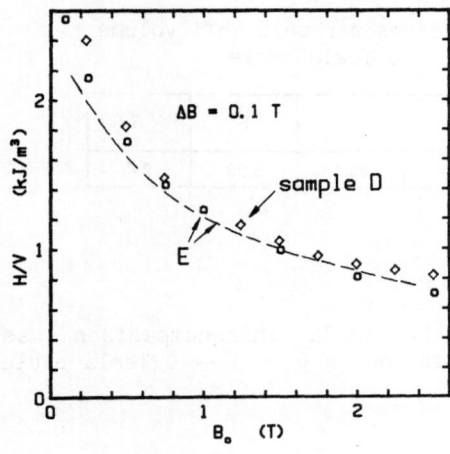

Fig. 3. Hysteresis losses as a function of the background field for the samples with "thick" filaments.

Fig. 3 and 4 show the hysteresis losses as a function of the background field for fixed alternating field amplitudes. The decreasing losses with increasing B_O reflect the $j_c(B_O)$ dependence of the samples. The dashed lines in Fig. 3 and 4 are calculated for samples E and F, using Eq. (1) and critical current values given in[8]. While the agreement for sample E is quite good, the losses of sample F are, in the whole field range, below the Bean model. For sample G, the Bean model gives values almost an order of magnitude to high.

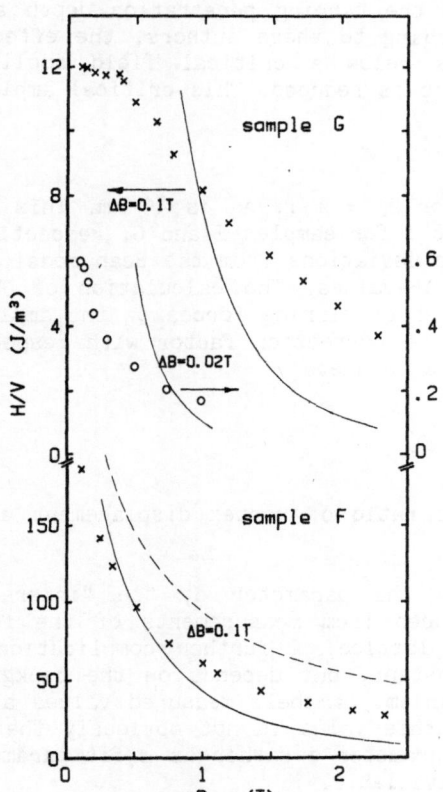

Fig. 4. Hysteresis losses vs. Bo for the very fine filament samples.

Table 3. Hysteresis losses per unit NbTi volume
for a 0 → 1 → 0 Tesla cycle

SAMPLE	D	E	F	G
H / V (kJ/m³)	31.4	26.1	3.09	2.4

A further requirement of the VAMAS intercomparison was the measurement of the hysteresis losses for an 0 → 1 → 0 Tesla cycle. The results are given in Table 3.

DISCUSSION

The preparation of superconductors with finer and finer filaments, in order to reduce the hysteresis losses, led, below a certain filament diameter, to a sharp increase of the losses.[9] It was understood that enhanced losses are due to the proximity effect, for which the critical parameter is not the filament diameter, but the spacing between filaments. The critical value for the filament spacing is about 150 nm.[1] If the condition of that minimal distance is fullfilled, however, a drastic reduction of losses below the expected values of the Bean model was found.[1,2]

Takács and Campbell[3] proposed a theory for the case where the filament diameter is smaller than the pinning penetration depth and a background field is present. According to these authors, the effect of reduced hysteresis losses appears below a critical field amplitude, which increases as the filament size is reduced. This critical amplitude is

$$\Delta B = 2 a_o B_o / r$$

where a_0 is the vortex spacing. For $B_0 = 1$ T, a_0 is 49 nm. This gives critical amplitudes of 0.2 and 0.56 T for samples F and G, respectively. The data of Fig. 2 show that major deviations from the Bean model occur indeed approximately below these ΔB-values. The calculation of Takács and Campbell assumes that the effect of pinning forces on the amplitude of flux line oscillation is small. A correction factor with respect to the Bean model is calculated using a parameter

$$c = \frac{1}{2} \Delta B r / (B_o \cdot d), \tag{2}$$

which physical significance is the ratio of vortex displacement amplitude to vortex spacing.

A problem in this theory is the parameter d, the "interaction distance", which can only be deduced from measurements of the force-displacement curve of the vortex lattice. A further complication is, that the parameter d is not a constant, but depends on the background field B_O, and on the pinning mechanism. Campbell measured values around 2.5 nm for a PbBi sample.[10] Since this value is not obviously the same for NbTi, we should consider the parameter d rather as a fit parameter, as long as no measured values are available.

The calculation of the correction factor, k, yields, according to [3]

$$k = 1 + (6/c^2)(1 - e^{-c}) - (2/c)(2 + e^{-c}).$$ (3)

The hysteresis losses expected from the Bean model (Eq. 1), must be multiplied with this factor. Eq. (3) is valid for a slab model, but according to Takacs and Campbell, the more complicated numerical calculation for a cylindrical geometry gives only a little difference between the correction factor for a cylinder and for a slab.

A calculation using d = 2.5 nm and the nominal filament radius (Table 1) was done for the $B_0 = 1$ T data (solid lines in Fig. 2). For sample E, the correction makes only a small change from the Bean model, but the corrected curve comes closer to the measured values.

For samples F and G, the calculation describes reasonably well the sharp decrease of the losses with decreasing ΔB, and the large deviation from the Bean model. For high ΔB values, where k approaches unity, the deviation between calculation and experiment can again be attributed to the difference between nominal and effective values of $r \cdot j_c$. Playing with the parameters r, j_c and d could improve the fit (a larger r would require a higher d in order to fit the data in the low ΔB range). It seemed however not useful to simulate, for a few measuring curves, a perfect agreement with the theory, which may perhaps not hold for other data.

The solid lines in Fig. 4 give the calculated dependence of H on the background field, using again d = 2.5 nm. The calculation for samples F and G gives, at least, values of the right order of magnitude, but the agreement with the measured dependence on B_0 is not perfect for sample F and very poor for sample G. It should be taken in mind, however, that the parameter d can vary with the background field. A dependence of $d \propto B_0^{-1}$, e.g., would remove most of the discrepancy between calculation and experiment of sample G. There is still some work to do on the theoretical side, before a better understanding of the hysteresis losses in very fine filaments is achieved.

ACKNOWLEDGMENT

The author greatfully acknowledges the help of E. Specht for sample preparation.

REFERENCES

1. F. Sumiyoshi et al., Jap. J. Appl. Phys., 25, L 148 (1986).
2. J. R. Cave, A. Fevrier, T. Verhaege, IEEE Trans. Magn. 25, 1945 (1989).
3. S. Takacs and A.M. Campbell, Supercond. Sci. Technol. 1, 53 (1988)
4. K. Kuroda, Cryogenics 26, 566 (1986).
5. C. Schmidt, Cryogenics 25, 492 (1985).
6. C. Schmidt, "Ac loss measurements on superconductors in the microwatt range", submitted to Rev. Sci. Instrum.
7. J. Lühning, Primärbericht No. 03.01.01.P46C, Kernforschungszentrum Karlsruhe (internal note, 1986). An almost similar expression was given by M.N. Wilson, "Superconducting Magnets", Clarendon Press, Oxford (1983) for $\Delta B > \Delta B_p$. For $\Delta B < \Delta B_p$ Wilson gives the result of a numerical calculation which is within ~ 7 % agreement with Eq. (1).

8. A. J. M. Roovers et al., this conference, paper AY6
9. J. R. Cave, A. Février, H.G. Ky and Y. Laumond, IEEE Trans. Magn, 23, 1732 (1987).
10. A. M. Campbell, J. Phys. C: Solid St. Phys., 4, 3186 (1971).

MAGNETIC CHARACTERISTICS AND MEASUREMENTS OF

FILAMENTARY Nb-Ti WIRE FOR THE SUPERCONDUCTING SUPER COLLIDER

R. B. Goldfarb and R. L. Spomer

Electromagnetic Technology Division
National Institute of Standards and Technology
Boulder, Colorado 80303

ABSTRACT

In synchrotron accelerator applications, such as the Superconducting Super Collider (SSC), superconducting magnets are cycled in magnetic field. Desirable properties of the magnets include field uniformity, field stability with time, small residual field, and fairly small energy losses upon cycling. This paper discusses potential sources of problems in achieving these goals, describes important magnetic characteristics to be considered, and reviews measurement techniques for magnetic evaluation of candidate SSC wires. Instrumentation that might be practical for use in a wire-fabrication environment is described. We report on magnetic measurements of prototype SSC wires and cables and speculate on causes for instability in multipole fields of dipole magnets constructed with such cables.

INTRODUCTION

A typical field cycle for the proposed Superconducting Super Collider (SSC) consists of an initial charge to a full field of 6.6 T, reduction of field to 50 mT, increase to 0.33 T for proton injection, and a slow increase of field to 6.6 T as the proton beam is accelerated.[1] The multifilamentary Nb-Ti cables used in the construction of the superconducting dipole magnets are themselves exposed to the field of adjacent windings, usually approximated as a transverse field. Electromagnetic characteristics of the wires and cables are potential sources of difficulty in meeting magnet specifications. Thus, requirements for the magnet imply design and performance specifications for the wires and cables. In this paper we discuss magnetic parameters useful for evaluating multifilar Nb-Ti superconductor wire and cable for the SSC.

MAGNETIC CHARACTERISTICS

Superconductors under steady-state conditions are lossless except for losses associated with thermally activated flux creep. With transient or ac currents and fields, there are several sources of energy dissipation. These ac losses may be classified according to their mechanism and localization within a wire composed of fine superconducting filaments in a normal-metal matrix.

Advances in Cryogenic Engineering (Materials), Vol. 36
Edited by R. P. Reed and F. R. Fickett
Plenum Press, New York, 1990

215

In this section we describe ac loss effects that may be detected with magnetic instrumentation. We discuss both time-independent and time-dependent phenomena for the wire and cable, not the dipole magnets made with these elements. Ideally, all of these ac losses should be minimized, subject to the often conflicting requirements of high critical current and stability against propagation of a normal zone (quench).

Time-Independent Effects

Hysteresis. Magnetic hysteresis upon field cycling is a major loss mechanism. Hysteresis loss per field cycle is frequency independent. It arises in type-II superconductors from irreversibility of the penetration of flux vortices and shielding currents resulting from flux pinning in the filament volume and at the filament surface. The energy dissipation *per se* might be viewed as a trivial problem in applications where field is only occasionally cycled. However, large hysteresis leads to large remanent magnetization from trapped flux in the filaments as the applied field is reduced to zero. This remanent magnetization is the source of residual field in a superconducting magnet. Even at low fields, such as the 0.33-T SSC injection field, trapped flux acts as an offset to the field expected from the magnet current. When it is predictable, the residual field may be compensated. Field uniformity is usually achieved by strategic magnet design. However, remanent magnetization due to trapped flux in the superconductor wires makes it difficult to obtain uniform fields at low currents.

Because the ability of a superconductor to pin flux is an essential requirement for high critical current, both small remanent magnetization and small hysteresis may be achieved, not by reducing flux pinning, but by reducing filament diameter, as predicted by the critical state model.[2] Hysteresis loss is generally higher for wire carrying transport current than for an open-circuited sample, with the extra energy provided by the current source, not the field.[3] Hysteresis loss is determined as the enclosed area in a plot of magnetization vs. field. Measurements often are made with field cycled from positive to negative values. The SSC cycle is such that fields are always positive. In typical multifilamentary wires measured in transverse field, positive-field hysteresis loops, with maximum applied fields of 1 T, have about 45% of the loss associated with complete hysteresis loops.

Self-field of transport current. When transport current changes, moving self-field lines dissipate energy.[4] The use of small-diameter wires reduces these virtually hysteretic losses. In fine-filament wires, the self-field loss may be greater than the magnetic hysteresis loss. In accelerator applications, cabling with fully transposed strands reduces self-field loss; simple twisted strands would still have a large self-field. Three methods of transposition are twisted rope, woven braid, and flattened twisted cable. The last is planned for the SSC, but it results in mechanical damage to the cable corners with a local reduction in critical current.[5]

Coupling between filaments. There are two time-independent sources of filament coupling. One is simply interfilament contact arising from metallurgical problems in processing. Coupled filaments act as a single filament of large diameter, with its associated problems. The second is interfilamentary coupling by the proximity effect when filament spacing is on the order of the coherence length. This type of coupling is important for wires with closely spaced fine filaments. The addition of impurities, such as Ni or Mn, to the matrix material is often effective in reducing the coupling.[6] In any event, proximity-effect coupling is disrupted when magnetic fields approach the effective critical field of the coupling medium. The losses associated with time-independent filament coupling are hysteretic.

Time-Dependent Effects

Coupling between filaments. An important coupling arises from eddy currents driven by voltages induced by a changing applied field. Coupling loss is caused by the transfer of current between filaments and dissipation within the matrix. This relaxation phenomenon leads to time-varying magnetization and field instability of the magnet. It may be reduced by transposing the filaments, approximated by twisting the wire, during manufacture.[7] This decreases the longitudinal distance over which the transverse coupling currents can flow. Other ways to reduce coupling are by increasing the resistivity of the Cu matrix and by increasing the distance between filaments. The former strategy is consistent with other wire requirements provided the stability of the conductor is not impaired.

Flux creep. Flux creep consists of thermally activated jumps of bundles of flux vortices between pinning sites at constant field. Flux creep causes slow changes in magnetization and, in a superconducting magnet, changes in magnet field. Flux creep is often ignored in strong-pinning superconducting materials. In particular, the critical state model assumes that there is no flux creep.[2] However, flux creep has been found to be a problem in accelerator dipole magnets.[8-10] The activation energy for flux creep is reduced by the Lorentz force of the applied field on vortex currents.[11] Flux flow results as a limiting case when flux vortices are no longer pinned at high fields.

Flux jumps. Flux jumps are sudden unpinning of flux vortices in response to instabilities, temperature increases, and breakdown in shielding currents as the applied field is changed. In wires with insufficient Cu or Cu-alloy stabilizer, flux jumps could lead to a quench. Flux jumps result in sudden drops in magnetization and could result in small changes in the field of a magnet.

Eddy currents. Eddy currents arise in the normal matrix material in response to a field change according to the classical mechanism dependent on the skin depth. The time constant of the eddy currents, a function of resistivity, is short. These eddy currents are differentiated from those that couple filaments, discussed above.

MAGNETIC MEASUREMENTS

The magnetic parameters of interest in evaluating multifilar Nb-Ti wire for the SSC may be obtained from measurements of magnetization as a function of field, time, and transport current. Ideally, measurements on cable samples should be also obtained. For a useful analysis, it is necessary to know the critical current density of the wire at several fields. In addition, the wire should be characterized by filament diameter, filament spacing, filament twist pitch, number of filaments, sample volume, and matrix-to-superconductor volume ratio. Magnetization values are usually reported per unit volume of superconductor or of total composite.

Magnetization in filamentary superconductors is the signal from superconductor shielding currents and other matrix currents discussed above. The magnetization-field cycle, or hysteresis loop, should have a variable cycle time, up to several hours, to extract time-dependent coupling information. The field should be transverse to the wire axis and cycle from zero to positive value. Additional information may be obtained from longitudinal-field measurements. Transport current could be controlled independently of the applied field, though in actual SSC operation the current would be approximately proportional to the field.

Total magnetic loss is the area enclosed by the loop in the magnetization-field plane. In the limit where coupling currents have decayed, the remaining area represents the magnetic hysteresis. The width of the loop at relatively high fields may be used to compute an "effective" filament diameter,[12] according to the Bean model,[13] if the critical current density is known. If coupling currents have not decayed or if filaments are coupled by the proximity effect, a large loop area will result.

Magnetization vs. Field

Several methods of magnetometry might be used for the magnetization measurements: integrated-flux, vibrating-sample (VSM), vibrating-coil (VCM), SQUID, and Hall-probe. We will discuss their advantages and disadvantages. To our knowledge, VCM and Hall-probe magnetometers have not been used for magnetic measurements of superconductors. They may be well suited for this task in a wire-fabrication environment.

Integrated-flux. This method[14,15] is good for measurements on wires carrying transport current. It detects flux jumps and frequency dependence. It requires large samples to increase the signal-to-noise ratio. Integration instrumentation limits the measurement to relatively fast field cycles.

VSM. Vibrating-sample magnetometer measurements are made with the vibration axis longitudinal[16] or transverse[17] to the field. This method is sensitive to small samples. It is a dc measurement when the field is stepped and the signal is allowed to stabilize. However, if synchronous detection is used or if the pick-up coils are well matched, data may be taken while sweeping the field. It is difficult to vibrate samples with current leads attached. We calibrate the pick-up coils with Ni wires, plates, cylinders, or spheres in the same configuration as the superconductor samples.

VCM. A vibrating-coil magnetometer[18,19] would be good for measurements on samples carrying transport current because the sample remains stationary. As with a VSM, the VCM may be used in stepped or swept fields. A wire sample could be formed into a coil, noninductively wound to avoid a magnetic signal from the transport current in the sample. As with a VSM, the field is supplied by an external magnet. The applied field should be uniform to avoid field-induced signals. Calibration would be similar as for a VSM.

SQUID. This method[20,21] is extremely sensitive and precise, suited to small samples. Any current to the sample would disrupt the SQUID circuitry. Field cycles are extremely slow.

Hall-probe. Two calibrated, cryogenic probes are used, one to measure the applied field, the other to measure the flux density at the sample surface.[22-24] The difference is the sample magnetization (after correcting for demagnetizing field, if necessary). The Hall probe could be positioned so the Hall element is parallel to the azimuthal magnetic field from the current, but perpendicular to the magnetic field and the magnetic moment from the superconducting shielding currents. This method would be appropriate for cable samples. Calibration may be achieved as for a VSM.

AC Susceptibility vs. Temperature

AC susceptibility is usually measured as a function of temperature in constant ac field, with or without a dc bias field. Measurements are made with a coaxial mutual-inductance system consisting of a primary excitation field coil, a secondary pick-up coil, and a secondary compensation coil.[25,26] Susceptibility is an excellent tool for determining critical temperatures and proximity-effect coupling in fine-filament superconductors.[27] Low frequencies are used to avoid eddy currents in the normal-metal matrix.

STUDY OF PROTOTYPE SSC WIRES AND CABLES

A disconcerting problem in accelerator magnets is field change over several hours at constant magnet current. This is often expressed as instability in multipole fields.[1,8,9] Possible mechanisms are flux creep and eddy-current coupling between filaments. If the mechanism is flux creep, there are two possibilities. One is flux creep intrinsic to the Nb-Ti superconductor filaments or their surface. The other is flux creep in the proximity-coupled matrix. (The proximity coupling itself is not time dependent.) The presence of proximity-effect coupling in filamentary superconducting wire may be deduced from measurements of magnetization vs. field or magnetic susceptibility vs. temperature.

Proximity-Effect Coupling

In hysteresis loops of magnetization vs. field, proximity coupling causes a magnetization peak centered near zero field.[28,29] (The exact position depends on a demagnetization correction of the field axis.) This peak is different from the peak in the second and fourth quadrants which is seen experimentally and predicted by the Kim model for critical current density.[30] The coupling peak arises from a large effective filament diameter when filaments are coupled at low field. The proximity coupling is destroyed at fields greater than about 0.2 T.

In ac susceptibility measurements, a large coupling peak in the imaginary part may be seen as a function of temperature. This peak represents hysteresis loss when the lower critical field of the proximity-coupled matrix becomes on the order of the measuring field as temperature increases. Thus, the peak temperature is a strong function of measuring field amplitude. We have used this technique to study intergranular coupling in high-temperature superconductors.[31]

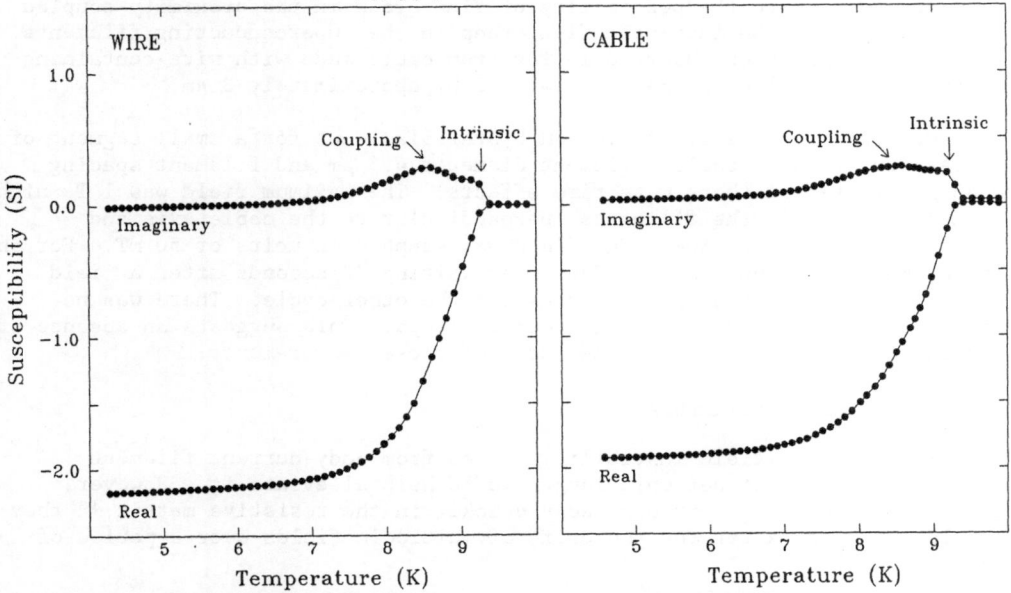

Fig. 1. AC susceptibility (uncorrected for demagnetization factor) vs. temperature measured at 0.1 mT rms at 10 Hz for wire and cable samples. The imaginary part shows intrinsic and coupling loss peaks. The real part shows a broad transition which includes intrinsic and coupling transitions.

Figure 1 shows the real and imaginary parts of external (not corrected for demagnetization factor) ac susceptibility as functions of temperature for samples consisting of small, sawed segments of prototype SSC wire and cable with 0.5% Mn in the Cu matrix. The filament diameter is 5.3 μm and the filament spacing is 0.53 μm. The susceptibility is plotted per unit volume of Nb-Ti. The measuring field was 0.1 mT rms at 10 Hz applied perpendicular to the wire and cable axes. The cable was the type used in the fabrication of the inner layer of dipole magnet D-15A-6 in Ref. 1.

Intrinsic and coupling loss peaks (partially overlapping) appear in the imaginary part of susceptibility. Correction for demagnetization factor would change the apparent shapes of the peaks. For a measuring field of 0.01 mT, the coupling peaks move to higher temperature. For a 1-mT field, the peaks are well separated as the coupling peaks move to lower temperature. Much less pronounced susceptibility peaks were seen for fields applied parallel to the wire and cable axes. But similar coupling peaks in the imaginary part were seen for wire and cable used for the outer layer of dipole magnet D-15A-6. The filaments are coupled at low temperature for low fields despite the Mn doping of the matrix. As expected, magnetization vs. field at 4 K showed characteristic coupling peaks near zero field.

Multipole-field instability occurs at the relatively high fields that would destroy proximity-effect coupling. However, there are portions of a dipole magnet which are exposed to very small fields where coupling is presumably intact. In multifilamentary wires that are proximity coupled, flux creep *may* occur in the interfilamentary matrix. We recently proposed this mechanism to explain subtle intergranular frequency effects in the ac susceptibility of high-temperature superconductors.[32] To inhibit low-field proximity coupling, more Mn could be added to the matrix, or a resistive Cu-Ni matrix could be used.

Intrinsic Flux Creep

In addition to the possibility of flux creep in the proximity-coupled matrix, there may be intrinsic flux creep in the superconducting filaments. Flux creep[10] has been observed in Tevatron cable made with wire containing presumably decoupled filaments separated by approximately 3 μm.

We measured positive-field VSM hysteresis loops for a small segment of D-15A-6 outer-layer cable, filament diameter 4.3 μm and filament spacing 0.43 μm, to see if there were time effects. The maximum field was 1 T, high enough to uncouple the filaments, perpendicular to the cable axis and parallel to the wide side. The field was stepped in units of 50 mT. For one cycle, magnetization was recorded after waiting 30 seconds after a field change. The wait time was 3 minutes for the other cycle. There was no significant difference in the hysteresis loops. This suggests an absence of intrinsic flux creep *on the time scale of these measurements*.

Eddy-Current Filament Coupling

If multipole-field instability arises from eddy-current filament coupling, more twist per unit length would help alleviate it. However, because the coupling currents decay quickly in the resistive matrix,[33] they should not be a factor for the drift of multipole fields over a period of hours.

To check for this time-dependent effect, we measured hysteresis loops, using the two field cycles described above, for two coiled samples of D-15A-6 wire used in the inner- and outer-layer cables. The field was applied along the axis of the coils, approximately transverse to the wire axis. These particular samples had 0.5 and 1.5 twists per centimeter, respectively, verified by etching in nitric acid the matrix of companion samples. Unlike the short cable segment described above, the length of wire for each coil sample was about 25 cm, long enough to contain many twists. There were no differences between long and short field cycles for the two coils. The result suggests that 0.5 twist per centimeter is adequate to inhibit eddy-current filament coupling for these wait times.

CONCLUSIONS

We have discussed several magnetic parameters to be considered in testing multifilar Nb-Ti superconductor for the SSC, with the goal of minimizing field nonuniformity, field instability, large residual fields, and large energy losses. These parameters may be extracted from measurements of magnetization vs. field. Several measurement methods, each with certain advantages, were described.

Measurements of ac susceptibility of candidate SSC wires and cables demonstrate proximity-effect coupling. Flux creep in the proximity-coupled matrix may be a source of time variations of multipole fields in prototype SSC magnets. Intrinsic flux creep was not observed over a period of minutes. Wires used for the inner and outer layers of the D-15A-6 prototype SSC dipole magnet did not show serious eddy-current filament coupling for these sample twist pitches.

ACKNOWLEDGMENTS

We had helpful discussions with C. E. Taylor, R. M. Scanlan, W. S. Gilbert, E. W. Collings, and D. C. Larbalestier. R. J. Loughran assisted with the VSM measurements. Sample wires and cables were generously provided by R. M. Scanlan. This work was supported by the Department of Energy, Division of High Energy Physics.

REFERENCES

1. W. S. Gilbert, R. F. Althaus, P. J. Barale, R. W. Benjegerdes, M. A. Green, M. I. Green, and R. M. Scanlan, Magnetic field decay in model SSC dipoles, *IEEE Trans. Magn.* 25:1459 (1989).
2. Y. B. Kim, C. F. Hempstead, and A. R. Strnad, Magnetization and critical supercurrents, *Phys. Rev.* 129:528 (1963).
3. M. N. Wilson, "Superconducting Magnets," Oxford University Press, Oxford, U.K. (1983), pp. 171-174.
4. *Ibid.*, pp. 139-140, pp. 194-197, p. 308.
5. L. F. Goodrich and S. L. Bray, Current capacity degradation in superconducting cable strands, *IEEE Trans. Magn.* 25:1949 (1989).
6. E. W. Collings, Stabilizer design considerations in fine-filament Cu/NbTi composites, *Adv. Cryo. Engr. (Materials)* 34:867 (1988).
7. G. H. Morgan, Theoretical behavior of twisted muticore superconducting wire in a time-varying uniform magnetic field, *J. Appl. Phys.* 41:3673 (1970).
8. D. A. Herrup, M. J. Syphers, D. E. Johnson, R. P. Johnson, A. V. Tollestrup, R. W. Hanft, B. C. Brown, M. J. Lamm, M. Kuchnir, and A. D. McInturff, Time variations of fields in superconducting magnets and their effects on accelerators, IEEE Trans. Magn. 25:1643.
9. R. W. Hanft, B. C. Brown, D. A. Herrup, M. J. Lamm, A. D. McInturff, and M. J. Syphers, Studies of time dependence of fields in Tevatron superconducting dipole magnets, *IEEE Trans. Magn.* 25:1647.
10. M. Kuchnir and A. V. Tollestrup, Flux creep in a Tevatron cable, *IEEE Trans. Magn.* 25:1839.

11. P. W. Anderson and Y. B. Kim, Hard superconductivity: Theory of the motion of Abrikosov flux lines, *Rev. Mod. Phys.* **36**:39 (1964).

12. S. S. Shen, Magnetic properties of multifilamentary Nb_3Sn composites, *in:* "Filamentary A15 Superconductors," M. Suenaga and A. F. Clark, eds., Plenum, New York (1980), pp. 309-320.

13. C. P. Bean, Magnetization of high-field superconductors, *Rev. Mod. Phys.* **36**:31 (1964).

14. W. A. Fietz, Electronic integration technique for measuring magnetization of hysteretic superconducting materials, *Rev. Sci. Instrum.* **36**:1621 (1965).

15. M. N. Wilson, *op. cit.*, pp. 243-245.

16. P. J. Flanders, Instrumentation for magnetic moment and hysteresis curve measurements, *in:* "Conference on Magnetism and Magnetic Materials," American Institute of Electrical Engineers, New York (1957), **T-91**, pp. 315-317.

17. S. Foner, Versatile and sensitive vibrating-sample magnetometer, *Rev. Sci. Instrum.* **30**:548 (1959).

18. D. O. Smith, Development of a vibrating-coil magnetometer, *Rev. Sci. Instrum.* **27**:261 (1956).

19. K. Dwight, N. Menyuk, and D. Smith, Further development of the vibrating-coil magnetometer, *J. Appl. Phys.* **29**:491 (1958).

20. E. J. Cukauskas, D. A. Vincent, and B. S. Deaver Jr., Magnetic susceptibility measurements using a superconducting magnetometer, *Rev. Sci. Instrum.* **45**:1 (1974).

21. J. A. Good, A variable temperature high sensitivity SQUID magnetometer, *in:* "SQUID: Superconducting Quantum Interference Devices and their Applications," H. D. Hahlbohm and H. Lübbig, eds., Walter de Gruyter, Berlin (1977), pp. 225-238.

22. D. A. Berkowitz and M. A. Schippert, Hall effect B-H loop recorder for thin magnetic films, *J. Sci. Instrum.* **43**:56 (1966).

23. D. J. Craik, The measurement of magnetization using Hall probes, *J. Phys. E: Sci. Instrum.* **1**:1193 (1968).

24. P. J. Flanders, A Hall sensing magnetometer for measuring magnetization, anisotropy, rotational loss and time effects, *IEEE Trans. Magn.* **21**:1584 (1985).

25. W. R. Abel, A. C. Anderson, and J. C. Wheatley, Temperature measurements using small quantities of cerium magnesium nitrate, *Rev. Sci. Instrum.* **35**:444 (1964).

26. R. B. Goldfarb and J. V. Minervini, Calibration of ac susceptometer for cylindrical specimens, *Rev. Sci. Instrum.* **55**:761 (1984).

27. J. R. Cave, A. Février, H. G. Ky, and Y. Laumond, Calculation of ac losses in ultra fine filamentary NbTi wires, *IEEE Trans. Magn.* **23**:1732 (1987).

28. A. K. Ghosh, W. B. Sampson, E. Gregory, and T. S. Kreilick, Anomalous low field magnetization in fine filament NbTi conductors, *IEEE Trans. Magn.* **23**:1724 (1987).

29. E. W. Collings, K. R. Marken Jr., M. D. Sumption, R. B. Goldfarb, and R. J. Loughran, AC loss measurements of two multifilamentary NbTi composite strands, paper AY-05, this conference.

30. D.-X. Chen and R. B. Goldfarb, Kim model for magnetization of type-II superconductors, *J. Appl. Phys.* **66**:2510 (1989).

31. R. B. Goldfarb, A. F. Clark, A. I. Braginski, and A. J. Panson, Evidence for two superconducting components in oxygen-annealed single-phase Y-Ba-Cu-O, *Cryogenics* **27**:475 (1987).

32. M. Nikolo and R. B. Goldfarb, Flux creep and activation energies at the grain boundaries of Y-Ba-Cu-O superconductors, *Phys. Rev. B* **39**:6615 (1989).

33. M. N. Wilson, *op. cit.*, pp. 176-181.

THE EFFECT OF FLUX CREEP ON THE MAGNETIZATION FIELD IN THE SSC

DIPOLE MAGNETS*

W.S. Gilbert, R. F. Althaus, P. J. Barale, R. W. Benjegerdes,
M. A. Green, M. I. Green, and R. M. Scanlan

Lawrence Berkeley Laboratory
1 Cyclotron Road
Berkeley, CA 94720

ABSTRACT

The sextupole fields of model SSC dipole magnets have been observed to change with time when the magnets are held at constant current under conditions similar to injection into the SSC accelerator. The changes in the sextupole component have close to a linear log time dependence, and is felt to be caused by flux creep decay of the magnetization currents in the superconductor filaments. Measurements of this decay have been made under various conditions. The conditions include various central field inductions and changes of field prior to when the decay was measured. The measured field decay in the dipole's sextupole is proportional to the magnitude and sign of the sextupole due to magnetization which was measured at the start of the decay. This suggests that the decay is a bulk superconductivity flux creep. Proximity coupling appears to play only a minor role in the flux creep according to recent LBL measurements with a stable power supply.

INTRODUCTION

At the 1988 Applied Superconductivity Conference, we presented data on the decay of magnetic field harmonics, at injection, of four model SSC dipole magnets.[1] One of these magnets, D15A-4F, had the power supply drifting at 5A/hr. during the decay. We have since shown that this drift causes an error in the measured field decay and that magnet has been remeasured with a low drift power supply. An additional five magnets have been measured in the past year and this report includes data on the sextupole field decay for all nine magnets. (The 12-pole field decay in the tested quadrupole). The other multipoles will be included in a more comprehensive LBL report.[2]

These measurements are extremely sensitive to details of set up cycles, ramp rates, power supply overshoot and stability. Some of these details are included here and others will be in the more complete report. The measurements with model dipoles are so time consuming and subject to unavoidable small variations in power supply repeatability that small changes in field decay in different magnets wound from superconductors of different designs are likely to be masked. Large variations in field decay were not observed.

---

*This work is supported by the Office of Energy Research, Office of High Energy and Nuclear Physics, High Energy Physics Division, Dept. of Energy under Contract No. DE-AC03-76SF00098.

Advances in Cryogenic Engineering (Materials), Vol. 36
Edited by R. P. Reed and F. R. Fickett
Plenum Press, New York, 1990

223

The association of the magnetization field decay with bulk flux creep is most strongly suggested by the linear log time behavior, but we are not aware of any theory that predicts this decay from first principles. Through the use of composite billets with different filament-matrix geometry, we have some data on the behavior of the composite decay with and without proximity coupling.

THE MAGNETIZATION PROBLEM IN SSC OPERATION

Fig. 1 shows the sextupole field at the reference radius of 1 cm as a function of magnet excitation. The current is ramped at approximately 6A/s from some low current, say 50A, to the injection current of 320A (approx. 0.33 tesla). The current is held constant from one to three hours while protons are injected into the two main rings. Then the ramp is resumed until the operating field is reached at 6600A. The stored beams interact for about a day, at which time the current is ramped down to near zero, the beams are dumped and the entire process is repeated. The reason the sextupole field changes with magnet current is the presence of magnetization currents in the superconducting filaments; otherwise the field shape would be constant and determined only by the transport currents flowing in the magnet coils. The observed slow decay of the magnetization sextupole can result in beam loss during the extended injection period. When ramping is resumed, the sextupole suddenly regains its pre-decay value, resulting in rapid beam loss.

Powered correction elements can correct for the magnetization sextupole if it is accurately known. The time decay of the field complicates this problem and if different magnets made from different superconductors were to have fields decay at different rates, the problem would be even more difficult. One of our goals was to see if there were differences in field decay for different conductors designs.

EFFECT OF POWER SUPPLY OVERSHOOT ON DECAY

When the current ramp is smoothly stopped at 320A, we get the sextupole decay curve shown in Fig. 2. The linear log time relationship indicates a flux creep behavior.[3] Current

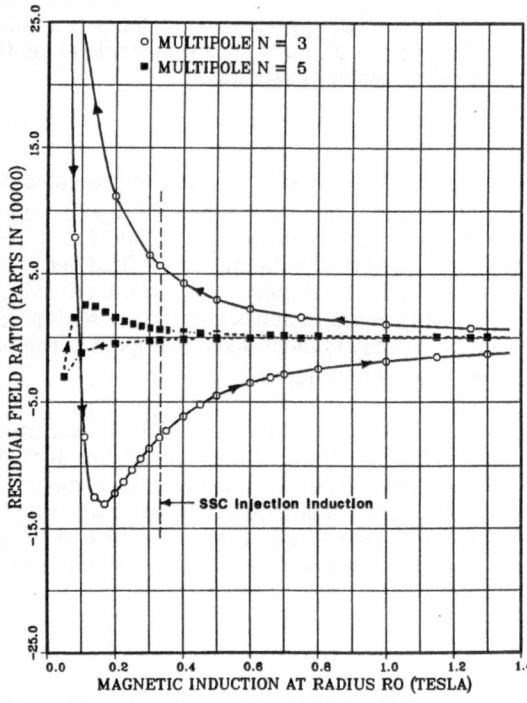

Fig. 1   The ratio of magnetization sextupole and decapole to
the transport current dipole as a function of dipole central induction.

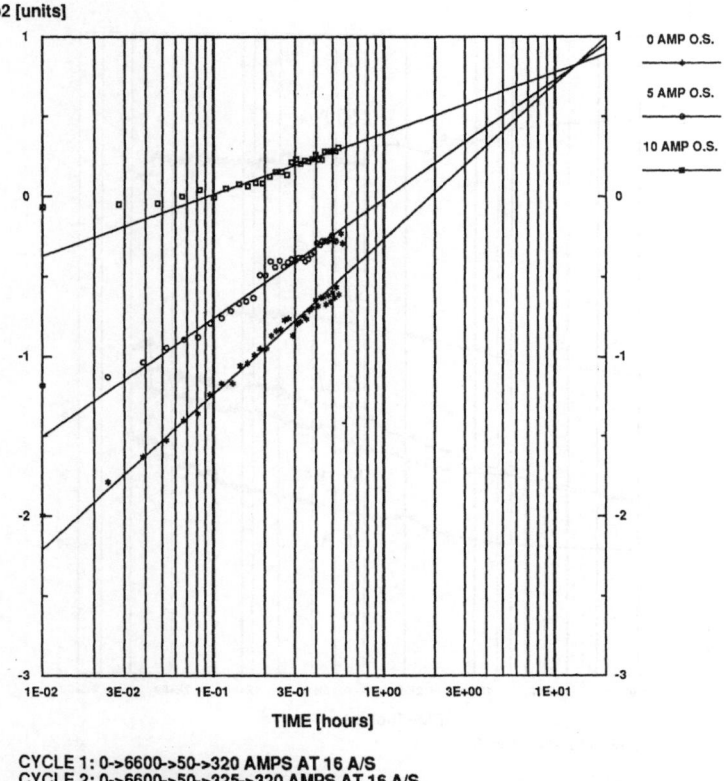

b2 [units]

CYCLE 1: 0->6600->50->320 AMPS AT 16 A/S
CYCLE 2: 0->6600->50->325->320 AMPS AT 16 A/S
CYCLE 3: 0->6600->50->330->320 AMPS AT 16 A/S

Fig. 2   LBL 1 Meter Model Magnet D-15A-5R3 320 A Decay @ 4.3K with overshoot.

overshoot was simulated in other runs by allowing the ramps to proceed to either 325A or 330A and then decreasing the currents to 320A before the decay data were taken. One can see that the overshoot reduced the initial sextupole fields and the subsequent decay rates.

EFFECT OF RAMP RATE ON DECAY

In Fig. 3 are shown the sextupole field decay curves at 320A for excitation ramp rates of 160, 50, 16, 6.6, and 1.6 A/s. The excitation cycle is from 0 to 6600A, 6600 to 50A, and 50 to 320A, which is then maintained for the one to three hour decay. Fig. 1 shows that the equilibrium sextupole field, in going from 50A to 320A, goes from more than positive 25 units (a unit is 10^{-4} of the dipole field) to a negative 7 units, going through a minimum of negative 12 units at 150A. It is clear that the magnetization currents take tens of seconds to stabilize. This could be a measure of the field diffusion time or an inward flux creep. Most of our data have been taken with a ramp rate of 16 A/s and the decay data are close to those taken at the projected SSC ramp rate of 6.6 A/s.

TEMPERATURE EFFECT ON DECAY

Fig. 4 shows sextupole decay at 4.3K and 1.8K for magnet D15A-5R2. The greater magnetization sextupole at injection field is expected as the conductor J_c is greater at the lower temperature. The 1.8K decay seems to be slightly slower. Similar data for magnet D15C-1 appear in Fig. 5. Here the 1.8K decay seems to be considerably slower than at 4.3K.

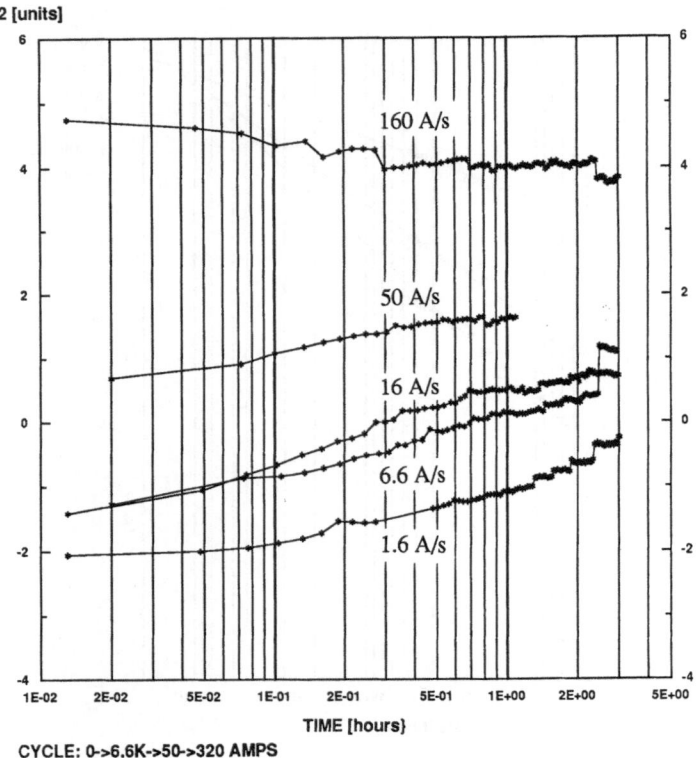

Fig. 3 D-15A-5R2 - Cold Measurements, 320 amp decay @ 4.3K various ramp rates.

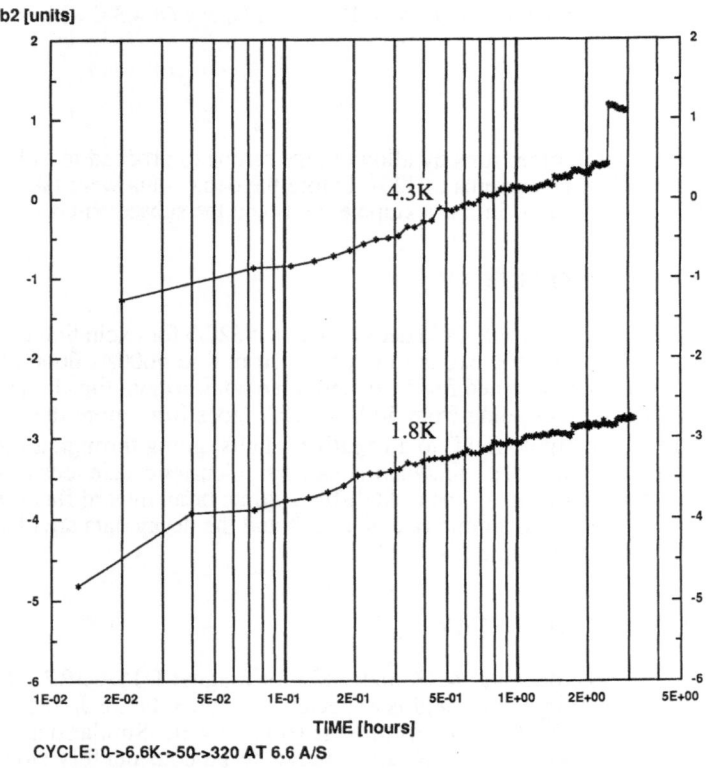

Fig. 4 D-15A-5R2 - Cold Measurements 320 amp decay @ 1.8K vs 4.3K comparison.

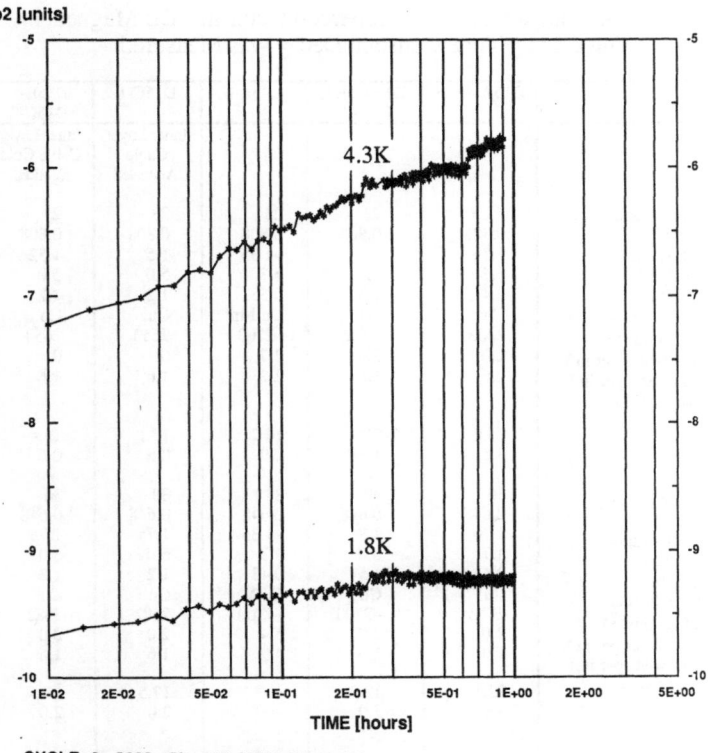

b2 [units]

TIME [hours]

CYCLE: 0->6600->50->320 AMPS AT 16 A/S

Fig. 5   D-15C-1 Cold Measurements 320 amp decay, 1.8K vs 4.3K comparison.

REPRODUCIBILITY OF DECAY RATES FOR SAME MAGNET

Magnet D15A-5R2 had six sextupole field decays measured under similar set up conditions. For each decay a straight line slope was fitted to the roughly linear log time data. The early, less that 0.1 hour, data usually lie above the fitted slope (slower decay). Often a sudden jump is observed. For these decays, the slopes yield 1.2 ± 0.1 units/decade. This spread is due not only to the data not lying on a perfectly straight line, but on different set up cycles. The 10,000 A power supply sometimes has a variation of a few amperes at the 50A turnaround and at the 320A levels. One wouldn't expect the reproducibility of the set up conditions to be any better for other magnet tests which are at least one month apart. Therefore, our present accuracy on sextupole decay slope is roughly ± 0.1 units/decade. Magnet D-15A-5 was run three different times in a five month period with slightly different pole shims. This should not influence the magnetization effects. The three different configurations are referred to as R1, R2, and R3. All the decay measurements compared below had the same set up cycles at a ramp rate of 16 A/s.

R1 has one decay with slope = 1.20 units/decade
R2 has six decays with slope = 1.18 ± 0.07"
R3 has three decays with slope = 0.98 ± 0.09"

PROXIMITY COUPLING - SMALL FILAMENT SPACING

Proximity coupling, which effectively increases the magnetization by coupling small diameter filaments together, occurs when the filament spacing is too small, less than 1μm. The conductor in magnet D-15A-4FR1 (see Table 1) has a spacing of only 0.4 μm, and has been measured to have a large magnetization at 0.3 tesla.4 The decay data, tentative at this time, show a sextupole decay rate of 1.00 unit/decade, which is about the same as that for other conductors. There is some evidence that the proximity coupling portion of the magnetization decays faster than the bulk property flux creep.

227

Table 1 - A Comparison of the Superconductor in LBL Magnets in
Which Long Time Constant Decay was Measured

Magnet--->	D15A-4FR1	D15A-5R1, R2, R3	D15A-6	D15C-1 Inner Layer Cable Annealed	D15B-1 D15C-2 Inner Layer Cable Cold Worked	Quadrupole QA-1R1
Inner Layer						
Number of Strands in Cable	23	23	23	23	23	30
Strand Diameter (mm)	0.808	0.808	0.808	0.808	0.808	0.648
Normal Metal to S/C Ratio	1.26	1.3	~1.35	1.52	1.52	1.69
Filament Diameter (μm)	4.7	6.0	5.3	5.0	5.0	5.0
Filament Spacing (μm)	0.4*	1.5	0.53	1.2	1.2	1.2
Material Between Filaments	Cu*	Cu	Cu-Mn**	Cu	Cu	Cu
J_c at 5 T and 4.2K (A mm^{-2})	2600	~2700	~2700	2650	2650	2743
Strand Twist Pitch (twists per in.)	2.0	2.0	2.7	0	0	2.0
Cable Twist Pitch (twists per in.)	2.0	1.6	2.2	1.6	1.6	1.6
Magnetization data (H=0.3T)						
Inner Layer						
2M (mT)	25.6	21.4	16.2	17.1	15.4	15.6
2Me (mT)	3.6	1.3	0.9	0.8	0.6	1.6
Outer Layer						
Number of Strands in Cable	30	30	30	30	30	30
Strand Diameter (mm)	0.648	0.648	0.648	0.648	0.648	0.648
Normal Metal to S/C Ratio	1.76	1.8	~1.35	1.75	1.75	1.69
Filament Diameter (μm)	4.7	6.0	4.3	6.0	6.0	5.0
Filament Spacing (μm)	0.4*	1.5	0.43	1.2	1.2	1.2
Material Between Fialments	Cu*	Cu	Cu-Mn**	Cu	Cu	Cu
J_c at 5 T and 4.2K (A mm^{-2})	2618	~2700	~2700	2582	2582	2743
Strand Twist Pitch (twists per in.)	2.0	2.0	5.4	2.0	2.0	2.0
Cable Twist Pitch (twists per in.)	2.0	1.6	4.9	1.6	1.6	1.6
Outer Layer Magnetization						
2M (mT)	22.3	19.2	-	17.5	17.5	15.6
2Me (mT)	3.1	2.0	-	2.0	2.0	1.6
2 M ≡ magnetization between upramp & downramp current sweeps						
2Me ≡ Excess magnetization due to eddy currents						

* & ** from page 2 of LBL-25139

Magnet D-15A-6 also has conductor with small filament spacing, 0.53 μm, but the normal copper is doped with Mn and doesn't show any measured increase in its 0.3 tesla magnetization.

DECAY RATES - DIFFERENT MAGNETS; DIFFERENT CONDUCTORS

In Table 1 are listed the conductor details for the various magnets in which field decay, at injection energy, was measured.

In Table 2 are listed the slopes of the various sextupole decays for similar set up cycles and ramp rate of 16 A/s. As discussed above, the data has enough scatter that one can't attribute the small differences in magnet decays to the conductor designs.

It is worth noting that a dipole magnet has conductor at various magnetic fields and effectively integrates the different magnetization cycles over the entire volume. Laboratory magnetization experiments on conductor at a single field possibly could more precisely probe the differences in field decay associated with different conductor designs.

QUADRUPOLE FIELD DECAY

QA-1R1 is a model SSC quadrupole built at LBL. The 12 pole magnetic field harmonic, called b5, is analgous to the sextupole field in the case of the dipoles already cited. The same set up cycle was used for the quadrupole and the decay of the 12 pole field is shown in Fig. 6. The magnetization offset at injection and the rate of decay are both about double that for the case of the dipoles.

Table 2 - b_2 decay @ 320 A, 4.3K

Magnet No.	Decay Slope (units/decade)
D-15A-5R1	1.20
D-15A-5R2	1.18 ± 0.07
D-15A-5R3	0.98 ± 0.09
D-15A-4FR1	1.00 (tentative)
D-15A-6	1.17
D-15B-1	0.85
D15C-1	0.80
QA-1R1 (Quad)	2.7 (b5)

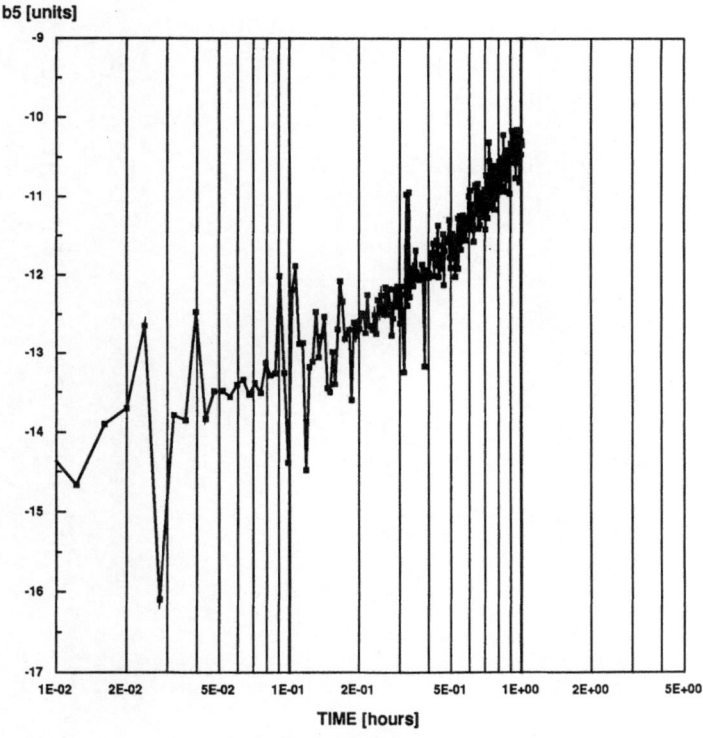

CYCLE: 3X(50->6600->50)->320 AMPS AT 16 A/S

Fig. 6   QA-1R1 - Cold Measurements - TBL 35, 320 amp decay @ 4.3K

CONCLUSIONS

The decay of magnetization currents as observed in the LBL-SSC model dipoles is roughly a linear log time relationship, suggesting a flux creep lasting over several hours. We have also measured a surprisingly long time to stablize these fields, some tens of seconds. The decay seems to be a bulk property effect and not particularly sensitive to details of conductor design.

REFERENCES

1. W. S. Gilbert, R. F. Althaus, P. J. Barale, R. W. Benjegerdes, M. A. Green, M. I. Green, and R. M. Scanlan, "Magnetic Field Decay in Model SSC Dipoles", paper

presented at the 1988 Applied Superconductivity Conference, San Francisco, CA, August 21-25, 1988, LBL-25139.

2. W. S. Gilbert, R. F. Althaus, P. J. Barale, R. W. Benjegerdes, M. A. Green, M. I. Green, and R. M. Scanlan, "The Effect of Flux Creep on the Magnetization Field in the SSC Dipole Magnets - Expanded Version", paper to be published, LBL-27488, SSC-MAG-246, August, 1989.
3. M. R. Beasley, et al, Physical Review 181, pp. 682-700, May, 1969.
4. A. Ghosh, BNL, private communication.

MAGNETIC STUDIES OF PROXIMITY-EFFECT COUPLING IN A VERY CLOSELY SPACED FINE-FILAMENT NbTi/CuMn COMPOSITE SUPERCONDUCTOR

E. W. Collings*, K. R. Marken Jr*., M. D. Sumption*†, E. Gregory**††, and T. S. Kreilick**

* Battelle Memorial Institute, Columbus, OH 43201
† Permanent address: Ohio University, Athens, OH 45701
** Supercon, Inc., Shrewsbury, MA 01545
†† Present address: IGC Advanced Superconductors, Waterbury, CT 06704

ABSTRACT

Magnetization studies have been conducted on a 23,000-filament composite (with a filament-spacing/filament-diameter ratio, s/d, of about 0.19) drawn down to d = 11.5 to 0.5 μm. Various techniques have been used to explore the occurrence and properties of proximity-effect coupling between the filaments across the Cu-0.5wt.%Mn matrix. This coupling, which sets in at d < 1.5 μm -- much smaller than 2.5 μm intended for superconducting supercollider (SSC) magnet applications -- is studied both at low fields (well below the H_{c1} of the NbTi) and at high fields (of up to 1.5 tesla (15 kG)).

INTRODUCTION

When a superconductor is subjected to a time-varying external magnetic field the "height" of the M(H) hysteresis loop is proportional to the critical current density, J_c, and the thickness, d, of the superconductor. Thus in order to minimize the residual magnetization retained in a partially de-energized coil, such as an SSC magnet following de-excitation from an operating field of 6.6 tesla to a beam-injection field of 0.3 tesla, it has been recommended that the filaments of the cable strands should be made as small as possible. The preservation of a high J_c under these conditions dictates the use of a filament-spacing/filament-diameter ratio, s/d, of some 0.13-0.17. A consequence of this is that filaments less than 3-4 μm in diameter are coupled by proximity effect (even at 0.3 tesla) which contributes an unwanted excess magnetization [1,2,3] to at least a portion of the M(H) hysteresis loop. Interfilamentary coupling can be suppressed by alloying the matrix with a high concentration of Ni (e.g. 30 wt.%) or, preferably, a low concentration of Mn (e.g. 0.5 wt.%)[1,2,3,4]. For the multifilamentary strands prepared for this study, an interfilamentary alloy of Cu-0.5wt.% Mn was used. As a result, filaments as small as d = 2.5 μm (hence s = 0.5 μm, at the design s/d of 0.19 in this case[3]) were successfully decoupled.

The purpose of the study was: (a) to determine for Cu-0.5wt.%Mn the coupling-threshold value of d (hence s); (b) to examine the manner in which coupling magnetization manifested itself both in the low-field regime below the H_{c1} of NbTi as well as in the high-field regime up to about 15 kG.

Advances in Cryogenic Engineering (Materials), Vol. 36
Edited by R. P. Reed and F. R. Fickett
Plenum Press, New York, 1990

231

Table 1. Specifications of 23,000-Filament Strands

Sample Code	Strand Diam. (nom.), mils	Strand Diam. (actual), μm	Avg. Fil. Diam.* μm	Cu/SC Ratio†	A_{tot}/A_{SC}
CMN-115	119	2998.7	11.556	1.941	2.940
CMN-25	25	635.0	2.459	1.912	2.912
CMN-21	21	529.4	2.051	1.909	2.909
CMN-15	15	385.0	1.495	1.897	2.897
CMN-10	10	274.5	1.086	1.885	2.885
CMN-5	5	127.2	0.4995	1.831	2.831

* Obtained by etching using measured density of bulk Nb-46.5Ti ($= 6.097$).
† Obtained from measured strand and filament diameters.

EXPERIMENTAL

Magnetization Measurements

Magnetization was measured as function of temperature up to the T_c of NbTi with field sweep amplitudes of from a few tens of gauss up to 15 kgauss on cylindrical bundles of multifilamentary strand. Measurements were taken with the applied field transverse to the sample axis. A computerized PAR-EG&G vibrating-sample magnetometer (VSM) was used, in association with a 17-kG iron-core electromagnet powered by a $\pm$ 65 A field-controlled bipolar power supply. In completing a full hysteresis loop, including the initial branch from the origin, the instrument records 1,023 data pairs. Thus the field resolution in any experiment is about $1/200^{th}$ of the field-sweep amplitude, which enables all fine structure associated with coupling magnetization to be fully recorded.

Sample Material

Samples were prepared from high-homogeneity Nb-46.5wt%Ti rods clad with a thin barrier-layer of Nb (whose presence was ignored in the data analyses) and enough Cu-0.5wt.%Mn to provide an s/d ratio of 0.19 [5]. The strand design called for an annulus of 22,902 filaments encased in Cu and surrounding a Cu core for a total matrix/superconductor ratio of about 1.9. Photomicrographs illustrating this configuration have been presented elsewhere[3] (see also Table 2). The strands under study, whose filament diameters ranged from 11.5 μm (CMN-115) down to 0.5 μm (CMN-5), had not been twisted or heat treated (except for a final 4h/225°C anneal). Some of their specifications are listed in Table 1. Some further critical dimensions of CMN-5 are listed in Table 2.

Magnetometer-Sample Preparation

The samples consisted of cylindrical bundles, about 3 mm in diameter and 6 mm in length, of parallel multifilamentary strands imbedded in epoxy. Depending on the strand diameter, the number of strands in the bundle varied from about 15 (2.5 μm filaments) to 200 (0.5 μm filaments) so as to keep the volume of superconductor roughly constant at about 0.01 cm³. Specifications of the magnetometer samples, including that of a sample prepared from unclad CMN-5 (i.e. CMN-5B, in which the Cu and Cu-Mn matrices had been removed by etching), are given in Table 3.

Table 2. Configuration of CMN-5

D_0	D_B	D_{IB} (μm)
128.0	98.04	39.62
126.5	101.35	39.62
125.5	98.04	38.10
125.0	97.79	36.58
126.3	98.81	38.48

Length of magnetization sample = 0.588 cm
Tot. Bundle Volume (incl. core) = 4.509×10^{-5} cm^3.
Tot. Bundle Vol./(n x fil. vol.) = 1.709

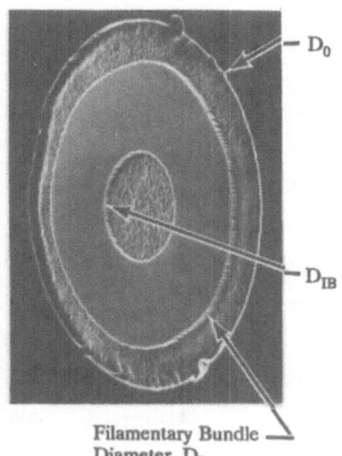

Filamentary Bundle
Diameter, D_B

Table 3. Magnetization Sample Specifications

Sample Code	Fil. Diam., μm	Number of Strands	Sample Length, mm	NbTi Filament Volume, 10^{-3}cm^3
CMN-115	11.556	1	5.892	14.15
CMN-25	2.459	15	5.64	9.201
CMN-21	2.051	21	6.15	9.772
CMN-15	1.495	33	5.60	7.429
CMN-10	1.068	58	5.736	6.826
CMN-5	0.4995	200	5.88	5.278
CMN-5B*	0.4995	200	5.734	5.147

* All the matrix material removed by etching -- 4.5804×10^6 independent filaments.

LOW-FIELD MAGNETIZATION

Provided that the field-sweep amplitude remains below the H_{c1} of NbTi, the magnetic susceptibility of a cylindrical composite or its filamentary components in a transverse magnetic field, is given by the Meissner (flux-exclusion) value: $\chi = dM/dH = (-1/2\pi)P$. Here P, the "flux-exclusion" volume fraction, is given by: $P = 1-(2/x)I_1(x)/I_0(x)$, where x represents the ratio of filament radius ($R = d/2$) to the penetration depth, (λ_L), and I_0 and I_1 are modified Bessel functions of the first kind of order 0 and 1, respectively (see Ref. 6 and the companion paper in this proceedings[7]). For large-diameter cylinders, $P = 1$; on the other hand significant temperature-dependent departures from unity are noted for very fine filaments when the radius becomes comparable to λ_L.

The low-field magnetization loops for the entire series of clad multifilamentary specimens are presented in Fig. 1. Two features are immediately noted: (a) a decrease in overall dM/dH slope in going from CMN-115 to CMN-5; (b) for CMN-10, a slight opening of the loop near the origin, and for CMN-5, a pronounced hysteresis. The susceptibility decrease is due to the increasing influence of field penetration as R/λ_L decreases with decreasing R (see Ref. 7 for a full account of this effect). The first appearance of magnetic hysteresis with decreasing R (hence decreasing s) signifies the onset of coupling.

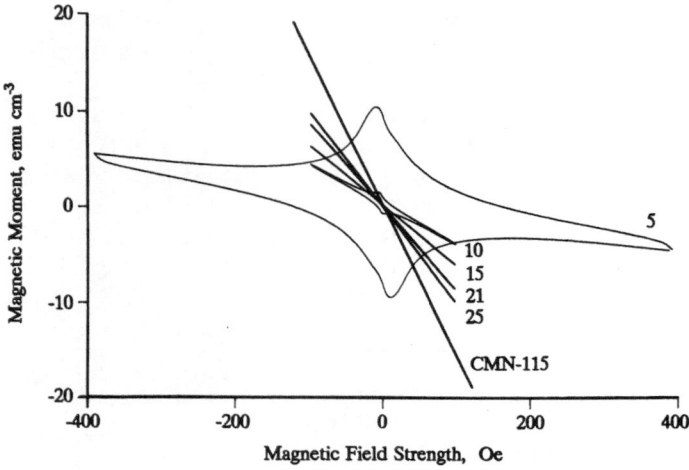

Fig. 1 Low-field ($<H_{c1}$) hysteresis loops for samples CMN-115 to CMN-5 at 4.2 K.

The initial magnetization for CMN-5, throughout the superconducting temperature range above 4.2 K, is shown in Fig. 2. The slope of the initial branch is -1.26×10^{-3} emu/G which, normalized to the exclusion-volume of the 200 filamentary bundles each of diameter 9.88×10^{-3} cm (see Table 2) that make up the sample, yields a volume susceptibility of -0.140. Comparing this with the ideal value of $-1/2\pi = -0.160$ for a cylinder in a transverse field, we note that in fields of up to about 8 or 9 G the filamentary bundle excludes (or screens out due to proximity-effect-permitted circulating supercurrents) about 88% of the applied field.

In CMN-5 at 4.2 K, departures from dM/dH linearity begin to occur as the applied field exceeds about 9 G. Above that field, flux begins to penetrate the bundle just as if it were a type-2 superconductor with a lower critical field, "H_{c1B}", of about 9 G. From the intercept of a plot of H_{c1B} versus temperature a T_c of 8.95 K for the fine-filamentary NbTi is deduced (a value agrees well with the 8.93 K obtained from the susceptibility temperature dependence of the bare filaments[7]). A plot of H_{c1B} versus t^2, where t = T/T_c, the reduced temperature, is presented in the inset to Fig. 2. We note that H_{c1B} has a quadratic temperature dependence and a zero-K value of 10.5 G.

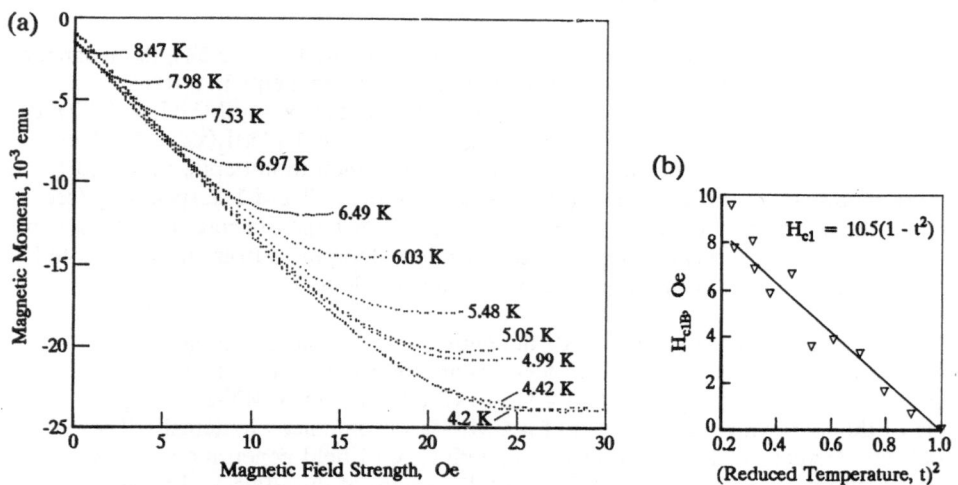

Fig. 2 Temperature dependence of (a) the initial magnetization of CMN-5; (b) the lower critical field, H_{c1B}, of the filamentary bundle.

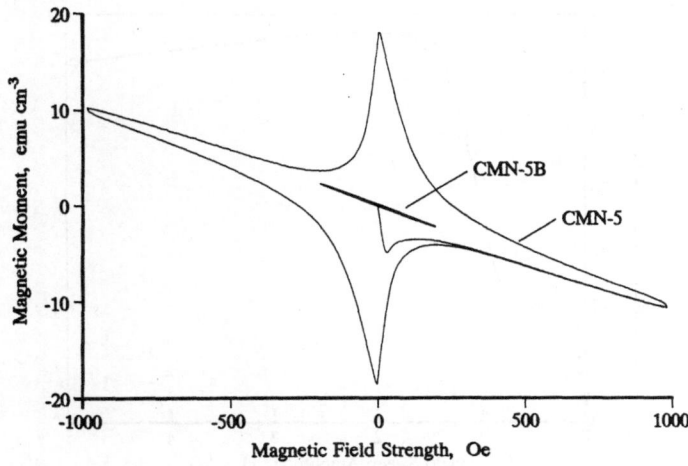

Fig. 3 Low-field ($<H_{c1}$) hysteresis loops for CMN-5 and the unclad CMN-5B at 4.2 K.

The full magnetization loop for CMN-5, at 4.2 K and with field amplitude less than the H_{c1} of the NbTi filaments[*], is shown in Fig. 3. Inserted is the magnetization loop (closed line) for the corresponding bare-filamentary sample CMN-5B, the slope of which approximates the between-wings slope of the Cu-clad sample. The susceptibility of the bare CMN-5B, measured as part of a companion study[7], was found to be -0.0108 (a value which is a factor of 14.74 lower than it would be in the absence of field penetration, and which yields a 4.2-K penetration depth[7] λ_L = R/x = 323 nm). The measured ratio of the clad to bare low-field (<9 G) susceptibilities[*] is 0.140/-0.0108 = 13.0, which in itself is spectacular evidence for the existence of coupling. This comparison of the magnetizations of clad and unclad materials is exploited further in the high-field regime.

With decreasing d (hence s), coupling also manifests itself as a development of the M(H) line into a hysteresis loop, as indicated in Fig. 3 for CMN-5. This excess magnetization, say M_{ex}, which decreases monotonically with increasing field, can be thought of as being superimposed on an imaginary inclined line (drawn between the wings of the M(H) loop) representing the Meissner diamagnetism of the filaments themselves.

TRANSITION FROM THE LOW-FIELD- TO THE HIGH-FIELD MAGNETIZATION REGIMES

As the field-sweep amplitude, H_m, increases three new features of the coupling emerge: (1) The peak of M_{ex} moves away from H = 0. (2) As depicted in Fig. 4, the peak value of M_{ex} increases with H_m in two stages -- stage-i, for $H_m < H_{c1}$, corresponding to filaments in the Meissner state, and stage-ii, for $H_m > H_{c1}$, corresponding to filaments in the mixed state. (3) As will be seen below, whereas at low values of H_m, a coupling M_{ex} appears during both the increasing and decreasing sweeps of the applied field (Fig. 3), when H_m exceeds H_{c1} the coupling occurs predominantly during the field-decreasing (so-called "trapping") segments of the M(H) loop.

---

[*]   As discussed in a companion paper[7], the critical field of fine filaments is strongly enhanced as a result of field-penetration effect. For example, the measured H_{c1} of CMN-5 at 4.2 K is 607 gauss[7].

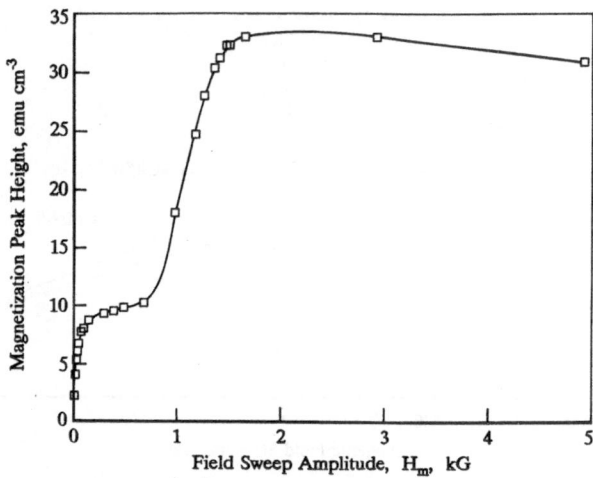

Fig. 4 Excess magnetization (clad-unclad, with respect to CMN-5 and CMN-5B) at the coupling peak in the positive M(H) quadrant, as function of the field-sweep amplitude.

HIGH-FIELD MAGNETIZATION

Clad-Filament/Critical Current Method

For a cylindrical superconducting composite in a transverse magnetic field sufficiently strong to place the filaments in the mixed state, the full height, ΔM, of the magnetic hysteresis loop (in emu/cm^3 based on the <u>strand</u> volume) is given by

$$\Delta M = (0.4/3\pi) \, L \, J_c \, d \qquad (1)$$

i.e.
$$\Delta M/(I_c/A_{strand}) = (0.4/3\pi) \, d \qquad (2)$$

where L is the filling factor ($= A_{SC}/A_{strand}$), J_c and I_c are the strand critical current density (A/cm^2) and critical current (A), respectively, and the strand diameter, d, is in cm. The conventional approach is thus to compare (as function of filament diameter, at constant applied field) the measured ΔM with the expected height based on Eqns. (1) or (2). General examples of the use of this approach have been presented elsewhere[1,4]. For the present material, the results of applying the Eqn.-(2) approach are given in Ref. [3]. From the plot of $\Delta M/(I_{SC}/A_{strand})$ versus d we noted that coupling was just beginning to appear in CMN-10, in agreement with the low-field conclusion (Fig. 1) and was more strongly present in CMN-5.

Clad-Filament/Bare-Filament Method

A second approach to the study of coupling uses magnetization only and eliminates the need for critical current measurement. It does, however, call for the preparation of a second set of samples in which all the matrix material has been removed by etching.

In analyzing the results of magnetization measurements performed on clad materials and their unclad counterparts, the left-hand side of Eqn. (2) may be essentially replaced by the height-ratio $(\Delta M)_{clad}/(\Delta M)_{bare}$. There are then two useful ways of dealing with this quotient: (1) Select a particular field strength, and plot the height-ratio versus d; (2)

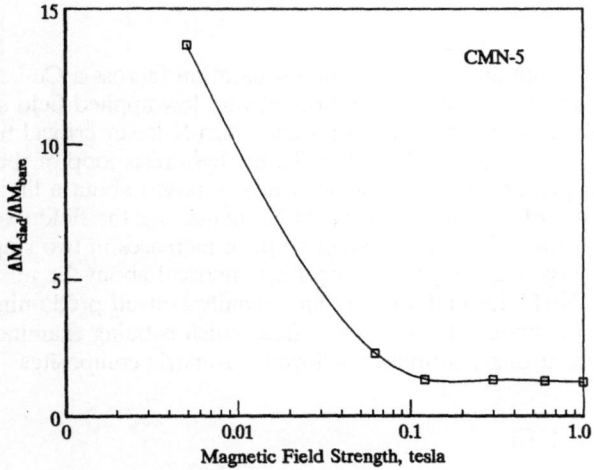

Fig. 5   Ratio of the hysteresis-loop height (clad/unclad) for CMN-5 and CMN-5B as function of the corresponding field.

select a particular strand, and plot the height-ratio versus applied field strength. The result of applying the second method is shown in Fig. 5 for the strand-pair CMN-5 and CMN-5B.

Finally, the availability of unclad-strand magnetization data sheds further light on the properties of the coupled strands. In Fig. 6 the high-field magnetization loops for CMN-5 and CMN-5B are superimposed. The former exhibits a tilt due to the paramagnetism of the Cu-0.5wt.%Mn interfilamentary matrix. But once this tilt is removed it can be seen that the curves superpose almost completely along the field-increasing (shielding) branches, while most of the coupling appears along the trapping segments of the M(H) loop.

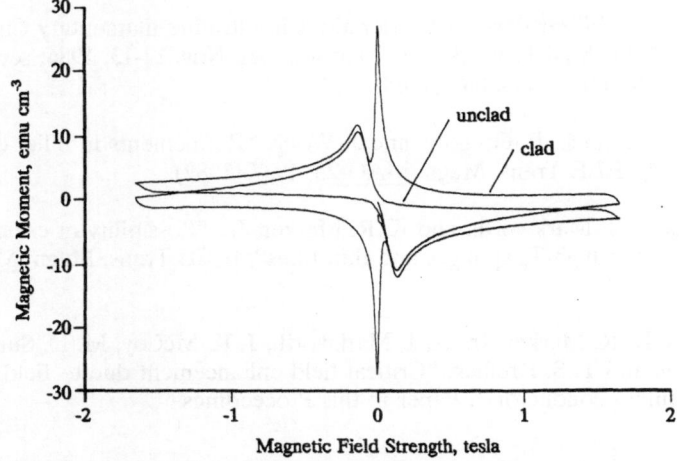

Fig. 6   High-field (mixed-state) hysteresis loops at 4.2 K for CMN-5 and CMN-5B, indicating the predominance of coupling along the trapping branches of the M(H) loop.

237

SUMMARY

In sample CMN-5, with an interfilamentary separation (across a Cu-0.5wt.%Mn matrix) of about 0.1 μm, the filamentary bundle at very low applied field strengths simulates a solid cylinder of superconductor with a zero-K lower critical field, H_{c1B}, of 10.5 G. Above H_{c1B}, but below the H_{c1} of NbTi, the hysteresis loop of the composite exhibits an excess magnetization, M_{ex}, symmetrically disposed about a line representing the Meissner diamagnetism of the uncoupled NbTi filaments. As the field sweep amplitude increases beyond H_{c1}, the height of the coupling peak increases in two steps separated by $H_m = H_{c1}$. During large H_m, M_{ex} is no longer symmetrical about the mixed-state magnetization of the NbTi. Instead, the coupling manifests itself predominantly during the "trapping" stroke of the hysteresis loop -- an effect which is being examined more extensively using more strongly coupled unalloyed-Cu-matrix composites.

ACKNOWLEDGEMENTS

The clad and unclad epoxy-potted magnetization samples were prepared by R. D. Smith, Battelle. The research was sponsored in full by the U.S. Department of Energy, Division of High-Energy Physics.

REFERENCES

1. A. K. Ghosh, W. B. Sampson, E. Gregory, and T. S. Kreilick, "Anomalous low field magnetization in fine filament NbTi conductors", IEEE Trans. Magn. MAG-23, 1724 (1987).

2. A. K. Ghosh, W. B. Sampson, E. Gregory, S. Kreilick, and J. Wong, "The effect of magnetic impurities and barriers on the magnetization and critical current of fine filament NbTi composites", Tenth Int. Conf. Magnet Tech., Boston, MA, Sept. 21-25 (1987).

3. E. Gregory, T. S. Kreilick, J. Wong, E. W. Collings, K. R. Marken Jr., R. M. Scanlan, and C. E. Taylor, "A conductor with uncoupled 2.5 μm diameter filaments designed for the outer cable of SSC dipole magnets", IEEE Trans. Magn. MAG-25, 1926 (1989).

4. E. W. Collings, "Stabilizer design considerations in ultrafine filamentary Cu/NbTi composites", Sixth NbTi Workshop, Madison, WI, Nov. 12-13, 1986; see also Adv. Cryo. Eng. (Materials) 34, 867 (1988).

5. P. Valaris, T. S. Kreilick, E. Gregory, and J. Wong, "Refinements in billet design for SSC strand", IEEE Trans. Magn. MAG-25, 1937 (1989).

6. E. W. Collings, A. J. Markworth, and K. R. Marken Jr., "Possibility of critical field enhancement in high-T_c sponges and thin films", IEEE Trans. Magn. MAG-25, 2491 (1989).

7. E. W. Collings, K. R. Marken Jr., A. J. Markworth, J. K. McCoy, M. D. Sumption, E. Gregory, and T. S. Kreilick, "Critical field enhancement due to field penetration in fine-filament conductors", Paper in this Proceedings.

LOW TEMPERATURE SPECIFIC HEAT AND MAGNETIC SUSCEPTIBILITY
OF NbTi AND NbTiMn ALLOYS

E. W. Collings*, R. D. Smith*, J. C. Ho†, and C. Y. Wu†

* Battelle, Columbus, OH 43201
† Wichita State University, Wichita KS 67208

ABSTRACT

We report on the results of a low temperature specific heat and magnetic susceptibility studies of a series of annealed-and-quenched binary NbTi alloys in the concentration range 41~53 wt.% Nb, which includes most compositions of technical interest. Also discussed is the influence of Mn on the calorimetrically measured properties of NbTi. Three series of alloys are considered: (i) a set of binary control alloys; (ii) alloys with Mn content fixed at ~0.5 wt.% and with Nb concentration between about 41 and 53 wt.%; (ii) alloys with an almost fixed Ti content (~46 wt.%) and with Mn concentration between 0 and ~5 wt.%. In alloys such as Ti-52Nb it is demonstrated that the addition of several percent of Mn causes a decrease in the electronic specific heat coefficient and a concomitant decrease in T_c. In that regard the effect of Mn on T_c (dT_c/dc = -0.22 K/at.%) is comparable to that of Cr, Mo, or Re (for which dT_c/dc = -0.2 K/at.%). It is concluded: (i) that Mn is a strong stabilizer of the bcc phase in Ti-base alloys; (ii) that Mn in bcc NbTi acts like any other nonmagnetic transition-element and influences T_c through its influence on the band density of states.

INTRODUCTION

Low temperature specific heat and magnetic susceptibility measurements have been made on three series of binary and ternary Ti-Nb-base alloys: (i) A set of binary control alloys, designated NTM43 to NTM53, with Nb concentrations in the range 43~53 wt.%; (ii) essentially the same set of alloys but with the inclusion of ~0.5 wt.%Mn -- these are designated NTM43/1 to NTM53/1; (iii) a series of alloys, designated NTM0 to NTM6, with an almost fixed Ti content (~46 wt.%), but with a Mn content that varies from 0 to ~5 wt.%. The exact compositions (as determined by electron-beam/x-ray wavelength analysis) are listed in Tables 1-3.

EXPERIMENTAL

Low temperature specific was measured at temperatures, T, from ~4 to ~14 K on samples weighing typically 20 g; they had been water quenched following an anneal in the bcc (β-phase) regime. Pulsed Joule heating and germanium thermometry were used. The heater and thermometer were attached to a Cu block to which the sample was thermally coupled via a layer of GE 7031 varnish; the heat capacity of these "addenda" was separately measured. The low temperature specific heat, C, of a metal in the normal state is usually represented by the relationship

$$C = \gamma T + \beta T^3 \tag{1}$$

Advances in Cryogenic Engineering (Materials), Vol. 36
Edited by R. P. Reed and F. R. Fickett
Plenum Press, New York, 1990

239

where the electronic specific heat coefficient, γ, is related to the Fermi density of states, and lattice specific heat coefficient, β, when expressed in the units mJ/mol K^4, yields the Debye temperature, θ_D, when inserted in the formula

$$\theta_D{}^3 = (1.944 \times 10^6)/\beta \tag{2}$$

If the metal is a superconductor, its second-order transition into the superconductive state at the transition temperature, T_c, is accompanied by a jump, ΔC, in the electronic specific heat. In BCS theory the relative height of this jump is $\Delta C/\gamma T_c = 1.43$. With further decrease in temperature the electronic heat capacity descends exponentially to zero at 0 K. These features are exhibited in subsequent figures.

Magnetic susceptibility, χ, in the temperature range ~298 K to ~77 K, was measured by the Curie technique using an electronic microbalance in association with an electromagnet fitted with 7-in.-diam. "constant-force" pole caps. Calibration was against pure Pt. At each temperature, magnetic force was measured at five fields, H, between 3.6 and 9.7 kgauss enabling the effect of ferromagnetic contamination (of saturation moment M) to be corrected for using the Honda-Owen method. In that method, the corrected susceptibility, χ_∞, was obtained as the intercept of a reciprocal-field "plot" based on the equation

$$\chi = \chi_\infty + M/H \tag{3}$$

All data reduction was performed numerically as described in Ref. 1. The susceptibility samples, weighing typically 100~150 mg, were measured in the bcc(β)-annealed-plus quenched condition except for members of the NTM0-NTM6 series which had received precipitation heat treatments (PHT) as for technical superconducting wire.

DATA SUMMARY

Calorimetric Data

The calorimetric data for the series of binary Ti-Nb alloys (NTM43, NTM47, NTM51 and NTM53) are presented in Fig. 1; the corresponding calorimetric parameters are listed

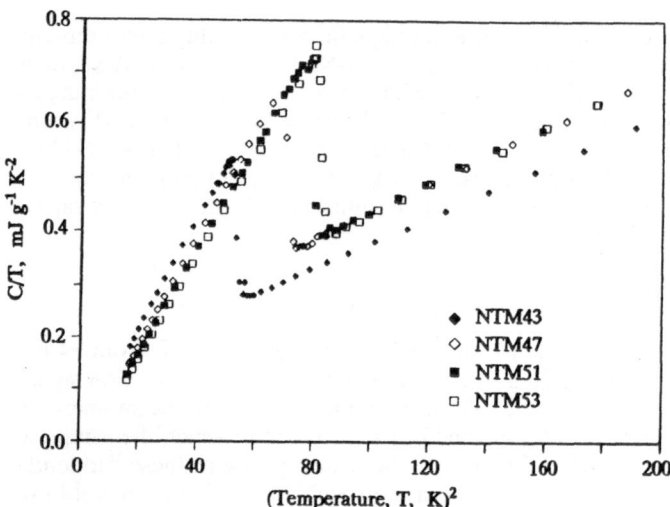

Fig. 1. Low temperature specific heat of the binary Ti-Nb alloys NTM43-NTM53 in the condition β-annealed (48h/1000°C) and quenched.

Table 1. Specific Heat of Binary Ti-Nb Alloys

Alloy Code		NTM43	NTM47	NTM51	NTM53
Wgt.% Nb		41.23	46.49	50.39	52.80
Wgt.% Ti		58.77	53.51	49.61	47.20
At.% Nb		26.56	30.94	34.37	36.58
At.% Ti		73.44	69.06	65.63	63.42
Molar Wgt, g		59.85	61.82	63.37	64.36
Electronic S.H. Coeff., γ, mJ/mol K^2		7.97	10.39	10.80	10.41
Lattice S.H. Coeff., β, mJ/mol K^4		0.146	0.165	0.170	0.176
Debye Temp., θ_D, K		237	228	225	223
Supercond.	onset	7.5	8.6	9.1	9.2
transition	middle	7.3	8.5	8.9	9.0
temp., K	tail (peak)	7.2	8.4	8.7	8.9

Table 2. Specific Heat of NTM43/1 to NTM53/1

Alloy Code		NTM43/1	NTM47/1	NTM51/1	NTM53/1
Wgt.% Nb		41.20	45.91	50.33	52.58
Wgt.% Ti		58.28	53.61	49.02	47.08
Wgt.% Mn		0.52	0.49	0.65	0.34
At.% Nb		26.56	30.46	34.35	36.39
At.% Ti		72.87	68.99	64.90	63.21
At.% Mn		0.57	0.55	0.75	0.40
Molar Wgt, g		59.89	61.65	63.41	64.31
Electr. S.H. Coeff., γ, mJ/mol K^2		8.46	10.07	9.57	10.34
Lattice S.H. Coeff., β, mJ/mol K^4		0.156	0.168	0.176	0.168
Debye Temp., θ_D, K		232	226	223	226
Supercond.	onset	7.8	8.6	8.9	9.1
transition	middle	7.6	8.5	8.7	8.9
temp., K	tail (peak)	7.5	8.4	8.6	8.8

Table 3. Specific Heat of NTM0-NTM6 and NTM53/1

Alloy Code		NTM0	NTM53/1	NTM2	NTM4	NTM6
Wgt.% Nb		53.3	52.6	52.1	51.9	50.2
Wgt.% Ti		46.7	47.1	46.1	44.7	44.7
Wgt.% Mn		----	0.3_4	1.8	3.4	5.2
At.% Nb		37.0	36.4	36.1	35.9	34.5
At.% Ti		63.0	63.2	61.8	60.1	59.5
At.% Mn		----	0.4	2.1	4.0	6.0
Molar Wgt, g		64.56	64.31	64.30	64.35	63.84
E.S.H. Coeff., γ, mJ/mol K^2		11.1	10.3	11.1	8.04	6.32
L.S.H. Coeff., β, mJ/mol K^4		0.168	0.168	0.135	0.135	0.128
Debye Temp., θ_D, K		226	226	243	243	248
Supercond.	onset	9.1	9.1	8.8	8.7	8.7
transition	middle	9.0	8.9	8.6	8.0	7.4
temp., K	tail (peak)	8.9	8.8	8.4	7.4	6.8

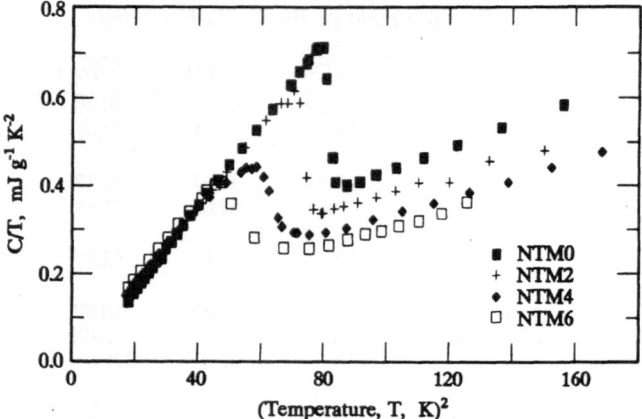

Fig. 2. Low temperature specific heat of the ternary Ti-Nb-Mn alloys NTM0-NTM6 in the condition β-annealed (24h/1000°C) and quenched.

in Table 1. Table 2, for the alloys NTM43/1, NTM47/1, NTM51/1 and NTM53/1, lists the effect of including ~0.5 wt.% of Mn in the formulations of the binary series.

The calorimetric data for the alloy series NTM0-NTM6, in which a varying level of Mn (0~5 wt.%) is added to an almost-constant Ti concentration (~46 wt.%) Ti-Nb base, are presented in Fig. 2. The corresponding calorimetric parameters are listed in Table 3. Also included in that table, for completeness, is one of the above-mentioned data sets.

Susceptibility Data

The susceptibility-temperature-dependence data for the binary Ti-Nb alloy series NTM43-NTM53 and the ternary Mn-containing series NTM43/1-NTM53/1 are presented in Fig. 3. Those for a representative member of the PHT series (viz. NTM6) are

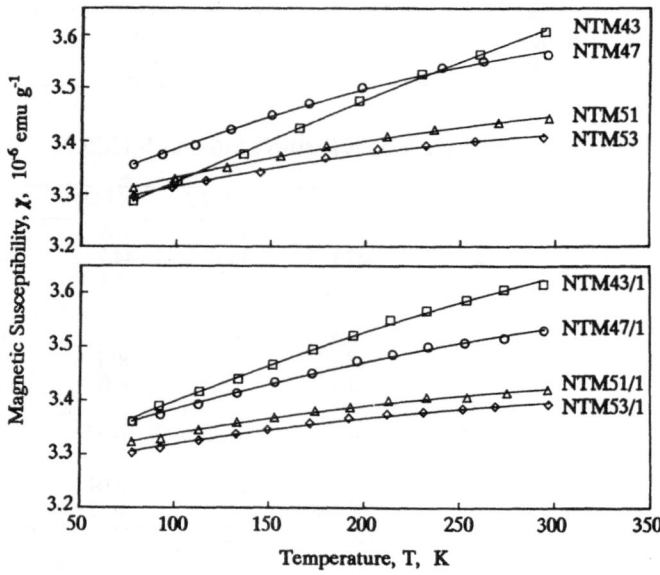

Fig. 3. Magnetic susceptibility temperature dependence of (a) the binary Ti-Nb alloys NTM43-NTM53, and (b) the ternary alloys NTM43/1-NTM53/1, all in the condition β-annealed (48h/1000°C) and quenched.

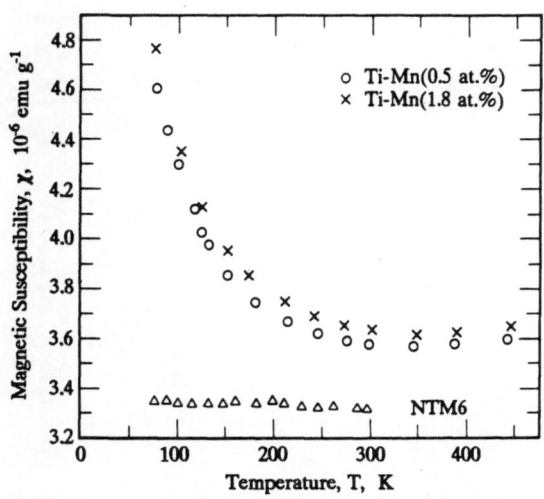

Fig. 4. Magnetic susceptibility temperature dependence of the binary Ti-Mn alloys Ti-Mn(0.5 at.%) and Ti-Mn(1.8 at.%) in the condition β-annealed (5h/1000°C) and quenched, and the ternary alloy NTM6 in the condition β-annealed (48h/1000°C) and quenched.

presented in Fig. 4. Also included in the latter figure are a set of data for two hcp-phase binary Ti-Mn alloys -- Ti-Mn(0.5 at.%) and Ti-Mn(1.8 at.%).

RESULTS AND DISCUSSION

Calorimetric Results -- Binary Alloys

The calorimetrically measured T_c of the present binary Ti-Nb alloys is plotted versus Nb content in Fig. 5 where a comparison can be made with other data that have been reported in the literature. Note the strong T_c composition dependence exhibited by the present alloys; they are evidently on the threshold of a regime of rapidly increasing (with decreasing Nb content) ω-phase volume-fraction[2,pp.100-110]. As also indicated in the figure, the ($\omega + \beta$)-phase region gives way (below ~20 at.% Nb[2,pp.78-79]) to a region of martensitically transformed structures indicated by α^m. The metastable structures mentioned refer to *quenched-from-bcc*, rather than equilibrium-phase, alloys.

The relative height of the specific heat jump at T_c is plotted versus Nb content in Fig. 6. We note that the data all lie above the "BCS-line" indicating that the alloys concerned are not of the "weak-coupling" kind. A similar conclusion was reached by Sasaki et al[3] in studies of Ti-Nb(21 at.%). With a $\Delta C/\gamma T_c$ of 1.86 (and other relevant properties), that alloy was claimed to exhibit, like pure Nb itself, "intermediate-coupling behavior".

Magnetic Results -- Binary and Ternary Alloys

Comparing Fig. 3 with the curves for Ti-(0.5 at.%) and Ti-Mn(0.8 at.%) in Fig. 4, we conclude (see also Ref. 8) that, within the regime of Ti-base alloys, only when Mn resides in an hcp environment does it exhibit a localized magnetic moment. Conversely, the absence of a Curie-Weiss 1/T-type susceptibility temperature dependence implies an immeasurably small level of so-called α-phase (hcp) precipitation. Applying this conclusion to the curve in Fig. 4 for NTM6, we deduce that PHT in that alloy did not produce measurable precipitation. The $\chi(T)$s of the other heat-treated NTM-series alloys were also flat[9]. The results indicate that Mn inhibits α-phase precipitation in Ti-Nb. In

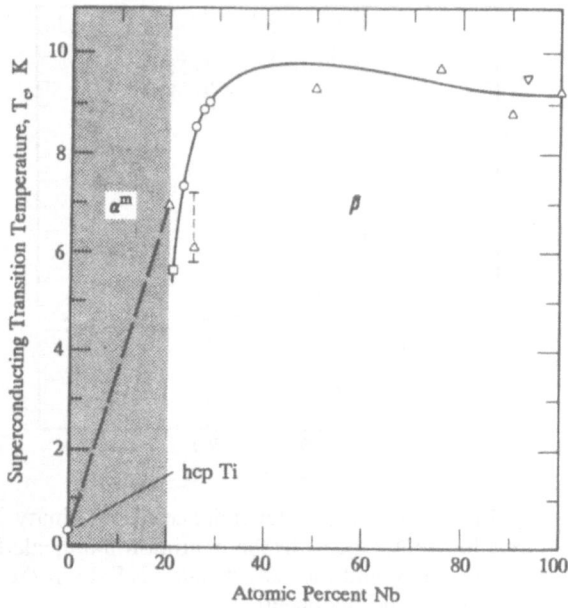

Fig. 5. Calorimetrically determined T_c of Ti-Nb alloys. Data sources: (□) Ref. 3; (△) Ref. 4 and Ref. 5; (▽) Ref. 6; (⊥) Ref. 7; (○) present data.

other words it acts as if to increase the equivalent Nb content; the rate at which it does so will be discussed below.

Calorimetric Results -- Ternary Alloys

According to Table 2, The inclusion of ~0.5 wt.% of Mn in Ti-Nb alloys influences their T_c in a manner that depends on the host's Nb content. The change of T_c changes sign in going from NTM43 (whose T_c is increased) to NTM53 (whose T_c is decreased). This seems to be the result of the following competing tendencies: (i) Mn increases the effective Nb content of the alloy (which will increase T_c in the low-Nb-content range, where T_c has a strong positive composition dependence -- see Fig. 5); (ii) the addition of transition elements other than Nb to Ti-Nb alloys generally tends to lower T_c[10].

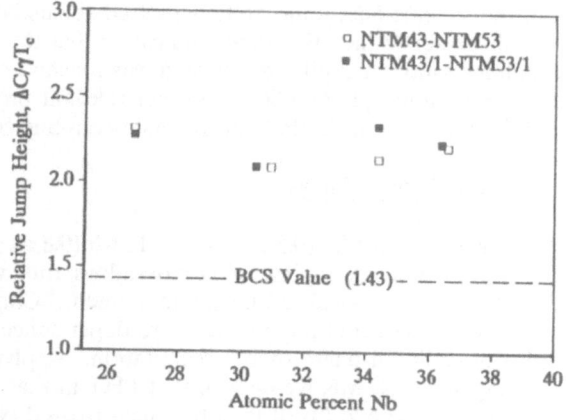

Fig. 6. Relative height of the specific heat jump at T_c.

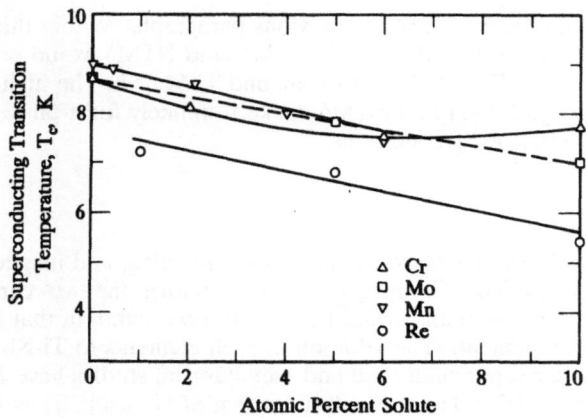

Fig. 7. Transition temperature of Ti-Nb alloyed with Cr, Mo, Mn, and Re.

To investigate further the influence of Mn on β-quenched Ti-Nb alloys the NTM0-NTM6 series of alloys was created. The data listed in Table 3 show that T_c scales with γ indicating that the decrease of T_c with increasing Mn content for these alloys is simply a band density of states effect. Fig. 7 indicates that the effect of Mn on calorimetrically measured T_c (viz. dT_c/dc = -0.22 K/at.%) is comparable to that of Cr, Mo, or Re (for which dT_c/dc = -0.2 K at.% -- based on the first three data points in each case).

Beta-Stabilizing Strength of Mn in Ti

The experimental evidence suggests that Mn acts to increase the equivalent Nb content of the Ti-Nb alloy -- i.e. it increases the stability of the bcc or β phase. The rate at which it does so can be described from both superconductive and metallurgical standpoints.

Superconductive Evidence: The superconductive viewpoint has been discussed previously[9]. After comparing the composition dependences of the critical fields of the NTM series of alloys with those of binary Ti-Nb (taken from the literature), it was asserted that NTM4 was equivalent to a Ti-Nb binary alloy with 65-70 at.% of Nb. This indicates that 1 wt.% (at.%) of Mn is equivalent to ~8.3 wt.% (~7.9 at.%) of Nb.

Metallurgical Evidence: From a metallurgical standpoint, the β-stabilizing strengths of transition-element (TM) additions to Ti can be gauged by the rates at which they lower the martensitic transus (i.e. the slope of transformation-start temperature, M_s, versus composition, c) hence to the degree to which they permit the retention of the β phase at some reference temperature. Molchanova[11] has presented a set of $M_s(c)$ curves for twelve Ti-base binary alloys; from the low-temperature intercepts of these curves with the c-axis a list of critical concentrations for the retention of the β phase in Ti-TM alloys has been constructed. Normalized to the critical concentration for Nb, these data have led to the construction of Eqn. (4) -- a formula for the Nb-equivalency (from a β-stabilizing standpoint) of a fictitious multicomponent alloy of Ti with various non-interacting transition elements in solid solution.

$$[Nb]_{eq.} = [Nb] + 9[Fe] + 6[Co] + 6[Mn] + 4.5[Cr] + 4.5[Ni] + 3.6[Mo]$$
$$+ 2.4[V] + 1.4[W] + 0.7[Ta] \tag{4}$$

This approach indicates that Mn is six times (in terms of wt.%) more potent than Nb in stabilizing the β phase, and hence in suppressing the tendency for α-phase precipitation --

in good agreement with the results of the previous paragraph. Within this framework, in binary-alloy terms, the alloys NTM0, NTM2, NTM4, and NTM6 would be equivalent, respectively, to Ti-53.3Nb, Ti-57.7Nb, Ti-61.8Nb, and Ti-64.6Nb. The high equivalent-Nb levels, especially in alloys NTM4 and NTM6, make it unlikely for α-phase precipitation to take place under the usual PHT conditions.

SUMMARY

The influence of Mn on the calorimetric, superconducting, and magnetic properties of Ti-Nb alloys has been studied. The magnetic results confirm that Mn carries no localized magnetic moment in bcc Ti-Nb alloys and hence is to be treated, in that environment, as a nonmagnetic transition element. The addition of such elements to Ti-Nb generally decreases T_c. But earlier superconductive and metallurgical studies have indicated that the β-stabilizing strength of Mn in Ti-Nb is 6~8 times that of Nb itself. Thus if Mn is added to Ti-Nb in the concentration range where dT_c/dc is steeply positive T_c tends to increase slightly with the addition of Mn; on the other hand, when Mn is added to higher concentration Ti-Nb alloys it lowers T_c. The β-stabilizing potency of Mn causes it to suppress α-phase precipitation when it is added to the usual technical superconducting alloys. Thus the T_c and the α-phase content of PHT alloys will be preserved only if a small amount of Mn is added to a sufficiently dilute Ti-Nb base.

ACKNOWLEDGEMENTS

The alloys were prepared under the supervision of F. A. Schmidt, Materials Preparation Center Ames Laboratory, and analyzed by J. J. Rayment-Rudolf, Battelle. The research was supported by the U.S. Department of Energy, Division of High Energy Physics.

REFERENCES

1. E. W. Collings and S. C. Hart, "Low temperature magnetic susceptibility and magnetization studies of some commercial austenitic stainless steels", Cryogenics 19, 521-530 (1979).
2. E. W. Collings, Physical Metallurgy of Titanium Alloys, American Society for Metals, Metals Park, OH, 1984.
3. T. Sasaki, K. Noto, N. Kobayashi, and Y. Muto, "Specific heat of super-conducting $Nb_{0.21}Ti_{0.79}$ alloy", Proc. LT-17 Part 2, Conf. Date: Aug. 15-22, 1984, North-Holland, Amsterdam, pp. 1295-1296.
4. B. I. Verkin, personal communication.
5. B. Ya. Sukharevskii, I. S. Shchetkin, and I. I. Fal'ko, "Investigation of the superconducting state of solid solutions of the niobium-titanium system", Sov. Phys. JETP 33, 152-155 (1971).
6. I. S. Shchetkin and T. N. Kharchenko, "Superconductivity and electron structure of a solid solution of titanium in niobium", Sov. Phys. JETP 37, 491-493 (1973).
7. B. Ya. Sukharevskii and A. V. Alapina, "Some features of the temperature dependence of the specific heat of a niobium-titanium alloy at the transition to the superconducting state", Sov. Phys. JETP 27, 897-899 (1968).
8. E. W. Collings, Applied Superconductivity, Metallurgy, and Physics of Titanium Alloys, Plenum Press, NY, 1986, pp. 292-294.
9. E. W. Collings, K. R. Marken, Jr., J. C. Ho, and T. S. Kreilick, "Effect of manganese additions on the physical- and superconducting properties of Nb-46.5Ti", in Cryogenic Materials '88, Volume 1. Superconductors, ICMC, Boulder, CO, 1988, pp. 343-355.
10. E. W. Collings, A Sourcebook of Titanium Alloy Superconductivity, Plenum Press, NY, 1983, pp. 396-404.
11. E. K. Molchanova, Phase Diagrams of Titanium Alloys, Israel Program for Scientific Translations, Jerusalem, 1965, p. 158.

DESIGN OF COUPLED OR UNCOUPLED MULTIFILAMENTARY SSC-TYPE

STRANDS WITH ALMOST ZERO RETAINED MAGNETIZATION AT FIELDS

NEAR 0.3 T

E. W. Collings*, K. R. Marken Jr*., and M. D. Sumption*†

* Battelle, Columbus, OH 43201
† Permanent address: Ohio University, Athens, OH 45701

ABSTRACT

Multifilamentary Cu-matrix strands with interfilamentary spacing as small as 0.2 μm can be almost fully decoupled by the addition of 0.5 wt.% Mn to the interfilamentary Cu. Decoupling in this way seems to be beneficial from a field-stability standpoint. On the other hand, the elimination of coupling does little to reduce residual strand-magnetization at the injection field of about 0.3 T when that field is approached, as usual, along the shielding branch of M(H). This residual diamagnetic magnetization (say M_R) of the winding material is responsible for unwanted distortion (multipole formation) of the dipolar field. It is demonstrated that M_R can be locally cancelled to zero by associating the strand with a small volume-fraction (less than 2%, depending on filament diameter) of pure Ni or any other low-field-saturable ferromagnetic material. The presence of the Ni has little effect on the shape of the M(H) hysteresis loop of the strand, other than to shift its wings uniformly in the $+M$ (when H is positive) and $-M$ directions, respectively. In practice, the Ni could be administered as: (a) additional filaments, (b) interfilamentary barriers, or (c) an electroplated layer on the outside of the strand.

INTRODUCTION

Helmholtz coils, or modifications of them (e.g. saddle-coils) are commonly used for producing dipolar magnetic fields. But if the coils are wound from superconducting strands, residual magnetization, M_R resident in the strand material itself is responsible for multipolar distortions of the desired field. It is well known that the height of the M(H) hysteresis loop -- $\Delta M(H) \equiv (M_{R+} - M_{R-})$, where the signs refer to the trapping (paramagnetic) and shielding (diamagnetic) branches, respectively, of M(H) -- is proportional to the product of filament diameter, w, and critical current density, $J_c(H)$[1]. Thus in an attempt to reduce strand magnetization (in the presence of high J_c) and the attendant field distortion, a strong effort has been under way to produce, on a commercial scale, multifilamentary strands with smaller and smaller filaments. In order to preserve filament quality (i.e. to prevent thickness undulations, or "sausaging") in small filaments, it has been suggested necessary to confine the ratio of filament spacing (s) to filament diameter (d) to s/d $\leq$ 0.15±0.02[2]. The combination of small d with low s/d results in interfilamentary spacings sufficiently close to proximity-effect-couple the filaments. For example, at an s/d of 0.13 Cu-matrix filaments that have been reduced to 5-1/2 μm in diameter are beginning to exhibit coupling; and the coupling becomes worse as d is still further reduced. But if the matrix is alloyed with ~0.5 wt.% Mn, coupling is barely perceptible even with 1 μm diameter filaments[2,3].

Advances in Cryogenic Engineering (Materials), Vol. 36
Edited by R. P. Reed and F. R. Fickett
Plenum Press, New York, 1990

247

But is this reduction of coupling meaningful within the context of Superconducting Supercollider (SSC) performance? To be sure, the $\Delta M(H)$ of closely spaced material has been reduced to its uncoupled value. But what has this achieved from the standpoint of strand magnetization at the beam-injection field of 0.33 tesla? This question can be answered by a glance at Fig. 1, which compares the M(H) loops for "coupled" multi-filamentary strands both with and without the presence of the Cu matrix. Evidently practically all of the coupling magnetization shows up along the trapping segment of the loop. Beam injection takes place after a demagnetization cycle that terminates in a field-increase to 0.33 tesla along the *shielding* segment of M(H).

MAGNETIZATION DECAY IN SSC-TYPE STRANDS

Possibly related to coupling is a second serious problem exhibited by SSC dipole magnets, viz. magnetic field decay following change of field[4]. We are presently investigating this phenomenon with vibrating-sample magnetometry (VSM) using samples that consist of bundles of strand some 6 mm in length (see Ref. 5 for additional magnetometry and sample details). Prior to launching on the drift study proper, an initial test sample was assembled from a length of strand based on RHIC (relativisitic heavy-ion collider) materials. The specifications of it (designated RHIC-009) and related sample materials are given in Table 1.

In order to simulate the magnetic state of the strand just prior to beam injection, magnetization studies were performed along the shielding branch of M(H). Since preliminary experiments indicated that the magnetization drift following field change was very small, the magnetometer was set to one of its high-sensitivity ranges (10^{-2} emu, in this case). Next, to permit full-precision data to be taken on this range, it was necessary to find some way of neutralizing most of the background magnetization (viz. about -0.1 emu). It was decided that background subtraction, free of noise and phase instability, could best be achieved by attaching to the sample holder a weighed amount of fine pure Ni wire. Some M(H) loops for sample RHIC-009 with Ni attached are given in Fig. 2. Note the almost complete diamagnetic moment cancellation in the vicinity of 0.3 tesla.

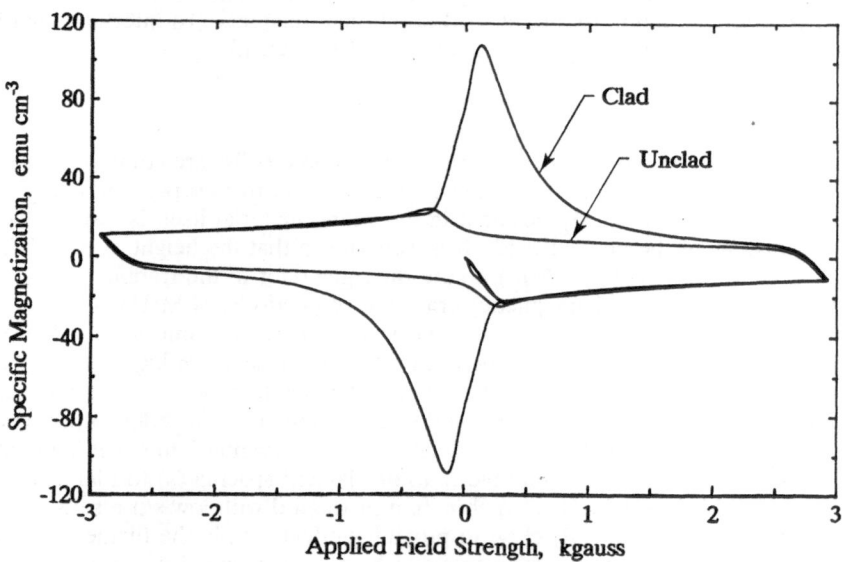

Fig. 1   Specific magnetization (NbTi volume) at 4.2 K of RHIC-009 material both with (clad) and without (unclad) the presence of the Cu matrix. Coupling magnetization is dominant only along the *trapping* branches of the loop.

Table 1 Specifications of Sample Material

Sample Code	Strand Diam.,D, 10^{-2} cm	Fil. Diam.,d*, μm	NbTi area, A_{SC}, 10^{-4} cm^2	Magnetization Test Sample	
				No. of Strands	Length, mm
Cu-Matrix Strands: 6,108 filaments (heat treated)					
RHIC-009	2.51	2.106	2.128	80	6.31
RHIC-013	3.3	2.890	4.007	48	6.31
RHIC-026	6.5	5.490	14.459	14	6.44
Cu-0.5wt.%Mn-Matrix Strands: 22,902 filaments (not heat treated)					
CMN-005	1.27	0.500	0.449	200	5.88
CMN-010	2.75	1.068	2.052	58	5.74
CMN-015	3.85	1.495	4.020	33	5.60

* Obtained by etching-and-weighing using separately measured density of bulk Nb-46.5Ti (= 6.097).

Having proceeded this far, it was recognized that a technique had been developed for fabricating SSC strand with close-to-zero residual magnetization at the injection field. On reflection, it was recognized that Ni barriers had been incorporated into multifilamentary strands to eliminate proximity-effect interfilamentary coupling[6], but the idea of using Ni to compensate for strand magnetization at injection is new. It may turn out to be convenient to add the Ni as an electroplated layer on the outside of the strand; but should internal Ni be preferred it will be possible to eliminate coupling and compensate for residual shielding magnetization in a single operation.

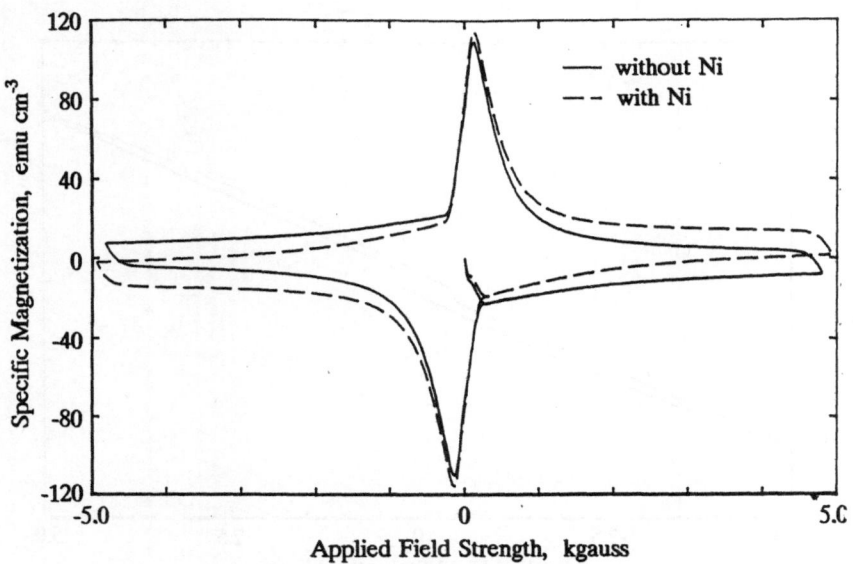

Fig. 2 Specific magnetization (NbTi volume) at 4.2 K of clad RHIC-009, and that of the same sample to which 1.84 mg of pure Ni wire has been attached.

In order to cancel the strong diamagnetic moment of a superconductor in the shielding mixed state a large positive moment is needed. Such a moment can be provided by either a paramagnetic or a ferromagnetic material.

Paramagnetic Compensation

Since Cu-Mn alloys have already been recommended as interfilamentary matrices for proximity-effect decoupling, it is natural to enquire into their potential for moment compensation. The 4.2-K hysteresis loop for a Cu-0.93 at.%Mn alloy is depicted in Fig. 3. (The offsets near the origin, the hysteresis, and the slight curvature at higher fields are a consequence of the alloy's mictomagnetism). Since the moment increases monotonically with field it is, in principle, possible to select, for a given superconductor/Cu-Mn volume-ratio, a field strength at which moment cancellation takes place. The disadvantage of a paramagnet is that even if moment-cancellation is practically feasible it would be strongly field-specific.

Ferromagnetic Compensation

Since the superconductor's diamagnetic moment decreases rather slowly with increasing applied field strength, near-compensation over a wide range can be achieved by a field-independent positive moment. An obvious candidate is the saturation moment of a soft ferromagnet (soft, because we would like the moment to drop to zero in zero field). The 10-K hysteresis loop for a sample of pure Ni wire (diameter, 0.1 mm) in what appears to be the as-drawn condition, is depicted in Fig. 4. Since the saturation moment of Ni increases by only 7.7 percent between 20 °C and absolute zero[7,p.5-144] it can be regarded as practically constant throughout the He-temperature range. In Fig. 4, the Ni moment is never fully saturated. The moment of *annealed* Ni saturates in fields less than 1 kgauss; special alloys such as 78 permalloy and deltamax saturate at even lower fields but have larger saturation magnetizations than Ni[8].

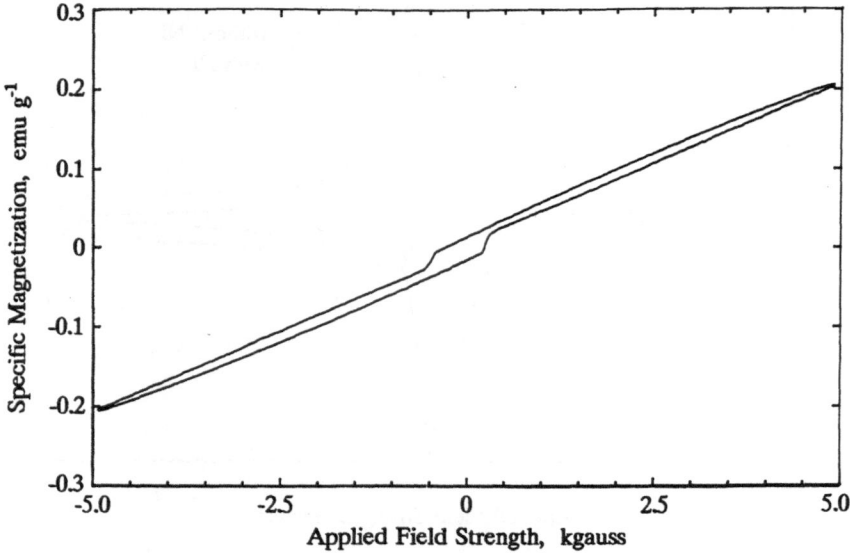

Fig. 3   Specific magnetization of Cu-Mn(0.93 at.%) at 4.2 K

Table 2 Specific Magnetization of Filamentary NbTi, Cu-Mn (both at 4.2 K) and Ni (at 10 K)

Sample Material	Strength of the Increasing Applied Field		
	0.30 T	0.33 T	0.40 T
Magnetization of NbTi, M_{SC}, emu/cm^3			
RHIC-009	-10.758	-10.176	- 9.124
RHIC-013	-13.073	-12.391	-11.185
RHIC-026	-20.881	-19.917	-18.164
CMN-005	- 6.366	- 6.036	- 5.178
CMN-010	- 6.223	- 5.796	- 5.048
CMN-015	- 6.865	- 6.472	- 5.685
Magnetization of Ni and Cu-Mn, M_{add}, emu/cm^3			
Cu-Mn(0.93 at.%)*	+ 1.136	+ 1.250	+ 1.500
Ni†	+ 467.4	+ 478.1	+ 496.8

* In normalization to unit volume, the density of pure Cu (8.95) was assumed[7,p.2-20].
† In normalization to unit volume, a density of 9.04 was taken[7,p.2-21].

DESIGN OF MAGNETIZATION-COMPENSATED STRANDS

The specific magnetizations of six samples of filamentary NbTi in composite-strand form, a sample of Cu-0.93 at.%Mn alloy (both measured at 4.2 K) and a piece of pure Ni wire (0.1 mm diameter; measured at 10 K) at three values of a steadily *increasing* magnetic field (shielding, in the case of the superconductor) are listed in Table 2. For convenience in strand design a per-unit-volume unit of magnetization has been selected.

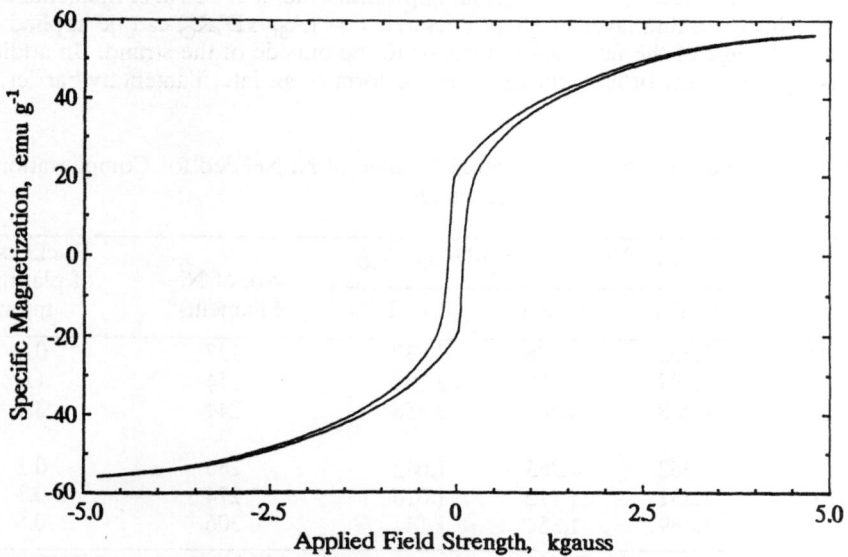

Fig. 4 Specific magnetization of (evidently) unannealed pure Ni wire (0.1 mm diameter) measured at 10 K (the actual sample was RHIC-009 plus Ni measured above the T_c of the NbTi).

Table 3 Volume Ratio of Cu-Mn(0.93 at.%) Needed for Compensation
at Various Fields

Sample Code	Volume Ratio, $R_C \equiv A_{add}/A_{SC}$		
	0.30 T	0.33 T	0.40 T
RHIC-009	9.5	8.1	6.1
RHIC-013	11.5	9.9	7.5
RHIC-026	18.4	15.9	12.1

If M represents a material's specific magnetization and A its cross-sectional area, while subscripts "SC" and "add" denote NbTi and the compensating addenda (Cu-Mn or Ni), then the fractional amounts of addenda material required for compensation are simply

$$R_C = A_{add}/A_{SC} = -M_{SC}/M_{add} \tag{1}$$

Compensation with Cu-Mn

For RHIC-type strands (the CMN series already contain Cu-Mn), the results of substituting the Cu-Mn magnetization data into Eqn. (1) are presented in Table 3. Clearly, compensation with the paramagnetic alloy requires an unacceptibly large volume fraction of additional material.

Compensation with Ni

The results of substituting the Ni magnetization data into Eqn. (1) are given in Table 4. Evidently strands with filament diameters between 1/2 and 5-1/2 μm require from 1-1/2 to 4 vol.% of Ni for compensation at fields near the injection field. These levels of Ni (which are based on the NbTi- rather than the total strand volume) can be introduced without significantly increasing the diameter, D, of the strand. Two methods of doing so are indicated in Table 4: (i) an appropriate number of NbTi filaments can be replaced by Ni; (ii) a thin layer of Ni, of thickness $t = (A_{SC}/\pi D)R_C$, can be applied (at any convenient stage of the fabrication process) to the outside of the strand. In addition to these is the possibility of introducing Ni in the form of an interfilamentary barrier.

Table 4 Volume Percentage and Actual Volume of Ni Needed for Compensation
at Various Fields

Sample Code	Vol. Pct. Ni, $100R_C \equiv 100A_{add}/A_{SC}$			No. of Ni Filaments*	Thickness of plating† t, μm
	0.30 T	0.33 T	0.40 T		
RHIC-009	2.302	2.128	1.837	127	0.6
RHIC-013	2.797	2.592	2.251	154	1.0
RHIC-026	4.468	4.166	3.656	244	3.0
CMN-005	1.362	1.263	1.042	286	0.1
CMN-010	1.331	1.212	1.016	274	0.3
CMN-015	1.469	1.354	1.144	306	0.5

* Number of Ni-replaced NbTi filaments for a total of 6,108 in the case of RHIC and 22,902 in the case of CMN.

† Plated layer applied to the outside of the strand (appropriate to 0.33 T operation) computed from the relationship $t = (A_{SC}/\pi D)R_C$.

CONCLUDING DISCUSSION

A few percent of Ni added to a superconducting strand can offset most of its shielding magnetization over a wide magnetic field range. Pure annealed Ni adds little to the existing magnetic hysteresis. Furthermore, since its magnetization saturates at fields below 0.1 tesla, its only significant effect on the shape of the M(H) loop is to shift the wings uniformly in the $+M$ direction when H is positive and in the $-M$ direction when it is negative.

The addition of Ni to the strand may relieve the SSC magnets' need for fine-filaments, the initial reason for which was to minimize winding magnetization over the operating field range.

We have indicated that the Ni can be introduced as an external coating at some convenient stage during strand processing, or may be incorporated into the strand in the form of replacement filaments. Other possible modes of deployment would be as: (i) a plating applied more-or-less directly onto the individual filaments (cf. Nb "diffusion-barriers"); (ii) an extended interfilamentary web throughout the strand (cf. "mixed-matrix" AC strands).

Diffusion Barrier Technology

In order to suppress intermetallic-compound node formation, it is customary to plate the NbTi elements with a few percent of Nb. In an extension of this procedure, the recommended few percent of Ni could be sandwiched between a pair of Nb layers. It is well known that the height of the magnetization loop, $\Delta M_v(H)$ (emu/cm^3) at any field H is related to the filament's critical current density, J_c (A/cm^2), and diameter, d (cm), by[1]

$$\Delta M_v(H) = (0.4/3\pi) J_c d \tag{2}$$

Next, taking $\Delta M_v = -2M_{SC}$, assuming a field of 0.33 tesla, and inserting $M_{add} = 478.1$ emu/cm^3 for Ni, it follows with the help of Eqn. (1) that

$$A_{add}/A_{SC} = (0.2/3\pi)(I_c/A_{SC})(d/478.1) \tag{3}$$

The thickness, t (cm), of a compensating Ni layer adjacent to a round filament is then given by

$$t = 1.41 \times 10^{-5} I_c(A) \tag{4}$$

Eqn. (4) shows that for a filament critical current at 0.33 tesla of 0.37 A (appropriate for RHIC-026) a Ni layer of thickness 0.052 μm would provide compensation. The corresponding Ni volume would of course be 3.8% of the filament volume. Eqn. (4) also shows that t is independent of filament diameter provided I_c is constant -- as it must be if the strand, whatever its diameter, is to carry a specified current at constant margin.

AC Strand Technology

In order to suppress eddy-current loss in AC applications, it is necessary to increase the transverse resistivity of the strand. One way of accomplishing this (and continuing to visualize the strand in cross-section) is to isolate the individual filaments within the meshes of a Cu-Ni net. To achieve the present goal, the Cu-Ni would be replaced by pure Ni. A step in this direction was taken several years ago by Curtis and MacDonald[9, pp.434-5] in connection with a Tevatron strand production program. NbTi-containing hexagonal-OD Cu tubes (3.05 mm across flats) were electroplated with about 20 μm of Ni prior to billet assembly. During extrusion, a well-preserved and continuous network of Ni barriers was developed. It is interesting to note that in this case, with a NbTi-rod diameter of 2.13 mm) the Ni/NbTi volume ratio was about 5.9%.

Ferromagnetic barriers will of course suppress proximity-effect coupling between the filaments. Accordingly if the Ni-barrier approach is adopted both coupling-elimination (if that still remains a goal) and moment-compensation may be accomplished in a single operation.

ACKNOWLEDGEMENTS

Samples for magnetization measurement were prepared by R. D. Smith, Battelle. The research stemmed from a magnetization-decay investigation initiated by R. M. Scanlan while with the SSC Design Group at the Lawrence Berkeley Laboratories. It was further stimulated by discussions with R. Steining of the SSC Laboratory in Dallas. The research was funded by the U.S. Department of Energy, Division of High-Energy Physics.

REFERENCES

1. W. J. Carr, Jr., and G. R. Wagner, "Hysteresis in a fine filament NbTi composite", Adv. Cryo. Eng. (Materials) 30, 923 (1984).
2. E. Gregory, T. S. Kreilick, J. Wong, E. W. Collings, K. R. Marken, Jr., R. M. Scanlan, and C. E. Taylor, "A conductor with uncoupled 2.5 μm diameter filaments designed for the outer cable of SSC dipole magnets", IEEE Trans. Magn. 25-2, 1926 (1989).
3. E. W. Collings, "Stabilizer design considerations in ultrafine filamentary Cu/NbTi composites", Sixth NbTi Workshop, Madison, WI, Nov. 12-13, 1986; see also Adv. Cryo. Eng. (Materials) 34, 867 (1988).
4. W. S. Gilbert, R. F. Althaus, P. J. Barale, R. W. Benjegerdes, M. A. Green, M. I. Green, and R. M. Scanlan, "Magnetic field decay in model SSC dipoles", IEEE Trans. Magn. 25-2, 1459 (1989).
5. E. W. Collings, K. R. Marken, Jr., M. D. Sumption, E. Gregory, and T. S. Kreilick, "Magnetic studies of proximity-effect coupling in a very closely spaced fine-filament NbTi/CuMn composite superconductor", Paper in this Proceedings.
6. T. S. Kreilick, E. Gregory, and J. Wong, "Geometric considerations in the design and fabrication of multifilamentary superconducting composites", IEEE Trans. Magn. MAG-23, 1344 (1987); see also A. K. Ghosh, W. B. Sampson, E. Gregory, S. Kreilick, and J. Wong, "The effect of magnetic impurities and barriers on the magnetization and critical current of fine filament NbTi composites", IEEE Trans. Magn. 24-2, 1145 (1988).
7. American Institute of Physics Handbook, Third Edition, McGraw-Hill, Inc., 1972.
8. R. J. Parker and R. J. Studders, Permanent Magnets and their Applications, John Wiley, New York, 1962, Fig. 4-30.
9. E. W. Collings, Applied Superconductivity, Metallurgy and Physics of Titanium Alloys, Volume 2, Plenum Press, 1986.

CRITICAL FIELD ENHANCEMENT DUE TO FIELD PENETRATION IN FINE-FILAMENT SUPERCONDUCTORS

E. W. Collings*, A. J. Markworth*, J. K. McCoy*, K. R. Marken Jr*.,
M. D. Sumption*†, E. Gregory**†† and T. S. Kreilick**

* Battelle Memorial Institute, Columbus, OH 43201
† Permanent address: Ohio University, Athens, OH 45701
** Supercon, Inc., Shrewsbury, MA 01545
†† Present address: IGC Advanced Superconductors, Waterbury, CT 06704

ABSTRACT

In samples in which at least one dimension, say 2R, is comparable to a penetration depth, λ_L, the flux-exclusion volume at fields below H_{c1} is less than the true sample volume. This "volume erosion" has two important consequences: (i) it causes the Meissner susceptibility to be less than the standard value -- for example, for a field-parallel plate, by a factor $1-(1/x)\tanh(x)$, where $x = R/\lambda_L$; (ii) it reduces the flux-exclusion energy density (based on the initial sample volume) by the square of that factor. In the latter case it calls for an enhanced applied field strength to terminate the Meissner state -- that enhancement factor being accordingly $1/\sqrt{1-(1/x)\tanh(x)}$. Both of these effects become more and more pronounced as T approaches T_c, since the field-penetration depth increases with temperature according to $1/\sqrt{1-t^4}$, where $t \equiv T/T_c$. The susceptibility-depletion and field-enhancement effects have been studied on a series of very fine filament composites (with NbTi filament diameters ranging from 0.5 to 11.6 μm), prepared from 23,000-filament NbTi/CuMn strands from which (to avoid unwanted proximity-effect coupling) the Cu has been removed by etching and replaced by epoxy. For example, at 4.2 K in the finest-filament material, susceptibility was found to be 1/15 of the Meissner value, and the sample exhibited a lower critical applied field of 607 gauss.

INTRODUCTION

In establishing the first experimentally-based phenomenological model for irreversible type-2 superconductivity, Bean[1] conducted a series of magnetization and critical-current density measurements on a synthetic filamentary superconductor (filament diameter, 10 nm or 30 nm) prepared by infiltrating molten Pb into porous Vycor glass. One perhaps surprising result of the study was the observation that, as a consequence of the filamentary subdivision, the 4.2-K critical field of Pb had become enhanced from 528 G, the bulk value, to a projected 26.0 kG. The enhancement was a size-effect function of the ratio of the filament radius (R) to the penetration depth (λ_L). In fact, for Pb, Tinkham[2] had already predicted that 1.2 nm thick foils should have a critical field of at least 25 kG.

Advances in Cryogenic Engineering (Materials), Vol. 36
Edited by R. P. Reed and F. R. Fickett
Plenum Press, New York, 1990

255

Table 1. Magnetization Sample Specifications

Sample Code	Fil. Diam., μm	Number of Strands*	Number of Individual Filaments	Sample Length, mm	Volume of NbTi, 10^{-3} cm^3
CMN-115	11.556	1		5.892	14.15
CMN-25	2.459	15		5.64	9.201
CMN-21B	2.051		4.8094×10^5	5.79	9.200
CMN-15B	1.459		8.2447×10^5	6.401	9.264
CMN-10B	1.068		1.6260×10^5	6.205	9.039
CMN-5B	0.4995		4.5804×10^6	5.734	5.147

* The strands each have 22,902 filaments.

An expression describing the effect of field penetration on the magnetization of spherical particles was published in 1950 by London[3]. The results of comparable calculations for cylinders and plates (of half-thickness R)[4] are presented here. Field penetration manifestly influences the energetics of the field-exclusion or Meissner state especially at small $x = R/\lambda_L$; hence, it enhances the critical field needed to terminate that state, which for type-1 superconductors such as Pb is H_c, the thermodynamic critical field. But with the recent availability of multifilamentary NbTi/Cu composites with very fine filament diameters[5], an opportunity arose to investigate the field enhancement effect as it might apply to the H_{c1} of a typical <u>type-2</u> superconductor, NbTi.

EXPERIMENTAL

Starting material for sample preparation was a series of multifilamentary NbTi/CuMn composite 22,902-filament strands with filament diameters between 11.6 and 0.5 μm, see Table 1 of Ref. 6. The strands were untwisted and had received no heat treatment except for a final 4h/225°C anneal. Samples for measurement consisted of cylindrical bundles, about 3 mm in diameter and 6 mm in length, of multifilamentary strands (fil. diams. of 11.6 and 2.5 μm) or parallel "bare" NbTi filaments (diams. of 2.1 μm and less) imbedded in epoxy. In the latter case, the filaments had been extracted by etching from the composite strands. In all cases, the filaments remained coated with a thin "diffusion-barrier" film of Nb, whose presence was ignored in the data analysis. Specifications of the samples are given in Table 1.

Magnetic susceptibility itself was measured as function of temperature up to the T_c of NbTi by vibrating-sample magnetometry in a transverse magnetic field that was swept between ±25 G. The lower critical field, H_{c1}, was measured at 4.2 K by observing the first departure from linearity of the initial magnetization curve.

MAGNETIC SUSCEPTIBILITY -- THEORY

The volume susceptibility, χ_v, of a macroscopic superconducting body in the flux-exclusion state below H_c (type-1) or H_{c1} (type-2) is a materials-independent function of its geometry and attitude. The susceptibilities of several such standard bodies are listed in the second column of Table 2. But when field-penetration cannot be neglected, the standard expressions must be modified by the appropriate shape-dependent f(x), as listed in the third column of Table 2. The <u>temperature dependence</u> of the susceptibility enters through that of λ_L, which is frequently taken to be given by $\lambda_L = \lambda_{L0}/\sqrt{(1-t^4)}$, where $t \equiv T/T_c$. By way of example, the susceptibility temperature dependences of four standard bodies/attitudes for two values of x, are given in Fig. 1. The zero-K penetration depth, λ_{L0}, is obtained by least-squares fitting the experimental $\chi_v(T)$ to the appropriate function of x.

Table 2. Magnetic Susceptibilities of Small Samples
with Various Shapes and Attitudes

Shape and Attitude	Large-Scale Shape-Dependent Susceptibility	Modification Due to Field Penetration, f(x)*	Ref.
Plate, field paral'l to faces	$-1/4\pi$	$1-(1/x)\tanh x$	[4]
Cylinder, field paral'l to axis	$-1/4\pi$	$1-(2/x)I_1(x)/I_0(x)$	[4]
Sphere	$-3/8\pi$	$1-(3/x)\coth x + 3/x^2$	[3]
Cylinder, field perp'r to axis	$-1/2\pi$	$1-(2/x)I_1(x)/I_0(x)$	[4]

* $x = R/\lambda$, where R = radius or half-thickness
I_0 and I_1 are modified Bessel functions of the first kind of order 0 and 1, respectively.

MAGNETIC SUSCEPTIBILITY -- EXPERIMENT

In order to eliminate proximity-effect coupling between the filaments at very small strand diameters (hence very small filament diameters and interfilamentary spacings)[6], which would interfere with the measurement of individual-filament susceptibility, the matrix material was removed by etching from all strands with filament diameters of 2.1 μm and below. The $\chi_v(T)$s of several representative sample groups are depicted in Fig. 2. We note the discrepancy between the temperatures at which the susceptibilities rise to zero, which suggests that T_c is diameter-dependent, at least in the finest filaments. The T_cs derived by extrapolating $\chi_v(T)$ to zero are displayed in Fig. 3 as function of filament size (expressed in terms of strain relative to the area of CMN-115B). The origin of the T_c size dependence is not yet clear. Since the strands had not received precipitation heat treatment during processing, proximity effect between the beta-phase matrix and alpha-phase precipitates (as in Ref. 7) could not be the mechanism. Likewise, proximity effect between the NbTi filaments and the Cu matrix (as in Ref. 8) must be discounted, since for the filaments concerned the matrix had been removed by etching. But the immediate requirement for accurate T_c values was to enable the experimental $\chi_v(T)$ data to be fitted to the computed susceptibility function, $\chi_v(t)$ [i.e. $\chi_v(x)$] in which $t \equiv T/T_c$. In Fig. 4, we display the experimental $\chi_v(t)$ data together with a set of least-squared fitted curves representing $\chi_v = (-1/2\pi)[1-(2/x)I_1(x)/I_0(x)]$. From the fitting variable, x, a penetration depth was derived for each sample. A least-squares fit to the data indicates that as R decreases below about 5 μm, λ_{L0} rises exponentially from a zero-K bulk value of 0.253 μm at a rate given by λ_{L0} (μm) $= 0.253 + 0.0725 \exp(-R)$.

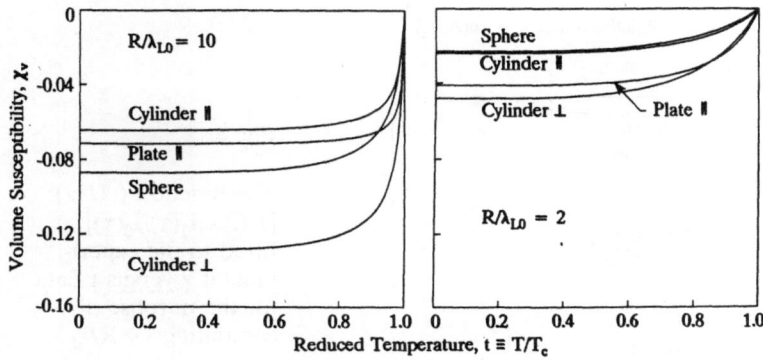

Fig. 1. Magnetic susceptibilities of four standard bodies/attitudes in the Meissner state for two values of $x = R/\lambda_{L0}$.

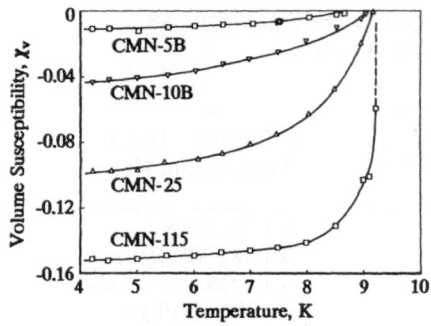

Fig. 2.

Magnetic susceptibility temperature dependences of four representative samples.

FIELD ENHANCEMENT -- THEORY

Field penetration causes the magnetic susceptibility to be reduced by a factor $f(x)$ below that of a similar macroscopic body. This effect is the result of a reduction in the "flux-exclusion volume". As a consequence of volume renormalization, the superconductor acquires energy density (averaged over the original volume) at the rate $P(H^2/8\pi)f(x) = P[H\sqrt{f(x)}]^2/8\pi$, where P is a shape-factor. It follows that the applied field must attain the value $H_{c1}/\sqrt{f(x)}$ before a type-2 superconductor can acquire sufficient energy to transform out of the Meissner state. In other words, the critical field is enhanced by the factor $1/\sqrt{f(x)}$, where the $f(x)$s are as listed in Table 2.

For a verification of this conclusion for the special case of a type-1 superconducting plate in a parallel magnetic field we cite the result of a direct calculation. According to London (Ref.3, pp.130-133), if field penetration is a factor, the unit-area free energy of a foil of thickness $d = 2R$ in the field-parallel attitude (otherwise equal to $-2RH^2/8\pi$) is reduced by an amount $(H^2/4\pi)\tanh(R/\lambda_L)/(1/\lambda_L)$. The energy balance equation in the Meissner state then becomes (on a unit-surface-area basis):

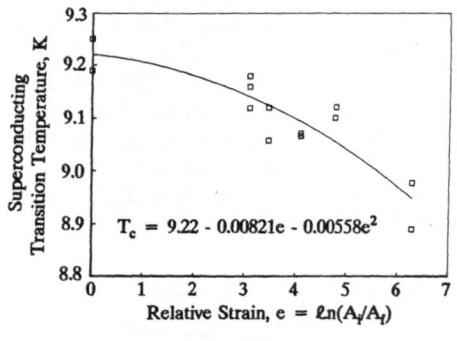

Fig. 3.

Superconducting transition temperature as function of relative strain for unheat-treated strands (except for a final 4h/225°C).

$$T_c = 9.22 - 0.00821e - 0.00558e^2$$

Relative Strain, $e = \ell n(A_i/A_f)$

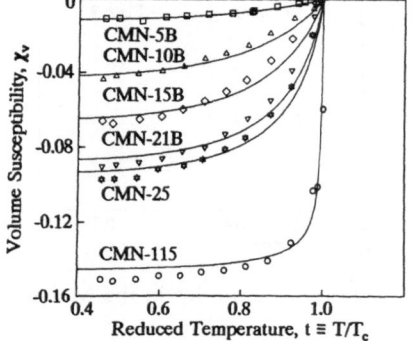

Fig. 4.

The function $(-1/2\pi)[1-(2/x)I_1(x)/I_0(x)]$ fitted to the experimental χ versus t data for the purpose of computing $x = R/\lambda_L$.

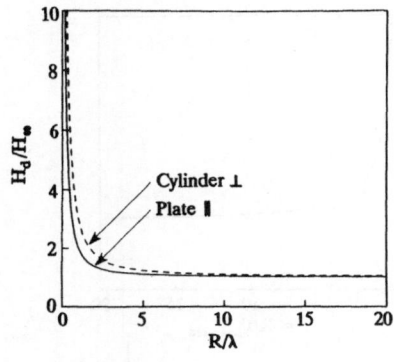

Fig. 5.

Calculated enhanced critical field, H_d relative to that of an infinitely thick cylinder (field perp.) or plate (field para.), respectively.

$$g_s(d) - g_n(d)$$

$$= -(d/8\pi)(H_c^2 - H^2) - (H^2/4\pi)\tanh(R/\lambda_L)/(1/\lambda_L) \tag{1}$$

$$= -(d/8\pi)[H_c^2 - H^2(1 - \tanh(R/\lambda_L)/(R/\lambda_L))]. \tag{2}$$

It follows that the applied field must be raised to an enhanced critical field, H_d, given by

$$H_d = H_c/\sqrt{1 - \tanh(x)/x} \qquad x = R/\lambda_L \tag{3}$$

before the superconductor transforms out of the Meissner state. This is in accord with the argument presented in the first paragraph of this section. Fig. 5 shows how H_d/H_c increases as R/λ_L decreases for a pair of standard bodies/attitudes.

FIELD ENHANCEMENT -- EXPERIMENTAL

Several methods have been recommended for the determination of H_{c1}, both in the past[9] and more recently[10,11]. H_{c1} is difficult to determine by direct measurement on an irreversible type-2 superconductor. However, the method adopted here, as in Ref. 12, is to accept as H_{c1} the applied field at which the the initial M(H) begins to deviate from linearity. Any deficiencies in this approach would tend to cancel out once the critical-field ratio (see below) is taken.

In Fig. 6, volume magnetization is plotted versus applied field strength for all the samples studied. In it we see that the H_{c1} of CMN-115, for example, is 165 G. Two

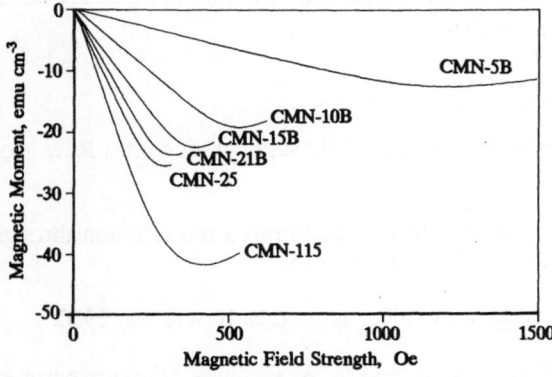

Fig. 6. Initial 4.2-K magnetization versus field. The slope of the linear portion is proportional to χ; the departure from linearity takes place at H_{c1} (i.e., H_d).

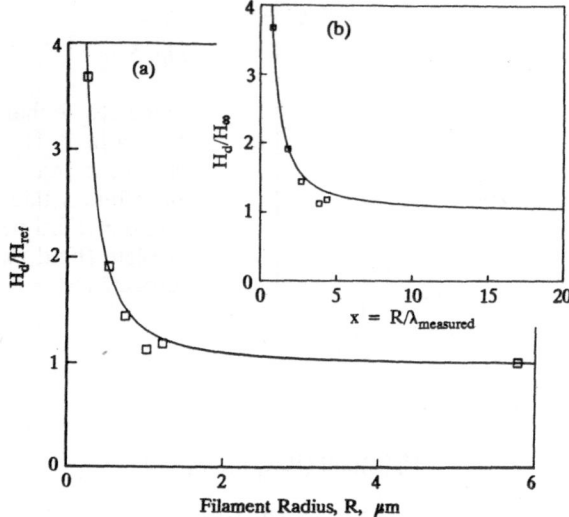

Fig. 7. (a) <u>Data points</u> -- measured relative critical field <u>versus</u> filament radius, R; <u>Curve</u> -- computed relative critical field; H_{ref} refers to the critical field of the largest (thickest) sample (CMN-115). (b) H_d/H_∞ plotted <u>versus</u> x, the ratio of filament radius to penetration depth.

consequences of field penetration are simultaneously displayed in Fig. 6: (1) the decrease in $\chi_v = dM/dH$ with decreasing filament diameter; (2) an accompanying increase in H_{c1}.

Using the H_{c1} of CMN-115B as reference (H_{ref}) the relative critical field (H_d/H_{ref}) is plotted versus filament radius, Fig. 7. On the same graph we also display $H_d/H_\infty = 1/\sqrt{1-(2/x)I_1(x)/I_0(x)}$ versus x (where $x = R/\lambda_{measured}$, and $H_\infty \equiv H_{c1, R=\infty}$). Note that we have returned to using the transverse-field enhancement factor. Although the curve is plotted independently of the data rather than fitted to it, the agreement between them is excellent, thereby confirming that lower critical field enhancement in a fine transverse-mounted filament can be described by the expected formula.

ACKNOWLEDGEMENTS

The clad and unclad epoxy-potted magnetization samples were prepared by R. D. Smith, Battelle. The research was sponsored in full by the U.S. Department of Energy, Division of High-Energy Physics.

REFERENCES

1. C. P. Bean, "Magnetization of high-field superconductors", Rev. Mod. Phys. <u>36</u>, 31-39 (1964).

2. D. M. Ginsberg and M. Tinkham, "Far infrared transmission through superconducting films", Phys. Rev. <u>118</u>, 990-1000 (1960).

3. F. London, <u>Superfluids</u>, John Wiley and Sons, New York, 1950.

4. A. J. Markworth, J. K. McCoy, and E. W. Collings, "Field penetration effects in variously shaped superconducting bodies below the critical field", in preparation.

5. E. Gregory, T. S. Kreilick, J. Wong, E. W. Collings, K. R. Marken Jr., R. M. Scanlan, and C. E. Taylor, "A conductor with uncoupled 2.5 μm diameter filaments designed for the outer cable of SSC dipole magnets", IEEE Trans. Magn. MAG-25, 1926-1929 (1989).

6. E. W. Collings, K. R. Marken Jr., M. D. Sumption, E. Gregory, and T. S. Kreilick, "Magnetic studies of proximity-effect coupling in a very closely spaced fine-filament NbTi/CuMn composite superconductor", paper in this Proceedings.

7. E. W. Collings, T. S. Kreilick, E. Gregory, P. J. Lee, and J. C. Ho, "Calorimetric studies of the superconducting transition as function of thermomechanical processing in fine-filament Cu/NbTi composites", Adv. Cryo. Eng (Materials) 34, 1027-1032 (1988).

8. K. Yasohama, K. Morita, and T. Ogasawara, "Superconducting properties of Cu-NbTi composite wires with fine filaments", IEEE Trans. Magn. MAG-23, 1728-1731 (1987).

9. E. W. Collings, Applied Superconductivity, Metallurgy and Physics of Titanium Alloys, Vol. 1, Plenum Press, New York, 1984, pp. 520-523.

10. A. Umezawa, G. W. Crabtree, J. Z. Liu, et al., "Anisotropy of the lower critical field, magnetic penetration depth, and equilibrium shielding current in single-crystal $YBa_2Cu_3O_{7-y}$", Phys. Rev. B 38, 2843-2846 (1988).

11. L. Krusin-Elbaum, A. P. Malozemoff, Y. Yeshurun, et al., "Temperature dependence of lower critical fields in YBaCuO crystals", Phys. Rev. B 39, 2936-2939 (1989).

12. S. S. P. Parkin, V. Y. Lee, and E. M. Engler, "Magnetic properties and critical fields of $RBa_2Cu_3O_{7-x}$ (R = Y, Pr, Eu, Gd, Dy, Ho)", Chemtronics 2, 133-139 (1987).

PROXIMITY EFFECT ON FLUX PINNING STRENGTH IN

SUPERCONDUCTING Nb-Ti WITH THIN α-Ti RIBBONS

Teruo Matsushita, Soji Otabe, and Tetsuya Matsuno

Department of Electronics
Kyushu University 36
Fukuoka 812, Japan

ABSTRACT

It has been found by Lee et al. that the critical temperature reduces monotonically, while the critical current density first increases and is followed by a decrease, according as the diameter of superconducting multifilamentary Nb-Ti wire is decreased. This behavior is mainly caused by the proximity effect between superconducting matrix and normal α-Ti ribbons that becomes more remarkable as the thicknesses of two regions become smaller. The critical temperature and the elementary pinning strength of α-Ti ribbons are theoretically estimated by solving the phenomenological Ginzburg-Landau equations for multilayered structure with superconducting and normal layers. The critical current density is calculated from a statistic summation of the elementary pinning forces. The obtained critical temperature decreases monotonically, while the critical current density increases, with decreasing thicknesses of the two layers. The decrease in J_c observed by Lee et al. is considered to result from constriction of superconducting layers by heavy cold work.

INTRODUCTION

Enhancement of critical current densities in superconducting Nb-Ti wires is required from the view of application. For a wire extensively cold-worked the critical current density of about 3,000 A/mm^2 was attained at B=5T. Larbalestier and his coworkers[1-3] obtained the maximum critical current densities for a final drawing strain ranging from 3 to 6. This indicates that the optimum condition is determined by other factors than the final strain. From the fact that precipitates of α-Ti phase are dominant pinning centers,[4] their morphology is considered as a key factor The average thickness of ribbon-shaped α-Ti phase in the cold worked wire was of the order of 1nm^3 and is much smaller than the coherence length in Nb-Ti matrix (β phase). Since the average thickness of β phase between adjacent α-Ti ribbons is also very small, the proximity effect is expected to be prominent. In fact, a reduction in the critical temperature was observed around the optimum condition for flux pinning. This means that the theoretical analysis taking account of the proximity effect is needed for a detailed investigation of the flux pinning. In this paper, this

Advances in Cryogenic Engineering (Materials), Vol. 36
Edited by R. P. Reed and F. R. Fickett
Plenum Press, New York, 1990

263

problem is treated in terms of the phenomenological Ginzburg–Landau theory
for an idealized superconducting and normal multilayers.

THEORY

We assume an idealized periodic multilayer shown in Fig. 1. The
thicknesses of superconducting and normal layers normalized by the coherence length ξ in the superconducting layers are denoted by $2d_s$ and $2d_n$,
respectively. From the periodicity, we have only to treat the region
$0 \leq \eta \leq d_s + d_n$ in Fig. 1, where η is the coordinate along the thickness
normalized by ξ. The phenomenological Ginzburg–Landau equations in the
superconducting layers $(d_n < \eta \leq d_s + d_n)$ are

$$\frac{1}{2m}(-i\hbar\nabla + 2e\vec{A})^2\Psi = -\alpha\Psi - \beta|\Psi|^2\Psi, \tag{1}$$

$$\vec{J} = -\frac{ie\hbar}{m}(\Psi*\nabla\Psi - \Psi\nabla\Psi*) - \frac{4e^2}{m}|\Psi|^2\vec{A}, \tag{2}$$

where $\vec{A}$ the vector potential, Ψ the order parameter, and $\vec{J}$ the current
density. In the normal region $(0 \leq \eta < d_n)$, the Schrödinger equation

$$\frac{1}{2m}(-i\hbar\nabla + 2e\vec{A})^2\Psi = -\alpha_n\Psi \tag{3}$$

is used instead of Eq. (1), where α_n is a positive parameter associated
with the repulsive potential for superconducting electron pairs. The
phenomenological theory can be approximately used even for a thin normal
layer, since the proximity effect from the superconducting layers is
prominent in it.

Critical Temperature

First, we shall investigate the proximity effect on the critical
temperature. In this case, the magnetic field and the current are not
applied and the problem is one-dimensional. Hence, the phase of Ψ can
be put as a constant. Thus, Eqs. (1) and (3) are reduced to[5]

$$\frac{d^2R}{d\eta^2} + R - R^3 = 0; \qquad d_n < \eta \leq d_s + d_n, \tag{4}$$

$$\frac{d^2R}{d\eta^2} - \theta R = 0; \qquad 0 \leq \eta < d_n. \tag{5}$$

In the above R is the absolute value of Ψ normalized by Abrikosov's unit
and the parameter θ is defined by

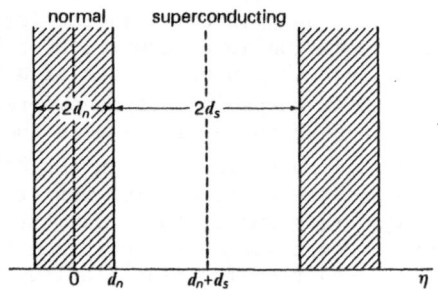

Fig. 1. Idealized periodic multilayer.

$$\theta = -\alpha_n/\alpha = \xi^2/\xi_n^{\ 2} \tag{6}$$

where $\xi_n = \hbar/(2m\alpha_n)^{1/2}$ is the coherence length in the normal layers. The solutions of Eqs. (4) and (5) are given in the form:

$$R = \sqrt{1-c}\ \text{sn}[\sqrt{\tfrac{1+c}{2}}\,(\eta-\eta_0)]; \qquad d_n < \eta \le d_s + d_n, \tag{7}$$

$$= R(0)\cosh(\theta^{1/2}\eta); \qquad 0 \le \eta < d_n, \tag{8}$$

where snz is Jacobi's elliptic function with a modulus

$$k = \sqrt{\frac{1-c}{1+c}} \tag{9}$$

and c and η_0 are the integral constants. Equation (8) was obtained from the condition of symmetry, $dR/d\eta=0$ at $\eta=0$, and $R(0)$ is a value of R at this point. Three constants, c, η_0, and $R(0)$ are determined from the boundary conditions at $\eta=d_n$ and $\eta=d_n+d_s$.

Here we assume that the thickness of the normal layer is much smaller than the electron mean free path. In this case, the transmission coefficient of superconducting electrons is not small and the Zaitsev's boundary conditions[6] are expected to be satisfied[5] at the S-N boundary $\eta=d_n$. If we assume that the state densities of electrons are the same in the two layers, the continuity of R and $dR/d\eta$ at the boundary is satisfied:[6]

$$R(0)\cosh(\theta^{1/2}d_n) = \sqrt{1-c}\ \text{sn}\ z, \tag{10}$$

$$R(0)\theta^{1/2}\sinh(\theta^{1/2}d_n) = \sqrt{\frac{1-c^2}{2}}\ \text{cn}\ z\ \text{dn}\ z, \tag{11}$$

where

$$z = \sqrt{\frac{1+c}{2}}\,(d_n-\eta_0). \tag{12}$$

From these equations we have

$$\theta^{1/2}\tanh(\theta^{1/2}d_n) = \sqrt{\frac{1+c}{2}}\ \frac{\sqrt{(1-\text{sn}^2 z)(1-k^2\text{sn}^2 z)}}{\text{sn}\ z}. \tag{13}$$

The condition of symmetry, $dR/d\eta=0$, at $\eta=d_s+d_n$ leads to

$$\sqrt{\frac{1+c}{2}}\,(d_N+d_s-\eta_0) = \int_0^1 \frac{dx}{\sqrt{(1-x^2)(1-k^2x^2)}}, \tag{14}$$

where the right-hand side is equal to the elliptic integral of the first kind K(k). From this equation we have

$$\text{sn}\ z = \frac{\text{cn}\ z'}{\text{dn}\ z'}, \tag{15}$$

where

$$z' = \sqrt{\frac{1+c}{2}}\ d_s. \tag{16}$$

Substitution of Eq. (15) into Eq. (13) leads to

$$\theta^{1/2}\tanh(\theta^{1/2}d_n) = \sqrt{\frac{2c^2}{1+c}} \ \frac{sn \ z'}{\sqrt{(1-sn^2z')(1-k^2sn^2z')}} . \qquad (17)$$

This is an equation only for c and can be solved numerically. Then, η_0 can be obtained from Eq. (15), and finally R(0) is obtained from Eq. (10). Examples of calculated R are shown in Figs. 2(a) and 2(b).

The condensation energy density F is given by

$$(\mu_0 H_c^2)^{-1}F = -R^2 + \frac{1}{2}R^4 + (\frac{dR}{d\eta})^2; \qquad d_n < \eta \le d_n + d_s, \qquad (18)$$

$$= \theta R^2 + (\frac{dR}{d\eta})^2; \qquad 0 \le \eta < d_n, \qquad (19)$$

where H_c is the thermodynamic critical field in the case where the proximity effect does not occur. At the center of the superconducting layer $\eta = d_n + d_s$, R takes the maximum value $\sqrt{1-c}$ and hence, F is minimum:

$$(\mu_0 H_c^2)^{-1}F(d_n + d_s) = -\frac{1}{2}(1-c^2). \qquad (20)$$

If the proximity effect does not occur, this energy density is –1/2. The increase in the condensation energy means a reduction in the effective thermodynamic critical field resulting in a reduction in the critical temperature. If we assume that the state density of electrons at the Fermi surface is not influenced appreciably by the proximity effect, the resultant critical temperature is proportional to the effective thermodynamic critical field. Thus, we have

$$T_c = T_{c0}\sqrt{1-c^2} , \qquad (21)$$

where T_{c0} is the critical temperature under no influence of the proximity effect.

Elementary Pinning Force

Elementary pinning force is estimated from a variation in the free energy when a fluxoid is virtually displaced. Here, we assume for simplicity an idealized situation shown in Fig. 3(a), where a fluxoid is normal to the multilayer. Since the Ginzburg–Landau parameter κ in Nb–Ti

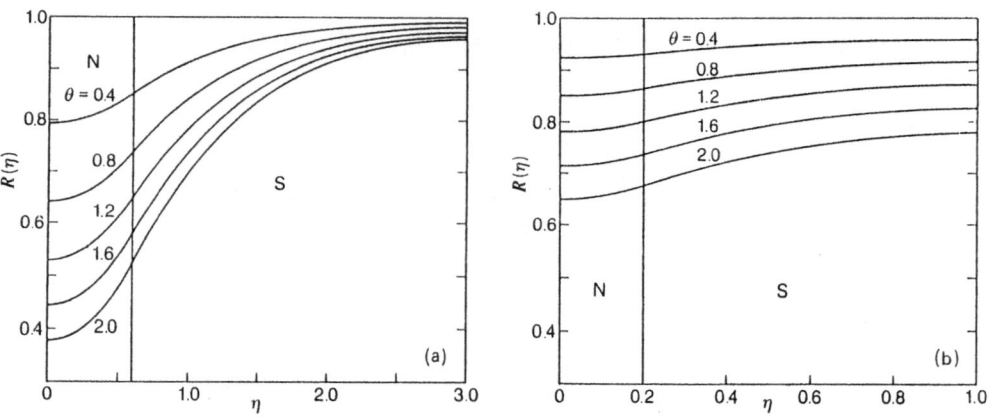

Fig. 2. Order parameter in the multilayer for (a) d_n=0.6 and d_s=2.4 and (b) d_n=0.2 and d_s=0.8.

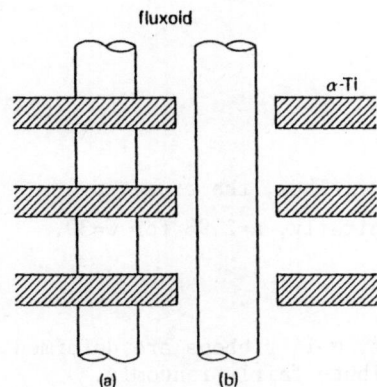

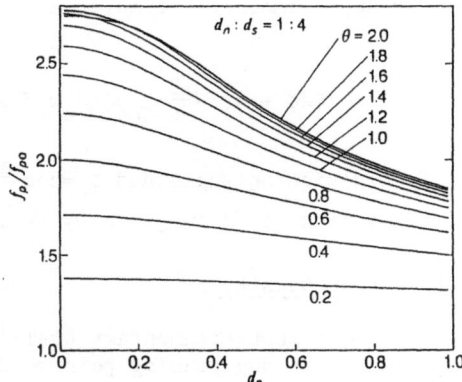

Fig. 3. Idealized arrangement of normal layers and fluxoid.

Fig. 4. Variation in f_p/f_{p0} when the thicknesses are reduced with the fixed ratio $d_n/d_s=0.25$.

is very large ($\kappa \simeq 70$), the magnetic field of each fluxoid is remarkably overlapped and is fairly uniformly distributed except in the very vicinity of the lower critical field. This means that the magnetic interaction is negligibly small in the practical field region for high κ materials. Hence, we concentrate only on the condensation energy interaction. Thus, the free energy can be approximated by Eqs. (18) and (19). In addition, we approximate that R=0 in the fluxoid core of diameter 2ξ. When the fluxoid is located at (a) in Fig. 3, its energy per periodic length $2(d_n+d_s)$ is given by

$$G = -2\pi\xi^3 \int_0^{d_n+d_s} F d\eta. \tag{22}$$

In the superconducting region, (b) in Fig. 3, R in the absence of the fluxoid can be approximated by a value $\langle R \rangle_s$ averaged in the superconducting layer. When the fluxoid exists in this region, the energy increase is

$$G' = 2\pi\mu_0 H_c^2 (d_n+d_s)\xi^3 (\langle R \rangle_s^2 - \frac{1}{2}\langle R \rangle_s^4). \tag{23}$$

Hence, the variation in the energy during the displacement of the fluxoid from (a) to (b) in Fig. 3 is obtained. We note that the edge of the normal layer works as a pinning center. Taking account of an overlap of normal cores at high fields, the elementary pinning force of the normal layer is approximately estimated from

$$f_p = \frac{\pi}{4} \cdot \frac{G'-G}{\xi} (1-\frac{B}{B_{c2}}). \tag{24}$$

In the idealized situation where the proximity effect does not occur, the elementary pinning force is reduced to

$$f_p = \frac{\pi^2}{4}\mu_0 H_c^2\xi^2 d_n (1-\frac{B}{B_{c2}}) \equiv f_{p0}. \tag{25}$$

Figure 4 shows a variation in f_p/f_{p0} when the thicknesses are reduced with keeping the ratio $d_n/d_s=0.25$.

In the practical geometry, the normal layers are not always perpendicular to the fluxoids. This means that the interaction volume is larger than $2\pi\xi^3 d_n$ in the above estimate in the simple case. The mean value of the volume overlapping between the fluxoid and the normal layer with a

267

width wξ is approximately estimated as

$$4\xi^3 d_n\int_0^{\zeta_c} \sec\zeta d\zeta + 4(1-\frac{2}{\pi}\zeta_c)\xi^3 d_n w \equiv 2\pi a\xi^3 d_n, \tag{26}$$

where ζ is a deviation angle and $\zeta_c=\cos^{-1}(2/w)$. Thus, the expected elementary pinning force is given by af_p. Typically, a=2.98 for w=30.

Critical Current Density

In practical multifilamentary Nb-Ti wires, α-Ti ribbons are deformed and their edges, strong pinning points, distribute fairly randomly. Hence, the net force on fluxoids is considered to be given by a statistical summation of randomly directed individual pinning forces. For very strongly pinned materials as commercial Nb-Ti wires, the transverse elastic correlation length ℓ_{66} of the fluxoid lattice is expected[7] to be less than the fluxoid spacing a_f and the longitudinal one ℓ_{44} seems to be given by a cut off length $4\pi a_f$. This suggests that each fluxoid behaves incoherently as in the amorphous state. If we denote the concentration of α-Ti ribbons in a transverse cross section of the superconducting filament by N, the number of ribbons that interact with the coherent region in a single fluxoid core is $n=8\pi\xi a_f N$. This value varies from 2 to 10 in the practical cases. Since n is not large enough, a deviation from the simple statistical behavior of Gaussian type is expected. In fact, the numerical simulation clarified that the total pinning strength due to randomly distributed n pins with the individual strength af_p is approximately expressed as $1.6n^{0.56}af_p$ in the range of $2\leq n\leq 10$. The exponent approaches asymptotically to 1/2 for large n and to 1 for small n. Thus, the macroscopic critical current density is obtained:

$$J_c = \frac{1.6n^{0.56}af_p}{4\pi a_f \phi_0}, \tag{27}$$

where ϕ_0 is the flux quantum.

DISCUSSION

Lee and Larbalestier[2] investigated in detail the correlation between J_c and the microstructure in Nb-46.5wt%Ti. The critical temperature T_c of the same specimens was reported by Meingast et al.[3] Here, we compare the present theoretical result with their experimental data. The starting specimen was a wire heat-treated after drawing by the true strain of 12 and had T_c=9.44K. This was drawn up to the strain ranging from 2.38 to 4.04. The content of α-Ti was kept constant during the drawing and was about 20%.[2] Hence, we can concentrate ourselves to the case of d_n/d_s=0.25. The average thickness of α-Ti ribbons in fine wires was close to the limit of measurement and contained ambiguity.[2] Hence, we assume the expected thickness, represented by the solid line in Fig. 14 in Ref. 2. The upper critical field of these specimens was about 10.3T and we have ξ=5.66nm. The starting specimen is also influenced by the proximity effect from α-Ti precipitates and we assume that the maximum T_c, 9.50K, for T_{c0}. The

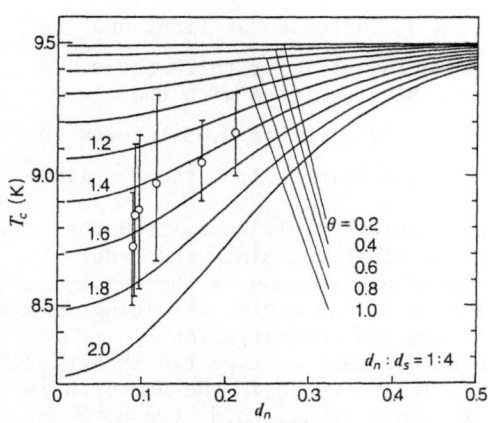

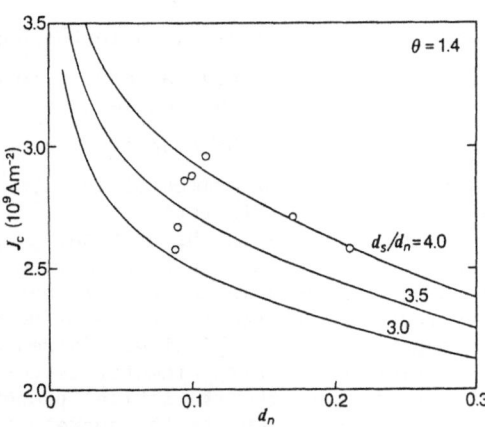

Fig. 5. Calculated critical temperature for d_n/d_s=0.25 and the
observed results (O) by
Meingast et al.[3]

Fig. 6. Calculated critical current
density for θ=1.4 and the
observed resluts (O) by Lee
and Larbalestier[2] at B=5T.

obtained theoretical result on the critical temperature is compared with
the experimental data in Fig. 5. Both critical temperatures decrease
monotonically with decreasing thicknesses. The observed T_c agrees with
the calculated result with θ=1.4 for d_n>0.1. For d_n<0.1, however, the
observed T_c drops sharply with decreasing thicknesses. The cause of this
deviation is discussed below.

If α-Ti phase is dirty, its coherence length is given by[8]

$$\xi_n = (8.34hk_B/e^2\rho_n\gamma_vT)^{1/2},\qquad(28)$$

where k_B the Boltzmann constant, ρ_n the normal resistivity, and γ_v the
electronic volume specific heat coefficient in SI units. Substitution of
$\rho_n\simeq4.0\times10^{-8}\Omega$m for Ti-1at%Nb[9] and $\gamma_v\simeq318$J/m^{3}K^2 for pure Ti[10] leads to
ξ_n=23.6nm. From an inverse proportionality between the electron mean free
path ℓ and ρ_n for Ti-Nb alloys,[11] we expect $\ell\sim10$nm. This value is much
larger than the thickness of α-Ti ribbons(0.9$\sim$2.1nm) and is sufficiently
samller than ξ_n. Hence, the required conditions are fulfilled. If the
effective mass of electron is the same between α-Ti and β-phase, we have
$\theta=(\xi/\xi_n)^2$=0.06. The value of θ obtained from the comparison on T_c is
fairly larger than this. This may be attributed to the neglect of higher
order term in the right-hand side of Eq. (3). That is, the effective
value of α_n is given by $\alpha_n+\beta_n|\Psi_n|^2/2$, where Ψ_n is the induced order
parameter in the normal layers. Thus, the effective θ is given by $\theta'=\theta+$
$\beta_n|\Psi_n|^2/2|\alpha|$. If we assume that β_n in α-Ti phase is comparable to β in
the matrix, the second term is of order unity under a significant prox-
imity effect and the large θ value obtained from comparison of T_c seems
to be reasonable.

Calculated critical current density at B=5T (a_f=21.9nm) is compared
with the experimental results in Fig. 6, where we used a=2.98 and
apprcximated as N=1/10wd$_n\xi^2\sim$1/350d$_n^2\xi^2$ from the results in Ref. 4. A
factor of 0.63 is multiplied to the calculated J_c. The calculated J_c

increases monotonically, while the observed J_c increases at first and then decreases sharply, according to a reduction of layer thicknesses. This deviation at small d_n is considered to be associated with the deviation in T_c. The quantitative difference by a factor 0.63 between the theoretical and experimental results seems to originate from the simplified treatment in this paper.

The increase in the critical current density mainly comes from the proximity effect as can be seen from Fig. 4. That is, since the order parameter induced in the normal layers causes an increase in the energy, its reduction by the existence of normal core is favorable, resulting in a stronger attractive pinning. Increase in the pin concentration by reduction of the wire diameter is not so significant as expected in Ref. 2. This comes from the statistical property. The increase in the number n is almost cancelled out by the decrease in f_p proportional to d_n (we note $n \propto 1/d_n^2$).

The degradation in J_c in Fig. 6 for small d_n value seems to be connected to the sudden drop of T_c shown in Fig. 4. Lee and Larbalestier[2] argued that the sausaging of filaments in the wire caused the decrease in J_c. However, the sudden drop of T_c cannot be explained by the mechanism of sausaging. Hence, some variation in microstructures in the wire is considered to affect the both degradations. For example, the heavy drawing may bring about a constriction of superconducting layers. If the thickness of the superconducting layers becomes relatively smaller, a significant reduction in the order parameter in these layers occurs. In fact, the critical current density decreases notably according to a reduction in the thickness of superconducting layers, as shown in Fig. 6. The decrease in T_c can also be explained qualitatively by assuming the constriction of superconducting layers. Such a constriction may be induced by sausaging of filaments.

The above agreement suggests a necessity of development of drawing technology that does not disorder the laminar structure inside the filament. If this is achieved, a further increase in J_c by reduction of wire diameter will be realized, as by the solid lines in Fig. 6.

REFERENCES

1. Li Chengren and D. C. Larbalestier, Cryogenics 27, 171 (1987).
2. P. J. Lee and D. C. Larbalestier, Acta metall. 35, 2523 (1987).
3. C. Meingast, M. Daeumling, P. J. Lee, and D. C. Larbalestier, Appl. Phys. Lett. 51, 688 (1987).
4. C. Meingast, Ph. Dr. Thesis (Univ. Wisconsin, 1988).
5. T. Matsushita, J. Appl. Phys. 54, 281 (1983).
6. R. O. Zaitsev, Sov. Phys. JETP 23, 702 (1966).
7. K. Yamafuji, T. Matsushita, T. Fujiyoshi and K. Toko, "Adv. Cryog. Eng. Mater.," Plenum, New York (1988) Vol. 34, p. 707.
8. E. W. Collings, "Applied Superconductivity - Metallurgy, and Physics of Titanium Alloys," Plenum, New York (1986) Vol. 1, p. 421.
9. R. R. Hake, D. H. Leslie and T. G. Berlincourt, Phys. Rev. 127, 170 (1962).
10. C. Kittel, "Introduction to Solid State Physics," 6th ed., John Wiley & Sons, New York (1986) p. 141.
11. E. W. Collings, "Applied Superconductivity - Metallurgy, and Physics of Titanium Alloys," Plenum, New York (1986) Vol. 1, p. 490.

THE TRANSITION TO THE NORMAL STATE

OF A STABILIZED COMPOSITE SUPERCONDUCTOR

O. Christianson and R. W. Boom

Applied Superconductivity Center, University of Wisconsin
Madison, WI 53706

ABSTRACT

The transition to the normal state of a stabilized composite
superconductor, induced by applying a thermal pulse directly to the
superconductor, is studied experimentally. The conductor is configured
such that heat transport is one dimensional and current diffusion times
are lengthened enabling measurement. The transport current capacity,
transport current distribution, time of the normal transition after
application of a thermal pulse, and recovery currents for transport
currents exceeding the critical current of the superconductor are
measured and explained.

INTRODUCTION

As large cryostabilized superconductors are designed for eventual
commercial use,[1,2] it is important to understand the physical processes
and their interrelationships affecting the stability of stabilized
superconductors. In this experiment the transition from the
superconducting state to the normal state is measured and explained.
This experiment is part of a larger study on the effects of current
diffusion on the stability of a stabilized superconductor.[3] Measurements
are made upon a uniquely configured conductor, see figure one, such that
current diffusion times are lengthened,[4] and the heat transport is one
dimensional. The conductor is a NbTi rectangular superconductor 0.1 cm
wide, 2×10^{-4} m thick, metallurgically bonded to a 75 residual
resistivity ratio (RRR) copper tab which is soldered to a 2 cm wide, 2.54
$\times 10^{-4}$ m thick 200 RRR copper stabilizer. The electrical resistance
between the superconductor and copper tab is 3.1×10^{-14} $\Omega\text{-m}^2$ (using the
current transfer method[3]), and, using the Lorentz ratio, the thermal
resistance is 4.7×10^{-7} $\text{m}^2\text{K/W}$. Cooling is provided by soldering the
stabilizer to a vertical, vacuum tight U shaped tube containing LHe at
atmospheric pressure. This configuration results in natural convective
cooling with the heat transfer being $q = h\Delta T$ where h is 10^4 $\text{W/m}^2\text{K}$ and ΔT
is the temperature of the stabilizer above the bath temperature.

RECOVERY CURRENT

The conductor exhibits typical V-I characteristics[5] with a critical
current, I_c, of 270 A, and a steady-state or transient recovery current,

Advances in Cryogenic Engineering (Materials), Vol. 36
Edited by R. P. Reed and F. R. Fickett
Plenum Press, New York, 1990

271

I_r, of 195 A. For total recovery to occur after the critical current is exceeded or after application of a thermal pulse, the power generated in the superconductor during current sharing must be less than the heat flux out of the superconductor. The heat flux out of the superconductor for a given current carrying capacity is found from the steady-state heat conduction equation, the one dimensional Poisson equation,

$$d(kdT/dx)/dx + q''' = 0 \qquad (1)$$

where T is the temperature, k is thermal conductivity, and q''' is the uniform volumetric heat generation equivalent to integrating over the temperature distribution. Because the thermal conductivity of the copper is large, the temperature difference across the stabilizer is small and is neglected. The heat conducted out of the superconductor is

$$q_w = 2kA(T_m - T_b)/w, \qquad (2)$$

where A is the heat transfer area to the copper stabilizer, T_m is the temperature on the outer edge of the superconductor, w is the width of the superconductor, and T_b is the bath temperature.

The current carrying capacity, I_{cc}, is calculated from the temperature distribution. The critical current density is approximately linear in temperature[6]

$$j_c(T) = j_{cb} [(T_c-T)/(T_c-T_b)], \qquad (3)$$

where j_{cb} is the critical current density at the bath temperature and T_c is the critical temperature of the superconductor. The current carrying capacity, for constant k and q''', is

$$I_{cc} = (I_{cb}/(T_c-T_b))[T_c-T_b-(2/3)(q'''w^2/2k)]. \qquad (4)$$

The average temperature of the superconductor is

$$\overline{T} = T_b + (2/3)(q'''w^2/2k). \qquad (5)$$

Rearranging Eq. 2 and 4 gives

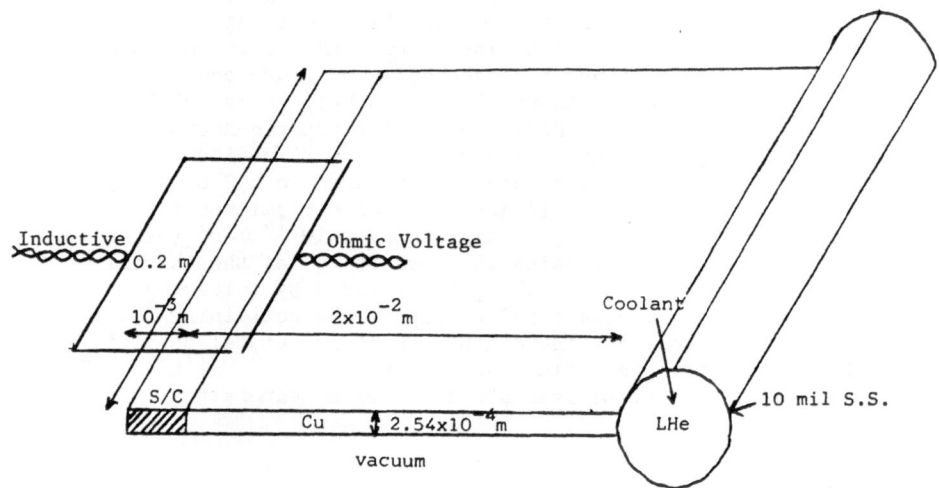

Inductive   0.2 m   Ohmic Voltage   Coolant

10^{-3} m 2×10^{-2} m   10 mil S.S.

S/C   Cu   2.54x10^{-4} m   LHe

vacuum

Figure 1.   The conductor geometrical configuration showing superconductor, stabilizer, voltage taps, and cooling.

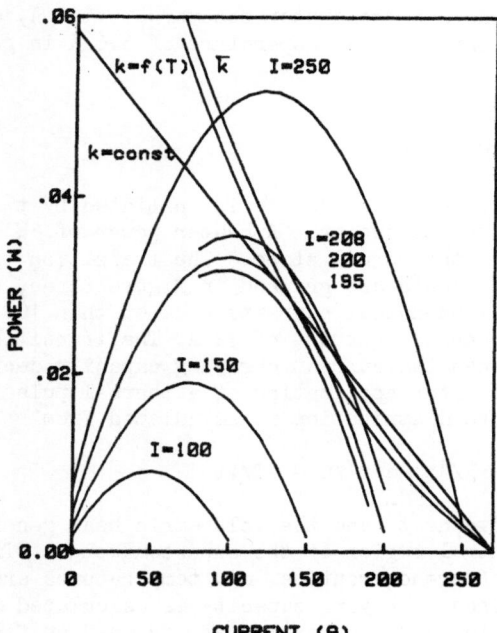

Figure 2.   Generated power in the superconductor at
             several transport currents and cooling curves.

$$T_m - T_b = (3/2)(\overline{T} - T_b).\tag{6}$$

The heat conducted away from the superconductor for a given current
carrying capacity with a constant thermal conductivity is

$$q_w = (2kA/w)(3/2)(T_c-T_b)(I_c - I_{cc})/I_c.\tag{7}$$

The power generated during current sharing must be less than the
heat flux out of the superconductor for recovery to proceed.   The
generated power is

$$q = V \cdot I = (I_t - I_{cc}) I_{cc} R\tag{8}$$

where V is the voltage along the conductor multipied by the current in
the stabilizer, $I_t - I_{cc}$, with I_t being the transport current.   The
condition for total recovery is

$$I_{cc}(I_t-I_{cc})R < (2kA/w)(3/2)(T_c-T_b)(I_c-I_{cc})/I_c.\tag{9}$$

The generated power of several transport currents is graphed versus
the current in the superconductor in figure two.   The generated power is
offset from zero due to joule heating at the contact between the current
leads and the sample.   Also graphed is the heat conducted out of the
superconductor for a particular current carrying capacity for a constant
thermal conductivity, an average thermal conductivity over the
temperature distribution in the superconductor, and a temperature
dependent thermal conductivity (solved numerically).   For transport
currents below 195 A, the recovery current, the generated power is less
than the cooling and recovery proceeds.   For currents greater than 195 A
partial recovery occurs to the point of the intersection of the power
generated curve and the cooling curve.   For 208 A the superconductor
should recover to 122 A and a temperature of 6.8 K.   It is found

experimentally that the superconductor recovers to 128 A, and thermocouple measurement gives a temperature of 6.6 K in good agreement with predictions.

TRANSITION TO THE NORMAL STATE

The transition to the normal state is initiated by two heaters, one on each face of the superconductor. A heater power of 8W is used to induce a transition to the normal state. The transition times for different transport currents are graphed in figure three, with times ranging from 0.5 milliseconds at currents greater than 100 A to 6 milliseconds for a transport current of 32 A. The transition to the normal state occurs when the current carrying capacity decreases below the transport current after application of a thermal pulse. The one dimensional temperature distribution is calculated from

$$\partial((k/C)\partial T/\partial x)/\partial x + q''' /C = \partial T/\partial t \qquad (10)$$

where C is the specific heat[7] and the volumetric heat generation, q''', is localized to a central region of the superconductor. The equation is solved by a finite difference routine, and temperatures are plotted in figure four. The current carrying capacity is calculated from this temperature distribution and the results are graphed on figure three. For transport currents less than 100 A the transition time at a given transport current agrees with the calculated current carrying capacity. As the transport current is increased above 100 A, the transition time does not decrease. The heaters are placed on the center of the face of the superconductor. Above 100 A, the current distribution extends into the superconductor beyond the heaters. As heat is applied, current beyond the heater is free to migrate to unoccupied, cooler regions of the

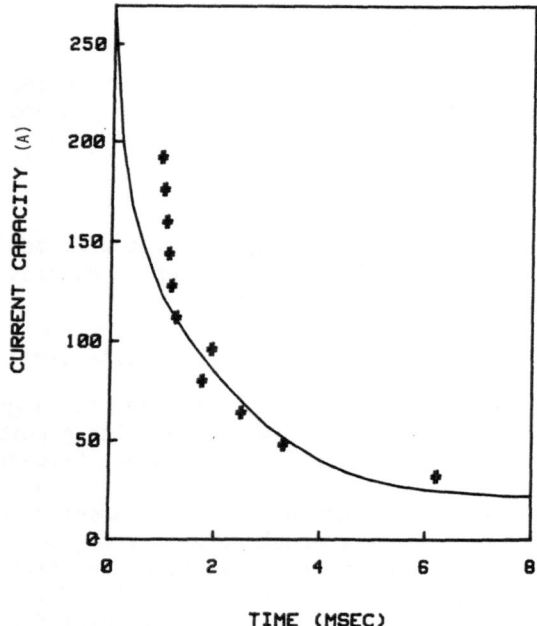

Figure 3. The current carrying capacity of the superconductor versus time after a heater pulse. The line is calculated, while '#' are measured transition times.

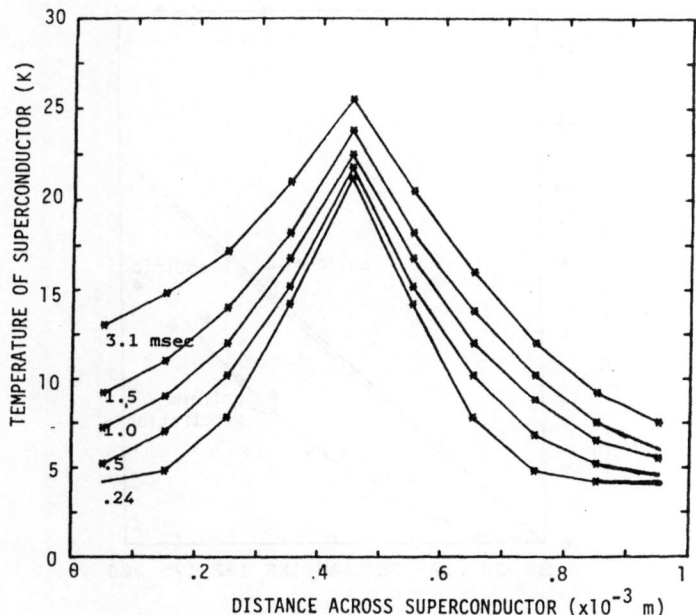

Figure 4. The calculated temperature distribution in the superconductor after a heater pulse.

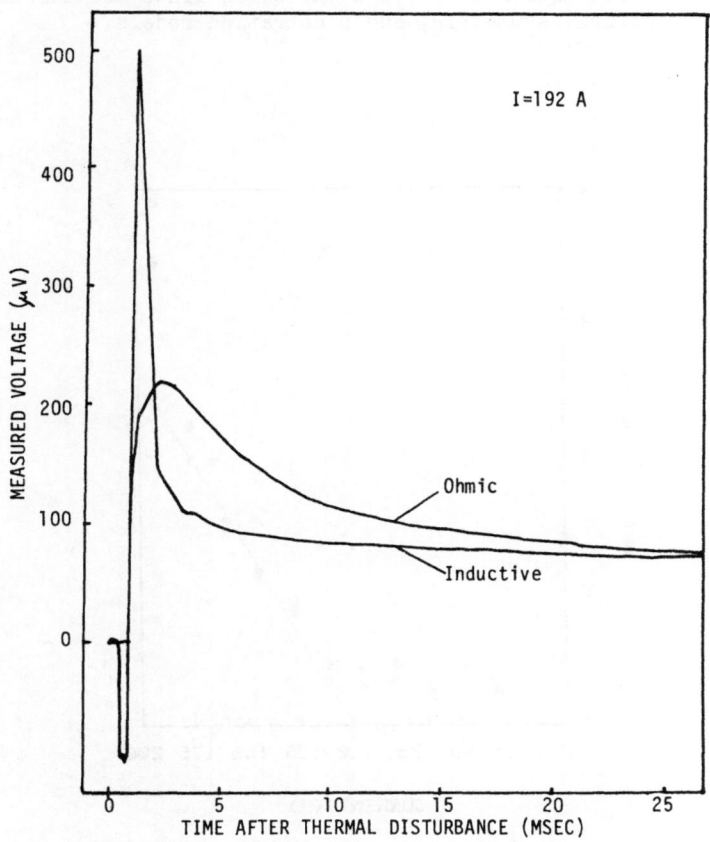

Figure 5. Voltages after an 8W thermal pulse from which the changing vector potential is calculated.

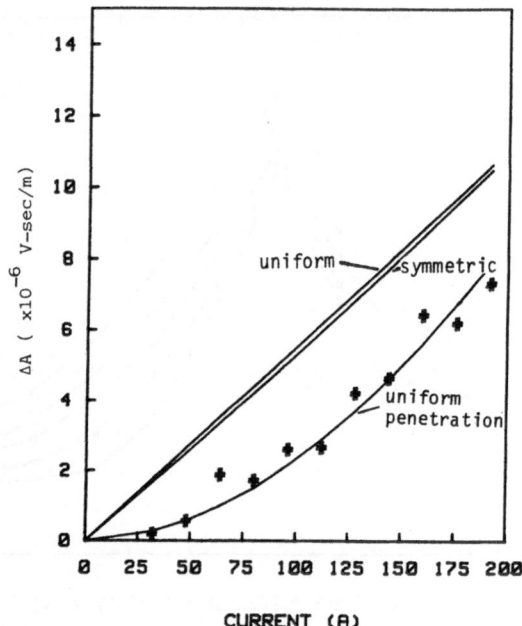

Figure 6. The change in vector potential due to current exiting the
superconductor after a normal transition versus transport current.
The experiment data are '#', and the solid lines are calculations
for the uniform, symmetric, and penetration models.

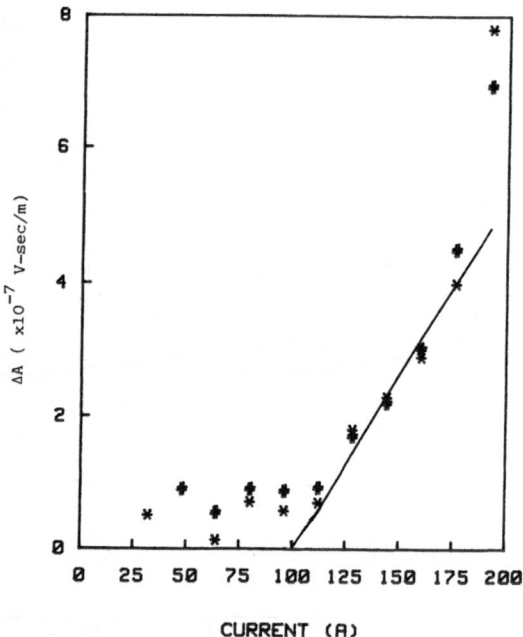

Figure 7. The experimental vector potential due to displacement of a
small amount of current before the transition to the normal state
versus transport current.

superconductor. The current close to the stabilizer does not migrate but remains constant. The normal transition occurs when the current carrying capacity decreases below the transport current in the region between the heater and stabilizer and is approximately independent of the transport current above 100 A.

CURRENT MIGRATION WITHIN AND OUT OF THE SUPERCONDUCTOR

The transition of the superconductor to the resistive normal state causes the transport current to exit from the superconductor into a small region of the stabilizer adjacent to the superconductor. This redistribution of current produces an inductive voltage, see figure five, which is related to the changing vector potential by

$$-V = d(\Delta A \cdot \ell)/dt \tag{11}$$

where $\Delta A = A_f(x) - A_i(x)$, ℓ is the separation of the voltage taps and f and i indicate the final and initial current distribution. The change in the vector potential is

$$\Delta A_{exp} = -(1/\ell) \int (V_{in} - V_o) \, dt, \tag{12}$$

where V_{in} is the inductive voltage and V_o is the ohmic voltage (see figure one for the placement of the voltage taps and figure five for V_{in} and V_o). Experimental values of ΔA are graphed in figure six.

The change in the vector potential can be calculated from the current distribution in the superconductor. The vector potential for a current I flowing in the z direction at a position x is

$$dA = (\mu_o \, I/4\pi) \, dz/r \tag{13}$$

and

$$A = (\mu_o I/4\pi) \ln (\ell+/\overline{\ell^2+x^2})/(-\ell+/\overline{\ell^2+x^2}) \tag{14}$$

For a given current distribution, the change in the vector potential is

$$A = A_f - (\mu_o/4\pi) \int dI(x) \ln (\ell+/\overline{\ell^2+x^2})/(-\ell+/\overline{\ell^2+x^2}), \tag{15}$$

which is evaluated numerically for several current distributions in the superconductor: 1) a uniform current across the width of the superconductor, 2) a symmetric configuration with current equally positioned on each edge, and 3) uniform current penetration into the superconductor at j_c, the critical current density. These values are graphed in figure six. Agreement between experimental data and the uniform penetration model is best, that is the initial current distribution in the superconductor is uniform at the critical current density up to a penetration distance given by the transport current.

An initial inductive voltage before the transition to the normal state occurs due to the heaters producing a small, hot region with the current within the region being displaced to the edge of the penetration of the transport current into the superconductor. In figure seven the experimental ΔA is graphed versus transport current, as well as calculations of the vector potenetial resulting from displacement of a 15 A normal region being produced by the heaters. This model predicts no inductive voltage until 112 A, while measured values are below the noise level for currents less than 128 A. The claculated ΔA agree with the measured values.

CONCLUSIONS

After application of a heater pulse, a normal zone is created at 0.5 milliseconds. Some current is displaced producing a small inductive voltage. As the superconductor warms up, the total current carrying capaciity decreases below the transport current inducing a normal transition at approximately 1 millisecond. For currents less than 100 A, the transition occurs at longer times, 6.3 milliseconds for 32 A. The current exits from the superconductor producing an inductive voltage when the transition to the normal state occurs. The current distribution, derived from this inductive voltage, within the superconductor is uniform at j_c extending into the superconductor to the value of the transport current. The recovery current is determined by the condition that heat produced inside the superconductor must be less than the cooling of the superconductor, which in this case is limited by the thermal conductivity of the superconductor.

REFERENCES

1. R. W. Boom, Ed., Wisconsin Superconducting Magnetic Energy Storage Project, Volumes I-IV, Engineering Experiment Station, University of Wisconsin, 1974-1979.
2. R. J. Loyd, T. E. Walsh, E. R. Kimmy, B. E. Dick, "An Overview of the SMES ETM Program: The Bechtel Team's Perspective," IEEE Trans. on Mag., Vol. 25, No. 2, pp. 1569-1575, March 1989.
3. O. Christianson, "Transition and recovery of a cryogenically stable superconductor," Ph.D. Dissertation, Dept. of Nuclear Eng., U. of Wisconsin-Madison, 1984.
4. O. Christianson and R. W. Boom, "Transition and recovery of a cryogenically stable superconductor," Adv. in Cryo. Eng., Vol. 31, pp. 207-214, 1985.
5. W. F. Gauster and J. B. Hendricks, "Flux Flow and Thermal Stability of Stabilized Superconductors," J Appl Phys 39(6), 2572-2578 (1978).
6. J. E. C. Williams, Superconductivity and Its Applications, Pion Limited, London, 1970.
7. D. C. Larbalestier, "Niobium-Titanium Superconducting Materials," in Superconducting Materials," Ed. S. Foner and B. B. Schwartz, Plenum Press, 1981.

A STUDY OF QUENCH CURRENT AND STABILITY OF HIGH-CURRENT

MULTI-STRAND CABLES HAVING A Cu OR A CuNi MATRIX

G.B.J. Mulder, H.J.G. Krooshoop, A. Nijhuis,
H.H.J. ten Kate and L.J.M. van de Klundert,

Applied Superconductivity Centre
University of Twente
Enschede, The Netherlands

ABSTRACT

This paper discusses the experimental results concerning maximum current and stability of two braided superconducting cables. The expected critical current of both conductors is 95 kA under self field conditions, at 4.2 K. An essential difference is that one of these conductors has a pure CuNi matrix, the other a Cu matrix. The maximum current of the cables was measured as a function of the temperature and the ramp rate of the current. We observed a remarkable decrease of the current-carrying capacity with increasing current rate in both cables, independent of the matrix material. Furthermore, the stability of the cables was investigated.

INTRODUCTION

Recently, a research programme was started in Twente to study superconducting cables that can be used in superconducting switches for currents from 50 kA up to 200 kA. The goal is to develop such cables and eventually apply them in the switches of high-current superconducting rectifiers. An important consequence of using conductors for switching purposes is that the matrix <u>must</u> have a high resistivity, otherwise it is not possible to attain the desired open state resistance of the switch with an acceptable length of conductor. This means that switch conductors are always poorly stabilized, usually having a matrix of CuNi or no matrix at all.

In many switches, constructed over the years in our laboratory, it was observed that reliable operation is only possible at currents far below the critical current. This undesirable phenomenon seems to grow worse as the current level increases. It is in fact the main problem that has to be solved in high-current cables for switching purposes. In order to determine the role the matrix material in this deterioration of the current-carrying capacity we compared the results of two cables, one with CuNi matrix, the other with Cu matrix. Obviously, the latter cable is unsuitable for superconducting switches. Our experiments were twofold. On the one hand, the quench current of each cable was measured as a function of parameters such as the temperature and the ramp rate of the current. On the other hand, the stability margin was determined, by measuring the response of the cable to a local heat input in one of the strands.

Advances in Cryogenic Engineering (Materials), Vol. 36
Edited by R. P. Reed and F. R. Fickett
Plenum Press, New York, 1990

279

Table 1. Specifications of the two cables

		Cable 1	Cable 2
strand:	materials	NbTi in CuNi	NbTi in Cu
	diameter	0.275 mm	0.255 mm
	filaments	576	367
	matrix/sc.	1.10	1.25
	twist pitch	12.5 mm	12.5 mm
	insulation	none	none
	manufacturer	MCA	MCA
sub-cable:	shape	6 strands around core	6 strands around core
	core	nichrome dummy wire	one strand
	twist pitch	20 mm	20 mm
cable:	type	round braided	round braided
	size	$\simeq 38 \times 6$ mm^2	$\simeq 35 \times 5$ mm^2
	sc. strands	2x24x3x6 = 864	2x24x3x7 = 1008
	braid pitch	300 mm	300 mm
	NbTi surface	24.4 mm^2	22.7 mm^2

THE TESTED CABLES

The parameters of the two investigated cables are summarized in Table 1. Except for the matrix material they are quite similar. From the measured critical current of one strand, we calculated a critical current of 95 kA for both cables, under self field conditions, at 4.2 K. Both cables were braided from 144 "six-around-one" sub-cables. In the CuNi sample the core of each sub-cable is non-superconducting so the total number of strands amounts to 864, contrary to the 1008 strands in the Cu sample.

A few remarks can be made about the switch cable :

- Superconducting switches mostly find their application in a relatively low background magnetic field. The self magnetic field of the cable is 1.7 T at 100 kA. In other words, the self field is usually dominant. Therefore the quench current <u>under self field conditions</u> is of main interest to us.
- In order to obtain switch conductors for several kA, multi-strand cables are required instead of monolithic conductors. This is a consequence of the self field stability that limits the usable diameter of a conductor to less than about 0.3 mm in the case of a CuNi matrix. In our samples the strands have a diameter of 0.25 mm, which means that they operate close to the limit predicted by the theory of self field stability.
- The cable is braided. The advantage of such a type of cable is that it can be bent easily, allowing a compact construction of the switch.
- Although the strands are uninsulated, their mutual electrical contact is poor. Therefore, the current in a strand can only be redistributed to the other strands via a soldered section of the cable, for example in the electrical joint.

THE TEST SET-UP

The experimental arrangement that was used to perform measurements in the range of 50 to 200 kA is described extensively in ref. 1. Essentially, it is a superconducting transformer where the secondary part consists of one short-circuited turn of the cable under test. Due to a large number of primary turns, a relatively small primary current is sufficient to induce the required transport current through the cable. The amplification factor of our transformer is 2300.

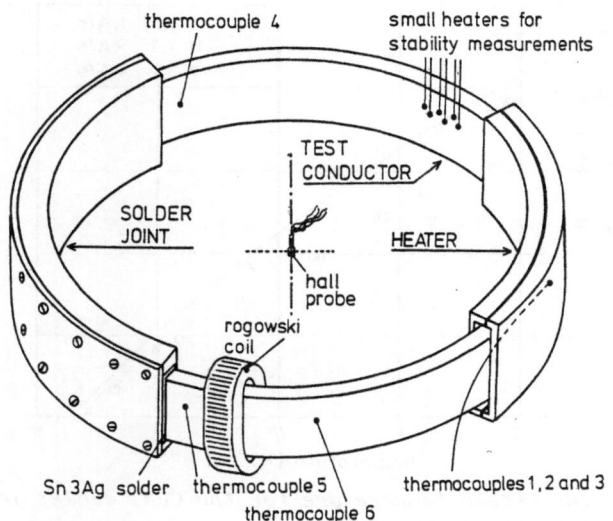

Fig. 1. *Lay-out of the samples, showing the conductor, the electrical joint and the heater. The diameter is 0.28 m.*

Figure 1 shows the lay-out of our samples. A length of 30 cm is used for the electrical connection. The remaining length, of about 60 cm, can partly be warmed up by means of a heater. The heatable sections are 13 and 21 cm long for samples 1 and 2 respectively. A few small heaters were fixed on the cable in order to enable stability measurements. The Hall sensor, shown in Fig. 1 in the center of the transformer, is used to determine the transport current in the sample. A Rogowski coil is included to check the current measurements.

RESULTS OF QUENCH CURRENT MEASUREMENTS

All quench currents were measured using a linear ramp current in the primary coil. The induced transport current through the sample also increases at a constant rate, because the measuring time is much shorter than the decay time of the secondary current. The decay time constants of the secondary circuit are over 5000 s, corresponding with joint resistances of less than 0.1 nΩ.

Maximum current as a function of temperature

The heater was used to raise the temperature T of the conductor and thus determine the maximum current I_{max} as a function of T. Figures 2 and 3 show the results. During the experiments it became clear that the CuNi sample suffers from training and the reproducibility of the measurements above 50 kA is poor, see the spread in the results in Fig. 2. This is an indication of the extreme sensitivity of conductors having a pure CuNi matrix to thermal disturbances, for example initiated by wire motion or epoxy cracking.[2,3]

In figures 2 and 3, the flat part of the curves below 5.5 K in the case of CuNi and below 5.0 K in the case of Cu is remarkable. It seems here that I_{max} does not depend on the temperature, which is actually not true. The flat part is caused by the electrical joint which has a lower maximum current than the rest of the sample. This was checked for the Cu sample by

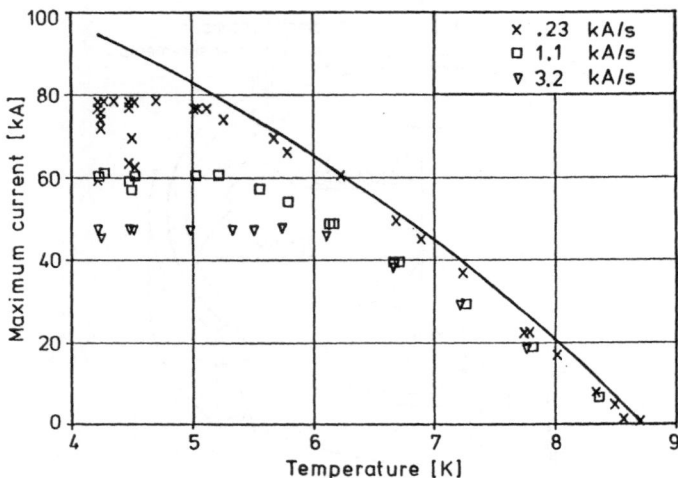

Fig. 2. *Maximum current versus temperature for the CuNi cable; measured at different values of dI_s/dt.*

reducing the bath temperature to 3.1 K and repeating the measurements above 4.2 K by means of the heater. By so doing, enhanced values of I_{max} were obtained at temperatures between 4.2 and 5.0 K. The flat part of the curves dissapears and instead we obtain the dashed lines indicated in Fig. 2.

The solid lines in figures 2 and 3 represent the critical currents for the cables as estimated from experimental data of the strands. Apparently, both of the cables can operate at currents close to the critical current provided :

1) the rate at which the current increases is sufficiently low, i.e. less than about 200 A/s,

2) the quench current of the joint section is sufficiently high,

3) there are no premature quenches due to instabilities, i.e. the current in the cable with CuNi matrix should not exceed 50 kA.

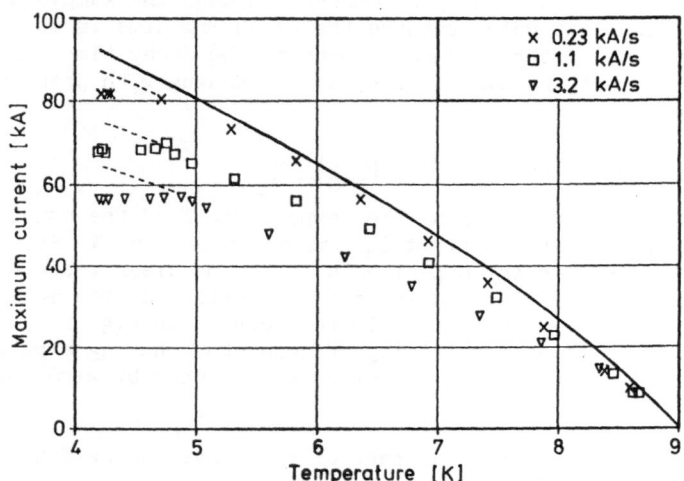

Fig. 3. *Maximum current versus temperature for the Cu cable; measured at different values of dI_s/dt. The dashed curves correspond with measurements where the joint was cooled to 3.1 K.*

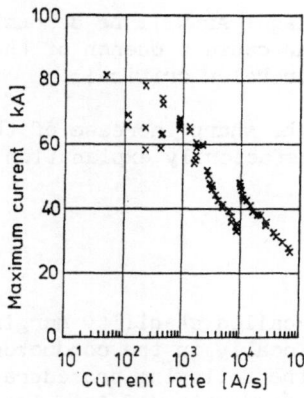

Fig. 4.
I_{max} versus current rate
for the CuNi cable at 4.2 K.

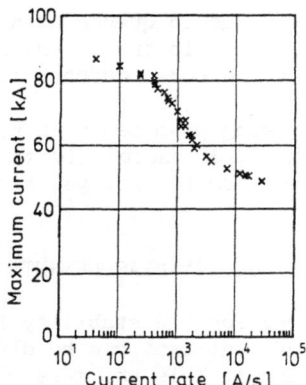

Fig. 5.
I_{max} versus current rate
for the Cu cable at 4.2 K.

Maximum current as a function of dI_s/dt

An important observation is that I_{max} depends rather strongly on the rate at which the current increases. Figures 4 and 5 show this dependence. The current-carrying capacity drops to approximately 50 % of the critical current if the rate exceeds 2 kA/s in CuNi cable or 10 kA/s in the Cu cable which corresponds with about 2 and 10 A/s per strand. In single strands a reduction of the transport current also exists, but this occurs at rates in the order of 5×10^4 A/s and can be attributed to a temperature rise of the conductor caused by AC losses. In the case of the cables, our thermocouples show for the measurements presented in Figs. 4 and 5 that the temperature rise is negligible, so the reduction of I_{max} is definitely not a thermal effect. A plausible explanation is that the magnetic field change at the cable induces an inhomogenous current distribution in the strands causing

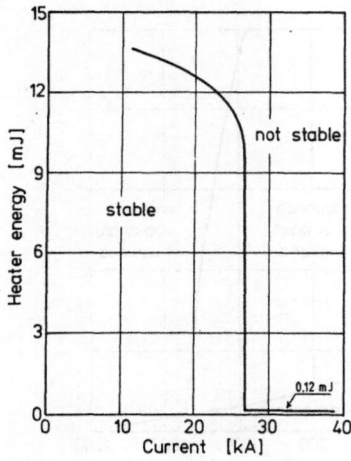

Fig. 6.
Quench sensitivity of the 864-
strands cable with CuNi matrix.

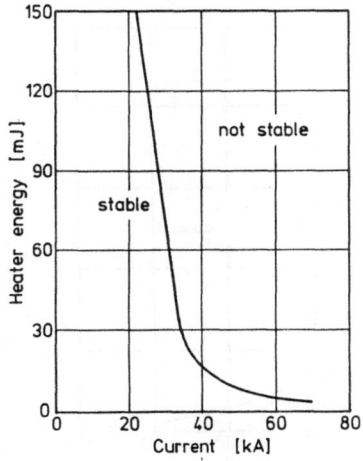

Fig. 7.
Quench sensitivity of the 1008-
strands cable with Cu matrix.

one of the strands to quench at an early stage. As will be demonstrated later on, a quench in one of the strands can cause a quench of the entire cable even if the total current is still far below critical.

A surprising phenomenon in Fig. 4 is the sharp increase of the quench current when dI_s/dt exceeds 10 kA/s. A satisfactorily explanation for this peculiar behaviour has not yet been found.

RESULTS OF STABILITY MEASUREMENTS

A measure for the stability is the so-called stability margin, i.e. the amount of heat that can be dissipated locally in the conductor without causing a quench. This was measured using the following procedure. At a stationary transport current, a pulse with a duration of 1 ms is supplied to a small heater, consisting of a thin manganine wire wound around one of the strands over a length of 2 mm. After supplying the heat pulse, there are three possibilities:

1) the heat pulse is too small to create a minimum propagating zone,

2) one strand quenches and its current is commutated to the other strands but the cable remains superconducting,

3) the entire cable quenches.

A Rogowski coil around the cable can be used to distinguish between these three cases, although the difference between cases 1 and 2 is sometimes difficult to detect. In case 2, the Rogowski voltage shows a small peak, corresponding with a few amperes current reduction during the process of current redistribution. By repeating the experiment for several combinations of current and heat pulse energy, it is possible to construct the stability plots such as Figs. 6, 7, 8 and 9.

Quench sensitivity of the cable

Figures 6 and 7 present the stability results for the cables. Clearly, the stability margin of the Cu cable is much larger than that of the CuNi

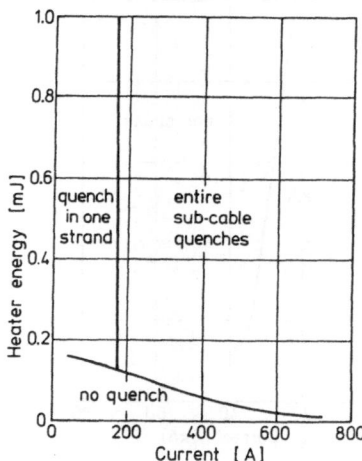

Fig. 8.
Quench sensitivity of a 6-strands sub-cable having a CuNi matrix.

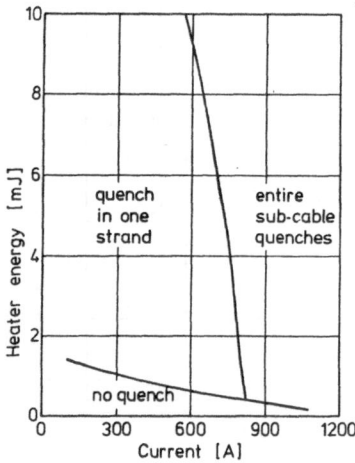

Fig. 9.
Quench sensitivity of a 7-strand sub-cable having a Cu matrix.

cable. The figures show that the difference is at least a factor of 10. For both cables, a remarkable phenomenon is observed. Above a certain value of the transport current there is a transition where the stability margin shows a sudden decrease by a factor of roughly 100. This happens at 27 kA for the cable with CuNi matrix and at 35 kA for the cable with Cu matrix.

The phenomenon can be explained as follows. A relatively small energy is sufficient to create a normal spot in just one of the strands, in which case its current will be transferred to the other strands. A calculation shows that almost 50 % of the current is transferred to each of the two neighbouring strands.[4,5] It should be noted that the redistribution involves large values of dI/dt and dB/dt since the whole process takes place within 100 μs for the CuNi cable and within 3 ms for the Cu cable. Depending on the current that was already present, the neighbours will or will not be capable of taking over the surplus current. Above a certain value of I_s, the surplus current cannot be taken over so the redistribution initiates a run-away effect driving the entire cable to the normal state. However, for lower values of I_s the redistribution takes place safely. In the latter case the stability margin is much larger because to quench the cable a thermal disturbance is necessary that creates normal spots in several strands simultaneously.

Quench sensitivity of the sub-cable

In a separate test arrangement we measured the stability of the sub-cables with CuNi and Cu matrix, see figures 8 and 9. These results have the same qualitative behaviour as the results of the full-size cables. Again, the difference between the stability margins of the two samples is at least a factor of 10. Here, it was actually possible to determine the region in the stability plot where current redistribution over the strands occurs without causing a quench of the sub-cable.

In the above experiments, a heater pulse duration of 1 ms was used. It appeared that a reduction of the pulse duration had no influence on the results. Below 1 ms, only the energy of the pulse is of importance. The Rogowski voltage shows that the redistribution takes place within 100 μs and 3 ms in the cases of CuNi and Cu respectively. If a heat pulse is supplied that is just sufficient to quench one strand, the peak of the Rogowski voltage, i.e. the quench, starts a few ms after the pulse.

CONCLUSIONS

Two cables having CuNi and Cu matrices were tested up to their maximum currents of 81 and 90 kA respectively. The cable with the CuNi matrix is typically suited for application in superconducting switches, due to its high resistance in the normal state. A serious disadvantage of the CuNi matrix is that the stability margin is at least ten times smaller than in a similar cable having a Cu matrix. Therefore, it is extremely sensitive to thermal disturbances, resulting in unreliable operation above 50 kA.

The experiments show the importance of making adequate electrical joints in such types of high-current cables. In our samples the electrical joints have resistances below 0.1 nΩ, which can be considered as of good quality, but nevertheless the quench current of the joint section proved to be significantly lower than the quench current of the conductor itself.

The multi-strand cables consist of numerous parallel strands that are electrically more or less insulated. Such a cable concept has an important consequence. If one of the strands quenches at an early stage, for example due to an inhomogeneous distribution of the currents or to a thermal disturbance, the current in this strand will be transferred to the other strands which can cause a quench of the entire cable. This partly explains two remarkable phenomena observed for both cables :

1) a drastic reduction of the maximum current with increasing values of the current rate,

2) a sudden decrease of the stability margin by approximately two orders of magnitude at about 30 % of the maximum current.

ACKNOWLEDGEMENTS

These investigations in the programme of the Foundation for Fundamental Research on Matter (FOM) have been supported by the Netherlands Technology Foundation (STW).

REFERENCES

1. G.B.J. Mulder, H.J.G. Krooshoop, A. Nijhuis, H.H.J. ten Kate and L.J.M. van de Klundert, A convenient method for testing high-current super-conducting cables, presented at CEC-89 paper BD-13, to be published in: "Advances in Cryogenic Engineering," Vol. 35, Plenum Press, New York (1990).

2. H.H.J. ten Kate, H. Pijper, A. Nijhuis and L.J.M. van de Klundert, Maximum current and quench sensitivity test of a 40 kA multistrand NbTi/CuNi conductor, in: "Proc. MT-9 Zurich 1985," Eds. C. Marinucci and P. Weymuth, SIN Switzerland (1985).

3. H.H.J. ten Kate, A.J.M. Roovers and L.J.M. vam de Klundert, Critical current and stability effects between 0 and 6 tesla in mono and multifilamentary NbTi conductors having a CuNi matrix, IEEE Transactions on Magnetics, MAG-21:362 (1985).

4. D. Faivre and B. Turck, Current sharing in an insulated multistrand cable in transient and steady state current conditions, IEEE Transactions on Magnetics, MAG-17:1048 (1981).

5. H.G. Knoopers, H.H.J. ten Kate and L.J.M. van de Klundert, Distribution of currents in a 6-strand superconducting cable, in: "Proc. MT-9 Zurich 1985," Eds. C. Marinucci and P. Weymuth, SIN Switzerland (1985).

RESTRICTED, NOVEL HEAT TREATMENTS FOR OBTAINING HIGH

J_c IN Nb46.5wt%Ti

P. J. Lee, J. C. McKinnell and D. C. Larbalestier

Applied Superconductivity Center, University of Wisconsin

Madison, WI 53706

ABSTRACT

The effect of greatly restricting heat treatment time and temperature, as well as the strain space between heat treatments was investigated for a high homogeneity Nb46.5wt% composite. The number of heat treatments was 3 or 6. It was found that the crucial heat treatment was the final one and that the initial ones could be considerably restricted. A strong linear relationship was found between J_c (5T, 8T) and the % α-Ti for J_c (5T) values ranging from $\sim$ 950–3050 A/mm^2 and α-Ti contents ranging from 2.5 to 21 vol%.

INTRODUCTION

The fabrication process for a high J_c Nb-Ti superconductor is illustrated in Figure 1 in terms of cold work strain. Strain will be defined here in terms of true strain, $\epsilon = ln\,(A_o/A)$, where A_o and A are the original last recrystallization and final cross-sectional areas of the Nb-Ti alloy. A cold work prestrain, ϵ_p, of approximately 5 is required before the first heat treatment of Nb-46.5wt%Ti alloy to ensure precipitation of α-Ti at β-Nb-Ti triple points. The exact prestrain depends on heat treatment temperature, alloy composition and alloy homogeneity.[1] Additional heat treatments are applied at strain intervals of 1.15; these heat treatments increase the amount of α-Ti and improve the uniformity of the microstructure.[2] Increasing the number of heat treatments and their duration increases the J_c and the quantity of precipitate.[3,4] After final heat treatment a large final drawing strain is applied. This strain distorts the α-Ti precipitates into a densely folded array of ribbons typically 1-2 nm thick, 4-8 nm apart. The final strain ranges from 4-5 with the higher strains being required to optimize composites having longer and/or higher temperature heat treatments (and higher peak J_c).[3]

In practice the total strain available for these operations is limited and is reduced by any warm or hot working process that reduces the stored work.[5] As normal commercial practice involves at least one warm extrusion using temperatures above 500°C, the loss in available strain space may be considerable. Methods of more fully utilizing the strain space include using hydrostatic extrusion in place of extrusion[6] or using NbTi rod annealed at a larger size. An alternative approach is to reduce the strain space between heat treatments. This experiment compares schedules using 3 heat treatments (HT) with one using 6 HT within a total heat treatment strain space of 3.4.

The duration of heat treatment is normally kept the same throughout the process and is typically 40 hrs or 80 hrs. Previous monofilament studies have shown that 80 hrs is superior to 40 hrs and that 160 hrs may also offer improved ultimate J_c.[3] Long heat treatments such as these, however, require excessive furnace time and may result in deleterious intermetallic formation at the filament surface if no diffusion barrier or one of insufficient thickness is used. Examining the amounts of α-Ti produced in previous heat treatment studies[1,2] there seems little advantage in using long initial heat treatments. For example, 10 hrs at 405°C produced 9 volume % α-Ti[2], whereas 80 hrs at 420°C produced only 11 volume % α-Ti at the first H.T.[1] Accordingly the principal thrust of the present

Advances in Cryogenic Engineering (Materials), Vol. 36
Edited by R. P. Reed and F. R. Fickett
Plenum Press, New York, 1990

287

work was to understand how the initial heat treatment(s) could be minimized, without unacceptably degrading the final J_c. The lowest heat treatment time and temperature used was 3 hrs at 300°C; this heat treatment, often used as a Cu anneal, has been found to only produce a thin grain boundary film of α-Ti in Nb-46.5wt%Ti.[7] The grain boundary film, typically less than 4 nm thick, is also produced at the higher temperature heat treatments.[8] By using a heat treatment of 3 hrs at 300°C in some schedules, the role of the grain boundary film in terms of triple-point α-Ti precipitate nucleation and microstructural refinement could be examined. Previous work suggested that considerable precipitate coarsening occurs in later heat treatments. If heat treatment times are minimized, a reduction in precipitate size would be expected. Monofilament studies have shown that long heat treatment times result in longer final strains being required for peak J_c.[3] If this is a consequence of the larger precipitate size at final HT size, then it might be speculated that ϵ_f can be reduced by minimizing heat treatment time.

EXPERIMENTAL DESIGN

Monofilament Manufacture

A Nb-diffusion barrier-clad high homogeneity grade monofilament was used throughout. The clad rod was supplied by Teledyne Wah Chang Albany at a size of 7.11 mm diameter with a total prestrain of 6.37. The heat treatment schedules are given in Table 1. The 3 heat treatment material received heat treatments at strain intervals of 1.39 (6 standard die passes) and the six heat treatment material at strain intervals of 0.682 (3 standard die passes). For direct comparison of wires, the 6 heat treatment series was given its first heat treatment at a prestrain of 6.37, compared with 7.00 for the 3 heat treatment series. The last heat treatment for both wires was thus applied at a total strain of 9.78.

Microstructural and Microanalytical Characterization

The local chemistry of the Nb-Ti billets was characterized on slices cut from the 5 3/4" diameter ingots using an ARL EMX 30 microprobe at 20 kV. Grain size was measured from polished and etched metallographic cross sections.

Transmission electron microscopy was performed on transverse cross-sections of the wires after final heat treatment in order to determine the quantity and size of the α-Ti produced. The α-Ti precipitates were identified by atomic number contrast. The atomic number contrast was enhanced

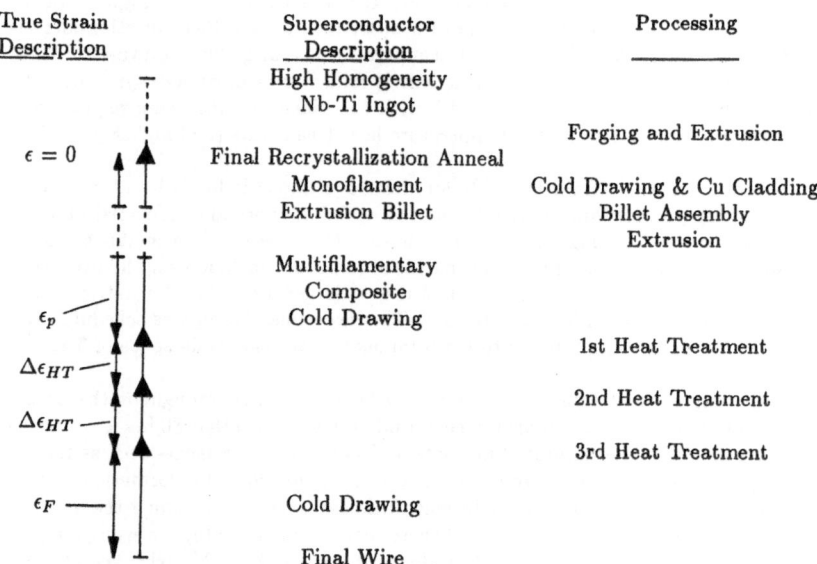

Figure 1: Strain space schematic illustration of a typical production route for a high J_c Nb-Ti composite.

Table 1: Heat Treatment Schedules for Monofilament Wires

Wire Designation	Heat Treatment (hrs and °)C Total Strain					
	6.37	7.00	7.70	8.39	9.09	9.78
UW210A	6h/405°C	6h/405°C	6h/405°C	20h/420°C	40h/420°C	80h/420°C
UW220A	3h/405°C	3h/405°C	3h/405°C	20h/420°C	40h/420°C	80h/420°C
UW250A	3h/300°C	3h/300°C	3h/300°C	3h/300°C	3h/300°C	3h/300°C
UW260A	3h/300°C	3h/300°C	3h/300°C	3h/300°C	3h/300°C	80h/420°C
UW270A	3h/300°C	3h/300°C	3h/300°C	20h/420°C	40h/420°C	80h/420°C
UW210B		6h/405°C		20h/420°C		80h/420°C
UW220B		3h/405°C		20h/420°C		80h/420°C
UW230B		6h/375°C		20h/420°C		80h/420°C
UW240B		6h/375°C		20h/405°C		80h/405°C
UW250B		3h/300°C		3h/300°C		3h/300°C
UW260B		3h/300°C		3h/300°C		80h/420°C
UW270B		3h/300°C		20h/420°C		80h/420°C

by combining images acquired at different specimen tilts using a Megavision image processor with a 1024 line x 1024 pixel resolution.[9] The same image processor was used to quantify both TEM and light microscopy images. In all but two cases, the total TEM image analysis area was 10 μm^2; in the other two cases (UW250A and B), a very fine scale precipitation was observed and the image analysis area was reduced to 1.8 μm^2.

Superconducting Critical Current Measurements

Critical current, I_c, measurements were performed on 0.6 m lengths of wire which were wound on barrels mounted coaxially inside the bore of a 12T solenoid. The current was determined at an overall wire resistivity of 10^{-14} Ωm ($\approx$ 10 μV m^{-1}). The critical current density, J_c, was obtained from I_c using Cu:NbTi ratios obtained by weighing wire samples before and after nitric acid removal of the copper matrix.

Table 2

Wire Designation (Heat Treatment)	Peak J_c		Microstructural Quantification			
	5T (ϵ_F)	8T (ϵ_F)	α-Ti Quantity	Mean Cross-Sectional Area		d^* α-Ti
	A/mm^2	A/mm^2	%	α-Ti nm^2	β-NbTi nm^2	nm
UW210A (3 × 6h/405°C+ 20,40,80h/420°C)	3223 (5.1)	1291 (5.6)	21.3	21,050	15,220	164
UW210B (6h/405°C+ 20,80h/420°C)	2973 (5.6)	1253 (5.6)	20.4	23,630	15,420	174
UW250A (6 × 3h/300°C)	2080 (3.3)	1040 (3.3)	12.8	1,010	4,830	36
UW250B (3 × 3h/300°C)	942 (4.2)	526 (3.3)	2.6	1,090	5,110	37
UW260A (5 × 3h/300°C+ 80h/420°C)	2973 (5.1)	1237 (5.1)	20.4	18,550	28,060	154
UW260B (2 × 3h/300°C+ 80h/420°C)	3044 (5.6)	1280 (5.6)	20.9	38,870	44,480	223

d* = mean diameter of α-Ti precipitates assuming circular cross-section

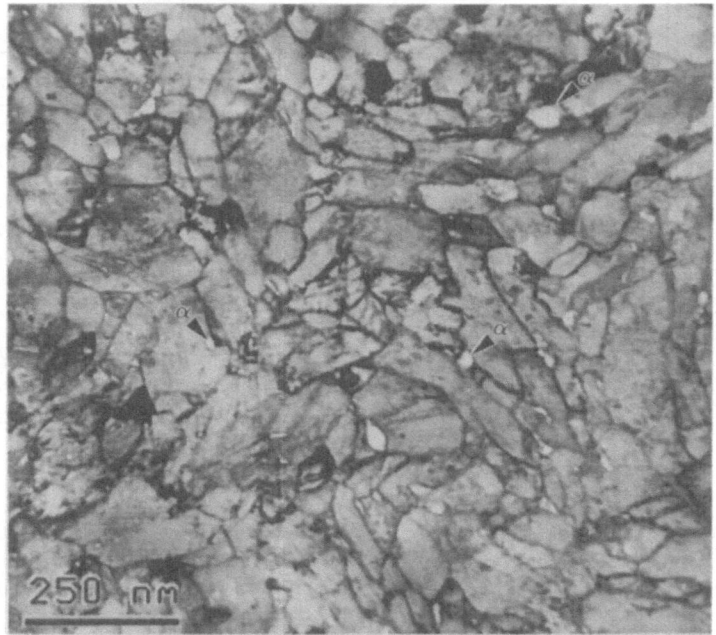

Figure 2: Image processed composite TEM micrograph of a transverse cross-section of wire UW250B after three heat treatments of 3 hrs at 300°C. α-Ti precipitates appear white.

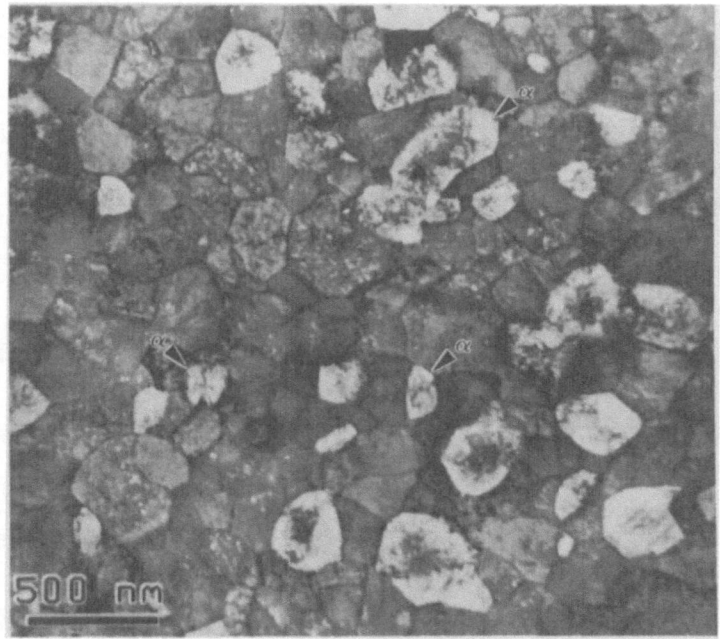

Figure 3: Image processed composite TEM micrograph of a transverse cross-section of wire UW260B after two heat treatments of 3 hrs at 300°C followed by one heat treatment for 80 hrs at 420°C. α-Ti precipitates appear white.

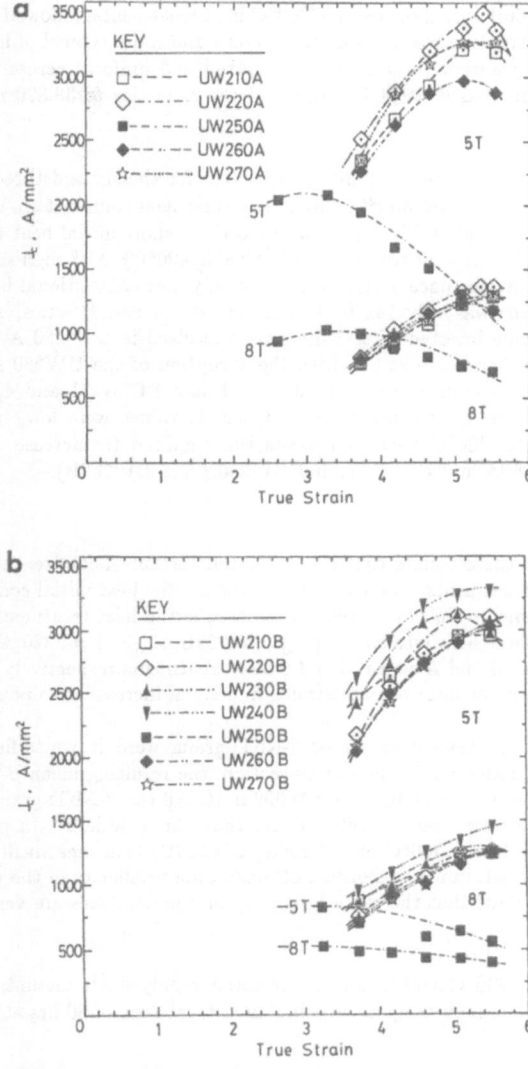

Figure 4: J_c versus final cold work strain at 5T and 8T, 4.2 K for (a) 6 heat treatment series, and (b) three heat treatment series wires.

RESULTS

The microprobe analysis performed on samples taken from the 5 3/4" diameter ingots revealed a very high level of chemical homogeneity. Local chemical variations were limited to ± 1/2 wt% over a 1 mm scale and ± 1 wt% over 6 mm. The mean transverse cross-sectional area of β-NbTi grains at 2/3 radius from the center of the billet was found to be 14,200 μm^2, corresponding to an ASTM grain size of 3.

The results of the microstructural quantification after final heat treatment are given in Table 2. The most striking results of the TEM analysis are those involving 3h/300°C heat treatments. For wire UW250B, where only three 3h/300°C heat treatments were given, less than 3% of the microstructure was α-Ti. The microstructure was on a very fine scale with a mean transverse cross-sectional β-grain area of only 5,110 nm². Most of the α-Ti occurred at β-NbTi grain boundary triple points (Figure 2). Comparing UW250B with UW250A (which received six 3h/300°C heat treatments), we found a large increase in α-Ti (12.8% of the cross-section) but the fine β-grain and α-Ti precipitate size was essentially identical. In Figure 3 we show the transverse cross-sectional TEM composite micrograph of the UW260B wire after two 3h/300°C heat treatments, followed by a heat treatment of 80h/420°C. The microstructure was then quite different and more typical of high J_c wires after final heat treatment.[2,4] The grains were much more equiaxed and uniform across the cross-section. The β-NbTi grain size had increased to 44,480 nm^2 and the α-Ti size to 38,870 nm^2 with a precipitate cross-section of 20.9%.

In Figure 4(a) and (b) we compare the J_c values for the six and three heat treatment series respectively. The highest J_c was obtained for a six heat treatment composite, UW220A, which achieved almost 3500 A/mm^2 at 5T and 4.2 K. This wire used three short initial heat treatments (3h/405°C) followed by increasingly longer heat treatments (20,40,80h/420°C). Although six heat treatments were used, they required the strain space normally used by only four conventional heat treatments and the total heat treatment time was only 149 h. Comparing the A and B series, we find that the use of additional heat treatments inserted after 3 die passes resulted in a $\sim$ 200 A/mm^2 increase in J_c at 5T and a 50–100 A/mm^2 increase at 8T, with the exception of the UW260 series where additional 3h/300°C heat treatments reduced both the J_c at 5T and 8T by 71 and 43 A/mm^2 respectively. Where only 3h/300°C heat treatments were used, the J_c values were low, but only one long heat treatment after initial 3h/300°C heat treatments was required to increase J_c at 5T beyond 3000 A/mm^2 (compare UW250A and UW260A, and UW250B and UW260B).

DISCUSSION

It is clear from these results that short low temperature heat treatments are very effective prior to long final heat treatments. For six heat treatments, the best initial conditions were 3h/405°C and for three heat treatments 6h/375°C. Using restricted initial heat treatments, high J_c values could be obtained with only three heat treatments [e. g. UW230B, J_c (5T) = 3320 A/mm^2, J_c (8T) = 1461 A/mm^2]. Comparing the B and A series (3 and 6 heat treatments respectively) we find that applying additional heat treatments at intermediate strains in general increases the peak J_c values.

If the original recrystallization anneal β-NbTi grains were drawn to final heat treatment size with only a geometrical reduction in their cross-section, the resulting mean β-NbTi transverse cross-sectional grain area would be approximately 800,000 nm^2. All the β-NbTi grain sizes found after final heat treatment were, however, considerably smaller than this (Table 2). In particular we note that even the smallest HT (3 hrs at 300°C in UW250A, UW250B) heat treatments produced a very fine grain size ($\sim$ 5000 nm^2), which are two orders of magnitude smaller than the geometrically expected one. Thus we must conclude that the grain boundary film precipitates are very effective in reducing the grain size.

Although 3 x 3h/300°C heat treatments resulted in only small amounts of precipitation (2.6% in UW250B), only one long high temperature final heat treatment of 80 hrs at 420°C was required to

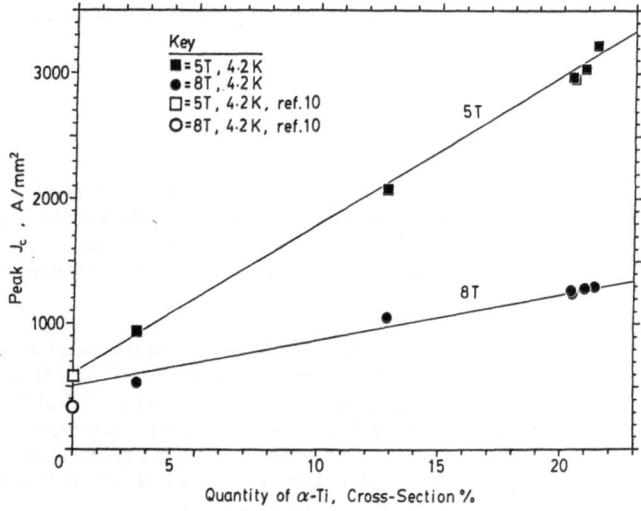

Figure 5: Peak J_c versus cross-section % α-Ti.

increase precipitate volume to amounts equivalent to three full heat treatments (UW260B, 21% α-Ti). Thus it appears that the crucial factor in developing high volume fractions of α-Ti is the nucleation of precipitate, rather than its subsequent growth.

Finally we present some initial results on the dependence of J_c on the % of α-Ti. In previous papers, we have shown that high J_c (e. g. $J_c > 3000$ A/mm^2 at 5T) requires of order 20% of α-Ti.[2,4] (We assume here that the elongated nature of our microstructure makes it reasonable to assume that the area fraction of α-Ti measured in cross-section is also the volume fraction. In general the aspect ratio of the precipitates at final heat treatment size is $>\sim$ 5:1 making this a reasonable assumption.) However, in a previous study we did not find that raising the % α-Ti above 20% produced a significant increase in transport current density (J_{ct})[4] and we were left to speculate that, although the intrinsic current density was still increasing with increasing α-Ti content, sausaging and other filament instabilities were limiting J_{ct}.[4] Figure 5 shows that over the range $\sim$ 2.5 to $\sim$ 21 cross-section % α-Ti, there appears to be a linear relationship to the J_c, the slopes being 123 and 36 A/mm^2 per % α-Ti at 5T and 8T respectively. The non-zero intercepts for zero % α-Ti provided an estimate of the pinning contribution of grain boundaries and other lattice defects. This linear dependence of J_c on % α-Ti again suggests the correctness of assuming that full summation of the elementary pinning force occurs in this system and reinforces the need to probe the way in which further increasing the α-Ti content can raise J_c. However, Gotoda et al[11] have found, using small angle neutron scattering measurements of the volume fraction at final wire size, that J_c is more closely proportional to the two thirds power of the α-Ti volume fraction.

Disappointingly none of the heat treatment variations significantly increased the slope of the J_c versus strain curves for the high J_c (J_c, 5T, 4.2 K > 3000 A/mm^2) wires. The peak J_c values for the high J_c wires occurred at final strains between 5 and 5.6. The higher the peak J_c, however, the lower the final strain required to surpass 3000 A/mm^2 at 5T; in the case of UW220A and UW240B this occurred at a final drawing strain of approximately 4.2.

SUMMARY

1. High J_c can be obtained within a restricted strain space by introducing additional heat treatments inserted after 3 dies. With one exception (UW260A), introducing intermediate heat treatments resulted in a $\sim$ 200 A/mm^2 increase in peak J_c at 5T and a 50–100 A/mm^2 increase at 8T.

2. Short low temperature initial HTs are very effective, providing the final HT is long.

3. Grain boundary α-Ti film nucleation of triple-point α-Ti has a major influence on the final quantity of α-Ti obtainable.

4. Repeated 3h/300°C heat treatments eventually produce triple-point α-Ti.

5. Precipitates of α-Ti even at very small levels produce very considerable β-NbTi grain size refinement.

6. 3h/300°C heat treatments show more grain size refinement than conventional heat treatments but this increase is lost when a final size long heat treatment is applied.

7. No significant increase in the slope of the J_c versus final drawing strain curve was obtained and all peak J_c values for the high J_c wires ($J_c > 3000$ A/mm^2) occurred at final strains between 5 and 5.6.

ACKNOWLEDGEMENTS

We are grateful to A. D. McInturff and Fermilab for supplying the NbTi alloy and to P. O'Larey and Teledyne Wah Chang Albany for carrying out initial processing of the monofilament. R. Remsbottom and W. Starch supervised wire drawing and A. Squitieri performed additional J_c characterization. Electron microprobe analysis was performed with the help of E. D. Glover. The work has been supported by the Department of Energy-Division of High Energy Physics and Fermilab.

REFERENCES

1. P. J. Lee, J. C. McKinnell, and D. C. Larbalestier, Microstructure control in high Ti NbTi alloys, <u>IEEE Trans. Mag.</u> MAG-25:1918 (1989).
2. P. J. Lee and D. C. Larbalestier, Development of nanometer scale structures in composites of Nb-Ti and their effect on the superconducting critical current density, <u>Acta Metall.</u> 35:2523 (1987).
3. Li Chengren and D. C. Larbalestier, Development of high critical current densities in niobium 46.5wt% titanium, <u>Cryogenics</u> 27:171 (1987).
4. P. J. Lee and D. C. Larbalestier, Determination of the flux pinning force of α-Ti ribbons in Nb46.5wt%Ti produced by heat treatments of varying temperature, duration and frequency, <u>J. Mat. Sci.</u> 23:3951 (1988).
5. P. J. Lee and D. C. Larbalestier, unpublished, Applied Superconductivity Center, Univ. of Wisconsin-Madison.
6. R. M. Scanlan, J. Royet, and R. Hannaford, Evaluation of various fabrication techniques for fabrication of fine filament NbTi superconductor, <u>IEEE Trans. Mag.</u> MAG-23:1719 (1987).
7. M. I. Buckett and D. C. Larbalestier, Precipitation at low strains in Nb46.5wt%Ti, <u>IEEE Trans. Mag.</u> MAG-23:1638 (1987).
8. A. W. West and D. C. Larbalestier, Microstructural changes produced in multi-filamentary Nb-Ti composite by cold work and heat treatment, <u>Met. Trans. A</u> 15:843 (1984).
9. P. J. Lee, Enhancement of atomic number contrast for image analysis of highly strained materials, <u>in</u>: "Proc. 45th Ann. Meeting of EMSA," ed. G. W. Bailey, San Francisco Press (1987) p. 358.
10. W. H. Warnes and D. C. Larbalestier, unpublished, Applied Superconductivity Center, Univ. of Wisconsin-Madison.
11. H. Gotoda, K. Osamura, M. Furusaka, M. Arai, J. Suzuki, P. J. Lee, D. C. Larbalestier, and Y. Monju, Influence of microstructure on the flux pinning in multifilamentary Nb-Ti wires, to be published in <u>Phil. Mag.</u> (1989).

INFLUENCE OF THERMOMECHANICAL WORKING SCHEDULES ON STRUCTURE

AND PROPERTIES OF HT-50 AND HT-55 SUPERCONDUCTOR ALLOYS

G. K. Zelenskiy, A. V. Arsent'ev, A. P. Golub',
V. E. Klepatsky, E. V. Nikulenkov, A. D. Nikulin,
V. Ya. Fil'kin, V. S. Titov,* P. P. Pashkov,*
V. A. Vasil'ev,** and A. I. Nikulin**

A. A. Bochvar's All Union Scientific Research Institute
of Inorganic Materials, Moscow, USSR

*All Union Scientific Research Institute of
Electromechanics
Moscow, USSR

**Institute of High Energy Physics, Serpukhov, USSR

ABSTRACT

The results of the staged critical current and structure
determination are given for HT-50 and HT-55 superconductor wire
specimens subjected to four intermediate anneals at 400°C. After
the first anneal, Jc of both the alloys grows by more than an
order of magnitude which is related to an intergranular and,
in the case of HT-55 alloy, also an intragranular α -phase
precipitation. Subsequent deformation cycles of thermomechanical
working (TMW) effect an increaseof Jc, a decrease of α -phase
and β -matrix subgrain sizes, while intermediate anneals have
an opposite effect. An increase of Jc is observed at the final
stage of drawing. The specimens of HT-50 and HT-55 alloys have,
respectively, Jc = 3.2×10^5 A/cm^2 and 3.9×10^5 A/cm^2 in a field
of 5T. The value is shown to depend on the size of β -matrix
subgrains, the boundaries of which are decorated with
α -particles.

INTRODUCTION

Recent progress in the reproduction technology of superconductor wires and, in particular, the introduction of a niobium
diffusion barrier between copper and niobium-titanium alloy
has lead to a considerable growth of their current carrying
capacity.[1,2] Progressive process developments lead to a
considerable extension of the temperature range (up to 400°-
450° C) and time (up to 28 h) and to an increase of the number
of intermediate anneals in the superconductor production.[3,4]
The correlation between the structural conditions and Jc of
wires indicates the importance of the β -matrix subgrain size
and the amount and morphology of α-precipitates.[5-8] However,
those papers do not contain data on structural and Jc variations

Advances in Cryogenic Engineering (Materials), Vol. 36
Edited by R. P. Reed and F. R. Fickett
Plenum Press, New York, 1990

295

Table 1. Characteristics of Ht-50 (numerator) and
HT-55 (denominator) wire specimen

Condition	β-subgrain size, nm	α-size, lxd, nm	Number α x10^18 cm^-3	Jc x10^4 5T, A/cm^2	Intensity (100)α	(110)α	(200)β
Cold worked prior (CWP) to 1-st intermediate anneal (IA)	165/173	-	-	0.2/0.1	0/0	0/0	45/65
After intermediate anneal (AIA) N 1	190/208	49x10 53x5/ 29x5	13 12/ 46	1.9/7.8*	-/6	-/5	-/63
CWP to 2-nd IA	119/137	105x5 53x6/ 20x1.5	11 2/ 60	3.7/14*	0/0	1.7/3	41/61
AIA N 2	131/140	41x8 33x5/ 8x5	12 24/ 80	3.0/8.8*	1/1.5	1.4/4	39/46
CWP to 3-d IA	80/79	112x5 40x4/ 14x4	1 -/ 100	7.8/11	0/0	2.5/5	45/45
AIA N 3	95/108	30x7 52x9/ 24x2	9 10/ 20	5.8/7.3	1/0.4	1/5	31/40
CWP to 4-th IA	78/53	118x5 34x4/ 25x4	7 -/ 20	8.4/14	0/0	3/6	48/63
AIA N 4	108/89	43x5 / 27x4/ 18x2	16 / 58 60	4.6/5.8	0/0.5	3/6	48/41
Final Drawing Strain							
ε =0.585	67/80	22x4 /23x4	20/19	-/8.5	-/0	-/4	-/76
ε =1.300	56/53	19x3.5/20x4	20/20	12/13	0/0	2.5/7	67/72
ε =1.970	49/46	16x3/ 8x4	23/25	16/17	0/0	2.2/5	59/55
ε =2.630				22/27			
ε =3.040				29/36			
ε =3.800	39/36	12x2.5/15x3	44/64	32/39	0/0	2/2	61/55

Note: The α-phase is along the boundaries and within the β-matrix.

during wire production. This does not always make it possible
to purposefully reach the wanted structural condition
corresponding to the high current carrying capacity of a
material. This work is a first attempt in this direction.

INVESTIGATIONS OF SUPERCONDUCTOR WIRES

Two assemblages in a copper matrix were produced with
traditional methods of extrusion and cold work of superconductor
HT-50 (Nb-48 mass % Ti) and HT-55 (Nb-55 mass % Ti) alloys.
The assemblages consisted of several tens of strands each
cotaining 55 superconductor filaments. The filaments are
surrounded with a niobium diffusion barrier that prevents
interaction between the Nb-Ti alloy and copper during wire

production.[1,9] Strands were removed from an extruded rod or
at different stages of wire production in order to measure the
critical current and perform the needed structural studies.

Nb-Ti filaments stripped of copper and niobium barrier
were investigated with an X-ray diffractometer with a continuous
counter motion using a graphite monochromator at the outlet
beam of the Cu K α-radiation. During the analysis the integral
intensity of the diffraction maxima (100)α , (110)α and (200)β
was measured from which one judged the amount of the phases
present.

The fine structure of the Nb-Ti filaments was studied by
transmission electron microscopy using light and dark field
(in α- phase reflections) images and microdiffractions (10).
The arithmetic mean values of the β-matrix subgrain width were
determined by the method of secants normal to the drawing axis
at 30-125 points of measurements. The α-particle size was
measured with a comparator CTM from the negatives of light and
dark field images. The critical current was measured at 4.2
K using an electric field criterion of E_o = 1 μV/cm. The
critical current density was determined using only Nb-Ti alloy.

Different TMW schedules were studied involving variation
in the number of intermediate anneals (up to 7), the anneal
temperature (350°-450° C), and duration of the anneal (3-50
hours). The comprehensive structural investigations were carried
out for specimens subjected to four intermediate anneals at
a temperature of 400° C for 24 hours. The results of the staged
measurements of Jc of HT-50 and HT-55 superconductor specimens
are tabulated in Table 1. The first intermediate anneal results
in more than an order of magnitude increase of Jc. The
subsequent anneals lead to a reduction in Jc as compared to
worked wires of the same diameter.

The authors' investigations show that, following an
intermediate anneal at 400° C for less than 24 hours, one can
observe Jc (5T) to increase in the HT-55 and HT-50 alloys up
to and including the third and second intermediate anneals,
respectively.

The dependence of Jc on the amount of deformation of an
annealed specimen is in the form of a curve with a maximum.
With an increase of the intermediate anneal temperature and
time, the position of a clearly defined maximum, or a maximum
in the form of a plateau, shifts to higher deformation regions
and depends on the schedule of prior TMW. Therefore, although
the higher cold work assists a more intensive α-phase
precipitation during the subsequent anneal, the defomation amount
must not exceed the values corresponding to the position of
the maximum of the optimized deformation. Such a deformation
arrangement of intermediate anneals is one of the important
requisites for production of high Jc superconductor wires.
In this work the selected deformations following the intermediate
heat treatment did not exceed the optimized ones.

The mesured Jc was higher for HT-55 alloy specimens than
for the HT-50 alloy, the condition of the alloys being similar.
The most drastic change of Jc with deformation of the specimens
annealed at 400° C for 24 hours was observed at the final
deformation after the fourth heat treatment (Fig. 1). In this
case Jc = 3.2×10^5 A/cm^2 and Jc = 3.90×10^5 A/cm^2 were measured
in a field of 5T for HT-50 and HT-55 alloys, respectively.

Initially as-extruded superconductor wire filaments reveal
a structure of slightly drawn (l:d=3:1) substrips (groups of
β'-grains) differently oriented to the drawing axis. In the
inner volumes of the groups the density of the dislocations
observed is $2 \times 10^{10} cm^{-2}$. The average cross-sectional size of
the substrips is 500 nm (HT-50) and 600 nm (HT55), but with
considerable scatter (100-1000 nm). After drawing, the size
of the -substrips decreased by a factor of 3 and the specimens
revealed a clearly defined deformation texture with the direction
[110] along the drawing axis. It corresponds to a Jc of
$2 \times 10^{3} A/cm^2$. The first intermediate anneal at 400° C for 24
hours leads to a growth of Jc, the substrip width, and to both
preferable inter- and intrasubgranular precipitations of disperse
α -particles in the HT-55 alloy, while in the HT-50 alloy the
precipitation is only intersubgranular. The α -phase amount
in the HT-55 alloy is larger than in the HT-50 alloy. The
appearance of the α-phase is evidenced by the microdiffraction
patterns and its reflections (100)α and (110)α on the X-ray
patterns (Fig. 2 a.b).

The drawing at the second stage of TMW again effects a
reduction in the cross-sectional size of the substrips and
imparts an elongated shape to the α -particles. In this case
the X-ray patterns show an oriented arrangement of (110)α -
particles. The anneal again increases the substrip width and
the amount of the α -phase precipitated (Table 1). The mode
of the structural changes identified at the subsequent stages
of TMW is generally retained. If the relation between the
intensity of (200)β and the intensity of (100)α or (100)α can
characterize the content of the α-phase, it follows from the
data of Table 1 that a significant increase of α-phase is observed after the third and fourth cycle of the TMW. The availability
of the (100) α-phase in the HT-50 alloy as annealed at 400°C
for 24 hours is detected up to and including the third anneal,
while in the HT55 alloy (Table 1). An increase in the average
cross-sectional size of the subgrains is less significant after
the third and fourth anneals, since substrips decorated with
the α -particles do not change their sizes. At the final
deformation stage (after the fourth anneal), the subgrain cross

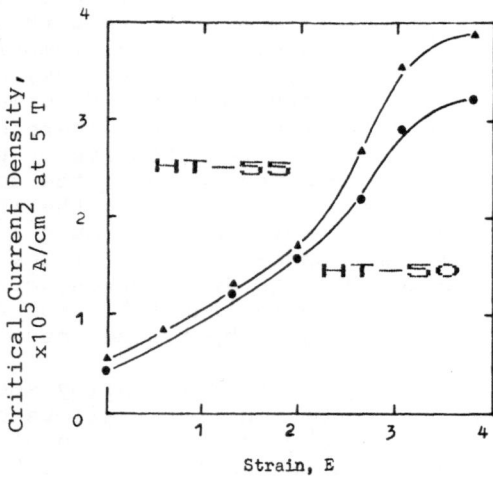

Fig. 1. Effect of final deformation amount of
         critical current density of HT-50 and HT-55.

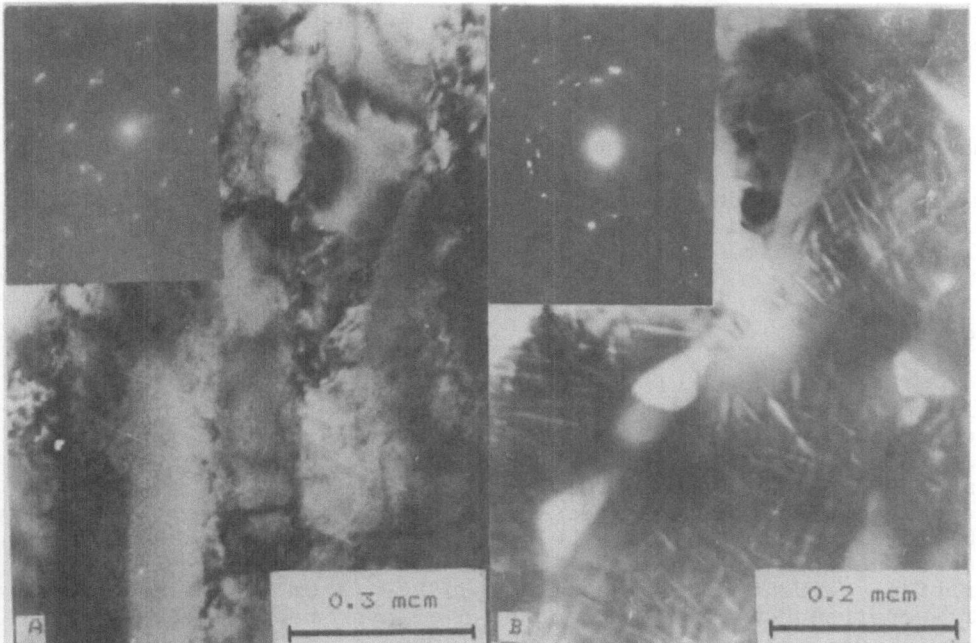

Fig. 2. Micrographs of filaments after first intermediate
         anneal of: HT-50(a)-longitudinal section and
         HT-55(b)-cross section.

sectional size is reduced, the available α-particles become
finer and the density of their precipitates increases. It can
be seen from Fig. 3 that in the case of the HT-50 alloy, where
α-phase precipitates were detected only at the subgrain
boundaries, a certain Jc-β-subgrain size relationship is
observed. In the case of the HT-55 alloy, where at the first
cycles of TMW the α-phase precipitates were observed both within
subgrains and along the subgrain boundaries, this β-subgrain
size dependence is more clearly revealed only at the final stages
of TMW. If there are also additional α-phase precipitates

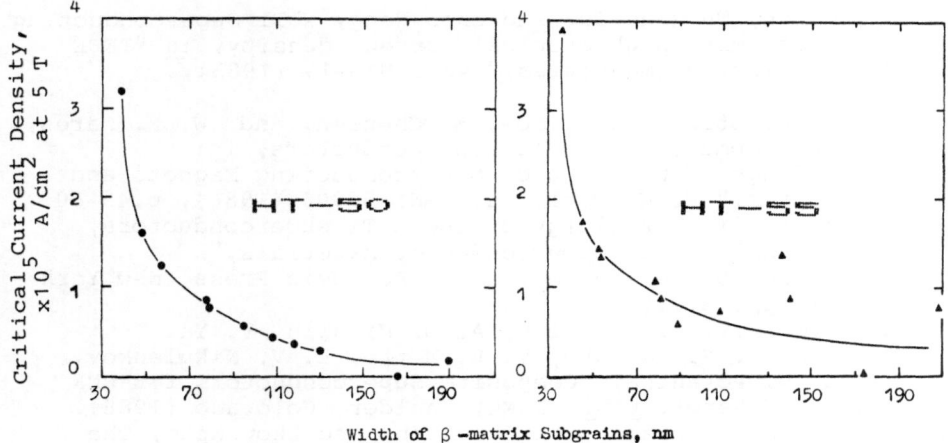

Fig. 3. Critical current density vs. width of β-matrix
         subgrains for HT-50.

within the β-matrix, higher Jc values are noted than for the case (as is observed for the HT-50) when α-precipitates are only intersubgranular.

Thus, the results make it possible to conclude that the main part in achieving the high Jc values as a result of TMW is played by the formed fine cellular substructure of the β-matrix, the boundaries of which are decorated with finely dispersed α-particles that provide for effective pinning of Abrikosov vortices.

CONCLUSIONS

The comprehensive investigations for the structure and critical current density changes at different stages of superconductor HT-50 and HT-55 alloy wire production made it possible to establish that the main part in achieving high Jc is played by the fine cellular structure of the β-matrix decorated with finely dispersed α-particles which result from multicycle TMW. Thus, in the case of wire specimens having superconductor filaments 4-6 mm dia and 35-40 ηm subcells critical current densities of 3.2-3.9×10^5 A/cm^2 (at field of 5T) were reached for HT-50 and HT-5 alloys, respectively. It is possible that with further developments of thermomechanical working and progressive methods of deformation^2 resulting in a finer celular substructure of the β-matrix, higher Jc would be attained.

REFERENCES

1. G. K. Zeleskiy, V. A. Vasi'liev, L. D. Bogdanova, E. V. Nikulenkov, A. P.Golub', "HT-50 composite superconductors fo UNK magnets," Report at the 24 International conference of countries-members of CMA on Physics and Techniques of Low Temperatures, 17-20.09, Berlin, DDR (1985).
2. G. Gregory, Recent advances in commercial multifilamentary Nb-Ti wires in the United states, in "Cryogenic Materials, 88," vol. 1, Superconductors, R. P. Reed, Z. S. Xing, E. W. Collings, eds., ICMC, Boulder, Colorado (1988), pp. 361-371.
3. Li Cheng-ren, Wu Xiao-Zu, and Zhon Nong, NbTi Superconducting composite with high critical current density, in "IEEE Transactions of Magnetics," vol. MAG-19 (1983), pp. 284-287.
4. D. C. Larbalestier, P. J. Lee, Li Chenren, and W. H. Wares, New developments in Nb-Ti superconductors, in: "Proceedings of Workshop on Superconducting Magnets and Cryogenics," P. G. Dahl, ed., BNL-52006 (1986), p.45-50.
5. A. D. McInturff, Metallurgy of the NbTi superconductors, in: "Metallurgy of Superconductor Materials," T. Luhman, D. Dew-Hughes, eds., Academic Press, New York (1979), pp. 67-96.
6. G. K. Zelenskiy, A. P. Golub, A. D. Nikulin, V. Ya. Fil'kin, V. P. Kosenko, V. L. Mette, E. V. Nikulenkov, and L. V. Potanina, "Composite Superconductors for UNK magnets," Report CB-3, ICMC, Boulder, Colorado (1988).
7. Zhang Tingju, Wu Ziaozu, Li Chengren, and Zhow Nong, The pining force and critical current in Nb-Ti superconducting wire with plate-like Ti precipitates, in: "Advances in Cryogenic Materials," vol. 32, N 4, (1986), pp. 903-909.

8. E. W. Collings, The Physical metallurgy of titanium alloys, ASM (1984).
9. V. Ya. Fil'kin, V. F. Gogulija, V. P. Kosenko, E. V. Nikulenkov, A. D. Nikulin, P. I. Slabodchikov, G. K. Zelenskiy, K. P. Myznikov, A. I. Nikulin, and V. A. Basil'iev, Composite Superconductors for UNK Magnets, in: "Proceedings of Workshop on Superconducting Magnets and Cryogenics," P. F. Dahl, ed., BNL52006 (1986). pp. 56-59.
10. V. S. Titov, G. N. Vlasov, Foils preparing method for electro-microscopic in investigation of superconducting wire, in: "Proceedings of the All Union Scientific Research Institute of Electromechanics," vol. 40, Moscow (1974), pp. 133-136.

EFFECTS OF HEAT-TREATMENTS AND OF PROCESSING PARAMETERS ON THE

MICROSTRUCTURE OF MULTIFILAMENTARY Nb-46.5 wt% Ti SUPERCONDUCTORS

Roland Taillard*, Christian-Eric Bruzek*, Jacques Foct*
Hoang Gia Ky**, Alain Lacaze**, Thierry Verhaege***

*   Laboratoire de Métallurgie Physique
Université de Lille I, Bâtiment C6
59655 Villeneuve d'Asq Cedex, France

**   Alsthom-Intermagnetics S.A.
90018 Belfort, France

* * * Laboratoires de Marcoussis,C.G.E.
route de Nozay
91460 Marcoussis, France

INTRODUCTION

The superconducting behaviour of the conventional Nb-Ti wires is governed by their internal microstructure [1,2,3,4,5,6] whereas that of the ultra-fine filaments seems to arise from surface pinning at the boundaries between the filaments and the matrix [7]. Therefore,it is particularly pertinent to examine the effects of the manufacturing process both on the internal and on the interfacial structure of the filaments. The authors have recently demonstrated that the Alsthom manufactured wires are highly polyphased [6,8] which can contain among other phases, significant amounts of interfacial intermetallic compounds and of internal α" martensite.

The aim of the present paper is threefold. Firstly, to supply further evidences of the α" existence at room temperature in this "Nb-46.5 wt% Ti" material. Secondly,to study the conditions of formation of the α" and intermetallic phases. And thirdly,to investigate the effects of the thermomechanical fabrication parameters and of a niobium diffusion barrier on the microstructure of the strands.

MATERIALS AND EXPERIMENTAL PROCEDURE

This study relates to Alsthom multifilamentary composites. These monolithic strands consist generally of 10^4 to 10^6 filaments which are embedded in a copper-based matrix. The filaments arise from Teledyne Wah Chang Albany billets with the Nb-46.5wt% Ti (Ti-36.8 at% Nb)standard composition. It is worthwhile to note that the filaments are either cladded or not with a niobium diffusion barrier. The manufacturing process is composed of a series of hot extrusion and room temperature drawing stages. Moreover,the fabrication encompasses some intermediate heat treatments of a few hours at 673 K and 523 K.It is worth emphasizing that in order to improve the wires quality,distinct fabrication schedules were followed which were differentiated by the possible cladding of

Advances in Cryogenic Engineering (Materials), Vol. 36
Edited by R. P. Reed and F. R. Fickett
Plenum Press, New York, 1990

Table 1: Heat treatments of some specimens

sample number	filament diameter ϕ (μm)	heat treatment (h/K)	quenching medium
0	22	none	none
1	22	100/573	water
2	22	500/623	water
3	22	120/823	water
4	22	3/1073+500/623	water+water
5	10^4	none	none
6	16	none	none
7	12.6	none	none
8	11.3	none	none
9	9.9	none	none
10,11,12	0.87	none	none

the filaments by a niobium diffusion barrier,by the extrusion preheating temperature over a 50K range and/or by the die-extrusion kinetics over a fivefold increase,by the filament twisting path or by the 293K deformation amplitude. The manufactured samples were investigated with or without subsequent isothermal heat treatments for up to 500 hours over the 573 to 1073K temperature extent. These treatments were given in silica tubes sealed under vacuum and finished by a water,nitrogen or helium quenching. Table 1 specifies the nature of some samples.

Structure analysis was performed by X-ray diffractometry,X-ray rotating-crystal method,optical,scanning backscattered and conventional transmission electron microscopies,S.T.E.M. E.D.X. and electron probe microanalysis. It should be noticed that the X-ray diffractometer is a very powerful tool which leads to the average volume fractions and to accurate lattice parameters measurements of the various phases. This later determination arises from a close analysis of the X-ray diffractograms and from the use of a high angle extrapolation method. The T.E.M. samples were either thin foils or chemically extracted filaments. The thin foils were prepared mainly in the longitudinal direction of the wires because of the difficulty encountered for obtaining large transparent areas in the transverse section .It worthwhile to note that the only jet-electropolished samples were wrapped in a thin matrix layer. In order to perform accurate S.T.E.M. E.D.X. analyses,this deposit was subsequently chemically destroyed by a few seconds immersion in 50% dilute HNO3.

RESULTS AND DISCUSSION

The two cases of the α'' martensite and of the intermetallic compounds are successively displayed.

α'' Martensite

Table 2 arises from a thorough examination of X-ray spectra. It suggests the existence of significant volume fractions "f" of an orthorhombic phase with the lattice parameters a,b and c in these "Nb-46.5 wt% Ti" samples that had never been at very low temperatures close to 0K. Similar results obtained in helium-quenched specimens have already been published [6,8] .Moreover,it is worth noting that the study of chemically extracted filaments by the rotating-crystal method may ascertain the well-established orientation relationship [9,10] : $[010]$ α'' 2° from $[1\bar{1}0]$ β. Furthermore and,as exemplified by Fig.1,the existence of α'' martensite at room temperature seems to be corroborated in transmission electron microscopy which shows possible α'' diffraction spots due to thin parallel platelets.

Table 2: X-ray data for samples that had never been at temperatures close to 0K

sample number	a(nm)	b(nm)	c(nm)	b/a	f(%)
0	0.3269	0.4554	0.4624	1.3931	0.25
1	0.3149	0.4940	0.4644	1.5680	9.10
2	0.3158	0.4905	0.4655	1.5532	11.2
3	0.3171	0.4845	0.4650	1.5269	17.3
4	0.3168	0.4858	0.4663	1.5333	13.6
5	0.3168	0.4846	0.4654	1.5290	traces
6					16.9
7	0.3187	0.4561	0.4672	1.4310	10.1
8					6.15
9					3.30
1 0	0.3155	0.4595	0.4863	1.4584	2.70
1 1	0.3147	0.4840	0.4676	1.5380	2.40
1 2	0.3140	0.4860	0.4659	1.5477	5.80

With regard to the influence of the manufacturing process, it is noteworthy,that α" formation takes place during the water-quenching of the hot extruded wires. This conclusion arises from the following results. The α" volume fraction is doubled by a 50K increase of the extrusion temperature. In the same way,Fig-2 shows that ,in water or more drastically quenched samples, f grows significantly with the applied tempering parameter. On the contrary,the air-cooled composites contain no α".Furthermore,Fig. 3 indicates that the α" proportion decreases markedly at the beginning of a cold drawing stage because of the α" metastability which leads to the α"--> β phase reversion.

All prior results are consistent with the occurrence of an α"-martensitic transformation above room temperature in these Ti-36.8 at% Nb mean compositionned-alloys. This observation is a priori astonishing because the linear extrapolation of the measured values of Ms obtained with the Ti-Nb alloys at Nb contents lower than 21.7 at% leads to a Ti-36.8 at% Nb-Ms estimate close to zero kelvin [11].This discrepancy seems to arise from the uneven alloys chemical composition. The 30nm probe sized investigations have assessed the local heterogeneities of composition to at least ±6.6 at% Nb from the nominal chemical analysis. It is worthwhile to note that,according to both the later measurements and the calculations of the To α-β equilibrium temperature,the upper bound of occurrence of the martensitic form of α : α" ,is close to room temperature in the present alloys . Accordingly, the α" lattice parameter measurements (compare for instance Table 1

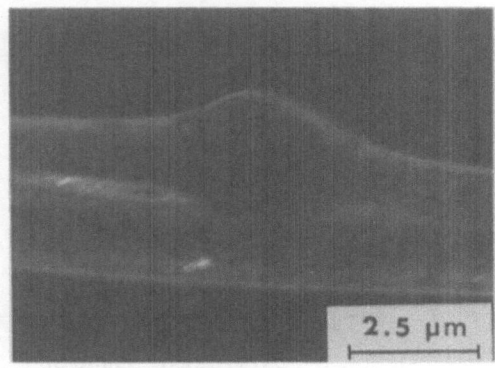

a) dark field image   b) associated diffraction pattern

Fig. 1. T.E.M. aspect of the α" martensite in a fine filamented-wire in its as-manufactured state.

and Fig.4 of ref.8) are generally consistent with the α" occurrence in the Nb solute lean zones. Furthermore, it should be emphasized that the α" volume fraction grows with the tempering parameter. This data suggests that the α" formation results from the lowering of the shear modulus [12] due to the Ti enrichment of the Nb-enriched areas. The efficency of a pipe diffusion mechanism is strongly guessed from the effect of the deformation amplitude on the heat treatment response. This result is made clear from the present $2 \, 10^{-13} \, cm^2 s^1$ estimate of the niobium diffusion coefficient in NbTi at 550°C. As matter of fact, this value is about 10^4 times higher that the intragranular coefficient of diffusion for Nb in a Nb-46.5 Ti alloy [13]. Neglecting the texture effect, f increases markedly with the tempering parameter in the samples 0 to 3 which have been previously cold worked with a true strain of 8. In contrast, after the same heat treatments, f remains at a trace level in 2.6 true hot strained specimens. Otherwise, and in accordance with our former proposal [8], the α" lattice parameter measurements may also locate the β/α" phase boundary at a niobium content higher than 36.8 at% (53.5 wt%).

<u>Intermetallic compounds</u>

As illustrated by Fig.4 and 5, the existence of large intermetallic compounds at the boundary between the filaments and the matrix can be very deleterious for applications where fine filaments are required. These particles, wich can be two to three times larger than their filament bearer bring about filament necking or breakage as well as proximity effects between the filaments. It should be noticed that we have observed a trend towards a linear decrease of the resistive transition index with the ratio: largest particle size over filament spacing.

However, it is worth emphasizing that the spatial distribution of the intermetallic particles is governed by the composite design as well as by the thermomechanical parameters of the fabrication process.

It is interesting to note that large intermetallic phases, such as that shown on Fig.4 and 5, were observed on niobium-sheathed filaments after a three hot extrusion process. For this Cu/Nb/NbTi composite design, the S.T.E.M-E.D.X analyses conclude more frequently to a $Cu_3(Ti_xNb_{1-x})$ particle composition, which agrees with its previous crystallographical identification [6,8]. In a more general way, we have just shown [14] that the choice of the materials for the diffusion barrier and the matrix is of prime importance as it determines the volume fraction and the chemical composition of the intermetallic particles. Returning to the Cu/Nb/NbTi composite case, the overall efficiency of the niobium diffusion barrier is proved by the following results. These observations were performed on solely hot-extruded wires. Electron probe traces have established that the

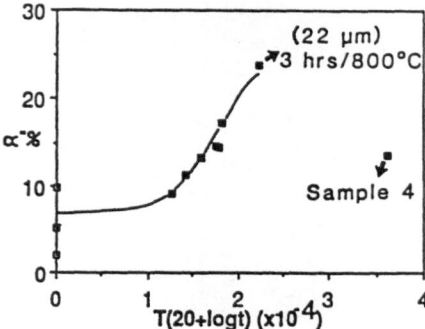

Fig-2. Effect of heat-treatments
on the α" volume fraction
in 22 μm-diametered filaments

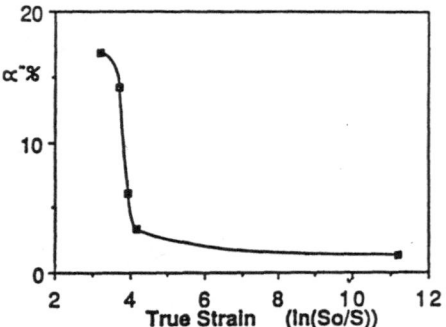

Fig-3. Relationship between cold-drawing
and the α"-volume fraction during
a stage II manufacturing process

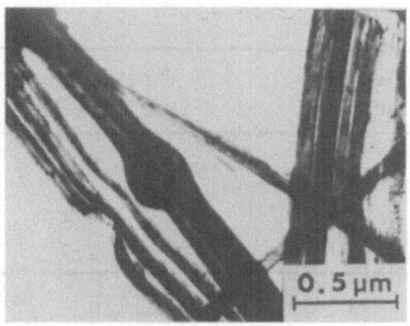
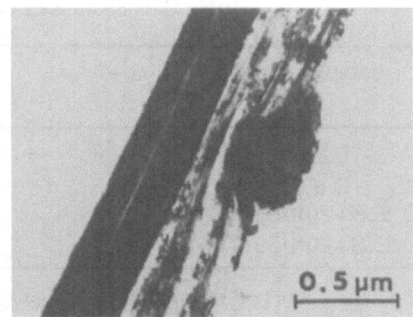

Fig-4. Impact effect of the intermetallic compounds in a composite with ultra-fine filaments

Fig-5. Irregular shape of the aggregates of intermetallic fragments

niobium cladding parts the Cu-and Ti-diffusion zones by the barrier thickness. The formation of the Cu_xTi_y particles is therefore delayed . For instance , during the first hot extrusion, the intermetallic volume fraction is divided by 6 in a monofilamentary specimen wrapped in a 0.9 mm thick niobium barrier. This amazing measurement of a 7.10^{-4} volume fraction of intermetallics in this filament-cladded specimen seems to arise from the intergranular diffusion of Ti and Cu through the barrier because after a further extrusion, the Cu-and Ti-diffusion zones extend respectively only over approximately 12 micrometers at the 1 atom percent level and 8.5 micrometers. As illustrated by Fig.6, the niobium barrier effect is self-evident after subsequent aging treatments for up to 120 h at 823 K. It should be noticed that such long treatment durations were chosen in order to promote the diffusion phenomena while allowing the comparison between the diffusion efficiencies respectively of a static aging treatment and of a dynamic hot extrusion. For the NbTi/Cu design,Fig.6a suggests that aging can give rise to at least three concentric intermetallic layers at the filaments-matrix boundaries. These various layers were only noticed at long aging times. They correspond to the two outermost plateaus of composition plus an innermost zone of variable composition which is an interdiffused region of Cu in the b.c.c. NbTi solid solution. The outermost intermetallic phase has a composition close to $Cu_3(Ti_xNb_{1-x})$ whereas the second intermetallic layer formula is of the $Cu(Ti_yNb_{1-y})_2$-type. These observations must be paralleled with the literature data .In aged samples, Larbalestier[15] has recently characterized three distinct layers of $Cu(Ti_xNb_{1-x})$ compounds at the NbTi/Cu interfaces, the microprobe analyses of the two outermost phases being respectively Ti_2Cu_7 and $(Ti_{0.5}Nb_{0.5})Cu$. Moveover, the Cu_4Ti ,

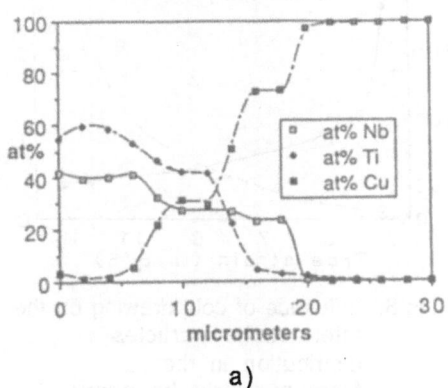
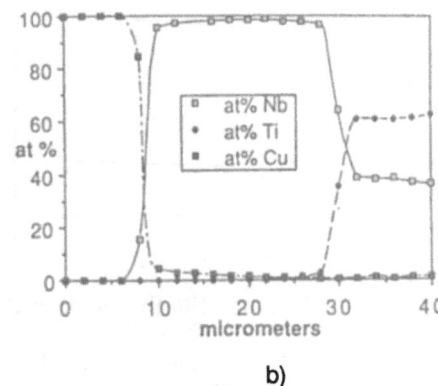

a)

b)

Fig-6. Electron probe trace across a matrix-filament interface in a

a) Cu/NbTi composite reacted 120h/550°C
b) Cu/Nb/NbTi composite reacted 20h/550°C

Table 3 Thicknesses of the Ti-and Cu-diffusion zones

sample nature	Ti case (µm) at the 0.2 at% level	Cu case (µm) at the 2 at% level
D.E.S.	5.1± 1.5	5.5
D.E.S+	8.5± 1.5	5
(D.E.S+20h/550°C)	10.5± 4.0	25
(D.E.S+20h/550°C)+	5.5± 0.5	7.7

D.E.S.:Double extruded sample   +:niobium sheathed filament

Cu_3Ti and Ti_2Cu intermetallic phases compositions have already been identified in NbTi/Cu composites [6,8,16,17].Furthermore, it is worth emphasizing that the 823 K kinetics of growth of the three layers taken all together is in perfect agreement with that of Larbalestier[15]. Based on the elementary $x= (Dt)^{1/2}$* relationship, it leads to a $3.65.10^{-12}$ $cm^2.s^{-1}$ D value instead of $1.07.10^{-12}$ $cm^2.s^{-1}$ [15] However, at a given aging time, the three layer thickness is three to five times larger than that measured by Larbalestier et al.[15] in solely cold drawn specimens. The high magnitude of this discrepancy suggests the major influence of the diffusion phenomena during hot extrusion. On the contrary, for the NbTi/Nb/Cu composite design, and as displayed by Fig.6b, the Cu_xTi_y compounds are scarcely electron probe-detected. Moreover, table 3 ascertains that the widths of the Ti-and Cu-diffusion zones are generally significantly reduced with a niobium diffusion barrier. It should be noticed that the Cu data takes account only of the concentration higher than 2 at % because of the very extensive penetration of the Cu at lower concentrations (see Fig.6b).

Furthermore,as suggested by the previous observations of a diffusion-controled intermetallic reaction,and as corroborated by Fig.7,the intermetallic formation is strongly temperature dependent. With regard to the manufacturing process,and in accordance with numerous results [15,16,18] this means that the reaction occurs the more,the higher the extrusion and preheating for extrusion temperatures. This conclusion explains why ,in a niobium-coated superconductor the intermetallic volume fraction decreases from 1.15 10^{-2} to 0.75 10^{-2} with a 50K reduction of the "extrusion temperature".

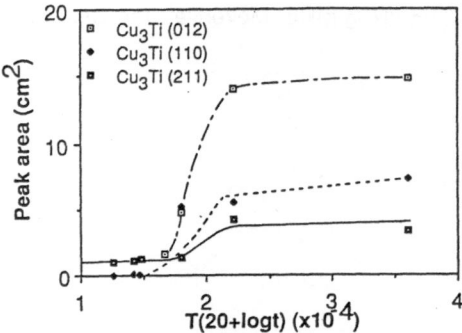

Fig-7. Cu/NbTi composite-Evolution of the Cu_3Ti X-Ray diffraction intensities with the isothermal heat treatments

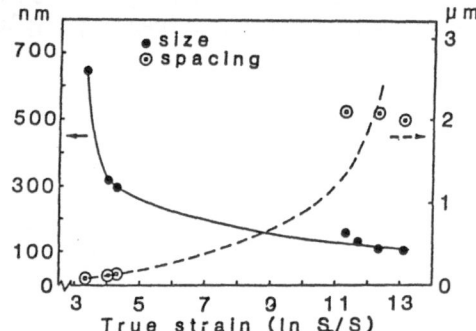

Fig-8. Influence of cold drawing on the intermetallic particles distribution in the filament-matrix boundary

(*): x is the overall thickness of the three layers, t the aging time, and D the diffusion coefficient.

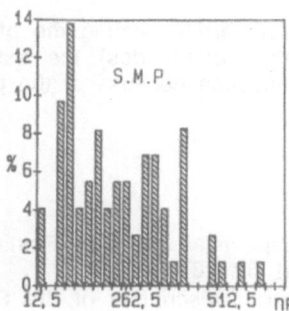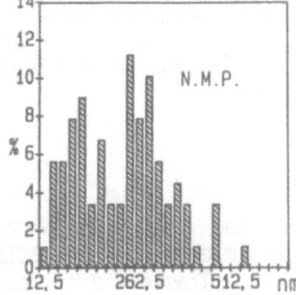

Fig-9. Effects of the temperature and of the kinetics of extrusion on the intermetallic particle size distribution on niobium-sheathed filaments.

S.M.P. Standard manufacturing process   N.M.P. New manufacturing process
(standard extrusion temperature - 50°C)
(low extrusion kinetics)

Otherwise,Fig. 8 exhibits the dependence of the particle size distribution on the extrusion kinetics. It is worth noting that the fivefold increase of the hot deformation velocity reduces the interparticle distance (from 3.7 μm to 2 μm) and leads to a bimodal distribution. These observations arise from the intrinsic brittleness of the intermetallic compounds that are the more prone to break,the higher the extrusion kinetics. The study of particular matrix compositions suggests that the high frequency of low-sized particles results from the embing of the intermetallic fragments into the filaments rather than from the initiation of preferential sites of intermetallic nucleation due to the tearing of the diffusion barrier by the intermetallic pieces [14].

Moreover, the intermetallic brittleness is also the cause of the decrease of the particle size during an isothermal hot deformation and of the evolution of the particle size distribution during a cold drawing step(see Fig.9).

On the whole,these results establish that it is of prime importance to limit the intermetallic compounds formation for the fine filaments applications. The best way to achieve this goal consists of interposing a judiciously chosen diffusion barrier (chemical composition and thickness) between the filaments and the matrix. The niobium barrier is shown to be rather efficient for the Cu/NbTi superconductor design. Moreover,it seems very well suited to look in a more thorough way at the significant effect of the kinetics of extrusion by a clear parting between the influences of the extrusion temperature and of the hot deformation velocity. In comparison with the previous fabrication parameters the intermetallic breakage during the cold drawing steps seems of secondary importance because of the continual increase of the ratio: largest particle size to filament diameter.

CONCLUSION

The formations of the α" martensite and of the intermetallic compounds are controlled by the manufacturing process.

The α" martensitic reaction occurs during the water-quench of the hot-extruded wires. Over the usual extrusion temperature range, the α" amount increases with the tempering parameter. These features are explained by the coring of these Nb-46.5 wt% Ti alloys and/or by an easier than expected α" occurrence.

The formation of the electrically and mechanically deleterious intermetallic compounds takes place mainly during the pre-extrusion and extrusion step. As illustrated for the Cu/NbTi composite design,the most powerful method for limiting the intermetallic content consists of a chemically suited diffusion barrier. Among the other parameters (extrusion temperature, hot or cold deformation amplitudes) the extrusion kinetics increment exerts a significant and favourable influence because of the breakage of the intermetallic particles.

REFERENCES

1. D.C.Larbalestier and A.W.West, "New Perspectives on Flux Pinning in Niobium-Titanium Composite Superconductors " Acta.Met. 32 :1871 (1984).
2. D.C.Larbalestier , " Towards a Microstructural Description of the Superconducting Properties ", I.E.E.E. Trans.Mag. 21:257 (1985).
3. I.Pfeiffer und H.Hillman, "Der Einfluss der Struktur auf die Supraleitungseigenschaften von NbTi 50 und NbTi 65", Acta.Met.16:1429(1968).
4. P.W Bach and A.C.A Van Wees, " High Critical Current Density with Low Current Sharing in a Multifilament Nb 55 wt% Ti Superconductor ", I.E.E.E. Trans.Mag. 21:351 (1985).
5. J.V.A.Somerkoski,D.P.Hampshire,H.Jones,R.O.Toivanen and V.K.Lindroos,"Structure and Superconducting Property Characterisation of MF Cu/Nb-46.5 wt% Ti Superconductors", I.E.E.E Trans. Mag. 23:1629 (1987)
6. R.Taillard,J.Foct and H.G.Ky,"Microstructural Study of Multifilamentary Nb-46.5 wt% Ti Superconducting wires", Adv. Cryog. Eng. 34:929 (1988)
7. K.Yasohama,K.Morita and T.Ogasawara,"Superconducting Properties of Cu-NbTi Composite Wires with Fine Filaments", I.E.E.E. Trans. Mag. 23:1728 (1987)
8. R.Taillard,J.Foct and H.G.Ky,"Effects of Heat Treatments on the Structure of Multifilamentary Ti-36.8 at% Nb Superconductors and Conductors", Jl. de Phys. (1989) to be published
9. B.A.Hatt and V.G.Rivlin,"Phase Transformations in Superconducting Ti-Nb Alloys", Brit. J. Appl. Phys. D.1:1145 (1968)
10. J.P.Morniroli et M.Gantois,"Etude des Conditions de Formation de la Phase Omega dans les Alliages Titane-Niobium et Titane-Molybdène",Mém. Sc. Rev. Met. 70:831 (1973)
11. D.L.Moffat and D.C. Larbalestier," The Competition between Martensite and Omega in Quenched Ti-Nb Alloys-Microstructure", Met. Trans. 19A: 1677 (1988)
12. C.C.Koch and D.S.Easton, "A Review of Mechanical Behaviour and Stress Effects in hard Superconductors", Cryog : 391 (1977)
13. D.L. Moffat and U.R.Kattner,"The Stable and Metastable Ti-Nb Phase Diagrams", Met.Trans. 19 A: 2389 (1988)
14. C.E.Bruzek and R.Taillard,Unpublished Work (1989)
15. D.C.Larbalestier,P.J.Lee and R.W.Samuel,"The Growth of Intermetallic Compounds at Copper Niobium Titanium Interface", Adv. Cryog.Eng. 32:   (1986)
16. H.Hillman,"Fabrication Technology of Superconducting Materials" in "Superconductor Materials Science",Plenum Publ. Corp. 275 (1985)
17. M.Garber,M.Suenaga,W.B. Sampson and R.L.Sabatini,"Effect of Cu4Ti Compound Formation in Fine Filament Nb-Ti Superconductors", I.E.E.E Trans. Nucl. 32:3681 (1985)
18. D.C.Larbalestier,Li Chengren,W.Starch and P.J.Lee,"Limitation of Critical Current Density by Intermetallic Formation in Fine Filament Nb-Ti Superconductors", I.E.E.E. Trans. Nucl. Sc. 32:3743 (1985)

NbTi SUPERCONDUCTORS WITH ARTIFICIAL PINNING STRUCTURES

Leszek R. Motowidlo , Hem C. Kanithi, and Bruce A.Zeitlin

IGC Advanced Superconductors Inc.
1875 Thomaston Ave.
Waterbury, CT, 06704

ABSTRACT

A multifilament NbTi superconductor having continuous non-random pinning centers has been fabricated by mechanical metallurgy. Specifically, present state of the art wire processing was utilized to manufacture a superconductor with controlled, uniform, pinning centers and predetermined spacing which match the flux line lattice (FLL) continuously. This is accomplished by designing the superconductor filaments to form continuous pinning centers in layers (laminar pinning centers) with spacing being predetermined and related to the fluxoid lattice. The use of drawing and extrusion techniques combined with this design yields a practical superconductor.
Preliminary current densities at low fields, significantly higher than in conventionally processed superconductors ,have been measured.

INTRODUCTION

Type II superconductors such as NbTi are characterized by two critical fields, H_{c1} and H_{c2}. Between these two critical fields we have the so called mixed state. The superconductive material in this state is now a composite of a triangular array of magnetic flux lines of equal spacing. The flux lines themselves are screened from the superconductor by circular super currents leaving most of the superconducting material free of field. The spacing between the fluxoids is uniquely determined by the applied magnetic field. When a current is passed through a type II superconductor in the mixed state, the flux lattice experiences a force called the Lorentz force.[1] This force tries to move the flux lines at right angles to both the current and magnetic field. When fluxoids move, energy is dissipated and an increase in temperature may ensue which can drive the superconductor normal. Fortunately, this motion of flux lines can be impeded or stopped by introducing metallurgical defects, such as voids, grain boundaries, alpha-Ti precipitates, etc. into the material. These normal state defects interact with the FLL to effectively pin the fluxoids. Recent developemental work on optimized NbTi has recognized that very fine alpha-Ti precipitates play a major role in pinning the FLL.[2] Improvements in the critical current density have been achieved thus , by combining several precipitation heat treatments with large degrees of cold work. Critical current densities as high as 3680 A/mm^2 at 5T and 4.2K in the optimized material have been achieived .[3] Despite the significant achievements in improving the J_c and recognizing the role of the alpha-Ti as effective flux pinners, further improvements beyond the 4000 A/mm^2 mark is uncertain. The flux line lattice may interact with the optimized microstructure more effectively than for example in the standard production wire, but the inherent randomness of the alpha-Ti clusters will not allow the interaction with the FLL in a continuous manner. As a result, there may be regions of superconducting material and normal state material interacting with the same flux line.

Advances in Cryogenic Engineering (Materials), Vol. 36
Edited by R. P. Reed and F. R. Fickett
Plenum Press, New York, 1990

311

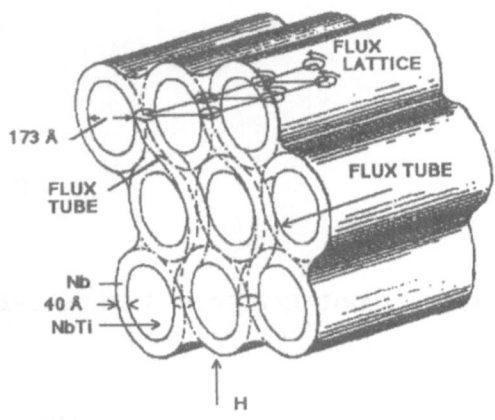

FIGURE 1. Schematic of the FLL threading the artificial pinning centers continuously.

In order to maximize the flux pinning characteristics and therefore the critical current density , the FLL should interact with the defect structure synchronously and continuously.[4] In this paper, we present a superconductor made by a process of successively drawing a group of Nb jacketed NbTi filaments until the ratio of core size to pinning layer thickness is dimensioned as desired. The objective in the dimensioning is to produce a pinning layer spacing comparable to the fluxoid spacing for the desired magnetic field strength, and to have the volume of the pinning layer shell approximate the volume of the fluxoids. Ideally, the structure would have a pinning thickness

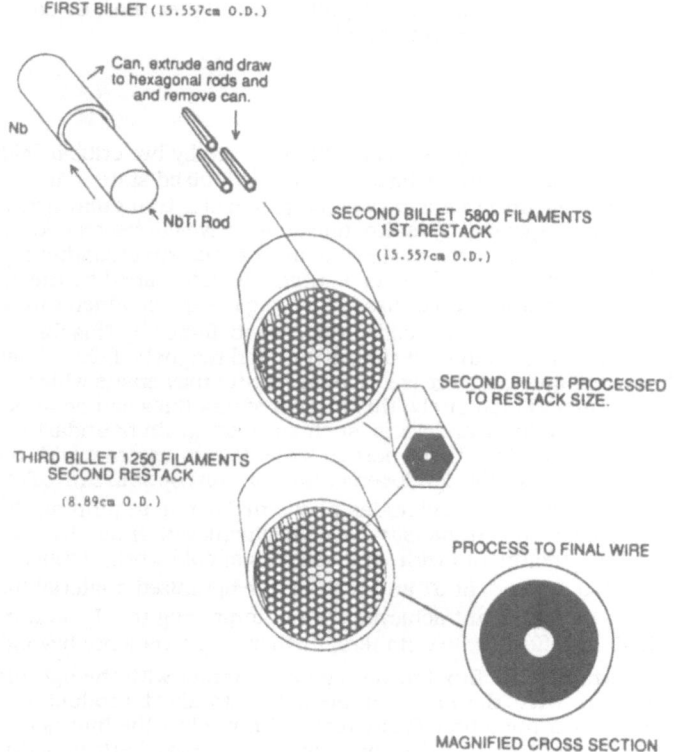

FIGURE 2. A schematic of the manufacturing process.

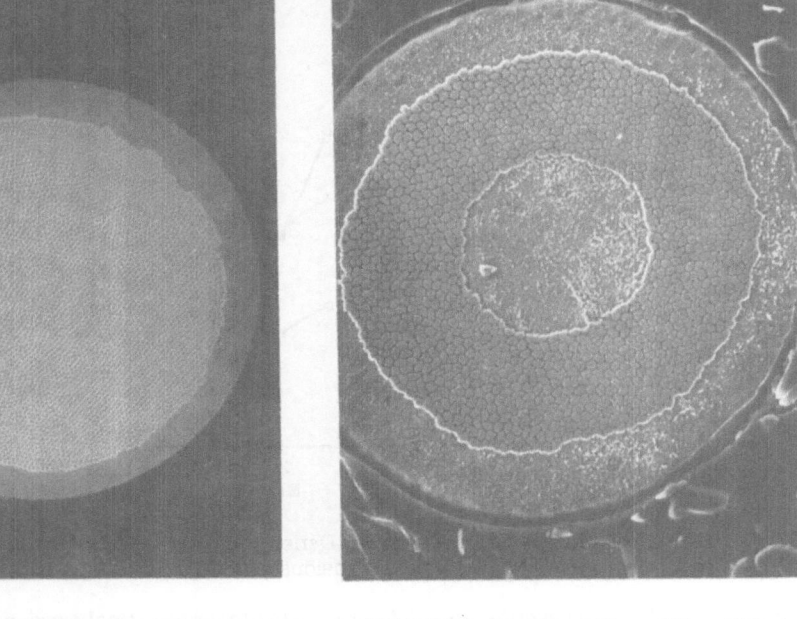

FIGURE 3. A cross section of a 1.8mm O.D. conductor filament with 5800 subfilaments.

FIGURE 4. A cross section of the APC with 1250 filaments, at final size of 0.152mm O.D.

comparable to the coherence length.[4] In figure 1 , we show a schematic view for the normal case of applied magnetic field. The hexagonal flux lines intersect the NbTi/Nb composite filaments perpendicularly so that the fluxoids themselves appear as transverse lines that distort around the hexagonal composite elements as they thread their way, in a continuous path provided by the Nb pinning material, across the composite lattice.

EXPERIMENTAL

To obtain the dimensions required for the artificial pinning structure, we start with a 9.96cm diameter Nb-46.5%Ti rod within a 2.16cm thick Nb jacket. This represents equal volume fraction of NbTi and Nb. The jacketed NbTi rod placed within a 15.56 cm O.D. copper billet and processed until it has an outer diameter of 0.18cm. The copper is then stripped off, leaving a 0.17cm diameter subfilament formed of NbTi surrounded with a uniform layer of Nb . A second restack billet is then assembled with approximately 5800 of these subfilaments placed in another copper can of 15.56cm O.D.. Again processed , this time to a copper-coated multifilament having a diameter of 0.18cm. A third and final restack billet is then assembled with 1250 of the second stage multifilaments. These are placed within an 8.89cm O.D. copper billet, and this time reduced to a final wire having a diameter of 0.152mm. A schematic of the process just outlined above is shown in figure 2.

The result is a superconductor, containing 1250 filaments each made up of 5800 subfilaments. Each subfilament is a superconducting NbTi core with an approximate 200 Angstrom diameter and embedded in a surrounding pinning layer of Nb approximately 40 Angstroms thick. These final dimensions correspond to a final wire diameter of 0.152mm. Figure 3 is a cross section of a 1.8mm O.D. restack rod containing 5800 NbTi subfilaments separated by Nb. Figure 4 is a cross section of the final product in its multifilament configuration. Samples were processed to various final sizes corresponding to the NbTi/Nb dimensions which may be optimal for the conductor under study. No precipitation heat treatments typical of conventional NbTi conductors have

313

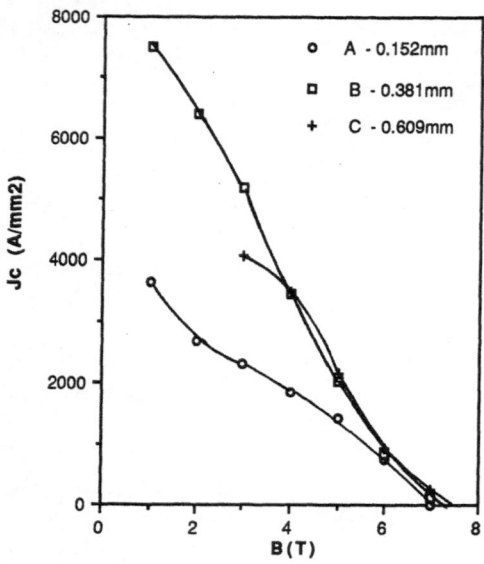

FIGURE 5. The Critical Current Density versus Applied Magnetic
Field for NbTi dimensions of 23.7nm, 59.2, and 96.8nm.

been given to these APC conductors. After drawing to various final sizes, samples of
approximately 200cm in length were cut for critical current measurements as a function
of external magnetic field. Measurements were taken at fields from 1 through 7 Tesla. A
standard four point probe on a helically wound sample at 4.2K was utilized. Voltage taps
were spaced 75cm apart and a resistivity criterion of 10^{-12} Ohm-cm was applied for I_c
results .

RESULTS AND DISCUSION

Critical current densities, J_c, in the filament as a function of applied magnetic field,
B, are illustrated in figure 5 for final NbTi filament diameters of 23.7nm, 59.2nm, and

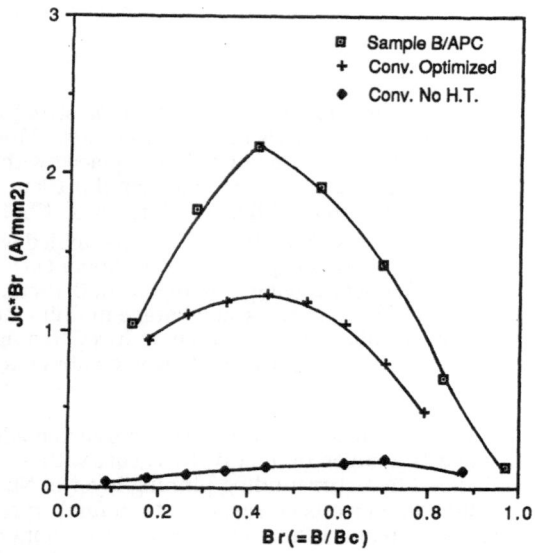

FIGURE 6.   A comparision of the normalized pinning strength versus
reduced field for the flux pinner (APC), conventionally optimized
wire, and cold worked wire without optimization (NHT).

96.8nm. The J_c in the NbTi is a factor of two higher then presented, as the NbTi represents 50% of the filament volume. These dimensions correspond to the final wire diameters of 0.152mm, 0.381, and 0.609. The critical current density for sample B in figure 5 is 3462 A/mm^2 at 4 T. The highest J_c obtained for a 0.508mm O.D. wire was 3700 A/mm^2 at 4 T. The results so far demonstrate clearly that the principle of conductor fabrication with artificially introduced flux pinning defects is feasable by mechanical metallurgy. The significance of this approach lies in the fact that no intermediate anneals are required to optimize the microstructure as currently required in present state of the art Nb-46.5%Ti superconductors. Furthermore, it should be possible to design a superconductor with a given current density independent of wire size.

The H_{c2} for samples A, B, and C exptrapolate to 7T, 7.2T, and 7.6T respectively . Thus the results for the present case however, show a dramatic decrease in the H_{c2} for all three dimensions investigated. Also, we find that the J_c varies with the wire diameter. The critical current for sample A in the range of 1 to 5 Tesla being the lowest, while sample B exhibiting the highest J_c. It is of interest to note that the J_c for sample C at the lower fields begins to fall below sample B, while at the higher fields near H_{c2} the J_c is larger than in both samples A or B. The shape of the J_c curve in sample A shows a slight increase "hump" centered about 4 Tesla where the flux lattice spacing interestingly corresponds to the NbTi dimension of 23.7 nm. In the case of sample B, the flux lattice matches the NbTi dimensions near 1 Tesla. However, in sample C no such FLL matching occurs.

At this time, it is our belief that the large H_{c2} deppresion experienced in these samples is due to the proximity effect. It has been shown both theoretically and experimentally that when two metals, a superconductor and normal metal are formed contiguously , the superconducting proporties, such as the H_{c2} and T_c are profoundly altered.[5,6] In particular, the dimensions for the artificial flux pinning in our samples are in the regime were the proximity effect would be important. This is especially evident in sample A were the H_{c2} is the lowest. The pinning shell Nb which surrounds the NbTi filament is 50% of the cross-section, thus at the smaller dimensions as in sample A , the change in the order parameter , due to proximity effect is more important. As we go to larger dimensions of the NbTi/Nb structure in sample C, the H_{c2} has already increased slightly. However, the FLL is no longer matching as well to the artifical pinning structure.

Despite the proximity effect on artifical flux pinning, we can nevertheless compare the relative pinning strength of the artificial pinning centers to the pinning characteristics of conventionally optimized Nb-46.5%Ti. This is shown in figure 6, where we plot the pinning force normalized by the reduced field versus the reduced field. In figure 6 we show conventional Nb-46.5%Ti , cold worked to final size without any intermitent anneals and conventional optimized wire [7]. We compare those results, to the artificial flux pinner which did not receive any intermediate anneals. As is clearly evident the relative magnitude of the flux pinner is nearly twice that of the optimized NbTi.

CONCLUSION

In principle the fabrication of artificial pinning centers by mechanical metallurgy has been demonstrated successfully. A first cut design including 50% Nb as the pinning centers however, resulted in a substantial decrease in the H_{c2} and J_c properties. Despite that, the critical current density at the low fields exceeds or equals the optimized Nb-46.5%Ti, and the relative normalized pinning strength of the APC conductor is nearly twice that of the conventionally optimized conductor, suggesting possible substantial further improvments in the superconducting properties. Future artificial pinning centers work, will focus on optimizing the volume fraction of pinning material to minimize the H_{c2} depression. TEM microstructural studies are currently underway and will be reported elsewhere as well as transition temperatures on the present APC conductor.

REFERENCES

1. J.Friedel, P.G. DeGennes, J.Matricon Appl. Phys. Lett., Vol.2, No.6, pp.119, (1963).

2. P.J. Lee, D.C. Larbalestier J. of Mat. Sci., 23 , pp.3951, (1988).

3. Li Chengren and D.C. Larbalestier IEEE MAG 23, No.2, March (1987).

4. B.A. Zeitlin, M. S. Walker, L. R. Motowidlo United States Patent, No. 4803,310, Feb.7, (1989).

5. N. R. Werthamer, Phys. Rew. , Vol. 132, No.6 , Dec.15, (1963).

6. P. G. DeGennes, Rev. Mod. Phys., p.225, Jan., (1964).

7. J.V.A. Somerkoski, D.P. Hampshire, H. Jones, IEEE,MAG.23, No. 2, March (1987).

THE PROPERTIES OF INDUSTRIAL SUPERCONDUCTING

COMPOSITE WIRES FOR THE UNK MAGNETS

V.Ya.Fil'kin[*], V.P.Kosenko[*], V.L.Mette[*], K.P.Myznikov,
A.D.Nikulin, V.A.Vasiliev, G.K.Zelensky[*], A.V.Zlobin

142284, Institute for High Energy Physics, Serpukhov, USSR

[*]All-Union Research Institute for Inorganic Materials
Moscow, USSR

ABSTRACT

The present note describes the properties of superconducting composite wires consisting of $6 \cdot 10^{-6}$ m Nb - 50% Ti filaments for the UNK magnets. The properties have been studied from the data on commercial batches of wires, 3000 km long and proceeding from the technology of twofold
billets (55 x 162). A strand is $0.85 \cdot 10^{-3}$ m in diameter and the copper-to-
noncopper area ratio is 1.385/1. The current density in the filaments is
$(2550+250) \cdot 10^6$ A/m^2 (B=5 T, T=4.23 K). The Residual Resistivity Ratio
(RRR) for the initial copper is $\geqslant 160$ and for the wire of the final size
it is $\geqslant 70$. The manufacturing technology for the wire with $6 \cdot 10^{-6}$ m filaments from 0.25-0.3 m billets has been developed. The current density obtained in the strands from large billets is 2800+290 A/mm^2. The current
rise parameter is $I_0 \simeq 11$ (n $\simeq 50$). Production of the wire for standard UNK
magnets has begun.

INTRODUCTION

The development and production of superconducting dipoles and quad-
rupoles for the IHEP 3 TeV Accelerating and Storage Complex (UNK) became
possible due to an appreciable progress in home industrial technology for
superconducting composite wires with thin nuobium-titanium filaments,
less than $10 \cdot 10^{-6}$ m in diameter. The current density in filaments may be
adjusted within a wide range up to 3.0×10^9 A/m^2. The chosen minimum current density, $2.3 \cdot 10^9$ A/m^2, for the UNK magnets is determined by the necessity to provide the 5 T operating field and a current reserve as
well as by the industrial process efficiency.

Within 1980-1986 large batches of superconducting strands SCNT-0.85-
-2970-0.42 with a current density of 2×10^9 A/m^2 in $10 \cdot 10^{-6}$ m filaments
for the 5 T warm-iron superconducting dipoles (T=300 K) were manufactured.
Simultaneously, the wire SCNT-0.85-7260-0.4 with filaments 6 m in diameter and less was also developed[1].Beginning with 1987 the industrial wire
SCNT-0.85-8910-0.42 for the UNK cold-iron magnets (T=4.5 K) has been under production. The basic characteristics of this wire are presented below.

Advances in Cryogenic Engineering (Materials), Vol. 36
Edited by R. P. Reed and F. R. Fickett
Plenum Press, New York, 1990

317

Table 1. Superconductor Parameters

1. Superconducting alloy	NT-50 (Nb - 50% Ti)
2. Admissible spread in Ti	4%
3. Matrix material copper	
4. Conductor diameter, m	$(0.85^{+0.03}_{-0.00}) \times 10^{-3}$
5. Filament diameter, m	$6 \cdot 10^{-6}$
6. Packing factor (copper-to-noncopper area ratio)	0.42 ± 0.02 $1.385 \pm 0.115/1$
7. Twist pitch, m	$10 \cdot 10^{-3}$
8. Residual Resistivity Ratio (RRR) of Matric	70
9. Critical current, A (B=5 T, T=4.23 K)	550
10. Current density, A/m^2 (specific resistivity, Ohm/m)	$2550 \cdot 10^6$ $2.55 \cdot 10^{-13}$

DIPOLE MAGNET FOR THE UNK

The dipole magnet for the UNK of the 1st version[2] was $90 \cdot 10^{-3}$ m in aperture and the iron shield was at room temperature. The two shells of the coil were wound from a flat transposed cable of uniform dimensions. The cable consisted of 23 strands SCNT-0.85-2970-0.42, each strand comprised 2970 $10 \cdot 10^{-6}$ m filaments. The cable was insulated by two layers of 2×10^{-5} m thick kapton tape and then it was coated by an epoxy impregnated glass tape 1×10^{-2} m wide, 1×10^{-3} m thick wound with a gap of $4 \cdot 10^{-3}$ m. Some magnets were manufactured the wire SCNT-0.85-7260-0.4 with $6.5 \cdot 10^{-6}$ m thick filaments. All in all, 50 short and 10 full-scale dipoles have been manufactured from the wires of these types. Strands were selected for the cable proceeding from the minimum current 310 A in the 7 T field (500 A in the 5 T field, T=4.23 K). The measurements showed that the full current in the cable is equal to the sum of the currents in single wires minus about 5% due to the deformation of the strands in the cable.

In 1986 a decision was taken at IHEP to change over to a cold-iron design with a view to simplify the cryostat design, to reduce the level of stationary heat leaks on the supports and to cut the amount of the superconductor used for the machine. The aperture of such a dipole was chosen to be $80 \cdot 10^{-3}$ m. The 0.18 m thick magnetic shield was placed in the helium cryostat around the collared coil. The working field was left unchanged, 5 T. For the new magnet there has been developed and manufactured a composite wire SCNT-0.85-8910-0.42 with $6 \cdot 10^{-6}$ m filaments and minimum current density of 2.3×10^9 A/m^2.

The operational features of the new dipole were simulated proceeding from two types of the superconducting cable, the 16- and 19-strand cable. The magnets produced from the 16-strand cable proved to be less stable from mechanical viewpoint. The 19-strand dipoles exhibited stable mechanical properties and the required field nonlinearities[3].

SUPERCONDUCTING WIRE SCNT-0.85-8910-0.42 FOR THE UNK

The commercial output of the superconducting wire SCNT-0.85-8910--0.42 for the UNK magnets was begun in 1987. Table 1 presents its tencnical parameters.

The production of this wire is based on traditional techniques of extrusion and cold deformation with intermediate annels. The total time

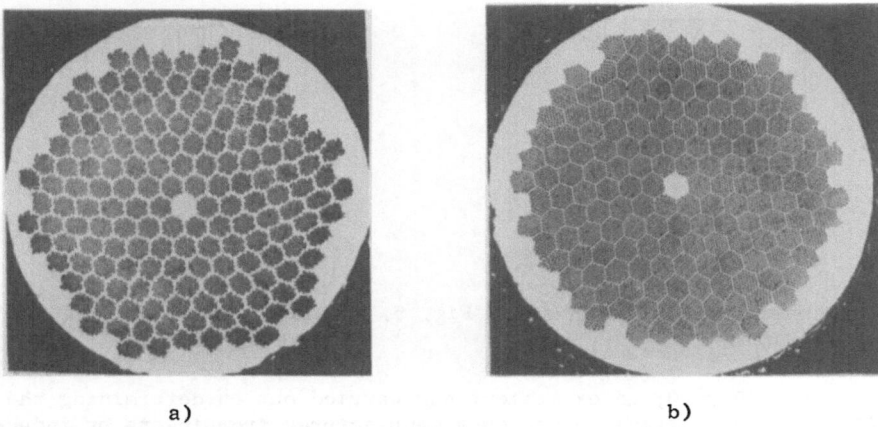

a)   b)

Figs. 1a,b. Cross section of composite wire of the UNK. 1a - wire from
a ∅ 0,13 m billet; 1b - wire from a ∅ 0,3 m billet.

of heat treatments was chosen depending on the diameter of the last billet and the required minimum current which was checked from its distribution in the industrial batch. Figures 1a,b shows the wire design under
development for billet diameters of 0.13 m and 0.3 m, respectively. Figure 2 presents the critical current versus field. The volt-ampere characteristic of the wire is described well by an exponential dependence
$E=E_0 \exp(I-I_c)/I_0$, where I_c is the critical transport current of the
wire measured for $E_0=10^{-6}$ V/cm, I_0 is the parameter of current ramp. Figure 3 shows the volt-ampere characteristics of the UNK wire from billets
0.13 m and 0.3 m in diameter. The stability of the filament cross sections
is determined primarily by the niobium diffusion barrier which has been
in use since 1980. The barrier between the Nb-Ti alloy and copper prevents the formation of intermetallic Cu-Ti particles during thermal and
mechanical treatment[4]. A high current density achieved was due to the
use of high-homogeneous ingots, multiple intermediate anneals and cold
extrusion at the final stage. The last annel on the finite-dimension wire
ensures the required value of the Residual Resistivity Ration (RRR) for
copper.

The parameter RRR ≳70 was chosen from the requirement to provide a
safe heat removal due to longitudinal heat conductivity of the cable and
limitation on the maximum coil temperature after a quench. To maintain

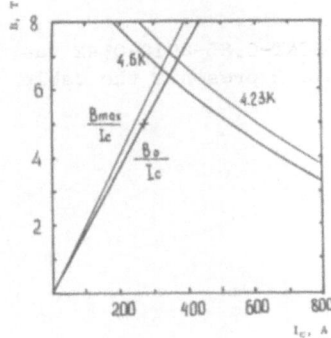

Fig. 2. The I_c(B)-depen-
dence for a single
UNK wire.

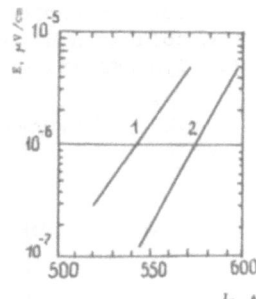

Fig. 3. V-I characteristics for a
UNK wire from ∅ 0.13 m (I),
∅ 0.3 m (2) billets.

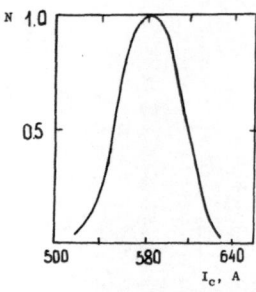

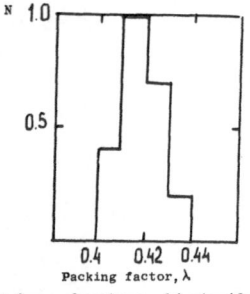

Fig. 4. Current distribution for
     UNK wires.

Fig. 5. Packing factor distribution
     for UNK wires.

the parameter RRR$\geqslant$70, an experiment was carried out on determining the
RRR effect of the initial copper cups manufactured from ingots by induction and electron-beam smelting. Electron-beam smulting was chosen for
all cups including the intermediate ones. Ingots are selected from the
quantity RRR$\geqslant$ 160.

Industrial production of the wire from two types of billets demonstrated that changing over to bigger billets, 0.25-0.3 m in diameter, is
more advantageous. With such billets used, some pieces of the wire became
longer, the end waste of wires also became less and the total anneal time required to attain the specified current density decreased about twice. The
quality factor of filaments increased 1.5-2 times. The UNK wires produced
from billets 0,13 m and 0.25-0.3 m in diameter possess about the same properties. The final confirmation of the adequate properties of strands will
come from experiments with future magnets.

The control of the basic parameters of the composite wire is carried
out at the plant with the agreed control method. The analysis of the commercial wire batches produced in 1988 showed that the technical tolerances
for the UNK superconductor are fulfilled. Figures 4,5,6 show the current
distribution, packing ratio and RRR, respectively. The spread in the wire
diameter is 2 times less than the tolerable one.

In the cource of the development of industrial technology the filament structure was investigated. The current density of about $3 \cdot 10^5$ A/cm^2
is provided by a fine-cell structure of the β-matrix decorated with $\measuredangle$-phase
particles, the β-matrix subcells are less than $50 \cdot 10^{-9}$ m. The study of the
filament structure is presented in detail in[5].

SUPERCONDUCTING CABLE

As was noted above, a cable of 19 strands SCNT-0.85-8910-0.42 was
manufactured for a cold-iron dipole magnet. Table 2 presents the cable
parameters and fig.7 shows its cross section.

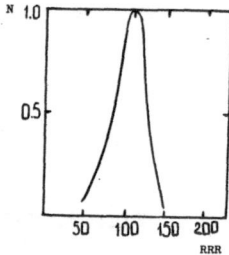

Fig. 6. RRR distribution for UNK
     wires.

Table 2. The Basic Parameters of the Cable

1. Number of strands	19
2. Cable type	ZEBRA
3. Coating of 9 strands	Sn + 5% Ar
10 strands	bare
4. Dimensions, mm	
- inner shell	(1.30 - 1.62) x 8.5
- outer shell	(1.33 - 1.59) x 8.5
5. Transposition pitch, mm	62
6. Critical current, A	
(B=5 T, T=4.23 K)	$\geqslant$ 9500

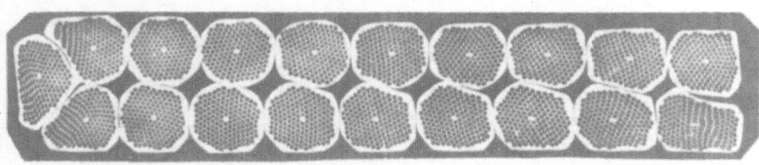

Fig. 7. Cross section of the cable for a dipole UNK magnet.

The decrease of the wire current due to the cable deformation into a trapezoid is 12% from the initial value (prior to transposition). The cable is produced and its current is checked at IHEP. The level of ac loss in the dipole coil determined by the sum of hysteresis and eddy current loss in the wires and cable is about 60 J/m of the magnet.

CONCLUSIONS

The industrial technology for the production of the NT-50 alloy (Nb-50% Ti) multifilamentary composite superconducting wire with a current density in filaments of $3.0 \cdot 10^9$ A/m^2 has been developed in the USSR. The maximum current density chosen for the UNK magnets is $2.3 \cdot 10^9$ A/m^2. The wire SCHT-0.85-8910-0.42 is produced from billets 0.130 m in diameter and single stage billets 0,3 m in diameter for the real magnets of the UNK.

REFERENCES

1. V.Y.Fil'kin et al. Composite Superconductors for UNK Magnets Proc. of Workshop on Superconducting Magnets and Cryogenics. BNL, 1986, p. 56.
2. N.I.Andreyev et al. Development and Study of the UNK Superconducting Dipole Models. Proc. of the XII Int. Conf. on High Energy Accelerators, Fermilab, USA, 1983.
3. A.I.Ageev et al. Study of a Superconducting Dipole Magnet for the UNK. Proc. XI Int. Conf. on Magnet Technology, Tsukuba, Japan, 1989.
4. L.D.Bogdanova et al. The Composite Superconductor from NT-50 Alloy for Impuls Magnets UNK. Proc. of Int. Conf. of Counsil of Mutual Assistance Member-States on Low-Temperature Physics, Berlin, GDR, 1985.
5. G.K.Zelensky et al. The Effects of Thermal and Mechanical Treatment Modes on the Structure and Properties of NT-50 and NT-55 Superconducting Alloys. Proc. XI Int. Conf. on Magnet Technology, Isukuba, Japan, 1989.

DEVELOPMENT OF LARGE KEYSTONE ANGLE CABLE

FOR DIPOLE MAGNET WITH IDEAL ARCH STRUCTURE

Takakazu Shintomi, Akio Terashima and Hiromi Hirabayashi

National Laboratory for High Energy Physics
Tsukuba-shi, Ibaraki, Japan

Masaru Ikeda and Hideki Ii

The Furukawa Electric Co., Ltd.
Tokyo, Japan

ABSTRACT

Compacted strand cables of NbTi with large keystone angle have been developed for applications to superconducting dipole magnets of big hadron collider accelerators. The trial-fabricated cables have the keystone angle of 1.6 to 3.0 degrees and the packing factor of 90 to 95 %. Strands of 0.808 mm in diameter with the filament diameters of 4.8 and 6 μm were used. The fabrication of those cables has not met with serious problems. The degradation measurements of the critical current have been performed and the degradation less than 3 % is observed up to the angle of 3.0 degrees for the cables with the packing factor of 90 %.

INTRODUCTION

A few projects of the high energy hadron collider accelerator have been planned and developed. Those accelerators require a large number of superconducting dipole magnets, for example about 8,000 for SSC and 4,000 for LHC. It is very important to control the field quality and to obtain high yield rate of the magnets for such huge accelerators. The superconducting dipole magnets of high energy colliders have a relatively small aperture in consideration of accelerated beam emittance and also the economical requirements. In the current designs, the inner diameters of the coil are 40 mm for the SSC dipole magnets[1] and 50 mm for LHC[2].

The dipole magnet for accelerator use is fabricated using a so called cos θ method with compacted strand cables. When the windings of the coil cross section are arranged to have an ideal arch structure, the cable should be made to have a proper keystone angle according to the arch structure. In this sense, a smaller inner coil diameter requires a larger keystone angle cable. As the angle of the existing cables has been limited below 2 degrees by the fabrication processes, the dipoles for high energy hadron colliders need some wedges inside the windings to fit the coil cross section to the arch structure. Those wedges sometimes cause difficulties of field quality control and fabrication processes.

We have developed large keystone angle cables up to 3 degrees. Using the cables it is possible to fabricate dipole magnets without any wedges inside the windings. The paper describes the development and the degradation test for the large keystone angle cables.

Advances in Cryogenic Engineering (Materials), Vol. 36
Edited by R. P. Reed and F. R. Fickett
Plenum Press, New York, 1990

323

Table 1. Parameters of the trial-fabricated cables

Strand	"A"	"B"
superconductor	NbTi	NbTi
diameter (mm)	0.808	0.808
Cu/SC ratio	1.3	1.5
number of filament	12,800	7,300
filament diameter (μm)	4.8	6.0
filament spacing (μm)	0.6	1.5
twist pitch (mm)	25	12.5
resistance ratio of Cu	> 110	> 110
J_c w/o Cu at 5 T, 4.2 K (A/mm^2)	2,590~2,740	2,630~2,660
Cable		
number of strands	23	
cabling pitch (mm)	79	
width (mm)	9.30	
keystone angle (degree)	1.6, 2.0, 2.4, 3.0	
packing factor (%)	90 ~ 95	

STRUCTURE OF DIPOLE MAGNETS

To obtain uniform dipole field, the cos θ method is adopted. As an approximation of cos θ, a double shell structure is usually used. The keystone angle cable is necessary to make up an ideal arch structure. If the required aperture of the dipole becomes small, the keystone angle of cable should be increased. Some wedges are set inside the windings to fit the keystone angle to the coil arch structure because of the small aperture and the limitation of the keystone angle. The wedges usually made of copper plate have different Young's modulus from that of the cable. This means non-uniform deformation under stresses at collaring and electromagnetic forces. It gives rise to the distortion of the field quality. Moreover the wedges will be causes of the difficulty of the coil fabrication processes.

RELATION OF KEYSTONE ANGLE AND COIL INNER DIAMETER

When a dipole magnet is wound to have an ideal arch structure, the coil inner diameter D is given as follows.

$$D = 2 (t / \theta - w)$$

where t and w are the thickness at the wide wedge and the width of a keystone cable with insulation, respectively. θ is the keystone angle. In case of the SSC dipole magnet, t and w for the inner coil are 1.888 mm and 9.600 mm including insulation, respectively. The keystone angle θ is 1.6 degrees. With those parameters the coil inner diameter will be 116 mm, if any wedge would not be used. To fabricate a dipole magnet of 40 mm inner diameter with the ideal arch structure, the keystone angle of the cable should be 3.6 degrees.

FABRICATION OF R & D CABLES

Cabling

Fabrication test has been performed for various keystone angles of compacted strand cables. To know the dependence of degradation of the critical current on the keystone angle and the packing factor, the cabling with several angles from 1.6 to 3.0 degrees and packing factors of 90 to 95 % has been tried. The trial-fabricated cables were stranded to have a width of 9.3 mm with 23 strands of which diameter is 0.808 mm. The parameters of the cables are shown in Table 1.

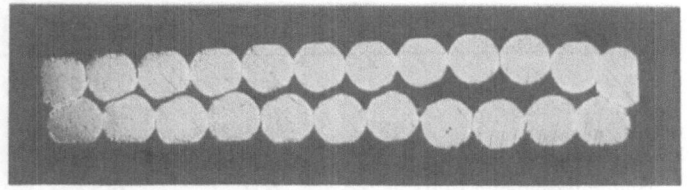

Fig. 1(a). The cross sections of the trial-fabricated cable
with keystone angle of 1.6 degrees.

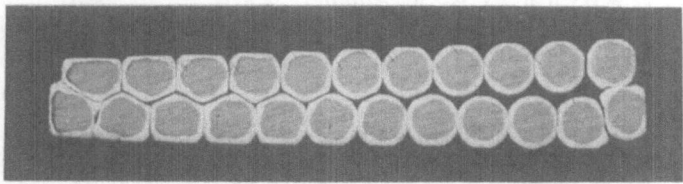

Fig. 1(b). The cross sections of the trial-fabricated cable
with keystone angle of 3.0 degrees.

Cabling has been easily performed for the large keystone angle up to 3 degrees. The examples of the cross sections of the fabricated cables are shown in Fig. 1. As shown in the pictures, heavy deformation is occurred at the narrow edge for the larger keystone angle cable and the deformed strands push other ones to the wide edge.

Degradation measurement

Critical currents of the fabricated cables were measured by sampling three strands from each compacted strand cable under the magnetic field of 5 T. Each sample of 2 m long is wound spirally on a sample holder. By comparing the measured values of the critical current of each sample to that of the same strand before cabling, the degradation due to cabling has been estimated. The degradation curves of critical current according to the keystone angles and the packing factors are shown in Fig. 2.

As in the figure, the degradation of the cable with the packing factor of 90 ~ 92 % is almost linearly increased with the keystone angle. As for the cable of the packing factor of 94 ~ 95 %, it juts out up to 5 % for 2.4 degrees but decreases down to 4 % for 3 degrees.

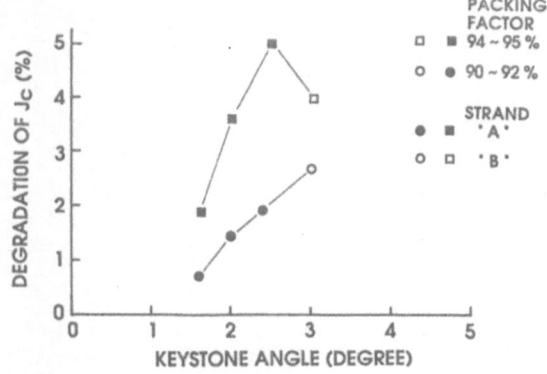

Fig. 2. Measured degradation of the critical current for the trial-fabricated cables
with the keystone angle and the packing factor of 90 ~ 95 at 5 T.

Table 2. Parameters of the large keystone angle cables for the R & D magnet

Strand	Inner	Outer
diameter (mm)	0.808	0.614
Cu/SC ratio	1.4	1.8
number of filament	7,600	3,750
filament diameter (μm)	6.0	6.5
filament spacing (μm)	> 1.0	> 1.0
twist pitch (mm)	13.0~13.5	13.0~13.5
resistance ratio of Cu	> 110	> 110
J_c w/o Cu at 5 T, 4.2 K (A/mm^2)	2,870	2,970
Cable		
number of strands	23	30
cabling pitch (mm)	79	74.5~75.0
narrow edge thickness (mm)	1.14	0.91
wide edge thickness (mm)	1.64	1.19
width (mm)	9.34	9.30
keystone angle (degree)	3.07	1.73
J_c w/o Cu at 5 T, 4.2 K (A/mm^2)	2,715	2,785
fabricated length (m)	300	300

On the whole the degradation of J_c becomes worse with the increase of the keystone angle and the packing factor, but more sensitive with the latter factor. Nevertheless, it is small and below 3 % for the packing factor of ~ 90 % which is a usually adopted value and the same level as one of the existing compacted strand cables.

To check the causes of the degradation, the break of the filaments has been also checked and only a small amount of breaks were observed for all of the trial-fabricated

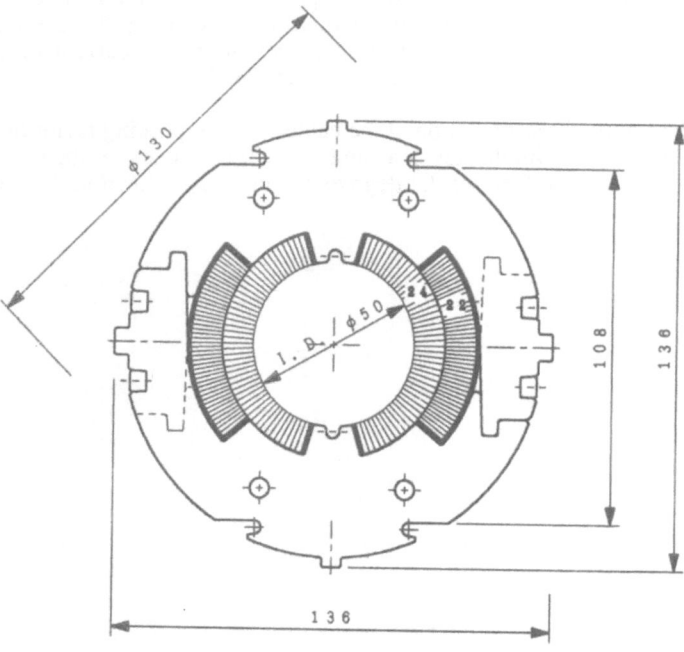

Fig. 3. The cross section of the R & D dipole magnet with ideal arch structure using large keystone angle of 3.1 degrees.

Table 3. Parameters of the R & D dipole magnet

Central magnetic field (T)	6.6
Rated current (A)	5,890
Coil length straight section (m)	1.0
overall (m)	1.35
Coil inner diameter (mm)	50.0
Number of turns inner	24
outer	22
Overall current density inner (A/mm^2)	390
outer (A/mm^2)	500

cables. From such a point of view, those degradations seem to be caused by the sausaging of filaments with heavy deformation at the narrow edge. The cable of the 2.4 degrees with the packing factor of ~ 95 % may be damaged with the heavy deformation. However, in general the serious degradation has not been observed and the large keystone angle cables are acceptable for dipole magnets.

R & D MAGNET WITH IDEAL ARCH STRUCTURE

To check the commercial production and applicability of those cables to the dipole magnets with a small aperture, a model magnet with the ideal arch structure is now being developed at KEK. The parameters of the used cable are shown in Table 2. The coil has an inner diameter of 50 mm and the length of the straight section is 1 m as shown in Table 3.

The coil cross section and the finished coil are shown in Figs. 3 & 4. As shown in the figure, there is no wedge inside the windings. The coil winding processes are very easy and have been finished in a short time. The fabricated dipole magnet will be tested in near future.

SUMMARY

Based on the cabling and the degradation measurement, the fabrication was very successful and there are no serious obstacle. The observed degradations of the large

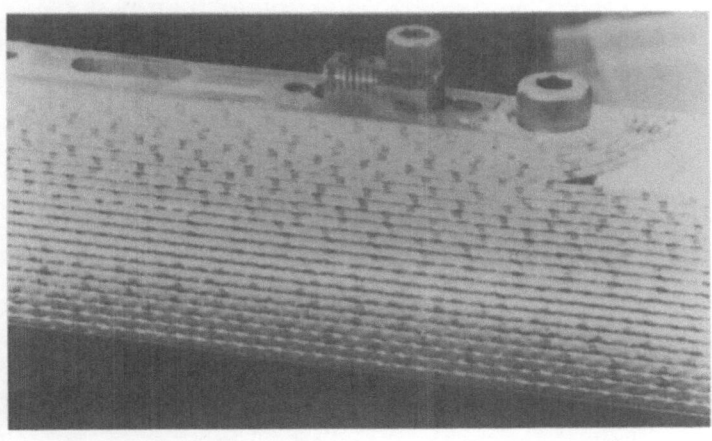

Fig. 4. The picture of the inner coil of the R & D dipole magnet.

keystone cables up to 3 degrees with the packing factor of ~ 90 % are below the value of 3 % which is almost the same level in the usually fabricated compacted strand cables. To check the commercial production of those cables, the cable which has the keystone angle of 3.1 degrees and length of 300 m was fabricated for a model magnet of 1 m long. The inner coil diameter of the magnet is 50 mm. The critical current density of the cable is more than 2,750 A/mm^2 without copper matrix. The value is above the level given in the SSC conceptual design report. The first model dipole magnet of 1 m long was fabricated using this cable. Those cables can be applied to the dipole magnets of big hadron colliders. Further development of larger keystone angle cables goes on to apply to a dipole magnet which has an inner coil diameter of 40 mm.

ACKNOWLEDGEMENT

The authors express their gratitude to Profs. T. Nishikawa and S. Ozaki for their continuous encouragement. They also express their thanks to the staff of the mechanical engineering center at KEK for their vigorous efforts for fabrication of the model dipole magnet.

REFERENCES

1. SSC Central Design Group, "Conceptual Design of the Superconducting Supercollider," Lawrence Berkeley Laboratory, CDG SSC-SR-2020, March 1986

2. D. Leroy, R. Perin, G. de Rijk and W. Thomi, "Design of a High-Field Twin Aperture Superconducting Dipole Model, IEEE Trans. on Magnetics, MAG-24, 1373 (1988)

CRITICAL CURRENT DENSITIES OF Sn-DOPED PbMo$_6$S$_8$ WIRES

Luc Le Lay, Philippe Rabiller, Roger Chevrel, Marcel
Sergent, Thierry Verhaege*, Jean-Claude Vallier**
and Pierre Genevey***

Université de Rennes, U.R.A. C.N.R.S. 254, F-35042 Rennes
Cédex, *C.R. C.G.E., Route de Nozay, F-91460 Marcoussis
**S.N.C.I., F-38042 Grenoble Cédex, ***C.E.N. Saclay
F-91190 Gif sur Yvette

ABSTRACT

PbMo$_6$S$_8$ wires were made using a cold powder metallurgy process.
The initial powder was a mixture of PbS, Mo, MoS$_2$ and Sn, so that the
stoichiometry was Pb:Mo:S:Sn = 1.05:6.2:8:x, where x varied from 0 to
0.6. We optimized the heat treatment conditions of small coils and
measured their transport critical current densities as a function of
x and the applied magnetic field B. The best result was 1.5 x 10^8 A/m^2
at 20 T and 5.5 x 10^7 A/m^2 at 27 T, achieved on a 1000 mm long wound
wire.

INTRODUCTION

Out of all the superconductors now known, the copper oxides recently
discovered possess the highest critical magnetic fields (Hc$_2$). However,
regarding the applications at high fields, they present so far the dual
disadvantage of being strongly anisotropic and having insufficient critical
current densities necessary to create strong magnetic fields.

These drawbacks in the new oxides have motivated continuing research
in the less recent compound, PbMo$_6$S$_8$ (PMS). The advantage of PMS is a
high critical magnetic field (Hc$_2$) of 54 T at 4.2 K, which remains the
most attractive for eventual applications beyond 20 T. Indeed,
regarding the transport critical current densities (Jc) in such filaments
(which, of course, is the form most practical for use), several research
teams in various countries |1-6| have already obtained encouraging
results, the best so far is on the order of 6 x 10^7 A/m^2 at 23 T which
was obtained in a small coil |6|.

We present here, essentially the new results of critical current
densities of such filaments : the first section will briefly discuss the
preparation of such filaments, the choice of compounds used as well as
the techniques used to measure Jc ; the second section consists of the
experimental results which will be discussed.

Advances in Cryogenic Engineering (Materials), Vol. 36
Edited by R. P. Reed and F. R. Fickett
Plenum Press, New York, 1990

TABLE I

configuration	distance between current leads	distance between voltage taps	applied field (T) and orientation
coil	1000 mm	100 mm	27 perp.
straight wire	20 mm	10 mm	18 perp.
straight wire	300 mm	50 mm	5 or 9 parall.

EXPERIMENTAL

The usual form of PMS is a powder ; thus, the transformation into wires is not a priori simple and requires expertise. This challenge has been met by most groups, at least for monofilaments.

The principle is to surround the powder with drawable material. This ensemble is then drawn through successive passes until a desired diameter is reached. The type of casing for the pure powder is determined by the type of drawing that is available, i.e. hot or cold. We opted for cold drawing and used copper for the casing. However, as we will see later, a final heat treatment is always necessary in order to make the powder superconducting ; but that allows a reaction between the copper and the PMS. This reaction destroys some of the superconducting properties expected : therefore, we inserted a niobium diffusion barrier between the two components. This precaution, in first approximation, generates the chemical and mechanical criteria demanded. Thus, typically, from initial billetts of o.d. 14 mm or 20 mm, we obtain o.d. 0.3 mm or 0.4 mm filaments. (These values are not the minima we can reach, but they allow an easy manipulation of the samples) |7|.

The cold drawing, as opposed to hot drawing, allows the use of two kinds of starting powders : PMS itself, or a mixture of its precursors (i.e. starting elements or binary components). We formerly showed |8-9| that the latter gives better Jc results in the case of a cold drawing : we have thus focused our present study on that kind of powder. It is composed of PbS, MoS_2 and Mo. The first two are synthesized from pure elements and Mo is reduced at 750° C under hydrogen flow. The mean size -measured on a Coulter LS 130 Laser granulometer- of each precursor is about 10 to 15 microns with sometimes the occurrence of coarser agregates. New routes are now under investigation in order to reduce both the mean value and the width of the grain size distribution. A doping agent can be added to improve the Jc characteristics of the wires. In this work, the doping agent is Sn(40 μm) or SnS(10 μm). The final stoichiometry is thus Pb:Mo:S:Sn = 1.05:6.2:8:x, where $0 < x < 0.6$.

For the present study, we principally stress on two experimental parameters : the doping agent concentration and the heat treatment conditions (of course others exist such as the mixing of the powders, their reactivity, the drawing conditions and so on). Besides their rather simple mastering, these are indeed important because we need to synthesize the superconductor in the wire itself after drawing. In order to evaluate and understand the influence of these two parameters, we had to run lots of Jc transport measurements. Unfortunately, it turned out that the coils, the most sensible configuration for such experiments |10|, are also the most difficult to achieve. This difficulty arises from the sample preparation as well as its manipulation. Hence, we have first tested our wires as straight samples. Table I gives the details of the different configurations we used.

The voltage taps are always symmetrically disposed with respect to the center of the wire.

Of course, straight wires give no absolute Jc values : nevertheless, they allow an easy comparison between the different filaments. Indeed, the measurements of the Jc properties of the coils showed a posteriori that this comparison was sensible. The general conditions of heat treatments were : 1 atm. of argon pressure ; the wire is put in the furnace at the heat treatment temperature, and pulled out directly to the room temperature (the heating times given are thus the actual duration of stay at the given temperature). The heat treatments varied from 850° C, 48 hours to 1050° C, 10 mn. Regarding the coils, because of the high annealing temperature needed, we selected barrels made of alumina rather than stainless steel, which reacts with copper sheath. The length of that barrel is 80 mm, its diameter 30 mm. In order to avoid measuring Jc at the ends of the wires which can be polluted (despite carefully done experiments), we cut the wires so that the total length of their measured part is only 1000 mm (compared to 1500 mm for the heating operation). All the critical current densities of the present work are calculated for the cross sectional area of the PMS core. The criterion for the critical current is always 1 µV/cm, as commonly reported in the literature.

RESULTS AND DISCUSSION

Straight wires

Figure 1a gives the results of the transport critical current measurements for the straight wires, as a function of the applied field B, up to 18 T. For every x value and for every form of the doping agent (i.e. Sn or SnS), only the best wire is shown (as it revealed experimental failures, 0.2 Sn (SnS) doped wire is not reported). Several pieces of information can be drawn from the curves : first, it is clear that almost all the heating temperatures are 950° C, 30 mn to 2 hours. Given the variety of the conditions tested, it thus appears that this temperature is the optimal one for these kind of samples. We already showed |11| that, for x = 0.6, a 1-to-2 µm-thick layer of Nb_3Sn is formed between PMS and the niobium diffusion barrier. Therefore, it is possible that the supercurrent is partially transported by that layer. On another hand, we can define, for practical purpose, a slope Jc(9 T)/Jc(18 T). In the wire x = 0.6 (Sn), this slope is about 6, -as it has been previously encountered in wires drawn from powder leading to the stoichiometry $SnMo_6S_8$ |12, 13|- whereas in almost all the others, it is closer to 3. That discrepancy can mean that, in such heavily tin doped wires, the dominating phase resembles $SnMo_6S_8$, whose critical field is half that of PMS. We cannot explain so far why the wire x = 0.6 (SnS) does not behave in the same way. So early conclusions could lead to misinterpretations.

Among the remaining wires, it appears that those whose x = 0.4 have better Jc's at high fields, without showing the evidence of any significant difference between Sn or SnS as a doping agent. It is also obvious that the wire x = 0 is worse than any other, showing the importance of the tin. Eventually, wire x = 0.4 can be regarded as the best candidate obtained so far, so it has been selected to build small coils to be tested in high field up to 27 T.

We also tested the reproducibility of the Jc as a function of both the heat treatment conditions and the doping agent : it turned out that for the same conditions Jc can vary by a factor of 3. One of the

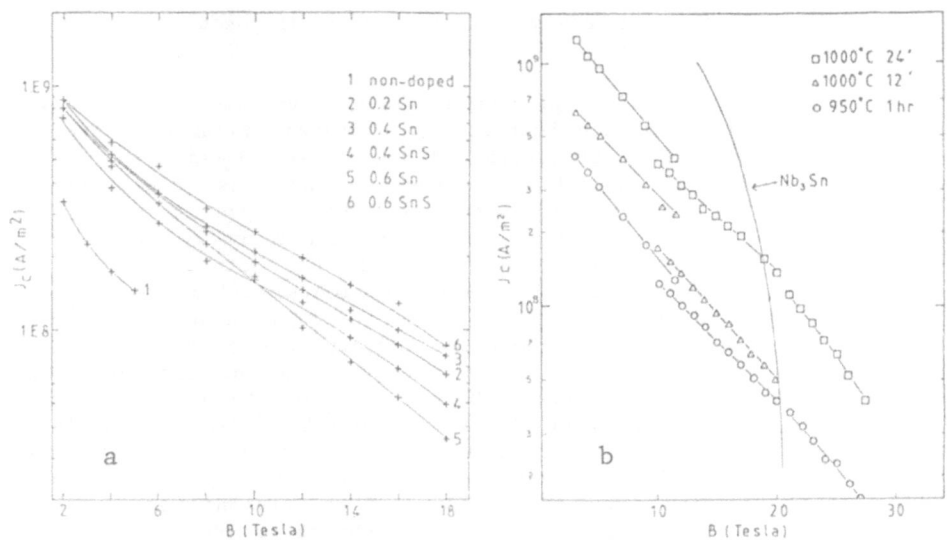

Fig. 1. Jc vs. B of Sn doped PMS wires (o.d. = 0.3, 0.4 mm). (a) on
straight wires as a function of Sn content, heat treatments
at 1) 950° C-30 mn 2) 950° C-2 h 3) 950° C-1 h 4) 950° C-30 mn
5) 1000° C-10 mn 6) 950° C-1 h; (b) on small coils for x = 0.4 Sn
at several heat treatment conditions. Breaks in the curves at
10 T and 20 T comes from changes in measurement apparatus.

possible reason to this unsatisfactory behaviour can be found in the
tin scattering in the wire after heat treatment as shown on picture 1.
Indeed, one can see that it is very inhomogeneous. It is unlikely that
tin scatters in the same way for every heat treatment (partly due, in
the case of Sn doping, to the large grain size) probably resulting in
the observed differences in the Jc.

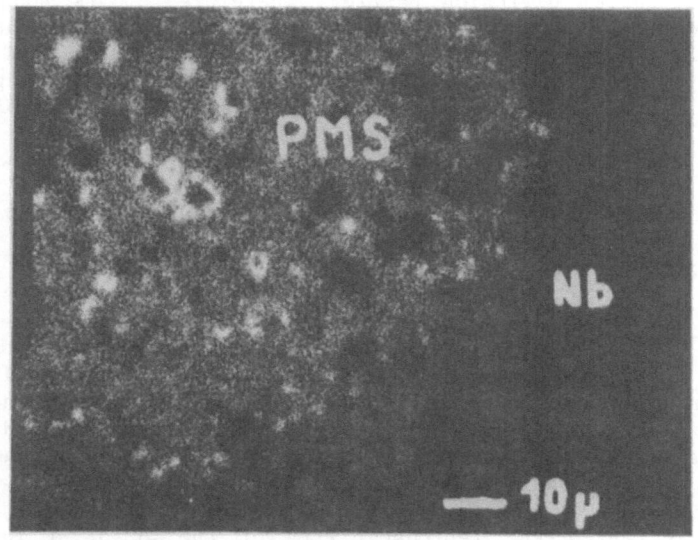

Picture 1. WDS elemental map of tin showing its inhomogeneity of distribution on a cross-sectional area of a 0.4 Sn doped PMS wire.

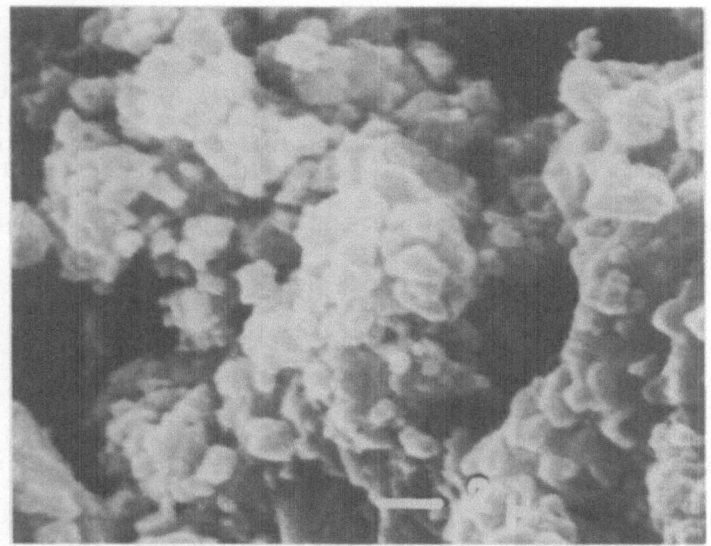

Picture 2. SEM micrograph of 0.4 Sn doped PMS wire. The large voids in the wire may reduce significantly the quality of grains' coupling.

Coils

As we mentioned in section II, the coil configuration is the less contestable one, as far as transport Jc's are concerned. For instance, we avoid the problem of the current transfert length, which sometimes is as large as 150 mm |13|. The results of the Jc transport measurements as a function of the applied field B, up to 27 T, are plotted on figure 1b. We can see that the coils generally behave as expected, i.e. the slope Jc(9 T)/Jc(18 T) is about 3. The best values achieved are close to 1.5×10^8 A/m^2 at 20 T. Although one can observe a slight improvement of the Jc, these agree fairly well with data recorded on short straight wires. However, the optimal heating temperature is now 1000° C, and the time close to 30 mn. These changes can be mainly attributed to technical improvement.

One can note that, in several cases, the Jc are larger than those obtained on straight wires. This seems logical as the large current transfer length observed on some straight samples allows only measurements of a Jc smaller than actual one.

These coils results are most certainly encouraging, but so far we still don't clearly understand why they were achieved. Indeed, we have almost no information about the microstructure of the PMS in the wires. A glance at the SEM picture of a cross section of a wire clearly shows that large voids exist between the grains (picture 2) : this is even amazing, in view of the relatively good Jc's achieved, for one can intuitively guess that the current path is closely related to the connections between the grains |14|. We do not have any clue yet to the precise role of the tin in our wires. Moreover, the other teams do not use tin and yet achieve good Jc results as well. In order to understand that, we have recently drawn wires in which tin is substituted to lead, instead of being added (i.e. the stoichiometry is (Pb+Sn):Mo:S = 1.05:6.2:8). These wires are now being studied.

SUMMARY and CONCLUSION

In order to improve the transport critical current densities of PMS wires, we have studied the effect of different concentrations of a doping agent (tin), as well as optimized their heat treatment conditions, as much for straight wires as for small coils. The main results are :
- despite the drawbacks of their shape, the straight wires are good samples for comparison.
- among all the trials carried-out so far, the overall initial stoichiometry appeared to be close to Pb:Mo:S:Sn = 1.05:6.2:8:0.4.
- although using SnS instead of Sn, as a doping agent, does not lead to significantly different Jc results, SnS may be preferred to pure Sn (hardly available in the form of small grains powder) because of its better efficiency during mixing operation.
- regarding the coils, the optical heat treatment is 1000° C, 10 mn to 30 mn. This slight shift in temperature is confirmed by the latest recorded data (not reported here).
- the best coil shows the following Jc's : 1.5×10^8 A/m^2 at 20 T and 5.10^7 A/m^2 at 27 T. To our knowledge, this is the best result obtained so far by transport measurements on Chevrel phase wires, regardless of the type of casing or fabrication process.

Several parameters are still to be understood, among which are the role of the tin, the connections between grains (i.e. the microstructure of the powder), the powder homogeneity. These are our next goals. Besides, we still expect to improve and master experimental parameters and therefore increase overall performances of our wires.

REFERENCES

1. H. Yamasaki and Y. Kimura, Proc. MRS Intern. Meeting on Advanced Materials, Ikebukuro, Tokyo, Japan, may 30-June 3, 1988
2. R. Chevrel, M. Sergent, L. Le Lay, J. Padiou, O. Pena, P. Dubots, P. Genevey, M. Couach and J.C. Vallier - Rev. Phys. Appl., 23:1777 (1988)
3. W. Goldacker, G. Rimikis, W. Specking, F. Weiss and R.Flukiger - Adv. Cryo. Eng., 34 (1989)
4. K. Hamasaki, K. Noto, K. Watanabe, T. Yamashita and T. Komata - Fifth US-Japan High Field Superconducting Materials Workshop, Fall (1987)
5. B. Seeber, P. Herrmann, J. Zuccone, D. Cattani, J. Cors, M. Decroux, Ø. Fischer, E. Kny and J.A.A.J. Perenboom - MRS Intern. Meeting on Advanced Materials, Tokyo, Japan, may 30-june 3 (1988)
6. Y. Kubo, K. Yoshizaki, F. Fujiwara, K. Noto and K. Watanabe - MRS Intern. Meeting on Advanced Materials, Tokyo, Japan, may 30-june 3 (1988)
7. M. Sergent, R. Chevrel, J. Padiou, O. Pena, R. Barathe, M. Hirrien, H. Massat, T. Pech, B. Turck, P. Dubots and M. Couach - Ann. Chim. Fr. 9:1069 (1984)
8. M. Hirrien - Thesis, University of Rennes, 1986
9. R. Chevrel, M. Hirrien, M. Sergent, M. Couach, P. Dubots and P. Genevey - Mat. Lett., 7:425 (1989)
10. L.F. Goodrich and F.R. Fickett - Cryogenics, 225 (1982)
11. L. Le Lay, P. Rabiller, M. Sergent, R. Chevrel and T. Verhaege - unpublished result
12. M. Sergent, M. Hirrien, O. Pena, J. Padiou, R. Chevrel, M. Couach, P. Genevey and P. Dubots - Adv. Cryog. Eng., 34:663 (1988)
13. L. Le Lay - Thesis, University of Rennes, 1988
14. H. Yamasaki, T.C. Willis, S.E. Babcock, D.C. Larbalestier and Y. Kimura - to be published in Adv. Cryog. Eng. 36 (1990)

EFFECTS OF HOT ISOSTATIC PRESSING ON SUPERCONDUCTING PROPERTIES

OF Nb/ss SHEATHED PbMo$_6$S$_8$ MONOFILAMENTARY WIRES

K.Hamasaki, Y.Shimizu, and *K.Watanabe

Department of Electronics, Nagaoka University of Technology
Nagaoka 940-21, Japan

*The Research Institute for Iron, Steel and Other Metals
Tohoku University, Sendai 980, Japan

ABSTRACT

The monofilamentary PbMo$_6$S$_8$ conductors with Nb diffusion barriers have been produced by the powder metallurgical preparation method. The influence of HIP treatment and Sn addition on the superconducting proper- ties are discussed on the basis of critical current densities (up to fields of 30.1 T), critical temperatures, as well as SEM analysis. The use of pure Pb starting powder with deoxidation is useful for increasing the packing density of the composites, and for increasing the superconducting properties of PbMo$_6$S$_8$ wires. The critical current density can be enhanced in the wires, HIP treated and Sn added. The best J$_c$ values for the wires with HIP treatment alone reached 10^4 A/cm^2 at a magnetic field of 20 T, and 2.7×10^3 A/cm^2 at 27 T, and for the wire with both HIP treatment and Sn addition were 5.4×10^3 A/cm^2 at 27 T and 2.5×10^3 A/cm^2 at 30.1 T. The behavior of the pinning force density F$_P$(B) was also analyzed with a modified flux-shear model.

INTRODUCTION

The interest of Chevrel-phase PbMo$_6$S$_8$ superconductor in practical applications originated in 1974, when Odermatt et al.[1] and Foner et al.[2] discovered the extremely high critical field. Since their discovery, this material has been the subject of intensive study and development for cables of high-field superconducting magnets. The upper critical field B$_{c2}$ of PbMo$_6$S$_8$ is larger than 60 T at 0 K. The critical temperature T$_c$ is also reported to be ~15 K. Since the 60 T critical field of PbMo$_6$S$_8$ is higher than those for all technological superconductors such as NbTi, Nb$_3$Sn, V$_3$Ga, and Nb$_3$Al, PbMo$_6$S$_8$ is today one of the most promising materials for the generation of high field (>20 T) superconducting magnets.

Needless to say, for the generation of high magnetic fields the conductor for a magnet must be in the form of a wire with constant physical properties and superconductivity of over a few km in length. For Chevrel-phase compounds, neither the conventional composite process nor the in-situ process, which have been successfully adoped to Nb$_3$Sn and V$_3$Ga, are applicable. Hence, alternative fabrication techniques must be developed for the production of PbMo$_6$S$_8$ filamentary conductors. Thus, the processing of PbMo$_6$S$_8$ filamentary conductors by the powder metallurgical technique has been studied.

Advances in Cryogenic Engineering (Materials), Vol. 36
Edited by R. P. Reed and F. R. Fickett
Plenum Press, New York, 1990

335

The production of $PbMo_6S_8$ wires by this technique is initiated by Luhman and Dew-Hughes.[3] At present many groups use powder metallurgical methods, and long $PbMo_6S_8$ monofilamentary wires with high J_c have been manufactured for several years.[4-8] The application of hot isostatic pressing(HIP treatment) on $PbMo_6S_8$ wires has been examined by Kubo et al., and Goldacker et al.[9-10] The critical current density of the HIP treated $PbMo_6S_8$ wires with Ta barrier has been 1.7×10^4 A/cm² at 12 T. Also the $PbMo_6S_8$ wires with Sn additions have been first produced by the group of the University of Rennes in France.[11-13] The best Jc values of $PbMo_6S_8$ wires with Sn additions were 1.7×10^8 A/m² at 14 T, and 4.2 K.[11] It is not clear whether or not the HIP treatment is useful in increasing the critical current of $PbMo_6S_8$ wires and the addition of the small amount of Sn in enhancing the flux pinning force in $PbMo_6S_8$ superconductors. However, these methods may be the most promising one for the improvement of the critical current density of $PbMo_6S_8$ wires.

This study was carried out to improve the critical current density of $PbMo_6S_8$ wires by the application of HIP treatment and Sn addition on the wires. The behavior of pinning force density in high fields is analyzed based on the modified flux-shear models.

PREPARATION OF MONOFILAMENTARY $PbMo_6S_8$ WIRES

In our powder metallurgical process, the starting powder is a mixture of powders of Pb/Sn/Mo/S.[14] The Pb(<120 μm), Sn(<240 μm), and Mo(3 μm) powders are deoxidized. The S powder is of less than 240 μm. The fine powders are thoroughly mixed, and then filled into a niobium tube which serves as matrix and diffusion barrier. After sealing, the composite is passed into a stainless steel tube which is neccessary for the adjustment of the thermal stress to the superconducting core.[15] The whole composite is sealed and fabricated into a wire through swaging and drawing at room temperature.

By the use of pure and soft Pb starting powder with deoxidization, we may get void-free samples. The void-free sample may result in the reduction of the residual oxide content in $PbMo_6S_8$ compounds which may influence T_c and B_{c2}.[16-17] The distribution of the void and Pb element in the composite wire before heat treating is studied with SEM and XMA. Figures 1(a), and 1(b) show the SEM micrograph and the x-ray emission Pb(L_{a1}) image on the longitudinal fractured section of a Pb/Mo/S filament before heat treating. As can be seen from Fig.1(a), the starting powders

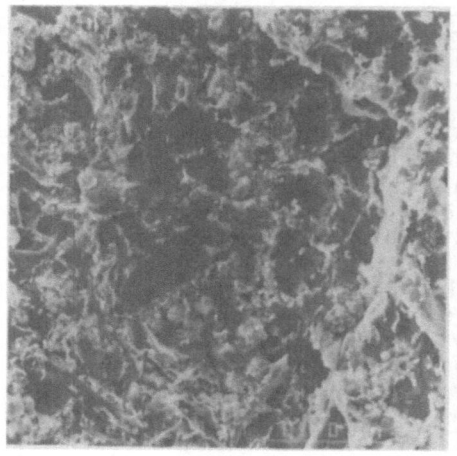

Fig.1 (a) SEM micrograph and (b) x-ray emission Pb(L_{a1}) image on longitudinal fractured section of a Pb/Mo/S filament before heat treating.

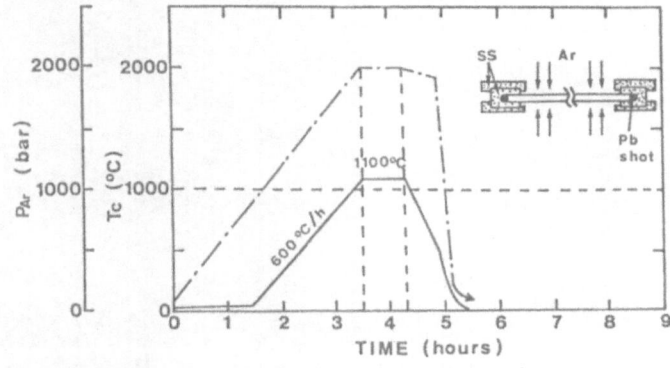

Fig.2 HIP parameters. The inset is a schematic structure of the
   composite.

are densely packed. Few voids are observed in the specimen. Note that Pb
element is almost uniformly distributed over the fractured section (Fig.1
(b)). This result suggests that the use of pure Pb powder is useful for
the increase of packing density of the composites.

   After the drawing, the wires were heat treated at 1030-1100 ℃ for
0.5-2 hours. Some of the cold drawn wires were HIP treated at 1100 ℃
for 45 minutes under 2 kbar argon gas atmosphere. Hydrostatic pressure
was applied before the temperature of the specimens was raised. The HIP
parameters are shown in Fig.2. The inset is a schematic structure of the
composite. The measurement of the critical currents I_c in the transverse
magnetic fields was carried out in hybrid magnets at HFLSM at Tohoku
University. The critical currents were determined by the standard four-
probe technique using the 2 μV/cm criterion.

RESULTS AND DISCUSSION

   Figure 3 shows the polished cross-section of a Nb/ss sheathed $PbMo_6S_8$
monofilamentary wire. The diameter of the $PbMo_6S_8$ layer is about 500 μm.
Figure 4(a) shows a typical fractograph(SEM) of a $PbMo_6S_8$ filament with
HIP treatment, and Fig.4(b) shows a high magnification view of fractured
surface. There are no large voids in the fractograph shown in Fig.4(a),
but some small voids are observed in the high magnification view. Contrary to the expectation, the voids in the specimen are not eliminated by
means of HIP treatment so far as we can see the fractograph of Fig.4(b).
Although this result suggests that the HIP treatment at 1100 ℃ for 45
minutes under 2 kbar argon(Ar) gas atomosphere is not suitable for the
improvement of the superconducting properties of $PbMo_6S_8$ wires, as

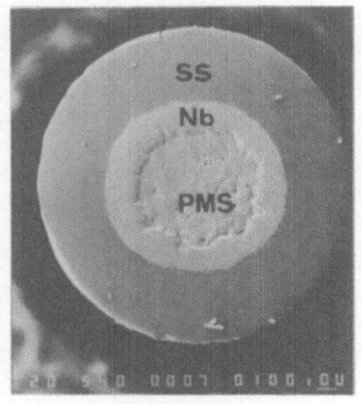

Fig.3 Polished cross-section
   of a Nb/ss sheathed
   $PbMo_6S_8$ monofilamentary
   wire.

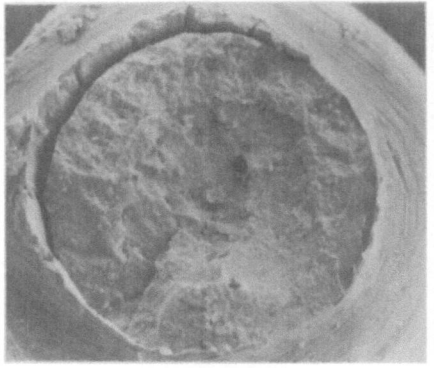 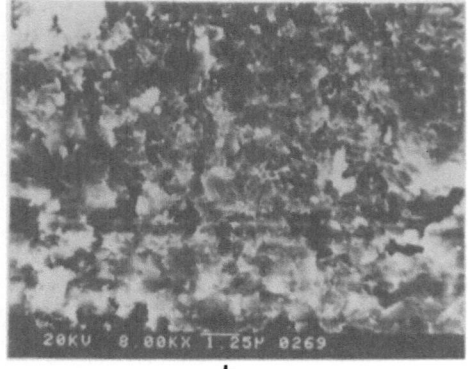

a   b

Fig.4 Typical fractographs of a PbMo$_6$S$_8$ wire with HIP treatment at
2 kbar, and at 1100 °C for 45 minutes. The wire was fractured in
a ductile manner, and hence the secondary crack can be seen in
the interface of Nb/PbMo$_6$S$_8$ (3(a)). (b) is a high magnification
view of fractured surface.

mentioned below, the critical current density of the wires is enhanced by
means of HIP treatment. It is difficult in the present situation to
evaluate the actual total-area of the voids in longitudinal fractured-
section of the wires with and without HIP treatment.

In order to estimate the grain size, SEM observations were made on
fractured section of the PMS wires chemically etched for 1-10 sec. in 5 %
nitric acid. Typical cross section of a monofilamentary wire is shown in
Fig.5. A finely divided structure is seen in this figure. The average
grain size for the PMS wires was about 0.2 μm. The pores (black areas)
partially resulting from the chemical etching in nitric acid are surrounded by PMS compound. The material etched in nitric acid may be lead
educed on cooling after the heat treatment at 1100 °C.

Fig.5 High magnification view of typical cross-section of a HIP treated
wire chemically etched for 5 sec. in 5 % nitric acid. A finely
divided structure of average grain size of 0.2 μm was observed.

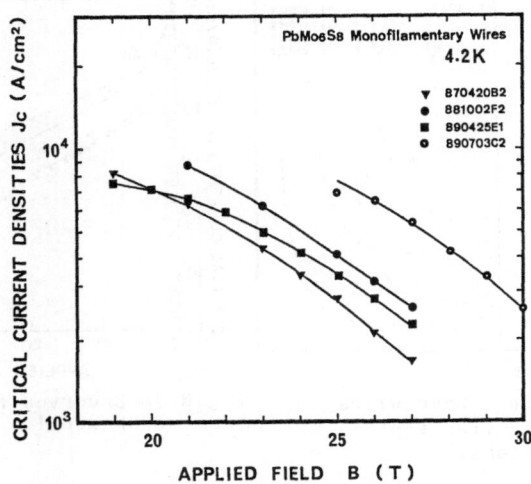

Fig.6 Critical current densities as a function of a transverse
magnetic field for several $PbMo_6S_8$ wires with different heat
treatment; ▼(870420B2, Pb:Mo:S=1:6:8) without HIP ; ●(881002F2,
Pb:Mo:S=1.1:6:8) with HIP ; ■(890425E1, Pb:Sn:Mo:S=1.1:0.2:6:8)
without HIP ; o(890703C2, Pb:Sn:Mo:S=1.1:0.2:6:8) with HIP.

In Fig.6 we show the J_c-B curves of some of the best $PbMo_6S_8$ monofilamentary wires being prepared with and without HIP treatment, and with
Sn addition. B is the transverse magnetic field up to 30.1 T. The best
J_c values of the wires without HIP and Sn additions are 7×10^3 A/cm^2 at
21 T, and 1.8×10^3 A/cm^2 at 27 T and 4.2 K. The J_c's of the wires with
HIP treatment alone and Sn addition are slightly increased compared with
the wires without HIP treatment and Sn additions.

More recently, we produced some wires with both Sn additions and HIP
treatment. Unfortunately, the current transfer phenomena are observed in
I-V curves as shown in Fig.7. This is due to the smaller diameter(d~24
mm) of the effective bore of the cryostat for 30 T Hybrid Magnet and to
the much shorter sample(~16 mm in length). The transfer current in Nb/ss
sheath, which estimated from the resistance of Nb/ss sheath, is less than
1 %. Therefore, most of the current flows in $PbMo_6S_8$ filament. The J_c's
estimated from Fig.7 is also shown in Fig.6 for the wire(o:890703C2).
The best J_c value of the wire with both HIP treatment and Sn addition is
much higher than those of wires with HIP alone, and with Sn addition.
These results indicate that HIP treatment and Sn addition is very useful
for increasing the critical current density of $PbMo_6S_8$ superconductors.

To increase J_c further, the flux pinning mechanism in $PbMo_6S_8$ should
be clarified. The critical current density is a microstructure sensitive
property which is not intrinsic in superconductors. From the present
knowledge of flux pinning and critical current density in A-15 compounds,
it is reasonable to assume that the J_c of Chevrel-phase $PbMo_6S_8$ compound
will increase as the flux pinning to the boundaries of the individual
crystallites is strengthened.

Based on Kramer's empirical model, flux pinning mechanism in $PbMo_6S_8$
has been extensively discussed in the past.[18-19] However, pinning
mechanism in $PbMo_6S_8$ is not yet made clear theoretically and
experimentally. We have to do further studies in order to clarify this
mechanism. Nevertheless, our interest is in finding out the highest value
of the critical current density as a function of the field. We will cal-

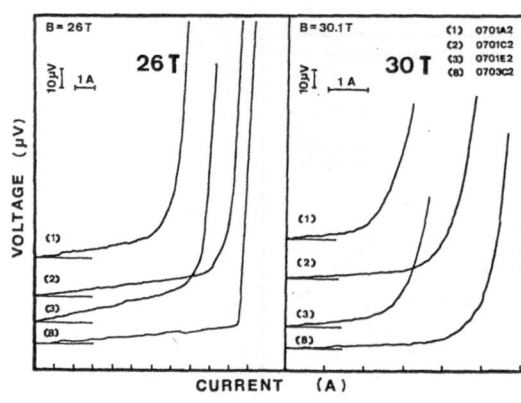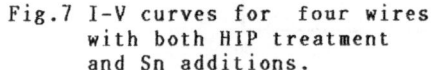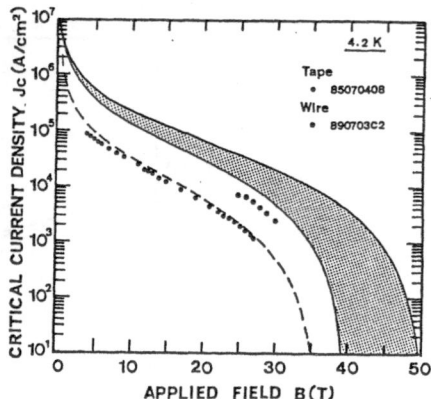

Fig.7 I-V curves for four wires with both HIP treatment and Sn additions.

Fig.8 J_c-B curves calculated using EPDH model.

culate here the maximum critical current density with the modified flux shear model(EPDH model)[20-21] which gives J_c as

$$J_c = (7.4 \times 10^4 / \pi \kappa^2 D) \cdot (B_{c2} - B)^2 / B$$

where κ is the G-L parameter, and D the grain size. A κ value of 130 was taken from Karasik et al.[22] Using this formula, the calculated J_c-B curves are drawn in Fig.8 together with the best experimental results for a $PbMo_6S_8$ tape[23] and wire with both HIP treatment and Sn addition. The full lines show the calculated values with above equation for B_{c2}=50 T and 40 T, and D=0.1 μm. The hatched area shows the effect of the anisotropy (20 %) of B_{c2} on the J_c of $PbMo_6S_8$ wires.[24] The dashed line shows for B_{c2}=36 T and D=0.3 μm. Regarding the field dependence, the experimental result for the tape is in good agreement with the J_c values calculated over a wide range of magnetic field. On the other hand, for the wire with both HIP treatment and Sn addition, we could not obtain the correct values of I_c in magnetic fields below 25 T due to the jouel heating in current terminals. Hence we can not discuss on J_c-B characteristics sufficiently, but the upper critical field of this wire is higher. The present results indicate that the Sn addition into $PbMo_6S_8$ can enhance the B_{c2} of this compounds, and hence the Sn atom may occupy the channel of Chevrel-phase crystal.

From Fig.8, if we obtain high B_{c2}(= 50 T) samples with D=0.1 μm, one is led to the conclusion that critical current density of the order 10^4 A/cm^2 at 30 T. Thus, the improvement of B_{c2}^* of $PbMo_6S_8$ wires remains one of the key problems in the fabrication of high J_c filamentary conductors of $PbMo_6S_8$. The effect of the anisotropy of B_{c2} on the J_c of $PbMo_6S_8$ wires is also still unclear and needs further studies.

CONCLUSION

Effects of HIP treatment and Sn addition on the $PbMo_6S_8$ mono-filamentary wires were investigated. Both HIP treatment alone and Sn addition can be slightly enhanced the critical current density of $PbMo_6S_8$ superconductors. The best J_c values for the wire with HIP treatment alone reached 10^4 A/cm^2 at 20 T, and 2.7×10^3 A/cm^2 at 27 T. Also, both HIP treatment and Sn addition is relatively improved the J_c, and the best J_c values were 5.4×10^3 A/cm^2 at 27 T and 2.5×10^3 A/cm^2 at 30.1 T. The present results indicate that the HIP treatment and Sn addition into $PbMo_6S_8$ is very useful for increasing the critical current density. The combination of improved Nb_3Sn, V_3Ga, and $PbMo_6S_8$ conductors may also make 30 T superconducting magnets possible in the near future.

ACKNOWLEDGMENT

This work was supported in part by a Grant-in-Aid for scientific Research of the Ministry of Education in Japan. We wish to thank Prof. K. Noto of Iwate University for his helpful discussion.

REFERENCES

1. R. Odermatt, ϕ. Fisher, H. Jones, and G. Bongi, Phys.C 7: L13 (1974).
2. S. Foner, E. J. McNiff,Jr., E. J. Alexander, Phys. Letters, 49A: 269 (1974).
3. L. Luhman, and D. Dew-Hughes, J. Appl. Phys., 49: 936 (1978).
4. B. Seeber, W. Glaetzle, D. Cattani, R. Baillif, and ϕ. Fisher, IEEE Trans. Magn., MAG-23: 1740 (1987).
5. H. Yamasaki, and Y. Kimura, IEEE Trans. Magn., MAG-23: 1756 (1987).
6. Y. Kubo, K. Yoshizaki, F. Fujiwara, and Y. Hashimoto, Adv. Cryog. Eng. Matrs., Vol.32: 1085 (1986).
7. T. Yamashita, K. Hamasaki, K. Noto, K. Watanabe, and T. Komata, Adv. Cryog. Eng. Matrs., Vol.34: 669 (1988).
8. W. Goldacker, W. Specking, F. Weiss, G. Rimikis, and R. Flükiger, Cryogrnics, (1989) (to be published).
9. Y. Kubo, K. Yoshizaki, F. Fujiwara, and Y. Hashimoto, Adv. Cryog. Eng. Matrs., Vol.32: 1085 (1986).
10. W. Goldacker, S. Miraglia, Y. Hariharan, T. Wolf, and R. Flükiger, Adv. Cryog. Eng. Matrs., Vol.34: 655 (1988).
11. R. Chevrel, M. Hirrien, and M. Sergent, Polyhedron 5: 87 (1986).
12. R. Chevrel, M. Sergent, L. Le Lay, J. Padiou, O. Pena, P. Dubots, P. Genevey, M. Couach, and J. C. Vallier, Revue Phys. Appl., 23: 1777 (1988).
13. M. Hirrien, Thesis PhD, University of Rennes(France) (1986).
14. K. Hamasaki, K. Noto, K. Watanabe, T. Yamashita, and T.Komata, Proc. of the MRS International Meeting on Advanced Materials, Tokyo, (to be published).
15. J. W. Ekin, T. Yamashita, and K. Hamasaki, IEEE Trans. Magn., MAG-21: 474 (1985).
16. S. Foner, E. J. McNiff, Jr., and D. G. Hinks, Phys. Rev. B, 31: 6108 (1985).
17. D. G. Hinks, J. D. Jorgensen, and H. C. Li, Solid State Commun., 49: 51 (1984).
18. S. Alterovitz, and J. Woollam, Cryogenics 19: 167 (1979).
19. B. Seeber, C. Rossel, and ϕ. Fisher, Ternary Superconductors, Shenoy, Dunlap and Fradin, eds., North Holland Inc., p. 119 (1981).
20. J. E. Evetts, C. J. G. Plummer, Proc. of International Symposium on Flux Pinning and Electromagnetic Properties in Superconductors, Fukuoka, Japan, p. 146 (1985).
21. D. Dew-Hughes, IEEE Trans. Magn., MAG-23: 1172 (1987).
22. V. R. Karasik, E. V. Karyaev, V. M. Zakosarenko, M. O. Rikel, and V. I. Tsebro, Sov. Phys., JETP 60: 1221 (1984).
23. T. Yamashita, K. Hamasaki, K. Noto, K. Watanabe, and T. Komata, Adv. Cryog. Eng. Matrs., Vol.34: 669 (1988).
24. M. Decroux, ϕ. Fisher, R. Flükiger, B. Seeber, R. Delesclefs, and M. Sergent, Solid State Commun. 25: 393 (1978).

MICROSTRUCTURE AND CRITICAL CURRENT DENSITIES OF PbMo$_6$S$_8$

IN HOT-WORKED Mo-SHEATHED WIRES

H. Yamasaki*, T. C. Willis, D. C. Larbalestier, and
Y. Kimura**

Applied Superconductivity Center, University of Wisconsin
Madison, WI 53706
*On leave from Electrotechnical Laboratory, Tsukuba-shi
Ibaraki 305, Japan
**Electrotechnical Laboratory

ABSTRACT

The critical current densities, Jc, of hot-worked Mo-sheathed PbMo$_6$S$_8$ wires are affected very much by their composition and heat treatments. For example, Jc for Pb$_{1+x}$Mo$_6$S$_{7.5}$ wires heat treated at 700°C took a maximum value at x = 0.02 (Jc = 2.1-2.9 x 10^8 A/m^2 at 8 T), and decreased monotonically with increasing x (Jc ≈ 0.8 x 10^8 A/m^2 for x = 0.05, and 0.3 x 10^8 A/m^2 for x = 0.1). We have investigated the Jc values for many Pb$_{1+x}$Mo$_6$S$_{8-y}$ sample wires, and microstructural analyses have been performed using SEM and TEM techniques. It has been demonstrated that Jc is not determined only by the grain size, which is only of order 0.1-0.3 μm at optimum, and that the connections between PbMo$_6$S$_8$ grains are in general not complete.

INTRODUCTION

The Chevrel phase compound, PbMo$_6$S$_8$, has a very high upper critical field, Hc$_2$ ≈ 50 T (4.2 K), and is one of the most promising materials for future high field superconducting magnets (>20 T). Lately, several groups have been making efforts to develop Chevrel phase PbMo$_6$S$_8$ (or SnMo$_6$S$_8$) wires with a powder metallurgical approach, and quite high Jc values have been reported.[1-3] Some groups use an *in-situ* technique to form the Chevrel phase after wire fabrication by cold-working operations.[2-3] Two of the present authors (HY and YK) have been doing fundamental studies on the development of hot-worked Mo-sheathed PbMo$_6$S$_8$ wires.[4] Although hot working operations are necessary to elongate Mo, unlike Nb (or Ta) which are used in the cold-working *in-situ* approaches, Mo does not react with PbMo$_6$S$_8$. Thus, there is a wider choice of heat treatment, and greater electromagnetic stability is expected, compared to the *in-situ* wires which have a possible high-resistivity interlayer (Nb sulfide) between the PbMo$_6$S$_8$ core and the Nb sheath. In this paper, Jc data on many wires with a wide variety of compositions are presented, and we also present microstructures of PbMo$_6$S$_8$ of some wires analyzed by SEM and TEM techniques. Magnetization measurements have also been performed to investigate the current carrying nature of the PbMo$_6$S$_8$ filament.

Advances in Cryogenic Engineering (Materials), Vol. 36
Edited by R. P. Reed and F. R. Fickett
Plenum Press, New York, 1990

343

EXPERIMENTAL PROCEDURES

Powder metallurgy processed $PbMo_6S_8$ wires with a Mo sheath and a stainless steel jacket were fabricated by hot-working operations, as described in a previous paper.[4] Sample wires (outer diameter = 1.0 mm, inner $PbMo_6S_8$ segment diameter ≈ 270 μm) were electroplated with Cu, and heat treated in a 54 cm long helical shape. 27 cm long helical wires were cut out, and the critical currents were measured with an FRP(G-10)-Cu specimen holder (outer diameter = 40 mm or 34 mm). The critical current was determined by a 1 μV/cm criterion, and the critical current density (Jct) was calculated for the segment of the $PbMo_6S_8$ core. DC magnetization was measured at 4.2 K for 3 mm long wires using a vibrating sample magnetometer, and Jcm (magnetization critical current density) was calculated by the critical state model (Bean model), using the equation:

$$Jcm = 3\Delta M/2R \quad .$$

EFFECT OF COMPOSITION ON THE CRITICAL CURRENT DENSITY

Critical current densities of $Pb_{1+x}Mo_6S_{7.5}$ wires which were heat treated at 700°C for 24 h are shown in Fig. 1. High critical current density is obtained at the composition with a slight amount of excess Pb (x = 0.02), and Jc decreases monotonically with x for x ≥ 0.05. Scanning electron micrographs of the fractured surface of $Pb_{1.1}Mo_6S_{7.5}$ wires are shown in Fig. 2(a). We can see that some of the $PbMo_6S_8$ grains in $Pb_{1.1}Mo_6S_{7.5}$ wires are "embedded" in excess lead. After the selective etching of interparticle Pb with an acetic acid/H_2O_2 solution, which dissolves Pb but not $PbMo_6S_8$, clear cuboidal $PbMo_6S_8$ grains appear (Fig. 2(b),(c)). Such interparticle Pb precipitates have been also observed in the fractured surface of $Pb_{1.05}Mo_6S_{7.5}$ and $PbMo_6S_{7.0}$ wires for which excess Pb is also expected. Apparently this interparticle Pb network shown in these SEM photographs degrades the good connection between the $PbMo_6S_8$ grains and brings lower Jc values (Fig. 1). In the previous paper,[4] it was demonstrated that Pb loss from regions near the ends of the wire caused a Tc reduction, resulting in the degraded critical current densities. A slight amount of excess lead is believed to be important in achieving a high Tc current path through $PbMo_6S_8$ grains.

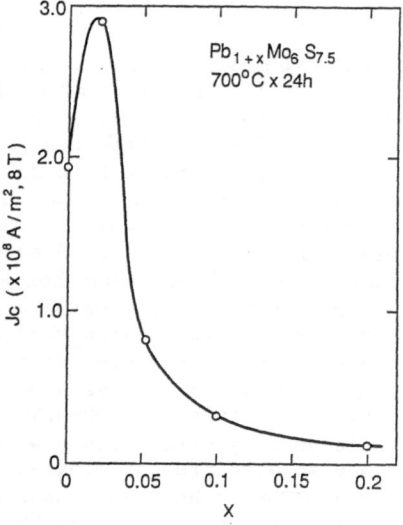

Fig. 1 Critical current densities at 8 T for $Pb_{1+x}Mo_6S_{7.5}$ wires heat treated at 700°C for 24 h.

(a)

(b)

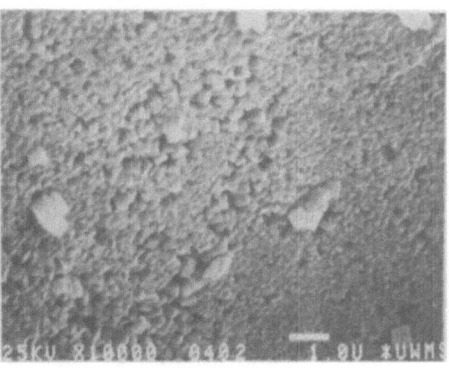

(c)

Fig. 2 Scanning electron
micrographs of the fractured
surface of $Pb_{1.1}Mo_6S_{7.5}$ wires heat
treated at 700°C for 24 h; (a) as
fractured, (b), (c) after
selective etching of Pb with an
acetic acid/H_2O_2 solution. White
bars in these pictures show the
length of 1 μm.

Critical current densities were measured for many wires with various
compositions, and the results are exhibited in Table 1. These wires were
heat treated at 700°C, because lower temperature heat treatment (~700°C)
usually gives higher critical current densities.[4] Jc and Tc results for
the $PbMo_6S_{7.0}$ sample wires heat treated at various temperatures (quoted from
ref. 4) are shown in Table 2. To our disappointment, high Jc values could
not be obtained in high sulfur content wires where higher Tc is expected.[5-7]
It may be caused by the second phase MoS_2 precipitate, which is commonly
observed in high sulfur content specimens.

On the contrary, high Jc values were obtained for the wires with lower
sulfur content, for which excess Mo phase is expected from the phase
relations.[5] Light microscope observations, together with wavelength-
dispersive X-ray microanalysis, have revealed that the excess Mo phase
exists as islands with the dimensions of 2-10 μm; therefore, excess Mo does
not degrade the connection between $PbMo_6S_8$ grains.

MICROSTRUCTURAL ANALYSES WITH HIGH-RESOLUTION SEM AND TEM TECHNIQUES

In order to get more insight on the Jc behavior of our hot-worked Mo
sheathed wires, microstructural analysis has been performed with a high-
resolution (low voltage) scanning electron microscope. Fig. 3 shows the
high-resolution SEM pictures of the fractured surface of the $Pb_{1.02}Mo_6S_{7.5}$
wire, corresponding to the high Jc composition shown in Fig. 1. The
microstructure is complicated, and there are many voids. Typical $PbMo_6S_8$
grain (or cluster) size is 0.1-0.3 μm, if the small nodules on the $PbMo_6S_8$
grains are neglected; the $PbMo_6S_8$ grains do not seem to be well-connected to
each other. It is surprizing that the material with such a far-from-dense

345

Table 1: Jc data for $Pb_{1+x}Mo_6S_{8-y}$ wires
heat treated at 700°C for 24 h

Composition	Jc ($\times 10^8$ A/m^2, 8 T)
$PbMo_6S_{7.0}$	2.1-2.6*
$PbMo_6S_{7.3}$	1.5-2.1
$PbMo_6S_{7.5}$	1.6-1.9
$PbMo_6S_{7.7}$	0.9-1.6
$PbMo_6S_{7.8}$	1.0**
$Pb_{1.02}Mo_6S_{7.5}$	2.1-2.9
$Pb_{1.05}Mo_6S_{7.5}$	0.8
$Pb_{1.1}Mo_6S_{7.5}$	0.3
$Pb_{1.2}Mo_6S_{7.5}$	0.1
$Pb_{1.05}Mo_6S_{7.7}$	1.0

* This shows the representative range of Jc observed for that
wire, based on the Ic data measured on not less than three
samples.
** The better value is shown when critical currents were
measured on only two samples.

microstructure can carry considerable current density (Jc $\approx 2.9 \times 10^8$ A/m^2
at 8 T). We suspect that the superconducting current must be taking a
percolating path between the $PbMo_6S_8$ grains. Better interconnectivity
should be important in achieving higher Jc.

The morphology of $PbMo_6S_8$ grains in $PbMo_6S_{7.0}$ wires heat treated at
700, 850 and 950°C is shown in Fig. 4-6. Fig. 4 corresponds to the high Jc
wire, Jc $\approx 2.3 \times 10^8$ A/m^2 at 8 T, heat treated at 700°C for 24 h. Typical

Table 2: Jc and inductively measured Tc for the $PbMo_6S_{7.0}$
sample wires heat treated at various temperatures
(quated from ref. 4)

Heat treatments	Jc at 8 T ($\times 10^7$ A/m^2)	Tc, ΔTc (K)
as drawn		9.2, 2.4
700°C x 12 h	22.4	11.8, 2.4
700°C x 24 h	27.9	12.1, 2.3
750°C x 12 h	17.0	11.7, 2.3
800°C x 12 h	9.3	12.2, 1.8
850°C x 12 h	7.7	13.0, 1.0
950°C x 12 h	6.4	12.9, 0.6

Fig. 3 High-resolution SEM pictures of the fractured surface of the $Pb_{1.02}Mo_6S_{7.5}$ wire heat treated at 700°C for 24 h. The scale is shown in the bottom row of the picture; 10 divisions show the length of 600 nm (left) and 300 nm (right). The operating voltage of the machine was 5 kV.

grain size is also 0.1-0.3 μm, but the intergranular connection seems to be better than that of Fig. 3. However, the microstructure is far from dense. At higher temperatures (850 and 950°C) the morphology of $PbMo_6S_8$ is more cuboidal and the microstructure is denser (Fig. 5 and 6). We can also recognize small nodules on $PbMo_6S_8$ grains in these pictures. The typical $PbMo_6S_8$ grain size is 0.15-0.4 μm (Fig. 5, 850°C heat treatment) and 0.2-0.5 μm (Fig. 6, 950°C heat treatment).

For those high temperature heat treated wires, the $PbMo_6S_8$ micro-structure has apparently become denser, and the grain size is about twice as large. However, the transport critical current densities, Jct, are much lower, 5-8 x 10^7 A/m² at 8 T, 3 to 5 times less than those for 700°C heat treated wires (Table 2).[4] This point is discussed in the next section.

Fig. 7 shows the transmission electron micrographs of the $PbMo_6S_{7.0}$ wire heat treated at 700°C for 24 h (the same sample as that of Fig. 4). We can recognize $PbMo_6S_8$ grains with the dimensions of 0.1-0.3 μm. Some are well-connected and others are not well-connected.

MAGNETIZATION MEASUREMENTS

DC magnetization has been measured for 3 mm long wires cut out from helical wires and straight wires, and the Jcm data are exhibited in Fig. 8 and 9. Fig. 8 shows the Jcm data for several wires heat treated at 700°C for 24 h, and the representative Jct data for those wires. In general, Jct is about 1.5-3 times higher than Jcm. However, the situation is different for the wires heat treated at 850-950°C (Fig. 9). Jcm is now 1.5-3 times higher than Jct, which is commonly observed for superconducting specimens with some decoupled areas. These results are perhaps not too surprizing in view of the HR-SEM pictures shown in Fig. 3-6, in which it is clearly shown that the connectivity of the grains changes very much with temperature.

The result for 700°C heat treated wires, Jct $\geq$ Jcm, seems rather strange. One complicating factor comes from the shape factor. We calculated Jcm on the assumption that the cross-section of the $PbMo_6S_8$ filament is a perfect circle with the radius calculated from the average cross-sectional area. However, the real cross-section is not a perfect circle and the effective radius should be smaller, resulting in higher Jcm values, for Jcm is inversely proportional to R. Further, the mechanical situation of the long Jct samples and the very short Jcm samples is very different; an effective compressive stress from the outer stainless steel jacket[8] is not expected for such short (3 mm) wires. More studies are needed to understand the Jcm/Jct behavior for these wires.

From these Jcm results and the microstructural analyses described in the previous section, we can make the following points:

1. The Jct decrease which is accompanied by the higher temperature heat treatment ($\geq$850°C, see Table 2)[4] is not explained only by the decreased pinning strength caused by the grain growth; the Jct decrease is a factor of 3-5, but the grain growth is only a factor of two.

Fig. 4 High-resolution SEM pictures of the fractured surface of the $PbMo_6S_{7.0}$ wire heat treated at 700°C for 24 h.

Fig. 5   HR-SEM pictures for the PbMo$_6$S$_{7.0}$ wire heat treated at 850°C for 12 h.   Note that the magnification set is different from Fig. 4.

Fig. 6   High-resolution SEM pictures of the fractured surface of the PbMo$_6$S$_{7.0}$ wire heat treated at 950°C for 12 h.

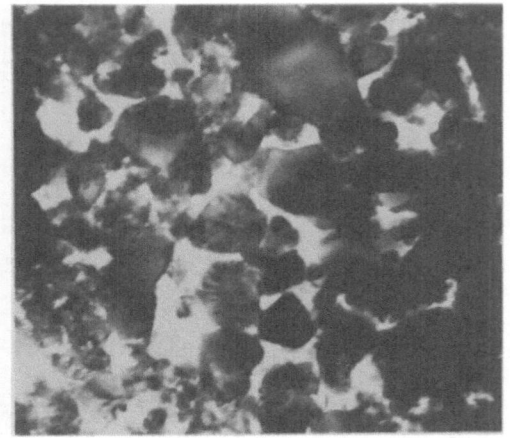

300 nm

Fig. 7   TEM pictures of the PbMo$_6$S$_{7.0}$ wire heat treated
at 700°C for 24 h, the same sample as that of Fig. 4.

2. For high temperature heat treated wires, the microstructure has become
   much denser, and Jcm is the same order or higher; moreover, the critical temperatures are higher with sharp transitions (Table 2).[4]   These
   facts make us consider another factor other than the reduced pinning
   force caused by the grain growth, for their poor transport Jc values.

3. The microstructure of PbMo$_6$S$_8$ grains has become denser as the
   sintering proceeds at higher temperatures.   During this densification
   process, the apparent volume should decrease, which is commonly
   observed in the sintering process.   However, in our Mo-sheathed PbMo$_6$S$_8$
   wires the total volume is determined by the Mo sheath and does not

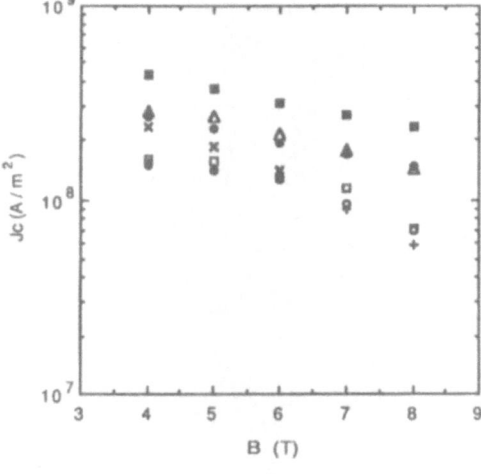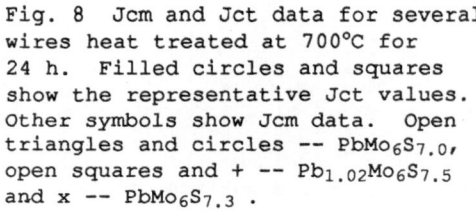

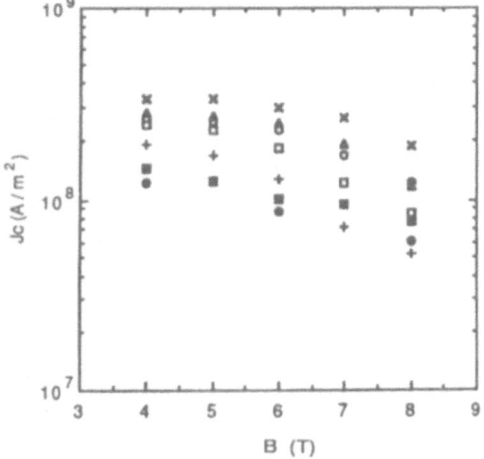

Fig. 8   Jcm and Jct data for several
wires heat treated at 700°C for
24 h.   Filled circles and squares
show the representative Jct values.
Other symbols show Jcm data.   Open
triangles and circles -- PbMo$_6$S$_{7.0}$,
open squares and + -- Pb$_{1.02}$Mo$_6$S$_{7.5}$
and x -- PbMo$_6$S$_{7.3}$.

Fig. 9   Jcm and Jct data for several
wires heat treated at 850 and 950°C
for 12 h.   Filled circles and
squares show the representative Jct
values.   Other symbols show Jcm
data.   Open circles and squares --
PbMo$_6$S$_{7.0}$ (850°C), open triangles --
Pb$_{1.02}$Mo$_6$S$_{7.5}$ (850°C), + -- PbMo$_6$S$_{7.0}$
(950°C) and x -- Pb$_{1.1}$Mo$_6$S$_{7.5}$ (950°C).

shrink during the sintering of $PbMo_6S_8$. This should result in the formation of micro-cracks in some area of the $PbMo_6S_8$ filament, bringing decreased transport critical current density. However, this would not depress the magnetization current density so much.

4. High temperature heat treatment is necessary to improve Hc_2 values.[4] If we can heat treat the wires at higher temperatures without formation of micro-cracks discussed above, higher Jc values are expected. A HIP (hot isostatic pressing) treatment seems to be promising to improve Jc.

CONCLUSIONS

We have investigated the Jc values for many $PbMo_6S_8$ wires with various compositions. Although the excess Mo phase does not degrade intergranular connectivity, the excess Pb precipitates between $PbMo_6S_8$ grains do degrade interconnectivity and bring lower Jc. High-resolution SEM observations have revealed that low temperature heat treated wires with higher Jct have a less dense microstructure, and that the microstructure becomes denser by high temperature heat treatments (850-950°C). The magnetization results suggest that the lower Jct values observed for high temperature heat treated wires are less likely to be caused by reduced pinning strength than by micro-cracks formed during the sintering process.

ACKNOWLEDGEMENTS

We would like to thank T. Udagawa of Tokyo Tungsten Co. Ltd. for his cooperation in wire processing, and M. W. Tengowski and J. Pawley of the Integrated Microscopy Resources for high-resolution SEM observations. The IMR in Madison is funded as an NIH Biomedical Research and Technology Resource (RR 570). Thanks are also to our colleagues of Wisconsin University; to J. McKinnell for magnetization measurements, to B. Starch, R. Noll and R. Casper for technical assistance, and to L. Le Lay for valuable discussions. This work has been supported by the Department of Energy-Office of Fusion Energy and the US-Japan Fusion Accords.

REFERENCES

1. B. Seeber, M. Decroux and Ø. Fischer, to be published in Physica B.

2. Y. Kubo, K. Yoshizaki, F. Fujiwara, K. Noto and K. Watanabe, Proceedings of the MRS meeting held in Tokyo, Japan, 1988.

3. R. Chevrel, M. Sergent, L. Le Lay, J. Padiou, O. Pena, P. Dubots, P. Genevey, M. Couach and J.C. Vallier, Revue Phys. Appl., 23, 1777 (1988).

4. H. Yamasaki and Y. Kimura, J. Appl. Phys., 64, 766 (1988); ibid., Proceedings of the MRS meeting held in Tokyo, Japan, 1988.

5. H. Yamasaki and Y. Kimura, Mat. Res. Bull., 21, 125 (1986).

6. H. Yamasaki and Y. Kimura, Solid State Commun., 61, 807 (1987).

7. H. Yamasaki, Y. Yamaguchi and Y. Kimura, Mat. Res. Bull., 23, 23 (1988).

8. B. Seeber, W. Glaetzle, D. Cattani, R. Baillif and Ø. Fischer, IEEE Trans. Magn., MAG-23, 1740 (1987).

EFFECT OF AXIAL TENSILE AND TRANSVERSE COMPRESSIVE STRESS ON J_c AND

B_{c2} OF PbMo6S8 AND SnMo6S8 WIRES

W. Goldacker, W. Specking, F. Weiss*, G. Rimikis
and R. Flükiger

Kernforschungszentrum Karlsruhe GmbH
Institut für Technische Physik
Postfach 3640, D-7500 Karlsruhe 1
Federal Republic of Germany

* ENSPG, F-38402 St. Martin d'Hères, France

ABSTRACT

The variation of the critical current density J_c of PbMo6S8 and SnMo6S8 monofilamentary wires with external transverse compressive stress σ_t and axial tensile stress σ_a at 22 and 13.5 T, respectively, is reported. Varying the reinforcing steel content in the wire matrix, different axial stress states of the filament were obtained: compressive, tensile and unstressed. The observed variation of both, J_c vs. σ_a and J_c vs. σ_t is significantly smaller compared with that of Nb3Sn wires, which is explained by the higher upper critical magnetic field B_{c2} of the Chevrel phase. After the appropriate steel reinforcement, a reversible strain range up to 0.6 -0.8 % (with 0.2 % compressive prestrain of the filament) is obtained, which fulfills the requirements for future technical applications of this kind of wire materials.

INTRODUCTION

Due to their high upper critical magnetic fields, e.g. B_{c2} (PMS, 4.2 K) = 51 T and B_{c2} (SMS, 4.2 K) = 31 T, [1,2] superconducting wires based on the Chevrel phase compounds PbMo6S8 (PMS) and SnMo6S8 (SMS) are promising for future technical applications at high magnetic fields (B > 20 T). In the last years various powdermetallurgical methods and wire configurations designs have been developed to produce monofilamentary wires reaching critical current densities of the order of $1 \cdot 10^4$ Acm-2 at B = 20 T.[3-8] For the future practical application at high magnetic fields (B > 20 T) with strong Lorentz forces acting on the wire, it is necessary to perform extended investigations on the effect of mechanical stresses on J_c, in analogy to those accomplished on Nb3Sn wires during the last decade. These investigations comprise the effect of axial tensile axial stress, σ_a, as well as that of transverse compressive stress, σ_t.[9,10,11,12]

The question of the sensitivity of Chevrel phase wires against transverse stresses has recently been studied by Goldacker et al.[11,12] For a direct correlation, the effect of the two kind of stresses, i.e. axial tensile and transverse compressive stress on J_c were measured on the same wires. The prestress state of the filaments, i.e. the starting

Advances in Cryogenic Engineering (Materials), Vol. 36
Edited by R. P. Reed and F. R. Fickett
Plenum Press, New York, 1990

353

condition before applying the external stress, was analyzed via the I_c vs. σ_a (axial tensile stress) relationship, which is also shortly reviewed in this paper. The use of experimental conditions identical to earlier investigations of Nb_3Sn wires[18] allows the comparison of both wire types with respect to transverse stress effects.

EXPERIMENTAL

Wire preparation

The presently analyzed wires were produced from Chevrel phase material, synthesized by HIP processing (Hot Isostatic Pressing) the starting powder mixtures of Pb, Mo, MoS_2 and Sn, Mo, MoS_2.[8] Cylindrical rods of PMS and SMS were machined from the HIP samples, packed into a Ta tube (barrier) and a Cu matrix and then cold worked by wire drawing to final diameters of 1.5 - 2 mm. The wires were then inserted into a stainless steel tube in order to obtain a precompression of the filament,[8] and subsequently cold drawn to the final diameter of 0.92 mm. A final heat treatment of 850°C / 30 hours was given in order to recover the superconducting properties, which are strongly affected by the deformation process. The compositions of the three wire types representing tensile, absent and compressive axial precompression of the filament, are given in Table 1.

I_c vs. σ_a, I_c vs σ_t The measurements of the critical currents I_c vs. the axial tensilestress σ_a were performed in the test facility previously described by Specking et al.[14] Chevrel phase wires of 120 mm length were soldered into the clamps of a 1 kN strain rig, which was inserted into the gap of a 13.5 T superconducting split coil magnet. The tensile stress was applied mechanically by a screw mechanism, the force was measured with a load cell and the strain by a capacitive strain gauge close to the sample. The stress values were obtained by the ratio of the applied force and the full cross section area of the wire. The I_c values were determined using a 1 µV/cm criterion. Transverse compressive stress was applied with a stress rig[17,18], being especially constructed for the use in a Bitter coil magnet providing magnetic fields up to 24 T.[19] In this case the length of the wire samples was restricted to 25 mm, and the length of the stress load zone being 4 mm. Force, current and magnetic field were perpendicular to each other. The stress values were calculated from the measured applied force, the affected cross section being defined by the wire diameter (0.92 mm) and the section of force load (4 mm). For the I_c determination we used a 2 µV/cm criterion, taking care that the voltage contacts were fixed as close as possible to the stress load zone.

RESULTS

Axial tensile stress

A full set of of the critical current measurements of the PMS wire with 73 % steel for various tensile strains ε at 13.5 K is plotted in Fig. 1. It illustrates how I_c increases with ε, the maximum being at $\varepsilon \sim 0.2$ % as a consequence of the high steel content. The normalized I_c vs. ε curves (ε is the measured applied axial strain) for the three PMS wires with different steel contents are shown in Ref. 12. Sample PMS-1, with the lowest steel content, exhibits a strong decrease of I_c from the first beginning of stress application, indicating that the filament was already in a tensile stress state (approx. 0.2 % tensile prestrain), thus resulting in a poor mechanical stabilization (0.1 % elastic strain range). The very flat I_c vs. ε dependence of wire PMS-2 for small strain

Table 1. Composition of PbMo$_6$S$_8$ and SnMo$_6$S$_8$ wires (% of cross section)

Sample	PbMo$_6$S$_8$	SnMo$_6$S$_8$	Ta	Cu	SS
PMS-1	13	–	25	11	51
PMS-2	10	–	17	9	64
PMS-3	7	–	14	6	73
SMS-1	–	16	27	12	45
SMS-2	–	10	19	8	63
SMS-3	–	8	12	7	73

values shows that this wire configuration has no significant axial prestress component in the filament. Compared to wire PMS-1, the reversible strain range is enhanced to $\Delta\varepsilon > 0.3$ %, irreversible cracks with sudden I_c drops occurring at strains > 0.5 %. The sample PMS-3 with the highest reinforcing steel content of $\approx$ 73 % has for I_c a maximum at $\varepsilon = 0.2$ %, and a reversible strain range of $\approx$ 0.7 % (including 0.2 % precompression). Within this strain range, full I_c recovery is observed after unloading the external stress. The good mechanical stability of this wire is due to the large content of reinforcing steel which enhances the stability of the filamentagainst cracks. A decrease of I_c with prestrain or externally applied axial strain results from the strain sensitivity of the peak pinning force density and the upper critical magnetic field B_{c2}. Combining previous results of J. Ekin[9] for tensile prestrained PMS wires and tapes with the change of B_{c2}^* in our wires (PMS-1, PMS-2 and PMS-3), we can obtain the normalized strain sensitivity of B_{c2}^* in both the compressive and tensile strain regime (see Fig. 3). The change of B_{c2}^* was determined from Kramer plots using B_{c2}^* (PMS-2) $\approx B_{c2m}^*$.

The relative strain sensitivity of B_{c2}^* of PMS, is given in Fig. 2. It is evident that the variation of $J_c(\varepsilon)$ in the field range between 20 and 25 T is significantly smaller in the case of the Chevrel

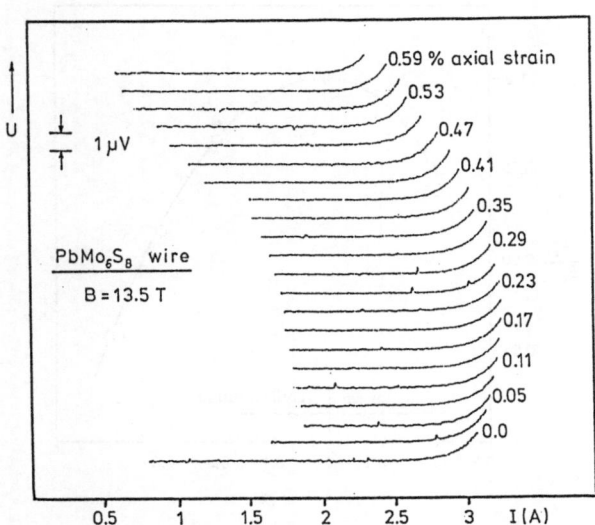

Fig. 1. I_c of the PbMo$_6$S$_8$ sample PMS-3 at 13.5 T at various axial strains ε.

phase wires than for Nb3Sn, whereas in the medium field range (B = 10 - 15 T) the effects are of similar size. For SnMo6S8 wires an analogous series of samples and investigations has been described in detail in Ref. 11 and 12. In principle the same behaviour as in PMS wires was observed, with slightly enhanced prestrain values and reduced elastic strain ranges.

<u>Transverse compressive stress</u>

For the investigations of the critical current as a function of transverse compressive stress σ_t, the Chevrel phase wires PMS-2 and SMS-2 were selected.[11,12] As was shown in the last section, this wire configuration has the minimum axial prestress component. Thus only an unknown small amount of hydrostatic compressive prestress has to be taken into account as the starting condition for the experiment. The existence of this hydrostatic prestress component can be concluded from observed T_C degradations in these wires, a consequence of the well known large sensitivity of T_C with respect to hydrostatic pressure[22]. In Fig. 3 the normalized critical current of sample PMS-2 as a function of transverse stress σ_t is given for different magnetic fields. For small stress values $\sigma_t < 50$ MPa, a slightly enhanced critical current, passing a flat maximum at $\sigma_t \approx 40$ MPa was observed, followed by a strong decrease of I_c at larger stress values ($\sigma_t \gg 50$ MPa). Usually the maximum vanished after cycling (stress load and unload), whereas the other parts of the I_c curve were nearly unchanged. It is remarkable that more than 90 % of the starting current is recovered after unloading the maximum applied stress σ_t (20 T) ≈ 270 MPa which causes 70 % I_c degradation (see Fig. 3).

The sensitivity of I_c against transverse stress is strongly depending on the magnetic field. It is found that I_c scales quite linearly with B, the slope and the starting values at B = 0 being a function of the applied stress. The degradation of I_c with B is about 2.5 times as strong for $\sigma_t = 264$ MPa compared to $\sigma_t = 115$ MPa. For the SMS-2 sample, quite analogous results as for the PMS wire were obtained.

From Fig. 3 it is seen that at $\sigma_t \sim 260$ MPa, I_c is lowered by $\sim 50\%$. Such high transverse stresses obviously exceed the limits of practical

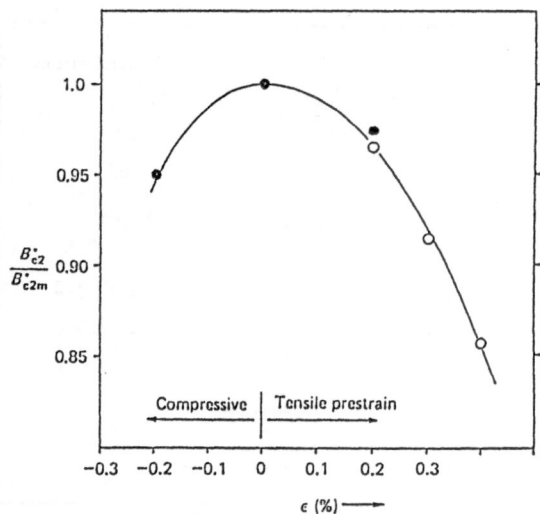

Fig. 2. Variation of B_{c2}^* /B_{c2m}^* with axial strain ε for PbMo6S8 wires (o: Ref. 9)

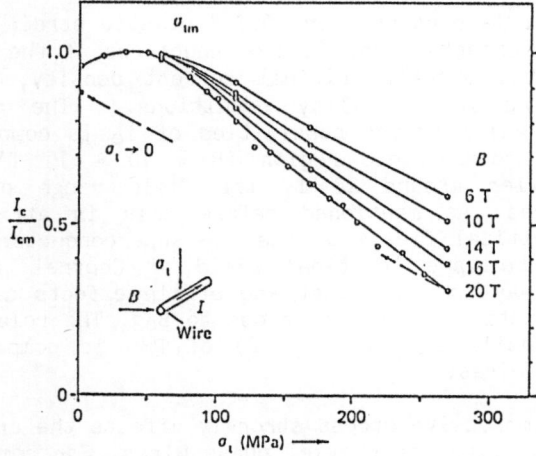

Fig. 3. Normalized I_c vs. σ_t for the PMS wires.

operation loads. Indeed, plastic flow of the stainless steel sheath has already occurred, as shown by Fig. 4. This figure shows the variation of the wire diameter of PMS-2 as a function of the angle α before and after 260 MPa stress load. This variation was measured at 300 K by a laser micrometer (Zygo, type 1200 B). The direction of compression corresponds to α= 90° in Fig. 4. It is seen that a difference of 13 μm was measured between D^+ and D^-, the error being ±1 μm (D^- and D^+ are the diameters parallel and perpendicular to the stress direction).

DISCUSSION

The investigations of I_c vs. tensile axial stress of Chevrel phase wires reinforced with various steel contents demonstrate that wires with a small compressive prestress possess sufficient mechanical and electri-

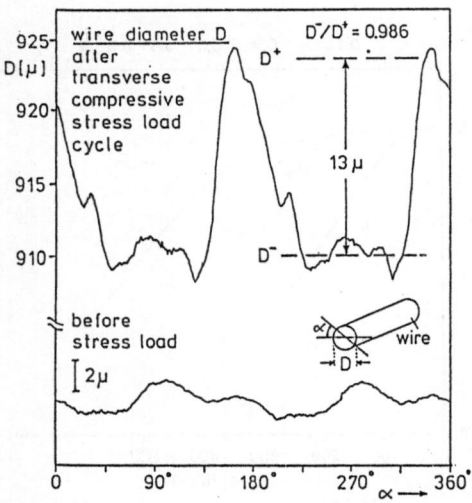

Fig. 4. Variation of the wire diameter of PMS-2 after 260 MPa compressive stress load cycle.

357

cal stability (full I_c recovery for 0.7 % elastic strain) for practical use. However it is expected that future reductions of the matrix to core ratio, enhancing the overall critical current density, will certainly affect these favourable stability conditions. The magnetic field dependence of the axial stress sensivities of I_c is comparable to that of Nb3Sn in the medium field region (B = 10 - 15 T), but becomes significantly smaller around 20 T, the field range of interest of Chevrel phase wires. As mentioned before this is attributed to the different upper critical fields of the two superconductor materials. At B = 20 - 25 T the upper critical field of Chevrel phase wires is sufficiently far away in order that the strain effects are thus not as dramatic as at magnetic fields very close to B_{c2}. The relative change of B_{c2} with axial tensile strain (Fig. 2) of PMS is comparable to that observed for Nb3Sn wires.

Transverse compressive stress strongly affects the critical current density I_c in both types of Chevrel phase wires. For comparison of the transverse and axial stress effects, both measurements were plotted in Fig. 6 for the PMS-2 sample at B = 13.5 T using a common stress axis. This figure clearly shows that transverse stress influences the critical current more dramatically. Indeed only 40 - 50 % of the axial tensile stress values σ_a are sufficient to cause the same I_c degradation via transverse stress application. In principle the same behaviour was recently observed for Nb3Sn wires, where the sensitivity against transverse stress in relation to the tensile stress is significantly larger; here the same I_c degradation can be obtained for $\sigma_t \approx$ (0.2 - 0.3) σ_a. This reflects that the Chevrel phase wires are relatively less sensitive to tranverse stresses than Nb3Sn wires, one reason being the σ_t sensitivity of B_{c2}^*.

This behaviour is quite different to the situation in Nb3Sn wires, where for σ_t a significantly stronger degradation of B_{c2}^* was measured. [15,17] Several reasons can be advanced for explaining this different behaviour. First the crystal structures of the two systems are quite different. The cubic A15 structure leads to an isotropic strain response of the material to hydrostatic stress application, while the rhombohedral Chevrel phase lattice has very anisotropic elastic properties

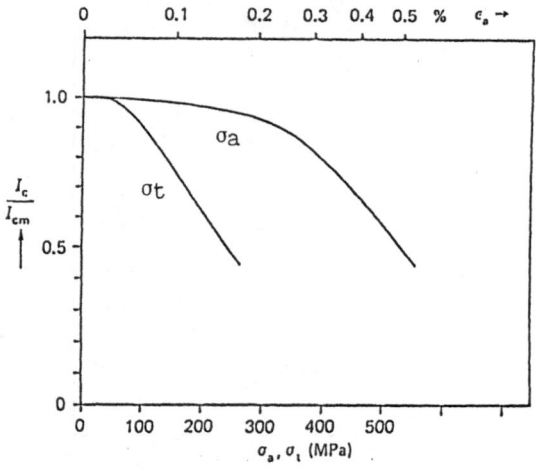

Fig. 5.   Effect of σ_a and σ_t, on I_c of a PMS wire at
T = 4.2 K and B = 13.5 T.

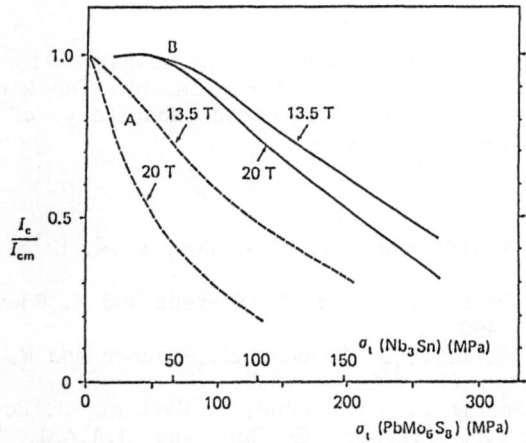

Fig. 6. Comparison of the influence of σ_t on the critical
current of PMS and Nb3Sn wires at B = 13.5 T and 20
T atT = 4.2 K.---:Nb3Sn wire,——:PbMo6S8 wire.

with respect to the different crystal axis.The compressibility of the a
axes is about 3 times stronger than that of the c axes, which leads to
anisotropic strains in the grains of polycrystalline samples under
hydrostatic pressure. The value of $B_{c2}{}^*$ for SMS wires seems to be
more sensitive to transverse stress compared to PMS [11], presumably due to
a somewhat different elastic properties of this compound, in agreement
with earlier observations made on I_c vs. σ_a measurements.[11,12]

A considerable field dependency of I_c vs. σ_t was obversed. From
B = 13.5 T to B = 20 T the I_c degradation doubles for the Nb3Sn wire,
while it enhances by only $\approx$ 20 % for the PMS wire. Therefore the
situation is quite similar to the case of axial tensile stress. Due to
the higher upper critical field, Chevrel phase wires are significantly
less sensitive to transverse stresses at fields around 20 T than Nb3Sn
wires.

CONCLUSIONS

The effect of transverse compressive stress has been measured for
the first time in PbMo6S8 and SnMo6S8 wires. Steel reinforced monofilamentary Chevrel phase wires have a good mechanical stability and
elastic behaviour, strongly depending on the degree of reinforcement and
thus of the stateof precompression of the filament. The sensitivity of
the critical current I_c (B $\geq$ 20 T)and the upper critical field B_{c2}
either to axial tensile stress or transverse compressive stress is
signicantly smaller than that of Nb3Sn wires. This is promising for
future applications of Chevrel phase wires at high magnetic fields (B
$\geq$ 20 T). In large superconducting coils transverse stresses in the range
of 50 - 100 MPa are expected which is quite well in the region of
tolerable degradation of both I_c and B_{c2} in Chevrel phase wires.

Our investigations also lead to the conclusion that the future
development of improved multifilamentary wires with possibly reduced
matrix to core ratios always will require a strong reinforcement of the
wire, favourably creating acompressive prestress in the filament. This
enhanced compressive prestress prevents the formation of cracks and is
thus a necessity to preserve the good current carrying capacity of
these wires and in particular an extended strain range of reversible I_c.

ACKNOWLEDGEMENT

The authors thank A. Kling for the collaboration in performing the I_C measurements and Dr. G. Maret und R. Klaschka for kindly supporting our experiments at the High Magnetic Field Laboratory, of the Max Planck Institut, CNRS in Grenoble, France.

REFERENCES

1. S. Foner, E.J. McNiff and E.J. Alexander, Phys. Lett., 49A, 269 (1974)
2. Ø.Fischer, H. Jones, G. Bongi, M. Sergent and R. Chevrel, H. Phys., C7, L 450 (1974).
3. D. Cattani, R. Baillif, B. Seeber, Ø. Fischer and W. Glätzle, Proc. MT 9, Zürich, P 560 (1985).
    B. Seeber, P. Herrmann, J. Zuccone, D. Cattani, J. Cors, M. Decroux, Ø. Fischer, E. Kny and J.A.A.J. Perenboom, in: Proceedings of the MRS International Meeting on Advanced Materials, Tokyo (1988), to be published.
4. Y. Kubo, K. Yoshizaki, F. Fujiwara and Y. Hashimoto, Adv. Cryog. Eng., 32, 1085 (1986).
5. M. Sergent, M. Hirrien, O. Pena, J. Padiou, R. Chevrel, M. Couach, P. Genevey, P. Dubots, Adv. Cryog. Eng. 34, 663 (1988).
    M.Hirrien, R. Chevrel, M. Sergent, P. Dubots, P. Genevey, Mat.Lett.5,173 (1987).
6. T. Yamashita, K. Hamasaki, K. Noto, K. Watanabe, T. Komata, Adv. Cryog. Eng. 34, 669 (1988).
7. H. Yamasaki and Y. Kimura, to be published in J. Appl. Phys. (1988).
8. W. Goldacker, S. Miraglia, Y. Hariharan, T. Wolf, R. Flükiger, Adv. Cryog. Eng., 34, 655 (1988);
9. J. E. Ekin, T. Yamashita and K. Hamasaki, IEEE Trans. Mag.,MAG-21, 474 1985).
10. B. Seeber, W. Glaetzle, D. Cattani, R. Baillif and Ø. Fischer,IEEE Trans. Mag., MAG-23, 17490 (1987).
11. W. Goldacker, G. Rimikis, W. Specking, F. Weiss andR. Flükiger, in: Proceedings of the ICMC Conference, Shenyang (1988), Eds. R.P. Reed, Z.S. Xing and E.W. Collings, Vol. 1, p.431.
12. W. Goldacker, W. Specking, F. Weiss, G. Rimikis andR. Flükiger, to appear in Cryogenics, Sept. 1989.
13. J. W. Ekin, "Filamentary A 15 Superconductors"., Eds. M. Suengaand A.F. Clark, Plenum Press, 187 (1980).
14. W. Specking, A. Nyilas and R. Flükiger, Proceedings of the 9th International Conference on Magnet Technology, Zürich (CH), 1985, Eds., C. Marinucci and P. Weymuth, printed by Swiss Inst. of Nuclear Res., p. 676.
15. J. W. Ekin, J. Appl. Phys., 62 (12), 4829 (1987).
16. J. W. Ekin, Adv. Cryog. Eng., Vol.34, 547 (1988).
17. W. Specking, W. Goldacker and R. Flükiger, Adv. Cryog. Eng.,Vol.34, 569 (19W. Specking, F. Weiss and R. Flükiger, Proc. 12th Symposium onFusion Engineering, Monterey, CA, USA, IEEE Catalog No. 87 CH 2507-2, p. 365.
19. High field magnet laboratory, MPI Grenoble, France.
20. W. Goldacker, R. Flükiger, IEEE Trans. Magn., MAG-21, 2,807 (1985).
21. W. Goldacker, R. Flükiger, Adv. Cryog. Eng., Vol.34, 561(1988).
22. D. W. Capone II, R.P. Guertin, S. Foner, D.G. Hinks and H.C. Li, Phys. Rev. B, 29, 6375 (1984).
23. J. D. Jorgensen, private communication.

IMPROVEMENTS IN HIGH FIELD PROPERTIES OF

CONTINUOUS ULTRAFINE Nb$_3$Al MF SUPERCONDUCTOR

T. Takeuchi, M. Kosuge, Y. Iijima, K. Inoue and [*]K. Watanabe

National Research Institute for Metals
1-2-1, Sengen, Tsukuba, Ibaraki 305, Japan

[*]Tohoku University
2-1-1, Katahira, Sendai, Miyagi 980, Japan

ABSTRACT

Attempts have been made to improve high field properties of continuous ultrafine Nb$_3$Al multifilamentary superconductors. The Al core, of which the hardness is a key to the excellent cold-workability of the Nb/Al composite, was hardened by being alloyed with Ge by the combination with Cu and Ag, resulting into an increase in T_c, H_{c2} and J_c of the Nb/Al composite at high fields. Very short heat treatments at temperatures above 1200°C were carried out by applying a large current pulse directly to the sample so that remarkable improvements in high field properties are attained.

INTRODUCTION

Recently, we have demonstrated[1-4] that a continuous ultrafine Nb$_3$Al multifilamentary superconductor could be fabricated by the improved composite diffusion-process. Alloying the Al core with Mg,[1,2,4] Ag,[4] Cu[3,4] and Zn[4] diminishes the difference in hardness between the Nb matrix and the Al core, and improves the workability of the Nb/Al multifilamentary composite, resulting in a successful fabrication of continuous ultrafine Al filaments embedded in the Nb matrix. Reacted wires at 700-950°C show superior properties to a commercial multifilamentary Nb$_3$Sn; (1) a higher critical current density J_c (for example, 1.5 x 10^9 A/m^2 at 4.2K and 10T) due to the strong flux pinning at the Nb$_3$Al-matrix and Nb$_3$Al-core interfaces,[3] and (2) the smaller degradation in J_c with strains and the larger irreversible strains(more than 1.3 %).[4,5] Furthermore, the multifilamentary structures with filament sizes less than 1 μm are realized, which is of great advantage to improving the stability against electromagnetic disturbances. If the Nb tubes used at bundling the elementary composite can be replaced by the resistive materials, the electric coupling between filaments will be reduced, making this conductor interesting as a low AC loss superconductor. Therefore, this Nb$_3$Al multifilamentary conductor has been promising as an alternative to the multifilamentary Nb$_3$Sn conductor.

However, the T_c(15.6K) and H_{c2}(21.5T) are lower than those of the stoichiometric Nb$_3$Al, resulting to remarkable degradation in overall J_c at high fields. Overall J_c becomes smaller than those of the commercial

Advances in Cryogenic Engineering (Materials), Vol. 36
Edited by R. P. Reed and F. R. Fickett
Plenum Press, New York, 1990

Table 1 Various properties of Al alloys and Nb/Al composites

Composition of Al alloy (at.%)	Precipitation (observed by OM)	Vickers hardness H_v $S_0/S = 2$	$S_0/S = 160$	Workability of Nb/Al composite
(Nb)	–	110	130	–
(pure Al)	no	30	34	very bad
5Ge	significant	60	94	bad
(5Mg)	no	90	138	very good
4Mg–2Ge	significant	53	60	very bad
0.8Mg–0.4Ge	significant	46	–	very bad
(2Cu)	no	94	135	very good
1.3Cu–1Ge	slight	90	–	–
1Cu–1Ge	no	90	117	very good
(3Ag)	no	63	80	not too bad
1.5Ag–1.5Ge	no	70	103	good
0.5Ag–2Ge	no	65	–	–
2Ag–2Ge	no	80	87	not too bad

S_0/S; area reduction ratio,   OM; optical microscope

(Nb,Ti)$_3$Sn at fields above 14 T. This study has been made to improve the high field properties of the multifilamentary Nb$_3$Al. The Al core has been alloyed with Ge, and the reaction has been carried out at temperatures above 1000 °C, resulting into an increase in T_c, H_{c2} and J_c at fields more than 14 T.

SAMPLE PREPARATION

    Table 1 shows the composition and the Vickers hardness H_v of the various Al alloys investigated in the present study, and the workability of Nb/Al composites when these Al alloys were used as Al cores. For the Al alloy, (1) the H_v comparable to that of the pure Nb matrix, (2) no second phase observed by the optical microscope, are required to get the good workability of the Nb/Al composite.[4] Since a maximum solubility of Ge in Al is about 2 at.%, Ge was precipitated along the grain boundary to deteriorate significantly the workability of the Al alloy even for the Al-2.5at.%Ge binary alloys. The Al-5at.%Ge, however, could be drawn into a wire when heat treated at 420 °C to spheroidize Ge. Nevertheless, the H_v of the Al-Ge binary alloys are too low to cold-work the composite. Ge can be added to Al cores by the combination with Cu and Ag so as to avoid not only decreasing the H_v but also precipitating any visible second phase. Small amounts of Ge added to Al-Cu and Al-Ag alloys rather increase the H_v to improve the workability of Nb/Al composite. G.P. zones of Al-Cu and Al-Ag compounds may disperse Ge homogeneously, although it is unclear whether the ternary compounds of Al-Cu-Ge and Al-Ag-Ge are formed or not. On the other hand, Ge added to the Al-Mg alloy precipitates the large second phase and deteriorates the workability of the composite.

    The details of the sample preparation have been already reported elsewhere.[4] An Al-2at.%Cu, Al-1at.%Cu-1at.%Ge, Al-3at.%Ag, Al-1.5at.%Ag-1.5at.%Ge or Al-2at.%Ag-2at.%Ge rod of 6.9 mmø was encased into a Nb tube of 7 mm i.d. and 14 mm o.d., where the atomic fraction of Al to Nb is 26.4%, if the Al alloy rod is replaced by the pure Al one. The resulting single-core Nb/Al composite was cold drawn by cassette-roller dies into a wire of 1.14 mmø and cut into short pieces. The 121 short, single-core wires were bundled in a Nb tube of 14 mm i.d. and 20 mm o.d., and the resulting 121-core Nb/Al composite was cold drawn into a wire and cut into short pieces again. These procedures were repeated two times more, and finally a 1.8 x 10^6 (121x121x121)-core Nb/Al composite with continuous

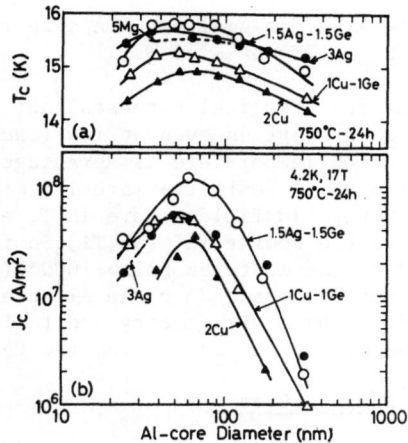

Fig. 1. Dependence of (a) T_c and (b) J_c (4.2K,17T) on the Al core diameter for the Nb/Al composite reacted at 750°C for 24 h.

ultrafine Al alloy cores was fabricated. The final composites with filaments 25 nm to 2.3 µm in diameter were heat treated to form Nb_3Al through the diffusion reaction between the Nb matrix and Al alloy cores.

SUPERCONDUCTING PROPERTIES

T_c and I_c were measured by a four-probe resistive methods. T_c was defined as the midpoint of the transition. I_c was defined as the current at which the sample showed the electric field of 1 µV/cm in steady transverse magnetic fields. J_c was defined as I_c/S. S was the cross-sectional area of the 1.8×10^6 Nb/Al single-core composite (i.e., excluding all Nb used for bundling). H_{c2} was determined by the extrapolation of a Kramer plot.

Heat Treatments at Low Temperatures

Figure 1 shows the dependence of (a) T_c and (b) J_c (4.2K,17T) on the Al core diameter for the composites reacted at 750°C for 24 h. The Al core diameter, which is calculated on the assumption that the filament shape is cylindrical one, represents roughly the inter-diffusion distance between Nb and Al. With decreasing Al core diameters, T_c and J_c initially increase, reach a maximum, and then decrease. The dependence of the Al core diameter on the critical values can be explained by (1) the relative stability of the Nb-Al compound phases(Nb_3Al, Nb_2Al and $NbAl_3$) in relation to the Al supply from the Al core(i.e., the instability of $Nb_2Al(NbAl_3)$ due to the exhaustion of the Al supply from the core is more remarkable at the smaller Al core.),[3,4,7] and (2) the proximity effect for extremely thin Nb_3Al filaments.[3,4] The poor workability of the Nb/Al-3at.%Ag composite(Table 1) is responsible for the fact that the critical values of the Nb/Al-3at.%Ag composite is insensitive to the calculated Al core diameter. Since the Al filament is deformed to the irregular shape, the calculated Al core diameter in the Nb/Al-3at.%Ag composite does not reflect exactly the inter-diffusion distance between Nb and Al. Ge added to the Al-Cu and Al-Ag alloy cores increases both the T_c and J_c(4.2K,17T); the highest T_c(15.9 K) and J_c(1.2×10^8 A/m^2) of the Nb/Al-1.5at.%Ag-1.5at.%Ge composite are higher than those of Nb/Al-5at.%Mg composite(T_c; 15.6 K, J_c; 0.96×10^8 A/m^2), respectively. Since the H_{c2}(4.2K) value is proportional to T_c, maximum T_c and H_{c2} values are obtained at the same filament size; the Nb/Al composite with the 50 nmø Al-1.5at%Ag-1.5at.%Ge core reacted at 750°C for 24 h showed the highest H_{c2}(4.2K) of 21.5 T.

It is very important for practical purposes that alloying Al with Ge can improve the high field properties even at low reaction temperatures (<1000°C). The low reaction temperature is advantage to the low AC loss conductor, because it requires a resistive material with a low melting point such as Cu-Ni. However, in fields above 15 T, even the Nb/Al-Ag-Ge composites are inferior to the commercial $(Nb,Ti)_3Sn$ in J_c properties, as long as they are reacted at temperatures below 1000°C(Fig. 5). This is probably because the amount of Ge added to the Al core is insufficient to suppress the deviation from the stoichiometry and to increase H_{c2}. Therefore, we tried to improve the H_{c2} by increasing the reaction temperature.

Heat Treatments at High Temperatures

J_c(4.2K) at 10, 15 and 20 T, and H_{c2}(4.2K) versus reaction temperature curves are shown in Figs. 2 (a) and (b) respectively, for the Nb/Al-1.5at.%Ag-1.5at.%Ge composite. The best superconducting properties are shown in this figure, where the numeral denoted beside each symbols represents the optimum reaction time at a given reaction temperature. Closed symbols represent the case of the two-stage reaction, where the samples were heat treated again at a low temperature of 700°C for 50 h to improve the long range order parameter(LRO). The optimum reaction time becomes shorter with increasing the reaction temperature. As long as the single-stage reaction is optimized, J_c does not decrease significantly with increasing reaction temperatures, in contrast to the other A15 compound conductors such as Nb_3Sn. This is because the major pinning center is not the grain boundary but the interface between Nb_3Al and matrix(or core). The second reaction at 700°C leads into an increase in J_c when the first reaction temperature is above 950°C. This increment in J_c due to the second reaction is pronounced by increasing both the first reaction temperature and the magnetic fields, since the increment in H_{c2} due to the second reaction becomes larger with increasing reaction temperature. It

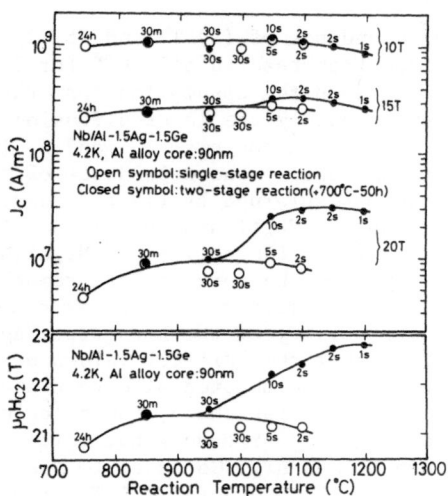

Fig. 2. (a) J_c(4.2K) at 10, 15, and 20 T, and (b) H_{c2}(4.2K) versus reaction temperature curves for the Nb/Al-1.5at.%Ag-1.5at.%Ge composite. The numeral denoted beside each symbols represents the optimum reaction time at a given reaction temperature.

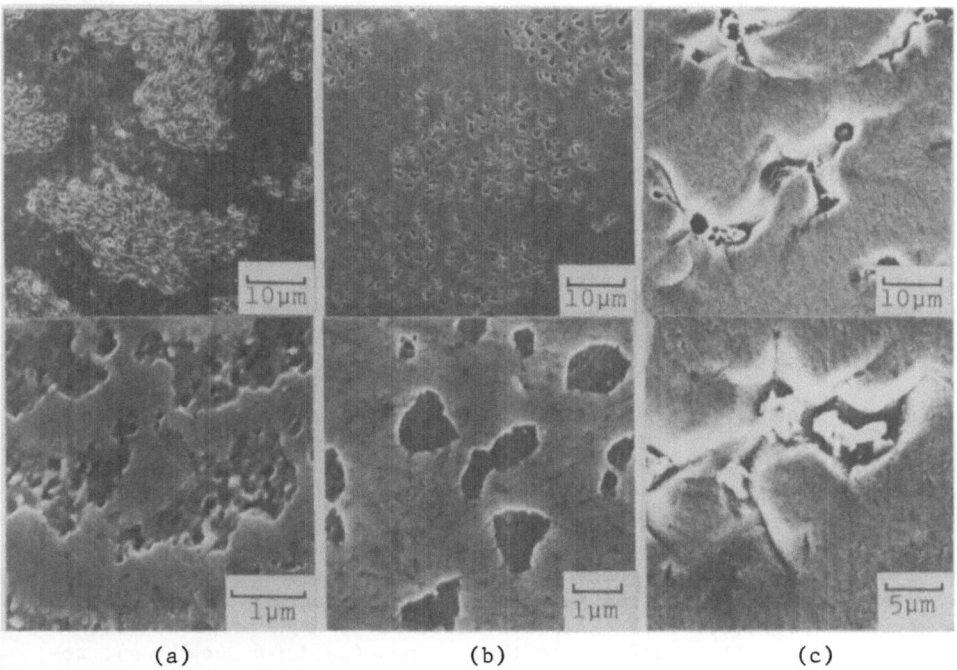

Fig. 3. Typical cross-sectional microstructures of the Nb/Al composite reacted by the pulse Joule heating; (a) the maximum electic resistivity ρ_{max} = 0.57 x 10^{-6} Ωm, (b) ρ_{max} = 0.75 x 10^{-6} Ωm, (c) ρ_{max} = 1.0 x 10^{-6} Ωm.

seems that the stoichiometric Nb_3Al forms easily at high temperatures and the second reaction improves the LRO, as reported for the powder processed Nb_3Al conductors.[8] However, to do the first reaction at further high temperatures, the optimum reaction time must be shorter than a second.

An attempt was made to do such a high-temperature reaction by applying a rectangular current pulse from a constant current source to the sample (0.1 m in length and 0.7 mm in diameter) immersed in liquid nitrogen. This technique is of great advantage to rapidly heating and cooling, so that it is easy to control the reaction time. The current pulse heating was controlled by varying both the pulse height from 35 to 75 A (0.9 - 2 x 10^8 A/m^2) and the duration time from 0.5 to 2 seconds. However, it is not so easy to measure the reaction temperature in such a short time. So, the electric resistivity ρ of the sample was monitored during the pulse Joule heating, since the resistivity is a measure of the sample temperature. The sample voltage which is proportional to the resistivity remained constant, when the applying current was at least below 35 A. In this case, a heat generation balanced with a heat cooling. On the other hand, the sample voltage came to increase rapidly with time when the sample current was at least over 39 A. This is because the boiling of liquid nitrogen changed from the nucleate boiling to the film one (the heat transfer coefficient became small), resulting in a decrease in the heat cooling. Furthermore, the applying current was constant during the pulse Joule heating so that the heat generation per unit volume($J^2\rho$) increased with increasing the resistivity ρ. In this study, the maximum electric resistivity ρ_{max} which was observed at the end point of pulse is adopted as the parameter representing the reaction temperature.

Figure 3 shows the typical cross-sectional microstructures of the Nb/Al composite reacted by the pulse heating. At ρ_{max} = 0.57 x 10^{-6} Ωm (Fig.3(a)), the individual Nb$_3$Al filament remains separated clearly, although the reached temperature was estimated at least above 1200 °C by using the thermocouples attached on the sample. With increasing ρ_{max}, however, the Nb$_3$Al filament comes to be joined with each other. It is hard to distinguish the individual filament at ρ_{max} = 0.75 x 10^{-6} Ωm (Fig.3(b)), at which the high temperature reaction rounds off the boundary between the 121 filaments part and the surrounding Nb part. At ρ_{max} = 1.0 x 10^{-6} Ωm (Fig.3(c)), it seems that (1) the melting occurs locally and (2) Al diffuses out of the 121 filament part and reacts with the surrounding Nb part. Therefore, the composition of Nb$_3$Al will deviate from the stoichiometry in the case of the excessive reaction at high temperature. When ρ_{max} reached 1.2 x 10^{-6} Ωm, the sample burnt out.

Figure 4 shows T_c of the various Nb/Al composite wires reacted by the pulse Joule heating as a function of ρ_{max}. Closed symbols represent the T_c of the sample subsequently reacted at 700 °C for 50 h. The second reaction increases the T_c. Higher T_c's more than 17 K are obtained around ρ_{max} = 0.57 x 10^{-6} Ωm where the reaction temperature is at least above 1200 °C. It seems that the pulse Joule heating diminishes the deviation from the stoichiometry and the subsequent second reaction at 700 °C improves the LRO, respectively. In this case, the multifilamentary structure remains unchanged as shown in Fig 3(a). The excessive reaction at high temperatures, however, causes the Al to diffuse to the Nb part(used for bundling), resulting into the deviation from stoichiometry and the degradation in T_c. It is noted that the highest T_c of 17.4 K is observed for the Nb/Al-Cu-Ge composite, in contrast to the case of the single stage reaction at low temperatures. At the single-stage reaction at 750 °C, the high T_c values are obtained in the sequence Nb/Al-Ag-Ge →Nb/Al-Ag→Nb/Al-Mg→Nb/Al-Cu-Ge→ Nb/Al-Cu→Nb/Al-Zn.[4] Ge added to the binary Al-Cu alloy is more effective in increasing T_c for the sample reacted by the pulse heating than those by the single-stage reaction. The T_c value of 16.5K is obtained, at least hitherto, for the Nb/Al-1.5at.%Ag-1.5at.%Ge composite, although the pulse Joule heating condition has not been optimized in this composite, yet.

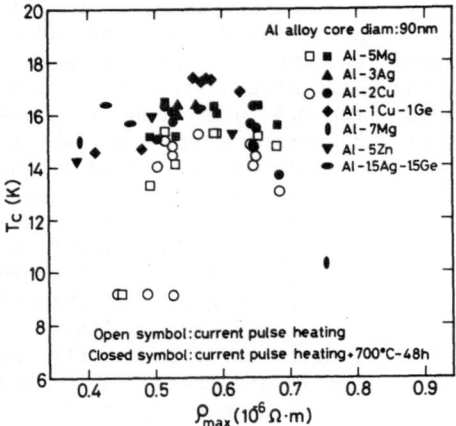

Fig. 4. T_c of the various Nb/Al composite reacted by the pulse heating. Maximum electric resistivity ρ_{max} is a parameter reflecting the reaction temperature.

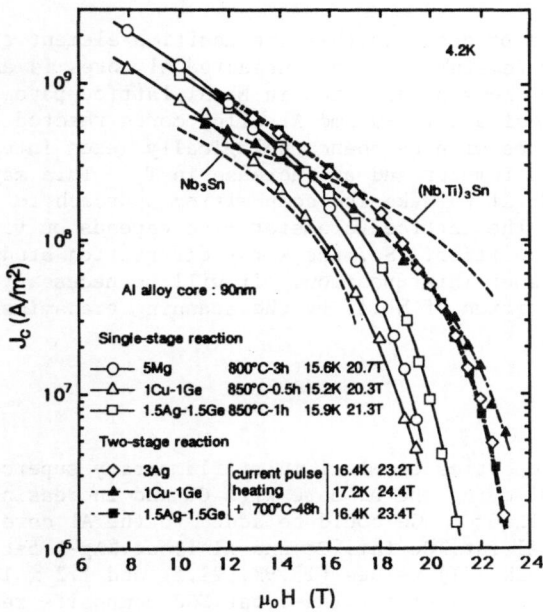

Fig. 5. Typical J_c versus $\mu_0 H$ curves for the various Nb/Al composites reacted by the single-stage reaction and the two-stage reaction (the pulse heating + 700°C-48 h). Non-Cu overall J_c of Nb_3Sn and $(Nb,Ti)_3Sn$ multifilamentary wires are shown in the figure for reference.

There is a linear relationship between $H_{c2}(4.2K)$ and T_c, irrespective of the reaction conditions and the Al alloys used as the core.[4] Therefore, an increase in T_c improves both H_{c2} and J_c at high fields, as shown in Fig. 5, by alloying Al core with Ge and increasing the reaction temperature. The highest J_c at 20 T is 9×10^7 A/m^2, which is obtained for the Nb/Al-3at.%Ag, Nb/Al-1.5at.%Ag-1.5at.%Ge and Nb/Al-1at.%Cu-1at.%Ge composites reacted by the pulse Joule heating. Furthermore, the most suitable pulse heating would improve J_c at high fields for the Nb/Al-1.5at.%Ag-1.5 at.%Ge composite, since the pulse heating condition has not been optimized yet. The improved J_c by the pulse heating is higher than that of the commercial $(Nb,Ti)_3Sn$ multifilamentary conductor, at least hitherto, up to 17 T.

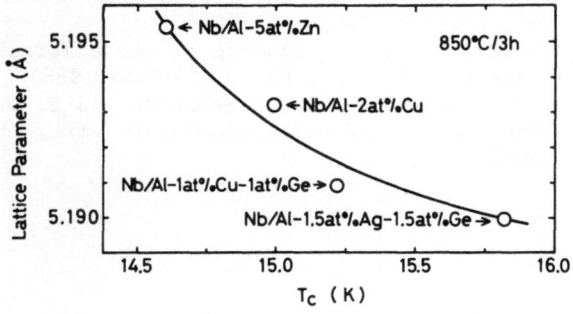

Fig. 6. Variation in the lattice parameter of Nb_3Al as a function of T_c.

In the present process, whether the additive element of Ge is soluble in Nb_3Al layers or remaining at the unreacted Al cores is an interesting problem. Figure 6 shows a variation in Nb_3Al lattice parameters for the various Nb/Al composite with 90 nmϕ Al alloy cores reacted at 850 °C for 3 h. Alloying Al cores with Ge phenomenologically leads into a decrease in the Nb_3Al lattice parameter and an increase in T_c. This may suggest that Ge is soluble in Nb_3Al to make the composition approach to the stoichiometric one. However, the lattice parameter also depends on various factors other than the composition. So, the x-ray diffraction study is insufficient to answer this question. It will be necessary to measure directly the composition of Nb_3Al by the scanning transmission electron microscopy, and so on.

CONCLUSIONS

High field properties of Nb_3Al multifilamentary superconductors have been improved by alloying the Al core with Ge and increasing the reaction temperature above 1200 °C. Ge could be added to the Al core by using the ternary alloys of Al-1at.%Cu-1at.%Ge and Al-1.5at.%Ag-1.5at.%Ge. The T_c, H_{c2}(4.2K) and J_c(4.2K,17T) values (15.9K, 21.5T and 1.2 x 10^8 A/m^2, respectively) of the Nb/Al-1.5at.%Ag-1.5at.%Ge composite reacted at 750 °C are the highest values among the various Nb/Al composites so far investigated. The pulse Joule heating makes it possible to react the sample above 1200 °C in a short time, resulting into an increase in the J_c at high fields(9 x 10^7 A/m^2 at 20 T) due to the improvement of H_{c2}(24.4 T), where the multifilamentary structure remains unchanged.

The authors are much indebted to Dr. T. Kiyosi at the NRIM and the staff of Tohoku University for operating 18T superconducting magnet and 23 T hybrid magnet, respectively.

REFERENCES

1. K. Inoue, Y. Iijima and T. Takeuchi, Superconducting properties of Nb_3Al multifilamentary wire, Appl. Phys. Lett., 52:1724(1988).
2. K. Inoue, Y. Iijima and T. Takeuchi, New superconducting Nb_3Al MF wire made by Nb/Al-Mg composite process, Cryogenics, 29:418(1989).
3. T. Takeuchi, Y. Iijima, K. Inoue, K. Watanabe and K. Noto, Pinning mechanism in a continuous ultrafine Nb_3Al multifilamentary superconductor, Appl. Phys. Lett., 53:2444(1988).
4. T. Takeuchi, Y. Iijima, M. Kosuge, T. Kuroda, M. Yuyama and K. Inoue, Effects of additive elements on continuous ultra-fine Nb_3Al MF superconductor, IEEE Trans. on Magnetics, 25:2068(1989).
5. T. Kuroda, H. Wada, Y. Iijima and K. Inoue, Strain effects on superconducting properties in Nb_3Al multifilamentary wires, J. Appl. Phys., 65:4445(1989).
6. Y. Im and J. W. Morris, Jr., Sequence of phase formation in Nb/Al multilayered samples, J. Appl. Phys., 64:3487(1988).
7. C. L. H. Thieme, S. Pourrahimi, B. B. Schwartz and S. Foner, Nb-Al powder metallurgy processed multifilamentary wire, IEEE Trans. on Magnetics, MAG-21:756(1985).

PHASE FORMATION AND CRITICAL CURRENTS IN PM Nb₃Al

MULTIFILAMENTARY WIRES

Klaus Heine[1] and René Flükiger[2]

[1]Vacuumschmelze GmbH, Grüner Weg 37, D-6450 Hanau, FRG
[2]Kernforschungszentrum Karlsruhe, Institut für
Techn. Physik, Postfach 3640, D-7500 Karlsruhe, FRG

ABSTRACT

It has been demonstrated that multicore Nb₃Al wires with stabilizing copper matrix can be prepared by powder metallurgy. High critical current densities of $j_c = 1.2 \cdot 10^5$ A/cm² at 10 T were achieved for single core wires with a high areal reduction ratio (ARR) of $q = 9 \cdot 10^5$. For high ARR the A15 phase is formed at relatively low temperatures of about 650°C, bypassing the σ phase. Phase formation has been studied by DSC, T_c measurements and XRD. The DSC results are quantitatively described by a layer model. Prior to the A15 phase NbAl₃ and bcc NbAl are formed. The amount and formation temperature of these preliminary stages depends on the ARR, indicating that the absolute layer thickness of Nb and Al influences the phase formation. Phase formation in Nb-Al multilayers of submicron scale does not follow the reaction sequence according to the equilibrium phase diagram.

INTRODUCTION

Nb₃Al is a promising superconductor for high field applications at 4.2 K and could be an alternative for Nb₃Sn. Very high critical current densities exceeding 10^5 A/cm² at 20 T have been reported for liquid quenched material[1] demonstrating the potential of this A15 phase superconductor. Unfortunately there is no "bronze process" for the formation of Nb₃Al wires as exists for Nb₃Sn, because of thermodynamical reasons[2] Nb₃Al is not stable in the presence of copper. At the moment there are two routes for the Nb₃Al formation which are of technological interest, e.g. the transformation of a supersaturated bcc phase[1,3] and the direct reaction of Nb and Al[4,5,6]. According to the binary phase diagram[7] a sequence of intermetallic compounds Nb₃Al, Nb₂Al and NbAl₃ exists in the Nb-Al system. All of this phases occur at high reaction temperatures and from thick reaction layers[8]. At low reaction temperatures which are convenient for the preparation of technical wires the formation rate of Nb₃Al is very slow making it difficult to obtain a substantial amount of A15 phase. This difficulty can be overcome by minimizing the diffusion length[3,6] and the thickness of Nb and Al layers, respectively.

Powder metallurgically (PM) prepared NbAl composites which were cold worked to an areal reduction ratio (ARR) of about $q=10^6$ consist of a struc-

Advances in Cryogenic Engineering (Materials), Vol. 36
Edited by R. P. Reed and F. R. Fickett
Plenum Press, New York, 1990

369

ture with Nb layers less than 100 nm in thickness. The phase formation in such PM wires[9] and in sputter deposited multilayer geometries with similar layer thicknesses[10,11] does not follow the reaction sequence according to the (high temperature) equilibrium phase diagram, e.g. bypasses the formation of the σ phase Nb_2Al. At low reaction temperatures of typically 700 - 900 °C the reaction kinetics has to be considered. In particular effects related to the extremely fine composite structure, e.g. interface energies and nucleation at interfaces might favour or suppress the formation of equilibrium phases. The layer thickness of PM processed NbAl is determined by the initial powder size and by the degree of cold working (ARR) of the composite. For different ARR it can be easily varied over a wide range, from some micrometers to less than 100 nm, thus giving the opportunity to study the phase formation at different length scales.

This work especially examines the reaction sequence prior to the formation of the A15 phase. The reactions were studied by X-ray diffraction (XRD), T_c measurements and differential scanning calorimetry (DSC). The DSC measures the heat flux from or to a sample while heating it up at a constant heating rate. Phase formation is associated with an exothermic heat release because it reduces the Gibbs' (free) energy of the sample. The reaction enthalpy ΔH related to the phase formation can be determined by integrating the corresponding DSC peak. A layer model was developed which quantitatively describes the DSC results as a function of layer thickness and areal reduction ratio, respectively.

The second part of this work deals with critical currents of single core and multicore PM Nb_3Al wires. High critical current densities j_c were obtained only for wires with high ARR. Studying the phase formation is therefore the basis for an understanding of this experimental result and might be helpful in finding optimized reaction treatments.

EXPERIMENTAL

Powder metallurgical (PM) NbAl composites were prepared from hydride-dehydride (HDH) Nb powders of 75 - 150 µm particle size with an oxygen content of 300 ppm and from low oxygen (<200 ppm) PREP Nb powder with spherical particles of 300 - 500 µm size. Al powder of 10 - 20 µm particle size was used. NbAl mixtures with a nominal composition of 8 wt.% Al (23 at.%) were cold pressed isostatically at 0.2 GPa, encapsulated in a copper can and hydrostatically extruded. Multicore wires were fabricated from PREP Nb by cold working the extruded Cu/NbAl rods to hexagonal geometry, bundling and a second hydrostatic extrusion. The extruded 18 core composites could be wire drawn to a final diameter of 1.2 mm which is equivalent to an ARR of $q = 6 \cdot 10^4$. Even higher deformation rates were achieved by additional rolling. Single core wires with ARR up to $q = 9 \cdot 10^5$ were prepared from HDH Nb. After hydrostatic extrusion and cold working the remaining copper can was removed. These wires were then bundled in a CuNi tube and further cold worked by swaging and rolling.

DSC measurements were performed on NbAl composites with ARR of $q = 2 \cdot 10^2 - 9 \cdot 10^5$. Samples of about 30 mg were sealed in Al crucibles and heated to a maximum temperature of 600 °C. Heating rates were varied from 1 - 40 °C/min. XRD was used to analyse the phases present in samples after DSC heat treatment and for samples isothermally heat treated at temperatures of 300 - 900 °C. The critical temperature T_c was determined inductively. T_c was defined as the midpoint of the superconducting to normal transition. Critical currents were measured in magnetic fields up to 20 T using a 1 µV/cm criterion. Critical current densities j_c are given with respect to the area of the NbAl cores.

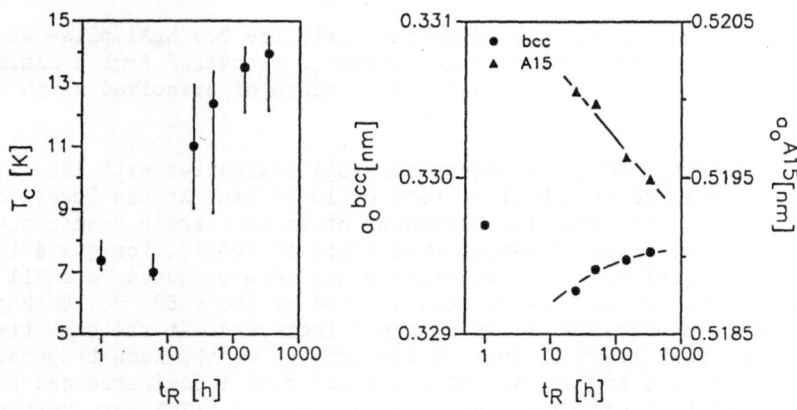

Fig. 1 a,b . Isothermal heat treatment of NbAl composites with q = 9·10⁵ at 650°C. Fig. 1a (left) critical temperature T_c. Fig. 1b (right) lattice parameter a_0^{bcc} and a_0^{A15} of bcc NbAl and A15 phase.

RESULTS

The kinetics of A15 phase formation was investigated in detail by iso-thermal heat treatments at 650 °C. For wires with q = 9·10⁵ the critical temperature T_c and the lattice parameter a_0^{bcc} of bcc Nb respectively a_0^{A15} of the A15 phase were measured as a function of the reaction time t_R (Fig. 1a,1b). The critical temperature of the as drawn wire was reduced to about 8.3 K due to the very high deformation rate[12]. After short reaction times T_c further decreased from this starting value to T_c = 6.5 K. This behaviour was associated with a decrease of the bcc lattice parameter a_0^{bcc}. Both T_c and a_0^{bcc} are consistent with a solid solution of Al in bcc NbAl of approx-imately 5 - 8 at.% Al. After further annealing T_c steeply increased indi-cating the formation of A15 phase. The formation of A15 phase was also con-firmed by XRD. At the beginning of the Nb_3Al formation an A15 phase with low T_c = 14 K and a_0^{A15} = 0.5200 nm was formed corresponding to an Al con-tent of approximately 18 at.% [7]. While the reaction proceeded the concen-tration of Al increased and approached the (low temperature) solubility

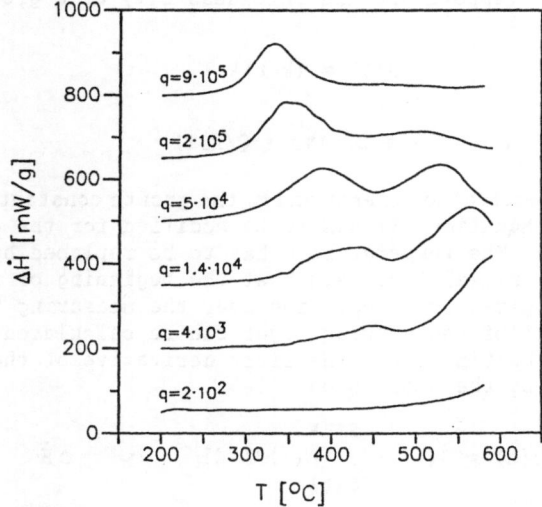

Fig. 2 . DSC mesurements for NbAl composites with ARR q = 2·10² - 9·10⁵ obtained at a heating rate of 10 °C/min. Baselines are offset for clarity.

limit which is about 21 at.% Al. Simultaneously the bcc NbAl phase was involved in the reaction. The lattice parameter recovered from a minimum value to a_0^{bcc} = 0.3295 nm indicating a decrease of dissolved Al in bcc NbAl.

Fig. 2 presents DSC measurements for NbAl composites with ARR q = $2 \cdot 10^2$ – $9 \cdot 10^5$ obtained at a heating rate of 10 °C/min. At the lowest deformation rate (q = $2 \cdot 10^2$) only the beginning of an exothermic reaction was observed at the experimental temperature limit of 600 °C. For q = $4 \cdot 10^3$ and q = $1.4 \cdot 10^4$, respectively two exothermic peaks were measured, a small one at about 380 – 450 °C and a large peak located at 580 – 600 °C. Both peaks were shifted to lower temperatures as q was increased. At the same time the low temperature peak (peak 1) grew at the expense of the high temperature peak (peak 2). At the highest ARR of q = $9 \cdot 10^5$ peak 1 dominated and peak 2 had almost completely vanished. The results for q = $2 \cdot 10^5$ were very similar to those obtained by Ref.[9] for PM wires with q = $1 \cdot 10^5$ and a finer Nb powder of 75 – 105 μm particle size.

According to Ref.[9] the origin of the two DSC peaks is related to the formation of NbAl$_3$ and bcc NbAl, respectively. The formation of Nb$_3$Al was detected near 800 °C and could therefore not be observed in the present DSC investigation. The identification of the first two peaks was in agreement with XRD and T$_c$ measurements on DSC samples as well as on samples which were heat treated isothermally at 300 °C, 450 °C and 550 °C, respectively. Prolonged anneals at 550 °C caused a solid solution of Al in bcc Nb, as revealed by a decrease of both, T$_c$ and a_0^{bcc}. The suppression of the bcc peak at very high ARR was surprising. As mentioned before the bcc solid solution is a preliminary stage of the A15 phase formation. This behaviour might be due to the extremely fine structure and will be discussed later.

The microstructure of cold worked NbAl composites was investigated by transmission electron microscopy (TEM) . It consisted of a curled layer like structure, which is due to the ⟨110⟩ deformation texture of niobium. Assuming a simplified layer structure with alternating planar Nb and Al layers a layer model was developed which quantitatively describes the DSC data[13]. In this case the interdiffusion and the formation of intermetallic compounds at the interface can be treated like a one dimensional layer growth. At constant temperature the thickness S(t) of a growing layer is given by

$$S(t) = (k \cdot t)^{1/2} \tag{1}$$

with

$$k = k_0 \cdot \exp \, (-Q/k_B T) \tag{2}$$

where Q denotes the activation energy and k_0 the growth constant of a diffusion controled process. Equation (1) has to be modified for the case of a constant heating rate α. The temperature T has to be replaced by T = T_0 + αT with T_0 beeing the initial temperature at the beginning of the DSC measurement. S(t) then is given by integration over the measuring time t. The volume fraction X(t) of the growing layer can be calculated as a function of temperature and heating rate. The first derivative of the volume fraction $\dot{X}(t)$ is given by the expression

$$\dot{X}(t) = \dot{S}(t) \cdot \int_{S(t)}^{S^{max}} \wp(1)/1 \, dl \quad \sim \quad \Delta\dot{H} \tag{3}$$

where S(t) and $\dot{S}(t)$ is the thickness and the growth velocity, respectively of the reaction product. The function $\wp(l)$ considers the layer thickness distribution of a real PM wire. Here the lognormal distribution was used. Setting the second derivative equal to zero one finds an expression which

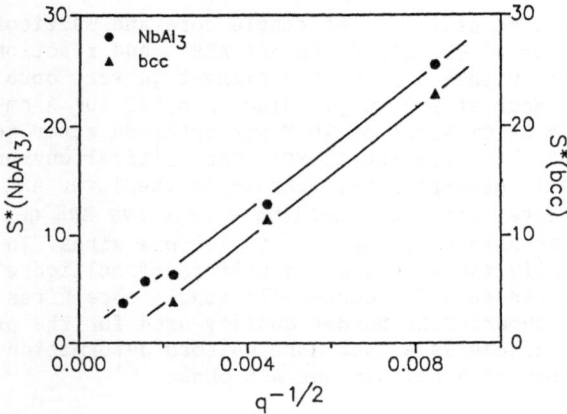

Fig. 3 . Calculated relative layer thickness S* for NbAl₃ and bcc NbAl
       vs. $q^{-1/2}$.

relates the maximum temperature T_m of the DSC peak to the activation energy:

$$\ln (\alpha z/T_m^2) = - 0.5\ Q/k_B \cdot 1/T_m + C \tag{4}$$

with

$$z = ((T_m-T_0)/\alpha)^{1/2}.$$

The activation energy Q can therefore be estimated by plotting the left
side of equation (4) vs. $1/T_m$ where T_m represents the peak temperature at
different heating rates α. Analysing the data for α = 1 to 40 °C/min yields
Q_1 = 2.2 eV for peak 1 (NbAl₃) and Q_2 = 1.95 eV for peak 2 (bcc NbAl).

   The estimated values for the activation energies Q_1 and Q_2 were taken
to calculate a relative layer thickness S* using equation (3). $S^* = S/k_0^{1/2}$
was obtained by fitting the mean value S and the standard deviation σ_L
which both characterize the layer thickness distribution to the measured
DSC peaks. In PM wires the mean layer thickness S depends on the areal
reduction ratio. Assuming constant volume and homogeneous deformation this
leads to the relation $S \sim q^{-1/2}$. Fig. 3 shows the calculated values of S*
for NbAl₃ and the bcc phase. A linear behaviour is observed for both reaction peaks in excellent agreement with the assumptions of this layer model.

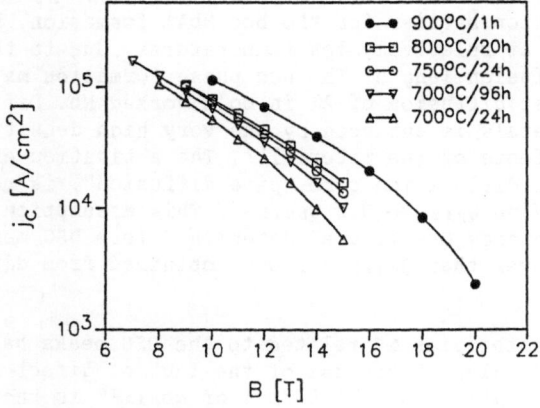

Fig. 4 . Critical current density for single core (closed symbols) and
       multicore (open symbols) PM Nb₃Al wires with different heat
       treatments.

Critical current densities j_c of single core and multicore PM Nb_3Al wires were measured on wires with different ARR[16] and reaction treatments. For single core wires with $q = 9 \cdot 10^5$ the highest j_c were obtained after a heat treatment of 1 hour at 900 °C yielding $j_c = 1.2 \cdot 10^5$ A/cm² at 10 T. For multicore wires $j_c = 8 \cdot 10^4$ A/cm² at 10 T was obtained after reaction treatments which were not optimized (8h/850°C). The critical current densities of these wires were in general lower because of the lower ARR $q = 5 \cdot 10^5$. Other powder sizes have been used yielding a relative ARR $q_{rel} = 3 \cdot 10^4$ with respect to the powder size used for the single core wires. In comparison with wires of nominally the same q_{rel} value[14] j_c of multicore wires is about a factor 3 higher than for comparable single core wires. This behaviour is due to the superior Nb powder quality used for the preparation of multicore wires which enables a much more uniform deformation of the composite and the formation of a homogeneous A15 phase.

The influence of different reaction temperatures on j_c is more pronounced at high magnetic fields (Fig. 4) where the upper critical field determines the critical current density. The extrapolated values[15] of B_{c2}^* systematically varied from $B_{c2}^* = 18$ T for low reaction temperatures and short times to $B_{c2}^* = 22$ T for high reaction temperatures and long times. The upper critical field of Nb_3Al is influenced by the stoichiometry and the degree of long range order[16] of the A15 phase. Both effects seem to determine j_c at high fields, especially for relatively low reaction temperatures of 700 – 750 °C.

DISCUSSION

This study demonstrates that the phase formation in PM NbAl composites depends on the absolute value of the layer thickness and on the diffusion length, respectively. At intermediate ARR's of $q = 10^4 - 10^5$ the phase formation followed the common thin film reaction sequence[9,10]. At low temperatures of about 300 – 450 °C the $NbAl_3$ phase is formed. At temperatures of 450 – 600 °C Al dissolves in a bcc <u>Nb</u>Al solid solution. The A15 phase is formed in a third step at higher temperatures. It nucleates with a low Al content of about 18 at.% and then approaches the (low temperature) equilibrium concentration of 21 at.% Al. The growth of Nb_3Al at the expense of bcc <u>Nb</u>Al and especially the bypassing of the σ phase does not follow the equilibrium phase diagram. The formation of bcc <u>Nb</u>Al and $NbAl_3$ seem to be two competitive processes. Both reactions are driven by the differences of the particular (free) enthalpy curves of the occurring phases[11] and are influenced by nucleation of $NbAl_3$ and growth kinetics. At moderate ARR there is no nucleation barrier for the bcc <u>Nb</u>Al formation, because Al penetrates into bcc Nb at relatively low temperatures, due to the high solubility and short diffusion length. The bcc phase formation may further be enhanced by the fast diffusion of Al in cold worked Nb. Diffusion in heavily cold worked metals is enhanced by the very high density of dislocations and point defects of the material[17]. The activation energy Q_{pipe} for diffusion along the dislocation core, "pipe diffusion", is reduced to about half of the bulk value $Q_{pipe} \approx 0.5 \cdot Q_{bulk}$[17]. This assumption is confirmed by the activation energy $Q = 1.95$ eV determined form DSC measurements which is significantly lower than $Q_{bulk} = 5.0$ eV obtained from diffusion experiments [18].

The reaction enthalpies ΔH related to the DSC peaks have to be compared with calculated values[19] because of the lack of direct measurements. Only the heat of formation $\Delta H = 41$ kJ/mol of $NbAl_3$[20] is known. For reaction temperatures less than about 600 °C only bcc <u>Nb</u>Al and the $NbAl_3$ phase have to be considered. The total energy release for the case that all Al of the Nb-8 wt.%Al composite reacts to form $NbAl_3$ + Nb or bcc <u>Nb</u>Al, respectively is about 15 – 20 kJ/mol[19]. This value was not found for one individual DSC peak. Only the sum of peak 1 and peak 2 at relatively low ARR ($q = 1.4 \cdot 10^4$) yields approximately 15 kJ/mol. The fact that this maximum formation enthalpy was not measured either for the $NbAl_3$ formation or for bcc $\underline{Nb}Al$ solid solution alone indicates that none of the reactions had consumed the total amount of Al or Nb. For higher ARR the formation enthalpy related to the first peak ($NbAl_3$) is approximately constant $\Delta H = 4 - 5$ kJ/mol while the second peak ($\underline{Nb}Al$) is suppressed. This DSC result implies that no bcc $\underline{Nb}Al$ is formed after the $NbAl_3$ formation. On the other hand the bcc phase has been found in competition with the A15 phase at 650 °C. At present this discrepancy is not well understood. There are only some speculations about this behaviour. For PM wires with such a fine layer structure the influence of phase boundaries and interface energies has to be considered.

CONCLUSION

It has been demonstrated that multicore Nb_3Al wires with stabilizing copper matrix can be prepared by powder metallurgy. The workability of the composite and the maximum degree of cold deformation is mainly determined by the quality of the Nb powder. In particular, a low oxygen content is essential for this fabrication process. In addition the work hardening and additional hardening at high areal reduction ratios (ARR) due to the high density of Nb-Al interfaces becomes important.

High critical current densities of $j_c = 1.2 \cdot 10^5$ A/cm^2 at 10 T were achieved for single core wires with a high ARR of $q = 9 \cdot 10^5$. For high ARR the A15 phase is formed at relatively low temperatures of about 650°C, bypassing the σ phase.

Phase formation has been studied by DSC, T_c measurements and XRD. The DSC results are quantitatively described by a layer model. Prior to the A15 phase $NbAl_3$ and bcc $\underline{Nb}Al$ are formed. The amount and formation temperature of these preliminary stages depends on the ARR, indicating that the absolute layer thickness of Nb and Al influences the phase formation. Phase formation in Nb-Al multilayers of submicron scale does not follow the reaction sequence according to the equilibrium phase diagram. The influence of phase boundaries and interface energies has to be taken into account in further investigations.

ACKNOWLEDGEMENT

The autors would like to thank R. Bormann for helpful discussions and additional DSC measurements. This work was supported by Bundesministerium für Forschung und Technologie (BMFT).

REFERENCES

1. T. Takeuchi, K. Togano, and K. Tachikawa, Nb$_3$Al and its ternary A15 compound conductors prepared by a continous liquid quenching technique, IEEE Trans. Magn., MAG-23:956(1987)
2. C. R. Hunt jr. and Aravamudhan Raman, Alloy Chemistry of σ(βU)-Related Phases, Z. Metallkde, 59:701(1968)
3. K. Lo, J. Bevk, and D. Turnbull, Critical Currents in Liquid-Quenched Nb$_3$Al, J. Appl. Phys.,48:2597(1977)
4. C. L. H. Thieme, S. Pourrahimi, B. B. Schwartz, and S. Foner,Improved high field perfomance of Nb-Al powder metallurgy processed superconducting wire,IEEE Trans. Magn.,MAG-21:756(1985)
5. K. Inoue, Y. Iijima, and T. Takeuchi, Superconducting Properties of Nb$_3$Al Multifilamentary Wire, Appl. Phys. Lett., 52:1724(1988)

6. R. Bruzzese, N. Sacchhetti, M. Spadoni, G. Barani, G. Donati, and S. Ceresara, Improved Critical Current Densities in Nb_3Al Based Conductors, IEEE Trans. Magn., MAG-23:653(1987)

J. L. Jorda, R. Flükiger, and J. Müller, A New Metallurgical Investigation of the Niobium-Aluminium System, J. Less-Common Met.,75:227(1980)

8. G. Salma and A. Vignes, Diffusion dans les Aluminiures de Niobium, J. Less Common Met.,29:189(1972)

9. K. R. Coffey, K. Barmak, and D. A. Rudman, Reaction Kinetics of Phase Formation in Nb-Al Powder Metallurgy Prcessed Wire, IEEE Trans. Magn. MAG-25:2093(1989)

10. J. M. Vandenburg, M. Hong, R. A. Hamm, and M. Gurvitch, Reactive Diffusion and Superconductivity of Nb_3Al Multilayer Films, J. Appl. Phys.,58:618(1985)

11. R. Bormann, H. U. Krebs, and A. O. Kent, The Formation of Metastable Phase Nb_3Al by a Solid State Reaction, Adv. Cryo. Eng., 32:1041(1986)

12. H. E. Cline, B. P. Stauss, R. M. Rose, and J. Wulff, Superconductivity of a Composite of Fine Niobium Wires in Copper, J. Appl. Phys., 37:5(1965)

13. K. Heine, to be published

14. R. Flükiger, W. Goldacker, and R. Isernhagen, Adv. Cryo. Eng., 32:925(19 86)

15. E. J. Kramer, Scaling Laws for Flux Pinning in Hard Superconductors, J. Appl. Phys., 44:1360(1973)

16. R. Flükiger, J. L. Jorda, A. Junod, and P. Fischer, Superconductivity, Atomic Ordering and Stoichiometry in the A15 Phase Nb_3, Appl. Physics Comm., 1:9(1981)

17. A. L. Ruoff and R. W. Balluffi, On Strain-Enhanced Diffusion in Metals. III. Interpretation of Recent Experiments, J. Appl. Phys., 34:2862(1963)

18. F. Brossa, G. Musso, and H. W. Schleicher, Studio della diffusione tra Al e Nb, La Metallurgia Italiana, 4:167(1969)

19. R. Bormann, Habilitation Thesis, Göttingen, 1988

20. A. Neckel and H. Nowotny, Zur Thermochemie von Aluminiden, 5. Intern. Leichtmetalltag, Düsseldorf 1968, p. 72 (1969)

THIN FILM PREPARATION AND DEVICE APPLICATIONS OF EPITAXIAL

MgO/NbN MULTILAYERS

K.Hamasaki, A.Irie, Z.Wang, T.Yamashita and [+]K.Watanabe

Nagaoka University of Technology, Nagaoka, 940-21 Japan

[+]The Research Institute for Iron, Steel and Other Metals,
Tohoku University, Sendai 980, Japan

INTRODUCTION

Many problems fundamental to the preparation of high quality Josephson junctions are strongly related to interface morphology in the superconducting electrodes and the insulator layer. One of the problems for NbN films is that the superconducting coherence length, ξ, of NbN is at least a factor of 2 lower than for Nb. For NbN and Nb, ξ (4.2 K) is 4-7 nm and 10-30 nm, respectively.[1] Another problem is that the NbN films are often inhomogeneous, and their properties are very sensitive to film preparation conditions. Variations in grain size, crystalline orientation, voids and columnar growth all contribute to changes in the electrical properties of the films.

Hence, reliable process technology is necessary in order to prepare the high T_c NbN layers with good surface and interface morphologies which are deposited first, i.e., layers adjacent to the insulator layer with a thickness characterized by ξ. Since one of the principle procedures permitting formation of ultrathin NbN layers with characteristics similar to those of bulk materials is heteroepitaxial growth, the preparation of highly oriented MgO films is of great scientific interest and of practical applications such as tunnel junction fabrication. For practical purposes, the textured MgO films can be used as underlayer to deposit other oriented films on top.

So far, epitaxial NbN films on highly oriented MgO films have been successfully prepared by several researchers using the technique of dc or rf magnetron sputtering.[2-7] However, in the present situation, for tunnel junction applications the deposition conditions remain still unoptimized. As mentioned by Talvacchio et al., tunnel properties are determined by the uniformity of the coverage of tunnel barriers, and the coverage is independent of crystal orientation of NbN.[6] Also Kerber et al. has found that the junctions with oriented MgO underlayers exhibited higher subgap leakage and only low current density junctions showed enhanced sum-gap voltage, and hence they concluded that for logic devices randomly oriented NbN layers should be used.[7]. However, it does not seem that the randomly oriented NbN is the most favorable one. It will not improve the tunnel property in future. If we get the perfect single crystal epitaxy or the fine column and very small voids, the coverage, i.e., tunnel property, may be improved. In the present situation, the study of surface and interface morphologies is still insufficient.

Advances in Cryogenic Engineering (Materials), Vol. 36
Edited by R. P. Reed and F. R. Fickett
Plenum Press, New York, 1990

377

Therefore, We will investigate the optimum deposition conditions and the surface and interface morpholgies of ultrathin(<10 nm) NbN films. The aim of this investigation is to achieve high transition temperature (low tensile strain), low resistivity(small voids) NbN films with good surface morpholgy under conditions suitable for use in all-NbN Josephson junction fabrication. In this paper we report on the results of a recent study on the deposition of oriented MgO films by rf sputtering from single crystal and hot-pressed MgO targets and of the epitaxial growth of MgO/NbN multilayers.

FILM PREPARATION

MgO films were deposited by rf sputtering in a multitarget sputtering system. The sputtering gases were high-purity(99.9995 %) Ar, N_2, and Ar+N_2 mixture. The substrate temperature was in the range of 150-210 ℃. The target was disks of single crystal MgO(85 mm diameter) and hot-pressed MgO(100 mm diameter) of 99.99 % purity. MgO and NbN film deposition rates were inferred from a Tncor surface profiler measurement on thicker films.

Following the MgO deposition, thin NbN layers were sequentially deposited onto predeposited MgO underlayers on silicon substrates without breaking the vacuum. NbN films were produced by reactive rf magnetron sputtering in 20 mTorr of Ar+N_2 mixture. The substrate temperature was about 210 ℃, and the sputtering rate for an rf power density of 5.1 W/cm^2 was about 80 nm/min in Ar+(~9 %)N_2 mixture. A calibrated silicon diode (Lake Shore Cryotronics, Inc) was used as a temperature sensor, with a measurement accuracy of ±0.1 K in the 4-20 K.

EXPERIMENTAL RESULTS AND DISCUSSION

XRD Analysis

The results of the preparation of highly oriented MgO films have already been reported elsewhere.[2,4] These films were deposited with hot-pressed MgO target and in Ar or Ar+N_2 mixture. In particular, the deposition in Ar+N_2 mixture resulted in the increased intensity of orientation of MgO films. The same phenomenon has been recently observed by Kerber et al.[7] Here we present the results of MgO films prepared with single crystal MgO target and in two different environments, N_2 gas and Ar+N_2 mixture.

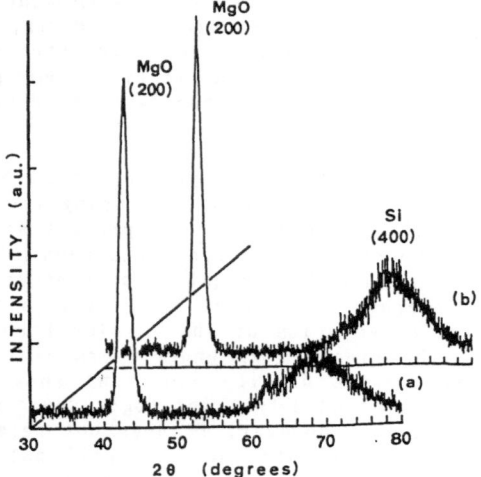

Fig.1 X-ray diffraction patterns of MgO films sputter-deposited with a substrate temperature of 150 ℃, and (a) in 9 mTorr of N_2 gas and (b) in 20 mTorr of Ar+(30%)N_2 mixture.

Fig.2 Surface morphology(SEM) of MgO films on silicon substrates
deposited with single crystal MgO target. The deposition
conditions are (a) in 9 mTorr of N_2 gas, and (b) in 20 mTorr
of Ar+(30%)N_2 mixture.

Figures 1(a) and 1(b) show the x-ray diffraction results for 120 nm
thick MgO films sputter-deposited with a single crystal MgO target. The
films were deposited in 9 mTorr of N_2 and in 20 mTorr of Ar+(30%)N_2,
respectively. One can clearly see the (200) prefered crystalline
orientation. It is notable that the single crystal MgO target without
containing the absorption gas produced these highly oriented films by rf
sputtering. On the subject of the effect of N_2 gas on the MgO crystal
orientation, the following mechanism can be proposed. MgO revealed
catalytic activity in the plasma synthesis of ammonia from H_2-N_2 mixed
gas.[8] Also, in N_2 plasma in which water vapor is mixed, NH* and N_2H_4 are
producted. The reaction occures on the surface of MgO. Excited nitrogen
molecules may be absorbed on MgO, and the MgO surface becomes a nitride
which is, at the same time, reduced by hydrogen atoms. In the results
reported by Sugiyama et al.,[9] the ESCA spectrum showed the presence of
magnesium nitride but the amount of the nitride was negligible. Although
the mechanism whereby the nitrogen ion responds to increase in the
intensity of line (200) of MgO is not fully understood, our results might
be explained by the generation of very small amounts of magnesium nitride,
or magnesium hydroxide.

Surface Morphology

Figures 2(a) and 2(b) show the surface morphology of MgO films. The
films were sputter-deposited in 9 mTorr of N_2(Fig.2(a)), and in 20 mTorr
of Ar+(30%)N_2 mixture(Fig.2(b)), respectively. Both the surfaces are
relatively smooth, and the surface shown in Fig.2(b) is somewhat smooth
compared with that in Fig.2(a). Figures 3(a) and 3(b) show high magnification views of fractured sections of the films as shown in Fig.2(a) and
2(b). The columnar growth of the microstructure is clearly visible. The
axis of the columns is perpendicular to the substrate surface. The width

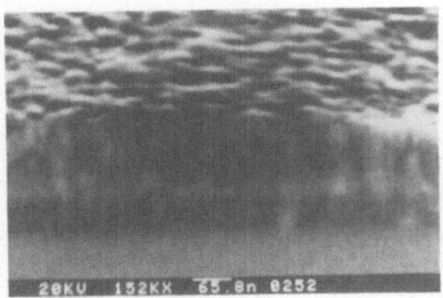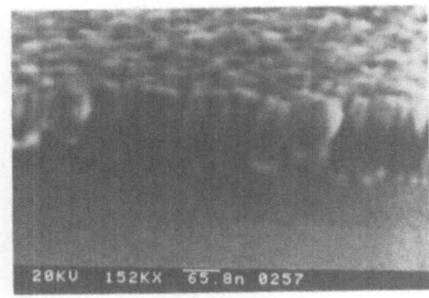

Fig.3 High magnification views of fractured MgO films as shown in
Fig.2(a) and 2(b).

379

Fig.4 High magnification view of
a fractured MgO film deposited
in 20 mTorr of Ar. The film
has no preferential orientation.
The grain size is 10-20 nm.

of the columns is 20-30 nm in Fig.3(a) and 40 to 50 nm in Fig.3(b), respectively. These results indicate that the microstructure of the MgO underlayer is controlled by the change of N_2 partial pressure.

In order to compare the surface morphology, we present the high magnification view of a fractured section of a MgO film with no preferential orientation, in Fig.4 which was deposited in 20 mTorr of Ar. Note no columnar growth in this figure. The coverage, in the present situation, when amorphous MgO shown in Fig.4 is used as tunnel barriers, is much better than that by oriented MgO underlayers. However, if we obtain MgO underlayers with the fine columns and small voids, the coverage of tunnel barrier(MgO) may be still better. The MgO films with fine columns and very small voids may improve the characteristics of Josephson tunnel junctions, especially the subgap leakage. This problem is still unsolved, but in future we should improve the surface and interface morphologies MgO underlayers and MgO tunnel barriers.

Film strain and Electrical Properties of NbN

The strain of sputter-deposited NbN films has been recently discussed by Kerber et al.[7] Here, we will also discuss the tensile strain of very

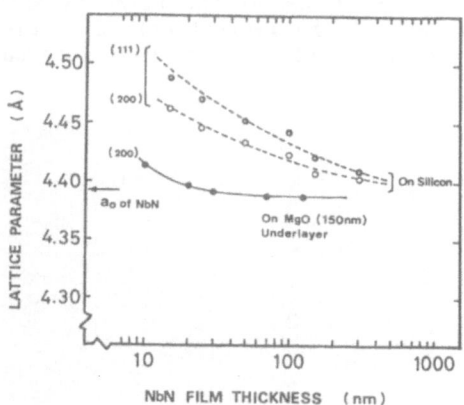

Fig.5 Variation of lattice constant of cubic NbN with the film
thickness.

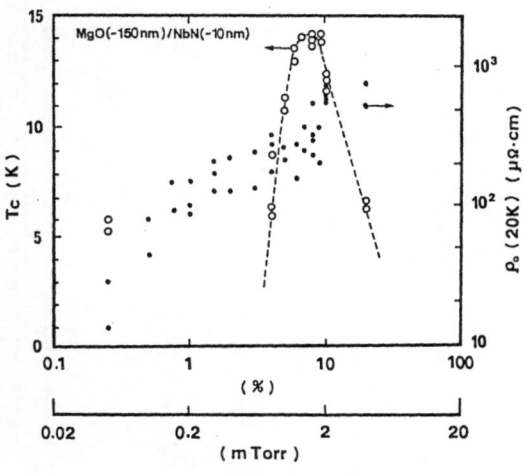

Fig.6 Critical temperature and
residual resistivity of
epitaxial NbN films as a
function of nitrogen
partial pressure.

thin NbN films with and without MgO underlayers. The standard value of
the lattice constant, a_0 is 0.421 nm for MgO and 0.439 nm for NbN. To test
the orientation effect of a_0 for NbN on the tensile strain, many MgO(150
nm)/NbN bilayers were prepared.

Figure 5 plots the a_0 of (100) oriented and randomly oriented NbN
films. The lattice constant of randomly oriented NbN films is much longer
than that for (100) oriented NbN in thinner films. For the 15 nm thick
NbN on silicon substrates, a_0 calculated from the (200) and (111)
diffraction peaks is 0.445 nm and 0.448 nm, respectively. Hence, grains
with (111) orientation in 15 nm thick NbN films without MgO underlayers
has about 1.4 % tensile strain while the NbN (100) orientation has about
2.0 % strain. This values are significantly large, and the strong strain
degrades the superconducting properties of NbN.[10]

On the other hand, for the 20 nm thick NbN on oriented MgO underlayers, a_0 is 0.440 nm which is calculated from the (200) diffraction
peak. This result shows that the NbN (200) orientation is almost
unstrained. Since the lattice constant of cubic NbN is related to the
critical temperature, the unstrained films may be useful for device
applications. However, in much thinner(<10 nm) NbN films the tensile
strain is still larger than 0.6 %. Hence we must further optimize the
film preparation conditions.

The data shown in Fig.6 is an indication of film quality(critical
temperature T_c, and film resistivity ρ_0) as a function of nitrogen
partial pressure, P_{N_2}. The film thickness of NbN is of ~10 nm, and MgO
underlayer is of about 150 nm. Variation in N_2 partial pressure strongly
affects T_c and ρ_N. Maximum T_c of about 14 K is obtained in the N_2
partial pressure of about 1.6 mTorr which is in good agreement with our
result for the thicker(~100 nm) NbN films. These results indicate that the
N_2 sticking coefficient is equal in both cases of thin and thick film
preparation. To obtain the order of magnitude of ξ (4.2 K), we measured
the critical current,I_c of oriented NbN(60 nm) films, prepared under the
same procedure, as a function of applied field. Figure 7 shows $I_c^{1/2}B^{1/4}$
vs B plots. The GL coherence length estimated from B_{c_2} was of 4-4.5 nm.

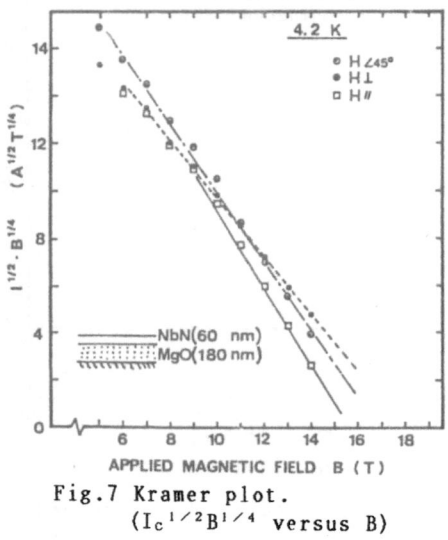

Fig.7 Kramer plot.
($I_c^{1/2}B^{1/4}$ versus B)

JUNCTION FABRICATIONS

Various types of Josephson junctions were fabricated with NbN/MgO multilayers deposited on silicon substrates. Junction fabrication is accomplished using a combination of photoresist patterning, reactive ion etching, and lift-off. The detailed fabrication process has been reported elesewhere.[11-14] Tunnel junctions were fabricated with MgO(underlayer)/ NbN/MgO(barrier)/NbN tetralayers. MgO underlayers were sputter-deposited with hot-pressed MgO target. The epitaxial junctions had high sum-gap voltage of 4 to 5 mV. However, the subgap leakage is somewhat large. The subgap leakage may be improved by using the MgO and NbN films with fine columns and small voids, and the improvement is now in progress.

The all-NbN nanobridges were successfully fabricated using the very thin NbN films with high T_c. The schematic structure of a nanobridge dc SQUIDs is ahown in Fig.8(a), and typical flux-modulation characteristics is shown in Fig.8(b). The maximum value of transfer function, $\partial V/\partial \phi$, and

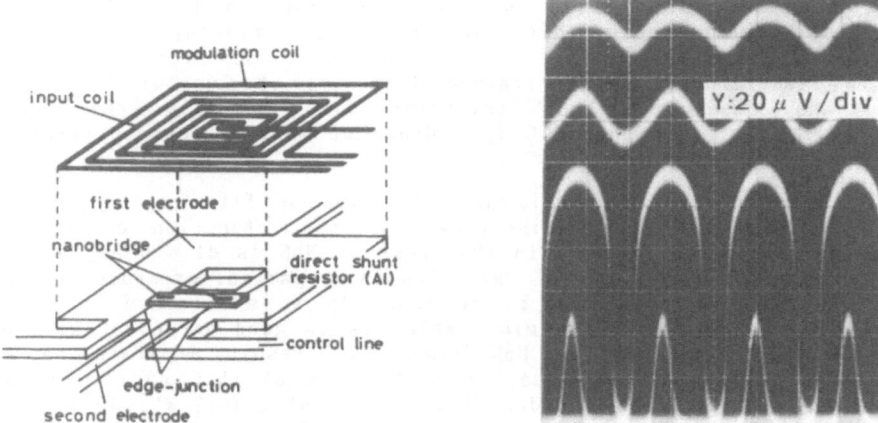

Fig.8 (a) Schematic structure of NbN nanobridge dc SQUID, and
(b) typical flux modulation(V-ϕ) characteristics.

the minimum energy sensitivity of a SQUID with $L_s=100$ pH is ~ 1 mV/ϕ_0 and about 20 h, respectively. These nanobridge also had good Josephson response on millimeter to submillimeter wave irradiations. Also, we fabricated the vertical-type NbN nanobridges by means of field evaporation technique. Using this technique, the nanobridge dc SQUIDs with low capacitance and desired L_sI_0/ϕ_0 values were in-situ produced at the operating temperature.

CONCLUSIONS

The thin film preparation and device applications of epitaxial MgO/NbN multilayers have been investigated. We found that (100) oriented films of MgO can be deposited with single crystal MgO target by the presence of N_2 in the sputtering gas. The surface morphology of MgO underlayers can be changed by the sputter gas conditions. Low film strain of NbN was observed in highly oriented MgO($\sim$150 nm)/NbN($\sim$10 nm) bilayers, which may be suitable for junction fabrication. The preparation conditions of the MgO/NbN multilayers are still unoptimized, but all-NbN nanobridges with good Josephson response on magnetic fields and millimeter to submillimeter wave irradiations were successfully fabricated. Also, the subgap leakage of tunnel junctions is still unimproved, and the MgO underlayers with fine columns and very small voids may improve the leakage of tunnel junctions.

ACKNOWLEDGMENTS

We wish to thank S.Nagaoka of Toso. Corp. for SEM analysis, and K. Yamaguchi and T. Nakata of Nagaoka University for their technical assistance.

REFERENCES

1. A. I. Braginski, J. R. Gavaler, M. A. Janocko, and J. Talvacchio, SQUID 85, H. D. Hahlbohm and H. Lubbig, eds., Walter de Glruyter, Berlin (1985) p.591.
2. T. Yamashita, K. Hamasaki, and T. Komata, Adv. in Cryog. Eng. Matrs., 32: 617 (1986).
3. S. Thakoor, H. G. Leduc, J. A. Stern, A. P. Thakoor, and S. K. Khanna, J. Vac. Sci. Technol., A5: 1721 (1987).
4. K. Ueda, K. Hamasaki, I. Yamada, T. Yamashita, and T. Komata,T., Sinku, 28: 796 (1985). (in Japanese)
5. J. Talvacchio, J. R. Gavaler, and A. I. Braginski, Proc. TMS-AIME: Metalic Multilayers and Epitaxy, H. Hong and S. A. Wolf, eds. (1987).
6. J. Talvacchio, and A. I. Braginski, IEEE Trans. Magn., MAG-23: 859 (1987).
7. G. I. Kerber, J. E. Cooper, R. S. Morris, J. W. Spargo, and A. G. Toth, IEEE Trans. Magn., MAG-25: 1294 (1989).
8. L. Hochard, A. Plain, and A. Ricard, Proc. 4th Int. Symp. Plasma Chem., S. Veprek, ed., University Zurich: 545 (1979).
9. K. Sugiyama, K. Akazawa, M. Oshima, H. Miura, T. Matsuda, and O. Nomura, Plasma Chem. & Plasma Processing 6: 179 (1986).
10. Y. M. Shy, L. E. Toth, and R. Somasundaram, J. Appl. Phys., 44: 5539 (1973)
11. K. Hamasaki, T. Yakihara, Z. Wang, T. Yamashita, and Y. Okabe, IEEE Trans. Magn., MAG-23: 1489 (1987).
12. K. Hamasaki, Y. Misaki, T. Yamashita, S. Nagaoka, and T. Komata, Proc. ICEC-11, p. 508, G. Klipping and I. Klipping, eds., (1986).
13. K. Hamasaki, S. Nagaoka, T. Komata, and T. Yamashita, Superconducting Materials, p. 10, J. Bevk and A. I. Braginski, eds., Material Res. Soc., Pittsburg (1986).
14. A. Irie, K. Hamasaki, and T. Yamashita, Extended Abstracts of the 1989 International Superconductivity Electronics Conference (1989).

STATIC AND TIME DEPENDENT MAGNETIZATION STUDIES OF

$Y_1Ba_2Cu_3O_z$ THIN FILMS*

S. T. Sekula, R. Feenstra, J. R. Thompson, Y. C. Kim,
D. K. Christen, H. R. Kerchner, H. A. Deeds, and L. A. Boatner

Solid State Division, Oak Ridge National Laboratory
P.O. Box 2008
Oak Ridge, Tennessee 37831-6061

ABSTRACT

Thin films of $Y_1Ba_2Cu_3O_z$ (YBCO) were deposited by electron-beam coevaporation of Y, Cu, and BaF_2 onto single-crystal substrates of $SrTiO_3$ and $KTaO_3$. Various oxygen annealing protocols produced different epitaxial alignments of the films, which were then studied using vibrating sample (VSM) and SQUID-based magnetometry. The magnetization behavior and the critical current density $J_c(H,T)$ deduced from the magnetic hysteresis is observed to be quite sensitive to the substrate orientation as well as to annealing procedures that result in variations of the film morphology as evidenced by x-ray diffraction and and SEM techniques. Flux creep effects are also observed over short time intervals in VSM studies and examined quantitatively over longer periods with a SQUID magnetometer.

INTRODUCTION

A large body of work has established that the superconducting properties of thin $Y_1Ba_2Cu_3O_z$ films prepared by electron-beam evaporation or sputtering techniques followed by annealing procedures are very sensitive to the substrate on which they are deposited. In some cases, a severe degradation in the superconducting properties is observed due to formation of reactive layers at the interface between the YBCO film and substrate (e.g., Al_2O_3, ZrO_2, and Si).[1,2] Such interfacial reaction effects are particularly severe in those formation methods requiring thermal anneals in the temperature range near 850°C. Interdiffusion across film-substrate interfaces presumably also occurs, but to a lesser degree, with the use of single-crystal $SrTiO_3$ and $KTaO_3$ substrates,[3] but in these cases the transition temperatures T_c of the YBCO films remain high at ~90 K and with a transport critical current J_{ct} that depends critically on the epitaxial growth that, in turn strongly depends on substrate orientation.[4]

---

*Research sponsored by the Department of Materials Sciences, U.S. Department of Energy under contract DE-AC05-84OR21400 with Martin Marietta Energy Systems, Inc.

Advances in Cryogenic Engineering (Materials), Vol. 36
Edited by R. P. Reed and F. R. Fickett
Plenum Press, New York, 1990

385

Accordingly, in the present work magnetic studies of high T_c films formed by coevaporation onto single-crystal $SrTiO_3$ and $KTaO_3$ substrates were undertaken in order to evaluate these materials for use in electronic devices and other possible applications.

EXPERIMENTAL TECHNIQUES AND APPARATUS

Thin films of YBCO ~350 nm thick were formed by e-beam coevaporation of Y, Cu, and BaF_2 onto single-crystal substrates that previously had been ultrasonically machined into circular discs ~6 mm in diameter. The $Y_1Ba_2Cu_3O_z$ phase was formed by a high-temperature reaction in wet oxygen. The substrate materials in this case were single-crystal $SrTiO_3$ with (001) and (110) surfaces and $KTaO_3$ with a (001) surface. Preparation of the samples has been reported elsewhere.[3,4] It is worth noting that the deposition of films on the (001) and (110) $SrTiO_3$ substrates was made in the same run and was followed by identical annealing procedures. The morphology of these films can be summarized as follows: (1) on a (001) $SrTiO_3$ substrate, the film consisted predominantly of domains with the c-axis perpendicular to the substrate surface and with the a- and b-axes approximately aligned with the in-plane [100] axes of the substrate[5] (sample A), (2) by suitable heat treatment, the film on (001) $KTaO_3$ consisted of comparable fractions of grains with either the a-axis or c-axis perpendicular to the surface (sample B), and (3) on a $SrTiO_3$ (110) substrate, a mixture of finely divided domains (~1 × 1 μm^2) was observed with [110]- and [103]-type orientations relative to the surface[4,6] (sample C).

Magnetic properties of these samples were investigated in the temperature range from 1.5 K to 100 K in a vibrating sample magnetometer (VSM) and a commercial SQUID magnetometer. The static magnetic field capability was 8.0 T and 5.0 T respectively for these studies. The circular disc samples were oriented with the plane of the disc normal to the applied field and to first order can be considered as oblate spheroids with demagnetization factor D ~ 1. Isothermal magnetization data were obtained using both types of magnetometry. With the VSM, the applied field was ramped at ~100 Oe/sec and periodically stopped to observe the early stages of flux relaxation which can be quite large. The same procedure was followed while ramping the magnetic field back to zero. Similar magnetization investigations were made on these samples with the SQUID magnetometer. While the earliest stages of flux creep could not be probed with this apparatus, it did offer the ability to study flux creep rates over extended periods of time. In addition to high-field magnetization measurements, constant low applied magnetic fields (~10 Oe) were used to obtain zero-field cooled data while varying the temperature in order to determine T_c. The 10% onset of the T_c values was found to be 89.1 K, 88.4 K, and 71.2 K, respectively, for samples A, B, and C. The low value of T_c for sample C is not understood but it may be related to the granularity of this film. At 10 Oe no Meissner flux exclusion was observed in field-cooled studies for any of the samples because of significant volume and boundary flux pinning.

MAGNETIZATION RESULTS

Figure 1 shows the isothermal magnetization curves at 4.2 K for the YBCO film A with the c-axis normal to the (001) surface and parallel to the

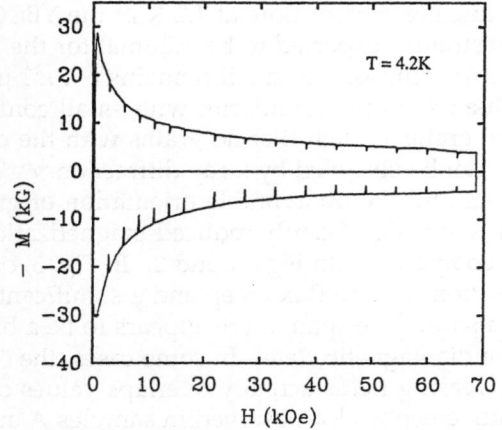

Fig. 1. Magnetization curves of a YBCO film
on a (001) SrTiO$_3$ substrate (sample A)

applied field. Values of the magnetization are quite large and indicate that
large shielding currents are being generated. This is to be expected since
optimal superconducting current flow is known to occur in the Cu-O planes
of YBCO parallel to the a-b axes. By periodically stopping the increasing
(decreasing) field ramp, one observes a relaxation to smaller (greater)
diamagnetic values. This relaxation is a manifestation of the flux creep
phenomenon and the end of the tick marks in the magnetization curves
represents "quasi-static" values of the magnetization, that are obtained after
20 seconds in the constant field. These points are taken to define the
envelope of the static irreversible magnetization curves. An intermediate
case is that of a film grown on a (001) KTaO$_3$ surface (sample B) which consists
of a mixture of a-axis and c-axis grains oriented normal to the surface.
Superconducting current can then flow normal to the a-b planes in the a $\perp$
grains, through grain boundaries between a$\perp$ and c$\perp$ grains, and through any
contiguous regions of c$\perp$ domains. The resultant superconducting current
flow is considerably reduced as reflected in the smaller values of the
magnetization hysteresis shown in Fig. 2 relative to those given in Fig. 1.

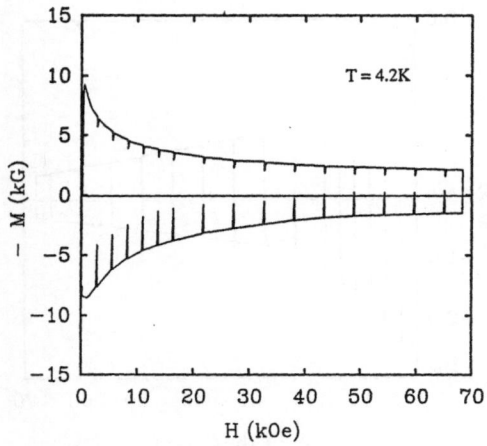

Fig. 2. Magnetization curves of a YBCO film
on a (001) KTaO$_3$ substrate (sample B)

The superconducting current flow at 4.2 K in the YBCO film prepared on a (110) $SrTiO_3$ substrate is expected to be minimal for the following reasons: (1) the film is composed of small domains (~1 × 1 μm^2 in area) that have numerous high-angle grain boundaries with small contact area, and (2) the orientation of the grains is such that no grains with the c-axis normal to the substrate surface can be observed by x-ray diffraction.[4,6] Thus, weak intergranular coupling and the unfavorable orientation of grains relative to the applied field leads to a significantly reduced magnetization behavior as is shown in Fig. 3 and compared with Figs. 1 and 2. In Fig. 3, one notes the relatively large relaxations due to flux creep and a significant asymmetry in the sense that, over a short time span, there appears to be a bias for the material to prefer the diamagnetic state. In some cases, the "quasi-static" magnetization in decreasing fields actually overlaps values observed in increasing fields. Flux creep is also observed in samples A and B, although proportionately less, as is an asymmetry in magnetic relaxation that requires further investigation.

TEMPORAL EFFECTS

For conventional applications, the motion of magnetic flux through a superconductor in an applied field is very undesirable due to the associated generation of heat. Such flux motion can be detected with great sensitivity by observing changes in the superconducting state magnetic moment due to the decay of "persistent" currents in a sample. The decay of such currents generally leads to a reduction in the absolute value of the magnetization with time and this is exactly the effect observed in Figs.1–3. These effects, which tend to be rather pronounced in high-temperature superconductors, are referred to as "giant flux creep"[7] and have been interpreted as arising from thermally activated motion of flux bundles over pinning barriers of height U_o. In the Kim-Anderson[8] model, the magnetization $M(t)$ has an asymptotic dependence on the logarithm of time t, with $M(t) \sim \ln(t)$. An example of this is shown in Fig. 4 which illustrates that a $\ln(t)$ dependence accurately

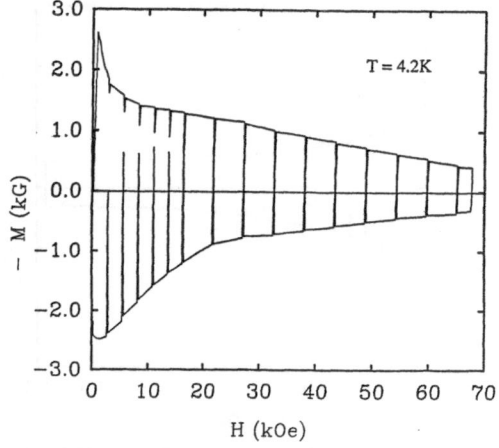

Fig. 3. Magnetization curves of a YBCO film on a (110) $SrTiO_3$ substrate (sample C)

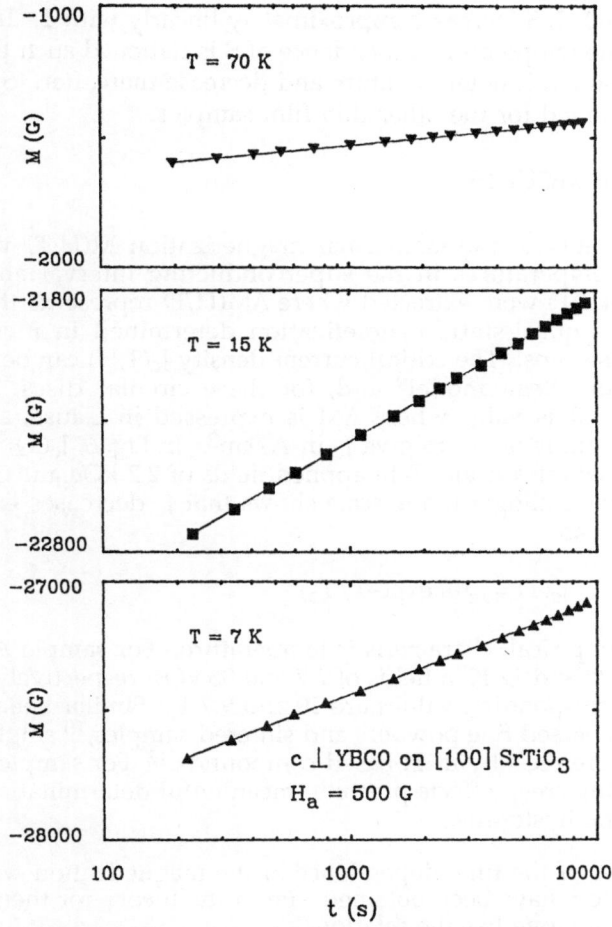

Fig. 4. The magnetization of a YBCO thin film ~350 nm in thickness on (001) SrTiO₃ vs the logarithm of time, following application of an external field increased to 500 G (sample A).

describes the experimental data on these films. To within experimental accuracy, all of the flux creep data reported here followed this logarithmic dependence on time after an initial transient.

The creep rate $S = dM/d[\ln(t)]$ was obtained from the slope of plots such as those in Fig. 4. In this figure, the vertical scale factors are identical so that the slopes can be compared directly; it is evident that the creep rate first increases with increasing temperature T and subsequently decreases. This behavior is confirmed in Fig. 5, a plot of the absolute value of the creep rate S versus T for sample A in 500 G after increasing field (squares) and decreasing field history (triangles), respectively. The solid line is the average of the two sets of data. It is evident that the creep rate exhibits a peak near 20 K, and that the measured rate is higher for decreasing field history, as is seen qualitatively in the VSM magnetization curves (Figs. 1–3). At lower temperature, Fig. 5

389

shows that, in 500 G, S increases approximately linearly with T. In higher applied fields, the temperature dependence of S is flattened such that the data are nearly constant at low temperature and decrease thereafter. Similar results were obtained for the other thin film samples.

DISCUSSION OF RESULTS

Measurements of the isothermal magnetization M(H,T) were carried out at various temperatures in the superconducting interval and values of the quantity $\Delta M(H,T)$ were extracted where $\Delta M(H,T)$ represents the hysteretic difference in the quasi-static magnetization determined in increasing and decreasing field sweeps. The critical current density $J_c(T,H)$ can be deduced by application of the Bean model[9] and, for these circular discs, the relation $J_c(T,H) = 15\,\Delta M/R$ is valid where ΔM is expressed in Gauss, and R is the sample radius in units of cm to give J_c in A/cm^2. In Fig. 6, $J_c(T)$ is plotted vs temperature for samples A and B in applied fields of 2.7 kOe and 35 kOe. The linear plot on the semilogarithmic scale shows that J_c decreases exponentially with temperature as

$$J_c(T) = J_c(0)\exp(-T/T_0) \tag{1}$$

where T_0 is an empirical, characteristic temperature. For sample A, the values of T_0 are 28 and 12 K in fields of 2.7 and 35 kOe, respectively; for sample B, the corresponding values are 20 and 9.7 K. Similar behavior has been noted in dispersed fine powders and sintered samples,[10] single crystals[11] of YBCO, and more recently in Tl-based compounds.[12] For sample C, the large observed flux creep effects preclude meaningful determinations of J_c from the magnetic hysteresis.

From studies of the time dependence of the magnetization, values for the flux creep rate S have been obtained. From the theory for thermally activated flux creep, one has the relation

$$S = dM/d[\ln(t)] = (RJ_{co}/3c)(k_BT/U_o) \tag{2}$$

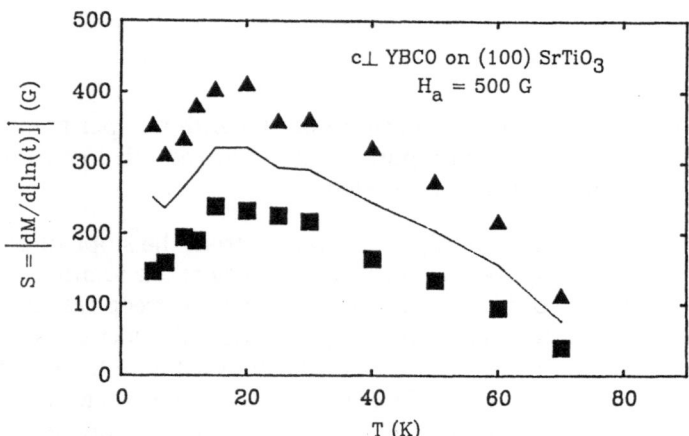

Fig. 5. The magnitude of the flux creep rate S vs temperature T for sample A in 500 G, for increasing (■) and decreasing (▲) field history, and their average (solid line).

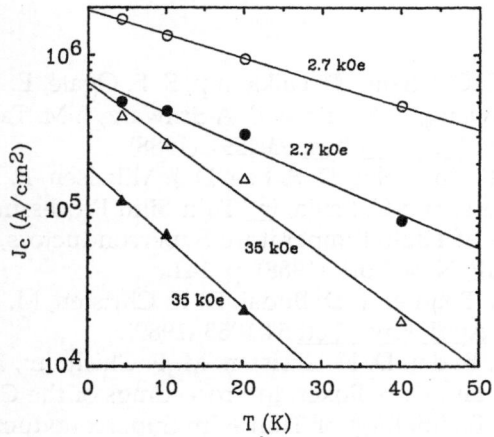

Fig. 6. Critical current density of $J_c(T)$ of sample A (open symbols) and sample B (solid symbols) vs temperature T in applied fields of 2.7 kOe and 35 kOe.

where the symbols follow standard notation, J_{co} explicitly denotes the critical current density at temperature T in the absence of flux creep, and all quantities are expressed in cgs units. Using experimentally obtained estimates for J_{co}, we find the initial linear increase in the quantity (S/J_{co}) with temperature that is expected from Eq. (2). From this analysis for sample A, we obtain the values for the effective pinning well depth U_0 of 0.16 eV (in 500 G applied field) and 0.12 eV (in 3 kG field). It must be noted, however, that the experimental results in Fig. 5 do not extrapolate to zero as T→0. This "nonthermal" component in flux motion has been observed previously in high-temperature superconductors, i.e., in Tl-based films,[13] but its origin is not understood at this time. Preliminary results from flux creep studies of the other films B and C appear qualitatively similar. The details of these investigations will be presented in a subsequent work.

CONCLUSIONS

Thin films of YBCO were formed on two different substrate materials and their magnetic properties were studied. For epitaxial films with the c-axis normal to the substrate, the results indicated a large critical current in the basal plane that decreased exponentially with increasing temperature. From measurement of flux creep in low applied fields, the average flux pinning well depth was determined. It was found that the flux creep rates depended on the magnetic history of the samples well. Lower J_c values, but similar overall results, were found in films with mixed a- and c-orientations. For triaxial films deposited on (110) $SrTiO_3$ surfaces, flux creep effects rapidly diminished the smaller magnetization, relative to the former films, that was observed under dynamic, field swept conditions. In conclusion, it is clear that the long-term behavior of supercurrents that is probed by these magnetic measurements is greatly dependent on the detailed morphology of the YBCO thin films.

REFERENCES

1. T. Venkatesan, C. C. Chang, D. Dijkkamp, S. B. Ogale, E. W. Chase, L. A. Farrow, D. M. Hwang, P. F. Miceli, S. A Schwarz, J. M. Tarascon, X. D. Wu, and A. Inam, J. Appl. Phys. 63:4591 (1988).
2. J. J. Cuomo, M. F. Chisholm, D. S. Lee, D. J. Mikalsen, P. B. Madakson, R. A. Roy, E. Giess, and G. Scilla, in "Thin Film Processing and Characterization of High-Temperature Superconductors," AIP Conf. Proc. No. 165, AIP, New York (1988), p. 141.
3. R. Feenstra, L. A. Boatner, J. D. Budai, D. K. Christen, M. D. Galloway, and D. B. Poker, Appl. Phys. Lett 54:1063 (1989).
4. R. Feenstra, J. D. Budai, D. K. Christen, M. F. Chisholm, L. A. Boatner, M. D. Galloway, and D. B. Poker, in Proceedings of the Conference on the Science and Technology of Thin Film Superconoductors, Colorado Springs, Colorado, 1988, ed. by R. McConnell and S. Wolf, Plenum, New York (1989)., in press.
5. J. D. Budai, R. Feenstra, and L. A. Boatner, Phys. Rev. B 39:12355 (1989).
6. T. Murakami, Y. Enomoto, and M. Suzuki, Physica C 153-155: 1690 (1988).
7. Y. Yeshurun and A. P. Malozemoff, Phys. Rev. Lett. 60:2202 (1988).
8. M. R. Beasley, R. Labusch, and W. W. Webb, Phys. Rev. 181:689 (1969).
9. W. A. Fietz and W. W. Webb, Phys. Rev. B 178:657 (1969).
10. S. T. Sekula, J. Brynestad, D. K. Christen, J. R. Thompson, and Y. C. Kim, IEEE Trans. Mag. 25:2266 (1989).
11. S. Senoussi, M. Oussena, G. Collin, and I. A. Campbell, Phys. Rev. B 37:9792 (1988).
12. J. R. Thompson, J. Brynestad, D. M. Kroeger, Y. C. Kim, S. T. Sekula, D. K. Christen, and E. D. Specht, Phys. Rev. B 39:6652 (1989).
13. E. L. Venturini, J. F. Kwak, D. S. Ginley, B. Morosin, and R. J. Baughman, in Proceedings of the Conference on the Science and Technology of Thin Film Superconductors, Colorado Springs, Colorado, 1988, ed. by R. McConnell and S. Wolf, Plenum, New York (1989), in press.

MAGNETIC STUDIES OF YBaCuO CRYSTAL PREPARED BY QUENCH AND MELT GROWTH

PROCESS

Masato Murakami, Satoshi Goton, Naoki Koshizuka, and Shoji Tanaka

Superconductivity Research Laboratory
International Superconductivity Technology Center
1-10-13, Shinonome, Koto-ku, Tokyo, 135 Japan

INTRODUCTION

Since the discovery of LaBaCuO with Tc exceeding 30K by Bednorz and Muller[1], tremendous effort has been conducted to raise critical temperature (Tc), leading to the discovery of LaSrCuO[2], YBaCuO[3], BiSrCaCuO[4], and TlBaCaCuO[5].

On the other hand, for practical applications, large critical current densities (Jc) of 10^4 - 10^6 A/cm^2 are required often in the presence of significant magnetic field. However, Jc values of bulk sintered specimens have remained low, typically three orders of magnitude lower than the required level. This is attributed to the presence of weak-links[6]. Extremely short coherence length and low carrier density, which are characteristic of high Tc oxides[7], seem to be the source of the weak-link behavior at boundaries[7]. But it has become clear that low Jc is not intrinsic to high Tc oxides by the fact that very high Jc values are obtained in single crystals[8] and thin films[9].

However, it should be noted that some kind of defects must exist in type II superconductors in order to achieve resistanceless supercurrents in the presence of magnetic field[10]. If single crystals were defect-free, their Jc would be zero in the magnetic field higher than H$_{c1}$. In the YBaCuO system, single crystals contain dense twin structures, therefore, they are not single crystals in a real sense. Since twin planes can work as pinning sites, high Jc values in single crystals can be partly explained by the presence of twin structures[11]. It has been also proposed that anisotropic crystal structure may contribute to intrinsic pinning in high Tc oxides[12].

On the other hand, in spite of large Jc, which implies high pinning potential, flux creep is reported to be fairly large in single crystals[13]. At non-zero temperature the motion of trapped magnetic flux is possible with the help of thermal energy. According to Anderson's thermally activated model[14], flux creep rate (v) can be given by

$$v = v_0 \exp(-U/kT)$$

where v_0 is the frequency with which a flux bundle attempts to escape from the pinning site, U is the pinning potential and kT is thermal energy. Since kT is fairly large at 77K, large flux creep is expected. It is also known that U scales with coherence length, therefore U must be small

Advances in Cryogenic Engineering (Materials), Vol. 36
Edited by R. P. Reed and F. R. Fickett
Plenum Press, New York, 1990

in high Tc oxides. A combination of high kT and low U results in so called "giant" flux creep. In fact, fairly large flux creep was observed in single crystals[13]. This has lead to a very pessimistic conclusion that their application is not feasible at 77K.

However, as mentioned above the presence of some defects which can work as pinning centers is required in order to obtain large Jc. In that sense, the pinning energy in single crystals, if they are really good in quality, should be the lowest for high Tc oxides.

Recently, large YBaCuO crystals that exhibit high Jc have been fabricated by quench and melt growth (QMG) process[15]. Their Jc values exceeded 2×10^4 A/cm^2 at 77K and 1T. Bean's critical state[16] was established in such samples[17]. It was found through microstructural analysis that fine Y_2BaCuO_5 particles are trapped in $YBa_2Cu_3O_7$ matrix in the QMG processed samples and they are expected to contribute to pinning.

In this paper, we report magnetic properties of the QMG processed YBaCuO crystals and show that their magnetization behavior can be understood in terms of the theory which has been developed for strongly pinned type II superconductors. We also report some flux creep data and show that flux creep rate is much smaller than that of single crystals.

EXPERIMENTAL

YBaCuO crystals were prepared by the quench and melt growth (QMG) process. Calcined $YBa_2Cu_3O_x$ powders were heated to 1400°C and quenched by using copper hammers. Then quenched plates were heated to 1100°C and held for 20 min, then cooled to 1000°C at the rate of 100°C/h followed by slow cooling at the rate of 5°C/h. Details of the process is described in ref. 15.

Rectangular samples were cut from the QMG processed plates. As reported previously[14], second phase layers and cracks were occasionally found along the c-plane. These defects affect the result of magnetization and complicate the analyses, and therefore in the present measurements, magnetic field was applied parallel to the c-axis. In this situation, effects of these defects can be neglected.

Magnetization measurements were made using a vibrating sample magnetometer, an AC susceptometer, and a SQUID magnetometer.

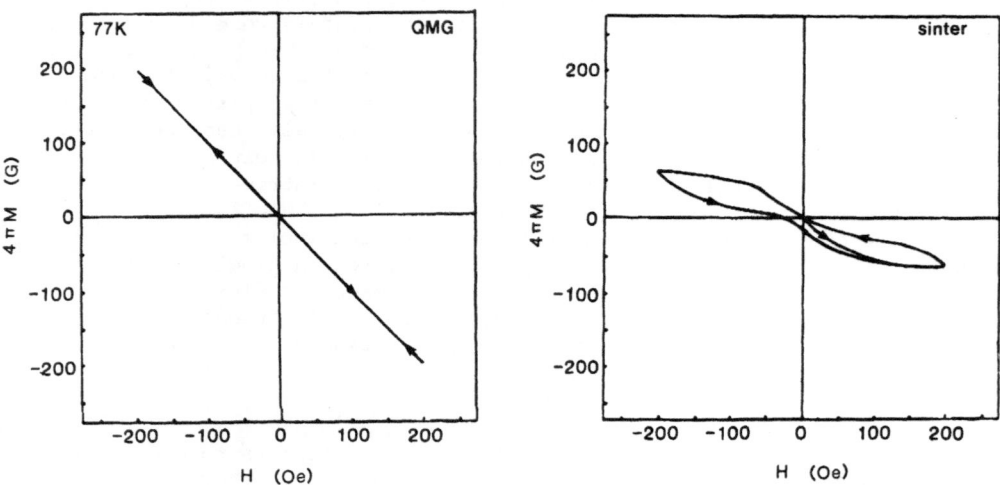

Fig. 1.   M-H relationships for QMG processed and sintered YBaCuO samples at 77K.

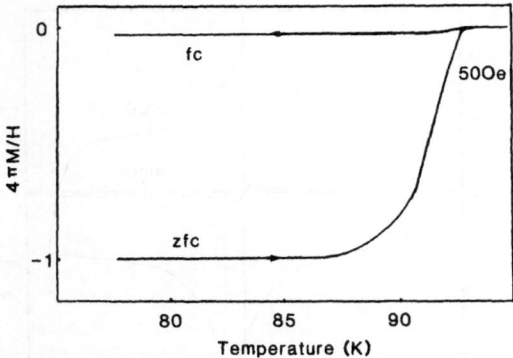

Fig. 2.   Temperature dependence of magnetization for the   QMG   processed
          YBaCuO under a magnetic field of 50Oe.

RESULTS AND DISCUSSION

Magnetization at low fields

    Magnetization (M) - field (H) properties at low-fields.   Figure 1
shows M-H relationships for a QMG processed YBaCuO crystal at   77K.   The
data for a sintered YBaCuO is also presented for comparison.   As shown   in
the figure, perfect diamagnetism is preserved during the sweep of external
magnetic field from -200 to 200 Oe in the QMG processed YBaCuO,   while   a
large hysteresis is observed in a sintered sample.   This magnetization
behavior of the QMG processed sample is the one   expected   for ordinary
superconductors   and   low-field   magnetization behavior   of   sintered
materials can be understood by the presence of weakly-coupled region
between grains. This result guarantees the quality of the QMG processed
sample.

    Temperature dependence of magnetization at low-fields. Figures 2   and
3 show temperature dependence of magnetization for the QMG processed   YBCO
in field-cooled (fc) and zero-field-cooled (zfc) processes under   magnetic
fields of 50 Oe and 200 Oe, respectively.   It is notable that   diamagnetic
signal is very small in the fc process, while perfect diamagnetism (4π M/H
= -1) is found up to certain temperatures in the zfc process.

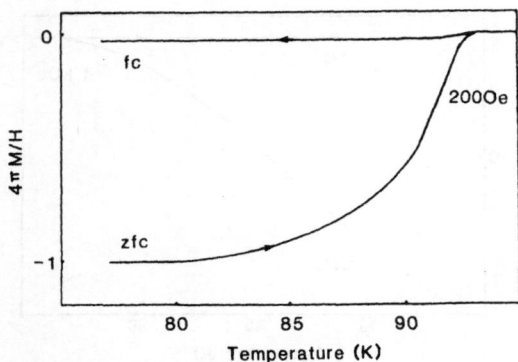

Fig. 3.   Temperature dependence of magnetization for the   QMG   processed
          YBaCuO under a magnetic field of 200Oe.

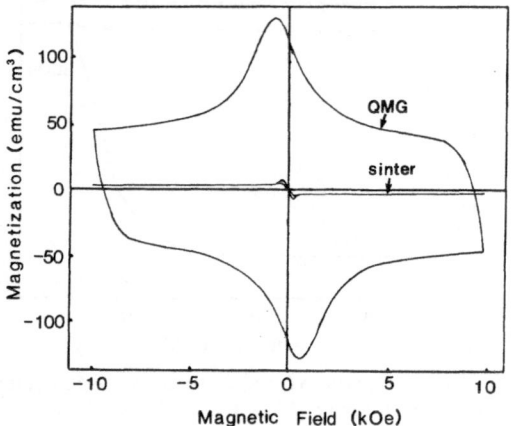

Fig. 4.   M-H loops for QMG processed and sintered YBaCuO samples at 77K.

It is believed that large Meissner effect or field exclusion effect
can be observed in the fc process if the sample quality is very good.   It
is also customary to obtain the superconducting volume fraction from the
fraction of M/H.   However, this M/H does not yield the superconducting
volume fraction in type II superconductors. When type II superconductors
are cooled in the presence of magnetic field, magnetic field will be
excluded from the superconducting region below Tc.   But, as is generally
recognized, magnetic flux can be pinned at crystal defects or normal
precipitates.   From technological view point, the presence of pinning
centers is critical, because resistanceless large current can be passed
only when the pinning force due to the pinning centers is larger than the
Lorentz force, J x B, acting on the flux vortices. Therefore, any Jc in
magnetic field indicates the presence of pinning centers.   If
superconductors contain any pinning centers, some magnetic flux will   be
pinned.   The higher the Jc is, the more magnetic flux will be pinned.
Small diamagnetic signal or small flux expulsion in the QMG processed
sample in the fc process is thus the indication of large pinning forces or
large Jc.
     While in the zfc process, type II superconductors will   resist the
penetration of magnetic field up to H_{c1} and perfect diamagnetism is
observed.   In external magnetic fields of 50 Oe and 200 Oe, perfect
diamagnetism is observed up to 85K and 80K, respectively. Although   more

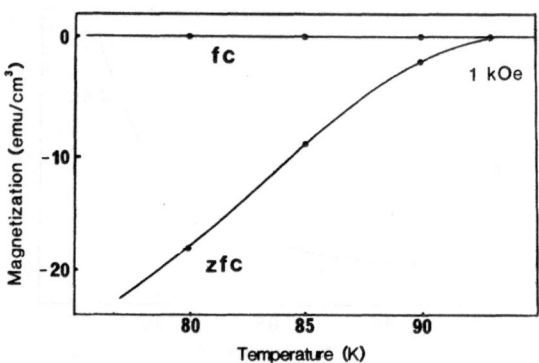

Fig. 5.   Temperature dependence of magnetization for the QMG processed
          YBaCuO under a magnetic field of 1kOe.

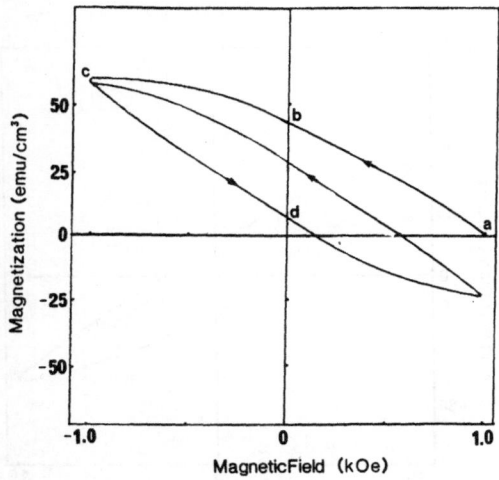

Fig. 6.   Magnetization of the QMG processed YBaCuO after field-cooled
under a magnetic field of 1kOe.

careful experiments are required for obtaining absolute values, this
result shows that H_{c1} is 50 Oe at 85K and 200 Oe at 80K. As the
temperature is raised higher than these temperatures, the external
magnetic field exceeds H_{c1}, and then can enter the sample.

Magnetization at high fields

M-H loops at high-fields.   Figure 4 shows M-H loops for QMG
processed and sintered YBaCuO samples at 77K. Magnetization hysteresis is
much larger in the QMG processed sample. As reported previously[17],
magnetization difference between increasing and decreasing field
branches ($M^+ - M^-$, ΔM) is proportional to sample thickness in the QMG
processed sample, while ΔM is almost independent of d in sintered samples.
This indicates that weak-links present in sintered samples are almost
absent in the QMG processed YBaCuO. It also indicates that Bean's
critical state is established in the QMG processed sample.

Temperature dependence of magnetization at high-fields.   Figure 5
shows temperature dependence of magnetization in the fc and zfc processes
under a magnetic field of 1kOe. In the fc process, no diamagnetic signal
is observed. This indicates that almost all magnetic flux is trapped in
the superconductor. In order to confirm this, we removed external
magnetic field at 77K after field-cooled. As shown in Fig. 6
magnetization appeared and increased as we decreased the external field.
This result demonstrates that magnetic field was trapped in the
superconductor below Tc in the fc process. For comparison, we also measured ordinary M-H loop at 77K after zero-field-cooled. Magnetization
hysteresis loops in the fc and zfc processes were the same in their shape
but differs such that the origin shifted upwards in the fc case compared
to the zfc case. These magnetization is understood based on Bean critical
state model. Magnetization in the fc and zfc can be schematically
illustrated as shown in Fig. 7. This figure is constructed referring to
the values which were obtained from the previous magnetization data of the
QMG processed sample[17]. In an initial state, magnetization is zero in
both cases. In the fc case, magnetic field was penetrating to the entire
sample. As the external field was decreased, magnetic flux in the
superficial layers was excluded. This exclusion continued down to 0 Oe.
Further decrease of magnetic field cause the penetration of magnetic field
of the opposite sign to the superficial layers down to -1 kOe. Then the

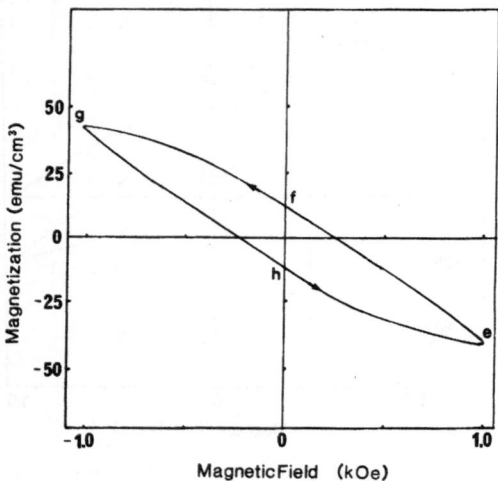

Fig. 7.   Magnetization of the QMG processed YBaCuO after zero-field-
          cooled down to 77K.

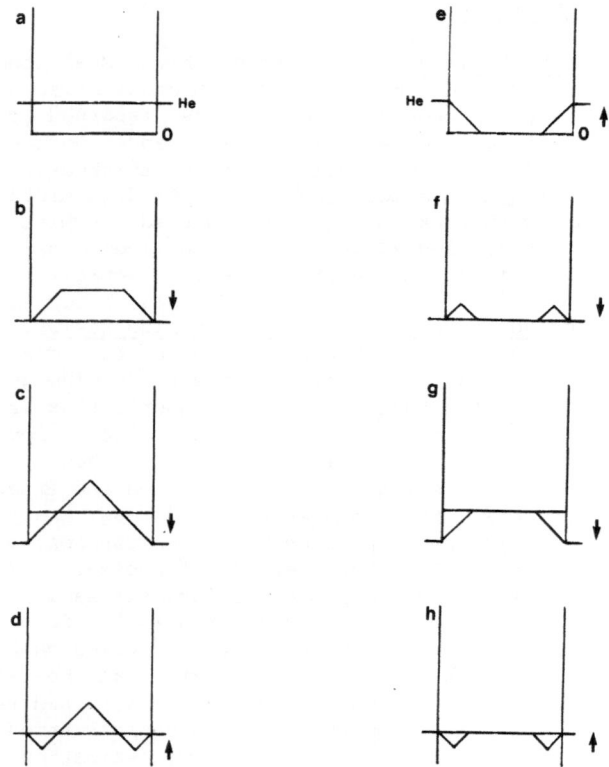

Fig. 8.   Magnetization processes corresponding to Figs. 6(left) and
          7(right).   The figure was constructed based on Bean's critical
          state model.

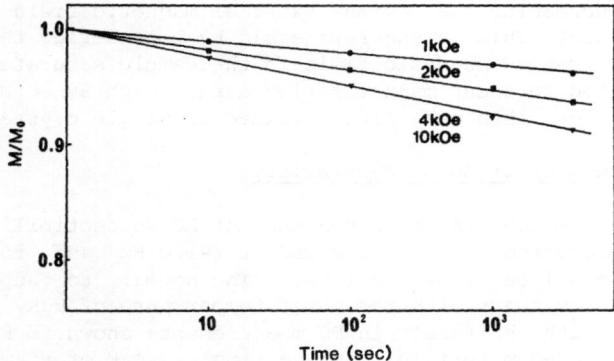

Fig. 9. Decay of remanent magnetization for the QMG processed YBaCuO after applied magnetic field up to 10kOe and removed.

sign of external field was reversed. In the following process, magnetic field enters the sample creating zigzag like distribution of magnetic field.

Magnetization in the zfc case is also consistently understood by the process based on Bean's critical state model, which is schematically shown in Fig. 8. Corresponding magnetization process is indicated by the same alphabet in the figure.

It is also notable that reversible region in the fc and zfc processes is not observed. Since the reversible region is believed to be evidence for large flux creep[13], this result indicates that flux creep is small in the QMG processed YBaCuO sample.

Magnetic relaxation

Figure 9 shows decay of remanent magnetization for the QMG processed sample. The sample was cooled down to 77K, and then magnetic field was applied up to 10kOe(1T) and removed at the rate of 100 Oe/min followed by measuring the relaxation of remanent magnetization. The relaxation rate

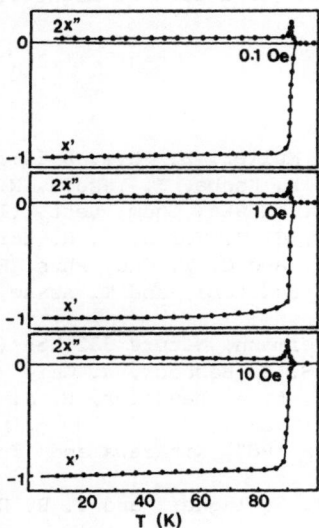

Fig. 10. Temperature dependence of AC susceptibility for the QMG processed sample measured at f=100Hz and for AC field amplitudes h_0=0.1, 1, and 10 Oe.

increased with increasing the maximum external magnetic field reached and saturated at 4 kOe. This is understandable by considering that remanent magnetization or trapped magnetic field in the sample saturates at 4 kOe. From the saturated remanent magnetization data, we obtained $U =$ 0.7 eV, which was much larger than the value obtained in single crystals[13].

Temperature dependence of AC susceptibility

Figure 10 shows temperature dependence of AC susceptibility (x'+ix") for the QMG processed sample measured at f=100 Hz and for AC field amplitudes h_0 = 0.1 Oe, 1 Oe and 10e. The normal to superconducting transition was very sharp with the onset temperature of 93K, which is in good agreement with the result in DC measurements shown in Fig. 2. As differed from sintered materials[18] only a single peak of x" was observed and the peak was not broadened even we increased h_0. This result also indicates high quality of the QMG processed YBaCuO.

CONCLUSIONS

Magnetic measurements were made on YBaCuO crystals fabricated by the quench and melt growth technique. Perfect diamagnetism was observed in the zfc process, while a small diamagnetic signal was observed in the fc process, indicating that the sample is high quality type II superconductors with large pinning forces.

Magnetic measurements at high fields show that magnetization behavior of the QMG processed sample can be understood in terms of Bean's critical state model, which indicates that the sample does not contain weak-links typically observed in sintered samples. Magnetic relaxation was found to be small in such samples and U value of 0.7 eV was obtained at 77K and 1kOe.

ACKNOWLEDGMENTS

We would like to express our gratitudes to K. Miyamoto and M. Morita of Nippon steel corporation for their help in sample preparation. We are also grateful to N. Koyama for a part of magnetic measurements in this paper.

REFERENCES

1. J. G. Bednorz and K. A. Muller, Z. Phys. B 64:189 (1986).
2. K. Kishio, K. kitazawa, S. Kanbe, I. Yasuda, N. Sugii, H. Takagi, S. Uchida, K. Fueki, and S. Tanaka, Chem. Lett. (1987) 429.
3. M. K. Wu, J. R. Ashburn, C. J. Torng, P. H. Hor, R. L. Meng, L. Gao, Z. J. Huang, Y. Q. Wang, and C. W. Chu, Phys. Rev. Lett 58:908 (1987).
4. H. Maeda, Y. Tanaka, M. Fukutomi, and T. Asano, Jpn. J. Appl. Phys. 27:L209 (1988).
5. Z. Z. Sheng and A. M. Hermann, Nature 332:L55 (1988).
6. D. C. Larbarestier, S. E. Babcock, X. Cai, M. Daeumling, D. P. Hampshire, T. F. Kelly, L. A. Lavanier, P. J. Lee, and J. Seuntjens, Physica C 153-155:1580 (1988).
7. G. Deutcher, "Proc. ISS '88", Kitazawa and Ishiguro ed., Elsevir, Tokyo, 1989 p. 383 .
8. T. K. Worthington, W. J. Gallagher, and T. R. Dinger, Phys. Rev. Lett. 59:1160 (1987).
9. S. Tanaka and H. Itozaki: Jpn. J. Appl. Phys. 27:L662 (1988).
10. M. Tinkam, "Introduction to Superconductivity" McGraw-Hill, New York (1975).

11. P. H. Kes, A. Pruymboom, J. van den Berg and J. A. Mydosh, Cryogenics 29:228 (1989).

12. M. Tachiki and S. Takahashi, to be published in Solid State Comm.

13. Y. Yeshrun and A. P. Malozemoff, Phys. Rev. Lett. 60:2202 (1988).

14. P. W. Anderson, Phys. Rev. Lett. 9:309 (1962).

15. M. Murakami, M. Morita, K. Doi, and K. Miyamoto, to be published in Jpn. J. Appl. Phys. 28 (1989) No. 7.

16. C. P. Bean, Phys. Rev. Lett. 8:250 (1962).

17. M. Murakami, M. Morita and N. Koyama, to be published in Jpn. J. Appl. Phys., 28 (1989) No. 7.

18. V. Calzona, M. R. Cimberle, C. Ferdeghini, M. Putti and A. S. Siri, Physica C, 157:425 (1989).

FLUX PINNING AND PERCOLATION IN HIGH-T$_c$ OXIDE SUPERCONDUCTORS

Teruo Matsushita, Baorong Ni, and Kaoru Yamafuji

Department of Electronics
Kyushu University 36
Fukuoka 812, Japan

ABSTRACT

Critical current characteristics in quench and melt growth (QMG) processed Y-Ba-Cu-O are investigated by ac inductive measurements. The critical current in these samples is percolative as is observed in sintered materials. However, this percolative behavior is not caused by weak-link grain boundaries but seems to be mainly attributed to layers of nonsuperconducting solidified melt. The experimental result of magnetization critical current density is compared with the theoretical estimate from the effective medium theory. It is also found that the shielding current with very high density flows locally inside the sample. Candidates for the dominant pinning centers in QMG processed samples are also discussed.

INTRODUCTION

Critical current densities obtained in bulk high-temperature super-conductors are fairly low, although very high values are attained in thin films prepared by various methods.[1-3] Low critical current densities in sintered materials are attributed to weak coupling between grains, since closed currents with fairly high density flow inside grains. In order to eliminate weak links in superconductors, efforts have been made on crystal-axis alignment and two methods have been tried. Mechanical treatments such as cold rolling or pressing of silver-sheathed tape wire make the c-axis alignment normal to the tape surface and this causes improvement of transport critical current density.[4-6] However, the weak links still remain and the critical current density falls down as the magnetic field higher than a few Tesla is applied. The other approach is elimination of weak-link grain boundaries by melt-process. Unidirectional solidification under a temperature gradient is effective for the purpose. Further improved critical current density with better field dependence has been obtained.[7-9] However, it is still far below the critical current density in single crystalline thin films. This seems to be caused by defective structure such as normal phase, cracks, and still remaining grain boundaries.

For a further improvement of the critical current density in oxide superconductors, detailed characterization of melt-processed materials is

Advances in Cryogenic Engineering (Materials), Vol. 36
Edited by R. P. Reed and F. R. Fickett
Plenum Press, New York, 1990

Fig. 1. Optical micrograph of the QMG processed sample.
The nonsuperconducting second phase remains as
thin layers along the growth direction. The
c-axis is directed vertical in this micrograph.

necessary. The critical current characteristics of melt-processed
Y-Ba-Cu-O specimen was investigated by the ac inductive method for this
purpose. In this paper, the experimental result is reported. Candidates
for dominant pinning centers and the percolative behavior due to defective
structure are discussed.

EXPERIMENT

The samples we used were quench and melt growth (QMG) processed
Y-Ba-Cu-O prepared by Nippon Steel Co. Precursor powders prepared by a
common method were heated to 1300-1500 ^{0}C in a platinum crucible and the
melt was quenched on a copper hearth. The quenched plates were again
heated to 1000-1300 ^{0}C and cooled down in flowing oxygen under a careful
temperature control. The plates were cut in the shape of slab of approximately $0.8 \times 3 \times 10$ mm^3 in dimension. The c-axis was roughly directed normal

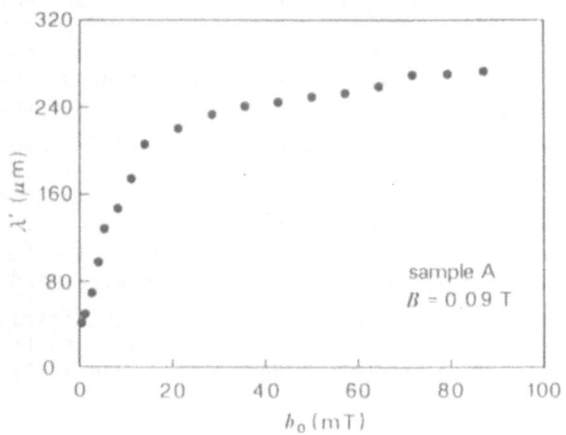

Fig. 2. Ac penetration depth λ' vs ac field
amplitude b_0 in sample A at B=90 mT.

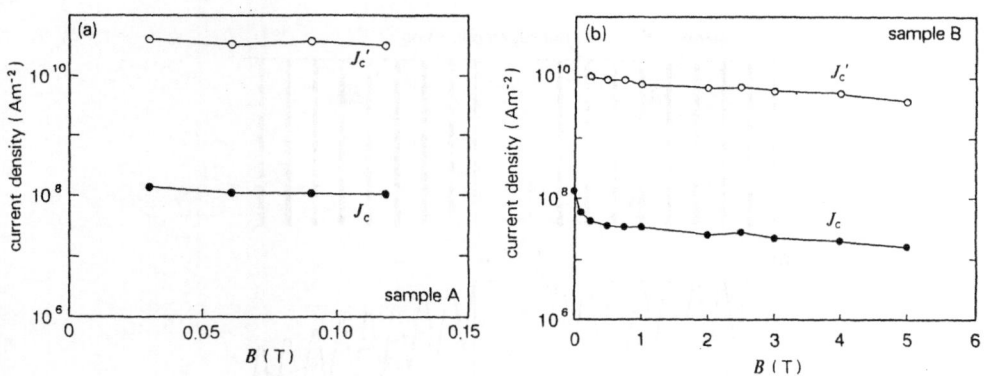

Fig. 3. Bulk critical current density J_c and local critical current density J_c' in (a) sample A and (b) sample B.

to the slab surface. Microstructure of the sample is shown in Fig. 1. Many layers of nonsuperconducting second phase are developed along the growth direction. Their mean separation is approximately 25 μm. These are solidified liquid phases that remained from the unidirectional growth process and are identified as a mixture of $BaCuO_2$ and CuO according to EDX analysis. It is found from transmission electron microscopic observation that fine precipitates of Y_2BaCuO_5 (211) phase with diameters of around 0.5 μm are distributed in a wide area of the samples. Coarse precipitates of a few micrometers are also found.

The critical current characteristic of the specimen was measured by the ac inductive method[10-12]. Dc and small ac magnetic fields were applied parallel to the wide surface of the sample, and hence normal to the c-axis. The measurements were carried out at the liquid nitrogen temperature. The dc bias magnetic field was changed up to 5 T. The frequency of the ac magnetic field was varied from 20 to 200 Hz. The obtained result did not depend on the frequency. This shows that the ohmic shielding current does not flow inside the sample. Figure 2 shows an example of the observed ac penetration depth λ' vs the ac field amplitude b_0 at f=40.2 Hz. This curve has a kink, suggesting that two kinds of the current, like inter- and intragrain currents in a sintered sample, flow inside the melt-processed sample.

The bulk critical current density can be estimated from the slope of the initial λ' vs b_0 curve in Fig. 2:

$$J_c = \left(\mu_0 \frac{d\lambda'}{db_0}\right)^{-1}. \tag{1}$$

The obtained results are shown in Figs. 3(a) and 3(b). The bulk critical current density in sample A takes 1.0×10^8 A/m^2 around B=0.1 T. This value is comparable to the estimate by Murakami et al.[9] from dc magnetization measurements. The critical current density in sample B is slightly smaller than that in A but has 1.6×10^7 A/m^2 even at B=5 T. This result is close to the result of transport current measurement by Murakami et al.[8] The outstanding points in the bulk critical current density in the QMG processed samples are: (i) value is very high, (ii) it is not degraded significantly by the magnetic field, and (iii) it does not depend on the increasing or decreasing process of the magnetic field (as for the history effect in a sintered sample, see Ref. 13). These facts show that the

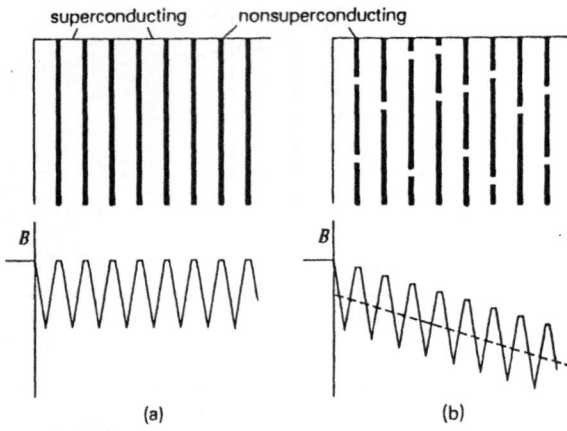

Fig. 4. Microstructure and corresponding magnetic flux
distribution in the specimen at a region sufficiently apart from the edge: (a) for the case
where superconducting regions are completely separated by nonsuperconducting layers and (b) for
the case where the separation is not complete.

weak-link nature of the grain boundary does not govern the transport
characteristics in the QMG processed samples.

However, the existence of the kink in the λ' vs b_0 curve insists that
the current does not flow uniformly inside the sample. From the microscopic observation shown in Fig. 1, the nonsuperconducting layers are the
dominant defects that prevent uniform current transportation. That is,
the current is forced to flow mainly parallel to the nonsuperconducting
layers. This means that the flux invasion into the sample occurs along
the nonsuperconducting layers. If superconducting regions are completely
separated by the nonsuperconducting layers as shown in Fig. 4(a), the kink
does not appear in the λ' vs b_0 curve and the observed magnetization does
not depend on the sample size, i.e., the thickness of the sample. In the
present samples, superconducting regions are not completely separated, as
illustrated in Fig. 4(b). The mean current density along the direction
normal to the nonsuperconducting layers is not zero and the magnetization
depends on the sample size. The broken line in the lower figure in Fig.
4(b) represents the coarsed magnetic flux distribution due to the
shielding by the bulk current.

The local shielding current corresponding to a steep slope in the
flux distribution shown in this figure is for flux invasion from nonsuperconducting layers into superconducting ones. The gradual increase in
the ac penetration depth at large ac field amplitude in Fig. 2 represents
the stronger local shielding currents. The ac penetration depth in Fig. 2
is defined for a bulk flux invasion into the sample and is obtained from

$$\lambda' = \frac{1}{2W} \frac{\partial \Phi}{\partial b_0} , \tag{2}$$

where W is the width of the sample and Φ is the amplitude of the ac
magnetic flux going in and out of the sample. In Eq. (2), 2W is the
length of the perimeter from which the magnetic flux invades the sample.
On the other hand, the pattern of the local flux invasion into the
superconducting layers is completely different from the above. The total

length of the perimeter for this flux invasion is approximately given by $2WD/d$, where D is the sample thickness and d is the mean interval of the nonsuperconducting layers. Hence, the local critical current density is estimated from

$$J_c' = \frac{D}{d}(\mu_0 \frac{d\lambda'}{db_0})^{-1} \tag{3}$$

in terms of the λ' vs b_0 curve in Fig. 2. In principle, existence of grain boundaries may affect the λ' vs b_0 characteristic. However, the mean distance of grain boundaries is much larger than d in the present samples and the influence on J_c' is considered to be negligibly small.

It is to be noted that this affects the bulk critical current density, as will be discussed later. The obtained local critical current densities are represented in Fig. 3. These are larger by two orders of magnitude than the bulk critical current densities. Here, we note that the ratio of the bulk critical current density to the local one J_c/J_c' is independent of the magnetic field. This supports the above-mentioned expectation that the percolative characteristic of the current in the QMG samples is mainly caused by other defects than weak-link grain boundaries.

DISCUSSION

The experiment on the QMG processed samples has clarified the following facts:

   (i) Two kinds of current flow inside the sample.
  (ii) The local critical current density is very high.
 (iii) The bulk critical current density smaller by two orders of magnitude
       than the local one is mainly attributed to the nonsuperconducting
       layers developed along the crystal growth direction.
The second point verifies a high quality of the QMG processed samples. The third point is concerned with the fact that the weak links inside the sample are greatly improved in comparison with the sintered samples.
Then, our interests are focused on the following two points:
   (i) A quantitative description of the bulk critical current density J_c
       in terms of the local critical current density J_c'.

  (ii) Explanation of the local critical current density J_c' from the flux
       pinning theory.

The first point is associated with the percolation problem and the transport phenomena in a percolative system can be treated by using the effective medium theory. For this purpose, we need to simplify the defective structure in the sample. The present sample contains parallel

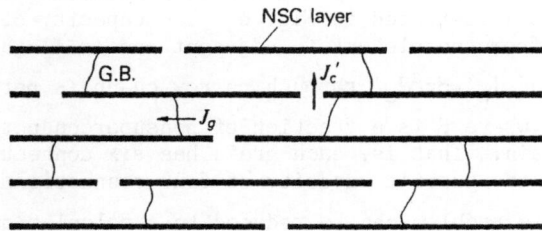

Fig. 5. Schematic illustration of microstructure of the QMG sample.

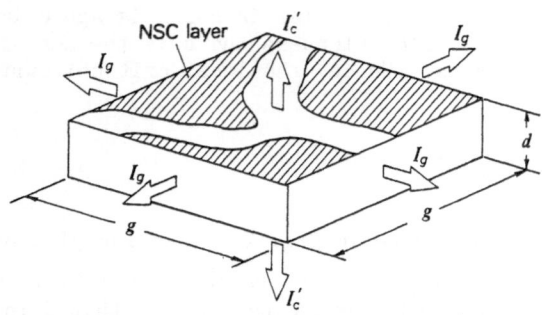

Fig. 6. Illustration of one element composed of one grain
for the effective medium theory. I_g and I_c' are
the capacities of current transportation to the
nearest neighbors.

nonsuperconducting layers roughly along the a-b plane and grain boundaries
of fairly low density, as schematically illustrated in Fig. 5. The
current can flow along the c-axis through narrow channels between
nonsuperconducting layers with the density J_c' (for simplicity we assume
that the local current density inside the superconducting regions is
isotropic). It can also flow through the grain boundaries with the
density J_g. Since the c-axes of two adjacent superconducting regions are
expected to be parallel in most cases, the ratio of J_g to J_c' is
considered to be given by the experimental result by Dimos et al.[14] We
denote the tilt angle in the a-b plane between two adjacent grains at the
grain boundary by θ. The observed dependence of the grain boundary
critical current density on the tilt angle can be approximated as[15]

$$j_g(\theta) = \frac{J_c'\sin\delta}{\sin(\theta+\delta)} ,\qquad(4)$$

where δ is a constant and the good fit with the experiments[14] is obtained
for $\delta=0.87\times10^{-2}$. We assume that J_g in our model is approximately given by
the simple average of Eq. (4) with respect to θ and we have

$$J_g = \frac{2}{\pi} J_c'\sin\delta \, \log[(1+\sec\delta)(1+\cosec\delta)] \equiv \alpha J_c'.\qquad(5)$$

The numerical factor α amounts to 0.074.

We suppose that the sample is simulated by a cubic array of elements:
each element composed of one grain has dimensions of grain size g in the
layer plane and mean separation of the nonsuperconducting layers d in the
third direction, as illustrated in Fig. 6. The capacity of nondissipative
transport current from one element to adjacent one is $I_g=gdJ_g$ through
grain boundaries and $I_c'=\beta g^2 J_c'$ through narrow channels between nonsuperconducting layers, where β is a fraction of nonsuperconducting-layer-free
surface of each grain. That is, each grain has six connections to the
nearest neighbors: the current capacity of four connections is I_g and that
of two is I_c'. This problem can be reduced to a calculation of the
electric conductivity of a bond percolation system composed of two kinds
of bond. According to the effective medium theory of Kirkpatrick,[16] the

mean conductivity σ_m of the mixture of two kinds of particles (the fractions of particles with the conductivity σ_1 and σ_2 are p and 1-p, respectively) is given by

$$\sigma_m = \left\{ (\frac{zp}{2} - 1)\sigma_1 + [\frac{z}{2} (1-p) - 1]\sigma_2 \right\} / (z-2)$$

$$+ \left[\left\{ (\frac{zp}{2} - 1)\sigma_1 + [\frac{z}{2} (1-p) - 1]\sigma_2 \right\}^2 + 2(z-2)\sigma_1\sigma_2 \right]^{1/2} / (z-2), \tag{6}$$

where z is a number of bonds at each node of the network. In the present case of cubic bond, z=6 and we can replace σ_1 and σ_2 by I_g and I_c' with p=2/3. Thus, the bulk critical current density along the layer plane, which corresponds to the transport critical current density, is given by

$$J_{c/\!/} = \frac{1}{4dg} [I_g + (I_g^2 + 8I_g I_c')^{1/2}] = \frac{J_g}{4} [1 + (1 + \frac{8\beta g}{\alpha d})^{1/2}]. \tag{7}$$

The mean bulk critical current density along the direction normal to nonsuperconducting layers is given $J_{c\perp} = (d/g)J_{c/\!/}$.

From optical micrograph observation, β is estimated to be around 0.09. If we assume d=25 μm, g=200 μm, we have $J_{c/\!/} = 0.112 J_c'$ along the layer plane and $J_{c\perp} = 0.014 J_c'$ in the normal direction to the layers. From a geometry of the sample, the magnetization critical current density J_c is related to $J_{c/\!/}$ as

$$J_c = \frac{WJ_{c/\!/}J_{c\perp}}{WJ_{c\perp} + DJ_{c/\!/}} = \frac{J_{c/\!/}}{1 + (Dd/Wg)} . \tag{8}$$

In the present sample, we have $J_c \approx 0.32 J_{c/\!/} \approx 0.036 J_c'$. On the other hand, the experimental value of J_c is smaller by a factor of 200 than J_c'. That is, the theoretical value is several times as large as the experimental result. This disagreement may be due to a coarse estimate of β. There may be thin nonsuperconducting layers invisible from optical micrograph of low magnification.

The above simple analysis shows that $J_{c/\!/}$, the transport critical current density in a common geometry, is larger than the magnetization critical current density J_c. This agrees qualitatively with the recent measurement by Murakami.[17] He observed the transport critical current density of 1×10^9 A/m^2 at zero field. This value is close to the present theoretical estimate with the observed J_c' value. The theoretical result, Eq. (7), insists that the transport critical current density can be improved more by increasing the grain size g.

The second point is the candidates for the causative pinning centers to explain the observed high value of local critical current density J_c'. Before we start this discussion, we wish to stress that such a high value of J_c' is not visionary. It is known that the intragrain currents with fairly high density flow in sintered samples (even higher than J_c in melt-processed samples).[10-12] It is difficult to suspect that the pinning characteristic in melt-processed samples is inferior to that in sintered samples. Such a high J_c' value is supported by the larger pinning potential U_0 estimated from the flux creep measurement[18] than that for intragrain currents in sintered samples.[19] We note that U_0 is proportional to $-1/2$ power of the pinning current density in the category of strong flux pinning.[19]

First, twinning planes with high density are seen in the samples. These planes are directed roughly perpendicular to the specimen surface and the direction of the most fluxoid motion is roughly parallel to the twinning plane in the present experimental situation. Hence, the contribution from these planar defects to the local critical current density is not so large as expected in Ref. 20. That is, the present experimental result cannot be explained by flux pinning by twinning planes.

The nonsuperconducting layers are effective for flux pinning in the present field geometry. The elementary pinning force of these layers per unit length of a fluxoid (S-N boundaries work as pinning centers) at low fields is

$$\hat{f}_p \simeq \frac{\pi B_c^2}{4\mu_0} \xi_{\parallel}, \tag{9}$$

where B_c is the thermodynamic critical field and $\xi_{\parallel}$ is the coherence length in the a-b plane. Thus, the linear summation of the elementary forces leads to

$$J_c' = \frac{\hat{f}_p}{a_f dB}, \tag{10}$$

where $a_f = (2\phi_0/\sqrt{3}B)^{1/2}$ is the fluxoid spacing with ϕ_0 denoting the flux quantum. At $B=0.12$ T ($a_f=141$ nm), we have $J_c'=7.76\times10^8$ A/m^2 for $d=25$ μm, where we used $\xi_{\parallel}=2.99$ nm ($B_{c2\perp}=36.8$ T) and $B_c=0.42$ T. If there are invisible thin nonsuperconducting layers, we have a larger contribution than the above estimate. However, the observed J_c' seems to be too large to explain by this flux pinning.

Fine precipitates of Y_2BaCuO_5 phase are distributed. These contribute to the flux pinning through the same condensation energy interaction as the nonsuperconducting layers. The critical current density is

$$J_c' = \frac{\pi B_c^2 S_V \xi_{\parallel}}{4\mu_0 B a_f}, \tag{11}$$

where S_V is the effective surface area of precipitates in a unit volume of the superconductor. If we denote the size of precipitates and their concentration by L and N_p, respectively, we have $S_V=N_p L^2$. The mean size is about 0.5 μm and the volume fraction of precipitates $N_p L^3$ is about 0.25. In this case, the critical current density is estimated as $J_c'=9.74\times10^9$ A/m^2 at $B=0.12$ T. This value is fairly large. But it is smaller by factor 4 than the observed value in sample A.

Finally we will discuss the possibility of flux pinning by point defects such as vacancies. According to Thuneberg et al.,[21] the pinning energy of a point defect with a size w through the electron scattering is $U_p \approx (B_c^2/2\mu_0)\xi_{\parallel}w^2$. Hence, the elementary pinning force is $f_p \approx U_p/\xi_{\parallel} \approx (B_c^2/2\mu_0)w^2$. This value is quite small because of small size w. These defects contribute appreciably only when their volume density is very high. We treat this case. We denote the longitudinal and transverse elastic correlation lengths of the fluxoid lattice by ℓ_{44} and ℓ_{66}, respectively. The fraction of the normal core in this volume $V=\ell_{44}\ell_{66}^2$ is approximately $(B/B_{c2\perp})V$. Hence, the number of effective point pins in this volume is $n=(B/B_{c2\perp})N_pV$. From the statistical nature, the collective force that these pins exert is approximately given by $\sqrt{n}f_p$. The resulting critical current density is of the order of

$$J_c' = (\frac{N_p}{\ell_{44}\ell_{66}^2})^{1/2} \frac{B_c^2 w^2}{2\mu_0(BB_{c2\perp})^{1/2}} . \qquad (12)$$

If we assume that the correlation lengths are given by the cut off lengths $\ell_{66} \sim a_f$, $\ell_{44} \sim 4\pi a_f$, the largest J_c' is obtained. Even in this case, we have $J_c'=5.0\times10^8$ A/m^2 at B=0.12 T for w=1 nm and $N_p=8\times10^{24}$ m^{-3}. In a strict sense, therefore, this does not belong to the category of strong flux pinning and the correlation lengths are much larger than the above assumptions, resulting in further small J_c'. Hence, the observed J_c' cannot be explained by the flux pinning by point defects.

In conclusion, the pinning centers sufficiently strong to explain the observed local critical current density cannot be found. This may suggest the possibility that all kinds of pinning centers contribute cooperatively. Hence, the problems of the dominant pinning centers and their pinning mechanism remain still open and further detailed investigation seems to be necessary.

The authors sincerely acknowledge Drs. M. Murakami and S. Matsuda for preparation of the QMG samples for this investigation.

SUMMARY

Critical current characteristics in QMG processed samples have been investigated and the following facts have been clarified:

(1) Two kinds of current flow inside the sample due to imperfections. Nonsuperconducting layers developed along the crystal growth direction and grain boundaries are responsible for the percolative current.
(2) The local critical current density is very high. The pinning centers that contribute to the high critical current density are not clear at this moment.
(3) The magnetization critical current density and the transport one are theoretically estimated for the percolation model system, where nonsuperconducting layers and grain boundaries are considered as obstacles. The agreement with the experimental result is not bad.

REFERENCES

1. S. Tanaka and H. Itozaki, Jpn. J. Appl. Phys. <u>27</u>, L622 (1988).
2. B. Roas, L. Schultz, and G. Endres, Appl. Phys. Lett. <u>53</u>, 1557 (1988).
3. K. Watanabe, H. Yamane, H. Kurosawa, T. Hirai, N. Kobayashi, H. Iwasaki, K. Noto, and Y. Muto, Appl. Phys. Lett. <u>54</u>, 575, (1989).
4. M. Okada, R. Nishiwaki, T. Kamo, T. Matsumoto, K. Aihara, S. Matsuda, and M. Seido, Jpn. J. Appl. Phys. <u>27</u>, L2345 (1988).
5. T. Hikata, K. Sato, and H. Hitotsuyanagi, Jpn. J. Appl. Phys. <u>28</u>, L82 (1989).
6. K. Osamura, Ext. Abstract of ISTEC Workshop on Superconductivity, Oiso, 1989, p. 107.
7. S. Jin, T. H. Tiefel, R. C. Sherwood, M. E. Davis, R. B. van Dover, G. W. Kammlott, R. A. Fastnacht, and H. D. Keith, Appl. Phys. Lett. <u>52</u>, 2074 (1988).
8. M. Murakami, S. Matsuda, K. Sawano, K. Miyamoto, A. Hayashi, M. Morita, K. Doi, H. Teshima, M. Sugiyama, M. Kimura, M. Fujinami, M. Saga, M. Matsuo, and H. Hamada, Adv. in Superconductivity (Springer-Verlag, Tokyo, 1989) p. 247.
9. M. Murakami, M. Morita, K. Miyamoto, and S. Matsuda, Prog. in High Temperature Superconductivity (World Scientific, Singapore, 1989) p. 95.
10. H. Küpfer, I. Apfelstedt, W. Schauer, R. Flükiger, R. Meier-Hirmer, and H. Wühl, Z. Phys. B <u>69</u>, 159 (1987).
11. B. Ni, T. Munakata, T. Matsushita, M. Iwakuma, K. Funaki, M. Takeo, and K. Yamafuji, Jpn. J. Appl. Phys. <u>27</u>, 1658 (1988).
12. H. Küpfer, I. Apfelstedt, R. Flükiger, C. Keller, R. Meier-Hirmer, B. Runtsch, A. Turowski, U. Wiech, and T. Wolf, Cryogenics <u>29</u>, 268 (1989).
13. T. Matsushita, B. Ni, K. Yamafuji, K. Watanabe, K. Noto, H. Morita, H. Fujimori, and Y. Muto, Adv. in Superconductivity (Springer-Verlag, Tokyo, 1989) p. 393.
14. D. Dimos, P. Chaudhari, J. Mannhart, and F. K. LeGoues, Phys. Rev. Lett. <u>61</u>, 219 (1988).
15. T. Matsushita, B. Ni, and K. Yamafuji, Cryogenics <u>29</u>, 384 (1989).
16. S. Kirkpatrick, Phys. Rev. Lett. <u>27</u>, 1722 (1971).
17. M. Murakami, private communication.
18. M. Murakami, M. Morita, and N. Koyama, to be published in Jpn. J. Appl. Phys.
19. T. Matsushita, S. Funaba, Y. Nagamatsu, B. Ni, K. Funaki, and K. Yamafuji, to be published in Jpn. J. Appl. Phys.
20. T. Matsushita and B. Ni, IEEE Trans. Magn. <u>MAG-25</u>, 2285 (1989).
21. E.V. Thuneberg, J. Kurkijärvi, and D. Rainer, Phys. Rev. B <u>29</u>, 3913 (1984).

POSSIBLE "PROXIMITY MATRIX" ROUTE TO HIGH CURRENT CONDUCTORS[†]

John Moreland, Y. K. Li[‡], J. W. Ekin, and L. F. Goodrich

National Institute of Standards and Technology
Electromagnetic Technology Division
Boulder, CO 80303

ABSTRACT - The conductance of point contacts between the surfaces of superconducting $YBa_2Cu_3O_{7-\delta}$ thin films is very low. This is probably due to a native insulating surface layer. The conductance of these point contacts can be markedly increased by vacuum depositing and subsequently annealing a thin layer of Ag into the surface of the films. We believe that what might be described as a normal-metal superconducting "proximity matrix", is formed at the surface of the Ag coated $YBa_2Cu_3O_{7-\delta}$ films. In this paper, we describe our efforts to adapt this method to $YBa_2Cu_3O_{7-\delta}$ powder. In particular, we have developed a procedure for vacuum deposition of very thin Ag coatings onto the surface of $YBa_2Cu_3O_x$ powder grains. The Ag-treated powder is then pelletized, sintered, annealed, and cut to form small conducting bars for electrical transport testing.

INTRODUCTION

The proximity effect has been observed at many superconductor-normal metal (SN) interfaces. In particular, several authors have studied superconductive coupling through SNS sandwich structures as a function of normal layer thickness, temperature, and applied magnetic field.[1-3] In general, the data are consistent with Ginzburg-Landau theory. For this reason, we think that observation and characterization of the proximity effect in high temperature superconducting CuO compounds may be key to understanding the fundamental mechanism of superconductivity in these materials. Proximity structures may also have important roles in the development of practical devices including Josephson junctions and high-critical-current wires based on high-temperature superconductors.

Recently, we observed a remarkable increase (a factor of 10^6) in the conductance of contacts between $YBa_2Cu_3O_{7-\delta}$ thin films when thin Ag coatings were annealed into their surfaces.[4] The current-versus-voltage (I-V) characteristics of the Ag-treated contacts also had a peak in conductance at zero voltage bias. A distinct zero bias-current (pair tunneling) was observed in one of the junctions with electrodes having 5 nm Ag coatings. In another experiment, Mankiewich et al.[5] detected Josephson

---

[†] Contribution of the National Institute of Standards and Technology, not subject to copyright.

[‡] Visiting scientist from the Institute of Physics, Chinese Academy of Science, Beijing, China.

Advances in Cryogenic Engineering (Materials), Vol. 36
Edited by R. P. Reed and F. R. Fickett
Plenum Press, New York, 1990

413

coupling through Au patches between $YBa_2Cu_3O_{7-\delta}$ films 1 μm apart. We believe that these results are evidence for a SNS proximity effect operating between the films, where the Ag or Au interlayer is induced into the superconducting state.

Work by Atoh et al.[6] and Nakayama et al.[7] on Nb (or Pb)-Au-$YBa_2Cu_3O_{7-\delta}$ thin film sandwiches also indicates that the proximity effect is operating between conventional superconductors and $YBa_2Cu_3O_{7-\delta}$ films. This implies that the type of pairing in conventional superconductors and $YBa_2Cu_3O_{7-\delta}$ is similar. Perhaps, then, high-temperature superconductors will lend themselves to some of the conventional applications of the proximity effect like interfilament supercurrent transfer in copper-matrix multifilamentary superconductors. The proximity effect might even be used to improve upon the weak link nature of sintered bulk high-temperature superconductor ceramics that apparently limits their high-field J_c.[8] Problems that stem from variations in stoichiometry and crystal structure near grain boundaries may be reduced using grains coated with a proximity layer. Since superconducting compounds typically have short coherence lengths (less than 2 nm for $YBa_2Cu_3O_{7-\delta}$; see ref. 9), the superconducting properties of their interfaces are easily affected by very short range material variations. Methods that improve stoichiometry and crystallography of the superconductor near the grain boundaries should improve the J_c of polycrystalline material. Alternatively, the electrical properties of the superconductor compound can be desensitized to the effects of short-range material variations by increasing the coherence length in the vicinity of the grain boundary. This can be done using a normal-metal proximity layer in the grain boundary. This paper is a progress report describing our initial efforts to adapt the thin-film Ag coating method for improving the electrical properties of $YBa_2Cu_3O_{7-\delta}$ thin-film surfaces to a method for improving the electrical properties of grain boundaries in sintered $YBa_2Cu_3O_{7-\delta}$ pellets. Simply, we have adapted vacuum evaporator and sputter deposition systems for evenly coating $YBa_2Cu_3O_{7-\delta}$ powder with very thin Ag layers. These powders are then pelletized and sintered to form samples for the transport measurements reported here.

Very low surface resistivity contacts between deposited Ag films on $YBa_2Cu_3O_{7-\delta}$ pellets have been reported in references 10 and 11. Is the proximity effect responsible for these experimental observations as well? If so, then it may be possible to optimize these powder mixture and sintering processes to develop a high-current conductor. According to Deutscher and de Gennes[1] there are three requirements for a proximity effect at an SN contact. First, electrical contact between S and N must be very good. Second, S and N must not migrate across the S-N interface. Third, the normal-metal coherence length, ξ_N, should be comparable to the thickness of the normal-metal layer, ℓ. Metallic Ag in grain boundaries may serve as high J_c links between grains. For very thin layers of Ag along grain boundaries, currents exceeding the BCS pair-breaking I_cR product of $\pi\Delta/2e$ are theoretically possible if $\xi_N \geq \ell$.[3]

For an ideal SNS junction, Likarev[3] shows that $\xi_N(T) = 0.58 \times (\hbar v_F\ell/6\pi k_BT)^{1/2}$ where v_F and ℓ are the Fermi velocity and mean free path of N. Based on normal-state resistivity data, he estimates that ξ_N is about 100 nm around 10 K in Ag. This is consistent with the results of Chaikin et al.[12] on Ag proximity junctions where they showed reduced electron coupling to Ag phonons for proximity layers thicker than 40 nm. If S has a T_c of 90 K then ξ_N in Ag would be reduced by a factor of 3 to about 33 nm (16 nm using the data of Chaiken et al.). We conclude that the Ag coatings on $YBa_2Cu_3O_{7-\delta}$ powder grains should be less than about 30 nm to use the proximity effect effectively for improving intergrain coupling. Standard thin-film deposition methods for Ag should provide the control necessary

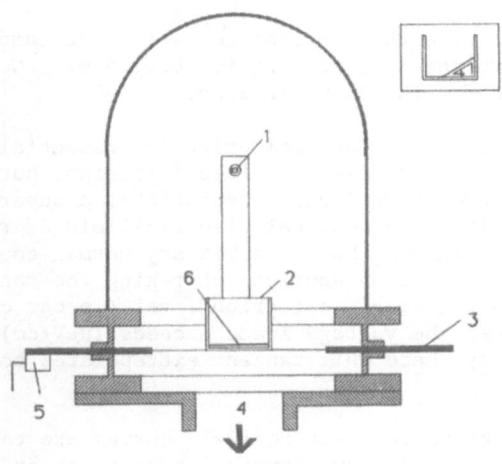

Fig. 1. Schematic showing the arrangement of the deposition source and powder holder in the high vacuum system for thin-film powder coating. The powder holder is mounted on a vibrating beam attached to an eccentric electric motor. Vibration of the powder holder insures an even coating of the powder during deposition. The parts of the diagram are as follows: 1. tungsten wire thermal source; 2. aluminum box for holding $YBa_2Cu_3O_{7-\delta}$ powder; 3. vibrating beam assembly; 4. vacuum pump port; 5. eccentric motor; 6. $YBa_2Cu_3O_{7-\delta}$ powder. The inset in the upper right hand corner shows the powder configuration during deposition.

for achieving very thin uniform Ag coatings on $YBa_2Cu_3O_{7-\delta}$ powder. If desired, it should be possible to coat each grain evenly with submonolayer coverage.

APPARATUS

The apparatus for Ag deposition is shown in Fig. 1. The $YBa_2Cu_3O_{7-\delta}$ powder is held in an aluminum box that is fastened to a stainless steel sheet ($30 \times 3 \times 0.1$ cm^3). The sheet has two 6.3 mm diameter stainless rods attached at either end. The rods protrude from the vacuum chamber through O-ring seals. An eccentric variable speed motor is attached to one of the rods. The motor provides vibration that stirs the powder in the aluminum box. Different normal modes of the powder holder assembly can be resonantly excited by changing the speed of the motor. Typically the motor speed is about 10 Hz. At this frequency the vibration of the powder holder results in a uniform rolling motion of the powder in one side of the aluminum box (see inset Fig. 1). At higher resonant frequencies the powder is ejected out of the holder into the vacuum chamber.

The transport characteristics of the $YBa_2Cu_3O_{7-\delta}$ pellets are measured using computer-controlled data acquisition systems. Electrical contact to the samples is made by In-2%Ag solder pads ultra-sonically bonded to the surface of the samples. This technique results in current contact resistances of about 0.01 to 0.9 Ω and corresponding surface resistivities of about 10^{-2} to 10^{-3} Ω-cm^2. The electric field versus current density (E-J) curves are measured point by point at many dc bias currents. The current set points are selected so that the resulting voltages have approximately equal spacing in log space. The sample voltage is measured with an analog nanovoltmeter, and the output is amplified and filtered before being fed into a digital voltmeter. Sample current is supplied using a 10 A battery powered current supply. The current is measured with a differential voltmeter across a precision resistor. Samples are dry-cut with a diamond wheel from the sintered pellets. The sample size is about $15 \times 1 \times 1.5$ mm^3. The voltage tap spacing is about 5 mm. No current-transfer corrections are made for the J_c measurements.

J_c is defined using an offset criterion.[12] Essentially the offset criterion is similar to the electric-field criterion, but it defines a J_c that goes to zero where the E-J characteristic of a superconductor becomes completely ohmic, unlike conventional electric-field or resistivity criteria where the defined J_c has an arbitrary normal-conduction component. Basically the offset criterion consists of taking the tangent to the E-J curve at a given electric-field criterion level (in our case where the electric field between the voltage leads exceeds 10μV/cm). J_c is defined as the current density where this tangent extrapolates to zero electric field.

The resistance-versus-temperature (R-T) curves are taken with a bathysphere cryostat.[14] The four-terminal resistance of the sample is measured using a lock-in amplifier to detect the voltage drop along the sample in response to a constant 60 μA rms, (137 Hz) current flowing through the sample. Temperature is measured using calibrated Pt (above 70 K) and C glass (below 70 K) resistance thermometers. T_{c0} is defined as the temperature the resistance of the sample falls into the noise of the instrument, which is about 1 mΩ at the 60 μA excitation current. T_{cm} is defined to be the point at which the resistance falls below half of the interpolated normal state value.

SAMPLE PREPARATION

The $YBa_2Cu_3O_{7-\delta}$ starting powder for this experiment was made from "123" stoichiometric mixtures of Y_2O_3, $BaCO_3$, and CuO ground together and calcined at 930°C in air for 10 h. The resulting powder was reground and annealed at 600°C for 10 h and cooled at a rate of 2.4°C/min to 450°C in oxygen. This powder was then ground and sectioned into 4 g lots for Ag deposition and further processing.

The 99.9% pure Ag coatings were either sputtered or evaporated onto the starting powder. Sputtering Ag should provide better contact to the samples because of the energetic deposition conditions. For this reason some of the evaporations were preceded by a brief Ag sputter treatment. The evaporation source was a tungsten wire spiral about 2 mm ID, 4 mm OD, and about 3 cm long. The spiral was suspended about 15 cm above the powder in the aluminum holder. Evaporation rates were typically about 1 nm/s. An rf diode sputter system was also adapted for $YBa_2Cu_3O_{7-\delta}$ powder coating. A 7 cm diameter Ag target, 20 cm above the powder holder, resulted in sputtering rates of 0.1 nm/s in an argon atmosphere of 2.6 Pa. Evaporation and sputtering rates were measured using quartz crystal monitors. However, the actual deposition rate and thickness at the surface of the powder could only be estimated from the ratios of the approximate surface area of the powder charge in the holder and the surface area of the powder exposed to the Ag at any time during the deposition. A rough estimate of the actual coating thickness, t, can be surmised from the radius, r, of the powder grains; the density, ρ, of $YBa_2Cu_3O_{7-\delta}$; the mass, m, of the powder charge; the area, A, exposed to the deposition source; and the thickness monitor reading, t_m; assuming uniform coverage of spherical powder grains. We find $t = rA\rho/3m \times t_m$. Our depositions used an estimated $r = 10$ μm, $A = 8$ cm^2, $\rho = 6.4$ g/cm^3, and m = 4 g so t = 0.004 t_m. We emphasize that this is only an estimate, probably accurate within a factor of 2. This means that the actual Ag thickness may be as little as one thousandth of the thickness monitor reading! If 90 nm of Ag is sputtered according to the thickness monitor, less than a monolayer of Ag is being deposited onto the aggregate

Table I

sample #	sputtered Ag thickness (nm)	evaporated Ag thickness * (nm)	sintering schedule (C)	post annealing schedule (C)	T_{om} (K)	T_{co} (K)	ρ_{100}/ρ_{300} (mΩ-cm)	J_c (at 77 K, 0 T) (A/cm^2)	J_c (at 77 K, 0.1 T) (A/cm^2)
29	90	1500	900 for 10 h 900–600 in 2 h 600 for 10 h	none	91.0	88.0	6.3/11 = 0.57	47	0.67
30	90	1000	800 for 10 h 800–600 in 1.4 h 600 for 10 h	none	86.5	60.0	480/470 = 1.0	0	0
51	0	0	900 for 10 h 900–450 in 3.1 h	600–450 in 3 h	91.5	89.0	3.8/7.7 = 0.49	136	1.06
52	0	500	(in Ag pouch) 900 for 10 h 900–450 in 3.1 h	"	91.5	79.0	25/36 = 0.69	0.40	0

* thickness monitor reading

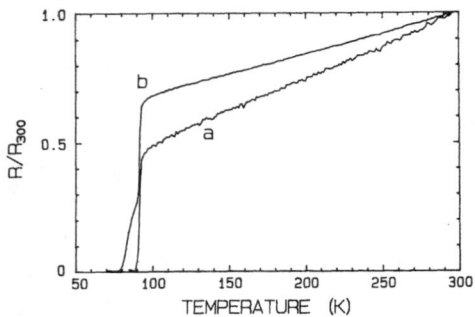

Fig. 2. Resistance versus temperature curves for samples made from (a) uncoated and (b) Ag coated powder.

surface of a typical 4 g $YBa_2Cu_3O_{7-\delta}$ powder charge. For this reason we would sometimes add Ag by following a sputter deposition with a faster subsequent evaporation of a thicker Ag layer.

The Ag-coated powder was pressed into 1.9 cm diameter pellets at a pressure of 20 MPa. The pellets were sintered at 900°C for 10 h in oxygen and slowly cooled at 2.4°C/min. Some of the pellets were wrapped in Ag pouches made from 0.127 mm thick Ag foil during the sintering step. Afterwards the pellets were removed from their pouches and subjected to a final low temperature oxygen anneal from 600°C to 450°C over 3 h.

RESULTS

Table I summarizes the processing, T_cs, residual resistivity ratios, and J_cs of some of the samples. These samples were chosen to represent trends in the data as a result of various processing methods. We caution that the data in table I are some of our first measurements, and there remains the need to gather further statistics given the large number of processing parameters involved. Typical R-T data are shown in Fig. 2. The

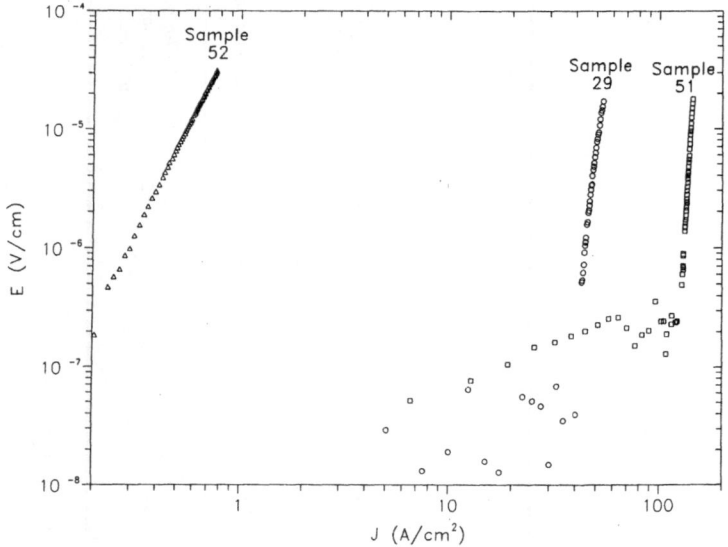

Fig. 3. E-J curves for samples 29, 51, and 52 in liquid nitrogen at 76 K (see table I for sample descriptions).

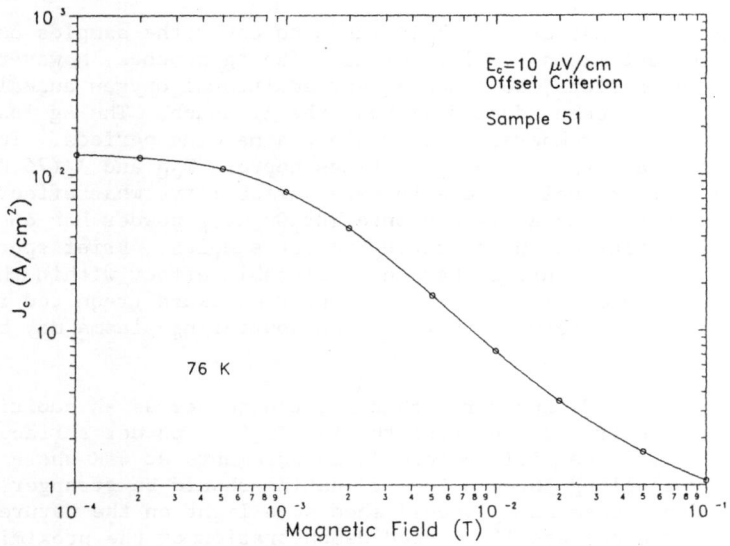

Fig. 4. J_c as a function of magnetic field applied normal to the direction of current in sample 51. The sample is immersed in liquid nitrogen at 76 K.

two curves in the figure are representative of the type of data observed for the Ag coated samples. Curve a shows the transition for a sample without Ag coating. The transition is sharp, with a T_{cm} of 91 K and a T_{c0} of 90 K. The Ag coated samples, on the other hand, show a foot in the transition (curve b). This foot typically starts as a break in the usual transition just below T_{cm} with a T_{c0} well below 90 K. We have seen T_{c0}s as low as 10 K for samples sintered at low temperatures (600°C). T_{c0} is preparation dependent whereas T_{cm} is not, remaining close to 90 K. A foot in the R-T data is an indication of a normal phase present at the grain boundaries.

The E-J curves of samples 29, 51, and 52 in table I are shown in Fig. 3. The data were taken without an applied field ($B < 10^{-4}$ T). The samples were immersed in liquid nitrogen at 76 K. Figure 4 shows the field dependence of J_c for sample 51. The data are typical of weakly linked high temperature superconductors showing a sharp decrease in J_c when a 10^{-2} T magnetic field is applied normal to the direction of the sample current.[15]

DISCUSSION

Sample 29 has much better transport characteristics than sample 30. Sample 30 is normal at 76 K. We think that low T_{c0}s are caused by relatively thick Ag boundaries between largely decoupled $YBa_2Cu_3O_{7-\delta}$ grains. The data for these samples indicate that at higher sintering temperatures the foot in the R-T data is substantially reduced even for relatively larger Ag deposition thicknesses. The improvement may be due to better oxygen stoichiometry achieved at higher sintering temperatures. In addition, Ag is driven off at the higher sintering temperatures. This is supported by the observation of a yellow-brown layer of AgO forming on the sample tube at 900°C. Unfortunately, the samples sintered at temperatures below 900°C have poor mechanical properties and often crumble during lead attachment for electrical measurements. These observations indicate that a key processing obstacle is achieving sintering temperatures high enough for sample integrity without driving the thin Ag coating from the powder. For this reason Ag pouches are now being used to cover the samples during 900°C anneals to prevent excessive loss of Ag. The Ag pouches, however, keep oxygen from the pellets, so a subsequent additional oxygen anneal from 600°C to 450°C has been adopted without the Ag pouch. The Ag loss should be minimal at these temperature over short annealing periods. In any case, Table I shows that evaporated Ag coatings depress T_{c0} and $J_c(76\ K)$. At this time, it is unclear, due to data variability, what effect sputtering of very thin Ag layers onto $YBa_2Cu_3O_{7-\delta}$ powder has on the transport properties of the resulting pellet samples. Brief sputtering periods of the powder surface have no noticeable effect within the variability of the results. However, longer exposure (required for thicker layers) at high temperatures in the argon sputtering plasma may be detrimental.

We plan to include sputter etching of the powder as an additional processing parameter for improving the $YBa_2Cu_3O_{7-\delta}$ powder surface before Ag depositions. Also, we will perform J_c measurements at 4 K where presumably the proximity coupling through the Ag coating should be stronger. The temperature dependence of J_c should shed some light on the nature of the weak links in the pellets.[16] A good demonstration of the proximity effect would be to show that there is a cross-over temperature where J_c for an uncoated weakly linked powder pellet would become less than that of a similarly prepared Ag-coated powder pellet.

ACKNOWLEDGEMENT

G. Reinacker, T. Larson, and T. Stauffer assisted with the critical current measurements. P. Rice assisted with the resistance measurements. This work was sponsored in part by the U. S. Department of the Navy under contract No. N615 3388F1365.

REFERENCES

1. G. Duetscher and P. G. de Gennes, in Superconductivity, R. Parks, ed. (Marcel Dekker, New York, 1969), p. 1005.
2. J. Clarke, Proc. Roy. Soc. A 308, 447 (1969).
3. K. K. Likharev, Rev. Mod. Phys. 51, 101 (1979).
4. J. Moreland, R. H. Ono, J. A. Beall, M. Madden, and A. J. Nelson, Appl. Phys. Lett. 54, 1477 (1989).
5. P. M. Manchiewich, D. B. Schwartz, R. E. Howard, L. D. Jackel, B. L. Straughn, E. G. Burkhardt, and A. H. Dayem, High T_c Superconductivity: Thin Films and Devices, edited by R. B. Van Dover and C. C. Chi, (International Society for Optical Engineering, Bellingham, 1988), Proc. SPIE vol. 948, p. 37.
6. H. Akoh, F. Shinoki, M. Takahashi, and S. Takada, IEEE Trans. Magnetics 25, 795 (1989).
7. A. Nakayama, A. Inoue, K. Takeuchi, H. Ito, and Y. Okabe, IEEE Trans. Magnetics 25, 799 (1989).
8. J. W. Ekin, A. I. Braginski, A. J. Panson, M. A. Janocko, D. W. Capone II, N. J. Zluzec, B. Flandermeyer, O. E. de Lima, M. Hang, J. Kwo, and S. H. Liou, J. Appl. Phys. 62, 4821 (1987).
9. T. P. Orlando, K. A. Delin, S. Foner, E. J. McNiff, Jr., J. M. Tarascon, L. H. Greene, W. R. McKinnon and G. W. Hull, Phys. Rev. B 35, 7249 (1988).
10. J. W. Ekin, T. M. Larson, N. F. Bergren, A. J. Nelson, A. B. Swartzlander, L. L. Kazmerski, A. J. Panson, and B. A. Blankenship, Appl. Phys. Lett. 54, 1819 (1988).
11. S. Jin, M. E. Davis, T. H. Tiefel, R. B. Van Dover, R. C. Sherwood, H. M. O'Bryan, G. W. Kammlott, and R. A. Fastnacht, to be published in Appl. Phys. Lett.

12. P. M. Chaikin, G. Arnold, and P. K. Hansma, J. Low Temp. Phys. <u>26</u>, 229 (1977).
13. J. W. Ekin, Appl. Phys. Lett. <u>55</u> (1989), to be published.
14. J. Moreland, Y. Li, R. Folsom, and T. E, Capobianco, Rev. Sci. Instrum. <u>59</u>, 2435 (1988).
15. R. L. Peterson and J. W. Ekin, Physica C <u>157</u>, 325 (1989).
16. M. Ashkin and M. R. Beasely, IEEE Trans. Magnetics <u>23</u>, 1367 (1987).

YBCO SUPERCONDUCTOR SUSCEPTIBILITY

IN AN ALTERNATING MAGNETIC FIELD

L. M. Fisher, N. V. Il'yn, I. F. Voloshin

All-union Electrotechnical Institute
111250, Moscow, Krasnokazarmennaja 12

INTRODUCTION

A great number of investigations concerning physical properties of high-Tc superconductors (HTSC) have been carried out on ceramics. Ceramics are essentially inhomogeneous media, for which various models have been proposed (see e.g. [1,2]). In the most widely used models, the ceramic are considered to consist of superconducting grains with high critical parameters, coupled by weak (Josephson) links. The macroscopic properties of such a medium appear to vary widely depending on grain dimension, orientation, relative volume, etc. As a result further investigations are needed to define the HTSC ceramics more precisely.

In this paper we report the results of our study of ac magnetic susceptibility (MS) of HTSX ceramics under various conditions. In particular, we have studied the change of MS as the superconducting state is destroyed by a dc transport current effect and we have studied the effect of ac magnetic field magnitude on MS. In both cases it is possible to estimate the volume fraction of superconducting grains in the ceramic. It is shown that the real and imaginary parts of the MS as a function of direct current have special features indicating that the current reaches its critical value. When the temperature is decreasing, one can examine the grain transition to the superconducting state, formation of weak intergranular links, and transition of a whole sample into a coherent state. For comparison, MS measurements of perfect epitaxial thin films have been carried out. We have obtained the temperature dependence of critical magnetic field B_{c1} from data for films. It is shown that ac magnetic properties of high-Tc samples can be explained in the framework of a generalized two-liquid model.

EXPERIMENTAL DETAILS

The complex dynamic susceptiblity $\chi = \chi' - i\chi''$ was determined by two methods. In the first we measured the complex emf voltage $\mathcal{E}$ induced by an ac magnetic field $B = B_0 \cos\omega t$, with a pick-up

Advances in Cryogenic Engineering (Materials), Vol. 36
Edited by R. P. Reed and F. R. Fickett
Plenum Press, New York, 1990

423

coil wound directly onto the sample $\mathcal{E}$= B$_o\omega$NS [χ''+i(χ'+1)], N
is the number of coil turns, and S the sample cross section
area. Voltage was detected by a wide-band, lock-in amplifier
adjusted so that only the in-phase or orthogonal components[4]
of the pick-up voltage were detected. In the second method
the coil was the inductor of an autogenerator resonant
LC-circuit.

We have studied a great number of ceramic Y-Ba-Cu-O
(1-2-3) samples. The size of the plate samples was about
10x10x1 mm^3 the cylindrical samples had diameters up to 10 mm.
The epitaxial films, deposited on SrTiO^3 substrates, were
approximately 0,2m in thickness. The film is made up of block
single crystals with preferential orientation of the c-axis
perpendicular to the substrate. A typical value of critical
current density was of order 10^6 A/cm^2 at T - 77 K; onset of
the superconducting transition was near 89 K, its width $\simeq$2 K.

RESULTS AND DISCUSSION

Critical Current Measurements

In this part we will consider the behavior of MS when a
current, I, flows through the ceramic sample. One can expect
the essential change of the function   (I) in the vicinity of
critical current Ic. The plots on fig. 1 demonstrate the current dependences of χ' and χ'' for the cylindrical sample in
zero external magnetic field. The probe ac field with magnitude
Bo $\simeq$ 10^{-6}T and frequency f =ω/2π = 300 Hz was parallel to the
current direction.

At first we shall consider a qualitative behavior of the
imaginary part of χ', which describes the energy losses, W,
per unit volume:

$$W = \chi''\omega Bo^2/2\mu_o \qquad (\mu_o = 4\pi\ 10^{-7}). \qquad (1)$$

At a small current, χ'' is unnoticeable. At I - Io = 18 A, the
function begins to increase strongly, reaches a maximum and
then decreases. The real part of MS, related to penetration
of magnetic flux, changes linearly with current and has a kink
at I = Io. At this point the derivative dχ'/dI increases
abruptly. Finally the function χ' (I) reaches the almost
constant value χ' = $\chi°$. The dashed curve on fig. 1 represents
the V-I-curve for the same sample. This curve was measured
by a coventional four-probe method. It is seen that a sharp
change of χ' (I) and χ'' (I) takes place in the vicinity of Io.
Notice that these changes become visible at a current which
is lower than the Io value determined from the V-I-curve. We
observed the same dependence of χ (I) for the ceramic plates.

All these experimental data may be explained easily in
a framework of recent HTSC ceramic models. As a result of
intergranular current weakness, a transport current penetrates
into the volume of the sample effectively. The probe ac field[5]
penetrates also into the sample following the transport current.
That is the reason for the linere charge of χ' (I) at I<Io.
The sharp changes of χ' (I) appear when the current begins to
destroy the weak links, enhancing the ac field penetration.
In this process the energy dissipation, W, appears and increases
quickly. As a result, this method allows us to distinguish the
boundary between the regions with non-dissipative current and
those with dissipative curents.

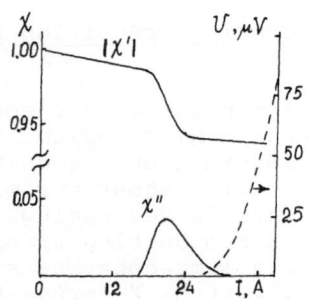

Fig. 1. The dependence of susceptibility on current through
a ceramic cylinder. Diameter - 10 mm, f = 300 Hz,
T = 7 K.

One can use the MS measurement not only for IC determination, but aso for estimating the volume fraction of superconducting grains in ceramics. Investigations carried out on
different samples show that in the low frequency region
(f<1kHz), the χ' value on a plateau of $\chi'(I)$ near Io correlates
with a value estimated on the basis of the static magnetization
curve. At higher frequencies, change of the function $\chi(I)$
becomes weaker and the maximum of $\chi''(I)$ takes place at higher
current. We emphasize that a kink position remains on curves
of $\chi'(I)$ and $\chi''(I)$ when frequency is ranged from several Hz
up to tens of kHz.

To explain the experimental curves '(I), "(I), we consider
the expression for dynamic susceptibility of a plate in the
two-liquid model[6]:

Magnetic Susceptiblity in a Two-liquid Model

$$\chi' = (2/d) \frac{k_2 Sh\ k_2 d + k_1 Sin\ k_1 d}{(k_1^2 + k_2^2)(Ch\ k_2 d + Cos\ k_1 d)} -1,\qquad (2)$$

$$\chi'' = (2/d) \frac{k_1 Sh\ k_2 d - k_2 Sin\ k_1 d}{(k_1^2 + k_2^2)(Ch\ k_2 d + Cos\ k_1 d)}\qquad (3)$$

where $k = k_1 + ik_2$,

$$k_{2,1} = (1/\sqrt{2})\{[\lambda^{-4} + (\omega\sigma\mu_0)^2]^{1/2} \pm \lambda^{-2}\}^{1/2}\qquad (4)$$

Here λ is a temperature dependent magnetic field penetration
depth and σ an effective dissipative conductivity. At
temperatures T > Tc, formulas (2)-(4) transform in the limit
$\lambda \to \infty$ to expressions describing the suscepiblity of a normal
metal plate. In accordance with (2)-(4), the frequency
dependence of χ is connected with the term $(\omega\sigma\mu_0)^2$ in (4).
At low frequencies, where this term is much smaller than λ^{-4},
the real part of k tends to zero. Because of such circumstances
as the undissipative character of current flow at I<Io as well
as the frequency independence of χ, we can conclude that the
term $(\omega\sigma\mu_0)^2$ in (4) is appreciably small.

Some Details of Superconducting Transition on Temperature in Ceramics

The features of $\chi(I)$ considered are concerned with weak links in the granular ceramic HTSC medium. In principle, the measurements of MS give information not only about the superconducting transition, but also about ceramic structure. In particular, when temperature is decreasing, it is possible to observe separately the superconducting transition of grains followed by creation of a volume-coherent state. Fig. 2 plots of the real and imaginary part of χ versus temperature are given for HTSC ceramics plates. One can see that the function $\chi'(T)$ is monotonic. On the contrary, there are two peculiar features on the curve of $\chi''(T)$: a maximum and a "step." Both of these features take place in the temperature range where dc resistance is finite and equals approximately 0.1 ρn (ρn = normal state resistance). The maximum in χ'' has been observed elsewhere (see e.g. [7]). We consider the peculiar features of $\chi(T)$ to be connected with an increase of the effective conductivity of the ceramic in the transition region. With regard to the step origin, we conclude that it is the result of a grain's superconducting transition. As temperature is decreased, the grains couple by weak links and finite superconducting clusters form. This leads to an increase of effective conductivity and a wave number k. The parameter kd for our samples is much smaller than unity in the normal state and at low frequency ($f < 10^5$ Hz) conditions. It increases strongly in the process of the superconducting transition. The function χ', in accordance with (3), has a maximum when $kd \simeq 1$. This maximum is observed on curve 2 of Fig. 3. The frequency dependence of its temperature position Tm confirms this suggestion. Actually, at a higher frequency, the condition $kd \simeq 1$ is reached at a lower conductivity, i.e. at a higher temperature. Using (3)-(4) it is possible to estimate the quantity σm corresponding to an experimental maximum $\chi''(T)$. The quantity σm exceeds the normal conductivity by many orders of magnitude. In other words, in spite of a finite conductivity, the ceramic medium with finite superconducting clusters may screen ac field effectively (see e.g. [8]). Not only does the maximum position, Tm, depend on frequency, but its height χ^m does also. The frequency dependence of χ^m is stronger for samples with a larger fraction of superconducting grains α. At the conditions $\alpha \simeq 0.8$, f = 10 Hz, the quantity χ^m is about 0.05. The quantity at a frequency on the order of 10^5 Hz reaches 0.3-0.4 corresponding to a case of normal skin effect. In contrast to χ^m, the height

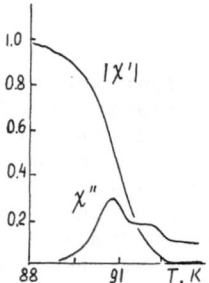

Fig. 2. The temperature dependence of ceramic plate susceptibility. Thickness d = 1.3 mm, f = 40 kHz.

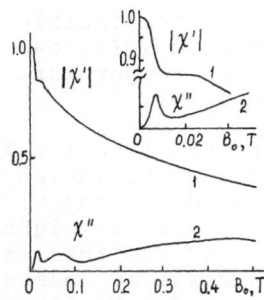

Fig. 3. Variation of susceptibility with ac magnetic field
          amplitude. Sample thickness 1 mm, f = 10 Hz,
          T = 77 K. The same dependence at low amplitudes is
          plotted on insert.

of the step increases with frequency and becomes comparable
with χ^m at f $\simeq 10^5$ Hz. The frequency dependence of step height
is approximately linear, that is in accordance with the behavior
of χ'' in the normal skin effect in a metal when kd<< 1.

A static field B as well as ac field amplitude B have
different influence on this feature of function χ. When the
magnitude of B or Bo is increasing the maximum position Tm
changes to a lower temperature. The position and form of the
step (but not its height) change with B, Bo and f weakly.

Magnetic Susceptibility in a Strong AC Field

In Fig. 3 is shown the typical dependence of χ' (curve 1)
and " (curve 2) on field amplitude, Bo, in the absence of a
static magnetic field. The form of these curves does not change
with frequency in the range of 10 - 10^5 Hz. The absolute value
of χ', which is very close to unity at Bo $\simeq 10^{-6}$T, decreases
with B. The abrupt fall of χ' takes place at Bo $\simeq$ 4 mT after
which the function χ' (Bo) has a plateau. (This part of '
[Bo] is sensitive to the ceramic quality.) At Bo=30 mT the
second abrupt fall of χ' is observed. The imaginary part of
MS at low Bo increases with Bo almost linearly so the losses,
W, are approximately proportional to Bo^3. At higher fields the
function χ'' (Bo) has three maxima differing in height. The
location of each maximum versus Bo corresponds to an abrupt
change of χ' (Bo). There is a correspondence between the
observed behavior of the ac and dc susceptibilities as functions
of field (Bo or B). It is well known that the main features
of χ(B) (dc susceptibility) are connected with ceramic and YBCO
material critical fields. In good ceramic samples, the ac and
dc MS at B $\simeq$ Bo are close to each other. Therefore, it is
possible to estimate the volume fraction of grains in ceramics
using the experimental results in ac field.

On Ms of Thin Films

Inhomogeneity of the ceramic medium permits us to consider
its magnetic properties only in a common sense. So one tries
to research a perfect single crystal. But it is difficult to
produce high quality samples. Many quantitative characteristics
of YBCO material can be obtained by measurements of epitaxial
properties. We have investigated MS of such films over the
frequency range 0.1-1.0 MHz. The value of ac MS of our films
is extremely small in the normal as well as in the superconduct-

ing states. Nevertheless, it is possible to observe some changes in $\chi''(T)$ near Tc. The function $\chi''(T)$ has a narrow single peak at some temperature Tm. This temperature has a weak sensitivity to the values of B, Bo and and f.

As was noticed earlier, the function χ'' has a maximum at kd $\simeq$ 1. We think that the same situation occurs in thin films. If it is true, we can estimate a conductivity at temperature Tm. It is eight orders of magnitude higher than the normal state value. It is known that two loss mechanisms are enhanced near Tc. At first, the increase of losses may be connected with conductivity enhancement due to fluctuation effects. The second explanation of the loss increase is based on the proposition that a flux melting near Tc is occuring.

The MS investigation of epitaxial films proves to be an effective method for determination of the first critic magnetic field. In Fig. 4 we present experimental plots of χ'' (B), taken at different temperatures in the case $B_o \perp B$. Both fields are in the film plane. One can see that there is a kink of the curves which occurs at a field B = B*. The temperature dependence of B* is in good agreement with the known data for the function B_{c1} (T). We find that B* = 500 $[1-(T//t)^2]$ Oe.

CONCLUSIONS

In this paper we report some results concerning magnetic properties of HTSC ceramics and thin films near the critical temperature. The main results of our study are the following:

Behavior of the dynamic magnetic susceptibility can be explained in the framework of a general two-liquid model. This is correct for such complicated objects as inhomogeneous ceramics and for perfect epitaxial films. For this purpose it is necessary to use the effective conductivity, depending on external parameters. In such a way one can explain the different experimental situations. Now the task of theory is to create an adequate model to describe the effective conductivity.

This work was supported by the Science Council on the HTSC problem and was carried out in the framework of project No. 134 of the State program "High-Temperature Super-conductivity."

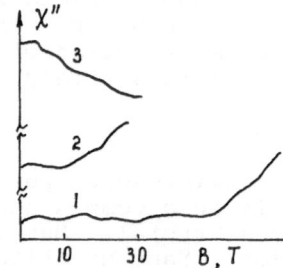

Fig. 4. The field dependence of χ'' (in arbitrary units)for an epitaxial film at f = 1 MHz. Curve 1 - T = 4, 2 K, 2 - 77 K, 3 - 85 K.

REFERENCES

1. C. Ebner, O. Stroud, Diamagnetic susceptibility of
   superconducting cluster: spin-glass behavior,
   Phys. Rev. B, 31:165 (1985).
2. H. Dersch, G. Blater, New critical-state model for critical
   currents in high-Tc superconductors, Phys. Rev. B,
   38:1391 (1988).
3. I. F. Voloshin, S. V. Kravchenko, L. M. Fisher,
   V. A. Yampolsky, Nonlinear anomalous skin effect in
   metals, Sov. Phys. JETP, 61:874 (1985).
4. A. Losche, "Kerninduction," Veb Deutscher Verlag Der
   Wissenschaften, Berlin, 1957.
5. P. Jinlin, Y. Shoesheung, Z. Hong, S. Yunzi, Li Guozhong,
   Z. Wenbin, W. Qingzhe, The effective penetration depth
   of a high-Tc superconductor $YBa_2Cu_3O_7$, Chinese Phys.
   Lett., 5:361 (1988).
6. I. F. Voloshin, S. V. Gaponov, N. V. Il'yn, M.D. Strikovsky,
   L. M. Fisher, Radio-frequency surface impedance of
   epitaxial films YBaCuO, Sov. Phys. SPEC, 2, n.5:94
   (1989).
7. J. R. Cave, M. Mautref, C. Agnous, A. Leriche and
   A. Fevrier, Electromagnetic properties of sintered YBCO
   superconductors: critical current densities, transport
   current and ac losses, Cryogenics, 29:341 (1989).
8. J. Rosenblatt, P, Peyral, Raboutou, Inhomogeneous super
   conductors, in "1979 AIP Conference Proceedings 58,"
   New York (1980).
9. T. K. Wortington, W. J. Gallaher, and T. R. Dinger,
   Anisotropic nature of single crystal YBCO, Phys. Rev.
   Lett., 54:1160 (1987).

MAGNETIC SHIELDING PROPERTIES AND PREPARATION

OF YBaCuO SUPERCONDUCTING TUBES

Wang Jingrong, Li Jianping, Xiu Peifei,
Teng Xinkang, Zhou Lian

Northwest Institute for Nonferrous Metal Research
P.O. Box 71, Baoji, Shaanxi, China

Qian Yongjia, Qiu Jingwu, Tang Zhiming

Dept. of Phys., Fudan University
Shanghai, China

INTRODUCTION

Superconducting magnetic shelding is very important for the application of superconducting devices such as dc and ac SQUID. The shields
made of high Tc superconductors become one of their most important
applications as it can work at liquid nitrogen temperature. Some results
of magnetic shielding on YBaCuO plates[1,2], disc and hollow cylinders[3,4,5,6] have been reported recently. But magnetic shielding fields
are generally in the range of 1-2 mT. In this paper, we reported a
YBaCuO superconducting tube with a bottom and its preparation, which can
shield dc magnetic field up to 4 mT, as well as some results of magnetic
flux trapping and penetration and saturation for such sintered 1,2,3 bulk
materials.

SAMPLE PREPARATION AND EXPERIMENTS

The preparation of YBaCuO superconducting magnetic shielding tubes
was worked out on the basis of a systematic study of the ordinary powder
sintering process for a series of bulk samples[7]. There are three kinds
of samples used in this paper:

L1,L2,L3—— bar samples with rectangular cross section, were cut
from the pellets sintered at different temperatures with finer powders
(~1μm) after two cycles of grinding and sintering from the well-mixed
powders of Y_2O_3 (99.95%), BaO(97%) and CuO(99%). The finer powders were
used in order to improve bulk homogenity and density. It is important
not only to enhance Jc but to avoid cracks for the bulk tubes of big
volume. Three temperatures of 900°C, 920°C and 950°C were chosen for the
final annealing to study its effect on cracks. The morphology and grain
size of samples were observed with SEM, and the resistivity at Tc and
the critical current density at 77K were measured with the standard 4-
probe technique as before[8], the criterion for Jc being 1μm/cm in this
paper.

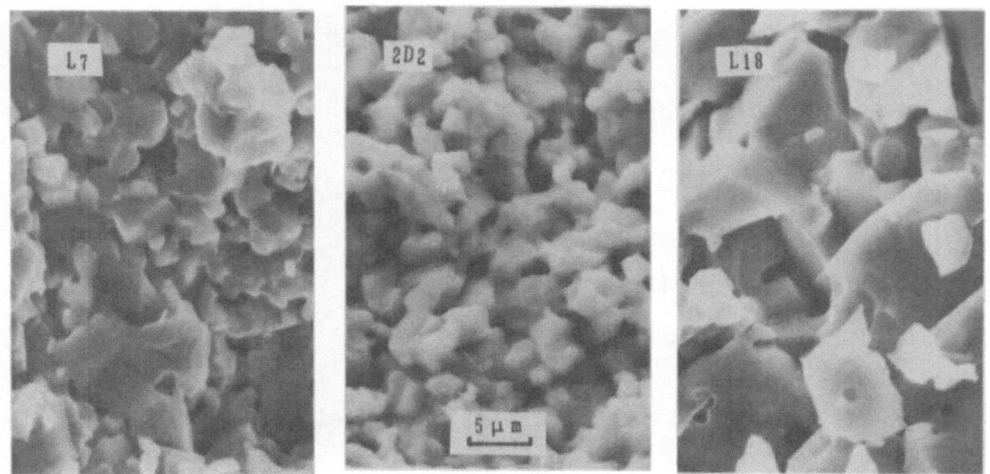

Fig.1.Micrographs of sample L7, L18 and 2D2.

L7, L18, 2D2——— bar samples, were cut from different pellets fabricated with different powders and for the investigations of magnetic flux
trapping and penetration in the bulk samples. L7 was sintered by different process but pressed with the same powders as that of L1; L18 was
prepared with co-precipitated powders by calcined at 900°C for 24h; 2D2
was prepared with the similar powder to that of L1, but calcined and
sintered for 4 cycles. The morphologies of fresh sections under SEM were
shown in Fig.1. Measurements were carried out in the device shown in
Fig.2, in which S is the sample, M, R,V and I are copper wire coil supplied with a cyclic power supply, a YEW Type 3036 X-Y recorder, the
voltage and current of sample, respectively. The sample and Cu coil were
immersed in liquid nitrogen of 77K. As pointed out by J.E.Evetts and
B.A.Glowacki[9] , trapped flux can be observed by measuring the critical
current irreversibility on cycling of applied field. In this work, the
variations of voltage (corresponding to Ic) with magnetic field at constant current were directly recorded by the X-Y recorder. The V(Ic)-B
curves in different maximum applied fields for L7 and at different constant currents for 2D2 have been drawn at 77K. The critical currents

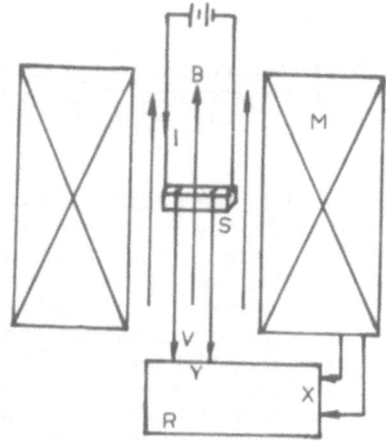

Fig.2.Experimental set up. S, sample, M, Cu coil, R, X-Y recorder,
V, voltage of sample, I, current of sample.

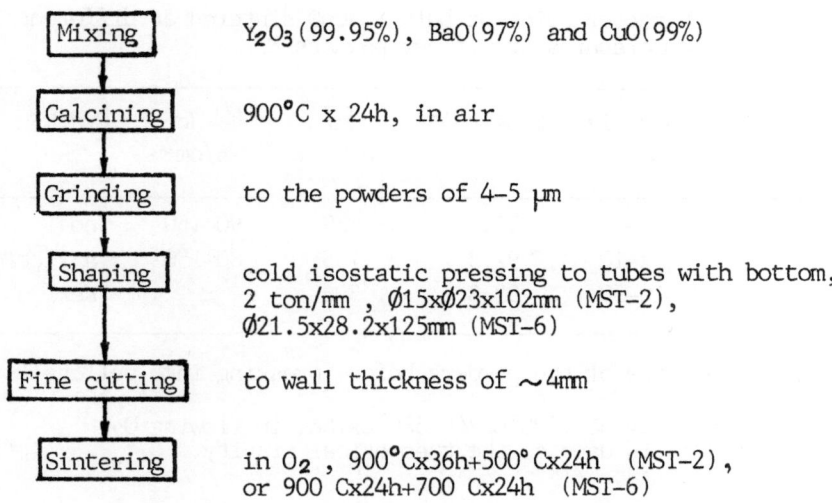

Mixing	Y_2O_3(99.95%), BaO(97%) and CuO(99%)
Calcining	900°C x 24h, in air
Grinding	to the powders of 4-5 μm
Shaping	cold isostatic pressing to tubes with bottom, 2 ton/mm , Ø15xØ23x102mm (MST-2), Ø21.5x28.2x125mm (MST-6)
Fine cutting	to wall thickness of ∼4mm
Sintering	in O_2 , 900°Cx36h+500°Cx24h (MST-2), or 900 Cx24h+700 Cx24h (MST-6)

Fig.3. Process for preparing the YBaCuO tubes with bottom.

were also measured just in remained flux after removing the applied field for 2D2 and L18, repeating the Ic measurements point by point during increasing the applied field. As a comparison, the reduced Ic as a function of the applied field has been measured by using the same criterion of 1μV/cm at 77K for 2D2.

MST-2, MST-6—— two sintered YBaCuO tubes with one end closed, have been prepared according to the process shown in Fig.3. The final anneal- ing temperature was 900°C in order to avoid micro and macro cracks. The magnetic shielding properties were measured with the same apparatus shown in Fig. 2, putting the tube in the position of sample. The magne- tic fields in the centre of tubes were measured at 77K with commercial Hall probes, Oxford Instrument Model 5200 for MST-2 and Lake Shore Type LHGA-321 for MST-6, respectively. MST-2 was also tested the variation and distribution along the axis of the internal field Bi by using an ac weak-signal modutation technique and lock-in amplifier.

RESULTS AND DISCUSSION

Effect of sintering temperature

Sintering temperature effects Jc of oxide superconductors in no small degree[8], especially for bulk materials of big volume, such as shielding tubes. The density, ρ_{Tc} , Jc and cracks of the three kinds of bulk samples sintered at 900, 920, and 950°C are listed in Table 1. As shown in Table 1, the density of bulk materials would increase with the increase of sintering temperature, but Jc of L3 sintered at 950°C couldn't be tested due to the serious micro and macro cracks, which come from the big difference of heat-contraction coefficient between a-b plane and c axis in the layer structure. As the another result, the resistivity near Tc increases by more than twice as large as that of L1, therefore, high sintering temperature is not suitable for big volume materials such as shielding tubes. At the same time, one must pay atten- tion to the cooling rate and the homogenity in the furnace in order to prevent fracture and deformation of tubes, otherwise magnetic flux will penetrate from the cracks, resulting in decreasing shielding property, although enhancement of Jc is also very important as generally speaking: the shielding field is proportional to Jc[1] .

Table 1. Characteristics of bulk YBaCuO sintered at different
temperatures from finer powders *

No	t_1** °C	Grain size μm	Density*** %	ρ_{Tc} μΩ-cm	Jc A/cm^2	Crack
L1	900	1–4	81.3	189	90–150	no
L2	920	2–10	92.3	155	80–150	a little
L3	950	10–50	93.3.	396		serious

* The mean size of the powders before pressing into pellets is
0.87μm.
**Treatment: t_1x36h+700°Cx24h+580°Cx16h, in flowing O_2.
*** 6.4g/cm is used as the theoretical gravity.

Magnetic flux trapping in bulk YBaCuO samples

Fig. 4 shows the critical current irreversibility of sintered
materials at constant transport current with cycles in applied field
drawn by the X–Y recorder. A series of maps in different maximum cycled
fields B_{max} for L7 at a constant transport current of 6A (0.5Ico) were
shown in fig.4a. No obvious irreversibility can be seen when $B_{max} < 10$mT.
This value can be regarded as Hc1 of L7 if such sintered YBaCuO materials were considered as non–idea type II superconductors. The trapped
flux increased with B_{max} in the range of 20–60mT. After 60mT, no big

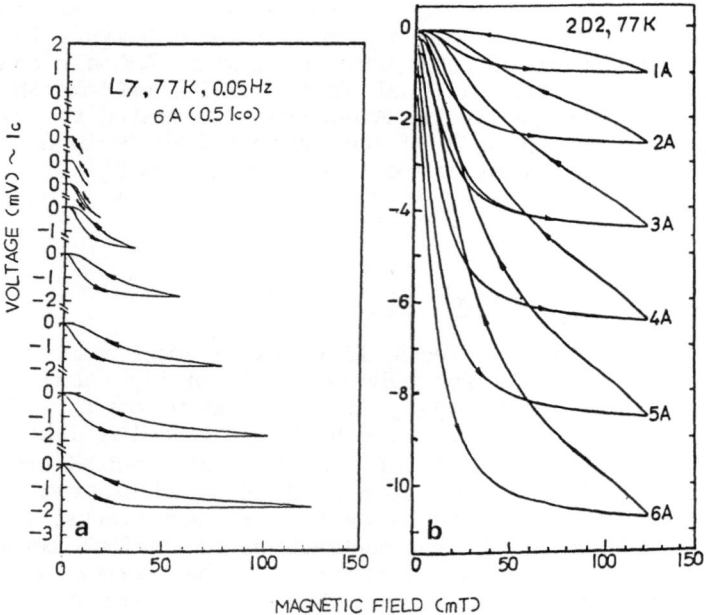

Fig.4 Observation of critical current irreversibility in bulk YBaCuO
materials, it directly reflects the flux trapping of intergrain and intragrain.
a,in different amplitudes of cycled field and at 6A(0.5 Ico)
for L7; b,at different constant currents and in the same
cycled field of 120mT for 2D2.

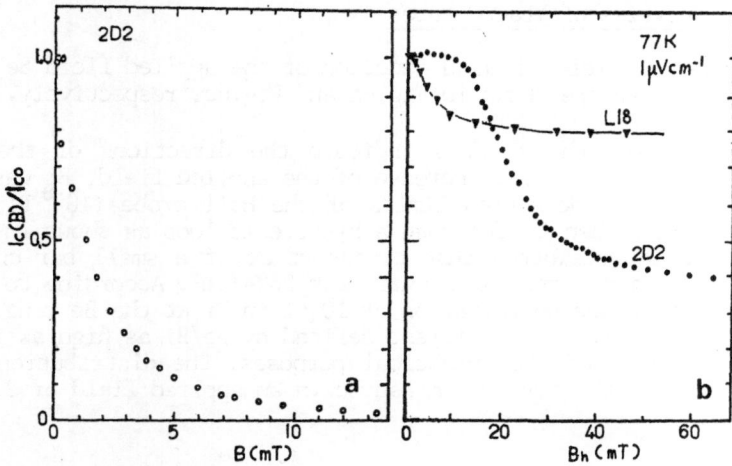

Fig.5 Observation of flux penetration and saturation field and Ic-B
curve for bulk YBaCuO materials. Ic was determined using an
electric-field crit rion of 1μV/cm.
a,reduced transport Ic vs applied field B for 2D2 at 77K;
b,reduced transport Ic vs magnetic field undergone historically
Bh for 2D2 (•) and L18(▲) at 77K.

difference of Ic can be seen according to Fig.4a, indicating the sample
began to entry the saturation state. Fig.4b shows different figures of
2D2 at different currents of 1-6A in the same B_{max} of 120mT of cycled
field. At every current, the initial voltage (i.e.Ic) decreased rapidly
with the increase of magnetic field due to the weak connection between
grains, then remained on a certain level. It is found that this voltage
increased with the constant current. The corresponding resistivity, V/I,
was about 80μV-cm at 1A, then increased to 144μV-cm at 6A. It is possibly due to the contact heating at the current connections. In spite of
this, the constant resistivity can be used to judge the intergrain connection as pointed out by us in another paper[10].

Flux penetration and saturation field

Fig.5a is the reduced transport Ic as a function of applied field
B for 2D2, it would correspond to the V-B curve of increasing field in
fig.4b if the constant current is near its Ico(8.5A).

Fig.5b shows the reduced transport critical currents in every historically undergone field Bh for 2D2 (•) and L18 (▲) at 77K, using a
criterion of 1μV/cm. There was no obvious change in Ic for 2D2 when Bh
was less than 10mT. It means no any trapped flux in such small magnetic
field region, corresponding to the upper curves of the same field region
in Fig.4a. When increasing the undergone field, flux began to be trapped
and Ic bean to decrease to a certain value of about 0.4 Ico as Bh was
about 60mT. It can be regarded as the saturation field of 2D2. The remanence was about 2mT according to 0.4 Ico from the Ic-B curve of 2D2
in Fig.5a. But the behaviour of L18 was different, its flux was trapped
from beginning and saturated in about 10mT. It is worthy to point out
that the density of L18 (~90%) was much high than that of 2D2 (only 67%)
and the grain size of L18 was bigger than that of 2D2 according to Fig.1,
although their critical currents were on the same level, (L18, 218A/cm²;
2D2, 241A/cm²), perhaps due to the cracks found in L18. It is obvious
that no cracks and high density and Jc are good for magnetic shields.

Shielding properties of YBaCuO tubes

The internal field Bi as a function of the applied field Be at 77K for MST-2 and MST-6 are shown in Fig.6a and Fig.6c, respectively.

The arrows in the figures indicate the directions of the field scanning. During the initial increase of the applied field, Bi was indetectable within the detection limits of the Hall probe (10^{-6} T) untill Be reached 3.9mT, then Bi followed a hysteresis loop as shown in Fig.6a for MST-2. It is noteworthy that transport Jc of a small bar cut from its open end with an area of 3.33mm was 262A/cm^2. According to the ac measurement, there was no variation of 10^{-7}T in Bi at the Be ranges less than 3.9mT. Therefore, a coefficient defined by Be/Bi as high as 10^5 for this tube was obtained. For practical purposes, the distribution of Bi along the axis of the tube was measured in an applied field of 2.8mT as

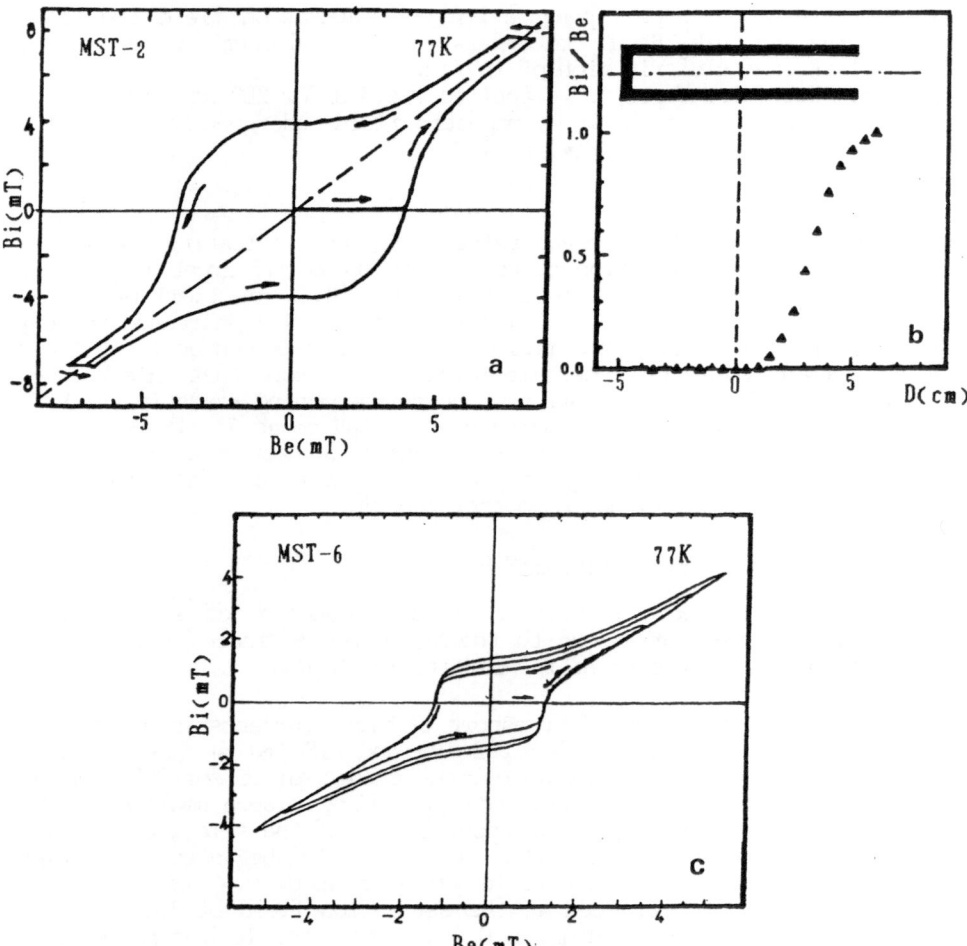

Fig.6 Shielding properties of YBaCuO tubes MST-2 and MST-6.
a.internal field Bi as a function of applied field Be at 77K for MST-2, the dc shielding field is 3.9mT; b,distribution of Bi along the axis in an applied field of 2.8mT at 77K for MST-2; c,magnetization loops in different Be ranges at 77K for MST-6, the saturation field is 12.8mT.

shown in fig.6b. It is found that the effective shielding space was greatly enhanced by the closed end design. For MST-6 tube, a series loops were drawn in different Be ranges of 3.5, 4.5 and 5.3mT as shown in Fig.6c. The maximum shielding field was 1.4mT, and the saturation field has also been measured to be 12.8mT. After saturation, a remanence of 2.01mT changed to 2.12mT after 9 hours, perhaps due to percolation of the screening currents.

SUMMARY

We have prepared bulk YBaCuO materials and two tubes with one end closed by using the ordinary powder sintering process and investigated the effects of sintering temperature on Jc and cracks and observed the behavious of magnetic flux trapping, penetrating and saturation in such bulk materials. On the basis of these investigations, a YBaCuO tube which can screen a dc magnetic field of 3.9mT at 77K has been prepared. It is very possible to increase the shielding fields of high Tc superconductors at 77K to the practical level for some applications such as SQUID according to their Hc1 and saturation field if optimizing the processes.

This work was supported by China R & D centre of high Tc superconductivity.

REFERENCES

1. Akihito Yahara, and Hironori Matsuba, Magnetic shielding effect on oxide superconducting plates, IEEE Trans. on Mag. 25: 2498 (1989)

2. T.Okada, K.Takahata, S.Nishijima, S.Yoshida, and T.Hanasaka, Applicability of oxide superconductor to magnetic shielding, IEEE Trans. on Mag. 25: 2270 (1989)

3. Y.J.Qian, J.W.Qiu, Z.M.Tang, J.R.Wang, J.P.Li, X.R.Teng, P.F.Xu and X.K.Teng, A magnetic shielding study of high Tc superconductor YBaCuO, proc. of ICEC12, Editor: R.G.Scurlock and C.A.Bailey, Butterworths, 998-1001 (1988)

4. J.O.Willis, M.E.McHenry, M.P.Maley, and H.Sheinberg, Magnetic shielding by superconducting Y-Ba-Cu-O hollow cylinders, IEEE Trans. on Mag. 25:2502 (1989)

5. J.W.Purpura and T.R.Clem, The fabrication characterization of high temperature superconducting magnetic shields, IEEE Trans. on Mag., 25:2506 (1989)

6. G.J.Cui, S.G.Wang, H.M.Jiang, J.Z.Li, C.Y.Li, C.D.Lin, R.Z.Liu, Q. L.Zheng, Y.S.Fu, Z.L.Luo and W.C.Qiao, A superconducting shielding can for high Tc SQUID, IEEE Trans. on Mag. 25: 2273 (1989)

7. J.R.Wang, J.P. Li, P.F.Xu, X.K.Teng, K.G.Wang, S.G.Wang and L.Zhou, Characteristics of transport Jc of bulk YBaCuO prepared with ordinary Powder sintering process, Proc. of ICEC12, Editor: R.G.Scurlock and C.A.Bailey, Butterworths, 978-982 (1988)

8. J.R.Wang, J.P.Li, P.F.Xu and L.Zhou, Limitation of transport Jc of bulk YBaCuO, Cryogenic Materials'88, Vol.1, Edited by R.P.Reed, Z.S.Xing, and E.W.Collings, 119 (1988)

9. J.E.Evetts and B.A.Glowacki, The relation of critical current irreversibility to trapped flux and microstructure in polycrystalline YBaCuO, Cryogenics, 28: 641 (1988)

10. Wang Jingrong, Xu Peifei, Teng Xingkang, Li Jianping, Wang Keguang, Tang Xiede, and Zhou Lian, Flux trapping and intergrain connection in bulk polycrystalline YBaCuO, submitted to The Beijing International Conference on High Tc Superconductivity (BHTSC'89)

INVESTIGATION OF THE INTERFACE BETWEEN A NORMAL METAL AND

HIGH TEMPERATURE SUPERCONDUCTOR

S.P.Ashworth, C.Beduz, K.Harrison*, R.G.Scurlock, Y.Yang

Institute of Cryogenics, University of Southampton
Southampton, UK
*Metco Ltd, Chobham, Surrey, UK

ABSTRACT

Data are presented on the contact resistance between YBCO and arc sprayed silver and copper. Contact resistivities of 3.4×10^{-7} Ωcm^2 (Ag on YBCO) and 6×10^{-3} Ωcm^2 (Cu on YBCO)) are obtained at 80K without annealing. With annealing, Ag on YBCO resistivities of 10^{-9} Ωcm^2 are obtained. A new technique for measuring these very low contact resistances is outlined. Measurements on Cu and Ag spray coatings on a 10 molar% Ag on YBCO matrix were also carried out, the addition of Ag to YBCO reduced the Cu contact resistivity. These results are interpreted in terms of a percolative Ag path allowing current to bypass a degraded surface region of YBCO.

INTRODUCTION

A low surface contact resistance is needed for the development of power applications of high T_c superconductors, not only as injection current terminals, but also in the formulation of composites for cryogenic stabilized conductors.

Previous studies[1-5] on the contact resistance of pressed, soldered, ultrasonically bonded, metal sprayed, sputtered and evaporated contacts have shown that pressed and soldered contacts have very high resistances ($0.01-10\Omega cm^2$). Although ultrasonic bonding produces better contacts, only metal sprayed sputtered and evaporated Ag and Cu contacts, with a resistivity in the region of $10^{-5}-10^{-10}\Omega cm^2$ offer the solution for high current applications. The observed contact resistance is the result of a degraded surface layer of the bulk superconductor. Based on photo-emission data, Egdell and Flavel[6] discuss the possibility that the YBCO surface is intrinsically non-metallic. Alternatively, Ekin et al[2] have presented strong evidence that a semiconducting oxide surface layer can be formed by oxygen uptake of the normal metal in contact with the superconductor.

In this paper we report results on the contact resistance between arc sprayed Cu and Ag and bulk sintered $YBa_2Cu_3O_{7-\delta}$ and $YBa_2Cu_3O_{7-\delta} + Ag_x$ with x = 0.1. The effect of low magnetic fields on the contact resistance was also studied.

Advances in Cryogenic Engineering (Materials), Vol. 36
Edited by R. P. Reed and F. R. Fickett
Plenum Press, New York, 1990

439

Table 1. Samples

Sample	Sintering temperature °C	pad	Post annealing	$\rho_c(80K)$ Ω cm^2
1.YBCO	990	Cu sprayed	–	$6.4\times10^{-3}(-)$
2.YBCO+Ag	990	Cu sprayed	–	$1.0\times10^{-5}(+)$
3.YBCO	960	Ag sprayed	–	$1.4\times10^{-6}(+)$
4.YBCO	970	Ag sprayed	–	$3.4\times10^{-7}(+)$
5.YBCO	980	Ag sprayed	–	$8.4\times10^{-6}(+)$
6.YBCO	990	Ag sprayed	–	$7.3\times10^{-7}(+)$
7.YBCO+Ag	990	Ag sprayed	–	$8.0\times10^{-7}(+)$
8.YBCO	990	Ag sprayed	(750°C,5 hours)	$<10^{-9}$

+ and – in column 6 indicate sign of $d\rho_c/dT$ for $T<T_c$. + sign corresponds to metallic type conductivity.

SAMPLE PREPARATION AND EXPERIMENTAL METHOD

The bulk specimens of YBCO were prepared by firing, cold pressing, sintering at the temperature indicated in Table I and annealing in an oxygen atmosphere. Several sintering temperatures were selected to study the effect of varying the hardness and density of the bulk superconductor on the contact resistivity ρ_c. Arc spraying was used instead of plasma spraying because of the lower surface heating involved during the metal spraying process. The surface of the samples was abraded with grade 1200 emery paper immediately before spraying. The thickness of the metal coatings was of the order 50-100μm. Samples 1 to 7 were measured as sprayed without post-annealing. Samples 2 and 7 were prepared by mixing silver oxide particles 3μm diameter with the fired powder.

To avoid errors arising from inhomogeneous current distribution at the normal metal-superconductor interface special attention was paid to the sample mounting. The usual four wire technique can produce erroneous results when measuring very low contact resistances. If the contact resistance is smaller than the normal metal film resistance between the

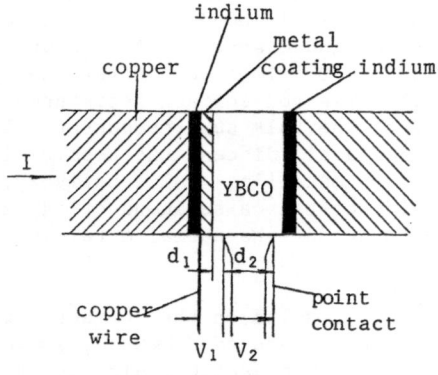

Fig.1. Sample Mounting

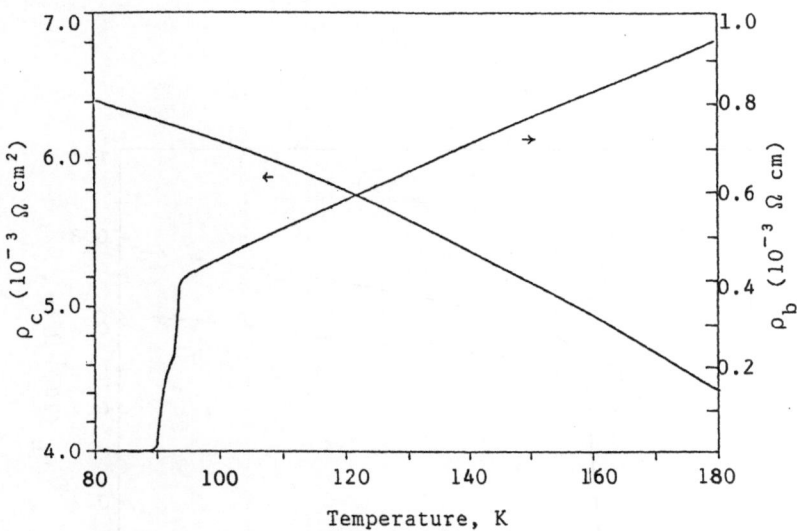

Fig.2. Contact resistivity (ρ_c) and bulk resistivity
(ρ_b) as function of temperature for a Cu sprayed
YBCO (Sample 1)

current and voltage pads, most of the current will be injected through the current lead directly into the superconductor. The effective surface is thus smaller than the film surface. A further consequence of non homogeneous current injection would be that the voltage reading could be greatly reduced. The samples were mounted as shown in Fig.1. A 2 x 2mm square section cut from the 2mm thick metal coated YBCO pellets was pressed between two spring loaded copper block of the same cross section, which were then used to inject current into the sample. To ensure good electrical contact across the whole surface, a thin indium sheet (~100µm) was placed between the sample and the copper blocks. The voltage was measured by a 25µm copper wire located between the indium and the normal metal coating and two pressure contact needles at different position on the sample. V_1 is the voltage across the interface plus some contribution from the bulk.[1] The net voltage across the interface could be obtained from the geometric ratio d_1/d_2 or from the measured jumps of V_1 and V_2 at the superconducting transition. The resistance per unit area of a Cu or Ag sprayed coating 50µm thick is in the range 1-5 x $10^{-9}\Omega cm^2$ at 80K. It is not possible to measure contact resistivities below this value using metal sprayed samples.

The resistance was measured with an a.c. method. The voltage generated by a 0.1A current, 31 Hz, was measured, using a lock-in amplifier, as a function of the temperature.

RESULTS AND DISCUSSION

The bulk resistivity ρ_b and net contact resistivity ρ_c of sample 1 (Cu on YBCO) as a function of temperature are shown in Fig.2. ρ_c shows a semiconductor type dependence with temperature above and below T_c, with a small reduction in the slope for $T < T_c$. The results for sample 2 (Cu on YBCO + Ag) are plotted in Fig.3. Although both samples were processed in the same   the addition of Ag produced a slightly different bulk

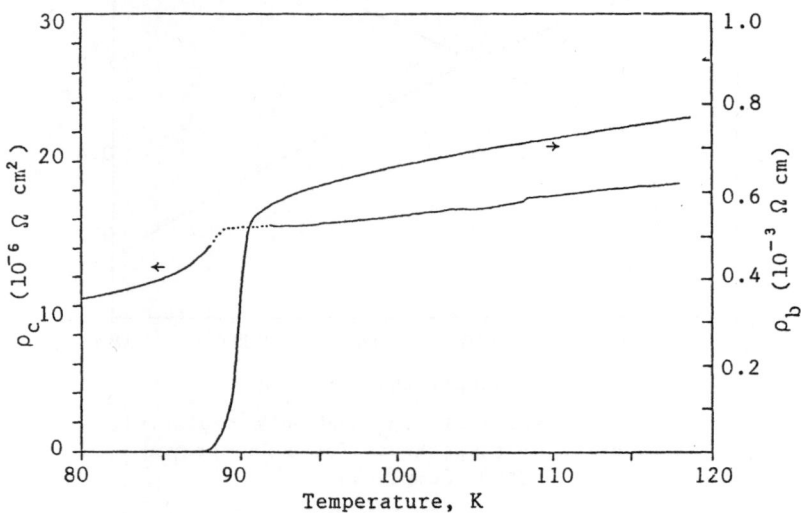

Fig.3. Contact resistivity (ρ_c) and bulk
resistivity (ρ_b) as function of temperature
for a Cu sprayed on YBCO + Ag (Sample 2)

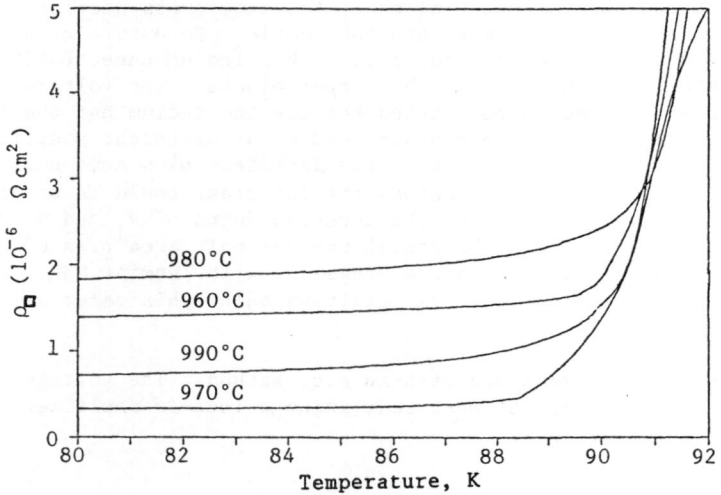

Fig. 4. Contact resistance per unit area of Ag
sprayed samples 3 to 6. Numbers indicate
sintering temperatures.

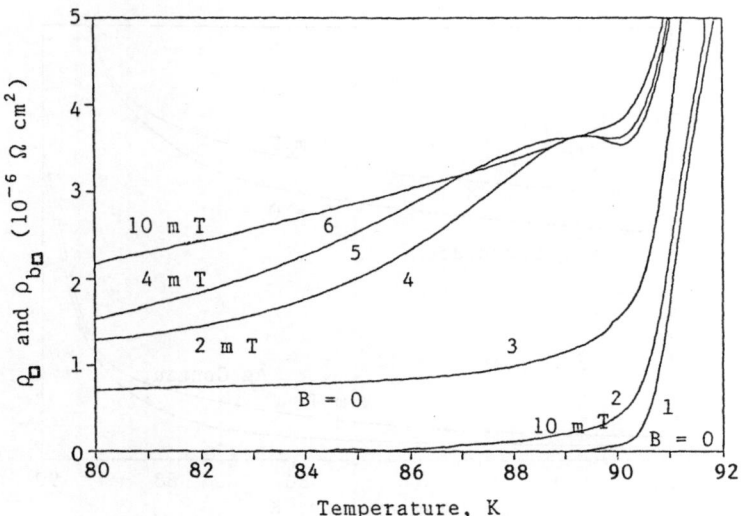

Fig. 5. Contact resistance per unit area $\rho_{\square}$
(curves 3-6) and bulk resistance per unit
area $\rho_{b\square}$ (curves 1,2) of sample 6 under
various magnet fields as a function of
temperature.

ρ_b-T curve. The contact resistivity of this sample is approximately 600
times smaller than the measured value of sample 1; also the temperature
dependence is now metal-like over the whole temperature range. Due to rapid
change of resistivities in the transition region the calculated contact
resistance is prone to large errors in the dotted region of the curve.
The microscopic analysis of sample 2 revealed that the silver agglomerated
into pockets of 5-10μm diameter. The temperature dependence and large
reduction in the contact resistivity of this sample suggests that the Ag
particles form a percolative path through the surface semiconductor layer.
We infer that the thickness of this degraded layer is smaller than the
percolative coherence length of the Ag particles in this sample, which was
estimated to be of the order of 10-25μm.

The change of ρ_c with T above T_c is consistent with an Ag matrix
percolating through a medium of high resistivity. The rapid decrease below
T_c is interpreted as the thinning of a semiconducting surface layer with
decreasing temperature. If the thickness of the semiconducting layer is
strongly dependent on temperature, the measured resistance can show a metal-
like behaviour.

Fig. 4 shows the results obtained from samples 3 to 6. The calculated
values of ρ_b for these samples are subject to large errors due to the
small values of ρ_c for these samples. Therefore, in this figure we have
plotted the resistance per unit area $\rho_{\square}$ as measured by V_1. The bulk
contribution to $\rho_{\square}$ can be neglected below 88K (where $\rho_b = 0$ and $\rho_{\square} = \rho_c$).
The difference in density and hardness resulting from varying the sinter-
ing temperature does not correlate with the values of the contact resistance. The contact resistance of all these samples show a metal-like
temperature dependence. The bulk characteristic of these samples are
very similar and the differences in ρ_c is probably due to variations
of the coatings.

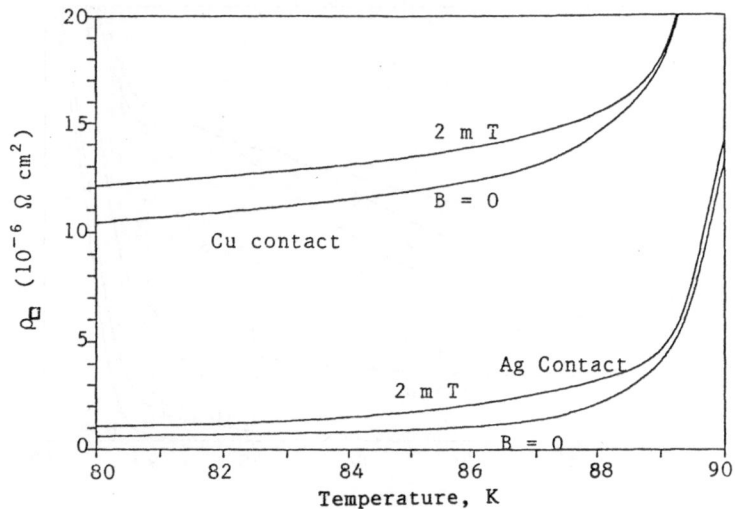

Fig.6. Effect of a 2 m T magnetic field on $\rho_\square$
for Cu and Ag sprayed contacts onto YBCO + Ag
composite (Samples 2 and 7)

In agreement with previous authors, the low values of ρ_c shown by the
Ag sprayed samples indicate that the use of metals with low oxygen affinity
reduces the extent of the degraded layer of superconductor at the interface
[2,7]. The effect of introducing Ag into the YBCO bulk is negligible when
an Ag coating is used, i.e. compare curve 1 in Fig. 6 with Fig.4.

Sample 8 was cut from the same pellet as sample 6 but was annealed in
oxygen after coating at 750°C for 5 hours. The measured value of $\rho_\square$
after annealing was comparable to the surface resistivity of the Ag pad
(i.e. $10^{-9} \Omega cm^2$) showing a reduction of at least two orders of magnitude
from the value before annealing. A similar decrease of $\rho_\square$ after the
annealing of sputtered contacts was obtained by Ekin et al[2], although Katz
et al[5] reported no change of $\rho_\square$ for plasma sprayed Ag contacts after
annealing at 500°C for two hours.

The effect of an external magnetic field parallel to the current on
$\rho_\square$ (sample 6) is shown in Fig.5. The bulk resistance per unit area
$\rho_{b\square}$ is also plotted for zero field and for the maximum field used
(10 mT). The increase of $\rho_\square$ with field strength is probably produced by
the increase in the thickness of the semiconducting layer due to the
gradual transition of the weak superconducting regions near the surface.
The small peak at 88.5K is not understood and may be the result of a small
contribution from the bulk magneto resistance. The increase of $\rho_\square$ for
samples 2 and 7 in a 2mT field is shown in Fig.6.

CONCLUSIONS

The metal spraying technique can be used to produce contact pads in
situ of any geometry and large areas with contact resistivity values in
the region of $10^{-7} \Omega cm^2$ without further processing of the superconductor.
In the formulation of composites for cryogenic stabilization a reduction
in cost or an increase in performance may be achieved by using a thin
Ag buffer layer between a thicker Cu cladding and the superconductor
or by the incorporation of Ag into the bulk superconductor.

Work is under way to extend this study to lower temperatures and higher magnetic fields.

If, in a particular application, it is possible to anneal the composite YBCO + contact then contact resistivities of lower than 10^{-9} Ωcm^2 may be obtained. This compares favourably with values achieved by vacuum techniques (eg sputtering etc).

When very low contact resistivities are being measured, care must be taken to ensure that a uniform current distribution is obtained, otherwise data must be treated with caution.

ACKNOWLEDGEMENTS

We wish to acknowledge helpful discussions with R.Webb. One of us (Y.Y.) acknowledge financial support from The British Council and the Chinese Government.

REFERENCES

1. J. W. Ekin, A. J. Panson and B. A.Blankenship, Method for making low-resistivity contacts to high T_c superconductors, Appl. Phys. Lett. 52, No.4: 331 (1988)

2. J. W. Ekin et al, High T_c superconductor/noble metal contacts with surface resistivities in the 10^{-10} Ω cm^2 range, Appl. Phys Lett. 52, No.21 : 1819 (1988)

3. Yoshihiko Suzuki et al, Low-resistance contacts on $YBa_2Cu_3O_{7-\delta}$ ceramics prepared by direct wire bonding methods, Appl. Phys. Lett 54, No.7 : 666 (1989)

4. S. Jin et al, Low-resistivity contacts to bulk high T_c superconductors, Appl. Phys. Lett 54, No.25 : 2605 (1989)

5. J. D. Katz et al, Low-resistivity, YBa Cu O -to-silver electrical contacts by plasma spraying, J.Appl.Phys. 65, No.4 : 1792 (1989)

6. R. G. Edgell and W. R. Flavell, Is the surface of $YBa_2Cu_3O_{7-\delta}$ intrinsically non-metallic?, Z.Phys.B 74 : 279 (1989)

7. S. P. Ashworth, C. Beduz, K. Harrison and Y. Yang, Electrical Contacts to YBCO using metal spray techniques. Cryogenics to be published.

EFFECTS OF ANNEALING TREATMENT OF SUPERCONDUCTIVITY

IN POWDER SINTERED YBCO

X. D. Tang, J. R. Wang, S. G. Xiong, J. R. Xu,
G. R. Guo and L. Zhou

Northwest Institute for Nonferrous Metal Research
P.O. Box 71, Baoji, Shaanxi, China

ABSTRACT

The effects of annealing treatment on superconducting properties of YBCO bulk samples prepared by powder sintering have been reported and discussed. In the annealing temperature (T_A range from 525 to 750°C, the sample with a single Tc component and the highest $Tc^{\circ n}$ and Tc° can be obtained only by annealing at 750°C. When T_A=750°C the bulk samples consist of two Tc components (higher and lower Tc components). $Tc^{\circ n}$ of the higher Tc component and Tc° of the lower Tc component increase with increase of T_A. Jc has a parabolic dependence on T_A and attains a peak when T_A=700°C. The effect of annealing time (in the range of 4-36 h) on Jc and Tc is not too obvious for a given T_A. A simple explanation for these effects is that the perfection of the one-dimensional (1-D) Cu-O chains and 2-D CuO networks is related to annealing. A good superconductivity can be attained as long as the sample has the perfect 2-D Cu-O networks under the preconditon that the perfection of 1-D Cu-O chains is above a given level.

INTRODUCTION

The report by Bednorz and Müller[1] of superconductivity above 30 K in the La-Ba-Cu-O system generated an unprecedented amount of research on high temperature superconducting oxides, including the discovery of a superconducting transition above 90K in the Y-Ba-Cu-O system [2,3]. The phase responsible for superconductivity in the Y-Ba-Cu-O system, YBCO, has been identified and its nominal crystal structure has been determined. The diffraction studies show that YBCO has an oxygen-deficient perovskite structure with ordering of the yttrium and barium ions. R. Beyers et al.[4] have shown by in-situ TEM and hot-stage X-ray and neutron diffraction that maximizing of the perfection of the 1-D Cu-O chains by oxygen vacancy ordering should give the best superconducting materials. In this paper the effects of annealing treatment on Tc and Jc in YBCO will be reported and discussed. Good superconductivity can be attained as long as the sample has perfect 2-D Cu-O

Advances in Cryogenic Engineering (Materials), Vol. 36
Edited by R. P. Reed and F. R. Fickett
Plenum Press, New York, 1990

447

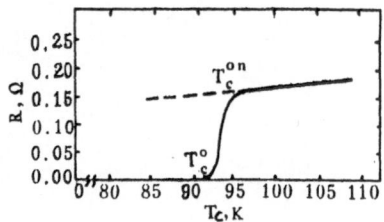

Fig. 1. The R-T curve of 750°C (x24 h) annealing sample,

networks under the precondition that the perfection of 1-D
Cu-O chains is above a given level.

EXPERIMENTS

        YBCO bulk samples were prepared by mixing Y_2O_3, $BaCO_3$
and CuO in a mortar, sintering the mixture at 920°C for
24 h (under flowing oxygen). Subsequent annealing treatments
at 525-750°C were accomplished in the process of cooling slowly
in the furnace to room temperature (1°C/min.). All samples
were determined to be single phase YBCO with an orthorhombic
structure by X-ray diffraction analysis. The resistance-
temperature curve of the samples was measured by the resistive
method with an accuracy of 0.1 K. The bulk samples were cut
into pieces with a cross-section area of $1.44mm^2$ and then Jc
was measured by a standard four-probe method (77 K, 0 T,
1 μV/cm criterion). The microstructure characteristics of some
samples were investigated by means of a SEM.

RESULTS AND DISCUSSION

        The R-T curves for the samples annealed for 24 h at
different temperatures show that the sample annealed at 750°C
consists of a single superconducting Tc component (see
Fig. 1), but there are two superconducting Tc components (higher
Tc and lower Tc) in other samples (see Fig. 2). The similar
evidence of two superconducting components in annealing samples
of YBCO has been reported.[5] The effect of annealing temperature
(T_A) on the onset transition temperature of the higher Tc
component (Tc^{on}) and zero resistance temperature (Tc^o) is shown
in Fig. 3. It is clearly illustrated Tc^{on} and Tc^o increase
with raising of the annealing temperature when T_A is in the
range of 525-750°C. For the sample annealed at 750°C, Tc^{on}

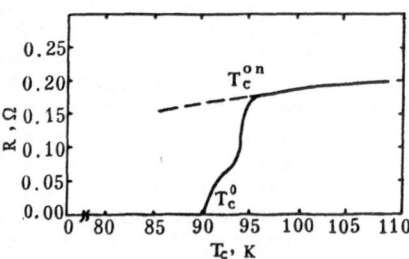

Fig. 2. The R-T curve of 525°C (x24 h) annealing sample.

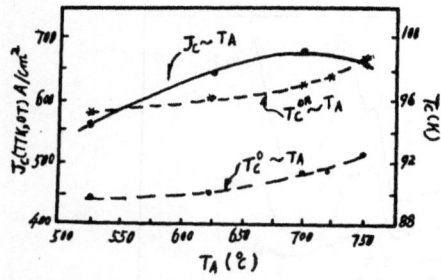

Fig. 3. The effects of annealing temperature (T_A)
on Tc^{on}, Tc° and Jc of YBCO bulk sample.

and Tc° are higher than those of all other samples. The R-T
curves are similar to the ones in Fig. 2 for the annealed
samples at 700°C for different times (t_A). The dependence of
Tc^{on} and Tc° increases slightly with increasing t_A. It is
specially worth noting that the R-T curve, the Tc^{on} and Tc°
of an un-annealed sample are the same as those of the sample
annealed at 525° C.

Fig. 5 shows the infrared ray (IR) absorption spectra
of the annealed-quenched samples.[6] In Fig. 5, the intensity
of the P1 and P2 peaks represents the perfection of 1-D
CuO chains and 2-D Cu-O networks respectively. The higher the
peak, the worse is the perfection of 1-D Cu-O chains or 2-D
Cu-O networks. Obviously, the perfection of 1-D Cu-O chains
increases with decreasing of T_A. For the samples annealed at
400-500°C, the perfection of 1-D Cu-O chains is the best.
From the point of view proposed by R. Beyers,[4] Tc should
increase with the decrease of T_A and the best Tc should be
given by annealing at 400-500°C. But, our experimental results
(Fig. 3) are contrary to it. Now, let's consider the relation
between the perfection of 2-D Cu-O networks and annealing treatments. From Fig. 5, it is found that the annealing at 750°C
can maximize the perfection of the 2-D Cu-O networks. The
perfection of the 2-D Cu-O networks of the sample annealed
at $T_A < 750$°C is worse than that at 750°C. Based on the results
of Fig. 3, we consider that slow cooling (1°C/min.) down to
room temperature is very important after annealing at 750°C.
During slow cooling, the perfection of 1-D Cu-O chains can
reach a certain level. This may be a precondition to get higher
Tc for the YBCO sintered samples. Thus the annealing treatment
for maximizing the perfection of the 2-D Cu-O networks (for
example, annealing at 750°C) should give the best Tc materials,
but the cooling rate after annealing treatment, as mentioned

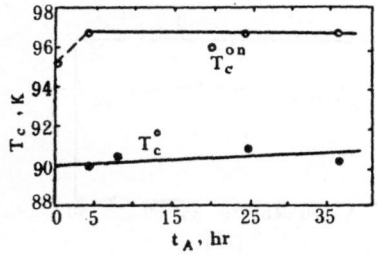

Fig. 4. The dependence of Tc^{on} and Tc°
on annealing time (t_A).

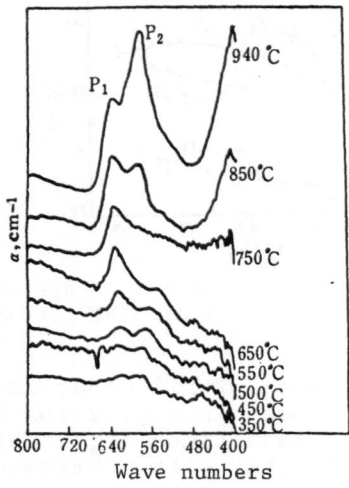

Fig.5 The infrared ray absorption spectra of
annealed-quenched samples.

above, has to be slow (e.g. 1°C/min.). When T_A=750°C, there
are two superconducting Tc components in the annealed samples.
This perhaps relates to the difference (in structure, compositon
and so on) between YBCO grains and intergranular zones. As
T_A=750°C, the difference between them is probably improved.
The R-T curve measured by the resistive method cannot show
the difference because of too small a current through the sample
when measuring Tc.

The effect of T_A and t_A on Jc is shown in Fig. 3 and
Fig. 6, respectively. When T_A is in the range of 525-750°C,
the dependence of Jc on T_A is parabolic and Jc reaches its
peak value when T_A=700°C. The change of Jc with t_A is very
small when t_A is in the range of 4-36 h (at 700°C). In order
to understand Jc-T_A and Jc-t_A relations in Fig. 3 and
Fig. 6, we have observed the microstructure annealed samples
by SEM and measured their density (ρ). For the samples
annealed at 520°C and 750°C, SEM observation shows that grain
size increases with increasing of T_A (see Fig. 7). The ρ values
of those two samples have only a little difference (respectively
5.75 and 5.83/cm3). For the samples annealed in the range of
4-36 h (at 700C) the grain size and ρ values are almost
unchanged. So, we consider that the increase of Jc with T_A
is due to the increase of Tc (Tc^{on} and Tc^{o}) of these samples,

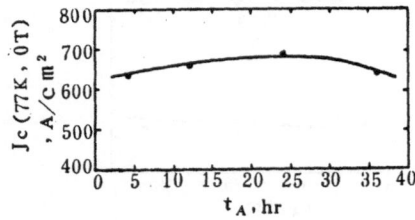

Fig.6 Jc of YBCO bulk sample as a function
of annealing time (t_A).

 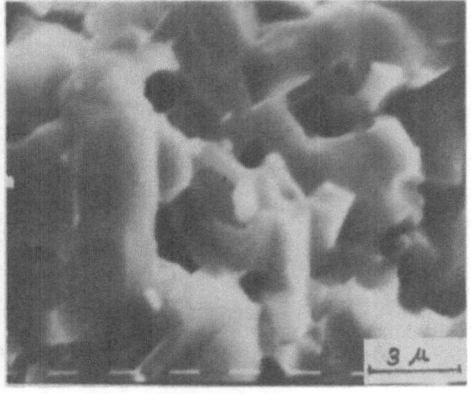

a   b

Fig.7 Photomicrographes
a.525°C annealing sample
b.750°C annealing sample

but raising T_A is unfavorable for increasing Jc because of
grain growth. The Jc-T_A curve in Fig. 3 and the Jc-t_A curve
in Fig. 6 result from simultaneous effect of those two factors
on Jc.

CONCLUSIONS

1. When YBCO bulk samples are prepared by a powder sintering
   process, it is very imporant for optimizing Tc and Jc of
   the materials to select a reasonable annealing treatment.
   In the annealing temperature (T_A) range of 525-750°C, the
   range of 525-750°C, the sample with a single Tc component
   and the highest Tc^{on} and Tc° can be obtained only by anneal
   ing at 750°C. When T_A<750°C, the samples consist of two
   Tc components. Tc^{on}_A of the higher Tc component and of
   the lower Tc component increase with increase of T_A. The
   dependence of Jc on T_A presents a parabolic pattern and
   the Jc peak value appears when T_A=700°C. At a given anneal
   ing temperature, the effects of an annealing period on
   Jc and Tc (Tc^{on}, Tc°) is very little in the range of
   4-36 h.
2. Under the condition of perfection of the 1-D Cu-O chains
   to a certain level, maximizing of the perfection of the
   2-D Cu-O networks is a decisive factor in obtaining the
   best Tc for the powder sintered YBCO bulk samples. Slow
   cooling after high temperature sintering and annealing
   (e.g. 1°C/min.) can satisfy perfection of 1-D Cu-O chains
   as requested. But the sample must be annealed at 750°C
   in order to maximize the perfection of 2-D Cu-O networks.

REFERENCES

1. J. G. Bednorz and K. A. Müller, Possible high Tc superconductivity in the Ba-La-Cu-O system, Z. Phys., B64,
   189 (1986).
2. M. K. Wu, J. R. Ashburn, C. J. Torng, P. H. Hor, R. L.
   Meng, L. Gao, Z. J. Huang, Y. Q. Wang and C. W. Chu,

Superconductivity at 93K in a new mixed phase Y-Ba-Cu-O compound system at ambient pressure, <u>Phys. Rev. Let.</u>, 58, 908 (1987).

3. Z. X. Zhao, L. Q. Chen, Q. S. Yang, Y. Z. Huang, G. H. Chen, R. M. Tang, G. R. Lin, C. G. Cui, L. Chen, L. Z. Wang, S. W. Guo, S. L. Li and J. Q. Bi, Superconductivity above liquid nitrogen temperature in Ba-Y-Cu oxides, <u>Kexue Tongbao</u>, 6, 412, (1987).

4. R. Beyers, G. Lim, E. M. Engler, V. Y.Lee, M. L. Ramirez, R. J. Savoy and R. D. Jacowitz, Annealing treatment effects on structure and superconductivity in Y Ba Cu O, <u>Appl. Phys. Lett.</u>, 51, 614 (1987).

5. R. B. Goldfarb, A. P. Clark, A. I. Braginski, and A. J. Panson, Evidence for two superconducting components in oxygen annealed single phase Y-Ba-Cu-O, <u>Cryogenics</u>, 27, 475 (1985).

6. T. S. Shi, Y. G. Zhao, P. X. Cai, H. F. Gu and and L. M. Xie, Infrared ray absorption spectra and oxygen deficient in YBCO, <u>in</u> "Annual report on high temperature superconductor," Shanghai Institute of Metallurgy Academia Sinica, Shanghai, China (1987).

PREPARATION OF YBa$_2$Cu$_3$O$_x$ THIN FILMS BY MULTISOURCE DEPOSITION*

M. Bhushan

Lincoln Laboratory, Massachusetts Institute of Technology
P.O. Box 73
Lexington, MA 02173

Abstract

Multisource evaporation and sputtering techniques for the preparation of
YBa$_2$Cu$_3$O$_x$ thin films are reviewed. The relative merits of different source materials,
processes for oxygen incorporation, and procedures for forming the superconducting
phase are discussed and the properties of films prepared by various techniques are
summarized. Methods for the monitoring of deposition rates, which is necessary for
achieving the desired film composition in a reproducible manner, are also described.

Introduction

For microwave device applications of superconducting YBa$_2$Cu$_3$O$_x$ films, it is
important to have a reliable process for producing high-quality films covering a
minimum area of the order of 20 cm^2 on low-dielectric-loss substrates. The electrical
properties required for device-quality films are a zero-resistance transition temperature T$_c$
of 90 K or above, a critical current density J$_c$ in the range of 10^6 A/cm^2 or higher at the
operating temperature, and a surface resistance at least an order of magnitude lower than
that of copper at the frequency and temperature of application. In addition, the film
surface must be sufficiently smooth to permit defining linewidths of a few micrometers or
less, and the film properties should not degraded by patterning and repeated thermal
cycling.

In stoichiometric YBa$_2$Cu$_3$O$_x$ films with x greater than 6.8, the superconductive
transition occurs at about 90 K or above. The value of J$_c$ is strongly influenced by the
crystallographic orientation of the films because of the two-dimensional nature of the
electronic conduction, which occurs mainly in the a-b plane. The highest values of J$_c$ are
obtained for epitaxial films oriented with the c axis perpendicular to the substrate. For

___

*This work is sponsored by the Department of the Air Force and the Defense Advanced
Research Projects Agency.

Advances in Cryogenic Engineering (Materials), Vol. 36
Edited by R. P. Reed and F. R. Fickett
Plenum Press, New York, 1990

453

low losses at microwave frequencies, it is important to have well-oriented films with a sharp film-substrate interface. A satisfactory technique for film preparation must therefore provide good control over the composition and crystal orientation, and the processing temperatures should be low enough to minimize film-substrate interactions.

$YBa_2Cu_3O_x$ (YBCO) films have been prepared by a variety of single-source or multisource vacuum deposition techniques. Epitaxial growth has been obtained on $SrTiO_3$ substrates, which have a good lattice match with YBCO. Laser ablation of stoichiometric material has produced well-oriented films with J_c's in the 10^6 -10^7 A/cm^2 range.[1] Films of equally good quality have been deposited by magnetron sputtering from a single composite target.[2,3] The target composition in this case is generally nonstoichiometric because of the differences in the sputtering yields and sticking probabilities of the three cations. The required target composition, which depends on the system geometry and sputtering parameters, has to be determined empirically. Resputtering effects due to negative electron bombardment of the substrate are minimized either by sputtering at high pressures (100 mTorr range) or by placing the substrate near the periphery of the target.

An attractive feature of multisource deposition techniques is that the arrival rate of each cation at the substrate can be controlled independently, allowing a greater degree of flexibility in achieving the desired film composition. The effects of a wide range of deposition conditions on the film properties can be explored without changing the source compositions. The success of these multisource methods is closely tied to the ability to monitor and control the deposition rates of each element. The growth of III-V semiconductor compounds by molecular beam epitaxy, utilizing as many as four independently controlled sources, has been highly successful. However, the difficulties in preparing YBCO are far greater as is apparent from the equilibrium phase diagram, which shows the existence of a number of ternary and quarternary oxides.[4] In addition to the correct cation ratios, the film must have an average oxygen content between 6.8 and 7 atoms per unit cell in order to obtain a T_c of 90 K (T_c decreases with decreasing oxygen content) and the crystal structure must be orthorhombic.

Deposition Methods

The multisource deposition methods currently employed for the preparation of YBCO films are broadly classified into two categories, evaporation and sputtering. Within each category, there are significant differences in the choice of source materials, the methods of achieving the desired crystal structure and oxygen stoichiometry by in-situ or post-annealing techniques, and the system configuration used to obtain large-area films. An attempt has been made here to cover the different methods, including their advantages and drawbacks, but reference is made to only a limited selection out of the large number of published reports.

Evaporation   A simple approach to YBCO film preparation utilizes thermal or electron-beam evaporation of Y, Ba and Cu metals followed by furnace annealing in O_2.[5] The as-deposited films are unstable in air because Ba metal is readily and completely oxidized to a white powder. Precautions must be taken in handling the films prior to annealing and in handling the Ba source in the vacuum chamber, especially in the

Table 1. Properties of YBCO films prepared by multisource evaporation

Cation Source Materials	Oxygen Source Type	Maximum Processing Temp. (^{0}C)	T_c (R=0) (K)	J_c at 4.2 K (A/cm^2)	Substrate	Reference Number
Y, Ba, Cu	None	800-900 Post-annealed	77	10^4	YSZ	5
Y, BaF$_2$, Cu	O$_2$ (10^{-5}Torr)	800-900 Post-annealed	90	10^6	SrTiO$_3$ LaAlO$_3$	6 9
Y, Ba, Cu	O$_2$ (10^{-3}Torr)	700 in-situ	81	2×10^6	SrTiO$_3$	10
Y, Ba, Cu	O	600 in-situ	87	2×10^5	SrTiO$_3$	12
Y, Ba, Cu	O$_3$	700 in-situ	82	6×10^5	SrTiO$_3$	13

absence of a vacuum load lock. These difficulties have been overcome by replacing Ba metal with BaF$_2$ which is chemically stable. Co-evaporation of Y, Cu and BaF$_2$ sources was first reported by Mankiewich et al.[6] and later by others [7, 8] who were successful in preparing high quality films on SrTiO$_3$ by this technique. Since BaF$_2$ evaporates as a molecule, the as-deposited films are chemically stable and can be patterned by standard photolithographic techniques even before annealing. The fluorine is removed from the films in the form of HF by annealing in wet oxygen at 850 ^{0}C. The superconducting phase is then formed by continuing the annealing process in dry oxygen. In comparison with YBCO films prepared from three metal sources, the annealed films are more resistant to moisture, and their superconducting properties do not degrade after exposure to ambient atmosphere. The stability of the films may be due to the presence of small quantities of fluorine. Recently, co-evaporation with BaF$_2$ has been used to prepare YBCO films with J_c's in the 10^6 A/cm^2 range on LaAlO$_3$ substrates, which are of great interest for microwave applications because of their excellent dielectric properties.[9]

A limitation of the BaF$_2$ process is set by the fluorine removal step, which requires annealing at temperatures greater than 800 ^{0}C. At these temperatures, interdiffusion between the substrate and the film can become significant. However, YBCO films with T_c's of over 80 K have been prepared on reactive substrates such as Si and Al$_2$O$_3$ by using a ZrO$_2$ diffusion barrier.[8]

The evaporation methods so far described all employ ex-situ annealing at temperatures of 800 ^{0}C or above. In order to reduce the processing temperature, as well as to avoid the problems resulting from the chemical instability of as-deposited films containing metallic Ba, a number of approaches have been investigated for preparing YBCO films from Y, Ba and Cu metal sources by in-situ processing. These approaches involve heating the substrate during deposition to temperature in the 600 to 700 ^{0}C range under conditions that result in the incorporation of enough oxygen to make as-deposited films superconducting. One procedure that has been used successfully is to establish a molecular oxygen partial pressure of a few millitorr adjacent to the substrate while keeping the partial pressure in the source region low enough to prevent oxidation of the metal sources.[10, 11] Successful in-situ preparation has also been achieved by injecting oxygen atoms provided by microwave plasma[12], oxygen ions generated by Kaufman ion source[12], or ozone molecules[13] into the substrate region, again maintaining a low oxygen partial pressure in the rest of the chamber. Superconducting YBCO films have also been prepared in-situ from Y, Cu and BaF_2 sources by injecting oxygen ions near the substrate.[14]

Table 1 summarizes the various evaporation techniques and the best YBCO film properties made by each.

Sputtering   Earlier difficulties with sputtering from a single source, together with the need to produce large-area films for electronic applications, have led to the development of several multitarget sputtering methods, all of which utilize post-annealing in O_2. One method employs co-sputtering of Y, Ba and Cu metal targets.[15, 16] Metal targets are preferred over oxides because of their higher sputtering yields. Cross-contamination of the targets is avoided by proper shielding and uniformity over large areas is achieved by increasing the target-to-substrate distance and by rotating the substrate.[17] Oxygen is introduced near the substrate to incorporate it in the growing film without oxidizing the targets. This process has yielded YBCO films with T_c of 90 K and J_c of 10^7 A/cm^2 at 4.2 K.

Because of the reactivity of Y and Ba, use of these metals as targets requires a vacuum load-locked sputtering chamber. Once the targets are exposed to air, pre-sputtering for several hours or days is necessary to remove the surface oxides and stabilize the sputtering rates. This problem can be alleviated by replacing Y and Ba metal targets with alloys such as YCu and Ba_2Cu_3.[18] Alternatively, as in the case of multisource evaporation, a BaF_2 target can be used. Magnetron sputtering of Y, BaF_2 and Cu has produced YBCO films on $SrTiO_3$ substrates with T_c of 91 K and J_c of 10^5 A/cm^2 at 77 K.[19]

Sequential sputtering, in which multilayer films are deposited by exposing the substrate to each of the targets in turn, permit the target-to-substrate distance to be reduced without sacrificing lateral uniformity, since the film composition is expected to be uniform over an area comparable to the target size. The targets can be spatially separated from each other to prevent cross-contamination. If the individual layer thicknesses are reduced and the substrate is heated, mixing of the elements can occur during deposition. A sequential rf diode sputtering method, using the chemically stable target materials Y_2O_3, BaF_2 and CuO, has been employed for YBCO film preparation.[20] Sputtering in an Ar-O_2 atmosphere yields as-deposited films containing very little fluorine since BaF_2 dissociates in the plasma and barium oxide is deposited. Post-annealing in dry O_2 at 800 ^{0}C or above has produced films on YSZ substrates with T_c of 90 K and J_c of 3×10^5 A/cm^2 at 4.2 K. Since the wet oxygen annealing step is not required to remove fluorine from the as-deposited films, this process has the potential for in-situ preparation of YBCO films.

Table II summarizes the various multisource sputtering techniques and the properties of the best YBCO films made by each.

Table 2. Properties of YBCO films prepared by multisource sputtering

Target Materials	Sputtering Techique	Post-Anneal Temperature (^{0}C)	T_c (R=0) (K)	J_c at 4.2 K (A/cm^2)	Substrate	Reference Number
Y, Ba, Cu	Magnetron Co-sputtering	850	90	1×10^7	SrTiO$_3$	15
YBa, BaCu	Magnetron Co-sputtering	950	75	500	Sapphire	16
Y, BaF$_2$, Cu	Magnetron Co-sputtering	800	91	10^5 (77 K)	SrTiO$_3$	19
Y$_2$O$_3$, BaF$_2$, CuO	RF Diode sequential sputtering	850	90	10^5	YSZ	20

Deposition Rate Control

For successful multisource deposition, it is necessary to monitor the arrival rate of each element at the substrate and to maintain the deposition rate constant by using a feedback loop to control the power applied to the source. The rate monitors have to be calibrated against the actual atomic concentration of elements in the YBCO film for a fixed set of deposition conditions such as system geometry, substrate temperature and oxygen partial pressure. Therefore, the ability of multisource methods to supply high-quality films on a routine basis requires reproducing these deposition conditions in each run.

Quartz-crystal thickness monitors are commonly used in both evaporation and sputtering systems. The crystals must be mounted in such a way as to avoid cross-talk between the different sources. Difficulties can arise since the monitors measure the total deposited mass. When film deposition from metal sources is carried out in the presence of oxygen, the deposited film could be a mixture of metal and oxide, with the metal fraction changing if the oxygen supply changes.

In an evaporation system, the individual evaporant fluxes can be monitored by electron impact emission spectrometers, by measuring the intensity of optical emission from elemental atoms produced by an electron beam. The Y, Ba and Cu emission lines at 407, 307, and 325 nm, respectively, can be monitored simultaneously without interference. Mass spectrometers can also be utilized to estimate the evaporation rate of each element independently. Since both of these techniques utilize hot filament sources, their use is limited to pressures below 10^{-5} Torr. Atomic absorption spectroscopy provides an alternative technique for monitoring the effluent flux at high background pressures.[13] In a sputtering plasma, the emission intensity of an excited atom gives a measure of the deposition rate.

Conclusion

Multisource evaporation and sputtering techniques have produced YBCO films that are comparable in quality to those prepared by single-source deposition methods such as laser ablation and magnetron sputtering. To date, films have been prepared in situ by evaporation, but the sputtered films require post-annealing in O_2. It is difficult to optimize the deposition conditions in a multisource sputtering method because of the large number of variables, not all of which can be measured directly. Improvment in rate monitoring techniques is required to achieve control over the film composition.

Acknowledgments

The author wished to thank A.J. Strauss for helpful discussions and assistance in the preparation of this manuscript.

References

1. T. Venkatesan, X.D. Wu, B. Dutta, A. Inam, M.S. Hagde, D.M. Hwang, C.C. Chang, L. Nazar and B. Wilkens, Appl.Phys. Lett. 54, 581 (1989).
2. Y. Enomoto, T. Murakami, M. Suzuki and K. Moriwaki, Jpn. J. Appl. Phys. 26, L1248 (1987).
3. X.X. Xi, J. Geerk, G. Linker, Q.Li and O. Meyer, Appl. Phys. Lett. 54, 2367 (1989).
4. G. Wang, S-J Hwu, S.N. Song, J.B. Ketterson, L.D. Marks, K.R. Poeppelmeier, and T.O. Mason, Adv. Ceramic Mater. 2, 313 (1987).
5. B-Y Tsaur, M. S. DiIorio and A. J. Strauss, Appl. Phys. Lett. 51, 858 (1987).
6. P. M. Mankiewich, J. H. Scofield, W. J. Skocpol, R. E. Howard, A. H. Dayem and E. Good, Appl. Phys. Lett. 51, 1753 (1987).
7. X. K. Wang, K. C. Sheng, S.J. Lee, Y. H. Shen, S. N. Song, D. X. Li, R.P.H. Chang and J. B. Ketterson, Appl. Phys. Lett. 54, 1573 (1989).
8. A. Mogro-Campero, L. G. Turner and G. Kendall, Appl. Phys. Lett. 53, 2566 (1988).
9. Martin C. Nuss, P. M. Mankiewich, R. E. Howard, B. L. Straughn, T. E. Harvey, C. D. Brandle, G. W. Berkstresser, K. W. Goossen and P. R. Smith, Appl. Phys. Lett. 54, 2265 (1989).
10. D. K. Lathrop, S.E. Russek, and R.A. Buhrman, Appl. Phys. Lett. 51, 1554 (1987).
11. Takahito Terashima, Kenji IIjima, Kazunuki Yamamoto, Yoshichika Bando and Hiromasa Mazaki, Jpn. J. Appl. Phys. 27, L91 (1988).
12. N. Missert, R. Hammond, J. E. Mooji, V. Matijasevic, P. Rosenthal, T. H. Geballe, A. Kapitulnik, M. R. Beasley, S.S. Laderman, C. Lu, E. Garwin, and R. Barton, IEEE Trans. Magn. MAG-25, 2418 (1989).
13. D.D. Berkley, B. R. Johnson, N. Anand, K.M. Beauchamp, L.E. Conroy, A. M. Goldman, J. Maps, K. Mauersberger, M. L Mecartney, J. Morton, M. Touminen and Y -J. Zhang, IEEE Trans. Magn. MAG-25, 2522 (1989).
14. Kazuyuki Moriwaki, Youichi Enomoto, Shugo Kubo and Toshiaki Murakami, Jpn. J. Appl. Phys. 27, L2075 (1988).
15. K. Char, M.R. Hahn, T.L. Hylton, M.R. Beasley, T.H. Geballe and A. Kapitulnik, IEEE Trans. Magn. MAG-25, 2422 (1989).
16. R.M. Silver, J. Talvacchio and A.L. deLozanne, Appl. Phys. Lett. 51, 2149 (1987).

17. R.W. Simon, C.E. Platt, K.P. Daly, A.E. Lee and M.K. Wagner, IEEE Trans. Magn. MAG-25, 2433 (1989).

18. M. Scheuermann, C.C. Chi, C.C. Tsuei, D.S. Yee, J.J. Cuomo, R.B. Laibowitz, R.H. Koch, B. Braren, R. Srinivasan and M. Plechaty, Appl. Phys. Lett. 51, 1951 (1987).

19. E. Wiener-Avnear, J.E. Cooper, G.L. Kerber, J.W. Spargo, A.G. Toth, J.Y. Josefowicz, D.B. Rensch, B.M. Clemens and A.T. Hunter, IEEE Trans. Magn. MAG-25, 935 (1989)

20. M. Bhushan and A.J. Strauss, this publication.

GROWTH AND SUPERCONDUCTING PROPERTIES OF $Y_1Ba_2Cu_3O_{7-\delta}$

THIN FILMS SPUTTERED ON FLEXIBLE YSZ SUBSTRATES

Satoshi Takano, Noriki Hayashi, Shigeru Okuda,
and Hajime Hitotsuyanagi

Osaka Research Laboratories, Sumitomo Electric Ind.
1-3, Shimaya 1-Chome, Konohana-Ku, Osaka, 554 Japan

Kiyoshi Hasegawa, Takuya Kisida, Jun Yamaguti

Research and Development Group
Kansai Electric Power Co., Ltd.
3-22, Nakanoshima 3-Chome, Kita-Ku, Osaka, 554 Japan

ABSTRACT

$Y_1Ba_2Cu_3O_{7-\delta}$ thin films were grown on polycrystalline, flexible YSZ substrates by RF magnetron sputtering. A Tc(R=0) of 90.3 K was obtained after annealing as the highest value. The films had c-axis orientation and were composed of islands even at a film thickness of 2.4 μm. The islands were partially impinged on each other. Jc largely depended on the degree of island impingement. A Jc value of 1.2×10^4 A/cm^2 was obtained on the film on which the islands mostly impinged on each other during annealing. By bending these films, the effect of strain on the superconducting properties was investigated. Tc(R=0) values raised as the compressive strain increased up to 0.3%, though they were lowered as the tensile strain increased. Jc did not degrade under a compressive strain of 0.3%, though it degraded by 33% under a tensile strain of 0.3%.

INTRODUCTION

There have been many processes proposed for obtaining high-Tc oxide superconductors which show a high Jc at 77.3 K. Many successful thin film depositions have given Jc above 10^6 A/cm^2. There are many reasons why the thin film had high Jc. The ability of thin film deposition to produce highly aligned materials must be one of those reasons.

Though all of the films which have been prepared to have high Jc by now are the films deposited on single crystals of oxides such as $SrTiO_3$ and MgO, it may also be possible to obtain high Jc thin films on polycrystalline materials.

Advances in Cryogenic Engineering (Materials), Vol. 36
Edited by R. P. Reed and F. R. Fickett
Plenum Press, New York, 1990

461

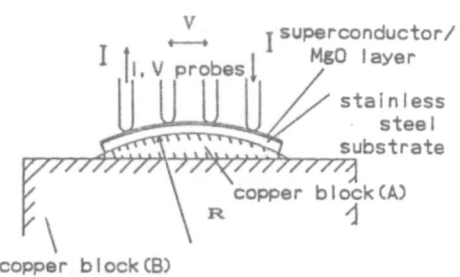

Fig. 1. Schematic diagram for measurement of the strain
effect on the superconducting properties.

There are many applications possible when films with high
Jc can be prepared on polycrystalline substrates. Especially
for superconducting wire applications, thin films grown on thin
substrates have the merit of flexibility.

$Y_1Ba_2Cu_3O_{7-\delta}$ thin films were grown on flexible YSZ (yittria
stabilized zirconia) substrates by RF magnetron sputtering.
The structural features and their effects on the superconducting
properties were investigated and compared with films grown on
(100) MgO. Strain effects on the superconducting properties
were also measured by bending these films.

EXPERIMENTAL PROCEDURE

We used polycrystalline, flexible YSZ substrates with
thicknesses of 100-200 μm.

The films were deposited on the substrates, heated to
700°C, from a composite single target with a composition of
$YBa_{2.4}Cu_{5.6}O_x$. Sputtering was carried out in an Ar+10% O_2
atmosphere with a pressure of 3×10^{-2}-8×10^{-2} torr. The films
were annealed in O_2 at 900-950°C for 2 h, cooled to 400°C at
a rate of 1°C/h and they annealed at 400°C for 3 h.

For all films, Jc was measured at 77.3 K by a four-probe
method using a criterion of 1 μV. Strain effects on the

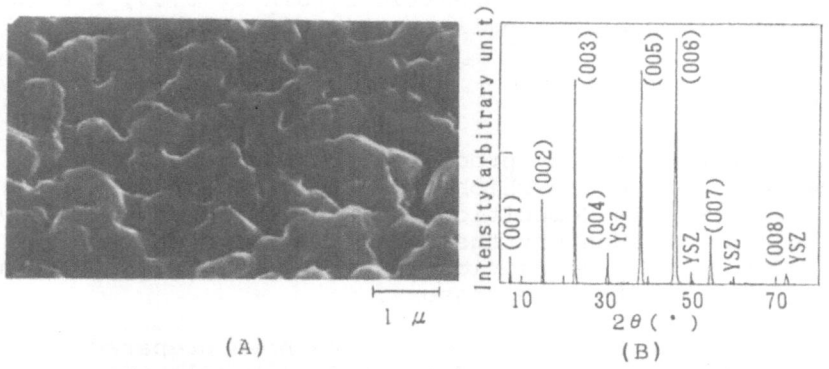

Fig. 2. (A) SEM micrographs; (B) X-ray patterns of
$YBa_2Cu_3)_{7-\delta}$ thin film grown on YSZ substrates.

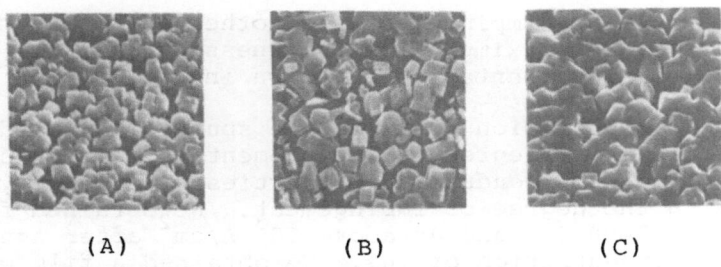

(A)   (B)   (C)

Fig. 3. SEM micrographs of the $YBa_2Cu_3O_{7-\delta}$ thin films grown on YSZ substrates under gas pressure of 8×10^{-2} torr with an O_2 concentration of (A) 50%, (B) 30%, and (C) 10%.

superconducting properties measured by bending the substrates along on a copper block with a part cylindrical surface (radius = R) as shown in Fig. 1. Several copper blocks with different radii were used to change the amount of strain, ϵ, applied to the superconducting films. ϵ was approximately calculated as follows:

$$\epsilon = (t_s + t_f) / 2R$$

where t_s and t_f are the thickness of the substrate and the film respectively.

Inductively coupled plasma atomic emission spectroscopy (ICP-AES) was used for the determination of composition. X-ray diffraction (XRD) was utilized for the characterization of structural features of the films.

RESULTS AND DISCUSSION

Bright black films were obtained on the substrates. Both as-grown films and post-annealed films were composed of islands

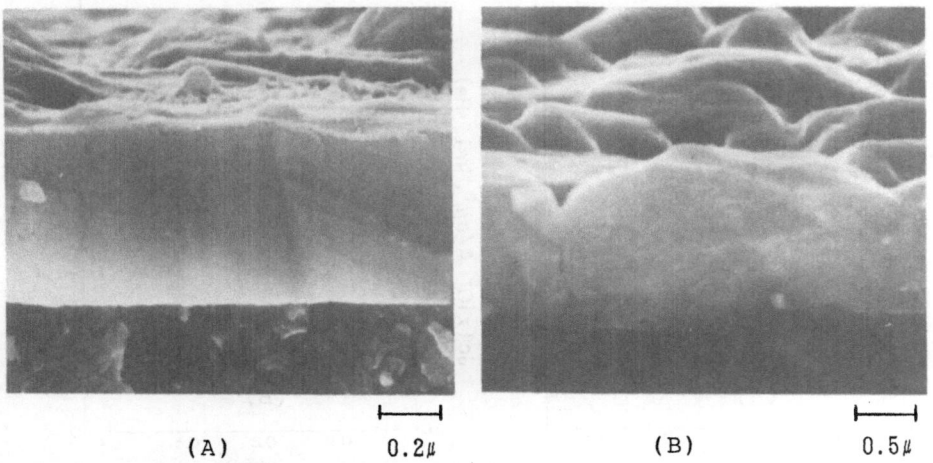

(A)   0.2μ (B)   0.5μ

Fig. 4. The cross-sectional SEM micrographs of (A) the film with Jc of 1.2×10^4 A/cm^2 grown on YSZ and (B) the film with Jc of 2.1×10^5 A/cm^2 grown on single crystal (100) MgO.

which were partially impinged on each other as shown in
Fig. 2, even at our maximum film thickness of 2.4 μm. Both
types had a c-axis orientation as shown in Fig. 2.

As the concentration of O_2 in the sputtering gas decreased
from 50% to 10%, the degree of impingement increased as shown
in Fig. 3. The superconducting properties were largely improved
by increasing the degree of impingement. We obtained films
with a Tc of 83-90.3 K and Jc above 10^3 A/cm^2 after annealing
under an O_2 concentration of 10%. We obtained a film with
a Jc of 1.2×10^4 A/cm^2 as our highest value of Jc.

Fig. 4 shows cross-sectional SEM micrographs of the film
with a Jc of 1.2×10^4 A/cm^2 grown on YSZ and a film with a Jc
of 2.1×10^5 A/cm^2 grown (100) MgO. The film with Jc of 1.2×10^4
A/cm^2 grown on YSZ is clearly more dense than the film with
Jc of 2.1×10^5 A/cm^2 grown on (100) MgO.

Fig. 5 shows the compressive strain effects on Tc(R=0),
Jc at 77.3 K and Jc at 4.2 K of the film with Jc of 5.3×10^3
A/cm^2 at 77.3 K before bending. Tc(R=0) was raised as the
compressive strain increased. Jc at 77.3 K and Jc at 4.2 K
of the film did not change up to a strain of 0.3%.

Fig. 6 shows the tensile strain effects on Tc(R=0) and
Jc at 77.3 of the film with Jc of 4.9×10^3 A/cm^2 at 77.3 K before
bending. Jc largely degraded as the tensile strain increased,
and was 80, 40 and 30% of the Jc before bending at a strain
of 0.1, 0.2 and 0.3% respectively.

The difference in the thermal expansion coefficients
between YSZ and $YBa_2Cu_3O_{7-\delta}$ seemed to result in the difference
in behavior of the superconductivity under compressive strain

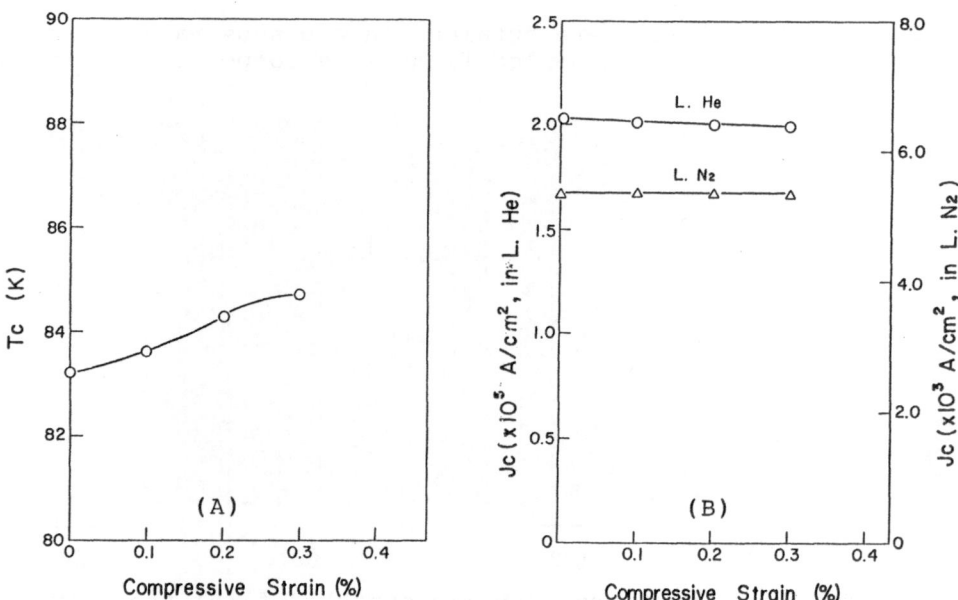

Fig. 5. The compressive strain effects on (A) Tc(R=0),
(B) Jc at 77.3 K and Jc at 4.2 K of the film
with Jc of 5.3×10^3 A/cm^2 at 77.3 K before bending.

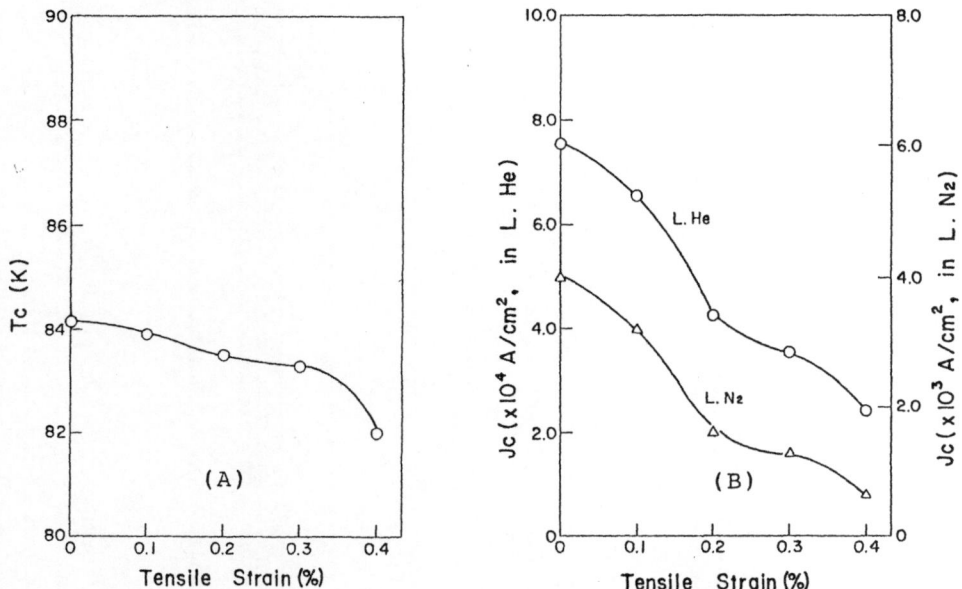

Fig. 6. The tensile strain effects on (A) Tc(R=O), (B) Jc
at 77.3 K and Jc at 4.2 K of the film with Jc of
4.9×10^3 A/cm^2 at 77.3 K before bending.

and under tensile strain. YSZ has a thermal expansion coefficient of 1.0×10^{-5}/°C during cooling from the anneal temperature
to 77.3 K. Therefore, the films grown on YSZ substrates
receive tensile stresses during cooling after annealing.

CONCLUSIONS

It has been confirmed that thin film deposition is one
of the most promising processes not only for preparing epitaxial
thin films for electronic devices, but also for manufacturing
superconductors for electric power application. Thin film
deposition can prepare high-Tc oxide superconducting wires
which are superior in flexibility to those manufactured by
other methods.

REFERENCES

1. M. K. Wu, J. R. Ashburn, C. W. Chu, Phys. Rev. Lett. 58:908
    (1987).
2. T. Nakahra, Proceedings of the First International Symposium
    on Superconductivity, Nagoya (1989).
3. H. Hitotsuyanagi, K. Satou, T. Takao, M. Nagata, Proceedings
    of 39th Electric Component Conference, Houston (1989).

THIN ALUMINUM OXIDE FILMS AS ELECTRICAL BARRIERS

BETWEEN SUPERCONDUCTING Y-Ba-Cu-O LAYERS

M. J. M. E. de Nivelle, B. Häuser, E. G. Keim, and H. Rogalla

Low Temperature Division
Faculty of Applied Physics
University of Twente
POB 217, 7500 AE Enschede
The Netherlands

ABSTRACT

Multilayer structures consisting of YBaCuO, Al_2O_3 and CuO have been sputtered on various substrates. Resistivity measurements show almost no deterioration of the electrical properties of a superconducting YBaCuO film, on top of which first an Al_2O_3 and then an YBaCuO film has been sputtered. First results of YBaCuO films deposited on the Al_2O_3 barrier show a T_c of 65 K.

Scanning electron microscopy (SEM) and scanning Auger microscopy (SAM) as applied on a cross section of a multilayer indicate no diffusion between the Al_2O_3 and YBaCuO layers.

INTRODUCTION

Since the discovery[1,2] of high T_c material many different ceramic superconductors have been found with superconducting transition temperatures (T_c) above liquid nitrogen temperature. Although compounds have been obtained with T_c's above 100 K, the $YBa_2Cu_3O_7-\delta$ compound with a T_c of about 93 K is still interesting, because of its relatively easy preparation and stability against the formation of other phases. Moreover, YBaCuO is free from the poisonous element thallium, which is essential for some of the oxidic superconductors with transition temperatures above 100 K.
Important for the application of the oxidic superconductors in electronic devices is the preparation of high quality thin films with as few grain boundaries as possible and high critical current densities. For applications like wiring or even the fabrication of Josephson junctions it would also be very interesting to separate the superconducting films by thin electrical insulating layers between them. Up to now it has always been very difficult to make such an insulating barrier between two superconducting layers due to diffusion between the different layers, which is accompanied by a loss of superconductivity.

We have investigated the applicabilty of aluminum oxide and copper oxide as an insulating barrier between YBaCuO superconducting thin layers. Because diffusion processes mostly take place along grain boundaries, least diffusion at the interfaces is expected when the oxide barrier layers are deposited in the amorphous state.

Advances in Cryogenic Engineering (Materials), Vol. 36
Edited by R. P. Reed and F. R. Fickett
Plenum Press, New York, 1990

467

For the preparation of the superconducting YBaCuO layers a nowadays, more or less, standard technique has been used. We deposited the layers at heated substrates by rf magnetron sputtering from a stoichiometric ceramic YBaCuO target. To ensure that there is enough oxygen load into the film during the deposition process, a mixture of argon and oxygen has been used as a sputtering gas. At the deposition temperatures used (between 600 and 750 ^{0}C) the film immediately grows in the correct crystalline phase, without need for a post-deposition anneal at higher temperatures (typically 920 ^{0}C).[3] The films prepared by this, so called, one step process have, compared with the films prepared by the three step process (with post anneal), less grain structure and a smoother surface. This is especially important for the preparation of a closed and homogeneous barrier layer on top of the YBaCuO film.

For the deposition of the insulating barrier layers we used both rf and dc magnetron reactive sputtering in a pure oxygen atmosphere. Metallic aluminum and copper targets were used. Solely oxygen was used to avoid sputtering of metallic Al or Cu, possibly leading to conducting barrier layers.
The actual preparation set-up is shown in Fig. 1. The sputtering chamber is evacuated by a turbo pump. Inside the chamber four targets can be mounted on four different cathodes. We use disk shaped targets with a diameter of 50 mm and a thickness of 3 mm. The YBaCuO target is prepared from metal oxides by sintering and annealing in oxygen. The ceramic target is glued to a copper mounting plate with silver conductive adhesive. Together they are clamped and pressed onto a water cooled cathode. Indium foil in between them ensures good thermal contact. Metallic aluminum and copper targets are mounted on two other cathodes.

The substrates are clamped onto a substrate holder. The holder is made of stainless steel and can be heated resistively. The substrate temperature is measured with a thermocouple, placed in a bore hole in one side of a sapphire dummy substrate. The holder is mounted on a rotatable shaft which can be turned from outside the vacuum chamber. Through this facility it is possible to change the target during the sputtering process. This allows the deposition of sandwich structures without affecting the vacuum. As a result there is no contamination with e.g. carbon dioxide at the interfaces of the different layers, leading to possible deterioration of the superconductivity. With this set-up, mono- and multi layered films have been prepared on yttrium-stabilized zirconia (YSZ), on strontium titanate ($SrTiO_3$) and on magnesium-oxide (MgO) single crystalline (100)-oriented substrates.

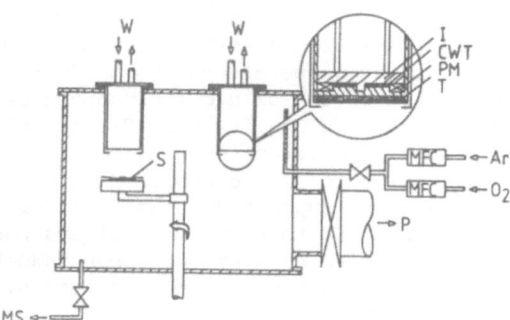

Fig. 1. Schematic cross-section of the sputtering chamber.
(W = cooling water, S = substrates, MS = mass spectrometer, I = iron core, CWT = cooling water tube, PM = permanent magnet, T = target, MFC = mass flow control, P = turbo pump)

Table 1. Sputtering parameters of an YBaCu-Al$_2$O$_3$-YBaCuO multilayer deposited on a (100)-oriented YSZ substrate [#456]

	YBaCuO	Al$_2$O$_3$	YBaCuO
sputtering time [min]	200	270	205
substrate temperature [^{o}C]	740	RT	670
argon partial pressure [Pa]	72	0	41
oxygen partial pressure [Pa]	48	80	27
distance target to substrate [mm]	43	44	43
rf power at 13,5 MHz [W]	100	100	100
cathode self-bias voltage [V]	50	110	50

PROPERTIES

Single YBaCuO films, prepared by the one-step sputtering technique, all show very smooth surfaces. They look shiny and from SEM pictures it appears that there are no grain boundaries. Depending on the substrates that have been used, the films have typical transition temperatures (zero resistivity) between 79 K (MgO) and 89 K (SrTiO$_3$) and residual resistivity ratios between 2.2 (MgO) and 3.1 (YSZ).

On top of these films we have deposited different additional layers. For instance, on an already characterized 110 nm thick YBaCuO film on YSZ substrate, with T$_c$ = 87 K and RRR = 3.1, an aluminum oxide layer of approximately 50 nm thickness and a second YBaCuO film were deposited (sputtering parameters listed in Table 1). By covering part of the first YBaCuO layer (two areas at the ends of the film) during sputtering of the next two layers, electrical contacts could be made to the first layer and the resistivity of the film beneath the alumina and YBaCuO layer could be measured (see Fig. 2). Resistivity measurements revealed that there is almost no degradation of the electrical properties of the first YBaCuO layer: T$_c$ is still 87 K, only RRR decreased from 3.1 to 2.7. This slight decrease of RRR could be due to some diffusion, or some oxygen loss out of the YBaCuO layer, especially at the contact areas on which no aluminum oxide was deposited, or due to the ion bombardment during the deposition of the upper layers.

Typical results of the electrical resistivity of the toplayer of an YBaCuO-Al$_2$O$_3$-YBaCuO sandwich show a superconducting onset at 90 K, a T$_c$ (zero resistivity) of 65 K and a RRR of 1. The reduced transition temperature and the low resistance ratio stem probably from the relative low deposition temperature of the top YBaCuO layer (see Table 1). A study to optimize the film deposition conditions for the top layer (especially the temperature) is in progress. In addition alternative methods like laser ablation[4] and indirect sputtering[5] will be used too, because the growth

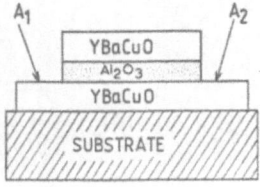

Fig. 2. Schematic cross-section of an YBaCuO-Al$_2$O$_3$-YBaCuO multilayer structure with two contact areas (A$_1$ and A$_2$) at the ends of the first deposited film. The contact areas have been covered during deposition of the next two layers.

Table 2. Sputtering parameters for a multilayer film deposited on (100)-oriented MgO substrate [#406]

	YBaCuO	Al$_2$O$_3$	YBaCuO	CuO	YBaCuO
sputtering time [min]	125	70	130	30	130
substrate temperature [$^{\circ}$C]	600	600	600	600	600
argon partial pressure [Pa]	6	–	6	–	6
oxygen partial pressure [Pa]	4	1,1	4	1,1	4
distance target to substrate [mm]	33	33	33	33	33
power [W]	80(rf)	40(dc)	80(rf)	25(dc)	80(rf)
cathode (self-bias) voltage [V]	-50	-250	-50	-360	-50

mechanism for YBaCuO films on the amorphous Al$_2$O$_3$ layer may be dependent on the deposition method.

Up to now there has been no clear characterization of the insulating properties of the alumina layer. From SEM pictures and Auger line scan measurements we expect good properties as an electrical barrier. Preliminary electrical measurements support this view.

The SEM picture and Auger line scan, shown in Fig. 3, have been recorded on a cross-section of an intentionally broken multilayer structure on a MgO substrate (sputtering parameters listed in Table 2). Experimental details are discussed elsewhere[6].

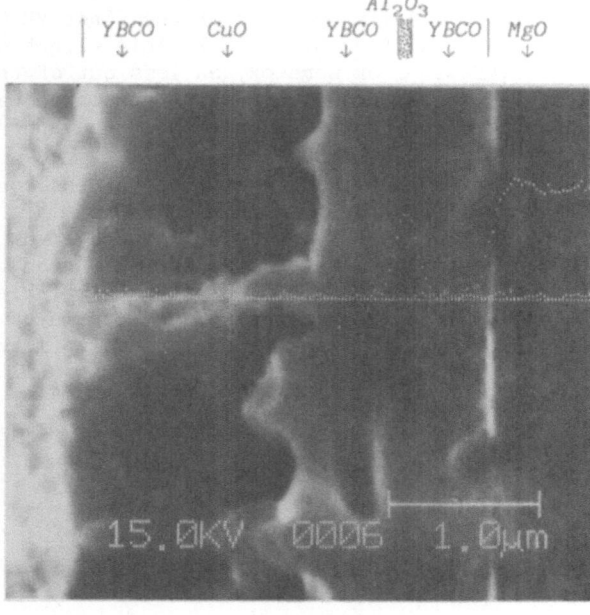

Fig. 3. SEM image of a broken MgO substrate with a YBaCuO-Al$_2$O$_3$-YBaCuO- -CuO-YBaCuO multilayer sputtered on top of it. The Auger Mg and Al line scans are superimposed on the SEM image (white dots). The aluminum oxide layer of about 170 nm thickness can be seen as a smooth dark band.

In contrast to the copper oxide layer, which has also been enclosed in this multilayer structure, the Al_2O_3 layer with a thickness of about 170 nm is smooth and has sharp boundaries. Within the experimental lateral resolution of the Auger line scan measurements[6] (about 90 nm) no diffusion between the alumina and YBaCuO layers can be observed, which is usually found between films and single crystal substrates.[7]

CONCLUSION

We have shown the possibility to deposit Al_2O_3 on an YBaCuO film without destroying the electrical properties of the superconductor. On top of such an amorphous alumina layer still superconducting YBaCuO films can be grown. Within the resolution limits, SEM pictures and Auger line scans indicate no diffusion between the alumina and Y3aCuO layers.
Our results show the potential of aluminum oxide as insulating barrier in high T_c superconducting electronics.

REFERENCES

1. J. G. Bednorz, and K. A. Müller, Possible high T_c superconductivity in the Ba-La-Cu-O system, Z. Phys. B 64:189 (1986).
2. M. K. Wu, J. R. Ashburn, C. J. Torng, P. H. Hor, R. L. Meng, L. Gao, Z. J. Huang, Y. Q. Wang, and C. W. Chu, Superconductivity at 93 K in a new mixed-phase Y-Ba-Cu-O compound system at ambient pressure, Phys. Rev. Lett. 58:908 (1987).
3. H. Adachi, K. Hirochi, K. Setsune, M. Kitabatake, and K. Wasa,. Low-temperature process for the preparation of high T_c superconducting thin films, Appl. Phys. Lett. 51 (26):2263 (1989).
4. D. H. A. Blank, D. J. Adelerhof J. Flokstra, and H. Rogalla, Parameter study of in-situ grown superconducting YBaCuO thin films prepared by laser ablation, proceedings M^2S-HTSC Stanford (1989).
5. J. Gao, B. Häuser, and H. Rogalla, Preparation of $YBa_2Cu_3O_x$ ultra thin (~15 Å) films by a modified rf-magnetron sputtering technique, submitted to Appl. Phys. Lett..
6. E. G. Keim, J. Halbritter, B. Häuser, and H. Rogalla, Scanning Auger microscopy as applied to the analysis of highly textured YBaCuO thin films, Journ. Less-Common Metals 151:23 (1989).
7. e.g. T. Venkatesan, C. C. Chang, D. Dijkkamp, S. B. Ogale, E. W. Chase, L. A. Farrow, D. M. Hwang, P. F. Miceli, S. A. Schwarz, J. M. Tarascon, X. D. Wu, and A. Inam, Substrate effects on the properties of YBaCuO superconducting films prepared by laser deposition, J. Appl. Phys. 63 (9):4591 (1988).

PREPARATION OF HIGH-Tc SUPERCONDUCTING THICK FILMS BY A LOW-PRESSURE

PLASMA SPRAYING

K. Tachikawa,
Y.Shimbo*, M.Ono*, M.Kabasawa* and S.Kosuge*

Department of Materials Science, Tokai University, Hiratuka
Kanagawa 259-12, Japan

*Steel Research Center, NKK Corporation, Kawasaki-ku
Kawasaki 210, Japan

ABSTRACT

Superconducting thick films of YBaCuO systems have been prepared by a low pressure plasma spraying technique. After the post-annealing in oxygen at 900~970℃, YBaCuO films showed zero resistance temperatures , T_c of ~90K. In order to improve the critical current density ,J_c of the films, a melt-reaction method has been newly developed. A flux layer richer in Ba and Cu was coated on a $YBa_2Cu_3O_x+Y_2BaCuO_x$ mixure layer by a low pressure plasma spraying, and then the double-layered film was annealed at 930~970℃, where the flux layer melted. The molten oxide acted as a flux and enhanced grain growth of $YBa_2Cu_3O_x$. The excess flux reacted with Y_2BaCuO_x and was fully converted to $YBa_2Cu_3O_x$. The grain growth of YBa_2Cu_3Ox and the reaction between the flux and Y_2BaCuO_x yielded a dense and uniform $YBa_2Cu_3O_x$ layer, so that the J_c of the films were improved up to 2000A/cm². This might be due to the dense structure which improves the coupling between the grains of the superconducting phase, and the grain growth which ruduces the grain boundary.

INTRODUCTION

Since the discovery of $YBa_2Cu_3O_{7-x}$[1] with a superconducting transition above the liquid nitrogen temperature, numerous efforts have been conducted to prepare thick films of the high-Tc superconductors. We have developed a low-pressure plasma spraying technique for the deposition of $YBa_2Cu_3O_{7-x}$ thick films[2]. By this technique, dense films can be deposited on the substrate with complicated shapes and large areas at a high deposition rate. Thus, from a practical point of view, low-pressure plasma spraying is much advantageous to fabricate thick films of the oxide superconductor.

Advances in Cryogenic Engineering (Materials), Vol. 36
Edited by R. P. Reed and F. R. Fickett
Plenum Press, New York, 1990

473

However, the films of $YBa_2Cu_3O_{7-x}$ prepared by plasma spraying contain some insulating or semiconducting phases such as $BaCuO_2$, Y_2BaCuO_5 or CuO, which were produced by the decomposition of $YBa_2Cu_3O_{7-x}$ during melting[3]. This second phase intrusion, which suppress the Jc of the plasma sprayed films, can not be avoided by optimizing the spraying or the post-annealing conditions.

Therefore, we have developed a new melt-reaction method to avoid the formation of the second phases. By this method, a plasma sprayed YBaCu oxide layer is converted into a dense and uniform $YBa_2Cu_3O_{7-x}$ layer, which includes little second phases. Jc of the layer is much higher than that of the $YBa_2Cu_3O_{7-x}$ film prepared by a conventional low pressure plasma spraying process.

In this report, firstly, we mention the second phase formation during the spraying and its effect on the Jc of the films, and secondly, we report the melt-reaction method and the improvement in Jc by this method.

EXPERIMENTAL PROCEDURE

Fig.1 shows the schematic drawing of present low pressure plasma spraying. The powders were injected into the plasma jet, then melted and blasted onto the substrate to form the film. Spraying was carried out in a chamber under 60 torr O_2.

The feedstock powder were prepared by the solid state reaction method. Mixed powder of Y_2O_3, $BaCO_3$ and CuO with nominal composition of Y:Ba:Cu=1:2:3 ($YBa_2Cu_3O_7$ powder) was calcined at 950°C and pulverlized. Powders with particle size ranging 26-44μm were used for spraying.

Meanwhile spraying powders with nominal composition of Y:Ba:Cu=3:3:4 (powder I) and Y:Ba:Cu=1:13:26 (powder II) were prepared for the melt-reaction method.

Electrical resistivity of the samples were measured by a standard four probe method. Transport critical current, Ic was obtain from I-V charactaristic curves at liquid nitrogen temperature. The microstructure

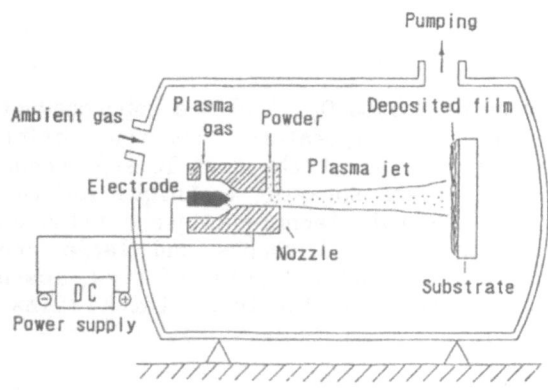

Fig.1 Schematic drawing of low pressure plasma spraying apparatus.

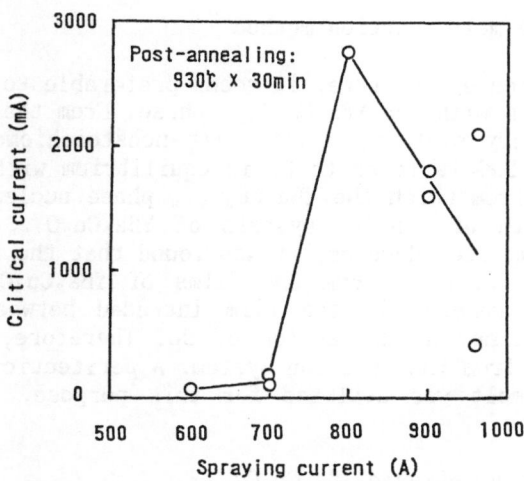

Fig.2 Effect of spraying current on critical current (at 77K) of films.

of the films were examined by an optical microscope and SEM. The chemical composition and the distribution of the elements in the films were studied by EDX analysis.

RESULTS AND DISCUSSION

Films prepared by a conventional low pressure plasma spraying.

The melting of the powder during spraying causes the decomposition of $YBa_2Cu_3O_{7-x}$ particles into Y_2BaCuO_5, $BaCuO_2$, and CuO as revealed by the ternary phase diagram[4], resulting in second phase formation in the as sprayed films. In order to obtain the desired $YBa_2Cu_3O_{7-x}$ phase in films, a post-annealing should be performed.

Spraying current affects the melting behavior of the powder, and conseqently the microstructure of films which is closely related to Jc.

The films with thickness of 80μm were deposited with $YBa_2Cu_3O_{7-x}$ powder on Ni-alloy substrates (Nimonic 80A), and subsequently annealed in oxygen at 930℃ for 30 minuites. Fig.2 shows the relationship between the spraying current and the critical current of the films. The film became denser as the spraying current increased, because the powder melted well at higher spraying currents. However, the critical current decreased when the spraying current exceeded 800A. Fig.3 shows the microstructure and distribution of elements of the films sprayed at 600A, 800A and 900A recpectively. A great amount of $BaCuO_2$ phase remained in the films sprayed at 900A even after post-annealing. This second phase might suppress Jc of the film. The formation of the second phases could be prevented by spraying at currents less than 800A. However, in that case, the powder did not fully melt, resulting in the formation of porous films and weak connectivity of grains. The muximum Jc obtained in the film was 700A/cm²

These results lead to the conclusion that a new process which produces dense films without second phase should be developed to achieve further enhancement in Jc.

Films prepared by a melt-reaction method

To produce a dense structure, it seems preferable to utilize a liquid phase in equilibrium with the $YBa_2Cu_3O_{7-x}$ phase. From the phase diagram of $YO_{1.5}$–BaO–CuO ternary system[4], a melt with nonstoichiometric composition richer in CuO and BaO is found to be in equilibrium with the $YBa_2Cu_3O_{7-x}$ phase. This melt, from which the $YBa_2Cu_3O_{7-x}$ phase nucleates directly, is used for the growth of single crystals of $YBa_2Cu_3O_{7-x}$. This method is known as the flux method. However, it was found that the flux method could not be applied directly to form the films of $YBa_2Cu_3O_{7-x}$, because the excess flux which remains in the film intruded between the grains of $YBa_2Cu_3O_{7-x}$, resulting in suppression of Jc. Therefore, the excess flux must be taken away from the reaction system. A peritectic reaction between Y_2BaCuO_5 and the melt was utilized for this purpose. This reaction is expressed as follows

$$Y_2BaCuO_5 + Liquid \rightarrow YBa_2Cu_3O_{7-x}$$

Through this reaction, the flux is converted into the $YBa_2Cu_3O_{7-x}$ phase.

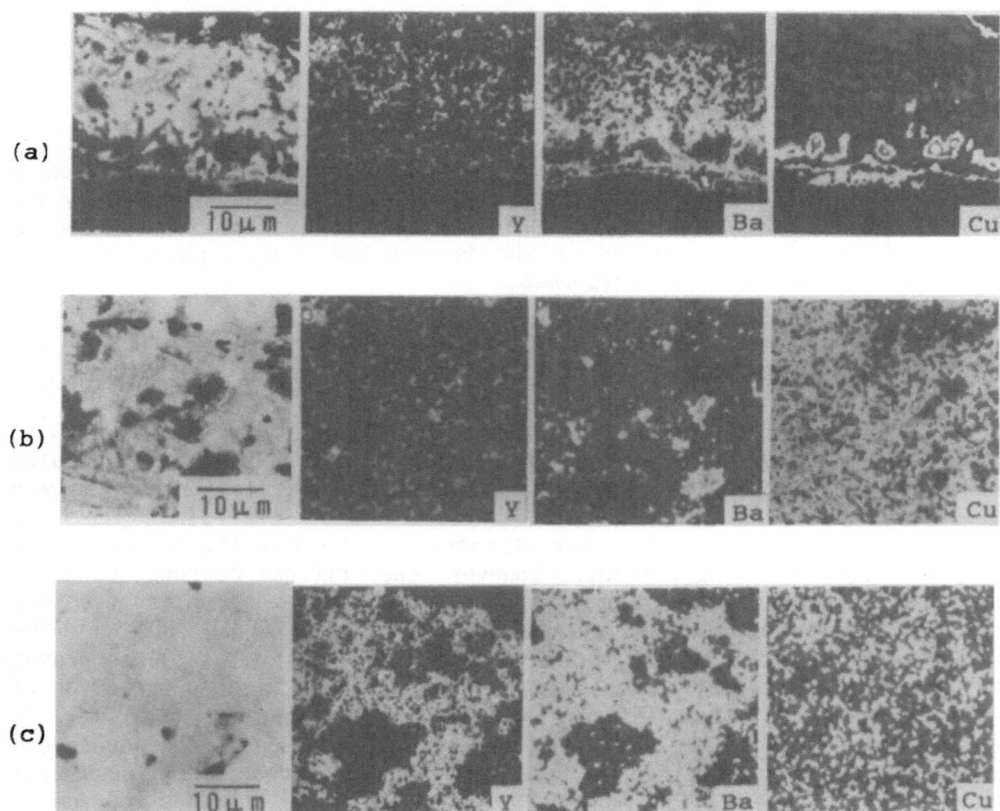

Fig.3 SEM images and distributions of Y, Ba and Cu in the cross section of annealed films.
Spraying current; (a) 600A, (b) 800A, (c) 900A.

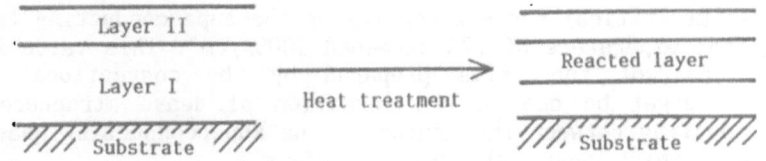

Fig.4 Schematic diagram of melt–reaction method.

The melt–reaction process consisted of two stages as shown in fig.4. Firstly, layer I composed of mixture of $YBa_2Cu_3O_{7-x}$ and Y_2BaCuO_5 was formed with the powder I on the Ni–alloy substrates by a plasma spraying. Subsequently, layer II composed of mixture of $BaCuO_2$, CuO and $YBa_2Cu_3O_{7-x}$, was overlayed on the layer I by a plasma spraying with the powder II. Then, the double–layered film was annealed in oxygen at 930℃ to 970℃, where the layer II melted. The molten oxide diffuses into the layer I and acted as a flux which enhances the grain growth of $YBa_2Cu_3O_{7-x}$ in the layer I. Simultaneously the excess flux reacts with Y_2BaCuO_5 to form $YBa_2Cu_3O_{7-x}$[5]. These reactions yield a dense uniform $YBa_2Cu_3O_{7-x}$ layer.

The microstructure of the oxide layer after annealing was studied by an optical microscope, SEM, and EDX analysis. Layer I and layer II reacted each other, and as a result, a dense and uniform layer with thickness of about 50μm grew between layer I and layer II. Fig.5 shows a SEM image and distribution of elements in the reacted layer. It consists of large rectangular grains about 20μm long and 5μm wide. EDX analysis revealed that the reacted layer mainly consisted of $YBa_2Cu_3O_{7-x}$ and contained a small amount of Y_2BaCuO_5 and CuO grains.

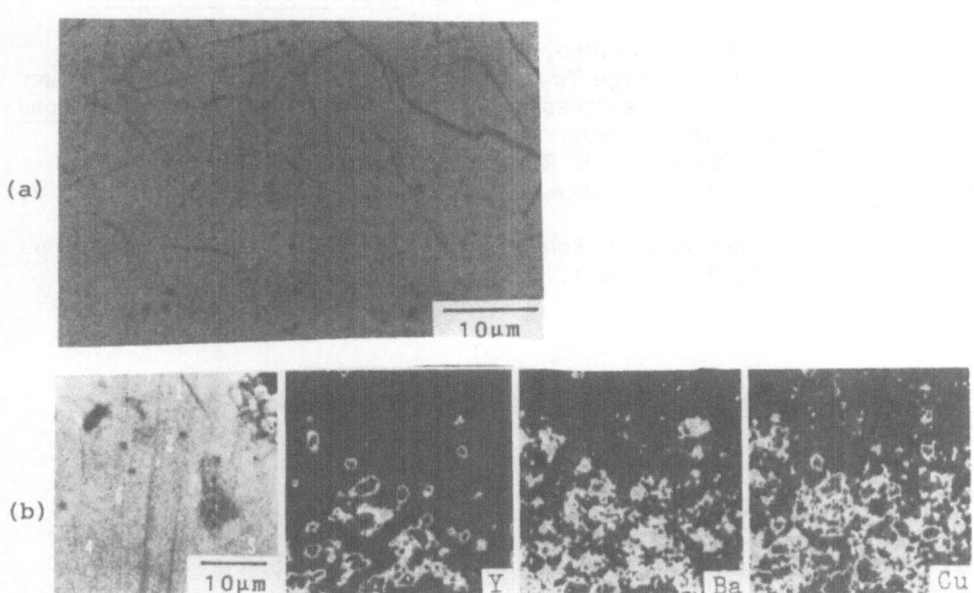

Fig.5 SEM images in a cross section of the oxide layer after annealing.
(a) A SEM image in a cross section of the reacted layer.
(b) A SEM image and distributions of Y, Ba and Cu in a cross section of the reacted layer and layer I (the boundary between two layers is about the middle of the photograph)

The transport critical current density of the superconducting layer at liquid nitrogen temperature of 77K exceeded 2000A/cm². This value is much higher than that of the films prepared by the conventional plasma spraying. This might be due to the formation of dense structure which improve the coupling between the grains of the superconducting phase, and the grain growth which reduce the grain boundary.

CONCLUSION

Superconducting thick films of $YBa_2Cu_3O_{7-x}$ were prepared by a low pressure plasma spraying. In order to improve the Jc of the sprayed films, a melt-reaction method, which utilized nonstoichiometric melt richer in Ba and Cu in equiblium with $YBa_2Cu_3O_{7-x}$, and the peritectic reaction between the melt and the Y_2BaCuO_5, has been developed. By this method, a dense and uniform $YBa_2Cu_3O_{7-x}$ films were formed, and as a result, the Jc of the film was significantly increased. The melt reaction method can be applied to products of complicated shape or large area, which is very advantageous for practical use of high-Tc oxide superconducting films.

REFERENCES

1. M. K. Wu, J. R. Ashburn, C. J. Torng, P. H. Hor, R. L. Meng, L. Gao, Z. J. Huang, Y. Q. Wang and C. W. Chu, Superconductivity at 93K in a New Mixed-Phase Y-Ba-Cu-O Compound System at Ambient Pressure. Physical Review Letters. 58; 908 (1987).
2. K. Tachikawa, I. Watanabe, S. Kosuge, M. Kabasawa, T. Suzuki, Y. Matsuda and Y.Shimbo, High-Tc Superconducting Films of Y-Ba-Cu Oxides Prepared by Low-pressure Plasma Spraying. Applied Physics Letters. 52; 1011 (1988).
3. K. Tachikawa, M. Ono, Y. Shimbo, T. Suzuki, M. Kabasawa and S. Kosuge, Preparation of High-Tc Superconducting Thick Films and Power Conducting Tubes by a Low-Pressure Plasma Spraying. IEEE Transactions on Magnetics. 25; 2029, (1989).
4. R. S. Roth, K. L. Davis and J. R. Dennis, Phase Equilibria and Crystal Chemistry in the System BaYCuO. Advanced Ceramic Materials. 2; 303 (1987).
5. N. Sadakata, M. Sugimoto, O. Kohno and K. Tachikawa. IEEE Transactions on Magnetics. 25; 2180 (1989).

NORMAL-STATE RADIATIVE PROPERTIES OF

THIN-FILM HIGH-TEMPERATURE SUPERCONDUCTORS

P.E. Phelan*, M.I. Flik§, and C.L. Tien†

Department of Mechanical Engineering
University of California
Irvine, CA 92717

ABSTRACT

One of the first applications of thin-film high-temperature superconductors
will be in the construction of liquid-nitrogen-temperature superconducting
bolometers. Successful design of a bolometer requires sufficient knowledge of
the radiative properties, and especially the absorptance, of the superconductor, if
the superconductor is to interact directly with the incident radiation. At present,
little quantitative information concerning the radiative properties of thin-film
high-temperature superconductors is available to the designer. Here, a
predictive model employing the Drude free-electron theory is applied to films of
the order of 1 μm thick. The only measured parameter required by the model is
the direct current electrical resistivity. Experimental data show good agreement
with the model's results for far-infrared normal-state properties for the
temperature range 100 K - 300 K.

INTRODUCTION

Thin-film high-temperature superconductors (HTSC) show promise for use in
bolometers capable of detecting infrared radiation.[1] Some applications of this
technology could be in surveillance systems, radio astronomy, imaging devices,
and the detection of "hot spots" in machinery or even in the human body.
Fundamental to the operation of a bolometer is a quantitative description of the
radiative properties of the absorbing material. If no optical coating is used, then
the absorber is identical with the bolometric element, i.e., the superconducting
film. Otherwise, the superconductor may still interact directly with the incident
radiation, provided the radiation is able to penetrate through the absorber and
reach the superconductor. Additionally, the quality of thin-film HTSC fabricated
by sputtering depends crucially on the radiative heat transfer from the film
surface during the sputtering process,[2] rendering a working knowledge of the
normal-state radiative properties essential to a more complete understanding of
the sputtering process.

---

* Also, Research Assistant in Mechanical Engineering, University of California
at Berkeley, Berkeley, CA 94720
§ Presently, Assistant Professor, Department of Mechanical Engineering,
Massachusetts Institute of Technology, Cambridge, MA 02139
† UCI Distinguished Professor

Advances in Cryogenic Engineering (Materials), Vol. 36
Edited by R. P. Reed and F. R. Fickett
Plenum Press, New York, 1990

479

Previous experimental studies of HTSC have primarily been concerned with their electron and phonon properties and have not provided a complete description of their radiative properties that is appropriate for engineering calculations. These studies can be separated into two categories: (1) bulk samples; and (2) epitaxial thin films and single crystals. Specular reflectance measurements of bulk samples have revealed a very complex reflectance spectrum. These investigators, for the most part, have fitted their data with a Drude-Lorentz model that assumes both free and bound charges interact with the incident radiation.[3-6] Contrarily, near-normal reflectance measurements of well-oriented thin films and single crystals do not indicate the presence of phonon contributions,[7-14] which implies a metallic behavior for the *ab*-plane—the plane parallel to the substrate in a *c*-axis-oriented thin film. Again, these authors have generally fitted a Drude-Lorentz model through their data.

What is lacking is a general model capable of *predicting* the radiative properties of thin-film HTSC; such a model would not have to rely on curve-fitting through measured reflectances, but would instead use a minimum of experimental data. Furthermore, this model would also consider substrate and film thickness effects. Since it appears that HTSC bolometers will be most useful for wavelengths greater than 20 µm,[1] an analytic model capable of predicting the radiative properties of HTSC in this wavelength range is required. The wavelengths of interest in the sputtering process, however, are less than 10 µm, and so the model should also apply to this wavelength region.

This study provides a general, predictive model capable of providing the far-infrared normally-incident specular reflectance and absorptance of normal-state thin-film HTSC of the order of 1 µm thickness. The present model relies on the simple Drude theory, which differs from the Drude-Lorentz theory in that it assumes only free charges interact with the incident radiation. The complex refractive index calculated from the Drude model is inserted into the electromagnetic equations governing an interface between two semi-infinite media. Results are generated for a $YBa_2Cu_3O_7$ film at temperatures ranging from 100 K to 300 K. Superconducting effects are briefly discussed for their impact on the radiative properties.

DRUDE THEORY

The observed metallic behavior of the *ab*-plane of thin-film $YBa_2Cu_3O_7$ suggests the use of the simpler Drude free-electron theory rather than the Drude-Lorentz theory. This model neglects all bound charges and assumes the free electrons are independent of one another. The effects of the lattice are represented by a viscous damping term, which serves to decelerate the electrons that have been accelerated by the electric field. The refractive index N, defined as $N \equiv n + ki$, derived from the Drude free electron theory where the magnetic permeability is assumed to be that of free space, is[15-18]

$$N = \left[1 - \frac{2\,\omega_p{}^2}{\omega(\omega + \beta i)} \right]^{1/2}$$

(1)

where ω is the angular frequency of the incident radiation and the plasma frequency ω_p and the damping coefficient β are given by

$$\omega_p \equiv \left[\frac{N_e\,e^2}{m\,\varepsilon_o} \right]^{1/2} \qquad \text{and} \qquad \beta = \frac{N_e\,e^2\,r_{dc}}{m}$$

(2a,b)

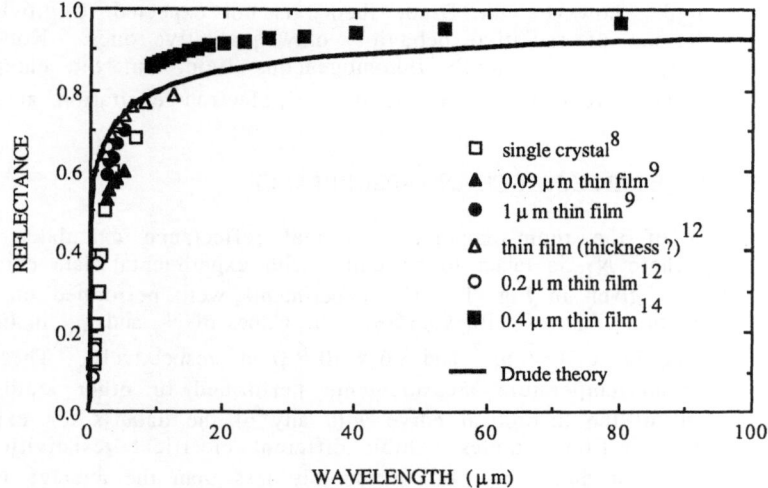

Fig. 1 A comparison of room-temperature experimental data with the normal specular reflectance calculated from the Drude free-electron theory. Two sets of data are for near-normal incidence,[8,14] two sets for normal incidence,[12] and for two sets the incidence angle is unknown.[9]

where N_e is the electron number density, e the absolute value of the electron's charge, m the electron rest mass, ε_0 the electrical permittivity of free space, and r_{dc} the direct current (dc) electrical resistivity. The value of N calculated from Eq. (1) can be inserted into the general expression governing the normal reflectance ρ at the interface between two *semi-infinite* media:[18]

$$\rho = \left| \frac{N_1 - N_2}{N_1 + N_2} \right|^2 \tag{3}$$

in which N_1 refers to the complex refractive index of the incident medium (typically vacuum or air, *i.e.*, $N_1 = 1$) and N_2 refers to that calculated from Eq. (1).

A tacit assumption of the Drude theory, and of the Drude-Lorentz theory as well, is the spatial homogeneity of the electric field. This assumption essentially requires the electron mean free path l to be much less than the penetration depth of the electric field, so that an individual electron experiences an electric field that varies only with time. The electron mean free path can be estimated from[20]

$$l \sim \frac{2\pi\hbar}{N_e^{2/3} e^2 r_{dc}} \tag{4}$$

where $\hbar$ is Planck's constant. Equation (4) is strictly applicable only for zero temperature. Calculations based on Eq. (4) and on measurements performed for the temperature range 100 K - 300 K[21] indicate $l \sim 0.02$ μm for 100 K and $l \sim 0.005$ μm for 300 K. As demonstrated in the following section, this magnitude of l is much less than the penetration depth for far-infrared wavelengths ($\lambda \geq 5$ μm).

For short wavelengths, however, the Drude theory is not expected to provide quantitative predictions of radiative behavior—only qualitative ones. Not only does an electron experience a spatially *in*homogeneous field, but the damping coefficient β becomes frequency-dependent due to electron excitation at higher frequencies.

ROOM-TEMPERATURE REFLECTANCE AND ABSORPTANCE

A comparison of the room-temperature normal reflectance calculated from Eqs. (1) and (3), where N_1 is taken to be unity, with experimental data extracted from other studies is given in Fig. 1. The experiments were performed on epitaxial, *c*-axis up, thin films of $YBa_2Cu_3O_7$. The values of N_e and r_{dc} in Eqs. (2a,b) are taken to be 7.4 x 10^{27} m^{-3} and 8.6 x 10^{-6} Ω-m, respectively. These values are averages of room-temperature measurements performed in other studies.[9,21-26] Exact agreement of the theoretical curve with any of the data is not expected since the different thin-film samples exhibit different electrical resistivities. In particular, for one set of data[14] r_{dc} is considerably less than the average value of r_{dc} used in the calculation, resulting in an underprediction of ρ (Fig. 1).

One could expect some discrepancy between ρ calculated for the interface between two semi-infinite media, as is Eq. (3), and the experimental results for thin films of finite thickness. However, if the radiation penetration depth δ, given by $\delta \sim \lambda/(4\pi k)$,[18] is much less than the thickness of the film, the film will behave like a semi-infinite medium. Figure 2 shows δ as a function of wavelength. The value of k is calculated from Eq. (1), which is applicable only for longer wavelengths since the Drude model fails at short wavelengths. It is apparent that for the wavelength range indicated in Fig. 1 (0 - 100 μm), a 1-μm-thick film should respond to incident radiation much like a semi-infinite medium. The data of the thinner films, of 0.09 μm, 0.2 μm, and 0.4 μm thicknesses, should show some sort of size effect, but that is not apparent in Fig. 1.

Figure 1 proves that the *ab*-plane room-temperature reflectance can be successfully modeled with the simple Drude free-electron theory employed here. The striking difference between this form of the Drude model and the Drude

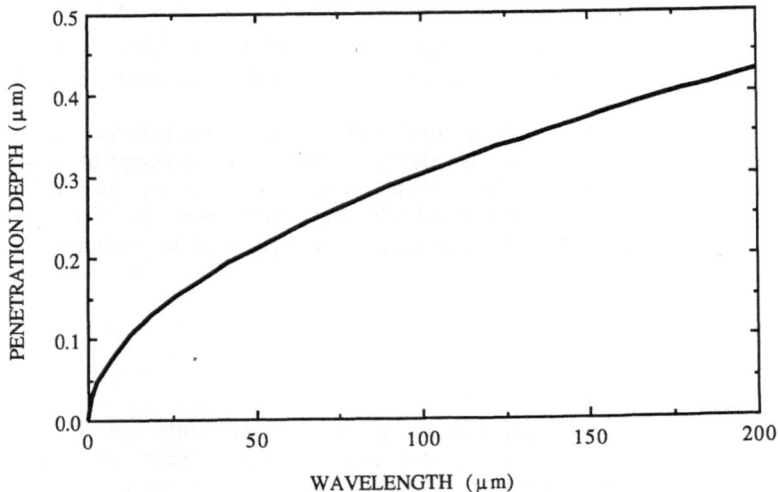

Fig. 2 The room-temperature penetration depth for $YBa_2Cu_3O_7$, as calculated from the Drude free-electron model.

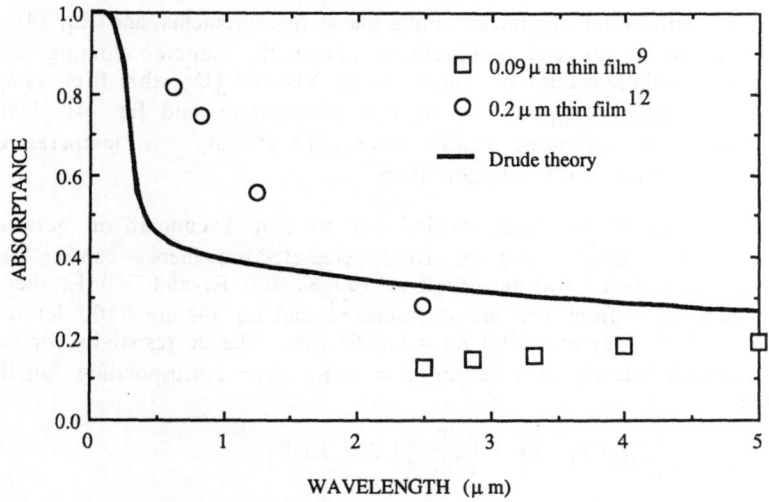

Fig. 3 A comparison of room-temperature absorptance data with that calculated from Eqs. (1) and (3) and the relation (1 - ρ).

relations others have used to fit through their data[3-14] is that this form requires only two experimentally-measured parameters: the electron number density, which can be deduced from Hall effect measurements, and the dc electrical resistivity. The electron rest mass is used instead of an "effective" mass. Most importantly, *no curve fitting is required.*

The absorptance of a thin film HTSC, which is the radiative property of most interest to bolometer designers, has been indirectly measured in two studies.[9,12] One study has measured the emittance, and hence the absorptance, of bulk $YBa_2Cu_3O_7$.[27] For the thin films, the reflectance and the transmittance were measured, and thus one can substract the sum of the two from unity and obtain the absorptance. Of course, any appreciable transmittance will occur only for very thin films—precisely where the semi-infinite media electromagnetic relations do not apply. Nevertheless, a comparison of the calculated absorptance, simply 1 - ρ, with the experimental measurements is useful and is presented in Fig. 3. It is seen that, as expected, there are large discrepancies between the Drude model prediction and the measured results, not only because of the size effect, but also because of the limitations of the Drude model at these short wavelengths. Even the experimental absorptance data show different trends; some increase with wavelength,[9] while others decrease.[12] The theoretical curve of Fig. 3, however, should be accurate for films on the order of 1 μm thickness and for $\lambda \geq 5$ μm.

TEMPERATURE EFFECTS

The two experimentally-measured parameters of the Drude theory, r_{dc} and N_e, are also the only two temperature-dependent parameters. Since r_{dc} is a function of N_e, r_{dc} and N_e exhibit similar temperature-dependent behavior: both decrease approximately linearly with temperature at least until the superconducting transition.[21,25,26] Ideally, both quantities would be measured experimentally and the results inserted into the Drude theory; however, measurement of N_e is generally not convenient since Hall effect experiments require a considerable magnetic field. It is possible, though, to relate N_e to r_{dc} through Eq. (4), which also includes the temperature-dependent electron mean free path l. One study[21] has measured both r_{dc} and N_e for thin films of

$YBa_2Cu_3O_7$ at various temperatures. From these measurements and Eq. (4) it is possible to calculate l for any temperature above the superconducting transition. If it is assumed that these values of l hold for all $YBa_2Cu_3O_7$ thin-film samples, then *a priori* knowledge of $r_{dc}(T)$, where T is temperature, and Eq. (4) yield $N_e(T)$. Alternatively, one can determine $N_e(T)$ and $\overset{\smile}{r}_{dc}(T)$ at only two temperatures and then connect the points with straight lines.

The above procedure has been carried out so that a comparison between measured reflectance data[14] and the Drude free-electron theory can be made. The normal-state reflectance was measured at 100 K, 200 K, and 300 K; the appropriate values of l from the measurements[21] and Eq. (4) are $l(100$ K$) = 0.0223$ μm, $l(200$ K$) = 0.0085$ μm, and $l(300$ K$) = 0.0051$ μm. The dc resistivity for the thin films[14] extrapolates linearly to zero for T = 0 K. The corresponding function $r_{dc}(T)$ is thus

$$r_{dc}(T) = \left(\frac{T}{T_{ref}}\right)\left[r_{dc}(T_{ref})\right] = \left(\frac{T}{200}\right)\left[1.2 \times 10^{-6}\right] \tag{5}$$

where r_{dc} is in Ω-m, T in K, and T_{ref} is the reference temperature at which r_{dc} is measured. Equation (5) and the values of l above yield $N_e(T)$: $N_e(100$ K$) = 4.1 \times 10^{28}$ m^{-3}, $N_e(200$ K$) = 6.2 \times 10^{28}$ m^{-3}, and $N_e(300$ K$) = 7.3 \times 10^{28}$ m^{-3}.

The results of the temperature-dependent Drude theory are presented in Fig. 4. Note that the theory tends to overpredict the reflectance. This discrepancy is due perhaps to the near-normal, rather than strictly normal, incidence angle during the measurements and to the general electromagnetic theory assumption that only specular reflectance occurs. In reality, some scattering of the incident radiation takes place at the surface of the film, resulting in a diffuse component of the reflectance which is not measured during the experiment.

SUPERCONDUCTING TRANSITION EFFECTS

The primary concern of a bolometer designer is not what happens in the normal state, but rather how do thin-film HTSC respond when they are in their

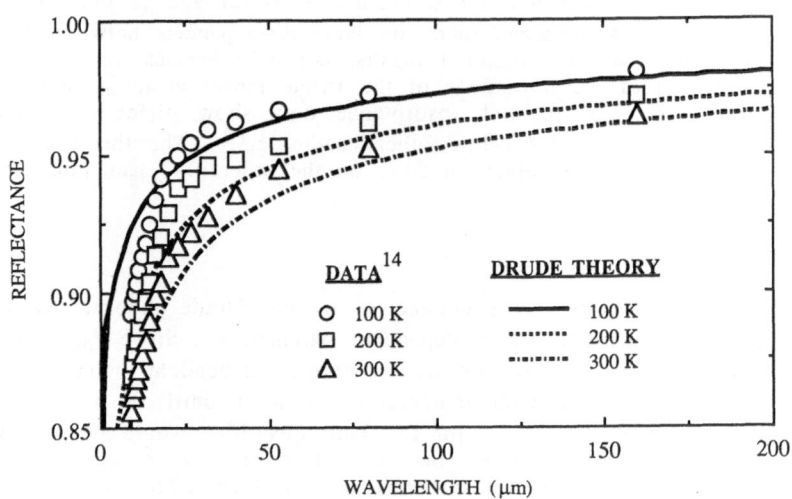

Fig. 4 A comparison of near-normal reflectance data, for 100 K, 200 K, and 300 K, with the normal reflectance as calculated by the temperature-dependent Drude free-electron theory.

superconducting transition region. Measurements on $YBa_2Cu_3O_7$ have indicated that there is not more than a 10% increase in reflectance when the film is in its superconducting state versus its normal state.[3,4,7,8,11,14,21] A corresponding decrease is expected for the absorptance. Future work will focus on the transition-region radiative properties.

CONCLUSIONS

A simple Drude free-electron model is capable of predicting the normal-state far-infrared radiative properties of 1-μm or thicker thin film $YBa_2Cu_3O_7$ for normal incidence. This model is successful for the temperature range 100 K to 300 K, and requires only the electrical resistivity r_{dc}, as a function of temperature, for input. Curve fitting through reflectance data is not required. The temperature-dependent electron number density is approximated from r_{dc}. Continuing work concentrates on application of the theory to very thin films, substrate effects, and on the radiative behavior of superconducting thin films in their transition region.

REFERENCES

1. P.L. Richards et al., "Feasibility of the high T_c superconducting bolometer," *Appl. Phys. Lett.* **54**, no. 3, pp. 283-285 (1989).

2. M.I. Flik and C.L. Tien, "Thermal phenomena in high-T_c thin-film superconductors," to appear in *Annual Review of Heat Transfer* **3** (1990).

3. T. Timusk and D.B. Tanner, "Infrared properties of high T_c superconductors," to appear in <u>The Physical Properties of High-Temperature Superconductors</u>, edited by D.M. Ginsberg, World Scientific Publishing Company, Singapore.

4. J.M. Wrobel et al., "Temperature dependence of the far-infrared reflectivity spectrum of the high-T_c superconductor $YBa_2Cu_3O_{7-y}$," *Phys. Rev. B* **36**, no. 4, pp. 2368-2370 (1987).

5. T.W. Noh, P.E. Sulewski, and A.J. Sievers, "Comparison of the electrodynamic properties of sintered $YBa_2Cu_3O_{7-y}$ and $La_{1.85}Sr_{0.15}CuO_{4-y}$," *Phys. Rev. B* **36**, no. 16, pp. 8866-8869 (1987).

6. D.A. Bonn et al., "Far-infrared properties of *ab*-plane oriented $YBa_2Cu_3O_{7-\delta}$," *Phys. Rev. B* **37**, no. 4, pp. 1574-1579 (1988).

7. R.T. Collins et al., "Comparative study of superconducting energy gaps in oriented films and polycrystalline bulk samples of Y-Ba-Cu-O," *Phys. Rev. Lett.* **59**, no. 6, pp. 704-707 (1987).

8. Z. Schlesinger et al., "Superconducting energy gap and normal-state reflectivity of single crystal Y-Ba-Cu-O," *Phys. Rev. Lett.* **59**, no. 17, pp. 1958-1961 (1987).

9. I. Bozovic et al., "Optical measurements on oriented thin $YBa_2Cu_3O_{7-\delta}$ Films: lack of evidence for excitonic superconductivity," *Phys. Rev. Lett.* **59**, no. 19, pp. 2219-2221 (1987).

10. X. Wang et al., "Optical reflectivity of single crystals of $YBa_2Cu_3O_{7-\delta}$ and $ErBa_2Cu_3O_{7-\delta}$," *Japan. J. Appl. Phys.* **26**, no. 12, pp. L2023-L2025 (1987).

11. T.W. Noh et al., "Far infrared measurements on single crystals, films and bulk sintered high temperature superconductors," *Materials Research Society Symp. Proc.* **99**, pp. 435-438 (1988).

12. I. Bozovic et al., "Optical anisotropy of $YBa_2Cu_3O_{7-x}$," *Phys. Rev. B* **38**, no. 7, pp. 5077-5080 (1988).

13. R.T. Collins et al., "Reflectivity and conductivity of $YBa_2Cu_3O_7$," *Phys. Rev. B* **39**, no. 10, pp. 6571-6574 (1989).

14. J. Schutzmann et al., "Far-infrared reflectivity and dynamical conductivity of an epitaxial $YBa_2Cu_3O_{7-\delta}$ thin film," *Europhys. Lett.* **8**, no. 7, pp. 679-684 (1989).

15. C. Kittel, <u>Introduction to Solid State Physics</u>, 6th ed., John Wiley & Sons, New York, pp. 256-257 (1986).

16. M. Born and E. Wolf, <u>Principles of Optics</u>, 4th ed., Pergamon Press, Oxford, pp. 624-627 (1970).

17. Y.K. Lim, <u>Introduction to Classical Electrodynamics</u>, World Scientific Publ. Co., Singapore, pp. 239-242 (1986).

18. C.F. Bohren and D.R. Huffman, <u>Absorption and Scattering of Light by Small Particles</u>, John Wiley & Sons, New York, pp. 251-257, 30-41, 29 (1983).

19. V.L. Newhouse, <u>Applied Superconductivity</u>, John Wiley & Sons, New York, pp. 62-64 (1964).

20. P. Chaudhari et al., "Properties of epitaxial films of $YBa_2Cu_3O_{7-\delta}$," *Phys. Rev. B* **36**, no. 16, pp. 8903-8906 (1987).

21. D. Hong-min et al., "Hall effect of the high T_c superconducting Y-Ba-Cu-O compound," *Solid State Comm.* **64**, no. 4, pp. 489-491 (1987).

22. U. Gottwick et al., "Transport properties of $YBa_2Cu_3O_7$: resistivity, thermal conductivity, thermopower and Hall effect," *Europhys. Lett.* **4**, no. 10, pp. 1183-1188 (1987).

23. M.R. Beasley, "Synthesis and properties of thin film high-T_c oxide superconductors," *Physica* **148B**, pp. 191-195 (1987).

24. J. Kwo et al., "Structural and superconducting properties of orientation-ordered $Y_1Ba_2Cu_3O_{7-x}$ films prepared by molecular-beam epitaxy," *Phys. Rev. B* **36**, no. 7, pp. 4039-4042 (1987).

25. B. Roas, L. Schultz, and G. Endres, "Epitaxial growth of $YBa_2Cu_3O_{7-x}$ thin films by a laser evaporation process," *Appl. Phys. Lett.* **53**, no. 16, pp. 1557-1559 (1988).

26. L.S. Fletcher et al., "Spectral properties of selected superconducting materials," Paper AIAA-1674, *AIAA 24th Thermophysics Conf.*, Buffalo, New York, June 12-14 (1989).

JOSEPHSON EFFECTS IN POINT CONTACTS AND BRIDGES

MADE OF $YBa_2Cu_3O_7$

R. Kleiner, P. Müller, H. Veith, J. Heise and
K. Andres

Walther-Meissner-Institute
D-8046 Garching, FRG

ABSTRACT

We report on investigations of the Josephson characteristics
of thin bridges of sintered $YBa_2Cu_3O_7$, both by four lead techniques as well as by observing the flux quantization properties
in ring structures containing a thin bridge.

INTRODUCTION

It is well known that in sintered samples of $YBa_2Cu_3O_7$,
the transport supercurrent density is limited by the intergrain
boundaries. At a temperature of 4 K, magnetic fields larger
than typically 40 Oe can penetrate along the grain boundaries
and reduce the transport supercurrent density drastically, the
reduction typically being a factor 10 in H = 100 Oe and a factor
25 in fields above 1000 Oe, staying then roughly constant up
to fields of 50000 Oe. This strongly suggests that i) most
of the intergrain boundaries behave like Josephson junctions
and that ii) some intergrain boundaries behave like weak links.
We must assume that the Josephson junctions possess a whole
spectrum of critical currents, varying with the quality of the
sinter. For a homogeneous sinter with well-connected grains,
these Josephson currents are higher and more uniform. This
is also manifested in the way that diamagnetic shielding currents
are reduced to zero when warming the sinter past its transition
temperature: a homogeneous and well-connected sinter shows a
sharp single transition in low applied fields (its width
increasing with increasing field), whereas a less well-connected
sinter often shows two transitions, the lower one being broader
and signalling the dying out of the Josephson currents before
the individual grains become normal conducting.

In the following, we report on investigations of the
Josephson properties of grain contacts by several means, namely
i) by directly observing the critical current, and its dependence
on applied magnetic field, of a thin bridge manufactured from
sintered material, ii) by doing the same thing on a broken bridge
pressed together again in situ ("point contact"), and also iii)
by incorporating such a bridge into a ring structure and
observing its flux quantization properties.

Advances in Cryogenic Engineering (Materials), Vol. 36
Edited by R. P. Reed and F. R. Fickett
Plenum Press, New York, 1990

487

$YBa_2Cu_3O_7$-sinter were prepared by the standard method of
first mixing and grinding the constituents (Y_2O_3, $BaCO_3$, and
CuO), pressing a pellet (2 kbar), reacting in O_2 at 950°C for
12 h, then grinding and pressing again for a sinter reaction
in O_2 at 950 °C for 24 h with a following slow cool-down in
flowing O_2 to room temperature (1). Highly phase-pure sinters
are obtained in this way, with grain sizes ranging between about
5 and 50 μm and with filling factors between 90 and 95%. Bridges
with diameters between 100 and 200 μm are then filed and sawed
into the material, and current and voltage leads are attached
on both sides of the bridge, using sintered silver paste. A
typical current-voltage characteristic at 77 K is shown in Figs.
1 (a and b).

At low supercurrents, the current flows uniformly through
all intergrain contacts. With increasing supercurrent, the
current distribution becomes increasingly non-uniform, since
it now has to percolate through the "better" grain to grain
contacts. At the current I_{c1}, no such percolation path can
be found any more and the bridge becomes resistive. However,
some isolated intergrain contacts still remain superconducting,
up to I_{c2}, where most of them suddenly revert to normal conduction. This sudden and hysteretic transition indicates that
ohmic heating, too, drives the contacts normal at this point.

If now small external magnetic fields are applied to the
bridge, we observe variations of the critical current I_{c1}. We
in fact use slightly overcritical bias currents in order to
observe the corresponding voltage variations. This also means
that we work in a range where there is no superconducting
percolation path through the bridge any more. The variations
with applied field, which are shown in Fig. 2a, are due to the
non-dissipative part of the current (I_{c1}) and are clear evidence,
that some of the remaining intergrain contacts behave like
Josephson junctions, whose critical current depends on the
magnetic flux in closed superconducting loops between them.

The Fourier transform (Fig. 2b) shows a broad range of
superimposed oscillation periods ranging from H = 5 to 100 mOe.
This corresponds to Josephson loop areas of 200 to 4000 (μm)^2.
Magnetic flux easily penetrates the bridge through its weakest
(and in part dissipative) intergrain contacts. The penetration
into single grains, however, occurs only to within the field
penetration depth, which is a fraction of a micrometer. As
the grain sizes range from about 5 to 50 μm, the active loops
are not around individual grains, but most probably around voids

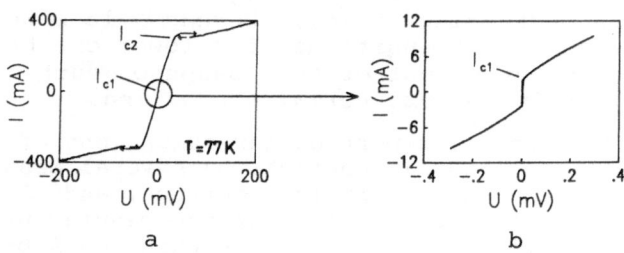

Fig. 1. Current-voltage characteristic of $YBa_2Cu_3O_7$ sinter
bridge at 77 K.

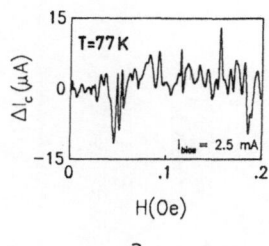
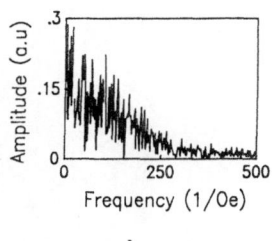

a   b

Fig. 2.   a)   Josephson oscillations of critical current of sinter
          bridge versus applied field.   b)   Fourier transform
          of a).

in the sinter which are surrounded by several grains. For
optimal interference effects in one such loop of area A_i, the
junctions around it should have critical currents of order
Φ_0/L_i, with Φ_0 the flux quantum (2×10^{-7} Maxwell) and L_i the
self-inductance of loop A_i. For an A_i of 100 $(\mu m)^2$, we obtain
$L_i \simeq 3 \times 10^{-11}$ H and $I_c \simeq 80 \mu A$ for one Josephson contact.
The actual maximum current carrying capacity of intergrain
contacts is estimated (from Fig. 1 and the percolative nature
of the supercurrent distribution) to be somewhat higher than
that.

     Increasing the bias current further past I_{c1} gradually
drives more junctions into the dissipative state and reduces
the number of remaining superconducting loops around voids.
This is shown in Figs. 3a and 3b (same bridge as Fig. 2 with
higher bias current), where it can be seen that the Fourier
spectrum of periods has narrowed.

PROPERTIES OF DOUBLE BRIDGES IN SINTERS

     We have also investigated the properties of double bridges
(shown in Fig. 4) and find that they show current-voltage as
well as current-field characeristics very similar to single
bridges.

     In particular we never observe the short critical current
oscillation periods versus field which would correspond to the
large area A between the two arms of the double bridge. This
can be understood by the poor match of the critical intergrain
currents to the inductance L of the area A: an estimate yields
$\Phi_0/L \simeq 1 \mu A$, whereas a critical intergrain current is of order
0.1 to 1 mA.

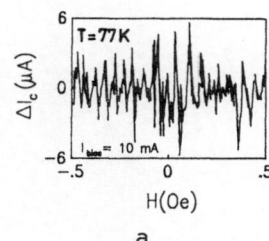
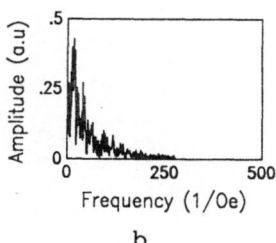

a   b

Fig. 3.   a)   Same as Fig. 2a, but at higher bias current (10mA).
          b)   Fourier transform of a).

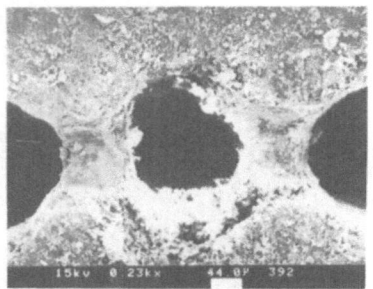

Fig. 4. Double bridge made out of $YBa_2Cu_3O_7$ sinter material.
The diameter of the bridge-arms is ~100 µm.

PROPERTIES OF BREAK CONTACTS

If a sinter bridge is broken and then pressed together
again, the number of "good" supercurrent paths across the bridge
is reduced, because only a finite number of "point-contacts"
will exist in the break. We observe moreover, that these point
contacts show significantly lower critical supercurrents,typical-
ly of order 100µA, (2, 3). This fact, then, automatically
locates the active Josephson junctions in the point contacts
of the break. Their reduced number leads to a reduced number
of areas between them, and hence a reduced number of oscillation
periods versus applied field. Again, by increasing the bias
current past I_{c1}, the number of active junctions further
decreases, leading to still fewer oscillation periods. This
behavior is demonstrated in Fig. 5 and 6 (a and b). Interestingly, there is is no evidence for Josephson interference between
individual point contacts, which would correspond to larger
areas and higher frequencies than those shown in Figs. 5 and
6. We therefore seem to observe interference effects only within
individual point contacts.

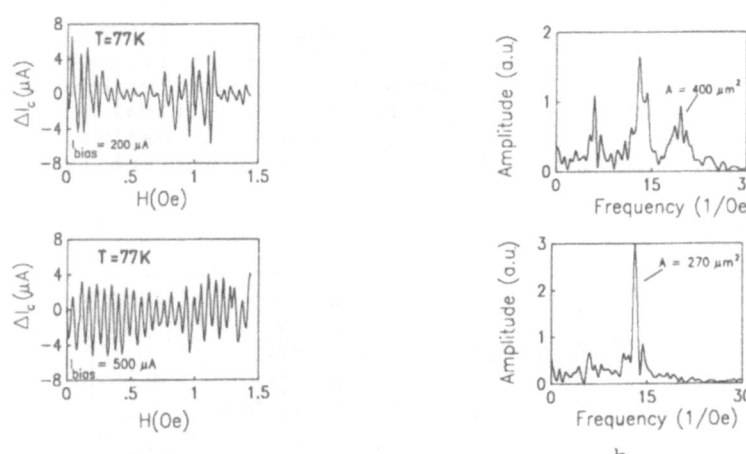

a

b

Figs. 5 and 6. Josephson oscillations of critical current of
break contacts (a) and their Fourier transforms (b)
for different bias currents. (See text.)

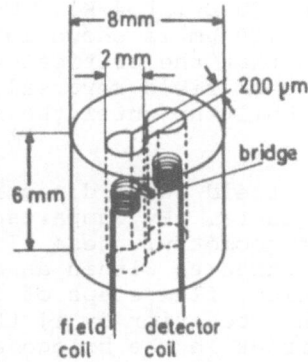

Fig. 7.   Geometry of two-hole r.f. SQUID made
out of $YBa_2Cu_3O_7$ sinter material.

Upon increasing the applied field further on break junctions, the amplitude of the Josephson oscillations also decreases, typically to 1% in a field of 100 Oe. This is as expected and is due to the well-known interference effects within individual junctions.

FLUX QUANTIZATION IN RING STRUCTURES CONTAINING A BRIDGE

To research the possibility of using $YB_2Cu_3O_7$ sinter material to manufacture r.f. SQUIDS, we have investigated the flux quantization properties of the classical two-hole geometry containing a bridge (Fig. 7).

In order for magnetic flux to be tansferred in quanta from one hole to the other, the condition $i_c \simeq \Phi_0/L$ should be satisfied (L = self inductance of the hole). This requires critical bridge currents of order 1 µA, which are difficult to obtain, as we have shown above. The flux quantization properties of the two-hole SQUID can be observed directly by sweeping a field in one hole and detecting the flux changes in the other hole by means of a second, superconducting niobium pick-up coil connected

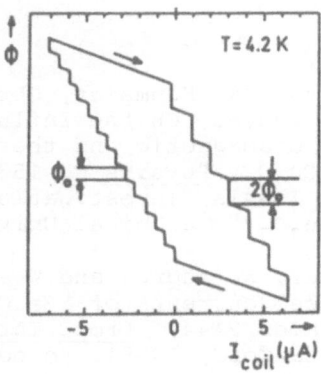

Fig. 8.   Observed flux changes in one SQUID
hole versus current through the
field coil located in the other hole.
(See text.)

to a conventional niobium SQUID (at ,4 K). The result for a
bridge of dimension 100 X 100 μm is shown in Fig. 8. The
observed hysteresis shows that the critical current of the bridge
is still too large: upon each field reversal, a flux change
of about 13 Φ_0 has to be built up until the quantized transfer
of flux begins.

However, only in one field sweep direction do we observe
flux transfer in single quanta. By comparison with the behavior
of a Nb-SQUID of the same geometry, these flux steps indeed
correspond to single Φ_0-quanta to within an accuracy of 5%.
In the other sweep direction, flux steps of 2 Φ_0 are observed.
we tentatively explain this behavior by i) the existence of
a frozen-in bias supercurrent in the bridge and ii) the existence
of a finite number of additional flux loops within the bridge.
When increasing the bias supercurrent, the flux is readily
transferred through the bridge loops, their flux state remaining
constant. When decreasing it, the flux changes can first be
stored in the bridge loops and only then released to the second
hole (in multiples of Φ_0).

CONCLUSIONS

By reducing the number of superconducting Josephson
junctions in a sinter bridge through increasing the current
above the current above the critical value, d.c. SQUID operation
can be observed up to temperatures close to Tc. The field
periodicities are consistent with the assumption that the SQUID
loops contain a few sinter grains around voids in the sinter.
The results on break junctions can be explained as arising from
similar, but fewer, SQUID loops located in the break area. By
properly adjusting the bias current, the number of active loops
can then often be reduced to one, resulting in only one
modulation frequency.

For proper operation of an r.f. SQUID made of sinter
material, the hysteresis in Fig. 8 has to be smaller than a
few Φ_0, corresponding to a rather small critical current in
the bridge. This can sometimes be obtained (presumably in
partially broken bridges), at the expense, however, of a poor
reversibility (especially on repeated warming and cooling
cycles).

REFERENCES

1. J. Heise, P. Gutsmiedl, K. Neumaier, Chr. Probst, H. Berndt,
   P. Müller and K. Andres, On the influence of the heat
   treatment on the diamagnetic and thermodynamic properties
   of sintered $YBa_2Cu_3O_7$, Physics C, 153-155, 1507 (1988).
2. R. Kleiner, Diploma Thesis "Investigations of Josephson
   Effects in $YBa_2Cu_3O_7$," Technical Univ. of Munich, Germany
   (1987).
3. R. Kleiner, P. Müller, K. Andres and W. Reith, DC Josephson
   effect and critical currents of $YBa_2Cu_3O_7$ and
   $Tl_2CaBa_2Cu_2O_8$, Paper 2A44, Proc. Int. Conf. on High
   Tc Supercond. (Stanford, 1989), to be published.

PREPARATION AND PATTERNING OF YBa$_2$Cu$_3$O$_x$ THIN FILMS*

M. Bhushan and A. J. Strauss

Lincoln Laboratory, Massachusetts Institute of Technology
P.O. Box 73, Lexington, Massachusetts 02173-0073

ABSTRACT

Superconducting YBa$_2$Cu$_3$O$_x$ thin films have been prepared by annealing multilayers deposited by sequential rf diode sputtering of Y$_2$O$_3$, BaF$_2$, and CuO targets. Reactive sputtering of BaF$_2$ in an Ar/O$_2$ atmosphere yields films containing primarily Ba and O. The F concentration in the as-deposited multilayers is so low that wet-O$_2$ annealing, which is necessary to remove F from films deposited from a BaF$_2$ source by other techniques, is not required. The superconducting phase is formed by post-annealing in dry O$_2$ for one hour at 850°C and two hours at 500°C. Films about 1 μm thick on yttria-stabilized zirconia substrates have a zero-resistance transition temperature of 90 K. Pattern definition is accomplished by using a photoresist mask followed by ion milling in an Ar$^+$ beam. Exposure to water is not detrimental to film properties, but some damage due to Ar$^+$ ions is observed after patterning.

INTRODUCTION

A variety of sputtering techniques have been used for the preparation of Y-Ba-Cu-O (YBCO) films. Although high-quality films have been produced by sputtering from a single composite target, experimental trials with a series of targets are generally required to determine the target composition needed to achieve the desired film stoichiometry, which is dependent on the sputtering parameters and system geometry. Methods that employ either simultaneous or sequential sputtering from three independent cation sources have the advantage of allowing better flexibility in controlling the film composition. The three sources may be the Y, Ba and Cu metals or their compounds. In order to incorporate O in the as-deposited film, it is convenient to use the metal oxides as target materials rather than the metals themselves. Unfortunately, BaO is not suitable as a sputtering source because it readily reacts with water vapor and CO$_2$ in the atmosphere. BaF$_2$, which is chemically stable, has been successfully used as the source of Ba in films prepared by evaporation [1-5] and magnetron sputtering.[6] When these methods are employed, BaF$_2$ is incorporated in the as-deposited films, and it is necessary to remove the

---

*This work was supported by the Department of the Air Force and the Defense Advanced Research Projects Agency

Advances in Cryogenic Engineering (Materials), Vol. 36
Edited by R. P. Reed and F. R. Fickett
Plenum Press, New York, 1990

493

F by annealing in wet O_2 at 700-800°C prior to forming the superconducting YBCO phase. This procedure is not compatible with in-situ preparation of YBCO films, which is desirable because it offers the possibility of lowering the processing temperature, thereby minimizing film-substrate interactions.

For this investigation we adopted a sequential rf diode sputtering technique using Y_2O_3, BaF_2, and CuO as target materials. We have found that reactively sputtering BaF_2 in an Ar + O_2 atmosphere yields films containing primarily Ba and O, with a F content of less than 5 at.%. (Unlike pure BaO, these films, which we shall refer to as BaOF films, are stable in air. The reactive sputtering of BaF_2 will be reported in more detail elsewhere.) Because the F content in the multilayer films is so low, wet-O_2 annealing is not necessary. In the experiments reported here, the superconducting YBCO phase was formed by annealing the as-deposited multilayers in dry O_2 at temperatures of 800°C or above. However, by raising the substrate temperature during deposition, the diode sputtering technique has the potential for in-situ preparation of superconducting YBCO films.

EXPERIMENTAL PROCEDURE

Film deposition was carried out in a commercial sputtering system equipped with a cryopump and three rf diode target assemblies. The base pressure in the vacuum chamber is 1×10^{-7} Torr. As shown schematically in Fig. 1, the substrate table can be rotated to position the substrate under each of the targets. A computer is used to control the motion of the table. Power from a single rf source is applied simultaneously to all three targets through separate tuning networks.

Single-crystal yttria-stabilized zirconia (YSZ) substrates with dimension up to 1.2 x 2.5 cm were used for most of the YBCO deposition runs, but some films were deposited on $SrTiO_3$ and MgO substrates. The sputtering targets were 12.5-cm-diam discs hot pressed from 99.9% pure powders of Y_2O_3, BaF_2 and CuO. The targets were presputtered for thirty minutes prior to commencing film deposition. The deposition rates obtained with each target were determined in separate experiments. For a fixed target-to-substrate distance and fixed partial pressures of Ar and O_2, the deposition rate was found to be proportional to V_{dc}^2, where V_{dc} is the dc self-bias potential of the target. During multilayer depositions, this relationship was exploited to correct for small fluctuations in the power distribution among the three targets. After the substrate was positioned under one of the targets, the computer read V_{dc} once every two seconds and calculated the instantaneous deposition rate. When a preset thickness was reached, the substrate table was moved to place the substrate under the next target. As many as ninety individual layers were deposited in a single run. The thickness of the layers ranged from as low as 20 nm for Y_2O_3 to as high as 60 nm for BaOF.

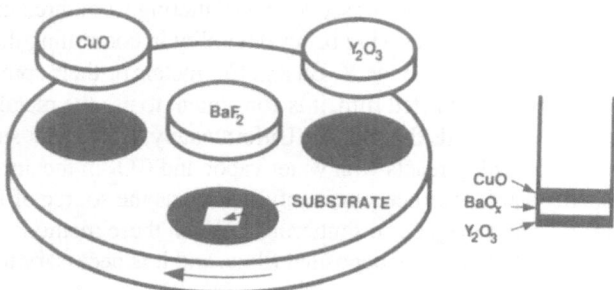

Fig 1. Configuration of targets and substrate in rf diode sputtering system.

Table 1. Properties of films deposited in Ar-O$_2$ Atmosphere

Target Material	Film Composition	Crystallinity Orientation	Optical Properties	Deposition Rate/ Target Voltage	Comments
Y$_2$O$_3$	Y$_2$O$_3$	Polycrystalline (100) on YSZ (111) on SrTiO$_3$ MgO	Refractive index=1.89	7.5 nm/min 1270 V	
BaF$_2$	BaO$_x$F$_y$ y<5%	Polycrystalline	Refractive index=1.5	20 nm/min 975 V	Stable in air
CuO	CuO	Polycrystalline (001) on YSZ SrTiO$_3$ MgO	Black-brown color	12.5 nm/min 1200 V	Resistivity ~ 1 Ω-cm at 300 K

The concentration of O in as-deposited multilayers of Y$_2$O$_3$, BaOF, and CuO is about 52 at.%. Formation of the superconducting YBCO phase requires interdiffusion of the individual layers as well as an increase in the O content to about 53 at.%. In this study the YBCO phase was obtained by annealing the as-deposited multilayers in a quartz tube in flowing dry O$_2$ at 850°C for one hour, reducing the furnace temperature to 500°C at the rate of ~1°C/min, annealing at 500°C for two hours to adjust the O stoichiometry, and then cooling to room temperature.

The YBCO films were patterned by using standard photolithographic techniques followed by Ar$^+$ ion beam milling. A 2-μm-thick masking layer of either a positive photoresist (Shipley 1350 J) or negative photoresist (Electrolux N-100) was used for defining the pattern. Ion milling was carried out with a Kaufman ion source at an Ar pressure of 3 x 10^{-4} Torr. The ion beam current density was 0.8 mA/cm^2, and the accelerating voltage was 500 V. The beam was incident on the sample at an angle of 90°. The sample was attached to a water-cooled table with thermal grease to keep the temperature low enough to prevent reaction between the photoresist and the YBCO. Under these milling conditions, the photoresist and YBCO were both removed at a rate of 30 nm/min. For satisfactory masking it was therefore necessary for the photoresist layer to be thicker than the YBCO film.

RESULTS AND DISCUSSION

The properties of Y$_2$O$_3$, BaOF, and CuO films deposited at a target-to-substrate distance of 4 cm and Ar and O$_2$ partial pressures of 22 and 5 mTorr, respectively, are listed in Table I. These operating parameters were selected to minimize resputtering of the films from the substrate due to bombardment by negative ions (F- or O-) and by electrons from the Y$_2$O$_3$ and BaF$_2$ targets, both of which have high secondary electron emission coefficients. The high-energy electron bombardment raised the substrate temperature to ~ 300°C during multilayer runs even though the substrate was not intentionally heated. The thicknesses of the Y$_2$O$_3$ and BaOF films, as measured by ellipsometry, were uniform within ± 1% over a 5-cm-diameter substrate placed in a position concentric with the target. The thickness variation in CuO films, as measured by a mechanical profilometer after patterning and etching the films in dilute nitric acid, was less than ± 5%.

The deposition rates listed in Table I are typical of the values used in preparing the YBCO films. The thickness ratios of the Y$_2$O$_3$, BaOF, and CuO layers were empirically adjusted to yield Y:Ba:Cu ratios of 1:2:3 in the annealed films, as determined by Auger

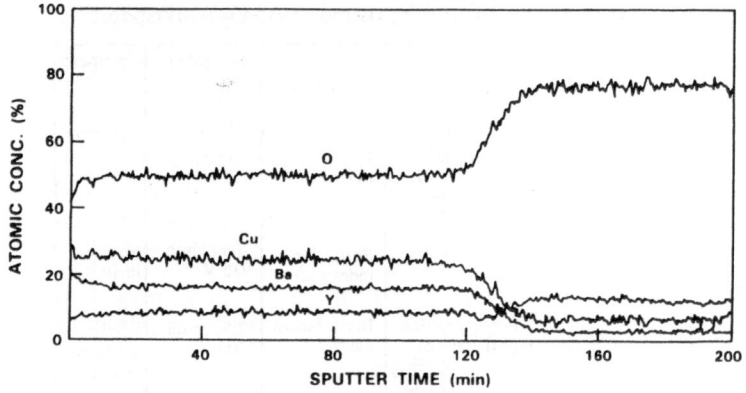

Fig. 2   Auger depth profile of annealed YBa$_2$Cu$_3$O$_x$ film.

analysis. The empirical thickness ratios were 1:10.8:4.1, compared to the ratios of 1:2.4:1.7 calculated from the literature values for the bulk densities of Y$_2$O$_3$, BaO, and CuO. Figure 2, which is an Auger depth profile of an annealed film with measured cation ratios of 1:2:3, shows that the composition is uniform through the thickness of the film.

The multilayer deposition and annealing conditions were optimized to obtain the highest zero-resistance transition temperature T$_c$ and maximum crystallographic texturing for YBCO films on (100) YSZ substrates. For such films about 1 μm thick, T$_c$ was 90 K, the transition width was 3 K, and the resistivity ratio ρ(293 K)/ρ(95 K) was in the range of 1 to 1.6. For films prepared under the same conditions on (100) SrTiO$_3$ and MgO substrates, T$_c$ was only in the 60 to 70 K range, and the transition width was 10 to 20 K.

An x-ray diffractometer scan of a 1-μm-thick YBCO film on a (100) YSZ substrate is shown in Fig. 3. The film is polycrystalline, but there is clear evidence of preferred orientation with the a and c axes perpendicular to the substrate, since the (h00) and (00l) diffraction peaks are much stronger in intensity than in a randomly oriented YBCO film. Films prepared under the same conditions on unoriented YSZ single-crystal substrates did not show any evidence of texturing. Thus the orientation of YSZ substrates can influence the orientation of the YBCO crystallites, even though there is a large lattice mismatch between YSZ and YBa$_2$Cu$_3$O$_x$. For patterned 1-μm-thick films on (100) YSZ substrates, the measured critical current density J$_c$ was ~ 4 x10^4 A/cm^2 at 4.2 K. Values of J$_c$ as high as 10^6 A/cm^2 at 75 K have been reported[7] for epitaxial YBCO films with the c axis perpendicular to the substrate. To measure the microwave surface resistance of YBCO

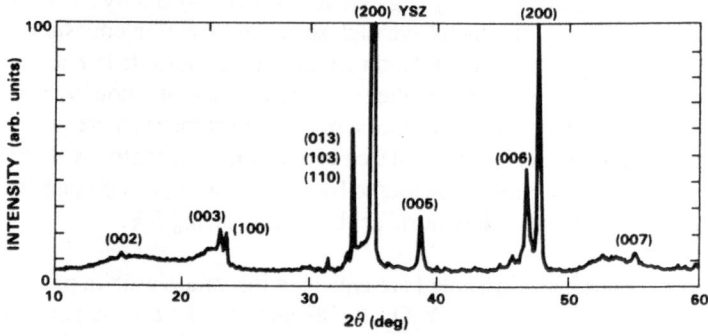

Fig. 3   X-ray diffractometer scan for 1-μm-thick YBCO film on (100) YSZ substrate.

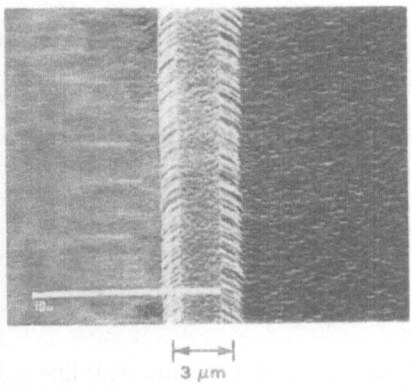

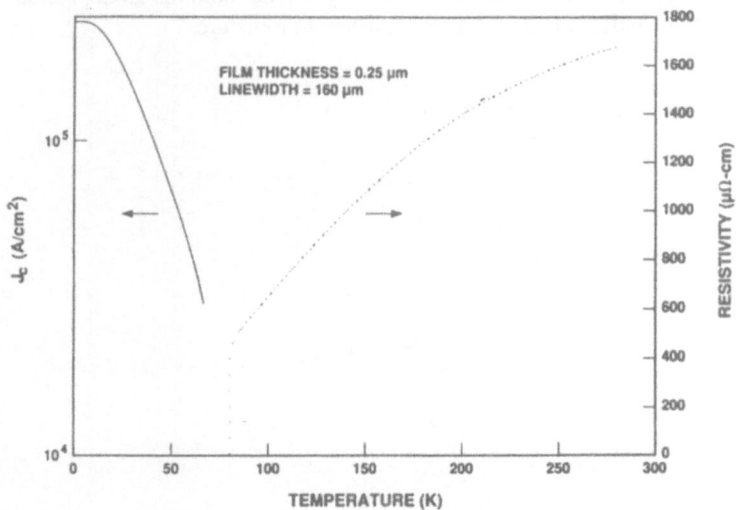

Fig. 4 Scanning micrograph of 0.25-μm-thick YBCO film.

films as a function of frequency, a film is substituted for the Nb ground plane in a stripline resonator.[8] For 1-μm-thick films on (100) YSZ substrates, the measured surface resistance was 0.2 mΩ/sq at 1.1 GHz and exhibited an $f^{1.8}$ dependence on frequency.

For YBCO films less than 0.3 μm thick that had been annealed in the usual manner, c-axis texturing was significantly greater than a-axis texturing. The c-axis texturing was further enhanced by pre-annealing the as-deposited multilayers in O_2 at 500°C for three hours to allow the layers to interdiffuse before raising the temperature to 850°C. These observations indicate that grain growth near the substrate occurs with the c axis perpendicular to the substrate. Flat platelets a few micrometers in size are present in films exhibiting c-axis texturing, and the platelet size is larger in films pre-annealed at 500°C. Figure 4 is a scanning electron micrograph showing the surface of a pre-annealed film 0.25 μm thick. For films of all thicknesses, the surface morphology was uniform over the entire area. In pre-annealed films, both $YBa_2Cu_3O_x$ and the superconducting $Y_2Ba_4Cu_8O_x$ phase[9] were present. For films with the $Y_2Ba_4Cu_8O_x$ phase predominant, T_c was in the range of 77 to 79 K with an onset at 80 K. For one such film 0.25 μm thick, the values of J_c and ρ are plotted vs temperature in Fig. 5. The J_c values are an order of magnitude higher than those for 1-μm-thick films.

Fig. 5   Plots of J_c and resistivity vs temperature for 0.25-μm-thick YBCO film.

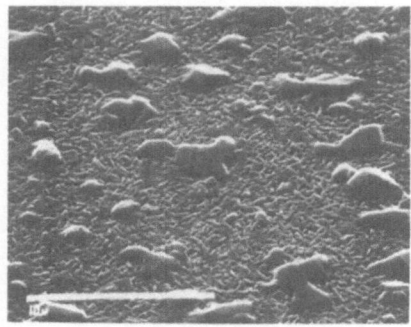

Fig. 6   Scanning electron micrograph of 3-μm-wide line in YBCO film.

Annealed YBCO films have been stored in room air for several weeks without any change in electrical properties or surface morphology. The film surface is sufficiently smooth to permit defining linewidths of the order of 1 μm. Figure 6 is a scanning electron micrograph of a 3-μm-wide line. The room-temperature resistivity and surface morphology were not affected by baking or photoresist development. After ion milling, however, an increase in room-temperature resistivity by a factor of 1.5 to 5 and a decrease of 5 to 10 K in T_c were observed for lines wider than 10 μm. Figure 7 compares the ρ vs T data for a 40-μm-wide line and the unpatterned film. For linewidths of less than 10 μm, there was a small temperature-independent residual resistivity, and the superconducting transition was not complete. The degradation is apparently due to damage caused by the Ar$^+$ ion beam.

CONCLUSION

The sequential rf diode sputtering technique described here is a unique approach to the preparation of YBCO films using a BaF$_2$ source. Sequential deposition permits the formation of films that are uniform in composition over a large area, while rf diode sputtering in an Ar/O$_2$ atmosphere offers the potential for in-situ preparation, since the F content in the films is so low that wet-O$_2$ annealing is not required. There is sufficient residual F, however, to reduce the chemical reactivity, so that both the as-deposited multilayers and the annealed YBCO films are resistant to moisture.

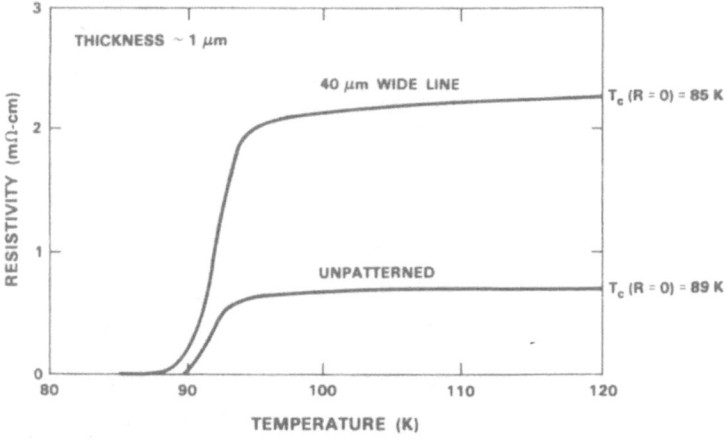

Fig. 7   Plots of resistivity vs temperature for unpatterned YBCO film and for a 40-μm-wide-line in this film.

ACKNOWLEDGMENTS

The authors wish to thank J. Hamer, G. L. Fitch and R. P. Konieczka for assisting in film deposition and electrical characterization and M. C. Finn for Auger analysis.

REFERENCES

1. P. M. Mankiewich, J. H. Scofield, W. J. Skocpol, R. E. Howard, A. H. Dayem, and E. Good, Appl. Phys. Lett. **51**, 1753 (1987).
2. M. Nastasi, J. R. Tesmer, M. G. Hollander, J. F. Smith, and C. J. Maggiore, Appl. Phys. Lett. **52**, 1729 (1988).
3. A. Mogro-Campero, L. G. Turner, E. L. Hall, and M. C. Burrell, Appl. Phys. Lett. **52**, 2068 (1988).
4. N. Hess, L. R. Tessler, U. Dai, and G. Deutscher, Appl. Phys. Lett. **53**, 698 (1988).
5. S-W. Chan, B. G. Bagley, L. H. Greene, M. Giroud, W. L. Feldmann, K. R. Jenkin, II, and B. J. Wilkins, Appl. Phys. Lett. **53**, 1443 (1988).
6. E. Wiener-Avnear, J. E. Cooper, G. L. Kerber, J. W. Spargo, A. G. Toth, J. Y. Josefowicz, D. B. Rensch, B. M. Clemens, and A. T. Hunter, IEEE Trans. Magn. **MAG-25**, 935 (1989).
7. M. C. Nuss, P. M. Mankiewich, R. E. Howard, B. L. Straughn, T. E. Harvey, C.D. Brandle, G. W. Berkstresser, K. W. Goossen, and P. R. Smith, Appl. Phys. Lett. **54**, 2265 (1989).
8. M. S. DiIorio, A. C. Anderson, and B-Y. Tsaur, Phys. Rev. B **38**, 7019 (1988).
9. A. F. Marshall, R. W. Barton, K. Char, A. Kapitulnik, B. Oh, R. Hammond, Phys. Rev. B **37**, 9353 (1988).

SUSPENSION SPINNING OF HIGH Tc OXIDE SUPERCONDUCTOR BY USING

POLYACRYLNITRILE

Tomoko Goto

Materials Science & Engineering
Nagoya Institute of Technology
Gokiso-cho, Showa-ku, Nagoya, Japan

ABSTRACT

The suspension spinning of a high Tc oxide superconductor by using PAN was studied to prepare a long filament with high Jc. The highest Jc of 1285 x10⁴ A/m² at 77 K was obtained for the Y-Ba-Cu-O filament spun through PAN-DMF suspension medium containing the oxide of 92 wt%.

A high-Tc phase superconducting Bi-Pb-Sr-Ca-Cu-O filament was successfully prepared using a combined technique of suspension spinning and densification of the pyrolyzed filament by two stages of pressing and sintering. Zero electrical resistivity was achieved at 90 K and the Jc of the filament at 77 K was 55 x10⁴ A/m².

INTRODUCTION

An application of the high Tc oxide superconductor for superconducting magnets requires the fabrication of the brittle ceramic materials into tapes or wires with high Jc.[1-3] We have studied the preparation of a oxide superconducting long filament using a suspension spinning method.[4-6] : The oxide powder was suspended in a polymer solution. The viscous suspension was extruded as a filament into a precipitating medium and coiled on a winding drum. The as-drawn filament was heated to remove volatile components and to generate the superconducting phase. It is important in this technique how to enhance the densification of the oxide in the filament and the crystallographic alignment of the high-Jc direction along the longitudinal direction of the filament and to form the fewer and cleaner grain boundaries. Although the polymer was removed as a volatile component by heating treatment, the dispersion behaviour of the oxide powder in the filament was dependent on the suspension medium, hence the microstructure and superconducting properties of the filament obtained are affected by the suspension medium. High Jc value of more than 1000 x10⁴ A/m² at 77 K, 0 T was attained for Ho-Ba-Cu-O, Bi(Pb)-Sr-Ca-Cu-O and Tl-Ba-Ca-Cu-O filament produced by this method through poly(vinyl alcohol) (PVA) suspension medium.[7] This paper deals with suspension spinning of Y-Ba-Cu-O and Bi-Pb-Sr-Ca-Cu-O superconductor by using polyacrylnitrile (PAN) suspension medium.

Advances in Cryogenic Engineering (Materials), Vol. 36
Edited by R. P. Reed and F. R. Fickett
Plenum Press, New York, 1990

501

EXPERIMENTAL

A fine mixed powder with nominal composition of $Y_1Ba_2Cu_3Ox$ was prepared by coprecipitating the oxalates of Y, Ba and Cu into ethyl alcohol. The powder was calcined at 1123 K for 7.2 ks and the particle size of the heated powder was up to 150 nm. The powder was suspended in PAN solution and the suspension dope was extruded as a filament into a precipitating medium of methyl alcohol and coiled on a winding drum. The as-drawn filament was sintered at 1253 K for 1800 s in oxygen gas flow, followed by furnace cooling at a cooling rate of 100 K/h. The effects of PAN nonaqueous suspension medium on Jc of the filament sintered were explored with PAN in N-N'-dimethyl formamide (DMF), N-N'-dimethyl acetamide (DMAc), N-N'-dimethy sulfoxide (DMSO), ethylene carbonate and malononitrile. PAN with molecular weight of Mv = 20 x10⁴ was kindly supplied from Asahi Chemical Industry Co Ltd.

Suspension spinning of high-Tc Bi-Pb-Sr-Ca-Cu-O was made by the same method for PVA medium.[7] Appropriate amounts of Bi_2O_3, PbO, $SrCO_3$, $CaCO_3$ and CuO powders with more than 3 N purity were mixed into the composition of $Bi_{0.96}Pb_{0.24}Sr_1Ca_{1.1}Cu_{1.8}Ox$, calcined at 1073 K for 54 ks and were pressed into pellet and then sintered at 1123 K for 432 ks in air to form the high-Tc phase of the Bi system. The resultant pellet was milled into a fine powder and was suspended in PAN-DMF solution. The as-drawn filament was heated at 773 K for 3.6 ks to remove volatile component. The filament was then cold pressed and sintered at 1113 K for 54 ks and then pressed and sinted at 1113 for 57.6 ks in low oxygen gas pressure of 1/13 x10⁵ Pa. The electrical resistivity (ρ) of the filament heated was measured by a standard four-probe method. Silver paint was used to connect the filament with silver electrodes of 50 μm in diameter. The specimens temperature was measured using a calibrated chromel-gold + 0.07 % iron thermocouple. The transport Jc measurement was performed at 77 K and 0 T with a criterion of 0.01 μV/m. The structure of the filament was examined with scanning electron micrography (SEM) and X-ray diffractometer.

RESULTS AND DISCUSSION

Y-Ba-Cu-O filaments

Superconducting $Y_1Ba_2Cu_3$ oxide long filaments with less than 100 μm in diameter were produced from various PAN suspension media and the electrical resistivity of the filaments sintered was measured. A rapid resistivity drop was observed around 90 K and zero resistivity was 85 K. The effect of the PAN-DMF and PAN-DMAc suspension media on Jc was examined. The Jc of the filament was dependent on the oxide powder content. Figure 1 shows the Jc of the filament spun through various powder contents of the spinning dope of PAN in DMF and DMAc, respectively. A maximum Jc of the filament was measured at the oxide content of 92 wt %. Previous works by using PVA in DMSO, hexamethyl phosphoric triamide (HMTA) and mixed solution of DMSO and HMTA (1:1) had shown the maximum Jc of the filament spun in DMSO, HMTA and the mixed solution was observed at the oxide content of 95, 93 and 96 wt %, respectively and These behaviours could be explained by overlapping the effect of the oxide densification and the critical molecular concentration (CMC) of PVA.[8] On the present case, the CMC of PAN needed to play the dispersant in the spinning dope and coagulator of the oxide through the precipitating medium is suggested to be at 8 wt % and the effect of PAN for the dispersant and coagulator was poor than that for PVA. The Jc of the filament spun in various other PAN solutions was measured and the Jc at 77 K and ρ at 100 K of the filament spun from the spinning dope of various PAN solutions containing the powder oxide of 92 wt % were listed in Table 1. The polarity of the solvent is considered to be related to the CMC, hence Jc of the filament obtained was varied with solvent. The maximum Jc value of 1285 A/cm² was

Table.1 Critical current density at 77 K and normal-state resistivity (ρ) of the $Y_1Ba_2Cu_3Ox$ filament spun through powder contents of 92 wt % in various PAN solutions

PAN solution	Jc at 77 K (10^4 A/m²)	ρ at 100 K (10^{-5} Ω・m)
N-N' dimethyl sulfoxide	109	0.38
N-N' dimethyl acetamide	448	0.31
N-N'dimethyl formamide	1285	0.15
ethylene carbonate	268	0.38
malononitrile	208	0.21

observed by using PAN-DMF solution. The surface and cross-section of the filament with the highest Jc was shown in Fig.2. The blade-type grains with a grain size of $9\mu m \times 2\mu m$ were packed densely on the surface. From the cross section of the filament the grains with larger grain size of $12\mu m \times 3\mu m$ and some voids are found out. It was reported that the maximum Jc of the $Y_1Ba_2Cu_3O$ filament spun by using PVA-DMSO solution was 680×10^4 A/m² and the oxide powder was drawn up the blade type grains with a grain size of $3\mu m \times 5\mu m \times 1\mu m$.[5] The filament with higher Jc and larger blade-type grains could be produced by using PAN-DMF solution.

Although a high tensile strength of 37 MPa and elongation of 1.2 % was observed for the filament produced by PVA medium, the filament by PAN medium was weak with tensile strength of 10 MPa and elongation of 0.5 %. Moreover, the as-drawn filament without sintering, which was formed from PAN-DMF solution by extruding into a precipitating medium of methyl alcohol was very brittle with elongation of less than 1 %. Then the mixed solution of DMF and H_2O (1:1) was used for the precipitating medium and the plastic deformation with tensile strength of 2 MPa and elongation of 1.8 % was found out for the as-drawn filament. Thus the Y-Ba-Cu-O filament through PAN suspension medium had a higher Jc compared with the filament by PVA and the high Jc is due to a coarse texture allowed of larger superconducting crystal growth.

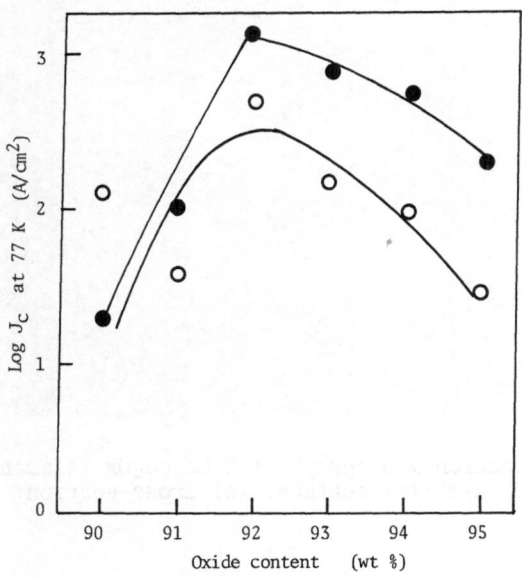

Fig.1 Relation between critical current density of the $Y_1Ba_2Cu_3Ox$ filament at 77 K and oxide powder content by using PAN-DMF and PAN-DMAc suspension medium. ● PAN-DMF medium and ○ PAN-DMAc medium

(a) surface

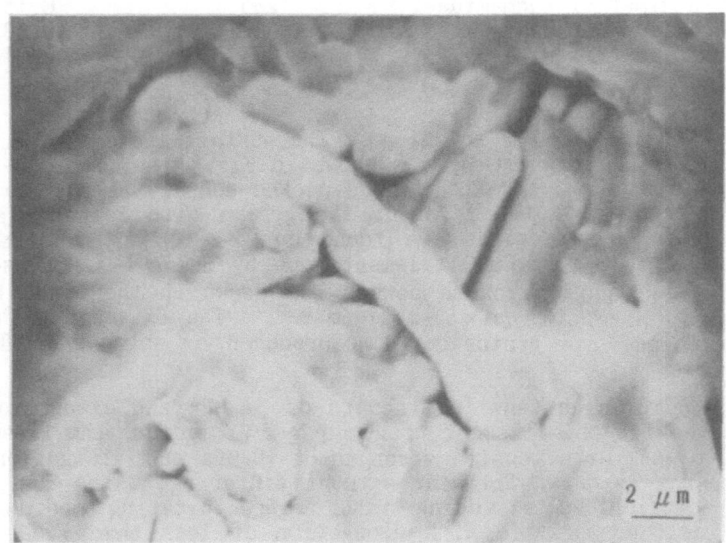

(b) cross-section

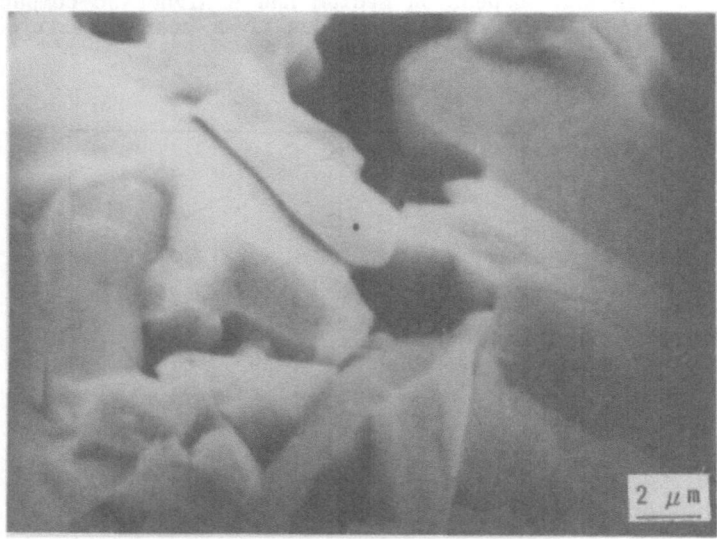

Fig.2. Scanning electron micrograph of $Y_1Ba_2Cu_3Ox$ filament exhibiting Jc of 1285×10^4 A/m^2:(a) surface, (b) cross-section.

Since the discovery of superconducting behaviour at an onset temperature above 100 K in Bi-Sr-Ca-Cu-O system, much effort has been made to prepare the single high-Tc phase. It is possible to fabricate the oxide with higher Jc by using the starting oxide of the single high-Tc phase. A long Bi-Pb-Sr-Ca-Cu-O superconducting filament was also produced as same as for the Y-Ba-Cu-O filament. However the low-Tc phase of 80 K still remained and high Jc was not attained for the Bi system filament. A high-Tc phase of Bi-Pb-Sr-Ca-Cu-O superconducting filament was successfully prepared by using a combined technique of suspension spinning through a mixed PVA solution of DMSO and HMTA, and densification of the pyrolyzed filament by pressing and sintering.[7] The suspension spinning of Bi-Pb-Sr-Ca-Cu-O was also made by using PAN-DMF solution. The as-drawn filament of 200 μm in diameter was heated at 773 K for 3.6 ks to remove volatile components. The filament was then cold pressed at 20 MPa and was sintered at 1113 K for 54 ks under the oxygen pressure of 1/13 x10^5 Pa in argon gas and was slowly cooled to 1023 K in the same atomosphere, then the flowing gas was switched to pure oxygen and the sample was cooled to room temperature. The densification was repeated to obtain higher Jc. Figure 3 shows the results of the temperature dependence of the resistivity for the filaments spun from PAN and PVA suspension media, respectively. The resistivity shows a sharp drop at about 115 K and becomes zero at 90 and 95 K for the filament from PAN and PVA respectively. A low Jc value of 55 x10^4 A/m^2 and high ρ value of 0.55 x10^{-5} $\Omega \cdot$ m were found out for the filaments spun from PAN medium comparing with the filament from PVA having Jc at 77 K of 1210 x10^4 A/m^2 and ρ value at 150 K of 0.24 x10^{-5} $\Omega \cdot$ m.

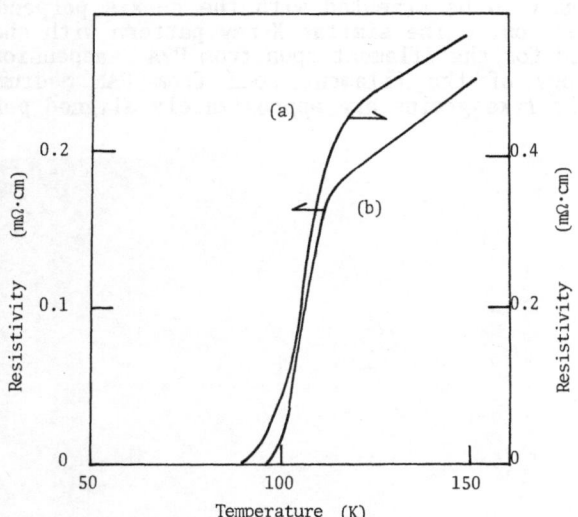

Fig.3. The temperature dependence of the electrical resistivity for the Bi-Pb-Sr-Ca-Cu-O filaments spun from PAN and PVA suspension medium. (a) The filament with Jc of 55 x10^4 A/m^2 at 77 K, which was spun through PAN suspension medium and two stage pressed with sintering at 1113 K under the low oxygen pressure of 1/13 x10^5 Pa for a total time of 111.6 ks. (b) The filament with Jc of 1210 x10^4 A/m^2 at 77 K, which was spun through PVA suspension medium and two stage pressed with sintering at 1113 K under the low oxygen pressure of 1/13 x10^5 Pa for a total time of 115.2 ks.

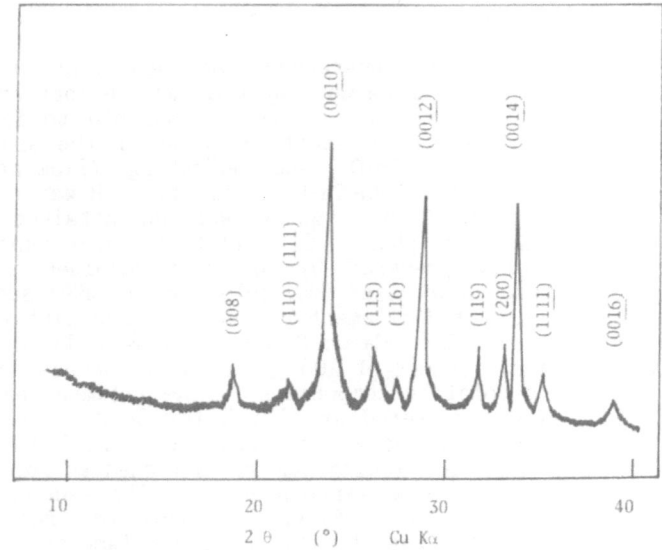

Fig.4. X-ray diffraction pattern of the Bi-Pb-Sr-Ca-Cu-O filament spun
through PAN suspension medium using Cu Kα radiation.   The X-ray
was applied perpendicular to the wide plane in the longitudinal
direction.

The X-ray diffraction pattern of the filament from PAN using Cu Kα
radiation is shown in Fig.4.   The X-ray was applied perpendicular to the
wide plane in the longitudinal direction.   This series of these strong
peaks is indexed at (00ℓ) of the high Tc phase having a tetragonal unit
cell with a = b = 0.5396 nm and c = 3.718 nm .   There is no indication of
the 80 K low-Tc phase or any other impurities.   The grains of the
filament were found to be oriented with the c-axis perpendicular to the
longitudinal direction.   The similar X-ray pattern with sharper peaks at
(00ℓ) was observed for the filament spun from PVA suspension medium.
Fracture morphology of the filament spun from PAN medium is shown in
Fig.5.   The flake-like grains are approximately aligned perpendicular to

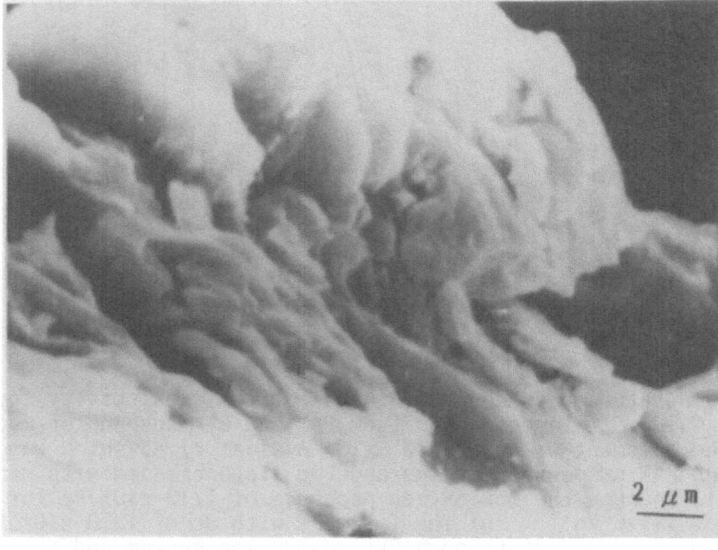

Fig.5. Fracture morphology of the Bi-Pb-Sr-Ca-Cu-O filament spun through
PAN suspension medium.

the longitudinal direction of the filament and the alignment is not perfect. The lower Jc for the Bi system filament spun from PAN suspension medium is related to the poor effect of PAN for the dispersant in the spinning dope and coagulator of the oxide through the precipitating medium, hence the poor densification of the oxide.

CONCLUSION

The suspension spinning of a high Tc oxide superconductor by using PAN was studied to prepare a long filament with high Jc. The spinning of $Y_1Ba_2Cu_3$ oxide was examined in various PAN solutions and the oxide powder contents of the spinning dope. The highest Jc of 1285 x10^4 A/m^2 at 77 K, 0 T was attained for the filament spun through PAN-DMF suspension containing the oxide of 92 wt %. The high-Jc is considered due to a coarse texture allowed of large superconducting crystal growth.

A high-Tc superconducting Bi-Pb-Sr-Ca-Cu-O filament was successfully prepared using a combined technique of suspension spinning and densification of the pyrolyzed filaments by two stages of pressing and sintering. Zero electrical resistivity was achieved at 90 K and the maximum Jc of the filament was 55 x10^4 A/m^2 at 77 K. The filament consisted of a single high-Tc phase and the grains were approximately oriented with the c-axis perpendicular to the longitudinal direction of the filament. The lower Jc for the Bi system filament spun from PAN suspension medium is considered due to the poor effect of PAN for the dispersant in the spinning dope and coagulator of the oxide through the precipitating medium, hence the poor densification of the oxide.

ACKNOWLEDGEMENT

This work was partly supported by a Grant-in-Aid for Special Project Research for the Ministry of Education, Science and Culture. (No. 01645511).

REFERENCES

1. K.Togano, H.Kumakura, H.Maeda, E.Yanagisawa, J.Shimoyama and T.Morimoto, Fabrication of flexible ribbons of high-Tc Bi(Pb)-Sr-Ca-Cu-O superconductors, Jpn.J.Appl.Phys., 28:L95 (1989).
2. M.Okada, R.Nishiwaki, T.Kamo, T.Matsumoto, K.Aihara, S.Matsuda and M.Seido, Ag-sheathed Tl-Ba-Ca-Cu-O superconductor tape with Tc=120 K, Jpn. J.Appl.Phys., 27:L2345 (1988).
3. T.Hikata, K.Sato and H.Hitotsuyanagi, Ag-Sheathed Bi-Pb-Sr-Ca-Cu-O superconducting wires with high critical current density, Jpn.J.Appl.Phys., 28:L82 (1989).
4. T.Goto and M.Kada, Preparation of high Tc Y-Ba-Cu-O superconducting filaments by suspension spinning method, Jpn.J.Appl.Phys., 26:L1527 (1987).
5. T.Goto, Nonaqueous suspension spinning of high-Tc Ba-Y-Cu-O superconductor, Jpn.J.Appl.Phys., 27:L680 (1988).
6. T.Goto and M.Kada, Critical current density of $Ba_2Y_1Cu_3O_{7-x}$ superconducting filaments produced by various suspension spinning conditions, J.Mater.Res.,3:1292 (1988).
7. T.Goto, Preparation of high-Tc superconducting Bi-Pb-Sr-Ca-Cu-O filament by the suspension spinning method, Jpn.J.Appl.Phys., to be published.
8. T.Goto and K.Yamada, Effects of poly(vinyl alcohol) suspension medium on critical current density of $Ba_2Y_1Cu_3$ oxide superconducting filaments produced by suspension spinning method, J.Appl.Poly.Sci., in submitting.

SPATIAL VARIATIONS IN THE TRANSPORT

PROPERTIES OF $YBa_2Cu_3O_{7-\delta}$ FIBERS

M.J. Neal and J.W. Halloran

CPS Superconductor
Milford, MA

INTRODUCTION

In this study fibers of dense, apparently homogeneous $YBa_2Cu_3O_{7-\delta}$ were used to study the spatial distribution of the critical current density in weak linked ceramic superconductors. It was discovered that for these samples not only does the critical current vary along the length of the fiber, but so does the transition temperature and the normal state resistivity.

FIBER PROCESSING

The fibers used in this study were produced in conjunction with the effort by CPS Superconductor to develop high Tc super-conducting wire. The starting material for the fiber was commercial YBCO powder (CPS Superconductor, lot #0602) with phase purity $\geq$ 99% and 1.69 micron average particle size. The powder is mixed with a polymer formulation and spun into fiber using a proprietary technique based on the melt spinning technology commonly used by the textile industries. Spools of polymer matrix ceramic composite with fiber diameters as fine as 50 microns have been fabricated. These green fibers are then subjected to a low temperature heat treatment for binder burnout. The fibers are sintered by placing them on a moving belt which passes at a constant rate through the hot zone of a furnace set at approximately 990°C, the fibers spend approximately 20 minutes at temperatures greater than 920°C. This sintering technique ensures that all points on a fiber have exactly the same sintering history. The final heat treatment is an oxygen anneal at 500°C for ten hours. The as-fired diameters of these fibers are approximately 80% of the green diameter. Most of the fibers used in this study had a nominal sintered diameter of 100 microns. Figure 1 shows the fracture surface of typical sintered fibers at different magnifications both before (A,C) and after (B,D) the oxygen anneal. These fibers are approximately 95-98% dense with little visible microcracking in either the tetragonal (unannealed) or orthorhombic (annealed) states. Critical current densities as high as 2600 A/cm^2 at 77K in self field have been measured in these fibers.

Advances in Cryogenic Engineering (Materials), Vol. 36
Edited by R. P. Reed and F. R. Fickett
Plenum Press, New York, 1990

509

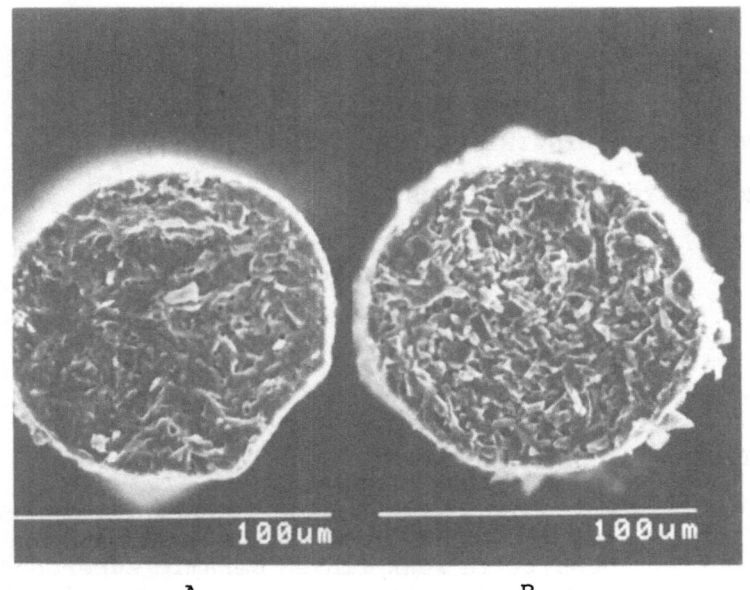

A B

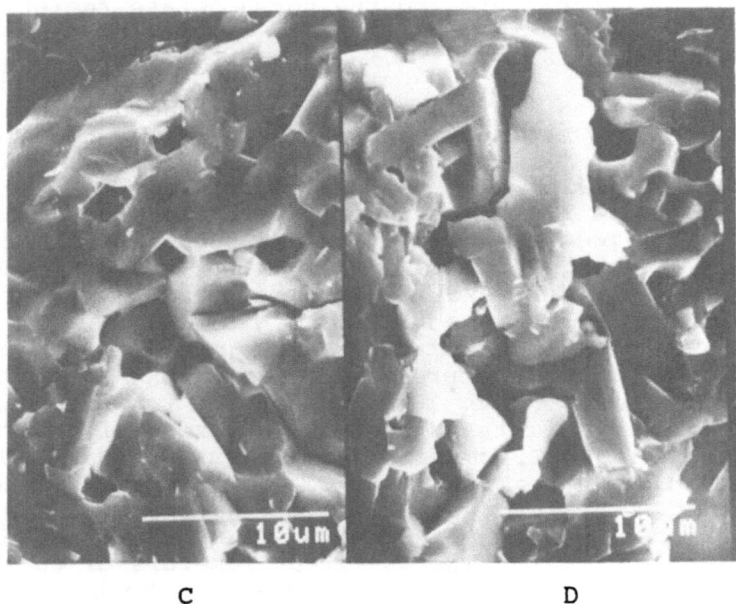

C D

Figure 1 SEM micrograph of the fracture surfaces of typical
YBCO fibers before (A,C) and after (B,D) oxygen anneal.

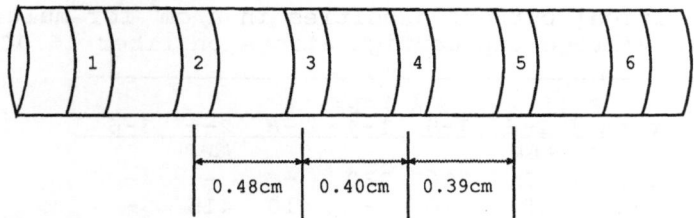

Figure 2 Schematic of the contact pad configuration on fiber 1670C.

EXPERIMENTAL PROCEDURE

Electrical contact to the ceramic fiber was made by making a silver contact pad using either sputter deposition or refiring the fiber after painting on contact pads with a silver paste. In the case of the refired fibers, a second 10 hour oxygen anneal was necessary in order to recover the orthorhombic ceramic phase. No subsequent heat treatments were administered to the sputter deposited contacts. Up to 10 silver contact pads, with a spacing of between 0.25 and 2.4 cm were patterned onto the bare fibers. The contact pads were connected to conductor lines on a test board using silver paint. In general, the contact resistivities of the refired contact pad fibers were lower than those of the sputter deposited contact pad fibers, but since the current required to test for Jc in these small diameter specimens was only 50 - 200 mA I^2R heating was not a problem. This was verified by measuring Jc across a particular section of a fiber as a function of the location of the current leads. It was assumed that since the contact resistance was somewhat variable the presence of contact heating effects would show up as a variation in the shape of the Jc curve with current contacts. This was not observed.

Measurements of the critical current densities of adjacent segments on a single fiber were performed. The tests were performed in liquid nitrogen using a DC current source (Keithley model 228), a digital multimeter (Keithley model 196) and an electric field criterion of 1 μV/cm for Jc. The instruments were controlled by a computer over an IEEE bus using an algorithm which increased the current in a stepwise fashion (step duration 6 seconds) and measured the voltage response at each step. All voltmeter filters were disabled during data acquisition and noise reduction was achieved with a discreet averaging routine for each current level. The noise level for the system using this routine was generally about 0.5 μV. The tests were performed in self field which for most samples was in the range of 2-5 Gauss. Tc measurements were by DC transport, with thermal voltages eliminated by reading the voltage signal under the influence of first positive current and then negative current.

RESULTS

There was a surprising variation in Jc between different segments of the same filament. In the most uniform specimen, the critical current densities varied only 7% between the segments. The least uniform specimen had up to 60% variation in critical current density. Figure 2 shows the contact geometry for fiber

Table I Critical current densities in A/cm^2 for multiple current and voltage tap configurations on fiber 1670C

V taps	I taps					
	1-6	1-5	1-4	2-6	2-5	3-6
2-3	310	310	330	–	–	–
3-4	395	410	–	410	415	–
4-5	500	–	–	500	–	510
2-4	315	325	–	–	–	–
3-5	420	–	–	430	–	–
2-5	330	–	–	–	–	–

1670C. The numeric labels assigned to each contact pads are in reference to Table 1 which shows the critical current density values in A/cm^2 for different segments of this fiber. There were six contact pads approximately 1 mm wide spaced 4 to 5 mm apart (center to center). This configuration allowed the testing of three adjacent segments of fiber. As Table 1 shows, Jc was independent of the choice of current contact pairs but highly dependent on which segment was measured. Note also that when two segments were measured together the Jc value corresponded to the lower of the two Jc's. The critical current densities for different segments of the same fiber are shown again in Figure 3 in a different format for two different fibers, 34008A and 33006C3. Again the variation in Jc is significant, in one case varying by more than a factor of 2. Figure 4 shows the resistive transitions for each of the three segments whose Jc's are shown in Figure 3b. The high Jc segment has a higher Tc(0) and a lower normal state resistivity than the other two segments. The difference in the normal state resistivities of the center and right segments is not significant due to uncertainties in the actual current path when the ceramic fiber and the silver contact pad are in parallel. The differences in Tc(0) and the normal state resistivity between those two segments and the left segment are certainly real.

DISCUSSION

The observed variations in electronic properties are surprising since the starting powder is quite uniform in phase homogeneity, and there has been no evidence in preliminary microstructural evaluations of inhomogeneity. For these reasons we believe that it is the fine scale on which we are making these measurements (a 3mm length of 0.1mm diameter fiber contains about 120 μg) is revealing these phenomena rather than any gross inhomogeneity in our fibers.

Since the Jc of filaments varies spatially, this will lead to specimen length effects in characterization. That is the value reported will depend upon the length over which the measurement was made. The length over which the generation of an electric field at a low current density by a particular segment will dominate the characteristics of the wire will depend upon the n-values of the local I-V characteristics and the frequency with which such segments exist. Extrapolating values from measurements on short lengths of wire to characterize the properties of long wires will require a quantitative description

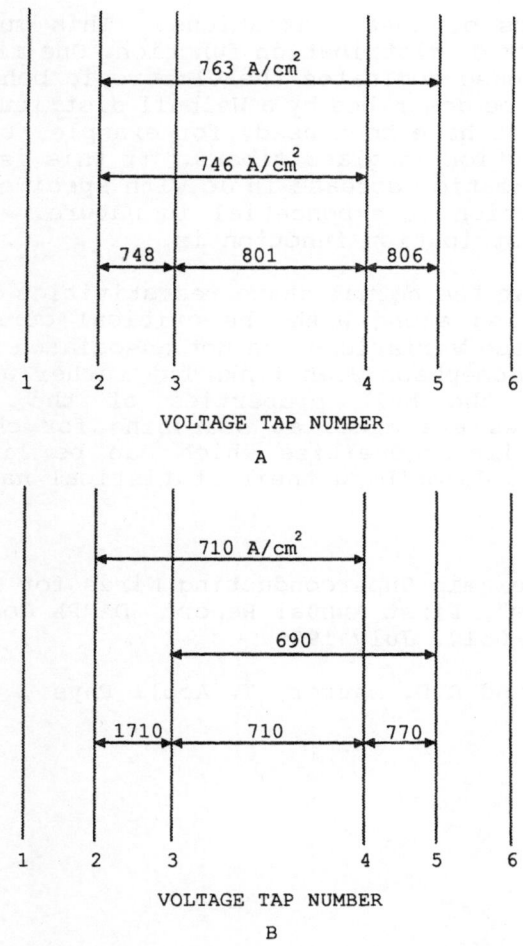

Figure 3 Critical current density for two similarly processed fibers(A) 34008, B) 33006C3) fibers as tested over different segment lengths.

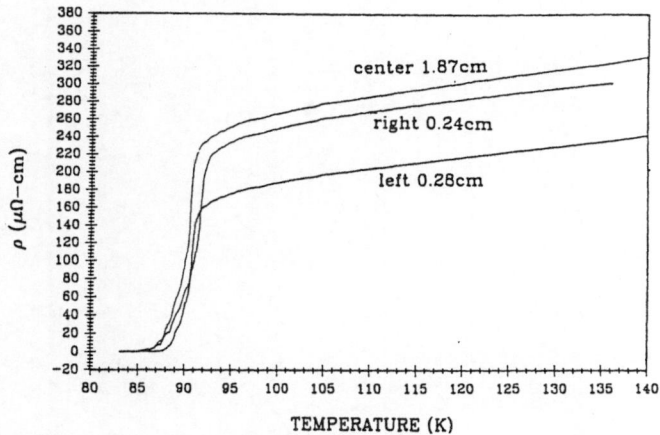

Figure 4 Temperature dependent resistivities for each of the three segments in fiber 33006C3.

of the statistics of these variations. This must be done in terms of some sort of distribution function. One might think that since the worst case dominates the electronic behavior that the statistics could be described by a Weibull distribution function. Weibull statistics have been used, for example, to describe the strength distribution in glass fibers.[2] If this is the case one must expect a dramatic decrease in Jc with specimen length since the Weibull function is exponential in nature. We do not yet know what the distribution function is.

The fact that the normal state resistivities and transition temperatures varied along with the critical current densities suggests that these variations are not associated with statistical grouping of Josephson weak links but rather with fine scale fluctuations in the bulk properties of the material. Our continuing studies are aimed at searching for chemical and/or microstructural inhomogeneities which can be linked to these variations and at describing their statistical nature.

REFERENCES

1. "Composite Ceramic Superconducting Wires for Electric Motor Applications", First Annual Report, DARPA Contract No. N00014-88-C-0512. July 1989.

2. R. Olshansky and R.D. Maurer, J. Appl. Phys., 46, 4497, (1976).

A STUDY OF TAPE AND WIRE PROCESSES FOR HIGH T$_C$ SUPERCONDUCTORS

L. R. Motowidlo, G. M. Ozeryansky, H. C. Kanithi, B. A. Zeitlin,
and B. C. Giessen*

IGC Advanced Superconductors Inc.
1875 Thomaston Ave.
Waterbury, CT 06704

*Northeastern University
Boston, MA 02115

ABSTRACT

Using the powder in tube method, tape and multifilament configurations were designed and fabricated using familiar methods of wire processing. The average estimated filament dimension in the multifilament wires at final wire diameter was 15 to 20 microns and the total reduction in area on a single filament was 99.99%. The total length of nearly 150 meters was acheived in the multifilament wire. Tape conductors were also fabricated from monofilaments by final rolling to a thickness of 0.152 mm. Preliminary transport current measurements at liquid nitrogen temperature show current densities on the order of 180 A/cm^2 for monofilaments while the tape shaped conductors gave a J$_c$ of approximately 1000 A/cm^2.

INTRODUCTION

The transition temperature,T$_c$, of the new metal oxide superconductors has been pushed to 122K[1]. Critical magnetic fields greater than 100 Tesla and intrinsic current densities on the order of 5×10^6 A/cm^2 at 1.5 Tesla in thin films, sufficient for any device have been demonstrated by magnetization[2] measurements. Transport current density however, has been disappointingly low in polycrystalline materials. The critical current density,J$_c$, is still some two to three orders of magnitude smaller than that of conventional metal alloy superconductors[3,4]. In addition to the fundamental T$_c$ properties, a technologically useful superconductor for power applications, must have a reasonable current density (10^4to10^5 A/cm^2) at magnetic working fields and be in a form that can be wound into a magnet.

The problem, in general, is to develop a process to manufacture long lengths of multi-filamentary metal oxide materials into practical material for device applications. In accordance with that, IGC Advanced Superconductors has initiated a program to explore fabricability of these materials via the technology now employed to process conventional multifilamentary low T$_c$ superconductors. In this paper, we present preliminary results on multifilamentary and tape fabrication.

Advances in Cryogenic Engineering (Materials), Vol. 36
Edited by R. P. Reed and F. R. Fickett
Plenum Press, New York, 1990

515

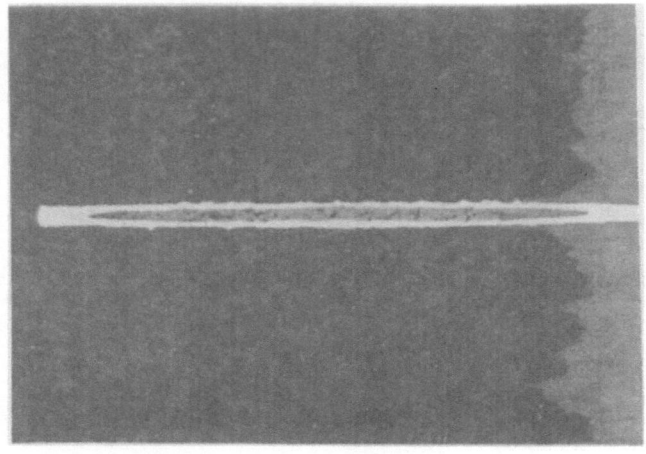

FIGURE 1. Cross section of rolled tape conductor in Ag sheath. Thickness 0.152mm.

EXPERIMENTS

Starting materials for these preliminary experiments included $Y_1Ba_2Cu_3O_x$ and silver tubing. The $Y_1Ba_2Cu_3O_x$ precursor powder was obtained from Rhone-Poulenc and consisted of powder particle sizes of two to four microns. Purity of this powder was at least 99.9%. The other key component for the fabrication of tape and wire was silver obtained from Johnson and Mathey in the form of tubing with a minimum purity of 99.9%.

Wire and Tape Processing

To prepare monofilament wire, typical starting billets with an overall length of 25 centimeters and outer diameter of 6.35mm and inner diameter of 4.45mm were prepared. To ensure complete filling of the tube space the powder was introduced gradually while tapping the tube. The ends were closed with silver plugs and lightly swaged. All monofilament billets were cold drawn

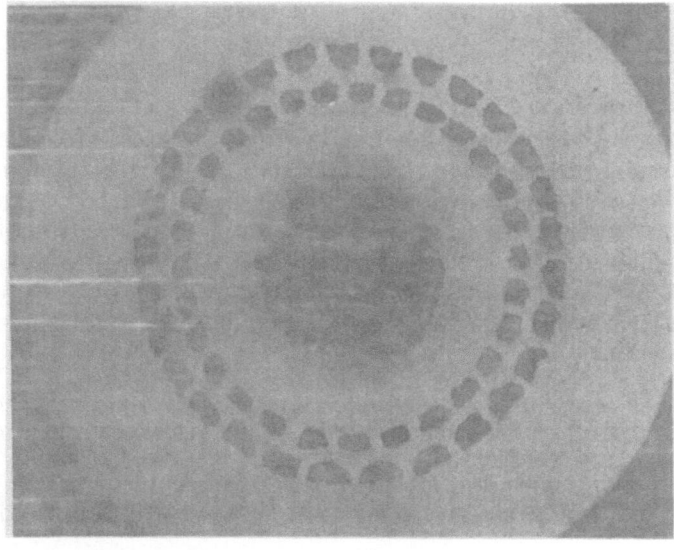

FIGURE 2. Cross section of a multifilament wire processed to 0.518mm diameter.

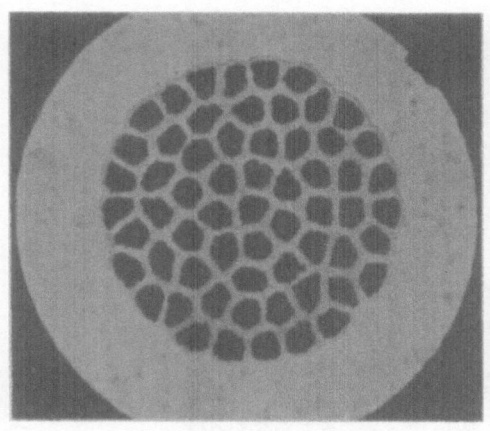

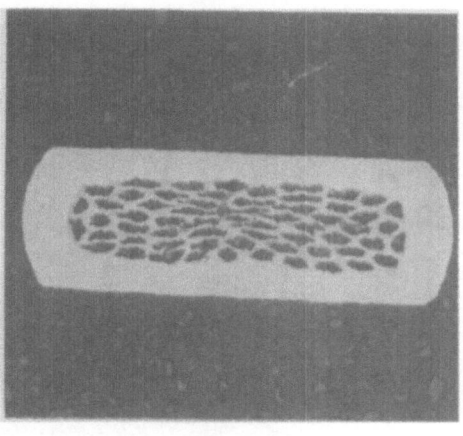

FIGURE 3. Cross section of a 61 element wire at 1.83mm diameter.

FIGURE 4. Cross section of a 0.254mm thick multifilament tape with 61 filaments.

first on a laboratory draw bench followed by wire drawing on conventional round captions. A total length of approximately 65 meters of wire at 0.381mm diameter was produced. Further drawing to 0.305mm diameter led to numerous wire breaks. Additional processing involved rolling the monofilament wire into tape form. A 1.625 mm diameter wire was rolled through 20 successive passes to a final thickness of 0.152-0.203 mm over the silver casing. A cross section of the rolled tape is shown in figure 1.

Several multifilament configurations were designed and processed using familiar methods of restacking and coprocessing. Monofilament composite wire was drawn to 0.737 mm and restacked into 58 elements with a large central core which also contained YBCO powder. The multifilament composite was processed to 0.381mm without any breaks. In figure 2 we show an example of a multifilament configuration. The average estimated filament dimension became roughly 15 to 20 microns. The total length acheived with this multifilament wire was nearly 150 meters, and the total reduction in area on a single filament was 99.99%. Another example of a multifilament configuration is shown in figure 3. In this case, we assembled a composite billet with 61 elements. Further drawing to 0.254 mm O.D., followed by rolling into tape shape is shown in figure 4.

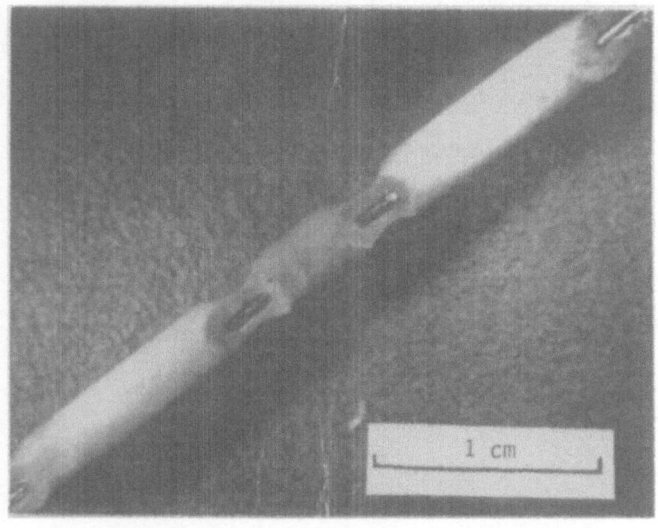

FIGURE 5. Ceramic holder for monofilament or multifilament wire.

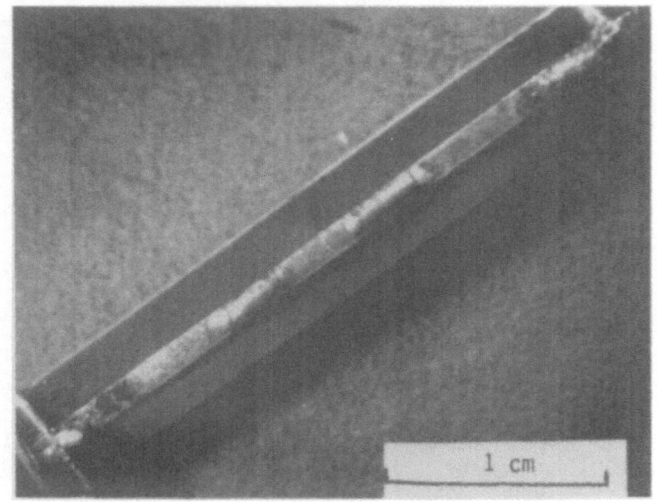

FIGURE 6   A flat ceramic holder for tape
conductors.

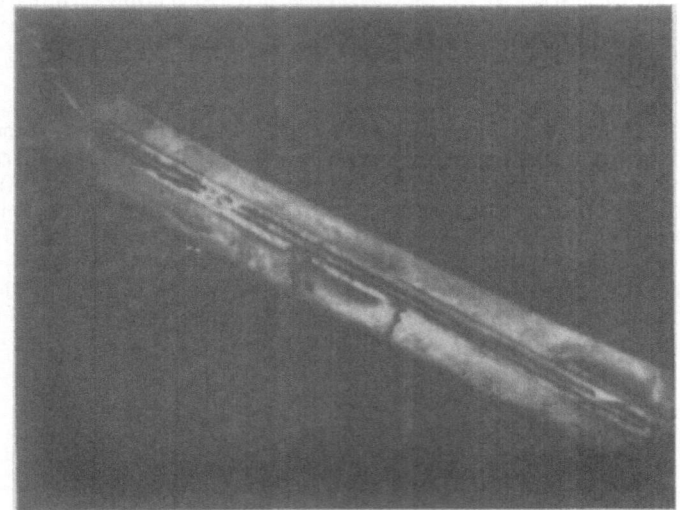

FIGURE 7a.   Tape conductor 0.178mm X 1.143mm.

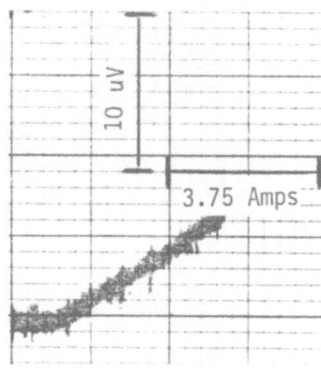

FIGURE 7b. Voltage versus current trace for the tape conductor

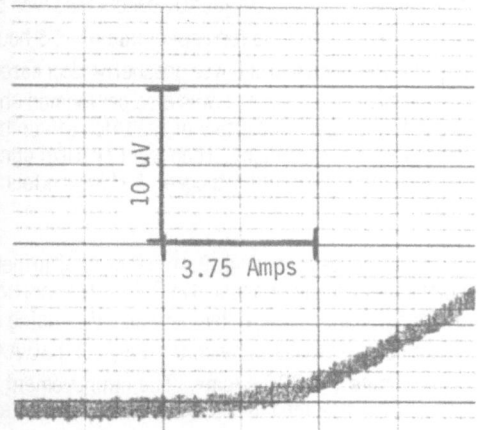

FIGURE 8a. A multitape configuration around a thin silver tube.

FIGURE 8b. The V–I trace for the multitape configuration.

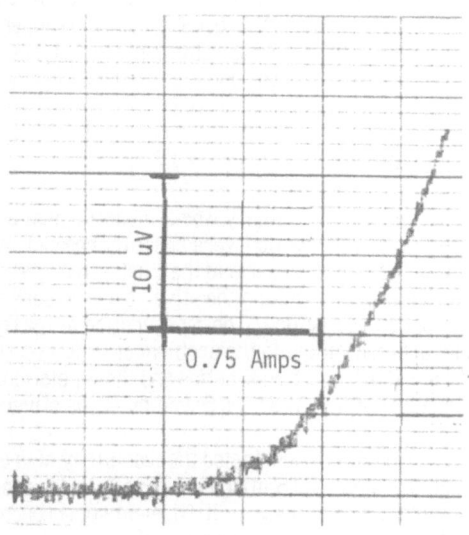

FIGURE 9a. A cable with three monofilaments.

FIGURE 9b. The voltage–current trace of the cable with three monofilaments.

Heat Treatment, Critical Transport and T_c Characterization

After final processing, samples were readied for heat treatment in special ceramic holders. In the case of the multifilament or monofilament wire, a thermocouple type ceramic tube with four holes, shown in figure 5, was fashioned to provide suitable access for voltage and current electrical leads. In this case four very thin multifilament samples of appoximately 0.305 mm diameter can be carefully and simultaneously handled for further measurements after heat treatment.

For tape type conductors, flat ceramic holders as shown in figure 6 were utilized. Each tape specimen was secured to the surface with Ag wire prior to heat treatment and characterization. Heat treatment of all $Y_1Ba_2Cu_3O_x$ conductors was performed under flowing oxygen. Samples were gradually heated and held for approximately 18 hours at 920°. After the high temperature sintering, the temperature was gradually decreased at 200°/hr to 450°C and left unchanged for 8 hours. Annealing at 450°C was also performed under flowing oxygen at a gas pressure of 2–3 psi. In order to minimize the effect of the thick silver barrier, we also thinned the silver chemically prior to the heat treatment. Transport critical current measurements on wire and tape samples were performed in liquid nitrogen. A standard four point resistance measurement technique was utilized.

In figures 7 through 9, we show preliminary transport current traces of the tapes and monofilament conductors tested at 77 K. The critical transport currents I_c varied between 1 and 7 amperes at liquid nitrogen temperature depending on the size and shape of the conductor. Monofilament current densities were on the order of 180 A/cm^2 while the tape shaped conductors gave a J_c of approximately 1000 A/cm^2. The improvement in the tape current density over the monofilament result is assumed to be a consequence of the texturing effect introduced by rolling. All measurements were taken at a resistivity criterion of approximately 10^{-10} to 10^{-11} ohm-cm. The transition temperature of the samples was found to be 89.6 K for samples heat treated without a Ag barrier. In samples with a Ag barrier, the T_c was 88.4 K as determined from susceptibility measurement.

In figure 8a we show a thin silver tube approximately 10 cm long and 6.35mm O.D. . The wall thickness was 0.255mm. Around the tube, several tape type samples approximately 1.5mm wide were held fixed by silver wire. The length of the samples was roughly 7.0 cm. A possible application for such a configuration could be superconducting down leads as in SMES devices. In figure 9a we also show a cable constructed from three 0.762mm monofilaments. The voltage-current trace of this configuration is shown in figure 9b.

CONCLUSIONS

In this preliminary work, monofilament and multifilament composites were prepared from precursor powders which were not fully reacted. As a result, initial difficulty was encountered in reacting the composites due to the thick silver barrier. For the case of the tape-type conductors this was alleviated by thinning the silver barrier such that the oxygen could penetrate more readily into the composite structure. The densification of the composites, such as is achieved in the multifilaments also contribute to the problem of fully reacting the structure. Further work is required to understand the reaction kinetics and their relationship on the density of the microstructure as well as the effect of thickness and geometry of the silver barrier.

We have demonstrated the feasibility of drawing multifilament wire with filament dimension on the order of 15 microns. The total reduction in area of a single filament was 99.99%. The total wire length of these multifilament superconductors was nearly 150 meters. Critical current densities on the order of 1000 A/cm^2 in the tape conductors were measured at liquid nitrogen temperature. Heat treatment optimization and characterzation of the multifilament wire are currently ongoing and will be reported elsewhere.

ACKNOWLEDGEMENTS

This work at IGC Advanced Superconductors is supported by the Department of Energy, Contract No. DE-AC02-88ER80607.

REFERENCES

1. P. Haldar, K. Chen, B. Maheswaran, A. Roig-Janicki, N. Jaggi, R. Markiewicz, B. C. Giessen, Science, 241 ,1198 (1988).
2. S. Nakahara et al, Journal of Crystal Growth, December,(1987).
3. B. A. Glowacki and J. E. Evetts, AA7.35, Materials Research Society meeting , Boston, MA, Nov30-Dec5, (1987).

INEXPENSIVE HIGH Tc SUPERCONDUCTING WIRES

P. Dubots, D. Legat

Laboratoires de Marcoussis – C.R – C.G.E Route de Nozay
91460 Marcoussis, France

F. Deslandes, B. Raveau

CRISMAT ISMRa Boulevard du Maréchal Juin
14032 Caen, France

INTRODUCTION

After the discovery of high Tc superconducting oxides, the
way to 77 K cryogeny based applications was open and worldwide
efforts have been undertaken to manufacture conductors, wires or
tapes, with or without matrices or substrates. More often than
not, these researches were carried out without taking into account
neither the specific superconducting (SC) properties of these
oxides nor those related to their potential utilization temperature. In fact, this leads to new conceptions of conductors[1]: for
instance, the filamentary aspect could be restricted to a.c
applications or when remanent magnetic field has to be avoided.
However, interest in making oxide filament based wires remains.

As SC oxides are not ductile, the only way to make these
composite wires is the "powder-in-tube" process, by analogy with
Chevrel phase based wires[2]. This process widely has been used with
these new materials, mainly YBa2Cu3O7 (YBCO)[3,4,5]. The main
problem encountered with this technique is the necessity to sinter
the powder in a final step, in order to establish a good electrical connection between the powder grains and thus to get the
current carrying capacity which is non-existant in the as-drawn
state.

In the case of YBCO, this problem is dramatically increased
by the strong temperature dependance of the oxygen content, and so
far, this only has been solved using a silver matrix. Since silver
is a very expensive and highly speculative material,thus limiting
its economical viability we have therefore tried to find other
materials for this purpose[6]. This paper presents results obtained
from wires manufactured with a copper based matrix.

Advances in Cryogenic Engineering (Materials), Vol. 36
Edited by R. P. Reed and F. R. Fickett
Plenum Press, New York, 1990

523

WIRE CONCEPTION AND MANUFACTURING PROCESS

The "With Oxygen Confinement"(WOC) wires

Silver has been shown to be a good matrix material because its high oxygen solubility allows exchange between YBCO and external oxygen, and the recovery of the oxygen stoichiometry of the oxide. In looking for a non silver matrix, our idea was to confine oxygen inside the wire during the heating cycle, so that oxygen remains available during the cooling, in order to recombine with the YBCO : this gives wires With Oxygen Confinement, WOC wires.

After having confirmed the validity of this principle using samples encapsulated in a sealed quartz tube[6], we have searched for a matrix material with the following requirements :
(i) ductility, to support the deformation of the wire process,
(ii) high melting temperature, which would not limit the sintering temperature,
(iii) reaction neither with YBCO, nor with oxygen, and with an oxygen solubility as low as possible,
(iv) cost as low as possible.

We have shown that it was not possible to fulfill completely point (iii)[6] ; this has been solved by inserting an oxide barrier between the metallic sheath and the oxide core. This oxide barrier has two main roles :
* it prevents reaction between YBCO and the metallic matrix,
* it constitutes an oxygen reservoir, which may compensate for small losses of oxygen.

We have thus reached the following wire composition, schematized on figure 1 (which also shows a view of a WOC wire cross section) :
- an inner powdered YBCO core,
- a powdered oxide barrier,
- an external metallic sheath.

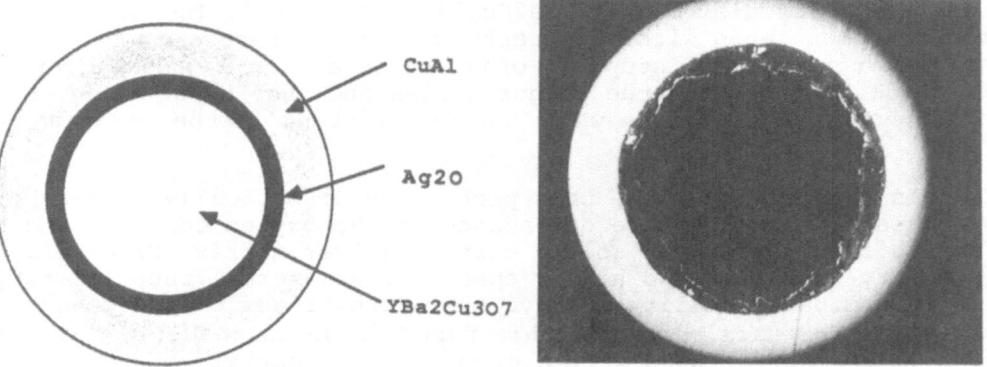

Figure 1 : Schematic and view (1mm diam. - as-drawn state)
        of a WOC wire cross-section.

Advantages of the WOC wires

WOC wires may present some specific advantages with regard to classical ceramic shaping.

We can expect that during the wire transformation, the oxide powder becomes more reactive, as grains are broken to smaller size, with fresh, unpolluted surfaces. Thus powder sintering may be favored.

We can expect some texturization of the powder, either during the transformation or during the sintering treatment, especially if the filament diameter is small.

Finally, as oxygen is confined inside the wire, during heating, the gas pressure increases, limiting the decrease of the oxygen content[7] ; we can thus hope that the orthorhombic - tetragonal transformation may be avoided during heating and cooling of the wire, removing the twin formation.

Manufacturing Process

The WOC wires were fabricated by the classical "powder-in-tube" technique which required the following steps :
(i) machining of a metallic tube (o.d $\sim$ 14 mm, i.d $\sim$ 10 mm),
(ii) isostatic pressing of a "normal" oxide powder tube (o.d $\sim$ 10 mm, i.d $\sim$ 8 mm),
(iii) isostatic pressing of a YBCO powder rod (diam. $\sim$ 8 mm), or filling the "normal" oxide tube with YBCO powder at a tap density,
(iv) sealing the metallic tube,
(v) drawing to diameter less than 1 mm.

This technique allowed us to obtain hundred meter long, mono- or multifilamentary YBCO based wires, the filament diameters of which were as low as 25 microns[8].

The studied materials

Sheath materials. In a first step, some experiments have been carried out on wires, the sheath of which was composed of concentric aluminum inner and copper outer tubes. This work allowed us to verify on wires the validity of the WOC conception, but the heat treatment temperature was limited to the melting temperature of aluminum, much too low to sinter the SC powder.

CuAl alloys were found to fulfill the sheath material requirements. Copper and aluminum present a large solubility field, up to about 10 wt% of aluminum (in copper), at room temperature. As the aluminum content increases, the oxidation rate of copper substantially decreases. But, at the same time, the alloy ductility becomes lower. Thus, the Al content has to be optimized in order to minimize oxidation and still yield a ductile product.

A good compromise has been found with an Al content around 8 wt%, which presents a very low oxidizability, as shown on figure 2 which gives TGA curves of a CuAl8 alloy and of copper heated up to 900°C under oxygen.

Table 1.Characteristics of the WOC wires

Reference	Metallic Sheath	Oxide Barrier	Outer Diam.	Core Diam.
WOC 3	Cu + Al	BaO2	1 mm	0.60 mm
WOC 15	CuAl 8	Ag2O	1 mm	0.69 mm
WOC 18	CuAl 8	Ag2O	1 mm	0.76 mm

Although the electrical resistivity of the CuAl alloy is obviously higher than that of copper, it has been shown that the stability of the SC oxides is intrinsically good and that a highly conductive matrix is not as necessary as for metallic super-conductors[1].Resistivity measurements have given 13.10^{-6} Ω .cm at room temperature and at 77 K.

Barrier material. As has been mentioned above, the oxide barrier must fulfill the following requirements :
* it must be inert with regard to the SC oxide and to the matrix,
* it may lose its oxygen at high temperature, but must not be more oxidizable than the SC oxide.

Preliminary experiments led to the choice of either BaO_2 or Ag_2O.However we showed subsequently that, at high temperature, barium oxide combined with YBCO and destroyed it[6]. Moreover silver oxide is more favorable for making electrical contacts[9].

Superconducting oxide. Experiments have been carried out on $YBa_2Cu_3O_7$ only. Figure 3 presents a typical grain size and a SEM view of the starting powder.

The characteristics of the different wires presented in this paper are given in table 1.

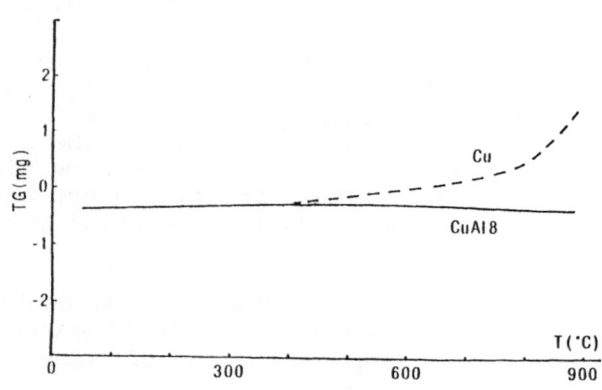

Figure 2 : TGA curves of Cu and CuAl alloy.

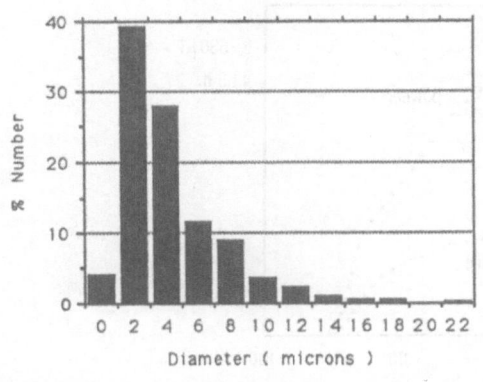

Figure 3 : Typical size and SEM view of the starting YBCO powder.

EXPERIMENTAL MEASUREMENTS

To characterize the SC properties of our wire samples, we have performed inductive (a.c susceptibility) and resistive (critical current) measurements.

A.c susceptibility measurements are rapid means of testing samples without the need of current leads. This method gives much information about the sample properties, but care must be taken in interpretation, especially in view of grain size, orientation distribution and weak interconnectivity of the powder[10].

For these measurements, the excitation frequency was 41.3 Hz and real K' and imaginary K" parts of the susceptibility were obtained mainly for an a.c field amplitude of 0.63 mT. No d.c bias field was added and the phase sensitive detector was set on broadband to obtain the full voltage waveform. The values of K' and K" obtained were normalized to the physical section of the inner core of the wire by calibration with a niobium sample of known cross-section.

Resistive measurements were made by the so-called four probe methods. Samples, 5 to 10 centimeter long, were measured at 77 K. Since it was very difficult to get good contacts directly on the wire, two kinds of samples were prepared :
- C samples : in that case, closed samples were heat treated without contacts. After heat treatment, the matrix was mechanically polished in order to make the filament visible and to supply contacts for the current and voltage leads,
- SO samples : in that case, the matrix was opened and contacts were made before the sintering treatment.
Similar results were obtained in both cases, but since the brittleness of C samples led to a lack of reproducibility, we report in this paper the results obtained on SO samples only.

RESULTS

As-drawn wire. As almost all the "powder-in-tube" processed wires[11],the YBCO based wires have no current carrying capacity in the as-drawn state, and it is necessary to sinter the powder to get a good electrical connectivity between grains.

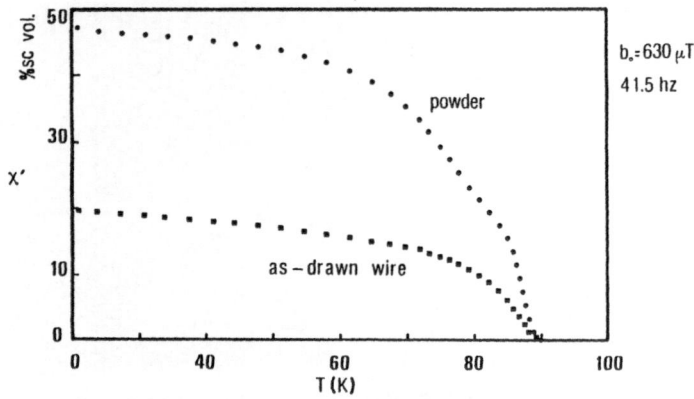

Figure 4 : K'(T) curves of starting powder and as-drawn wire.

However, even if the as-drawn wires do not carry any SC current, the intrinsic SC properties of the YBCO powder are not degraded in the fabrication process.The K'(T) curves of figure 4 show that the Tc is the same before and after drawing, although the SC volume has decreased by a factor of about two. This decrease is due to the reduction of the powder grain size during drawing, and is related to the ratio between the penetration depth and the grain size[10].

Control of the WOC wire conception on wire samples. Some C samples with concentric copper and aluminum sheaths displayed cracks in these sheaths after the sintering treatments, probably due to the high pressure generated inside the wire.In that case, we observe a large decrease of the SC volume probably due to the loss of oxygen by the SC oxide (figure 5). By increasing the thickness of the external copper tube in order to improve its mechanical resistance at high temperature, we succeeded in

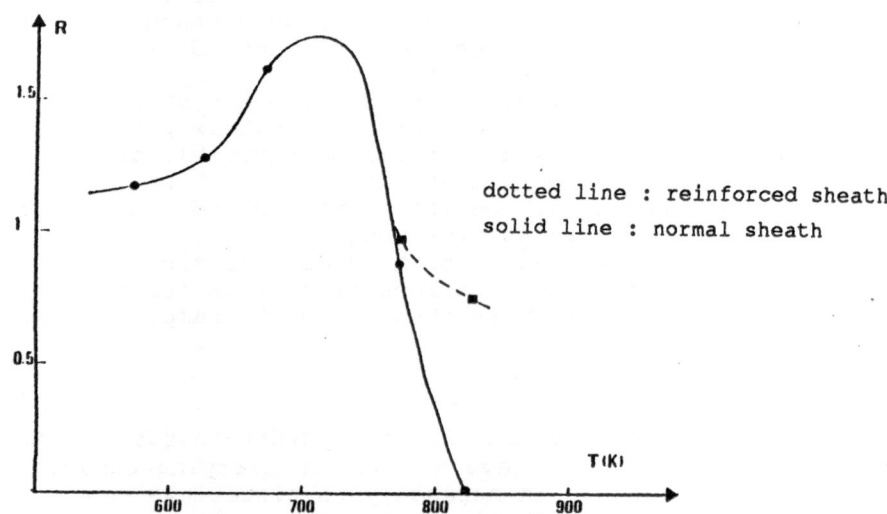

dotted line : reinforced sheath
solid line : normal sheath

Figure 5 : Sintering temperature dependence of the SC volume of WOC 3 wires.

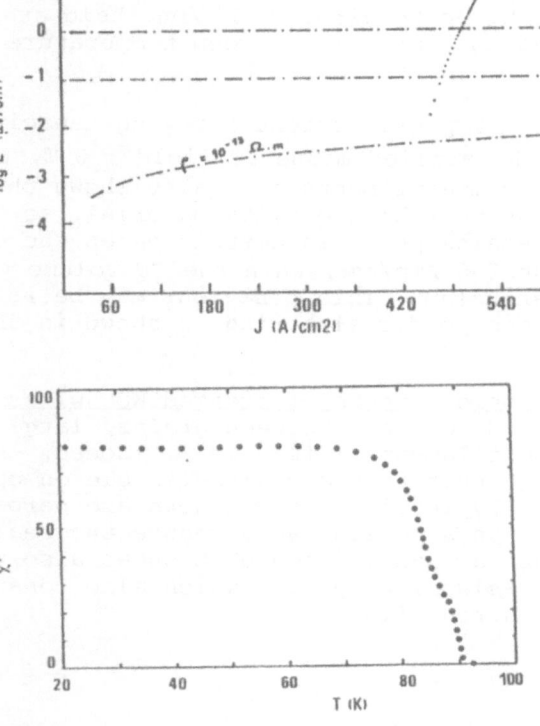

Figure 6 : Typical resistive and inductive SO WOC wire
          characteristics.

suppressing cracks and maintaining the SC volume up to 550°C. At
higher temperatures, reactions between barium oxide and YBCO, and
oxidation of the aluminum sheath, became too important not to
degrade YBCO properties. This result attested that no oxygen
escaped from the wire, when the treatment was performed under
vacuum.

    SC properties of treated samples. Figure 6 gives typical
results of resistive and inductive measurements of treated SO WOC

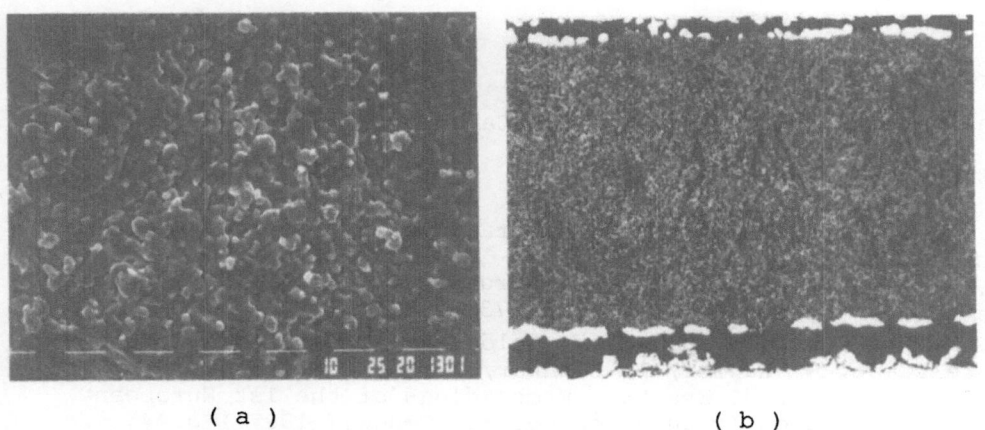

(a)   (b)

Figure 7 : Metallographic aspects of treated WOC wires.

wires. Good conditions of sintering have been found to be the following : heating up to 925°C at 1°C/mn, held at 925°C for one to two hours, and cooling down to room temperature at 1°C/mn, all under oxygen.

This leads to typical current carrying capacities of about 500 Amp/cm^2 (77 K, applied magnetic field : 0 T, E = 1μv/cm). A.c susceptibility measurements have also shown that this does not always corresponds to a high quality material, as the connectivity between grains remains poor. In certain cases,the critical current density was about 200 Amp/cm^2, when the SC volume was as low as 10% of the powder volume. This behaviour may be related to the poor quality of the powder sintering as shown in the following part.

Metallographic aspects of treated WOC wires.Although the SEM views show good contacts between grains, large cracks are also evident in the SC filaments (fig. 7a). Indeed, these cracks are sufficiently large that they are also visible on optical micrograph (fig.7b). Typically, these cracks are perpendicular to the current axis, and thus considerably reduce the real cross-section where the current may pass. Figure 7b shows also that there remains a large residual porosity, which also constitutes an obstacle to the current flow.

DISCUSSION

Although we have shown that inexpensive materials could be used as a matrix of SC oxide based wires, it appears that the original conception of the WOC wires leads to very different behaviours of the SC powder when the performances are about the same as those of Ag sheathed wires[3] or as those of sintered ceramics[12]. More particularly, it seems that satisfactory current carrying capacities may be obtained, even when the SC filament is dramatically porous or cracked.
That raises questions regarding the real or effective cross section of the wire and of the real current carrying capacity of the powder. Since no texturization has so far been observed on our filaments, we have no simple explanation for our critical current results which are comparable to those obtained on sintered ceramics. Work is in progress in that field and we hope that reducing the defects will lead us to large improvements of the WOC wire performances.

ACKNOWLEDGEMENTS

The authors are grateful to J.R.Cave for fruitful discussions.

REFERENCES

1. A. Fevrier, Journées SEE " Céramiques supraconductrices à haute Tc " Caen, sept. 1988.
2. M. Hirrien, R. Chevrel, M. Sergent, P. Dubots, P. Genevey, Materials Letters 5 (1987) p.173.
3. R. Flükiger, T. Müller, W. Goldacker, T. Wolf, E. Seibt, I. Apfelstedt, H. Küpfer, W. Schauer, Physica C 153 (1988) p. 1574.
4. R. Glovacki, J. Evetts, Proceedings of the 1st European workshop on high Tc superconductors. Gènes (1987) p.447.

5. H. Sekine, K. Inoue, H. Maeda, K. Numata, K. Mori, H. Yamamoto, Applied Phys. Lett. 52 (1988) p.2261.

6. F. Deslandes, B. Raveau, P. Dubots, D. Legat, This conference presentation BX 04.

7. J. Karpinski, E. Kaldis, S. Rusiecki, J. of Less Common Metals 150 (1989) p.207.

8. P. Dubots, J. Cave, D. Legat, C. Michel, B. Raveau, ibid 4 p.133.

9. J. Ekin, A. Panson, B. Blankenship Applied Phys. Lett. 52 (1988) p.331.

10. J.R.Cave, M. Mautref, C. Agnoux, A. Leriche, A. Fevrier, Int. Conf. on Critical Currents in high Tc superconductors Snowmass Colorado August 1988.

11. P. Dubots, J. Cave, Cryogenics 28 (1988) p. 661.

12. N. Chen, K. Goretta, M. Lanagan, D. Shi, M. Patel, I. Bloom, M. Hash, B. Tani, D. Capone II, Supercond. Science and Technology 1 (1988) p.177.

PRELIMINARY CHEMICAL STUDY FOR THE MANUFACTURE OF HIGH Tc SUPERCONDUCTING

WIRES

F. Deslandes, B. Raveau

CRISMAT ISMRa Boulevard du Maréchal Juin 14032 Caen (France)

P. Dubots, D. Legat

Laboratoires de Marcoussis Route de Nozay 91460 Marcoussis
(France)

INTRODUCTION

Superconducting composite wires containing both the superconductor and a metal sheath are required for many applications. This design, used for industrial superconducting wires is desirable in order to stabilize the superconducting state minimizing local heating and energy dissipation induced by magnetic flux changes, to carry current in case of local break down of superconductivity and of course to provide mechanical and environmental protection.

In the case of the new superconducting oxides, a powder-in-tube technique is considered because of the brittle material. However, this technique is complicated by a number of factors such as the control of oxygen stoichiometry changes during the sintering of powders, the anisotropy of $YBa_2Cu_3O_7$, the possible reactions with the matrix, etc...

Actually the best results are obtained by drawing a silver tube packed with $YBa_2Cu_3O_7$ superconducting powder.[1,2] Nevertheless the use of silver is industrially impractical for two reasons : silver is too soft and too expensive. These considerations have led us to look for an original approach that could use an inexpensive, relatively non-oxidizable matrix which can be isolated from the superconductor by a non-reactive sheath. To achieve the manufacture of superconducting wires using $YBa_2Cu_3O_7$, we studied the following essential points : the behaviour of the superconductor in an enclosed atmosphere was tested by measurement of variation of stoichiometry and changes in the diamagnetic response as a function of thermal treatment. We have tried to synthetize the 123 compound by solid state reaction in a sealed tube and we have shown the feasibility of using reagents like $Y_2Cu_2O_5$ and $BaCuO_2$. The reactivity of $YBa_2Cu_3O_7$ with different matrices and oxide sheaths was studied and the results obtained have been utilized in the manufacture of superconducting monofilamentary wires.[3]

BEHAVIOR OF THE $YBa_2Cu_3O_7$ SUPERCONDUCTOR IN AN ENCLOSED ATMOSPHERE

Before starting investigations on non-reactive and impermeable matrices, one of the crucial aspects of the work was to study $YBa_2Cu_3O_7$ in a sealed tube because if the compound is damaged during the thermal

Advances in Cryogenic Engineering (Materials), Vol. 36
Edited by R. P. Reed and F. R. Fickett
Plenum Press, New York, 1990

533

Table 1 . Heat treatment conditions of samples in sealed
quartz tube. Samples were held at the temperature indicated for two hours. Oxygen stoichiometry of the different
samples as a function of thermal treatment is indicated

Sample	Temperature (°C) of sintering	oxygen stoichiometry
starting powder	-	6.95
bars after sealing	-	6.98
a	700	6.98
b	800	6.97
c	900	6.94
d	950	6.94
e	950	6.64

treatments of sintering and annealing in an enclosed atmosphere, we have to
find another route to draw wires. We chose quartz tube to do the experiment
because metallic tubes are oxidizable or permeable to gas and especially
oxygen.

Experimental

$YBa_2Cu_3O_7$ powder was placed in a die body and pressed at 490 Mpa. The
bars obtained ($\approx$ 12X2X1.5 mm^3) were sealed in a quartz tube. The samples
were heated (300°C/h) to different temperatures (see Table 1), held at
these temperatures for two hours, cooled (100°C/h) to 450°C and finally
furnace cooled to room temperature after 5 hours at 450°C except for sample
e which was quenched from 950°C by immersion of the quartz tube in water.

Results

One remarkable feature of $YBa_2Cu_3O_x$ is that the oxygen stoichiometry
varies over a large range and superconducting properties are rapidly affected when x moves away from 7. The oxygen content was determined by high
resolution thermobalance measurements. Samples were slowly heated (60°C/h)
in a flow of argon / hydrogen 10% with a simultaneous symmetrical thermo-
analyser SETARAM TAG24. Under these conditions, only copper is reduced to
the oxidation state 0. The reduction of copper is complete at about 870°C.
The global formulation of the reacted mixture can be written "$YBa_2Cu_3O_{3.5}$"
and the weight loss corresponds to the loss of (x-3.5) atoms of oxygen.
Table 1 summarizes the results obtained with the different samples (oxygen
content of the starting powder and bars after pressing and sealing in
quartz tube are given as reference).

Oxygen content for each sample is quite similar to each other except
the quenched sample. If we consider the estimated precision of the measurement ($\pm$ 0.03) we can conclude that $YBa_2Cu_3O_7$ recovers the oxygen loss during
thermal treatment at high temperature. The oxygen stoichiometry of the
quenched sample seems to be higher than expected according Karpinski et
al.[4] but this difference can be due to the fact that during quenching, the
quartz tube limits the cooling speed and the compound may reabsorb a part
of the oxygen. X-ray diffraction investigations were performed on the different samples. All of them, except sample e, exhibit the $YBa_2Cu_3O_7$ orthorhombic structure without impurities. Sample e shows a diffraction pattern
(fig. 1) typical of the 123 phase in an orthorhombic form but with a significant loss of oxygen. These results show that $YBa_2Cu_3O_7$ is stable in a

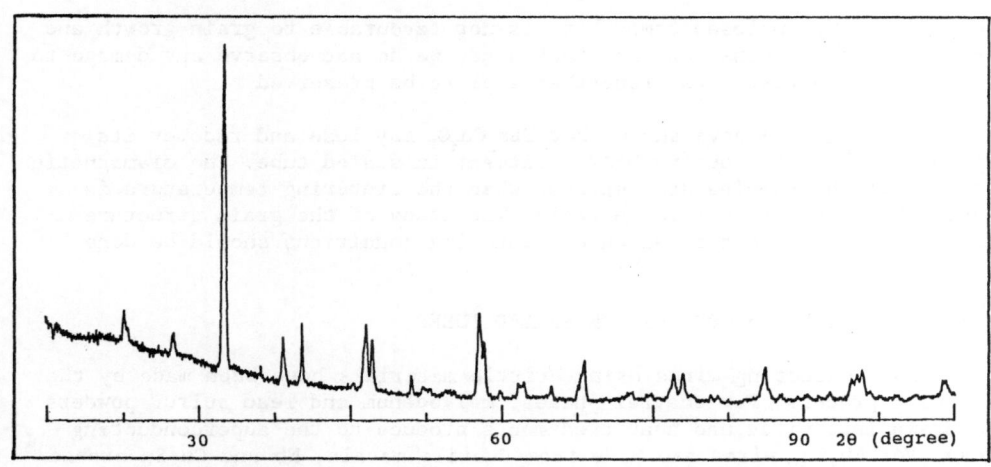

Figure 1. X-ray diffraction pattern of sample e.

confined atmosphere and even if we have a phase evolution at high temperature, it is reversible.

To complete the study of $YBa_2Cu_3O_7$ in sealed quartz tube, the diamagnetism response of the samples was measured using a Quantum Design MPMS SQUID Magnetometer. Fig. 2 shows the curves obtained. The basic curve corresponding to the "as compacted" powder is in good agreement with the results reported by Dubots and Cave[5]. The grain size of the $YBa_2Cu_3O_7$ powder is irregular and ranges from less than 1 μm to about 5 μm. At low thermal treatment temperatures, the total flux screened out of the sample is not affected. The samples have not begun to sinter and the grains are not electromagnetically coupled. When the temperature of thermal treatment increases, the screened volume is more important but not as much as expected. In order to link the magnetic observations to the microstructure of the sample, SEM studies were performed. Samples a and b do not exhibit sintering ; very small grains are still found and the grain size is not noticeably modified. Sample c is not significantly sintered but the very small grains have disappeared. A temperature of 950°C (sample d) is necessary to reach a beginning of significant sintering. SEM observations indicate that, even if we are not in optimal sintering conditions (heating rate

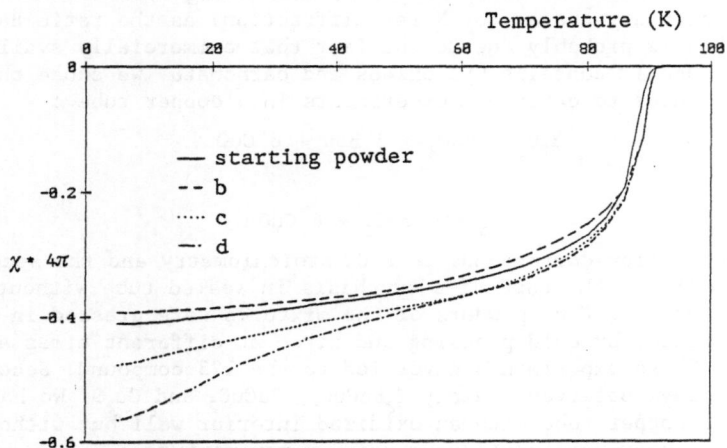

Figure 2. Screening graph for the different samples as a function of temperature (the applied magnetic field is 5 mT).

too rapid), the enclosed atmosphere is not favourable to grain growth and contact between grains. On the other hand, we do not observe any damage to the material in that, its properties seem to be preserved.

In summary we have shown that $YBa_2Cu_3O_7$ may lose and recover its oxygen stoichiometry during heat treatment in sealed tube. The diamagnetic response of the samples are improved when the sintering temperature is above 900°C but not on a large scale. The study of the grain structure indicates that further research of sintering conditions should be done.

SYNTHESIS OF THE 123 COMPOUND IN SEALED TUBES

Superconducting wires using brittle materials have been made by the following process. For Chevrel phases, molybdenum and lead sulfur powders are drawn into wires and heat treatments produce to the superconducting phase. For Nb_3Sn, wires are made from ductile metals, Nb and Cu-Sn bronze and a diffusionnal process is used to synthesize the superconducting Nb_3Sn material.

A similar approach was tried with $YBa_2Cu_3O_7$. However the problem is much more complicated due to the number of components of the ceramic, the secondary phases that can be formed and the need to control the oxygen level. Different ways were attempted in order to synthetize the 123 compound. Syntheses were, first, tried in a copper tube and second, in a non-reactive, sealed quartz tube subsequently stainless steel tubes were also tested.

Use of copper tube

Prior to the start of this study, we tested the reactivity of the needed oxides for the "in tube" synthetis because most of the time, the 123 compound is prepared by solid state reaction of CuO, Y_2O_3, $BaCO_3$ in air or oxygen or by using the nitrate precursor method. In our case, these preparations are not suitable because of the gas release during synthesis. Hence $BaCO_3$ must be replaced by BaO and/or BaO_2. Thus, mixtures of CuO, Y_2O_3, BaO and/or BaO_2 were fired at about 900°C in platinum crucibles in the ratio Cu : 3, Y : 1, Ba : 2. In all cases we obtained the 123 compound but we found that longer synthesis times were necessary. If BaO_2 is used as the source of barium, the pure 123 phase is obtained. On the other hand, if a mixture of BaO and BaO_2 is used, one finds increasing evidence of the presence of secondary phases (by X-ray diffraction) as the ratio BaO/BaO_2 increases. This is probably due to the fact that commercially available BaO is never pure and it contains hydroxides and carbonate. We chose the following two mixtures to carry out experiments in a copper tube :

$$Y_2O_3 + BaO_2 + 3\ BaO + 6\ CuO \tag{1}$$

$$Y_2O_3 + 4\ BaO_2 + 6\ CuO \tag{2}$$

The first composition corresponds to a O_7 stoichiometry and the second a $O_{8.5}$ stoichiometry in the case of a synthesis in sealed tube without reaction with the matrix. Fine powders of the mixtures were pressed in small copper tubes closed by cold pressing and fired at different times and temperatures. These experiments never led to the 123 compound. Secondary phases were always observed, mainly Y_2BaCuO_5, $BaCuO_2$ and Cu_2O. We have also tried to use a copper tube with an oxidized interior wall but without success. A last attempt was carried out using a new mixture :

$$Y_2O_3 + 4\ BaO_2 + 3\ CuO \tag{3}$$

Table 2. Results obtained in sealed quartz tubes. The oxygen pressure was obtained by decomposition of silver oxide. The estimated pressure at 930°C is about 700 KPa.
* 123 : $YBa_2Cu_3O_x$; 211 : Y_2BaCuO_5 ; 202 : $Y_2Cu_2O_5$.
# we observe $BaCO_3$ because the X-ray diffraction analysis is done in air and the carbonatation of BaO is very rapid

mixture	heat treatment		X-ray diffraction analysis*
	heating and cooling rate	temperature and time	
(4)	1°C/mn	930°C 36h	123 + 211 ($\approx$ 50/50) + ϵ CuO
(5)	1°C/mn	930°C 30h	123 + ϵ (CuO + 211)
(5) + O_2 pressure	1°C/mn	930°C 40h	211 + 202 + $BaCO_3$# + CuO

The idea was to use the matrix as a reactant but the experiment was unsuccessful.

We conclude therefore that the use of copper tube for the in situ synthesis of $YBa_2Cu_3O_7$ material does not lend itself to the technique used with Nb_3Sn. Another thing is that $YBa_2Cu_3O_7$ is very oxidizing, consequently the use of potentially reducing matrix is not desirable. This point will be reexamined next chapter.

Note : a study similar to that with copper tubes was done with INOX 316L tubes. One problem with stainless steel is to close the tube by cold pressing and after unsuccessful attempts, we decided to put the stainless steel tube containing the powders up in the sealed quartz tube. The 123 compound had never been obtained starting from oxide or peroxide powders in this matrix.

Use of sealed quartz tubes

The main question after these poor results was of course : is $YBa_2Cu_3O_7$ synthesis possible in a confined atmosphere ? To find out, the following experimental procedure was used : platinum crucibles filled with mixed, ground starting reagents were introduced into a quartz tube, sealed in air and fired for similar times and conditions listed above but the 123 phase was not obtained.

A new approach was then considered. If we do not observe any 123 compound formation it is probably due to the fact that the choice of the starting reagents was not appropriate. Two other mixtures were then tested:

$$(4)\ Y_2Cu_2O_5 + 3\ BaCuO_2 + CuO + BaO_2$$

$$(5)\ Y_2Cu_2O_5 + 4\ BaCuO_2$$

The main results are given in table 2. Numerous observations can be made. We have shown that the 123 compound can be synthesized in an enclosed atmosphere, nevertheless it is neither orthorhombic $YBa_2Cu_3O_7$ nor tetragonal $YBa_2Cu_3O_6$ but a new phase (see the X-ray diffraction pattern fig. 3). Similar phases have been reported[6,7] but the synthesis method was completly different. This 123 compound is found to have an oxygen content of about 6.7, and is not superconducting. An interesting observation is that the sintering temperature is rather low. We notice that the synthesis does not occur with an oxygen pressure.

The conclusion of this study is that a "draw and react" process does not appear to be feasible and, if a 123 compound can be formed by synthesis in an enclosed atmosphere, it will probably not be superconducting.

REACTIVITY OF $YBa_2Cu_3O_7$ WITH DIFFERENT POTENTIAL MATRIX AND OXIDE SHEATHS

Since the "in-situ reaction" process does not appear to be applicable for the fabrication of 123 wires, the "powder in tube" method was considered. The problem to be solved is to find a matrix which does not react with the superconducting compound, is sufficiently ductile to be drawn, is not oxidizable and is impermeable to oxygen, and of course is as inexpensive as possible for commercial applications. The first approach was carried out using Cu and CuNi. These two compounds are oxidizable but they are easy to draw and widely used in the fabrication of 4.2K superconducting wires. Subsequently we studied the use of stainless steel and silver.

Reactivity with Copper

The results obtained with copper and copper nickel alloy are quite similar so we will present only the results obtained with copper. To do the experiments, $YBa_2Cu_3O_7$ powder was pressed and introduced in copper tube closed by cold pressing. The samples, after thermal treatment were analysed by X-ray diffraction. Table 3 shows the results. $YBa_2Cu_3O_7$ quickly loses its oxygen and is destroyed when the temperature reaches 900°C. With CuNi, the destruction of $YBa_2Cu_3O_7$ is not as fast but occurs at the same temperature.

Reactivity with Stainless Steel

Thermal stability of $YBa_2Cu_3O_7$ is more important in this case. Secondary phases appear at 950°C (mainly Y_2BaCuO_5) but the material which ap-

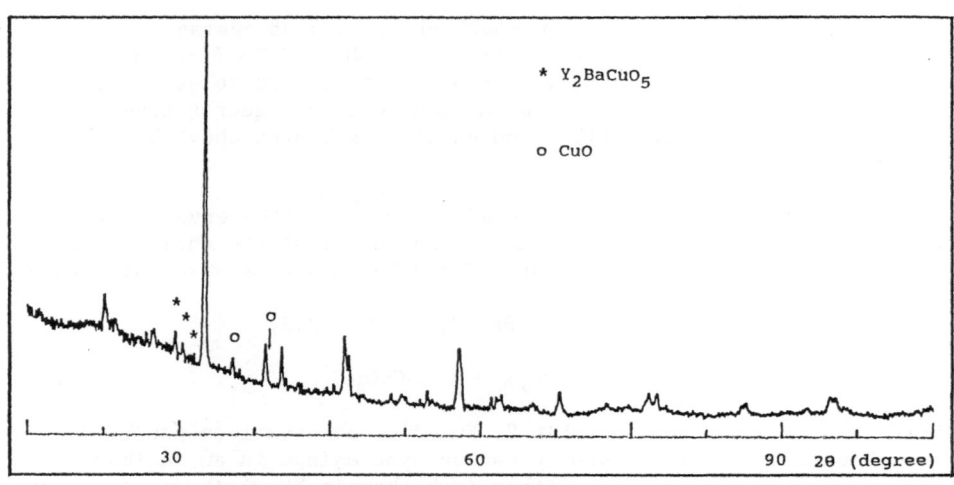

Figure 3 . X-ray diffraction pattern of the 123 compound obtained in sealed quartz tube.

Table 3 . Behavior of $YBa_2Cu_3O_7$ in a copper tube
as a function of temperature

thermal treatment	Results
800°C 1h	tetragonal 123
850°C 1h	tetragonal 123
900°C 1h	tetragonal 123 + 211
950°C 1h	211 + Cu_2O + other phases

peared to be pure tetragonal by X-ray diffraction, was shown by a Transmission Electron Microscope study to have a non-negligible proportion (10 to 15%) of modulated crystals characteristic of the superconducting material.[8]

Reactivity with Silver

A complete study[9] was done previously which shows that silver does not diffuse in the $YBa_2Cu_3O_7$ phase but that reaction may occur as a result of local deviation from stoichiometry of the 123 phase.

These results with the different matrices suggest that the use of a protective sheath (coating) between the superconductor and matrix may be advantageous. Silver appears to be a good candidate for this purpose even if it has the serious drawbacks which we enumerated previously. The role of this sheath must be of two orders : to prevent contamination by the matrix and to supply the oxygen which could be absorbed by the matrix. Two oxides were considered : barium peroxide, which could supply oxygen at high temperature (>800°C) and silver oxide which decomposes at temperature below 400°C.

Reactivity with BaO_2 and Ag_2O

The following experimental procedure was used : little bars of $YBa_2Cu_3O_7$ material were embedded in an oxide sheath by isostatic pressure (300 MPa). The samples were introduced into a stainless steel tube and sealed in quartz tube. X-ray diffraction analysis, oxygen stoichiometry and diamagnetic percentage measurements at 4.2K were performed. Table 4 shows these results. We observe with BaO_2 the appearance of secondary phases at temperatures as low as 800°C although, and maybe because of, the sintering seems to be better in comparison with the sintering of pure $YBa_2Cu_3O_7$ at 800°C. However BaO_2 does not supply oxygen and does not prevent oxygen loss of $YBa_2Cu_3O_7$. With silver oxide, no secondary phases appear, even at high temperature and in decomposing, silver oxide leaves a thin silver layer all around the superconductor and prevents its contact with the matrix.

We must keep in mind that the results are a bit distorted because of the stainless steel which is not completly unoxidizable and at 800°C when BaO_2 loses its oxygen, it may be taken up by the stainless steel. Silver oxide leads to a different behaviour and its use as a protective sheath may be feasible.

One of the main problems which is the matrix, has been partly solved by using a non oxidizable aluminium copper alloy and superconducting mono-filamentary wires have been obtained.[3]

Table 4 . Reactivity of $YBa_2Cu_3O_7$ with oxide sheath. The
heating and cooling rates are 5°C/mn.
O : orthorhombic ; T : tetragonal ; Ø : other phases.

oxide sheath	thermal treatment	oxygen content	diamagnetism % at 4.2K	Xray
BaO_2	600°C 1h	6.95	49	O
BaO_2	700°C 1h	6.68	24	O + T
BaO_2	800°C 1h	6.30	2	T + Ø
Ag_2O	700°C 1h	6.91	47	O
Ag_2O	800°C 1h	6.77	24	O
Ag_2O	930°C 1h	6.49	6	T

CONCLUSION

Chemical investigations of the systems superconductor / matrix and
superconductor / oxide sheath / matrix in the manufacture of superconducting wires have been studied. The following important results have been
found :

- $YBa_2Cu_3O_7$ may loose and recover its oxygen stoichiometry during
thermal treatment in enclosed atmosphere. Its superconducting properties
are not damaged.

- The synthesis of $YBa_2Cu_3O_7$ in enclosed atmosphere is possible using
$Y_2Cu_2O_5$ and $BaCuO_2$ but the obtained compound is not the superconducting
oxide and the synthesis does not occur under oxygen pressure. A "draw and
react" route is not applicable for making wires of this compound.

- A silver oxide sheath separating matrix and superconductor is found
to preserve the superconductor and to supply oxygen loss.

REFERENCES

1. S. Jin, R. C. Sherwood, R. B. Van Dover, T. H. Tiefel, D. W. Johnson,
Jr, Appl. Phys. Lett. 51, 203 (1987).
2. R. Flükiger, T. Müller, W. Goldacker, T. Wolf, E. Seibt, I. Apfelstedt,
H. Küpfer, W. Schauer, Physica C 153 - 155, 1574 (1988).
3. P. Dubots, D. Legat, F. Deslandes, B. Raveau, presentation BX-06 this
conference.
4. J. Karpinski, E. Kaldis, S. Rusiecki, J. Less Common Metals 150, 207
(1989).
5. P. Dubots, J. R. Cave, Cryogenics 28, 661 (1988).
6. Y. Nakazawa, M. Ishikawa, T. Takabakake, K. Koga, K. Terakura, Jap. J.
Appl. Phys. 26, L796 (1987).
7. F. Boterel, J. Wang, J.M. Haussonne, G. Desgardin, B. Raveau, Journées
SEE "Céramiques supraconductrices à haute Tc", Caen, sept. 1988.
8. M. Hervieu, B. Domenges, C. Michel, G. Heger, J. Provost, B. Raveau,
Phys. Rev. Let. B, 36, 3920 (1987).
9. F. Deslandes, B. Raveau, P. Dubots, D. Legat, Solid State Commun. 71,
407 (1989).

SOME STUDIES OF Ag/YBa$_2$Cu$_3$O$_7$ SUPERCONDUCTING COMPOSITE WIRE AND TAPE

Yao-xian Fu, Ying Pen, Yong-xiang Hu, Hong-chuan Yang,
Yue Sun, Chun-si Zhang, Su-hui Hu

Shanghai Institute of Metallurgy, Academia Sinica
Shanghai 200050, China

ABSTRACT

The Ag/YBCO superconducting composite wire has been fabricated by powder-in-tube technique. The samples of composite wire were heat-treated at 915°C for 4h, then quenched to 500°C and held for 4h in flowing oxygen and followed by furnace cooling. The distribution of Ag and O on cross section of Ag/YBCO superconducting composite wire and depth profile of Ag in surface of YBCO grain have been measured by AES. It was found that the Ag diffused into YBCO core and distributed in the region around grain boundary about 200Å.

The Ag/YBCO composite tape was heat-treated at 900°C for 24h. Then, the Ag sheath was dissolved with dilute nitric acid. After the etched YBCO core was heat-treated at 900°C×1h+500°C×2h and followed by furnace cooling in flowing O$_2$, the highest J$_c$ of 4,040A/cm^2 (1μV/cm) was achieved.

THE PREPARATION AND SUPERCONDUCTING CHARACTERISTICS OF Ag/YBa$_2$Cu$_3$O$_7$ COMPOSITE WIRE

The Y-Ba-Cu-O oxide superconductor discovered 2 years ago showed excellent high T$_c$ ($\geq$90K) and H$_{c2}$ (80-120T)[1,2]. The scientists in the world have a thirst for achieving the Y-Ba-Cu-O oxide superconducting composite wire or tape that could be used in liquid nitrogen conveniently. The different metal materials such as Cu, CuNi, Nb, stainless steel, Pt have been experienced as a jacket or diffusion barrier for fabricating superconducting YBa$_2$Cu$_3$O$_7$ oxide composite wire. All of these metal materials were found to be corroded by YBa$_2$Cu$_3$O$_7$ core. However, using Ag, Jin et al[3] have successfully fabricated Ag/YBa$_2$Cu$_3$O$_7$ superconducting composite wire. Afterwards some research groups have successively reported the processing technique and superconducting characteristics of the Ag/YBa$_2$Cu$_3$O$_7$ composite wire or tape[4-7].

The authors have made up of the Ag/YBa$_2$Cu$_3$O$_7$ superconducting wire by the powder-in-tube technique. The sintered and pulverized YBa$_2$Cu$_3$O$_7$ powder was packed into a silver tube of dia. 10/7.5mm, and codrawn to the dia. of 0.5-2mm.

Advances in Cryogenic Engineering (Materials), Vol. 36
Edited by R. P. Reed and F. R. Fickett
Plenum Press, New York, 1990

541

According to the results of neutron powder diffraction measurements of Jorgenson et al[8], the oxygen content x for samples of $YBa_2Cu_3O_x$ is 6.3 at 915°C, and 6.8 at 500°C in flowing 100% O_2. On the other hand, the solubility of oxygen in Ag increases along with the raising of temperature, therefore the exessive oxygen in Ag would release as the temperature of Ag decreases[9]. If the sample of an $Ag/YBa_2Cu_3O_7$ composite is fast cooled down to 500°C from sintered temperature 915°C, and kept at 500°C for several hours with furnace cooling, the $YBa_2Cu_3O_7$ core may continuously absorb oxygen from Ag sheath or through Ag sheath from the circumference of flowing oxygen, then the x could reach a value ≥ 6.8. So we selected the heat-treatment condition of 915°C for 4h, then quenched to 500°C and held for 4h in flowing oxygen and followed by furnace cooling.

The diamagnetism of the sample of the composite wire heat treated as above mentioned was obviously exhibited at 95K with AC susceptibility measurement. The zero resistance T_c was measured to be 89K by the standard four probe technique, and the four leads were contacted directively to the $YBa_2Cu_3O_7$ core with indium. The critical current density J_c (77K, 0T) is $145A/cm^2$ (1μv/cm) for the composite wire sample of dia. 2.2mm.

THE DISTRIBUTION OF Ag AND O IN $Ag/YBa_2Cu_3O_7$ SUPERCONDUCTING COMPOSITE WIRE

The knowledge of distribution of Ag and O in $Ag/YBa_2Cu_3O_7$ superconducting composite wire, no doubt, is extremely significant for improving the fabrication process and increasing critical current density. Here we give the results about this.

The JCXA-733 electronic microprobe (EP) and AMRAY-100B scanning microscopy (with TN-5400 EDS energy spectroscopy) were used for observing the topomicrography of cross section of sample and detecting the Ag in $YBa_2Cu_3O_7$ core. The distribution of Ag and O elements on surface of cross section of composite wire was measured by LAS-2000 Auger electron spectroscopy (AES).

In most cases here, the observing and measuring were performed after stripping the Ag sheath of composite wire heat treated as mentioned in part 1, and then breaking the $YBa_2Cu_3O_7$ core to eliminate the contamination from Ag tube.

It was found, by EP line scanning, that a small amount of Ag existed on the fracture face and reduced gradually from the edge to center of the core. Almost no Ag could be detected by EDS when its scanning region was on the center region of the $YBa_2Cu_3O_7$ core, however a little Ag was detected while the scanning region of EDS covered the whole section of core. Presummably a small amount of Ag has diffused into core and stayed at the interface of core with Ag tube or in the grain boundary of $YBa_2Cu_3O_7$. Therefore surface analysis was required for further investigation.

A superconducting composite wire with diameter 1mm was cut by a sharp blade, and then immediately put into UHV chamber of AES to measure the distribution of Ag and O on cross-section of composite wire. The SEM photography showed the typical intercrystalline fracture of the breaking face of the $YBa_2Cu_3O_7$ core.

The contimination of carbon was not found on the surface of sample as shown in Fig.1. The energy of primary electron beam of AES was 3 KeV, and the diameter of beam was about 1μm. The Auger spectrum was taken at 14 points on cross-section from the edge to center of the core in radial direction. The atomic concentration of Ag and O was calculated from peak to peak height of AES (Fig.2). It was shown in Fig.2 that Ag diffused from

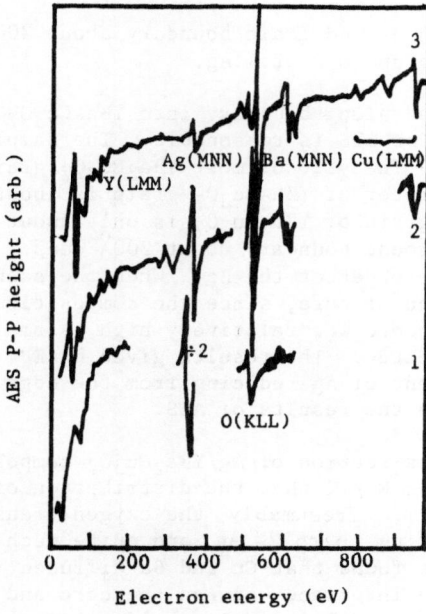

Fig. 1. The AES spectrum
at three points.

tube into YBa$_2$Cu$_3$O$_7$ core, suddenly reducing at interface between Ag tube
and core, and uniformly existing at fracture face of core with atomic composition about 2%. The results given from the sample broken after stripping
Ag tube was the same as Fig.2, implying there was no contamination of Ag
during cutting the composite wire by sharp blade. The Ag depth profile was
performed by Ar$^+$ ion sputtering on the cross-section of core as shown in
Fig.3. Before sputtering, the composition of Ag was 2 at.%, sputtering for
3 min. (about 100Å deep) reducing to 1.2 at.% and for 6 min. (about 200Å
deep) no Ag was detected. Because of intercrystalline fracture, the fracture face observed should be the grain boundary, the results here indicate

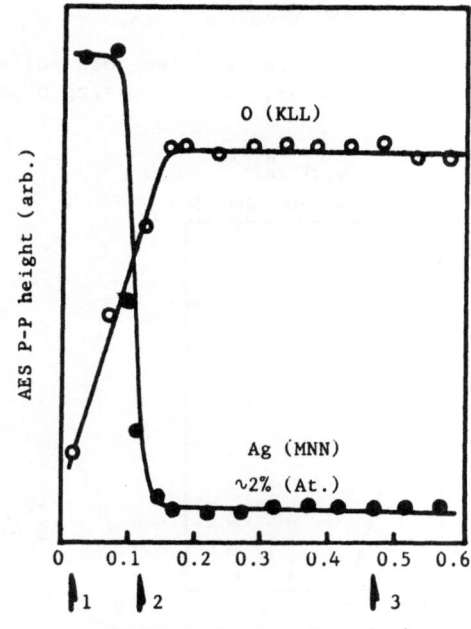

Fig. 2. The distribution
of Ag and O in radial direction of the cross-section
of Ag/YBCO composite wire.

that the Ag only existed in the region around grain boundary about 200Å depth, forming a thin shell containing up to 2 at.% Ag.

As the Ag diffused and distributed along boundary into $YBa_2Cu_3O_7$ core, the difficulty to detect the Ag by EP or EDS is reasonable. The sampling depth was about 1-3μm for EP or EDS. The size of most $YBa_2Cu_3O_7$ grain is less than 5μm. Assuming that the diameter of $YBa_2Cu_3O_7$ grain is about 1μm, the average composition of Ag in the grain of $YBa_2Cu_3O_7$ is only about 0.01 at.%, as the Ag exist in the region around boundary about 200Å deep. This is why EP and EDS was quite difficult to detect the Ag. When the scanning region of EDS covered the whole section of core, since the composition of Ag at the region near the edge of the core was relatively high, a small amount of Ag, therefore, could be detected. The results given by EP, which showed that there is a very small amount of Ag reducing from the edge to center of the core, is consistent with the results of AES.

The distribution of O on the cross-section of $Ag/YBa_2Cu_3O_7$ composite wire was measured too. It was shown in Fig.2 that the distribution of O at cross section of the core is uniform. Presumably the oxygen transported from Ag tube through the shell containing up to 2% Ag and quite much O surrounding the $YBa_2Cu_3O_7$ grain. It was found that Cu and Ba diffused into Ag tube and the composition around the interface between the core and the tube has deviated from the stoichiometry of 1-2-3. It would be necessary to do some further investigation of this phenomenon in the future.

SUPERCONDUCTING CHARACTERISTICS OF $YBa_2Cu_3O_7$ WIRE OR TAPE ETCHED Ag SHEATH

In order to make the $YBa_2Cu_3O_7$ core absorb the oxygen sufficiently from the ambient flowing oxygen, the sheath of $Ag/YBa_2Cu_3O_7$ composite wire of dia. 1/0.7mm presintered at $880^{\circ}C$ for 4h was dissolved by dilute nitric acid. Then the $YBa_2Cu_3O_7$ core wire of dia. 0.7mm was re-heat-treated at $900^{\circ}C$ for 1h and $500^{\circ}C$ for 2h in flowing oxygen and with furnace cooling to room temperature. The four measuring leads were attached to the core wire using silver paste, and the cross section of the core wire was determined from SEM photograph. The critical current density J_c is up to $885A/cm^2$ (1μm/cm) at 77K and 0T.

The composite wire was also cold rolled to tapes of 0.1-0.25mm in thickness. Afterwards, the processing procedure for the tape is about the

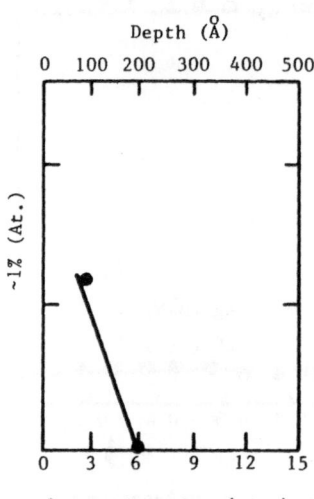

Depth (Å)

Sputtering time (min.)

Fig. 3. The depth profile of Ag at boundary of YBCO grain.

Fig. 4. Cross section after etching
by nitric acid.

same as that for the wire, but pre-sintered at $900^{\circ}C$ for 24h. The zero
resistance T_c of the core tape is 90.2K. The diamagnetism is obviously
showed at 91K on the AC susceptibility-temperature plot. The values of J_c
(77K, 0T) varied considerably from sample to sample, generally ranged from
800 to 1,200A/cm^2.

As the Ag sheath of Ag/$YBa_2Cu_3O_7$ composite tape was etched off by
emersing the tape in nitric acid, about 10μm depth of surface layer of
$YBa_2Cu_3O_7$ core was damaged by nitric acid as shown in Fig.4. The XRD results showed that the etched layer was composed of $Ba(NO_3)_2$, Y_2O_3, CuO,
$AgNO_3$, Ag_2O, Ag and some $YBa_2Cu_3O_7$ (Fig.5). In this case neither the
T_c (R=0) nor J_c could be measured by four probe technique over 77K. But a
sharp diamagnetic shift was detected by AC susceptibility measurement,
because there was still mainly the superconductive $YBa_2Cu_3O_7$ orthorhombic
phase inside the etched layer of the tape. After the etched sample was
properly re-heat-treated in flowing oxygen, the structure uniformity of the
sample was recovered, and it was composed of superconductive $YBa_2Cu_3O_7$
orthorhombic phase and some Ag as shown in the XRD pattern (Fig.6).
Although the measured values are varied severely with samples, the corresponding higher critical current densities of the re-heat-treated samples

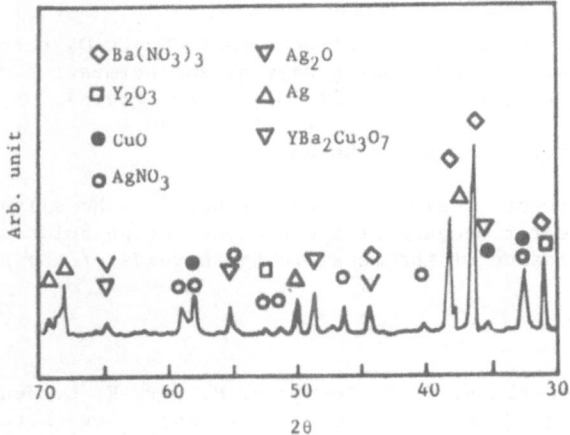

Fig. 5. XRD pattern of the corroded
surface of the $YBa_2Cu_3O_7$ core.

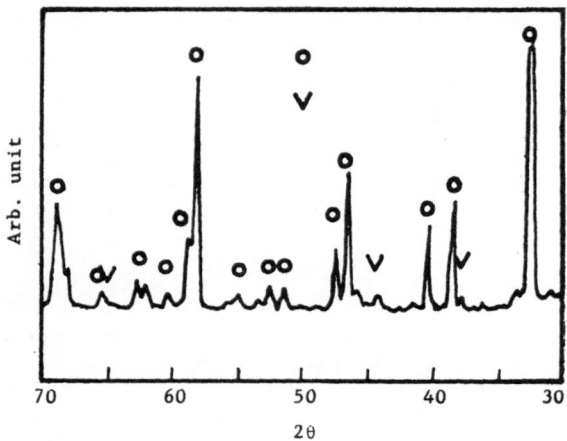

Fig. 6. XRD pattern for the re-heat-
treated specimen.

than that of the un-etched composite tape were gained. One of them reaches
the achieved highest J_c (77K, 0T) value of 4,040 A/cm^2 at a tape thickness
of 0.14mm. Yang et al[10] have reported that certain Ag addition to
$YBa_2Cu_3O_7$ superconductor could apparently improve its current carrying
capacity, and the addition of Ag could make the $YBa_2Cu_3O_7$ grain become
smaller and the content of C, which has been found at the grain boundary in
the form of $BaCO_3$ and BaC_2[11], decrease. Meanwhile we have observed that the
Ag only existed in the region around $YBa_2Cu_3O_7$ grain boundary about 200Å
depth (see part 2 of this paper). And it is well known that the solubility
of oxygen in Ag increases intensely at elevated temperature. Presumbly the
Ag at the grain boundary acts as a carrier of oxygen, and this oxygen could
promote the formation of appropriate $YBa_2Cu_3O_7$ structure and the decomposition of the barium carbonate and carbide along grain boundary, consequently
it can purify the grain boundary and to constrain the effect of week link.

CONCLUSION

    Among the various possible causes, besides the densification of the
$YBa_2Cu_3O_7$ core after drawing and rolling, the authors would rather believe
that the higher J_c of $YBa_2Cu_3O_7$ core that the Ag sheath was etched in dilute
nitric acid may be related to:

    (1) More perfect orthorhombic structure of $YBa_2Cu_3O_7$ through exposition
of the surface of core to the flowing oxygen, and permeation of oxygen to
each grain through the O-rich Ag shell coating the grains.

    (2) The purification of grain boundary.

    (3) The improvement of electric contact between the silver paste and
$YBa_2Cu_3O_7$ superconductor because of the presence of Ag soluted in the grain
boundary and the presence of the metallic Ag in voids of the sample.

REFERENCES

1. M. K. Wu, J. R. Ashburn, C. J. Torng, P. H. Hor, R. L. Meng, L. Gao,
    Z. H. Huang, Y. Q. Wang, and C. W. Chu, Phys. Rev. Lett. 58, 908
    (1987).
2. Z. X. Zhao, L. Q. Chen, Q. S. Yang, Y. Z. Huang, G. H. Chen, R. M. Tang,
    G. R. Liu, C. G. Cui, D. Chen, L. Z. Wang, S. Q. Guo, S. L. Li, and

J. Q. Bi, Ke Xue Tong Bao, 1987, No.6, 412 (in Chinese).

3. S. Jin, R. C. Sherwood, R. B. van Dover, T. H. Tiefel, and D. W. Johnson, Jr., Appl. Phys. Lett. $\underline{51}$, 203 (1987).

4. O. Kohno, Y. Ikeno, N. Sadakata, and K. Goto, Jpn. J. Appl. Phys., $\underline{27}$, L77 (1988).

5. M. Okada, A. Okayama, T. Morimoto, T. Matsumoto, K. Aihara, and S. Matsuda, Jpn. J. Appl. Phys., $\underline{27}$, L185 (1988).

6. Y. X. Fu, Y. Peng, Y. X. Hu, H. C. Yang, W. Q. Liu, Y. Sun, P. Y. Liang, and S. H. Hu, Proc. Chinese Symp. of High T_c Superconductor, p.63, April, 1988, Baoji, Shanxi.

7. H. Sekine, K. Inoue, H. Maeda, K. Numata, K. Mori, and H. Yamamoto, Appl. Phys. Lett. $\underline{52}$, 2261 (1988).

8. J. D. Jorgenson, M. A. Beno, D. G. Hinks, L. Soderholm, K. J. Voling, R. L. Hittermen, J. D. Grace, Ivan K. Schuller, C. U. Segre, K. Zhang, and M. S. Kleefisch, Phys. Rev. $\underline{B36}$, 3608 (1987).

9. C. J. Smithells (Editor), Metals Reference Book, 5th ed. London, Butterorths, 1976, p.852.

10. H. C. Yang, Y. Sun, M. Y. Wu, Y. X. Fu, Presented at MRS Spring Meeting, San Diego (1989).

11. K. G. Wang, T. J. Zhang, P. X. Zhang, S. Q. Wang, L. Zhou, and H. L. Huang, High T_c Superconductor, Baoji, 1988, p.101.

547

EFFECTS OF BENDING STRAIN ON THE CRITICAL CURRENT DENSITY IN SUPERCONDUCTING

YBa$_2$Cu$_3$O$_x$ MULTICORE WIRES

Jitsuo Chikaba

Faculty of Engineering
Kinki University
Iizuka-shi, Japan

ABSTRACT

Deterioration of the critical current density, Jc, as the result of bending strain is measured for silver-sheathed ceramic high Tc superconducting wires. The multicore composite wires (3-85 cores) are fabricated by a conventional powder-in-tube method using YBa$_2$Cu$_3$O$_x$ compound. Results concerning ceramic core diameters, thermal stress, and bending strain are discussed, comparing the multicore wires with single core wires. The value of Jc in the multicore wires cannot reach a higher value than in the single core wire (Jc=1600A/cm^2 in zero magnetic field) because of cracks and ruptures in the sheath which are attributed to unequal stress between cores. Multicore composite wires, however, have the advantage of suitable mechanical properties, especially flexibility due to good contact between the ceramic and the silver sheath.

INTRODUCTION

Concerning the practical use of ceramic high Tc superconductors, many problems associated with the improvement of the critical current density as well as mechanical fragilities have been accumulated. Experiments on coils fabricated from YBa$_2$Cu$_3$O$_x$ ceramic high Tc superconductor have been done and suggest that stabilization by silver(Ag), used for a mandrell, is effective for quench protection.[1] It is generally known that Ag-sheathed ceramic superconducting wires can be easily fabricated[2] by the powder-in-tube method. Although Ag is a suitable material in having a high solubility for oxygen, there is a problem with the considerable high cost for practical use. From this point of view, Sen et al. have proposed that Al(1100 grade) and stainless steel(304 grade) can be used as a substitute for the Ag sheath by adding a small amount of Ag$_2$O or Ag to the YBa$_2$Cu$_3$O$_x$ powder.[3] The mechanical properties were improved as a result. The Ag doping appears to be an effective method for both decreasing coupling resistance between grains and for enhancement of grain growth.[4] It has been also reported that, since many cracks come into existence in the ceramic core owing to the difference of thermal expansion between the metal sheath and the core, the Jc of wire which is sintered after removing the metal sheath is improved.[5]

Meanwhile, fine fibers were prepared by the spinning method[6,7,8] and the alginate method[9] to obtain further ductility. These methods have the advantage of manufacturing continuous long length wires of high tensile

Advances in Cryogenic Engineering (Materials), Vol. 36
Edited by R. P. Reed and F. R. Fickett
Plenum Press, New York, 1990

549

Table 1. Final Dimensions of Ag-Sheathed Superconducting Wires

Type	o.d (mm)	b.d (mm)	c.d (μm)	example of a cross section
Single Core	2.0–0.3		600–100	
3 cores	1.0	0.64	250	
7 cores	1.86	1.07	240	
14 cores	1.71	1.14	100	
19 cores	0.68	0.39	120	
65 cores	0.83	0.51	44	
85 cores	0.84	0.49	48	

o.d : outer diameter of the composite wire
b.d : bundle diameter of the core
c.d : core diameter

strength. However, it is necessary to improve the Jc value. As the wire-rolling progresses towards smaller diameter in the usual powder-in-tube method, the Jc value is improved because of the increasing densificaticn. Then, in order to further increase the densification, Ag-sheathed superconducting tapes were fabricated by pressing after wire drawing.[10] The results show that the Jc value is improved and also that distribution of the crystal orientation measured by neutron diffraction has an effect on the Jc value.[11]

This paper presents bending characteristics of Ag-sheathed superconducting $YBa_2Cu_3O_x$ multicore wires. Although the Jc value of the multicore wires is smaller than that of the tape so far, it will be worth studying because of the good superconducting stabilization and mechanical properties. The relation between bending strain and Jc values for various types of multicore wires is discussed.

EXPERIMENTAL ARRANGEMENT

A starting mixture of Y_2O_3 (99.9%), $BaCO_3$ (99.9%) and CuO (99.9%) was calcined at 950 °C for 5h in air and furnace cooled to room temperature at a rate of 80 °C/h. The calcining was repeated four times. The Ag sheath (6.0 mm o.d, 4.0 mm i.d) was filled with powder which was well pulverized after calcining and then rolled cylindrically. To produce the multicore wires, the required number (3, 7, 14, 19, 65 and 85 cores) of single core wires were packed in the Ag sheath mentioned above and rolled or swaged again to the test diameters. After rolling or swaging, they were sintered at 910 °C for 20h in an oxygen atmosphere and furnace cooled at a rate of 60 °C/h. The final dimensions of the sample wires and an illustration of the cross section are given in table 1. The measurements of critical temperature, Tc, and critical current density, Jc, were performed by a conventional four probe method where the sample length was 4 cm, the length between voltage terminals was 2 cm, and the Jc criterion was 0.1 μV/cm. The values of Jc were measured with various bending radii (R = 24cm – 1cm) and external magnetic flux densities, B, at 77K.

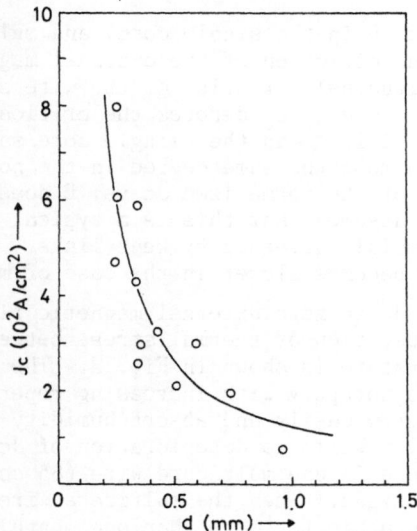

Fig. 1. Critical current densities vs. core diameters of the silver sheathed YBa$_2$Cu$_3$O$_x$ single core wire at 77K in zero external magnetic field (Tc = 85 K).

RESULTS AND DISCUSSION

Fig. 1 shows the relation between the value of Jc and the diameter of the YBa$_2$Cu$_3$O$_x$ single core wires at zero external magnetic field which were sampled during the wire rolling process. The value of Jc is improved as the wire diameter decreases. The core diameter dependence of Jc shows the same tendency as in the tape sample.[10] It is difficult to make the wire smaller with a circular cross section because of cracks in both the sheath and core before sintering. Consequently, the value of Jc has reached 1600 A/cm^2 (d=0.12 mm) at zero external magnetic field by reinforcing the sheath (double sheathed) in the present work.

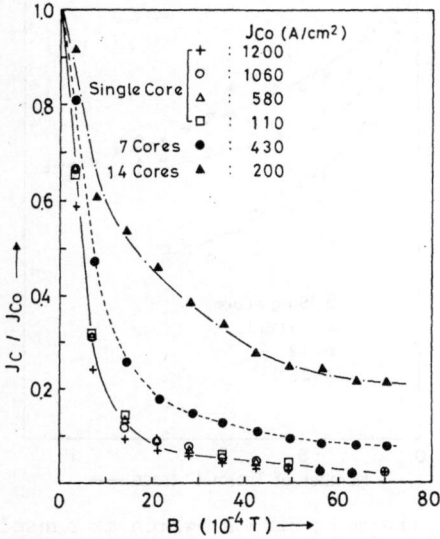

Fig. 2. External magnetic field dependence of normalized Jc at 77K with single core, 7-core and 14-core composite wires. Jco denotes the Jc in zero external magnetic field.

The behavior of Jc vs. B in the single core and multicore samples is shown in Fig. 2, where the direction of the external magnetic field is perpendicular to the longitudinal direction of the wire sample. In spite of the improvement of Jc, where Jco denotes the critical current density in zero external magnetic field, in the single core samples, the abrupt decrease of Jc with B is almost the same degree in the normalized ordinate Jc/Jco, that is, the plot of the normalized Jc vs. B does not depend on the value of Jco. It can be assumed that this is a typical behaviour of the sintered sample which is mainly governed by weak links. On the other hand, the decrease of Jc with B becomes slower in the case of multicore wires.

The deterioration of Jc in zero external magnetic field caused by subjecting the wire to repetition of thermal stress between room temperature and liquid nitrogen temperature is shown in Fig. 3. The value of Jc in the single core wire decreases abruptly with increasing repetition. It can be assumed that many cracks grow easily and absorb humidity because of low densification and porosity. While no deterioration of Jc can be observed for more than 15 repetitions in the multicore wire (85 cores) which is proof against wet and thermal fatigue. When the multicore wire was exposed to air and daylight at room temperature for more than one month, however, the Jc deterioration was about 10 %.

To evaluate the flexibility of the composite wires as well as the electrical and mechanical coupling between superconducting grains, the value of Jc was measured under various bending radii (R=24cm – 1cm) and external magnetic fields. Fig. 4 shows Jc/Jco vs. 1/R for various types of the wires where Jco is the value of Jc for R = ∞ . It is found that when the curvature is less than about 0.05 cm^{-1}, the deterioration of Jc occurs abruptly and then the value of Jc is improved with increasing the numbers cores, namely, the degree of Jc deterioration by bending clearly depends on the core dia-meter. It has been reported that the Jc deterioration occurs at the curvature of about 1.2 cm^{-1} – 0.6 cm^{-1} in the composite wire with silver added.[3]

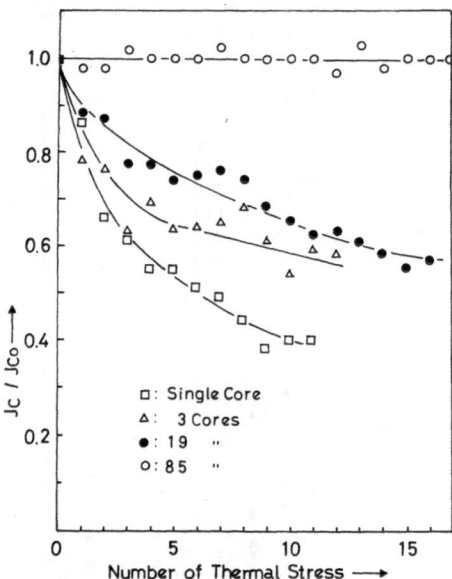

Fig. 3. Effects of the thermal stress which is caused by the difference between room temperature and liquid nitrogen temperature on the normalized Jc with various numbers of cores. The value of Jco is the initial value of Jc at B=0 T.

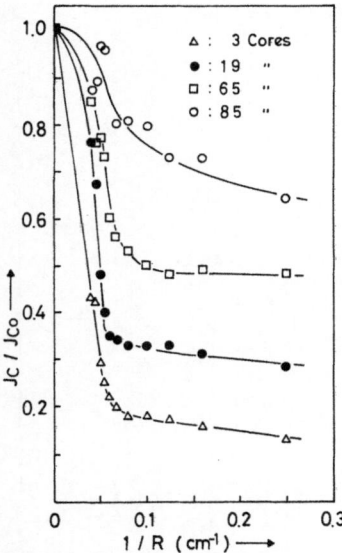

Fig. 4. Curvature dependence of normalized Jc at 77K. The value of Jco is
the Jc without both the bending and external magnetic field.
Jco=250 A/cm² for 85 cores, Jco=250 A/cm² for 19 cores, Jco=200 A/cm²
for 65 cores and Jco=200 A/cm² for 85 cores.

The effect of bending strain, r/R, on the Jc is shown in Fig. 5, where
r is a core or bundle radius and R is a bending radius, respectively. When
the average radius per one core is used as the value of r, the plot of Jc/Jco
vs. r/R in both the 85 core and 7 core wires is similar to the single core
characteristics as shown in Fig. 6. The magnetic field dependence of these
wires has been already shown in Fig. 2. On the other hand, using the bundle
radius as the value of the r, the deterioration of Jc seems to be smaller

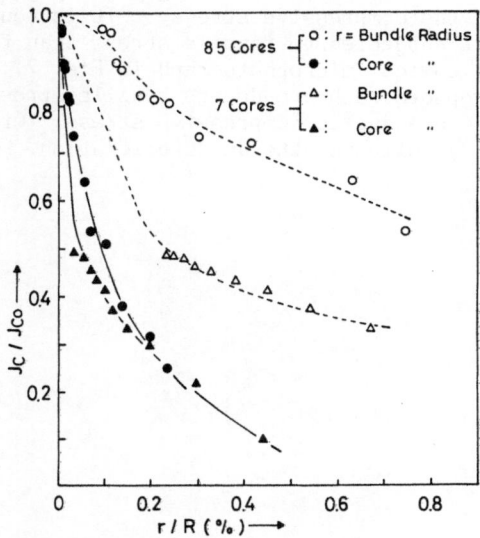

Fig. 5. Bending strain dependence of normalized Jc at 77K.
Jco=430 A/cm² for 7 cores and Jco=200 A/cm² for 85 cores,
respectively.

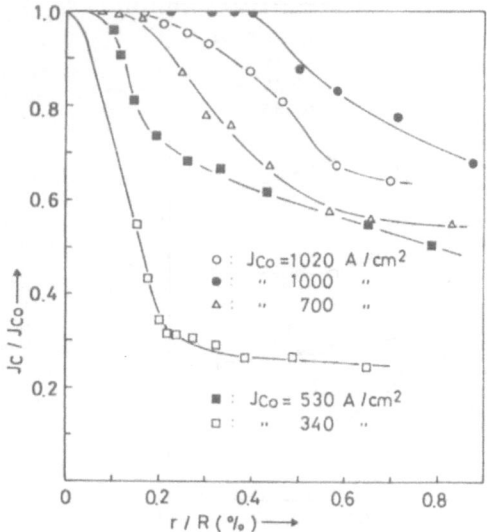

Fig. 6. Bending strain dependence of normalized Jc with various values
of the Jco in single core wires at 77K.

than in single core wires in spite of its thick diameter.   This indicates
that since the $YBa_2Cu_3O_x$ grains contact well with the silver sheath,   the
characteristics of bending strain in multicore wires behave as in a fairly
good composite wire with ductility.   According to the work reported by
Singh et al.[12], fractures were observed at the value of 0.1 - 0.2% bending
strain in the tape sample without silver addition. As compared with their
results, the Jc deterioration is about 10 - 20% at the same bending strain
(0.1 - 0.2%) without fractures in the multicore wire (85 cores).

Since both tensile and compressive stress act on the wire in the case
of bending, the wire does not fracture perfectly, but the deterioration of
Jc progresses rapidly because of cracks mainly caused by tensile stress in
the initial stage (1/R < 0.05 cm^{-1}) and then gradually goes with breakage
of the core texture by the compressive stress.   It is found that there are
more cracks in the part subjected to tensile stress than to compressive
stress as shown by the optical microphotograph in Fig. 7.   Cracks on the
upper side of the microphotograph are due to tensile stress and the broken
parts on the lower side are due to compressive stress.   Cracks generated
by tensile stress mostly influence the Jc deterioration.

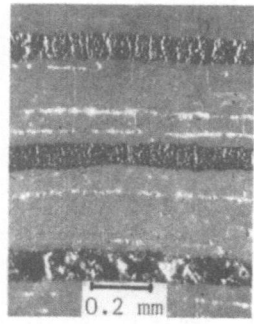

Fig. 7. Microphotograph of a longitudinal section of the 7-core composite
wire after bending (R=1cm).

CONCLUSION

The value of Jc in silver-sheathed $YBa_2Cu_3O_x$ superconducting single core wire increases with reduction of core diameter. With further thinning in multicore composite wires, the value of Jc has been improved to some extent and then decreased sharply (down to Jc=200 A/cm^2) by about 40 μm core diameter, contrary to our expectation. However, it is clear that the ceramic multicore composite wire is sufficiently flexible for practical use in magnets. It is considered that, since the stress between single core wires is not uniform during rolling processes, the cross section of the sheaths do not remain circular, but cracks and ruptures occur easily. To improve the Jc value of the multicore wire, it is important that first, no crack generates in the sheath at all during rolling or swaging, secondly the grain size of the powder and sintering conditions should be controlled. Moreover, the generation of cracks is inevitable as the volume of the ceramic decreases with increasing densification during the sintering process and also because of the difference of thermal expansion coefficients between the silver sheath and the $YBa_2Cu_3O_x$ core. It will be necessary to develop a malleable and tough material for the sheath which is stable at high temperature as well as new fabrication methods for ceramic cores.

REFERENCES

1. Yasuzo Tanaka, Kiyoshi Yamada and Takayuki Sano, YBCO Superconducting Coils Operated at Nitrogen Temperature, Jpn. J. Appl. Phys. Vol.27, No.5:799 (1988).
2. S. Jin, R.C. Sherwood, R.B.van Dover, T.H. Tiefel and D.W. Johnson, Jr., High Tc superconductors-composite wire fabrication, Appl. Phys. Lett.51(3):203 (1987).
3. S. Sen, In-Gann Chen, C.H. Chen and D.M. Stefanescu, Fabrication of stable superconductive wires with $YBa_2Cu_3O_x/Ag_2O$ composite core, Appl. Phys. Lett. 54(8):766 (1989).
4. P.N. Peters, R.C. Sisk, E.W. Urban, C.Y. Huang, and M.K. Wu, Observation of enhanced properties in samples of silver oxide doped $YBa_2Cu_3O_x$, Appl. Phys. Lett. 52(24):2066 (1988).
5. Osamu Kohno, Yoshimitsu Ikeno, Nobuyuki Sadakata and Kenji Goto, High Critical Current Density of Y-Ba-Cu Oxide Wire without a Metal Sheath, Jpn. J. Appl. Phys. Vol.27, No.1:L77 (1988).
6. Tomoko Goto and Masahiro Kada, Proparation of High-Tc Y-Ba-Cu-O Superconducting Filaments by Suspension Spinning Method, Jpn. J. Appl. Phys. Vol.26, No.9:L1527 (1987).
7. Tomoko Goto and Iwao Horiba, A Method for Producing High-Tc Y-Ba-Cu-O Superconducting Filaments by Die-Free Spinning, Jpn. J. Appl. Phys. Vol.26, No.12:L1970 (1987).
8. Tomoko Goto, Nonaqueous Suspension Spinning of High-Tc Ba-Y-Cu-O Superconductor, Jpn. J. Appl. Phys. Vol.27, No.4:L680 (1988).
9. Hajime Konishi, Takumi Takamura, Hisashi Kaga and Keiichi Katsuse, A New Fabrication Process for High-Tc Superconducting Oxide Ceramic Fibers, Jpn. J. Appl. Phys. Vol.28, No.2:L241 (1989).
10. Michiya Okada, Akira Okayama, Tadaoki Morimoto, Toshimi Matsumoto, Katsuzo Aihara and Shinpei Matsuda, Fabrication of Ag-Sheathed Ba-Y-Cu Oxide Superconductor Tape, Jpn. J. Appl. Phys. Vol.27, No.2 : L185 (1988).
11. Michiya Okada, Akira Okayama, Toshimi Matsumoto, Katsuzo Aihara, Shinpei Matsuda, Kunio Ozawa, Yukio Morii and Satoru Funahashi, Neutron Diffraction Study on Preferred Orientaion of Ag-Sheathed Y-Ba-Cu-O Superconductor Tape with Jc=1000-3000 A/cm^2, Jpn.J.Appl.Phys. Vol.27, No.9: L1715 (1988).
12. J. P. Singh, D. Shi and D.W. Capone, ll, Mechanical and superconducting properties of sintered composite $YBa_2Cu_3O_{7-\delta}$ tape on a silver substrate, Appl. Phys. Lett. 53(3):237 (1988).

PRELIMINARY DESIGN OF A HIGH-TEMPERATURE

SUPERCONDUCTING TRANSMISSION LINE

G. B. Andeen[1] and R.L. Provost[2]

[1] SRI International, Menlo Park, California 94025

[2] E.I. Du Pont de Nemours & Company, Inc., Wilmington, Delaware 19898

INTRODUCTION

High-temperature superconductors offer the potential advantage of reduced refrigeration costs, leading to reconsideration of transmission-line applications. However, because the properties of high-temperature superconductors differ considerably from those of low-temperature superconductors, we have begun with a preliminary design based on limited knowledge of the new materials. The result is a configuration that is quite different from the low-temperature superconducting transmission lines previously designed and demonstrated. The primary emphasis of the design has been to consider materials other than the superconductor itself, i.e., the construction materials. In particular, the low-temperature strength and favorable dielectric properties of aramid fibers have been used.

SUPERCONDUCTIVE TRANSMISSION LINE DESIGNS

History and Design Choices

Low-temperature superconducting transmission lines were considered by many investigators in the 1960s, and demonstrated and reported on in the 1970s and early 1980s. Cooling to liquid-nitrogen (LN2) temperature to reduce electrical resistance, although not superconductivity, was also considered.[1] Key developments and demonstrations are noted in References 2-11.

The various transmission line schemes chosen by investigators have been amazingly diverse. Some of the choices are alternating or direct current, high or low voltage (typically 60 to 138 kV), two- or three-phase alternating current, and flexible cable, a cable in a rigid conduit, or a rigid, bus-bar-like conductor. Other details such as whether wires or tapes are used, how the cooling fluid is carried, and material choice make for further differences. Figure 1 shows a diagram[12] of the design choices made by different low-temperature transmission-line teams.

Utility Requirements

Although technical feasibility of superconducting transmission lines has been demonstrated many times, there are no commercial applications. There are two key reasons: Since the technology became available, demand for transmission lines has dropped, owing to the slow growth of utilities together with an increase in dispersed cogeneration sources. Also, opportunities that were both economically and operationally attractive have been lacking.

Superconducting transmission lines are more expensive to build and operate than long-distance high-voltage dc lines in rural areas. For shorter distances, especially in and near urban areas, superconducting lines become attractive at high capacities where on-line performance requirements are very high. In these situations, the utilities would prefer to have multiple, lower-capacity lines.

Advances in Cryogenic Engineering (Materials), Vol. 36
Edited by R. P. Reed and F. R. Fickett
Plenum Press, New York, 1990

557

Property	Low-Temperature Nb-Ti *	High-Temperature 1-2-3 Material†
Critical current density (J_c)	1.5×10^9	2×10^6 (A/m^2)
Density (kg/m^3)	6.2×10^3	1.6×10^3
Specific heat (J/kg-K)	0.89	$\sim 10^2$
Thermal conductivity (W/m-K)[13]	0.05	3
Normal resistance ($\mu\Omega$-cm)	0.2	3-15
Critical strain (%)	0.5, tough	0.05, brittle

* At 4.2 K   † At 77 K

We have focused our design on the shorter urban application, and on compatibility with existing utility practices and preferences.

DIFFERENCES BETWEEN HIGH- AND LOW-TEMPERATURE DESIGN CONSIDERATIONS

Property Differences

Examples of properties of a low-temperature and a high-temperature superconductor are listed in Table 1. Very different properties are the result of operating at different temperatures as well as very different materials. The properties are at the respective operating temperatures.

The different properties shown in Table 1 inevitably lead to different configurations through considerations of stability of operation and other considerations in transmission line design.

Stability

Superconducting components have usually taken the form of fine filaments of superconducting material imbedded in a metal matrix, or a ribbon attached to or imbedded within another piece of metal. The reason has been to stabilize the superconductor, i.e., to prevent it from going normal in response to a disturbance such as a movement, a change in current, or an externally imposed magnetic field. Unfortunately, the specific heat of metals is vanishing near absolute zero, so that even a small amount of heat can cause an increase in temperature, triggering instability.

Since the new superconductors are ceramic and function at high temperature, the specific heat is not vanishing. The question is what form the superconductors should take. The "stability parameter" for static stability is given by

$$\frac{\mu_0 J_c^2 a^2}{2\gamma} C(T_c - T_o) < 3$$

where μ_0 is the permeability of free space, J_c is the critical current density, a is the radius, γ is the density, C is the specific heat, T_c is the critical temperature, and T_o is the operating temperature.[14] If we assume a difference between critical temperature and operating temperature of 2.3 K in both cases, the radius to give stability is 115 μm for the niobium compound and 0.47 m for the new materials. Finely divided fibers are not necessary; bus-bar configurations can be used.

Dynamic Stability and Graceful Failure

The concern is the dissipation of energy in the transmission line if some portion goes normal, as when a refrigeration system fails. When a low-temperature superconductor goes normal, relatively low electrical resistance is encountered, particularly if the superconductor is imbedded in an electrically conducting metal. The adjacent metal permits the current to bypass normal parts of the superconductor or to shunt to parallel superconducting filaments with low resistance. It is desirable to dissipate the energy slowly so that heat can be removed, dynamically stabilizing the superconductor. The new superconductors, however, jump to relatively high values of resistance. Catastrophic, rather than graceful, failure of bus-bar construction in our design is expected. Further consideration of shunts or parallel paths is desired.

Shielding

The superconducting transmission lines demonstrated to date have all used a coaxial arrangement of conductors, i.e.,the power-carrying superconductor is shielded by a second superconductor. A current is induced in the shielding that cancels the magnetic field outside the shield and eliminates the inductive loss associated with ac transmission lines. This also prevents cables from being attracted or repelled from each other, which could give rise to mechanical losses as a result of movement. As has been noted, the zero external field makes it possible to locate the cables close to one another, reducing trenching costs.[8] The zero external field could also be an advantage, especially in urban areas, if the concern over transmission line fields[15] proves valid. The coaxial design only applies to superconducting lines where the current can flow in the shield without resistive loss; it has not been used with cryoresistive lines (i.e., reduced in temperature to lower resistance).

A coaxial design is considerably more difficult to achieve with ceramic materials because of required high-temperature sintering. An insulating layer that can tolerate the high temperatures and will shrink with the superconductor is required. The ceramic superconductors are finished by firing at 900°C, a temperature incompatible with most materials that one would consider for spacers. Although it is possible to imagine green material being wound into appropriate shapes, it is difficult to imagine materials of construction compatible with the firing. Joining poses further difficulties. Both the inner conductor and the shield must be joined at every connection. Because of these difficulties, our design does not use coaxial shielding.

The question is whether shielding is necessary: Near absolute zero, shielding is essential. Losses add heat at the low temperatures that is expensive to remove.and a small loss cantrigger a thermal runaway. Such problems do not apply to high-temperature superconductors. Thus, although we would like to have a coaxial superconductor for electrical reasons, it is not a necessity.

Insulation

Vacuum with superinsulation is preferred at helium temperatures because the refrigeration costs are so high. At LN_2 temperature, the cost of refrigeration is greatly reduced, but the difficulty of maintaining a vacuum increases because there is less cryopumping. We believe vacuum insulation is still preferable, but the choice is not nearly so clear.

Construction Features

The rigid ceramic material suggests a pipeline-like construction. Although one might prefer a flexible cable and housing, assembly from rigid components is not the hurdle one might imagine.

Aramids are the choice of construction materials because of their combination of high strength and low thermal conductivity.[16-19] Figure 2 illustrates the relative relationship between strength and conductivity for various materials and how the relationship changes as the temperature is reduced (arrows). The solid lines show construction value in terms of constant heat lost through the a support of given strength. Aramids also show good electrical insulating properties,[20-21] which is a desired feature of transmission line internals. In addition to good functional properties, the aramids can easily be fabricated into a variety of shapes [22] They can be made in paper form with various thicknesses and porosity. The composite can be filled to make a wall that will hold a vacuum.

CHOICES

The first application of superconducting transmission lines is most likely to be underground lines moving power in an urban area for distances of perhaps 20 miles. Such distances imply compatibility with the existing distribution system; converting ac to dc and back again is not desired. Direct current has lower superconducting transmission losses (losses are only on transients) but requires conversion equipment to mate with existing distribution equipment. Three-phase ac is the choice of convenience, but has inductive and other cyclic losses not yet well defined in the high-critical-temperature materials.

We have chosen a three-phase system of 12 MVA operating at 20 kV. Our choice of voltage is somewhat arbitrary, but consistent with lines of 20-mile length. Higher voltage was selected because high-current densities characteristic of low-temperature superconductors are not expected. Wolsky et al.[23] suggest that current densities of at least 2000 kA/cm^2 are necessary for a superconducting transmission line. These values seem to be determined by comparison with underground power cables and demonstration cryogenic lines rather than through a design study. In any case, the desired value is far

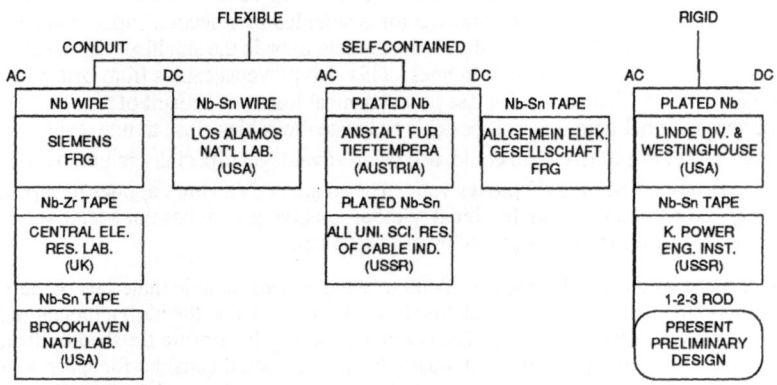

FIGURE 1   TECHNICAL OPTIONS TAKEN BY MAJOR SUPERCONDUCTOR TRANSMISSION PROJECTS.

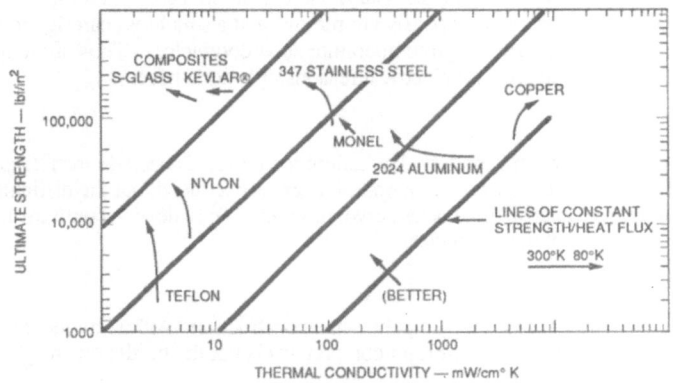

FIGURE 2   STRENGTH AS A FUNCTION OF CONDUCTIVITY FOR VARIOUS CONSTRUCTION MATERIALS.

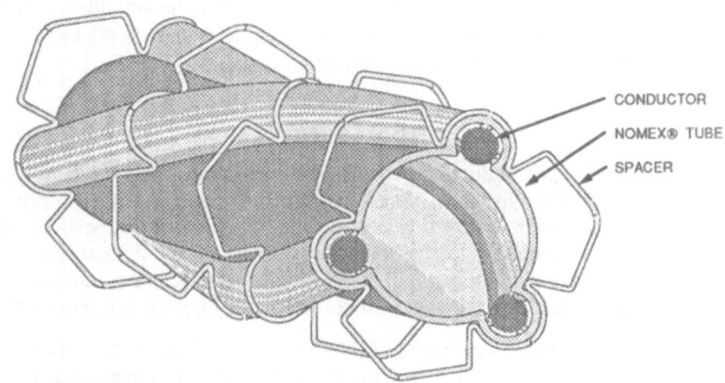

FIGURE 3   CARRIER TUBE, SHOWING TWIST.

higher than the 200 to 300 A/cm^2 usually obtained on high-temperature superconductors, or the 2000 to 3000 A/cm^2 that researchers consider excellent in laboratory samples. We do not believe the high value[24] quoted by Wolsky should be taken as the sign of the end of the superconducting transmission line concept. The relatively low superconducting current density simply means that lower capacities and higher voltages will be required.

PHYSICAL DESIGN AND MATERIALS

Description and Rationale

Liquid Carrier Tube, Conductor, Spacers—The current is carried in three solid conductors, a little over 1 cm in diameter. (We have assumed solid rods, but thin-film-coated carriers, or maybe bundles of filaments could also be used.) These current carriers are relatively rigid. The conductors are configured as rods that twist in a long-wavelength spiral so that the rods are constantly changing place with respect to the vertical along the transmission line, as shown in Figure 3. The primary purpose of the twist is to accommodate thermal changes with little strain. The conductors are individually insulated by Nomex® paper and carried in the three lobes of a porous Nomex® tube. The tube contains a flow of liquid nitrogen (LN$_2$) that completely fills the tube interior, surrounding the conductors and saturating the insulation. Nitrogen that vaporizes passes through the porous tube wall to an outer annulus, which is the passageway for flowing nitrogen gas. The inner porous tube is suspended within this tube by occasional spacers to the outer wall.

The Nomex® tube serves several functions: The first is to support the conductors during operation, but also during transportation from manufacture to the field site. The second is to separate the liquid from the gas. Two-phase flow typically has a larger pressure drop than single-phase flow, and it is to be avoided for that reason alone. One also would like to have the conductors in a liquid bath to increase the heat transfer coefficient and to increase the electrical breakdown resistance.

The attractive (and repulsive) forces between conductors are a factor to be considered. The attraction (or repulsion) between conductors is expressed by

$$\frac{F}{L} = \frac{\mu_0 \, i_a \, i_b}{2\pi d}$$

where F/L is the force per unit length, μ_0 is the permeability of free space, i_a and i_b are the currents on conductors a and b, and d is the separation distance. A preliminary look shows that the attractive (repulsive) forces will be on the order of 0.1 N/m. Extensive structural materials will not be required. Support spacing for the conductors will have to be considered carefully to avoid resonance.

Outer Vessel—The outer vessel (Figure 4) is a part of the protection and insulation package. This outer double-wall-construction tube comes in sections of convenient shipping lengths (6-meter lengths normally, but lengths up to 15 m are possible). The sections of the outer wall telescope at the joints. The inner surface, adjacent to the flowing nitrogen gas, is made of filled Nomex® pressboard, or epoxy-Kevlar® composite. The outer shell could be plain steel for structural purposes, but stainless steel is chosen because the environment could be corrosive. The space between is permanently evacuated (maintained by a getter) and insulated with superinsulation consisting of layers of aluminized Mylar®. We chose superinsulation with aluminized Mylar® to keep the diameter small.

The outer vessel is the size of an 8-inch, Schedule-20 pipe. The pressure in the pipe is slightly higher than atmospheric. The slight pressurization of the vessel will prevent ingestion of condensable gases (water vapor and carbon dioxide). Absolutely tight seals between the nitrogen annulus and the atmosphere are not a requirement.

Cooling—A two-phase cooling system is used, in which liquid nitrogen will be evaporated to the gas phase, thereby exploiting the latent heat of vaporization. The Nomex® tube accomplishes the phase separation. The liquid would flow inside the tube and evaporate through the tube to the outer annulus. Preliminary tests show that the concept is feasible. Refrigeration stations could be spaced at intervals of 10 km or greater. The liquid and vapor have been shown flowing in the same direction to reduce pressure differences across the Nomex® tube, but that is not necessary.

The heat removal requirement is expected to be much less than 600 W/km. This was the value of the Brookhaven line,[8-10] and it was about one-third due to heat leak with the rest due to electrical losses

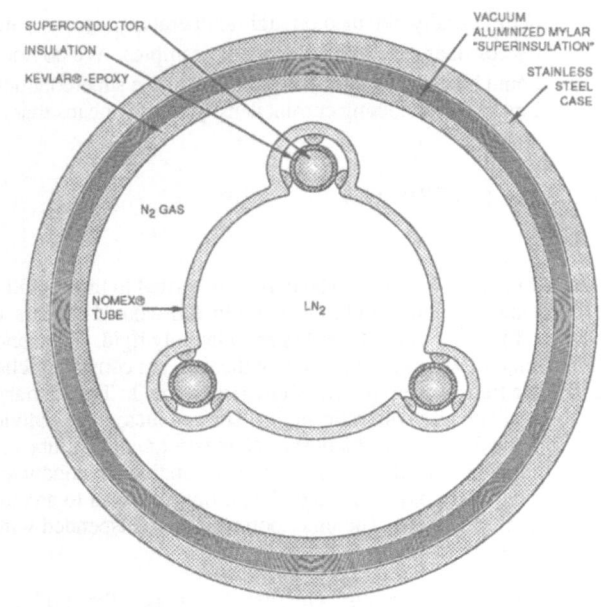

FIGURE 4 CROSS SECTION, SHOWING OUTER SHELL .

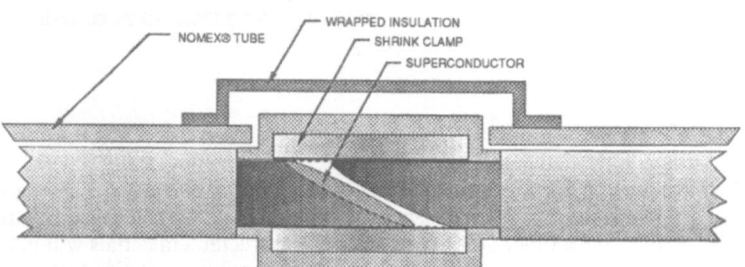

FIGURE 5 CARRIER CONNECTION .

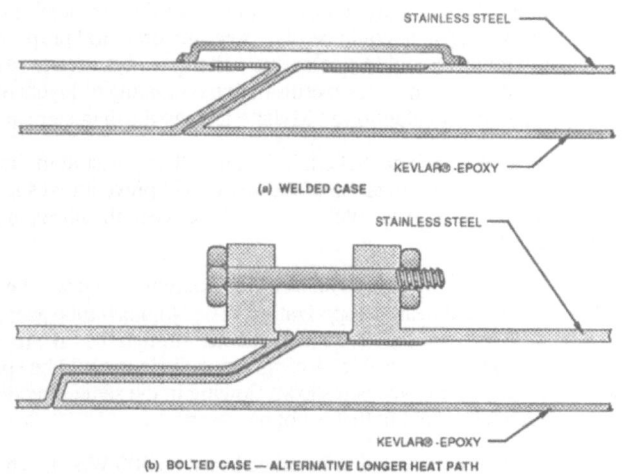

FIGURE 6 OUTER SHELL CONNECTION METHODS .

related to design, voltage, and current. The Los Alamos line[11] had 250 W/km—almost all heat leak because the line was dc. We expect the heat leak in our design to be considerably less than either of these because the diameter of the line is only about one-third that of the Brookhaven design. The electrical losses will also be much less. Some electrical losses are proportional to voltage (dielectric) and others are proportional to current squared (losses resulting from flux development and decay). We expect the electrical losses at our reduced voltages and currents to be less than one-tenth of those reported by Brookhaven. The total heat removal requirement is expected to be around 200 W/km. Nevertheless, we have used the 600-W/km value for our calculations, considering the unknowns with the new superconductors.

The expected gas velocity is about 0.06 m/s per kilometer or a reasonable 6 m/s at the end of 10 km. The pressure drop and viscous heating are negligible. Similar conclusions apply to the liquid.

Method of Assembly

The inner tube with the three conductors is shipped in sections to protect the conductors during shipping. The new section is brought up to the end of the most recently placed section, within one meter of the final location. The conductors are pulled out of the cloverleaf tube, and the diagonal flats of the superconductor are mated and secured by a clamp, as shown in Figure 5. The clamp has a high coefficient of thermal expansion and will tighten as the assembly is cooled. The connections between the conductors are then insulated, and the cloverleaf tube is slid to mate with the adjoining tube and glued. Spacers are then placed and attached to the cloverleaf tube and the double wall tube is then slid in place and mated to the previous section. A welded and bolted design for connecting the double-wall tube is shown in Figure 6.

Bends are somewhat more complex and will require special sections of all components. The construction method is similar to the one described above.

COMMENTS

Several features of the design require further investigation in the laboratory. For example, the connection of the conductors needs to be verified, as does the phase separation design involving a structure of porous Nomex®. Further design and component examination are also justified.

REFERENCES

1. Graneau, P., H. C. Parish, and J. L. Smith, Jr., "Refrigeration Requirements of the LN_2 Cryo-Cable," *Trans. ASME, J. Eng. Industry,* Vol. 93, Series B, No. 4, pp. 1161-1165 (1971).
2. Mauser, S. F., R. R. Burghardt, M. L. Fenger, T. W. Dakin, and R. W. Meyerhoff, "Development of a 180 KV Superconducting Cable Termination," *IEEE Trans Power Apparatus and Systems,*" Vol. PAS-95, No. 3, pp. 909-914 (1976).
3. Janocko, M. A., "Lattice Braided Superconductors," *IEEE Trans Magnetics*, Vol. MAG-15, No. 1, pp. 797-799 (1979).
4. Dolgosheyev, P. I., I. B. Peshkov, G. G. Svalov, I. M. Bortnik, V. L. Karapazyuk, L. P. Kubarev, A. A. Panov, Y. V. Petrovsky, and V. A. Turkot , "Design and First Stage Test of 50-Meter Flexible Superconducting Cable," *IEEE Trans. Magnetics,* Vol. MAG-15, No. 1, pp. 150-154 (1979).
5. Meshchanov, G. I., I. B. Peshkov, and G. G. Svalov, "The Results of Work Carried Out in the USSR on Creation of Superconducting and Cryoresistive Cables for Electric Power Lines," *IEEE Trans. Magnetics,* Vol. MAG-19, No. 3, pp. 662-667 (1983).
6. Klaudy, P. A., and J. Gerhold (1983, "Practical Conclusions from Field Trials of a Superconducting Cable," *IEEE Trans. Magnetics,* Vol. MAG-19, No. 3, pp. 656-661.
7. Klaudy, P., J. Gerhold, A. Beck, P. Rohner, E. Scheffler, and G. Ziemek (1981), "First Field Trials of a Superconducting Power Cable within the Power Grid of a Public Utility," *IEEE Trans. Magnetics*, Vol. MAG-17, No.1, pp. 153-156.
8. Morgan, G. H., F. Schauer, and R. A. Thomas (1981), "An Improved 60 Hz Superconducting Power Transmission Line," *IEEE Trans. Magnetics,* Vol. MAG-17, No. 1, pp. 157-160.
9. Thomas, R. A. (1985), "Operational Characteristics of a 1000 MVA Superconducting Power Transmission System," *IEEE Trans. Magnetics,* Vol. MAG-21, No. 2, pp. 795-798.
10. Thomas, R. A., and E. B. Forsyth (1987), "Preliminary Economic Analysis of a High T_C Super-

conducting Power Transmission System," Informal Report, Brookhaven National Laboratories, BNL 39973, July 1

11. Edeskuty, F. J., R. J. Bartlett, and J. W. Dean (1981), "Current Test of a dc Superconducting Power Transmission Line," *IEEE Trans. Magnetics*, Vol. MAG-17, No. 1, pp. 161-164.

12. Forsyth, E. B. (1987), "The Status of R&D for Superconducting Power Transmission Systems," BNL 40420, Workshop on Utility Applications of Superconductivity, Washington, D.C. (22-23 October).

13. Uher, C., and A. B. Kaiser (1987), "Thermal Transport Properties of $YBa_2Cu_3O_7$ Superconductors," *Physical Review B*, Vol. 36, No. 10, pp. 5680-5683.

14. Wilson, M. N. (1983), *Superconducting Magnets* (Oxford University Press, Oxford).

15. Slesin, L. (1987), "Power Lines and Cancer: The Evidence Grows," *Technol. Review*, pp. 53-59 (October).

16. Kasen, M. B. (1975), "Mechanical and Thermal Properties of Filamentary-Reinforced Structural Composites at Cryogenic Temperatures, 1: Glass-Reinforced Composites," *Cryogenics*, Vol. 15, No. 6, pp. 327-349.

17. Kasen, M. B. (1975), "Mechanical and Thermal Properties of Filamentary-Reinforced Structural Composites at Cryogenic Temperatures, 2: Advanced Composites," *Cryogenics*, Vol. 15, No. 12, pp. 701-722.

18. Kasen, M. B. (1975), "Properties of Filamentary-Reinforced Composites at Cryogenic Temperatures," in *Composite Reliability*, ASTM STP 580, ASTM, pp. 586-611.

19. Kasen, M. B. (1981), "Mechanical and Thermal Properties of Filamentary-Reinforced Structural Composites at Cryogenic Temperatures: An Update," *Cryogenics*, Vol. 21, No. 6, pp. 323-340.

20. Reed, R. P., R. E. Schramm, and A. F. Clark (1973), "Mechanical, Thermal, and Electrical Properties of Selected Polymers," *Cryogenics*, Vol. 13, No.2, pp. 67-82.

21. Weedy, B. M., and S. G. Swingler (1987), "Review of Tape Materials for Cable Insulation at Liquid Nitrogen Temperatures," *Cryogenics*, Vol. 27, December, pp. 667-672.

22. Maguire, F. J., J. D. Ramsden, and D. E. Wolman (1988), "Novel Approach to Supercritical Helium Flight Cryostat Support Structures," *Cryogenics*, Vol. 28, No. 2, pp. 142-146.

23. Wolsky, A. M., E. J. Daniels, R. F. Giese, J. B.L. Harkness, L. R. Johnson, D. M. Rote, and S. A. Zwick (1988), "Advances in Applied Superconductivity; A Preliminary Evaluation of Goals and Impacts," ANL/CNSC-64, Argonne National Laboratory.

Preparation of Bi(Pb)SrCaCuO Thin Film with High-Tc Phase

Tsutom Yotsuya, Yoshihiko Suzuki and Soichi Ogawa

Osaka Prefectural Industrial Technology Research
Institute, Nishi-ku, Enokojima, Osaka, 550, Japan

Hirofumi Imokawa

Tatuta Electric Wire and Cable Co., Iwata-cho
Higashiosaka, Osaka, 577

Masahiro Yoshikawa and Makio Naito

Hosokawa Micron Co. Ltd., Shoudai-Tajika, Hirakata
573

Kohei Otani

Hitachi Zosen Technical Research Laboratory Inc.
Corp., Sakurajima, Konohana-ku, Osaka, 554

ABSTRACT

The lead added BiSrCaCuO thin film of high-Tc phase has
been successfully prepared by using rf-magnetron sputtering
and post annealing process. The content of lead in the thin
film was easily decreased during the ordinary annealing process. In order to minimize the decrease of lead, the Pb-doped
BSCCO thin film was placed on the bulk ceramics and annealed
for over 170 hours at 855°C in air. As the result, the content of Pb was almost unchanged and the film showed superconducting transition at 104 K.

INTRODUCTION

After the discovery of high-Tc oxide superconductor of
BiSrCaCuO(BSCCO) compound by Maeda[1], many groups have made
great many efforts to obtain the high-Tc single phase of the
Bi system. Tarascon et al.[2] showed that BSCCO included at
least three superconducting phases: the high-Tc phase (ideal
composition was $Bi_2Sr_2Ca_2Cu_3O_{10}$) with Tc=110 K, the low-Tc
phase ($Bi_2Sr_2Ca_1Cu_2O_8$) with Tc=80 K, and the semiconducting
phase($Bi_2Sr_2CuO_6$) with Tc=20 K. The low-Tc phase was quite

Advances in Cryogenic Engineering (Materials), Vol. 36
Edited by R. P. Reed and F. R. Fickett
Plenum Press, New York, 1990

stable and can be easily formed. However the high-Tc phase is difficult to form. In order to assist forming the high-Tc phase, Pb was added by substituting a portion of Bi[3][4] and/or it was synthesized in low oxygen pressure[5].

It is important to prepare the BSCCO thin film with the high-Tc phase for many applications as well as fundamental studies. In the case of the BSCCO thin film, heat treatment caused Pb decrease in the BSCCO film[6]. Hakuraku et al.[7] succeeded in forming the high-Tc phase of BSCCO thin film by using dc-magnetron sputtering and post annealing process. However the metal compositions of their virgin film was Bi and Pb rich compared with the ideal composition of BSCCO high-Tc phase. Because these elements were gradually released during annealing, the annealing conditions were restricted by annealing time and annealing temperature.

The purpose of this study was to find a relatively easy method and conditions for producing the BSCCO thin film with the high-Tc phase by keeping the metal contents unchanged during annealing.

EXPERIMENTAL

A Pb-doped BSCCO thin film was deposited by a conventional rf-magnetron sputtering system with a single target. The target was prepared by mixing powders of Bi_2O_3, $SrCO_3$, $CaCO_3$, CuO and metal Pb powder then being heated for 10 hours at 840°C. The target composition was: Bi:Pb:Sr:Ca:Cu=2.1:0.7:2:2.1:3.0.

The polished MgO(100) single crystal was used as a substrate. Four substrates were located 5 cm above the target. The substrate temperature was elevated to 500°C during sputtering. Incident rf-power was fixed at 200 W and a mixed gas of oxygen(1×10^{-2} Torr) and argon(5×10^{-2} Torr) was used for a sputtering gas. Film thickness was about 1 μm. Under these conditions, reproducibility of the film composition was very good (within 8 %). The metal contents in a thin film was measured by the method of an inductive coupled plasma atomic emission spectroscopy(ICPAES). The accuracy of this method was within 5 %.

The resistance of the film was measured by the ordinary four-probe method. The electrodes were connected on the film surface with silver paste. The distance between potential leads was about 2 mm. The crystalline structure was determined by the X-ray diffraction analysis(Cu Kα radiation).

RESULTS and DISCUSSIONS

The long time annealing process was needed to obtain a BSCCO ceramics with the high-Tc phase as shown by Nobumasa et al[8]. In the case of a ceramics composition ratio of the metals including Pb element remained constant during synthesis. However in the case of the Pb-doped thin film, Pb content in the film was easily decreased during heat-treatment.

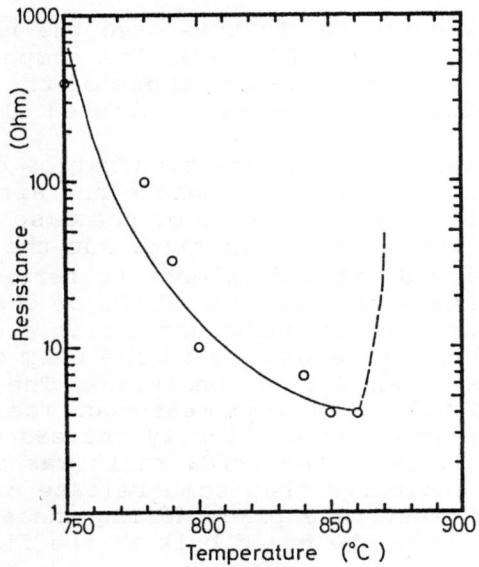

Fig. 1. Resistance of thin film
as a function of anneal-
ing temperature. The re-
sistance was measured at
300 K.

To confirm the annealing conditions, deposited films
were heat-treated in air for 1 hour. The typical metal composition ratio of virgin film was:
Bi:Pb:Sr:Ca:Cu=1.33:0.60:1.98:2.00:3.15. Before annealing,
crystalline structure of the film was almost amorphous. The
annealing temperature was changed from 700 to 860°C. Figure 1
shows the resistance measured at room temperature as a function of annealing temperature.

In the case of relative low annealing temperature(below
750°C), any resistance change due to superconductive transition was not observed above 10 K.

In the case of high annealing temperature(750~850°C),
the temperature dependence of the film's resistance became
metallic and superconductive transition was observed at a-
round 60 K. The critical temperature(Tc) gradually increased
as annealing temperature increased. The crystalline structure
of the low-Tc phase was observed by X-ray diffraction measurement. The high-Tc phase could not be formed in this annealing condition.

In the case of annealing temperature more that 860°C,
the thin film melted down and resistance of the film became
extremely large.

To obtain the high-Tc phase, the Pb added BSCCO thin
film was annealed for 50 hours at 855°C . Two annealing
methods were tried. One was an ordinary annealing method. The
film surface was continuously exposed to air(referred to as
Case-1 hereafter). Another was one in which the film was

located on the bulk ceramics. In this case the film surface contacted with a BSCCO pellet(Case-2). The composition ratio of the base pellet was the same as those of the virgin film. Detailed experimental procedure was published elsewhere[9].

As shown in Fig. 2, the composition ratios of the two films and base bulk were measured before and after annealing. The atomic ratios of Sr/Ca and Bi/Ca of the Case-1 film were slightly decreased from the virgin film. But the ratio of Pb/Ca was drastically decreased, almost to zero. On the other hand, the atomic ratios of Bi/Ca and Pb/Ca of the film annealed in Case 2 were almost unchanged. It is clear that the reduction of the Pb and Bi elements in the film did not take place under the Case-2 annealing conditions. The composition ratios of the based bulk were also measured. The atomic ratios of Cu/Ca and Sr/Ca were slightly increased by the annealing process. However, the Pb/Ca ratio was almost constant. This result indicates that some release of Ca and Pb from the based bulk must take place during annealing. The Pb atom might migrate from the based bulk to the film.

The X-ray diffraction analysis was performed for these two films, and obtained results are shown in Fig. 3. Diffraction lines due to the low-Tc phase were observed, but those of the high-Tc were not observed. This result indicates that the metal elements such as Bi and Pb in the Pb-doped film were easily released, and this reduction makes it difficult to form the high-Tc phase. On the other hand, the X-ray diffraction lines due to high- and low-Tc mixture phase were observed for the film annealed in Case 2. The intensity ratio of the X-ray diffraction lines due to the $\langle 002 \rangle (\langle 002 \rangle_H / \langle 002 \rangle_L)$ was about 4/1. This radio was consistent with the volume fraction of the high- and low-Tc phases measured by SQUID.

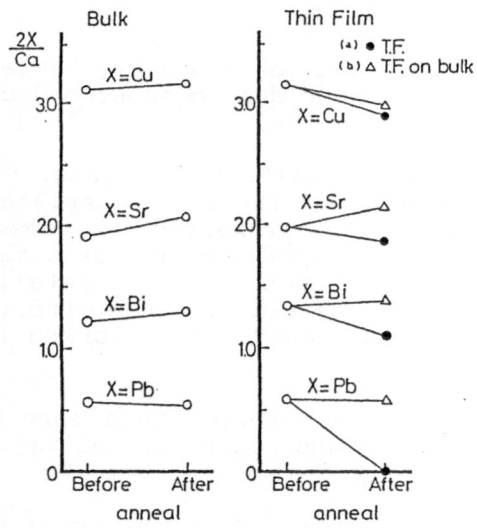

Fig. 2. Composition ratios before and after annealing for two films and based ceramics. The metal elements of the film annealed on a bulk ceramics was unchanged during annealing.

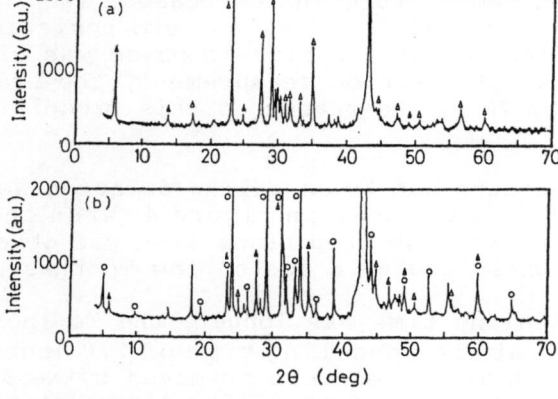

Fig. 3. X-ray diffraction analysis for two films.
Diffraction lines due to the low-Tc
phase(△) for the Case-1 annealing film(a),
and mixture phase (low-Tc phase(△) and
high-Tc phase(○)) for the Case-2 annealing film(b) were observed.

The temperature dependence of the resistance was measured in temperature range between 10 and 300 K for both
annealed films. The resistivity of the Case-1 film at 300 K
was about 20 mΩ cm, and it increased as the temperature decreased and showed maximum resistance at near 75 K. There was
no resistance change due to high-Tc phase near 100 K.
The resistance was rapidly decreased at around 75 K due to
the low-Tc phase, but it was not disappeared above 10 K.

The resistivity of the Case-2 film was also measured and
was 3 mΩ cm at 300 K. It showed metallic characteristics and

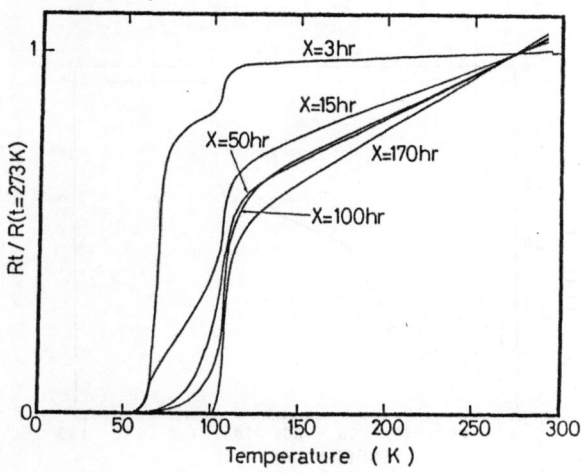

Fig. 4. Resistance changed as a function
of temperature. Annealing was
conducted at 855°C in air.

was decreased as temperature was decreased. The resistance began to drop rapidly at 110 K with zero resistance at 90 K. Although the existence of the high-Tc phase was clearly observed by X-ray diffraction measurement, the transition width was nearly 20 K. The reason of this broad transition is not clear.

Next the effects due to annealing-duration were examined for the Case 2 films as shown in figure 4. When the annealing time was three hours, two step transition was observed; 100 K due to high-Tc phase and 60 K due to low-Tc phase.

As the annealing time was longer, the Tc increased. The Tc_0 become 104 K at the annealing time of 170 hours. However both low- and high-Tc phases were observed by X-ray diffraction measurement. Studies of the influence of atomic composition ratios of the virgin film on the formation the high-Tc single phase are in progress.

The susceptibility of the film was measured by a SQUID for this low-and high-Tc mixed phase film with Tc=104 K. The external magnetic field was fixed to 100 Oe. As shown in Fig. 5, the susceptibility of the film began to change at 110 K, and gradually increased as the temperature decreased. Below 60 K, the susceptibility due to low-Tc phase was observed. However in this temperature region, it was extremely noisy. The surface morphology of the film was observed by a SEM. The large and flat plate-like crystalline grains were observed. This morphology was very similar to the bulk ceramics with high-Tc single phase. Above results show that the domains of the high-Tc phase was large and tightly contacted each other. On the contrary, the crystalline domains of the low-Tc phase was small and dispersed in several places. So that trapped flux in the low-Tc area easily moved.

The film with Tc=104 K was re-annealed at 850°C for one hour in pure oxygen atmosphere. In this case the film surface was exposed for pure oxygen gas. After the heat-treatment, X-ray diffraction analysis was conducted. As shown in Fig. 6,

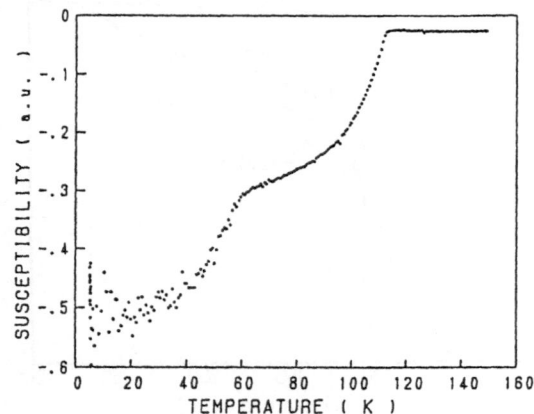

Fig. 5. Susceptibility change measured by SQUID for the BSCCO thin film with Tc=104 K.

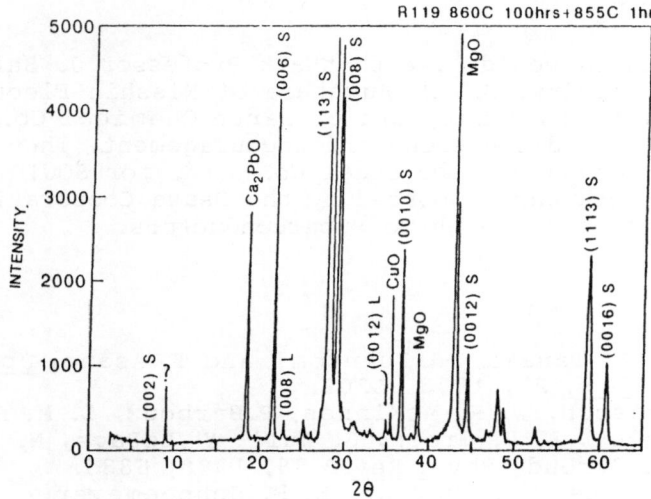

Fig. 6. X-ray diffraction pattern of the re-annealed
film. Diffraction lines due to semiconducting
phase and Ca_2PbO_4 were observed.

the high-and low-Tc phases were diminished and diffraction
lines due to apparent semiconducting phase and Ca_2PbO_4 were
observed. This result indicated that the high-Tc phase was
not stable, since Ca and Pb atoms in the high-Tc phase would
react each other and became Ca_2PbO_4. Since Ca atoms in the
high-Tc phase were removed, the high-Tc phase was changed to
semiconducting phase.

CONCLUSIONS

We have succeeded to prepare the Pb added BSCCO thin
film with high-Tc phase by using rf-magnetron sputtering
method and followed proper post annealing process. To realized the high-Tc phase of the lead added BSCCO thin film, we
annealed the film on the bulk ceramics for over 170 hours at
around 850°C in air. Obtained results are summarized as
follows.

1) Long-time annealing was needed to obtain the high-Tc
phase of BSCCO thin film. It is noted that the atomic composition ratios of the Pb-doped BSCCO thin film were almost
kept constant during annealing in which the film surface
contacted with the bulk surface. The film prepared by this
process showed a superconducting transition at 104 K. However
the film was a mixture of the high- and low-Tc phases, and
the intensity ratio of the X-ray diffraction lines due to the
(002) was about 4/1.

2) Simple annealing method reduced Pb in the film drastically. This reduction of the Pb made the formation of the high-
Tc phase difficult.

3) The high-Tc phase of BiPbSrCaCuO was not stable for re-
annealing at 850°C.

ACKNOWLEDGMENT

The authors would like to thank Professor J. Shirafuji of Osaka University, Dr. H. Kuwahara of Nisshin Electric Co.,Ltd. and Mr. M. Okabayashi of Earth Chemical Co., Ltd. for their useful discussions and encouragement. They also thank Dr. M. Ogawa of Kobe Steel Co., Ltd. for SQUID measurements. This work was supported by the Osaka Cooperative Research Project for High-Tc Superconductors.

REFERENCES

1) H. Maeda, Y. Tanaka, M. Fukutomi and T. Asano, Jpn. J. Appl. Phys. 27, 1988, L209.
2) J. M. Tarascon, W. R. McKinnon, P.Barboux, D. M. Hwang, B. G. Bagley, L. H. Green, G. Hull, Y .LePage, N. Stoffel and M. Giroud, Phys. Rev.B 38, 1988, 8885.
3) S. A. Sunshine, T. Siegrist, L. F. Schneemeyer, D. W. Murphy, R. J. Cava, B. Batlogg, R. B. van Dover, R. M. Fleming, S. H. Glarum, S. Nakahara, R. Ferrow, J. J. Krajewski, S. M. Zahurak, J. V. Waszczak, J. H. Marshall, P. Marsh, L. W. Rupp, Jr. and W. F. Peck, Phys. Rev. B36, 1988, 893.
4) M. Takano, H. Kumahara, J. Takada, K. Oda, H. Kitaguchi, Y. Miura, Y. Ikeda, Y. Tomii and H. Mazaki, Jpn. J. Appl. Phys., 27, 1988, L1041.
5) U. Endo, S. Koyama and T. Kawai ; Jpn. J. Appl.Phys.,27, 1988, L1476.
6) S. K. Dew, N. R. Osborne, P. J. Muhern and R. R. Parsons, Appl. Phys. Lett., 54, 1989, 1929.
7) Y. Hakuraku, Y. Aridome and T.Ogushi, Jpn. J. Appl. Phys.,27, 1988, L2091.
8) H. Nobumasa, K. Shimizu, Y. Kitano and T. Kawai, Jpn. J. Appl. Phys., 27, 1988, L846.
9) T. Yotsuya, Y. Suzuki, S. Ogawa, H. Imokawa, M. Yoshikawa, M. Naito, R. Takahata and K. Otani, Jpn. J. Appl. Phys., 28, 1989, L972.

STABILIZATION OF HIGH-Tc PHASE IN SPUTTERED

BiPbSrCaCuO FILMS BY POST ANNEAL

H. Yamamoto, M. Katoh and M. Tanaka

College of Sci. & Technol.
Nihon University
Funabashi-shi, Chiba, Japan

Y. Aoki, K. Maeda, and T. Shiono

Showa Electric Wire and Cable Co. Ltd.
Kawasaki-ku, Kawasaki, Japan

ABSTRACT

The post anneal was investigated for amorphous $(Bi,Pb)_2Sr_2Ca_2Cu_3O_x$ films deposited on (100) planes of MgO at about 400°C. The films were annealed in a closed Pt crucible together with a sintered bulk including Pb to supply enough constituent metal atoms in vapor. Also the film surface was covered by a MgO plate to suppress a change of the film composition during anneal. The film which contained the high-Tc phase above 70 % was obtained and the maximum Tc of zero resistance attained up 105 K. The results obtained revealed that the addition of Pb and/or the reduced pressure of oxygen were effective to promote the growth of the high-Tc phase under the conditions ensuring the stoichiometry during anneal.

INTRODUCTION

An oxide superconductor of Bi system discovered by Maeda et al.[1] has attracted a large interest because it reveals a high Tc above 110 K or characteristic crystallographic dynamics at the time of syntheses. There were found several crystal phases in the Bi system corresponding to the composition, $Bi_2Sr_2Ca_nCu_{n+1}O_x$ (n=0,1,2,3,4)[2]. So it is very difficult to isolate the phase with n=2, which is called as a high-Tc phase. Takano et al.[3] reported that a small amount of Pb of about 10 % to the Bi concentration effectively stabilize the high-Tc phase, though strict conditions for sintering are needed. Furthermore, Kawai et al.[4] suggested that the high-Tc phase can be obtained with comparatively ease at reduced pressure of oxygen.

The purposes of this work are to investigate the feasibility of the previous methods for the synthesis of the high-Tc phase in thin films and at the same time to develop heat treatment techniques peculiar to thin films. Concretely the atmosphere of the heat treatment was changed and optimized for amorphous thin films deposited on MgO substrates at about 400°C. The specimen film was annealed in a closed Pt crucible

Advances in Cryogenic Engineering (Materials), Vol. 36
Edited by R. P. Reed and F. R. Fickett
Plenum Press, New York, 1990

573

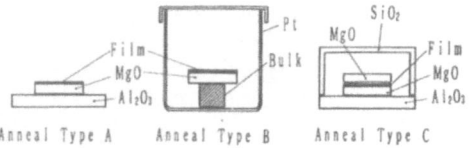

Fig. 1. Three types of anneal methods investigated in this work.

together with a bulk to supply enough constituent atoms in vapor. And also the film surface was covered by a MgO plate in order to suppress a change of the film composition during high temperature anneal. The maximum Tc of zero resistance of the annealed film attained 105 K. These experiments revealed that the addition of Pb and/or the reduced pressure of oxygen were quite available to promote the growth of the high-Tc phase also in a thin film process. Details of annealling conditions and properties of the obtained films will be stated in the following.

EXPERIMENTAL METHOD

Different several types of reactive sputtering: a conventional diode, a magnetron, and a facing targets sputtering[5] were adopted to prepare specimen films in order to grasp essential parameters being independent on a sputtering apparatus. The targets used were sintered powder or disks which were made by calcine at 820°C for 24 hrs and firing at 860°C for 24 hrs from mixed powder. The composition of the target was adjusted peculiarly for each the sputtering method to obtain the stoichiometry, $(Bi,Pb)_2Sr_2Ca_2Cu_3O_x$. The exact value of the film composition was $Bi_{0.9}Pb_{0.2}Sr_{1.0}Ca_{1.1}Cu_{1.6}O_x$, according to the experiment of a bulk[6]. The change of the concentration of copper after deposition was about 20 % in the case of a diode or a magnetron sputtering, and less than 10 % in the case of a facing targets sputtering. A substrate used was a (100) plane of a single-crystal MgO. Typical sputtering conditions were as following: discharge power P_D of 50 W, pressure of argon P_{Ar}/ oxygen P_{O2} of 1.5 mTorr/1.5 mTorr, deposition rate R_D of about 2.5 nm/min in the magnetron sputtering, and P_D of 100 W, P_{Ar}/P_{O2} of 1.5 mTorr/1.5 mTorr, R_D of about 1.0 nm/min for the facing targets sputtering.

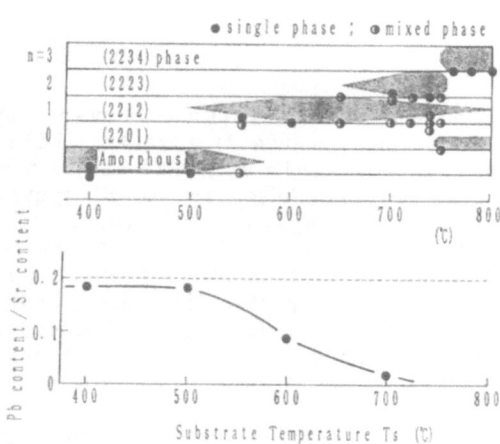

Fig. 2. Crystal phases observed in as-sputtered BiPbSrCaCuO films as a function of substrate temperature Ts. The concentration of Pb in the films changes corresponding to Ts.

Typical methods of anneal investigated are schematically illustrated in Fig.1. Newly invented methods were tried to ensure enough metallic vapor during high temperature anneal, not only the case in a conventional open system (type A). One method was anneal in a closed Pt crucible together with the specimen films and a sintered bulk. The weight of the bulk was about 2 g and its composition was $Bi_{0.9}Pb_{0.2}Sr_{1.0}Ca_{1.1}Cu_{1.7}O_x$ (type B). The specimen films including with and without Pb were investigated. In another method, the film surface was simply covered by a MgO plate to suppress a change of the film composition during anneal (type C). The anneal was done in a furnace heated by infrared rays. The temerature was raised in the rate of 60 °C/min and cooled down in the rate of 5°C/min. The atmosphere was a mixture gas of oxygen and nitrogen. The partial pressure of oxygen was 1, 1/5 and 1/13 atm.

The films composition was analized by Ion Coupled Plasma spectroscopy. The structure and the morphology of the film were observed by reflected x-ray diffraction (XRD) method and secondary electron microscope (SEM), respectively. The electric resistivity was measured by a four probe method.

RESULTS AND DISCUSSION

As-deposited Films

The crystal phases observed in the films deposited on MgO at substrate temperature Ts are summarized in Fig.2. At the comparatively low Ts, the composition of the film was controlled with good reproducibility, though the concentration of copper increased as $P_{Ar}+P_{O2}$ increasing. The composition of all the specimen films was adjusted to approximately the stoichiometry of the high-Tc phase by changing the composition of the target. However, the concentration of Pb in the film decreased as Ts increasing in the range above 500°C, as being shown in Fig.2. So it is difficult to discuss the detail with respects to the effect of addition of Pb in the as-deposited films on hot substrates.

The cystal structure of the film at Ts of 400°C was amorphous. The XRD pattern of the film is shown in Fig.3. The phase with n=1, which is called as a low-Tc (2212) phase appeared in the wide range of Ts above 550°C. On the contrary, the high-Tc (2223) phase was grown in a strict range of Ts from about 700°C to 750°C. At the higher Ts, the phases observed were very complexed: a semiconducting (2202) phase with n=0 or the phases with n=1 and 3, strongly depending on each the sputtering method. Superconductivity of all the films were not so good and the maximum Tc of zero resistivity was at most about 50 K.

Post Annealed Films

Open system

In the conventional open system, the amorphous films were annealed. The mixture phase with the low-Tc phase and the semiconducting phase appeared in the film which was deposited without Pb. On the contrary, the growth of the high-Tc phase took place in the film including Pb which was annealed in the oxygen atmosphere of 1/13 atm. The film annealed in the oxygen pressure of 1 or 1/5 atm. revealed only the low-Tc phase. Typical XRD patterns of the films are shown in Fig. 3. The reduced pressure of oxygen was effective to get the high-Tc phase as pointed out in the bulk process. A preferential growth of c-axis perpendicular to the film plane was observed.

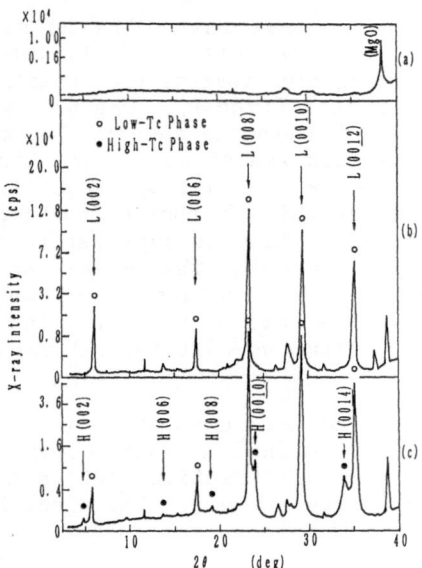

Fig. 3. Typical XRD patterns of the as-deposited BiPbSrCaCuO film on MgO (100) at 400°C (a), the annealed film in P_{O_2} of 1/5 atm. at 860°C for 1 hr (b), and the annealed film in P_{O_2} of 1/13 atm. at 860°C for 1 hr (c).

In Fig. 4 the volumetric fraction of the high-Tc phase estimated from the XRD peak ratio are shown as a function of anneal temperature. Though there was the scatter of the data caused by the difference of the composition of the specimen films, the promoted growth of the high-Tc phase was observed at temperature of about 850 °C in the oxygen atmosphere of 1/13 atm.

The typical ρ-T curves of the films annealed in the different three atmosphere are shown in Fig.5. The conduction of the film became more metallic as the oxygen pressure decreasing. The resistivity drop around 115 K became large corresponding to the growth of the high Tc phase. It was confirmed from these results that both the addition of Pb and the reduced pressure of oxygen are effective to promote the growth of the high-Tc phase. Its growth rate was, however, not still enough probably because the stoichiometry was not strictly satisfied.

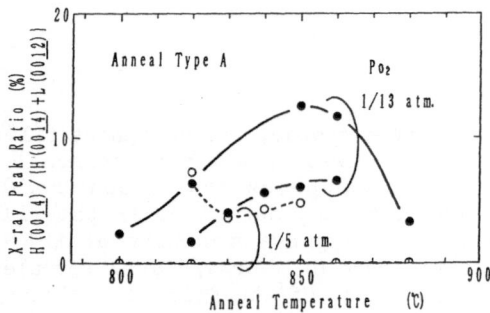

Fig. 4. The change of X-ray peak ratio of the high-Tc phase in the films annealed in the reduced pressure of oxygen, 1/5 and 1/13 atm. as a function of anneal temperature.

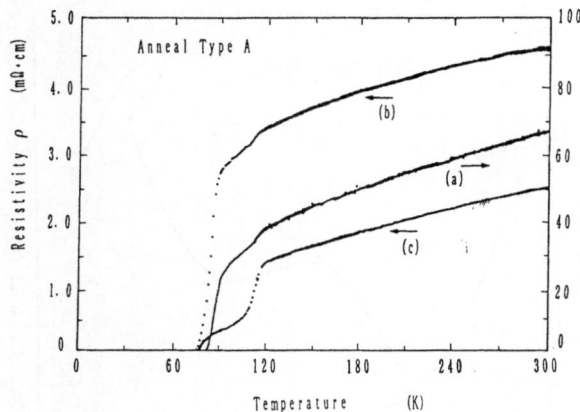

Fig. 5. Temperature dependence of resistivity for the films annealed in P_{O2} of 1/5 atm. at 880°C for 1 hr (a), P_{O2} of 1/5 atm. at 860°C for 1 hr (b), and P_{O2} of 1/13 atm. at 860°C for 1 hr (c).

Closed system

The closed system was effective to ensure the desired metallic vapor during anneal. Typical XRD patterns are shown in Fig.6. The volumetric fraction of the high-Tc phase which was observed in the film annealed in the Pt crucible is shown as a function of anneal temperature Ta in Fig.7. The film which was deposited with Pb showed the growth of the high-Tc phase even in the comparatively low Ta. In the range of temperature above 850°C, the high-Tc phase transformed into the low-Tc

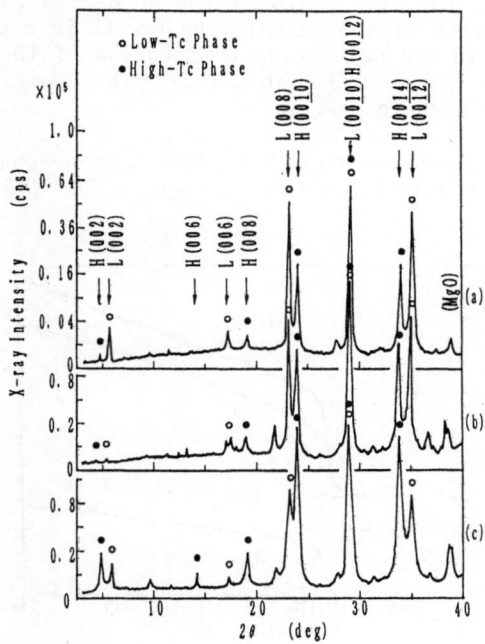

Fig. 6. Typical XRD patterns of the film without Pb annealed at 850°C for 15 hrs (a), the film with Pb at 835°C for 15 hrs (b) by the type B, and the film with Pb annealed at 848°C for 0.5 hr by the type C (c).

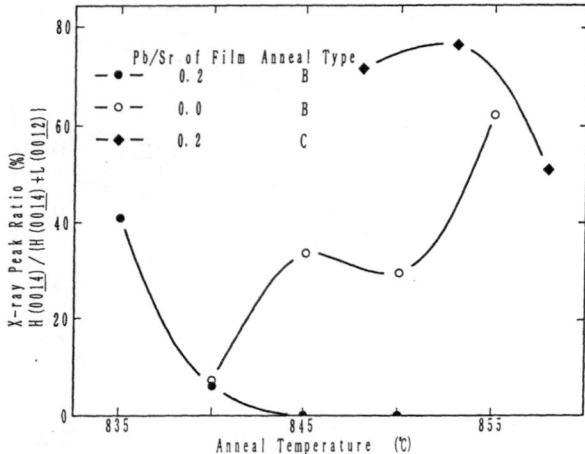

Fig. 7. The change of X-ray peak ratio of the high-Tc phase in the
films with and without Pb by the anneal type of B and the
film with Pb by the anneal type C as a function of Ta.

phase. For the film without Pb, the higher Ta resulted in much growth
of the high-Tc phase untill the melt down of the film took place. The
high-Tc phase was more stable for a long time anneal in the film
deposited with Pb, while the high-Tc phase was destructed by the anneal
for about 20 hrs in the films without Pb.

Before and after the anneal the composition of the film did not
change except of Pb. The amount of Pb in the annealed film was
constant, about 0.2 to that of Sr whether the film was deposited with Pb
or not. Incidentally the crystalline films deposited on hot substrates
were not hardly affected by the anneal type B. It is concluded that the
Pb atoms which come in and out during the process of the crystalization
from the amorphous phase take an important role for growth and
stabilization of the high-Tc phase.

The specimen film, on which a MgO plate was covered, was annealed
by the type C at the reduced pressure of oxygen. The XRD pattern of the

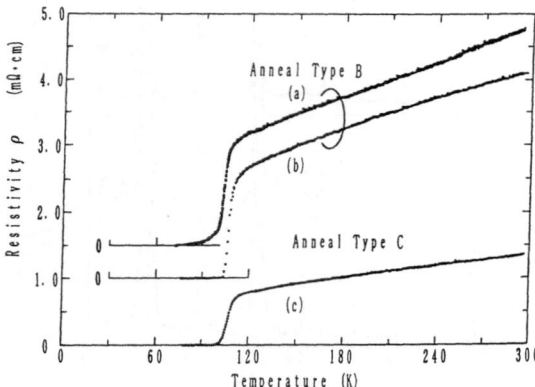

Fig. 8. Temperature dependence of resistivity for the film without Pb
annealed at 850°C for 15 hrs (a), the film with Pb at 835°C for
15 hrs (b) by the type B, and the film with Pb annealed at 848
°C for 0.5 hr by the type C (c).

film is shown in Fig.6. The volumeric fraction of the high-Tc phase observed in the films are also shown in Fig.7, as compared with the results by the previous anneal method. The high volumeric fraction of above 70 % was obtained at Ta of about 850°C for a short time less than 1 hr.

The typical ρ-T curves of the films annealed in the closed system are shown in Fig.8. All the films revealed good metallic conduction with the resistivity at room temperature of about 1.3 mΩcm and the resistivity ratio of about 2. The maximum Tc of zero resistivity attained up about 105 K and the transition width was about 10 K.

The surface morphology of the annealed film generally was like a scale, though the surface of the as-deposited film was very smooth. Such the structure may be caused from a partial melt during anneal. From considering the results of XRD and SEM, it was concluded that the ab plane was grown parallel to the film plane in the process of crystallization. The crystal growth with the anisotropy is favorable to improve the electric conductivity in the film plane and enables to achieve a high critical current density Jc.

SUMMARY

The specimen films were prepared by reactive sputtering. The relationship between substrate temperature and crystal phases was investigated. The high-Tc phase was grown at the very stricted range of Ts of 700°C ~ 750°C, while the low-Tc phase was easily obtained in the wide range of Ts above 550°C. The post anneal conditions were examined to promote and stabilize the growth of the high-Tc phase. It was found that the addition of Pb and/or the reduced pressure of oxygen were effective when the change of the composition of the film was suppressed as possible during anneal. As a result, the film which contained the high-Tc phase above 70 % was prepared, and it revealed the maximum Tc of about 105 K.

REFERENCES

1. H. Maeda, Y. Tanaka, M. Fukutomi and T. Asano, A New High-Tc Oxide superconductor without a Rare Earth Element, Jpn. J. Appl. Phys. 27: L209 (1988).
2. S. Ikeda, H. Ichinose, T. Kimura, T. Matsumoto, H. Maeda, Y. Ishida and K. Ogawa, Transmission Electron Microscope Studies of Intergrowth in BiSrCaCu2Ox and High-Tc Superconducting Phase, Jpn. J. Appl. Phys. 27: L999 (1988).
3. M. Takano, J. Takada, K. Oda, H. Kitaguchi, Y. Miura, Y. Ikeda, Y. Tomii and H. Mazaki, High-Tc Phase Promoted and Stabilized in the Bi,Pb-Sr-Ca-Cu-O System, Jpn. J. Appl. Phys. 27: L1041 (1988).
4. U. Endo, S. Koyama and T. Kawai, Preparation of the High-Tc Phase of Bi-Sr-Ca-Cu-O Superconductor, Jpn. J. Appl. Phys. 27:L1476 (1988).
5. Y. Hoshi, M. Naoe and S. Yamanaka, Facing Targets Type of High Rate and Low Temperature Sputtering Method and Its Application for Deposition of Magnetic Films, Trans. I. E. C. E.Jpn. J65-C: 490 (1982). (in Japanese)
6. S. Koyama, U. Endo and T. Kawai, Preparation of Single 110 K phase of the Bi-Pb-Sr-Ca-Cu-O Superconductor, Jpn. J. Appl. Phys. 27: L1861 (1988).

SUPERCONDUCTING PROPERTIES OF SINTERED Pb-DOPED Bi OXIDE

SUPERCONDUCTOR PREPARED BY AN INTERMEDIATE REPRESSING PROCESS

Toshihisa Asano, Yoshiaki Tanaka, Masao Fukutomi,
Kazunori Jikihara* and Hiroshi Maeda

National Research Institute for Metals, Tsukuba
Laboratories, Sengen 1-2-1, Tsukuba-City, Ibaraki 305,Japan

*Sumitomo Heavy Industries, Ltd. Hiratsuka Research
Laboratory, 63-30 Yuhigaoka, Hiratsuka, Kanagawa 254, Japan

ABSTRACT

Recently, we have shown that highly oriented microstructure was achieved by introducing intermediate repressing on bulk sample by powder method. The sintering has been done with the procedure of sintering of the pellet, re-cold-pressing(i.e. uniaxial compaction at room temperature) during the interrupted sintering and then resintering. We have also shown that the intermediate repressing accelerate the atomic diffusion rate, consequently, the growth rate of the high-T_c phase. By this method it becomes very easy to get almost single high-T_c phase by short sintering duration. However, in the higher fields above 0.1 T J_c decreases rapidly though those in the lower field were much improved by introducing this method. We studied the effects of Ag addition to the Bi system for the high Tc phase growth rate and superconducting characteristics on the bulk samples prepared by the powder method combined with the intermediate pressing process. It was found that the J_c decreases with the appearance of Ag rich phase, in particular, excess amount of Ag degrades the layer continuity of the high T_c phase though slight improvement in the growth rate of high T_c phase was observed.

INTRODUCTION

The high T_c superconductor in the Bi-Sr-Ca-Cu-O (BSCCO) system found by Maeda et al[1]. has been received considerable attention as this system shows the much higher T_c than the boiling temperature of liquid nitrogen and may realize the practical applications at this temperature. In the BSCCO system, however, the superconductor has at least two phases with different transition temperatures : $T_c \sim 80$ K (low T_c phase) and 110 K (high T_c phase). In addition, as the low T_c phase is apt to occupy a large volume portion of the sample it was difficult to make a high T_c single phase superconductor and attain zero resistance temperature of over 100 K. Though there were people who showed the effectiveness of Pb to increase the volume fraction of the high T_c phase[2-4], we must have waited for the study of optimization of the nominal composition and the heat

Advances in Cryogenic Engineering (Materials), Vol. 36
Edited by R. P. Reed and F. R. Fickett
Plenum Press, New York, 1990

581

treatment condition including the atmosphere of reduced oxygen partial pressure to get high T_c single phase BSCCO system.[5,6] Nevertheless these advances in getting high T_c single phase, the transport currrent density, J_c at 77 K and in 0 T has made poor progress and remained as the lower J_c than that of $YBa_2Cu_3O_{7-y}$.

Recently, we reported that the J_c of the (Bi,Pb)SCCO bulk sample can be improved by intermediate pressing between the sintering.[7-9] This intermediate pressing process, not only enhances the growth rate of the high T_c phase but improve electromagnetic characteristic as well with forming a compacted layered crystal structure.

In this paper, we shall show the effects of Ag addition for T_c and J_c together with sample morphology on the (Bi,Pb)SCCO system.[10] It is reported that the high T_c phase is produced by the partial melting. We expected an expansion of the temperature region forming the high T_c phase, an improvement in J_c in magnetic fields by doping Ag and the stabilization of the superconductor by good thermal conductance of Ag.

EXPERIMENTAL

Powder reagents of Bi_2O_3, PbO, $SrCO_3$, $CaCO_3$, CuO and AgO were mixed and ground with a cationic proportion of $Bi:Pb:Sr:Ca:Cu:Ag_x$ = 0.8:0.2:1:1.1:1.6:(x=0 - 2). The mixture was calcined in silver boat at 1023 K for 36 ksec in air. The compound thus obtained was thoroughly ground and pressed into a disk form of 20 mm in diameter and about 1 mm in thickness at room temperature under the pressure of 5×10^8 Pa. The samples sintered at 1108 K for 432 ksec, cooled to room temperature and re-pressed under the same pressure, then sintered again at same temperature for 72 ksec and cooled in the furnace to room temperature. All sintering were carried out in the atmosphere of Ar : O_2 = 12:1. Fractured cross section of the sample was observed by scanning electron microscopy (SEM), back-scattered electron image (BEI) was also observed on the polished cross section. X-ray diffraction analysis of CuK_α radiation was performed on the bulk and powder samples. Critical temperature, T_c were measured on bar-shaped specimens cut from the sintered disk using a four point probe technique in zero field and 0.8 T generated by water-cooled copper/iron (Cu/Fe) magnet. The specimen temperature was measured by calibrated Pt film resister. Critical current I_c was measured on bridge-shaped specimens at 77.3 K using Cu/Fe magnet and at 4.2 K using superconducting magnet. J_c was calculated by dividing I_c with the cross-sectional area of the specimen.

RESULTS AND DISCUSSION

Figure 1 shows SEM image of the samples of x=0, 0.4, 2 together with Ag map taken for the sample of x=2. The SEI (top) were taken on the fractured cross section and BEI (bottom) were on the polished cross section. Layered structure consisting of plate-like grains are observed not only in the surface region but also in the middle position of thickness in SEI pictures of Ag added samples up to x=2.

In the sample, x=0, at least three phase are recognized by BEI; high T_c layers (dark gray) are separated by Bi rich lamina (bright gray) in parallel and black islands are scattered among them . It is reported that the bright gray phase composition corresponds to the mixture of $(Bi,Pb)_2Sr_2CuO_y$ and the low T_c phase, and that the black phase corresponds to $(Ca,Sr)_3Cu_5O_8$. On the other hand, in the Ag added samples, There exists an other phase, Ag rich phase. This phase is shown by label A in the pictures of BEI. This phase corresponds to the white

582

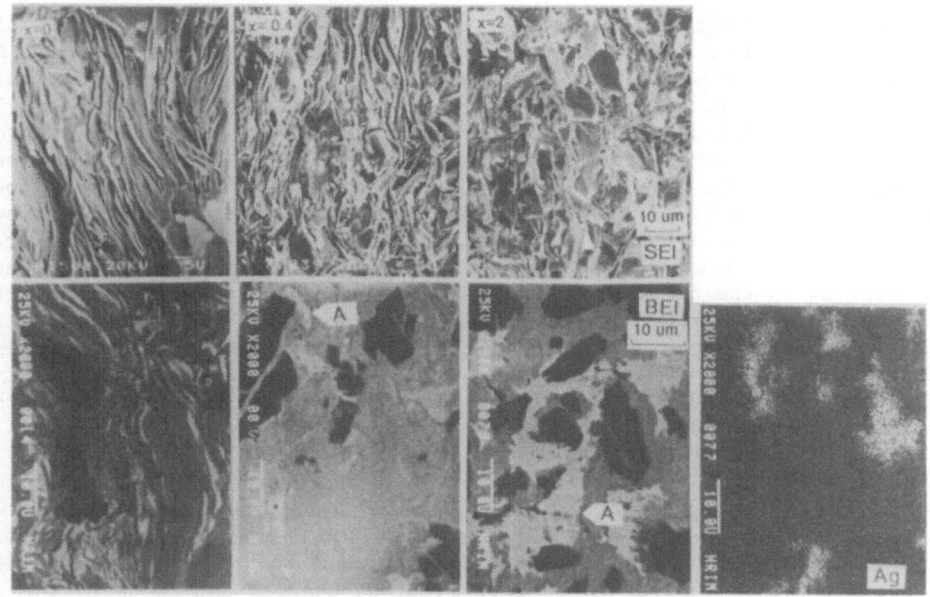

Fig. 1. SEI (top) and BEI (bottom) pictures for the fractured (SEI) and
polished (BEI) cross section of the bulk sample
$Bi_{0.8}Pb_{0.2}Sr_1Ca_{1.1}Cu_{1.6}Ag_xO_y$ (x=0,0.4,2) sintered at 1108 K
for 432 ksec, pressed then sintered again at 1108 K for 72 ksec.
Sintering were performed under the atmosphere of reduced oxygen
partial pressure; $Ar:O_2$=12:1. All pictures were taken on the
middle position in thickness. Label A shows the Ag rich phase
corresponds to white area on the Ag map.

area on the Ag map and increases with the amount of Ag addition, while the
phase so-called Bi rich phase decrease. However, as the non
superconducting phase dose not become flat by only single intermediate
pressing , the continuity of the high T_c phase degrades remarkably by the
Ag amount of x>1.

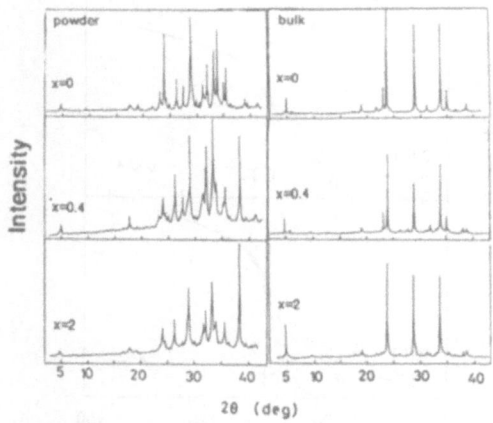

Fig. 2. X-ray spectra of powder sample (left) and bulk sample (right) of
$Bi_{0.8}Pb_{0.2}Sr_1Ca_{1.1}Cu_{1.6}Ag_xO_y$ prepared by the same way as those
in Fig. 1. In the bulk sample, spectra were taken on the surface
of the sintered disk.

Figure 2 shows XRD patterns. The XRD patterns for powder sample (left) shows that the peak from the low T_c phase becomes weak by Ag addition. On the other hand, one peak ($2\theta \cong 38$ deg) becomes remarkably strong with the increase of Ag addition. The (002) reflection of low T_c phase ($2\theta=5.8$ deg) disappears by Ag addition of x=0.4-1 in the samples prepared with intermediate repressing. The required resintering duration, in the sample x=1, to reduce the low T_c phase untill this peak almost disappear is less than half of Ag free samples as reported in a previous paper. The high T_c phase crystals, in the Ag added sample, are layered even though coexisting nonsuperconducting phase are not flattened as can be observed in the pictures of Fig. 1. In such case, it seems as if the bulk sample is almost consisting of the crystals of the superconducting High T_c phase as shown in Fig. 2 (right). In fact the nonsuperconducting phases coexist in the disk surface as well. But note that the relative intensity of the (002) reflection of the high T_c phase increases with the amount of Ag addition.

The magnetic susceptibility $x^{\bullet}$ versus temperature curve measured for the bulk sample shows single sharp transition. The transition curve shifts to the lower side from 108 K (x=0) to 104 K (x=2) with the increase of Ag addition but the sharpness remains almost the same. However, the transition curve measured by resistive method changes the shape by the Ag addition. Fig. 3 shows R-T curves measured with the current density of the order of 1 mA/cm^2. In the sample x=0, the resistance reaches zero (< 0.1μV) at 109 K in zero magnetic field with the tail scarcely recognized. The addition of Ag degrades this sharpness in the transition by causing tail throughout one third of the transition region and lower the zero resistance temperature down to 101 K independent of Ag amount. In the applied magnetic field, 0.8 T, the tailing occupies considerable resistance change of the transition and broadened in temperature width. The broadening takes place on more large scale in the Ag added samples

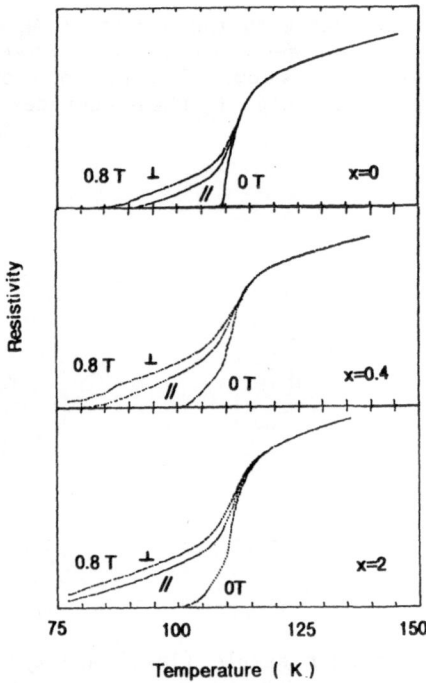

Fig. 3. Resistance-Temperature curves measured in the magnetic fields 0 and 0.8 T by four prove method.

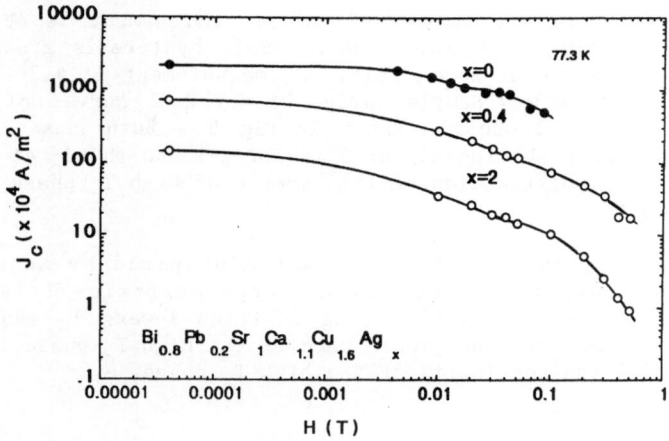

Fig. 4. H – J_c curves measured at 77.3 K. J_c were measured in static fields parallel to the pressed plane of the sample.

than that of Ag free samples at least the region $x \leq 0.4$. Though the zero resistance temperatures of the samples x=0.4 and 2 are approximately same in zero field, that of x=2 is more lower in the field 0.8 T corresponding to the result of x^a-T curves measured in the alternative magnetic field of 1.7 G (peak to peak). The resistance change with temperature is measured in two different field direction, one is perpendicular to the pressed plane and the other is parallel. In both case current flows perpendicular to the field. Broadenings are caused more greatly in the former case independent of Ag addition. Magnetic field induced broadening is activated by the Ag addition. Furthermore the broadening is not simple one but accompanies curve shift to the lower temperature side. This shift becomes significantly with the increase of Ag addition as observed in the Fig. 3. This suggests that the addition of Ag to the Bi system may have a possibility to provide voltex pinning centers though excess Ag degrades the continuity of the high T_c phase.

Critical current density, J_c was not improved in this study. Those of Ag added samples are lower than that of Ag free sample as shown in Fig. 4. The J_c measured at 77.3 K show rapid decrease above the fields 0.2 T.

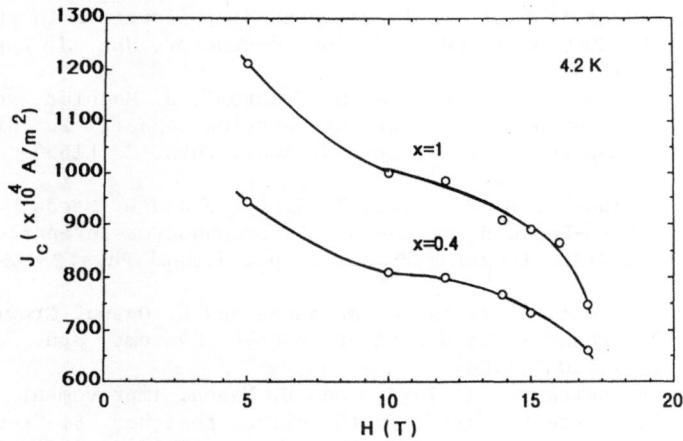

Fig. 5. H – J_c curves measured at 4.2 K in static fields parallel to the pressed plane of the sample.

It is more apparent in the sample x=2 and no improvement is observed in pinning characteristic. All sample shows small hysteresis around 0.01 T and relatively large one at zero field. J_c measurement at 4.2 K was also performed but as there was sample damage we got H-J$_c$ curve only for the samples x=0.4 and 1. Those are shown in Fig 5. Both show same field dependence. The sample showing higher J_c is of x=1 and the J_c at 15 T is 9 x 10^6 A/m^2. The volume fraction of this sample of high T_c phase is larger than that of x=0.4.

In conclusion, the Ag added Bi system prepared by intermediate pressing and resintering method shows large magnetic field induced broadening in resistive transition. Ag addition lowers T_c and J_c. Ag addition slightly enhances the growth rate of the high T_c phase though the excess Ag degrades the continuity of the high T_c phase.

REFERENCES

1. H. Maeda, Y. Tanaka, M. Fukutomi and T. Asano, A new High-T_c Oxide Superconductor without Rear Earth Element, Jpn. J. Appl. Phys. 27:L209(1988).
2. R. J. Cava, B.Batlogg, S. A. Sunshine, T. Siegrist, R. M. Fleming, K. Rabe, L. F. Schneemeyer, D. W. Murphy, R. B. van Dover, P. K. Gallaghers, S. H. Glarum, S. Nakahara, R. C. Farrow, J. J. Krajewski, S. M. Zahurak, J. V. Waszczak, J. H. Marshall, P. Marsh, L. W. Rupp,Jr., W. F. Peck and E. A. Rietman, Studies of Oxygen-Deficient $Ba_2YCu_3O_7$ and Superconductivity Bi(Pb)-Sr-Ca-Cu-O, Physica C 153-155:560(1988)
3. S. Green, C. Jiang, Yu Mei, H. L. Luo and C. Politis, Zero resistance at 107 K in the (Bi,Pb)-Ca-Sr-Cu oxide system, Phys. Rev. B 38:5016(1988)
4. M. Takano, J. Takada, K. Oda, H. Kitaguchi, Y. Miura, Y. Ikeda, Y. Tomii and H. Mazaki, High-T_c phase Promoted and Stabilized in the Bi,Pb-Sr-Ca-Cu-O System, Jpn.J.Appl. Phys. 27:L1041(1988)
5. U. Endo, S. Koyama and T. Kawai, Preparation of the High-T_c phase of Bi-Sr-Ca-Cu-O Superconductor, Jpn. J. Appl. Phys. 27:L1476(1988)
6. S. Koyama, U. Endo and T. Kawai, Preparation of single 110 K phase of Bi-Sr-Ca-Cu-O Superconductor, Jpn. J. Appl. Phys. 27: L1861(1988)
7. T. Asano, Y. Tanaka, M. Fukutomi, K. Jikihara, J. Machida and H. Maeda, Preparation of Highly Oriented Microstructure in the (Bi,Pb)-Sr-Ca-Cu-O sintered Oxide Superconductor, Jpn. J. Appl. Phys. 27:L1652(1988)
8. Y. Tanaka, T. Asano. K. Jikihara, M. Fukutomi, J. Machida and H. Maeda, Improvement in the Current Carrying Capacity in High-T_c BiSrCaCuO Superconductors, Jpn. J. Appl. Phys. 27:L1655 (1988)
9. T. Asano, Y. Tanaka, M. Fukutomi, K. Jikihara and H. Maeda, Properties of Pb-Doped Bi-Sr-Ca-Cu-O Superconductors prepared by the Intermediate Pressing Process, Jpn. J. Appl.Phys.28:L595 (1988)
10. T. Hatano, K. Aota, S. Ikeda, K. Nakamura and K. Ogawa, Growth of the 2223 Phase in Leaded Bi-Sr-Ca-Cu-O System, Jpn. J. Appl. Phys. 27:L2055(1988)
11. M. Mimura, H. Kumakura, K. Togano and H. Maeda, Improvement of the critical current density in the silver sheathed Bi-Sr-Ca-Cu-O Superconducting Tape, Appl. Phys. Lett. 54:1582(1988)

FABRICATION AND PROPERTIES OF FLEXIBLE TAPES OF HIGH-Tc

Bi-(Pb)-Sr-Ca-Cu-O SUPERCONDUCTORS

Kazumasa Togano, Hiroaki Kumakura, D.R.Dietderich,
Hiroshi Maeda, Eiji Yanagisawa[+] and Tsuyoshi Morimoto[+]

National Research Institute for Metals,
1-2-1, Sengen, Tsukuba-city, Ibaraki 305, Japan
[+]Asahi Glass Research Center, Hazawa, Yokohama 221, Japan

INTRODUCTION

The discovery of high-T_c oxide superconductors, especially of the
$Ba_2YCu_3O_7$,[1] Bi-Sr-Ca-Cu-O[2] and Tl-Ba-Ca-Cu-O[3] compounds, which show
superconductivity above liquid nitrogen temperature, has had great impact
not only basic physics but also on the field of superconductor
application. The application of a compound superconductor to a
superconducting magnet requires the winding of tape or wire conductors
with sufficient flexibility into a solenoid. Among the various
superconducting compounds having intrinsic brittleness, Nb_3Sn and V_3Ga
compounds with A15 structure have been successfully fabricated into a tape
or wire form by the so-called 'bronze process' which utilizes the solid
state reaction. The oxide high-T_c superconductors are also intrinsically
brittle, and special techniques must be developed for making tape or wire
conductors. The studies on stress dependence of the high-T_c oxide
superconductors are also practical interest, because the superconductor
must withstand various stresses caused by coil winding, Lorentz force etc.
Standard ceramic processes of mixing powders and organic formulations are
applicable for preparing of high-T_c oxide superconductor tape or wire, as
demonstrated for $Ba_2YCu_3O_7$ compound.[4,5] However, these tapes or wires
became rather brittle after the final sintering process. In a previous
paper,[6] we reported that the combination of the processes of doctor blade
casting, cold rolling and sintering is effective for the densification of
the tape, and hence to improve the flexibility of the tape as well as
superconducting properties. In this paper, we report some results on the
effects of processing conditions on microstructure and properties.
Preliminary results of bending test are also presented here.

EXPERIMENTALS

The appropriate amounts of Bi_2O_3, Pb_3O_4, $SrCO_3$, $CaCO_3$ and CuO powders
were mixed in the molecular ratio of $Bi_{0.7}Pb_{0.3}SrCaCu_{1.8}O_x$, calcined at
800°C for 15h and heat treated at 845°C for 120h to form high-T_c phase of
the Bi-Sr-Ca-Cu-O system. This was milled into a fine powder for
subsequent processing. Organic formulation consisting of solvent
(trichloroehylene), binder (polyvinyl butyral) and dispersant
(sorditantrioleate) was then added and again milled together. The
resulting slurry was cast under a doctor blade into a 125mm-wide and

Advances in Cryogenic Engineering (Materials), Vol. 36
Edited by R. P. Reed and F. R. Fickett
Plenum Press, New York, 1990

587

100–150μm-thick green tape on the carrier sheet (polyethylene terephthalate). Ribbon samples, typically 3mm in width and 100mm in length, were cut from the tape and heat treated at 500°C for 1h to decompose the organic formulation. The ribbon was then subjected to the various combinations of rollings and sinterings. The rolling was done by sandwiching the ribbon between stainless steel sheets and rolling them together. A small reduction of total cross section area results in an effective densification of the oxide ribbon, the thickness of the ribbon being reduced to 30–40%.

The superconducting transition temperature T_c was measured by employing a standard four-probe resistive method. Current and voltage lead wires were attached to the surface of the tapes using ultrasonic solder. The temperature was measured by an Fe-0.07at%Au-chromel thermocouple. The critical current I_c of the tape was also measured resistively in liquid nitrogen(77K). I_c was defined as the current which induced 1μV across a 10mm length of the tape and critical current density J_c was calculated by dividing I_c by the cross sectional area. The effect of bending strain on I_c was measured by using two fulcrums and one push rod.

RESULTS AND DISCUSSIONS

Although the initial ribbon composed of unsintered powder mixed with organic binder has good flexibility, the ribbon becomes extremely porous and brittle after the preheat treatment at 500°C due to the evaporation of the organic binder. However, the rolling and subsequent sintering greatly improves the packing of the grains. The density of the ribbon processed by adding rolling process is about 5.8g/cm^3, which is almost twice of the value of about 3g/cm^3 for the tape processed without rolling. The observation by optical and scanning electron micrographs shows that the rolling produces an alignment of plate-like grains parallel to the ribbon surface as observed for the pellet sample.[7] Figure 1 shows a scanning electron micrograph on the fractured surface after the rolling and sintering at 845°C for 4h. The grain alignment is also confirmed by

Fig. 1 Scanning electron micrograph on the fractured surface after the rolling and sintering at 845°C for 4h for the doctor-blade processed Bi(Pb)-Sr-Ca-Cu-O tape.

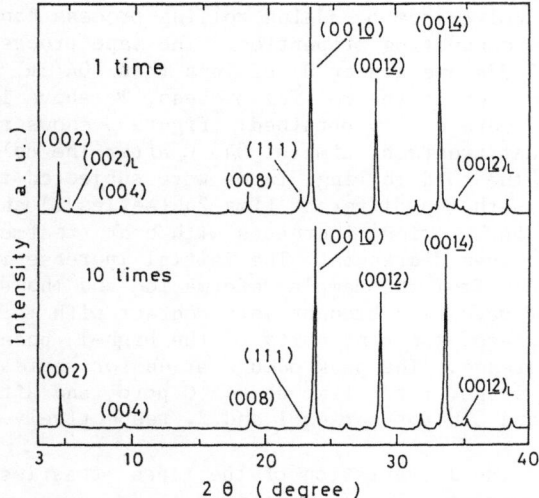

Fig. 2   X-ray diffraction patterns from the surface of
Bi(Pb)-Sr-Ca-Cu-O tape after the one time combination
of rolling and sintering and after the 10 times
repetition of this combination.

the x-ray diffraction data from the surface of the sample, in which the
dominant peaks are identified as (00ℓ) reflection of the high-T_c
phase(110K) indicating strong texturing of c-axis pependicular to the tape
surface. In the x-ray diffraction some of the peaks of the low-T_c
phase(80K) are observed, however they can be decreased by the repetitions
of the rollings and sinterings as shown in Fig. 2. However, small peaks
of nonsuperconducting phases, e.g., Ca_2PbO_4 and Ca-Sr-Cu-O, still exist.
It should be possible to increase the J_c of the sample if these
nonsuperconducting imprurity phases are eliminated by altering the
composition and heat treatment conditions.

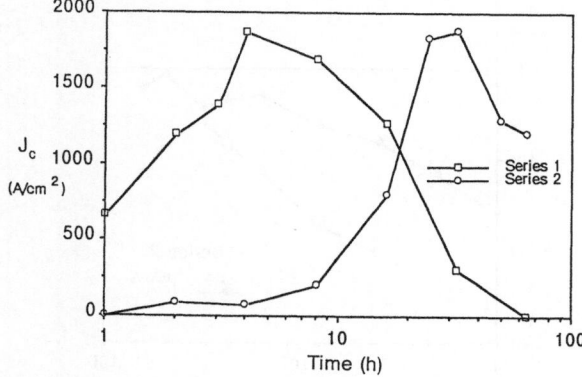

Fig. 3 J_c variation as a function of heat treatment time at 845 C
after the rolling process. Prior to the cold rolling,
the tapes were subjected to heat treatment with two different
times of 24h(series 1) and 2h(series 2) at 845 C.

Such a structural change by adding rolling process can result in an improvement of superconducting properties. The tape processed without rolling has a T_c of 97K and a poor J_c of less than 10A/cm^2 at 77K and 0T. However, by the addition of the rolling process, T_c above 100K and J_c in excess of 1000A/cm^2 were easily obtained. Figure 3 shows the J_c variation as a function of heat treatment time at 845 C after the cold rolling process. Prior to the cold rolling, tapes were subjected to the heat treatment at 845°C with two different time 24h(series 1) and 2h(series 2). The J_c of tapes of both series' increases with heat treatment time reaching a peak and then decreases. The initial increase in J_c is due to the healing of damage from the sample deformation and the sintering of high-T_c grains that have been brought into contact with this deformation. During this early stage, the continuity of the high-T_c phase and its volume fraction increase. The peak occurs at 4h for series 1 and 30h for series 2. At this stage, total time at 845°C prior and after rolling heat treatment are 28h and 32h for series 1 and 2, respectively.

Figure 4 shows the T_c variation of the tapes of series 2 for two different probing current. The T_c continues to increase with heat treatment time when measured by a 1mA probing current, while the T_c determined using a 100mA probing current has a similar shape and peak position as its J_c curve in Fig. 3. This shows there is a loss of continuity between high-T_c regions at prolonged heat treatment time, eventhough the intrinsic T_c does not degrade. This is probably caused by the decomposition of the high-T_c phase at grain boundary region.

Figure 5 shows the effect of applied bending strain on I_c of the sample cold rolled and sintered at 845°C for 4h. The bending strain was estimated by assuming that the sample deforms into a circular arc passing through the points A and B. The I_c was normalized by the value obtained at zero strain(I_{c0}). The tape can be bent up to the value of bending strain $\sim 0.12\%$, which is much larger than that of YBa$_2$Cu$_3$O$_7$ reported by Jin et al.[8] This means the tape with thickness of 40μm can be bent into a diameter of less than $\sim$ 40mm(Fig. 6). The I_c is almost constant up to the bending strain $\sim$0.11% and then small drop is observed in the last 0.01% up to fracture. It should be noted that the strain in Fig. 5 might be underestimated. Elastic deformation theory predicts that the strain at midway between A and B is larger than that determined in this experiment. However, the fracture strain is still smaller compared to that of the metallic Nb$_3$Sn compound superconductors.

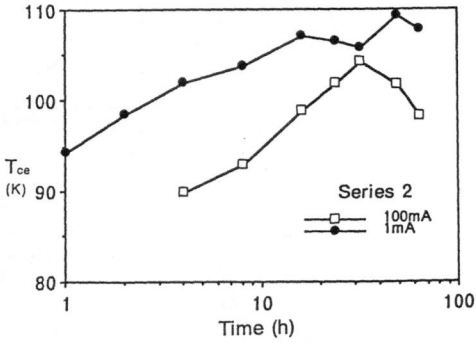

Fig. 4 T_c variation as a function of heat treatment time for the tape of series 2 in Fig. 3. The measurement was done by two different probing currents of 100mA and 1mA.

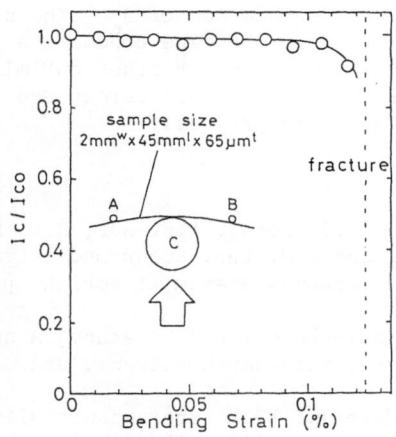

Fig. 5 Effect of applied bending strain for the doctor-blade processed
Bi(Pb)-Sr-Ca-Cu-O tape. The bending strain was estimated by
assuming that the sample deforms into a circular arc.

The bending of the tape occurs elastically and the elastic behaviour
seems to be maintained up to failure. The fracture strain in the elastic
regime for these brittle materials is affected by various metallographic
features such as intergranular bonding and porosity as well as the
presence of cracks and surface flaws. The addition of the rolling process
is considered to improve the packing density, intergranular bonding and
surface condition.

CONCLUSIONS

Flexible tapes of high-T_c Bi(Pb)-Sr-Ca-Cu-O superconductor were
successfully fabricated by the combination of doctor blading and rolling
processes. A zero-resistance state was achieved as high as 107K. The
highest J_c obtained so far is about 2000A/cm^2 at 77K and and in zero
magnetic filed. The tape has sufficient flexibility to be bent into a
small diameter corresponding to the bore diameter of small superconducting

Fig. 6 Doctor-blade processed Bi(Pb)-Sr-Ca-Cu-O tape bent on a 38mm
diameter cylinder.

magnet without breakage and degradation of J_c. The process is very promising for the fabrication of the tape conductors for future application in superconducting coils. Further optimization of fabrication processes, such as the heat treatment conditions and attempts at scaling-up production, is now in progress.

REFERENCES

1. M.K. Wu, J.R. Ashburn, C.J. Torng, P.H. Hor, R.L. Meng, L. Gao, Z.J. Huang, Y.Q. Wang and C.W. Chu, Superconductivity at 93K in a new mixed phase Y-Ba-Cu-O compound system at ambient pressure, Phys. Rev. Lett., 58:908(1987).
2. H. Maeda, T. Tanaka, M. Fukutomi and T. Asano, A new high-T_c oxide superconductor without a rare earth element, Jpn. J. Appl. Phys., 27 : L209(1988).
3. Z.Z. Sheng and A.M. Hermann, Bulk superconductivity at 120K in the Tl-Ca/Ba-Cu-O system, Nature, 332 : 138(1988).
4. D.W. Johnson,Jr., E.M. Gyorgy, W.W. Rhodes, R.J. Cava, L.C. Feldman and R.B. Van Dover, Fabrication of ceramic articles from high-T_c superconducting oxides, MRS Symposium Proc, High Temperature Superconductors(1987, Anaheim), p.193.
5. T. Goto, Nonaqueous suspension spinning of high-T_c Ba-Y-Cu-O superconductor, Jpn. J. Appl. Phys, 27 : L680(1988).
6. K. Togano, H. Kumakura, H. Maeda, E. Yanagisawa, N. Irisawa, J. Simoyama and T. Morimoto, Fabrication of flexible ribbons of high-T_c Bi(Pb)-Sr-Ca-Cu-O superconductor, Jpn J. Appl. Phys., 28 : L95(1989).
7. T. Asano, Y. Tanaka, M. Fukutomi, K. Jikihara, J. Machida and H. Maeda, Preparation of highly oriented microstructure in the (Bi,Pb)-Sr-Ca-Cu-O sintered oxide superconductor, Jpn J. Appl. Phys., 27 : L1652(1988).
8. S. Jin, R.C. Sherwood, T.H. Tiefel, R.B. Van Dover, D.W. Johnson,Jr., and G.S. Grader, Stress and field dependence of critical current in $Ba_2YCu_3O_{7-y}$ superconductor, Appl. Phys. Lett., 51 : 855(1987).

STUDIES ON Bi-(Pb)-Sr-Ca-Cu-O SUPERCONDUCTING TAPES

Hisashi Sekine, Kiyoshi Inoue,
and Hiroshi Maeda

National Research Institute for Metals
1-2-1, Segen, Tsukuba City, Ibaraki 305, Japan

Kouichi Numata

Advanced Technology Research Center
Mitsubishi Heavy Industries, Ltd.
1-8-1, Sachiura, Kanazawa-ku, Yokohama 236, Japan

ABSTRACT

Bi-(Pb)-Sr-Ca-Cu-O superconductor was fabricated into tapes without a sheath and into multifilamentary tapes with a Ag sheath. These specimens were prepared by combination and repetition of cold pressing or cold rolling and sintering. In these specimens, the c axis tended to align well. Measurements of magnetization and critical current density, J_c, in magnetic fields up to 23T were made. The magnetization measurement revealed that, at 4.2K, these tape specimens had excellent J_c-H characteristics while, at 77K, the flux pinning force in the tape specimens was reduced to almost zero above 0.3T due to thermal activation, i.e., flux creep. In the resistive J_c measurement at 4.2K these tape specimens showed J_c of $\sim 10^4$ A/cm^2 at 18T, and 7000 A/cm^2 at 23T. The J_c-H curves for some of these specimens showed a slight peak effect. The J_c's scarcely depended on H at 20-30T. These results indicate that the Bi-(Pb)-Sr-Ca-Cu-O tapes could be used, for example, for the innermost coil of an extremely high-field superconducting magnet at 4.2K in the near future.

INTRODUCTION

The Bi-Sr-Ca-Cu-O system material (BSCCO) shows a very high T_c (critical temperature, ~ 107K) and included neither rare earth elements nor poisonous elements.[1] This indicates that this material should be the most promising candidate for practical use among the oxide superconductors. Recently, the formation of the single high-T_c phase of this material has been made possible by partial substitution of Bi by Pb in the Bi-Sr-Ca-

Advances in Cryogenic Engineering (Materials), Vol. 36
Edited by R. P. Reed and F. R. Fickett
Plenum Press, New York, 1990

593

Cu-O system.[2-6] Therefore, development of wire (or tape)
fabrication processes, study of the magnetic behavior, and
improvement of critical current density, J_c, in magnetic fields
are most important subjects for application research on this
material.

In this study, tapes of Bi-(Pb)-Sr-Ca-Cu-O superconductor
(BPSCCO) with a Cu sheath were fabricated and sintered after
the Cu sheath was removed. BPSCCO multifilamentary tapes with
a Ag sheath were also fabricated. They were sintered with the
Ag sheath attached. Magnetization measurements were made at
various temperatures and J_c was measured in magnetic fields
up to 23T in order to study the J_c-H characteristics and the
flux pinning mechanism of this material.

EXPERIMENTAL PROCEDURE

The BPSCCO powder prepared by a co-precipitation method
was packed in a metal (CU and Ag) tube of 10 mm o.d. (outer
diameter) and 6.7 mm i.d. (inner diameter), and was cold worked
into a wire of 2.5 mm o.d. (Cu sheath) and 0.7 mm o.d. (Ag
sheath). The 2.5 mm o.d. (Cu sheath) wire was then rolled into
a tape 4 mm x 1.5 mm in size. The nominal cation ratio for
this power was Bi:Pb:Sr:Ca:Cu=0.8:0.8:1.0:1.4. (Here, this
compositon is indicated by composition (i)). We used this nominal
ratio following ref 5. We also used a slightly different cation
ratio of Bi:Pb:Sr:Ca:Cu=0.92:0.17:1.05:1.12:1.5 (composition
(ii)), according to ref 6. Composition (ii), however, was
attempted only for the tapes with a Cu sheath. These powders
were calcined at 800°C for 14h in air before packing.

For fabrication of a multifilamentary wire, 46 pieces were
cut from the monofilamentary Ag-sheath wire of 0.7 mm o.d.,
packed into a Ag tube of 7.5 mm o.d. and 5 mm i.d., and cold
worked into a tape of 1.8 mm x 0.16 mm.

The tapes from which the Cu sheath was removed with nitric
acid after cold working were cut into short samples. They were
sintered at 820-850°C for 20-100h in a mixed gas (Ar:O$_2$ = 12:1)
and then they were pressed and sintered again under the same
conditions as outlined for the first sintering at 820-850°C
for 20-50h in air, cold worked and sintered again under the
same conditions as outlined for the first sintering.

In order to study the effect of impurity element addition
on the flux pinning mechanism at 77K, BPSCCO pellet samples,
to which Ag particles 1μm in diameter were added, were also
prepared.

The transition temperature, T_c, was measured both by the
standard induction method and the resistive method. The
temperature was measured by a Au-Fe/chromel thermocouple.
X-ray diffraction analysis was performed on these specimens.
The magnetization measurements were performed with a vibrating
sample magnetometer in magnetic fields up to 8T at various
temperatures. Rectangular specimens 0.5 x 3 x 4 mm were cut
from the sintered tape. Resistive J_c measurements were also
made on these specimens at 4.2K in magnetic fields up to 30T
with a hybrid magnet at Tohoku University.

RESULTS AND DISCUSSION

Figure 1 shows an X-ray diffraction pattern for the BPSCCO tape specimen which has been sintered at 845°C for 36h, pressed, and sintered again at 845°C for 36h. In Fig.1, an X-ray diffraction pattern for the powder specimen into which the tape specimen has been ground is also shown. Comparison of these patterns indicates that the process in which pressing and sintering are repeated produces strong orientation of the c axis in the tape specimen. We reported in our earlier paper[7] that the weak link problem has been almost solved by the effect of strong orientation.

T_c measurements showed that the tape specimens without the sheath and the multifilamentary tape specimens with a Ag sheath have T_c values of 105K and 102K, respectively.

Figure 2 shows magnetization versus magnetic field (M-H) curves (a) at 4.2K and (b) at 77K for the BPSCCO tape specimen. At 4.2K, the width of the hysteresis in the M-H curve was broad and it was not dependent on the magnetic field, as shown in Fig. 2 (a). This means that J_c at 4.2K produced by flux pinning is sufficiently high and is not dependent on magnetic field. At 77K, however the width of the hysteresis became almost zero above 0.3T. This indicates that the mechanism of flux pinning does not work above 0.3T at 77K. This clear difference between the hysteresis at 77K and that at 4.2K is considered to result from the thermal activation of fluxoids, i.e., flux creep.

According to the theory of flux creep, the relaxation of the magnetization dM/d(1nt) is as follows,

$$dM/d(1nt) = (d.J_c/30)(kT/U_0) \tag{1}$$

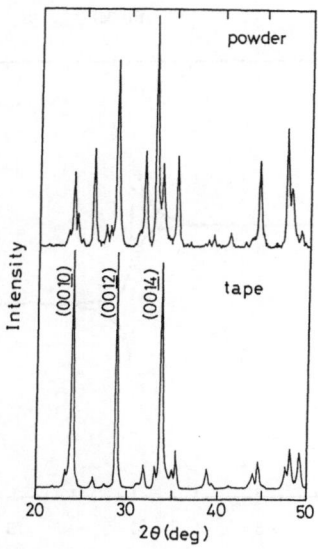

Fig. 1. X-ray diffraction patterns for the BPSCCO tape specimen and for the powder specimen into which the tape specimen has been ground (composition (ii)).

where d, k, T and U_0 are the dimension of the sample, Boltzmann constant, temperature and the flux pinning potential, respectively.

Figure 3 shows the relaxation of the normalized magnetization at various temperatures. In Fig. 3, magnetization decays almost linearly with lnt, indicating that this decay of the magnetization is caused by flux creep. At 4.2K, the relaxation of the magnetization was not observable. Figure 3 shows that the relaxation rate at 20K is much smaller than that at 50K or that at 60K. We obtained the pinning potential U_0 with the results shown in Fig. 3 using formula (1). U_0 values at 20K, 50K and 60K were 0.04eV, 0.054eV and 0.054eV, respectively. It is noteworthy that U_0 was almost constant in this temperature range. This result contradicts the theory of flux melting.[8] It is reported in ref 8 that the flux line lattice, FLL, becomes

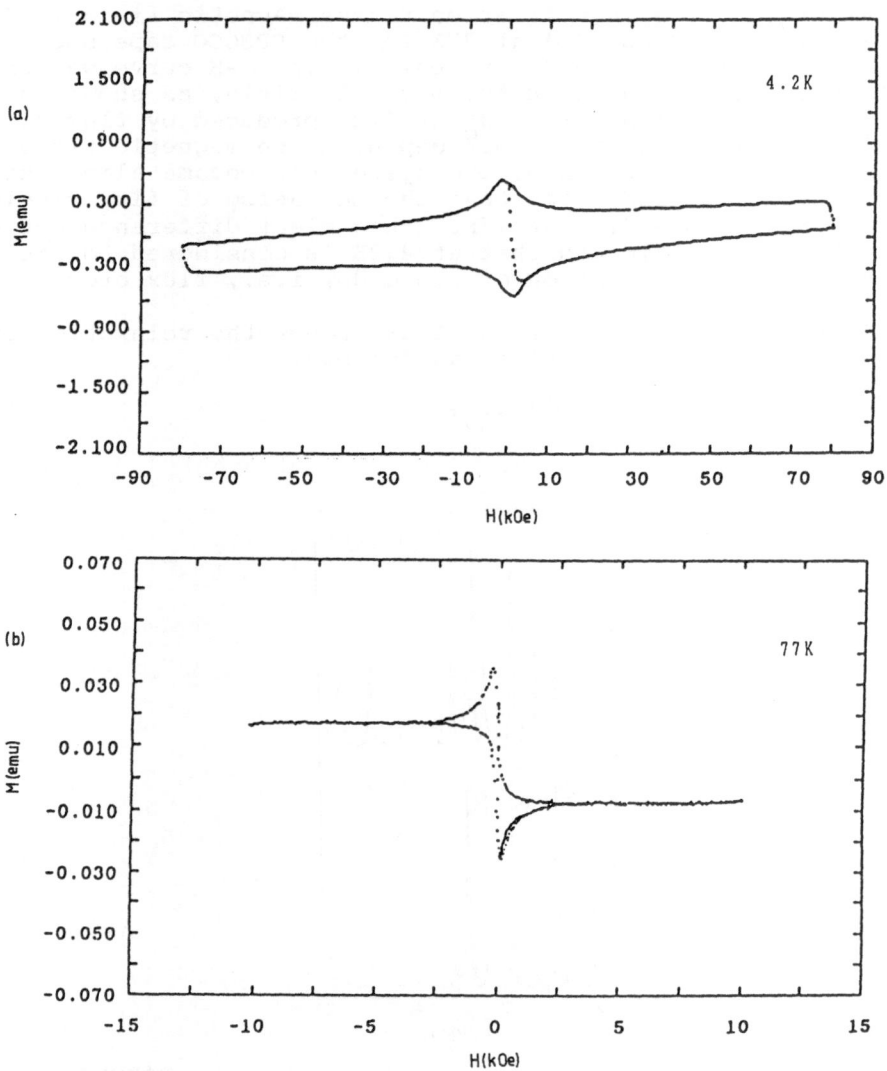

Fig. 2. Magnetization versus magnetic field curves at (a) 4.2K and (b) at 77K for the same BPSCCO tape specimen as shown in Fig. 1.

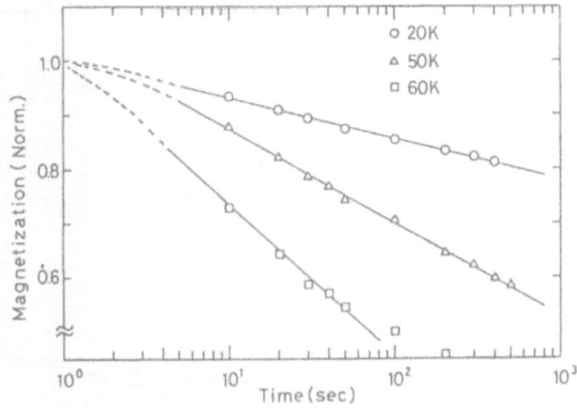

Fig. 3.   The relaxation of the normalized magnetization
          at various temperatures.

irregular as in an amorphous material at a temperature of T_m
(T_m = 40K in the BSCCO system).   If flux melting really occurs
at a temperature around 40K, U_o should undergo a sudden transition
also.   In our results, however, such a sudden change in U_o was
not observed.   At any rate, U_o must be enhanced by doping or
otherwise producing stronger pinning centers in order to prevent
flux creep.

     Figure 4 shows concentration maps of Ag and Bi for the
pellet samples to which 3 wt% Ag particles have been added.
Figure 4 shows that Ag particles have not reacted with the BPSCCO
matrix.   The Bi concentration map in Fig. 4 shows that, in addition to the Ag particles, the BPSCCO matrix includes inhomogeneous (non-superconducting) regions which are Bi poor.   We
expected these non-superconducting regions, i.e., Ag particles
and Bi-poor regions, to work as flux pinning centers.   The result
of the magnetization measurement at 77K, however, was totally
negative.   The magnetization curve for the BPSCCO pellet to
which Ag particles had been added indicated no sign of improvement in the flux pinning force.   There could be three reasons

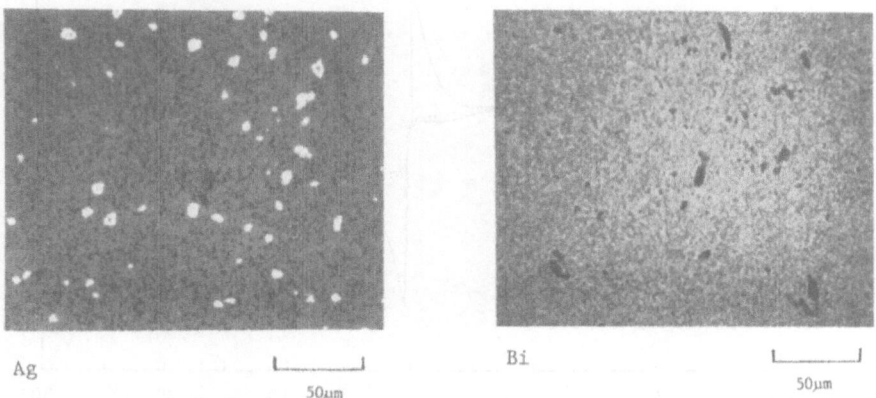

Fig. 4.   Concentration maps of Ag and Bi for the pellet samples
          to which 3 wt% Ag particles have been added
          (composition (ii)).

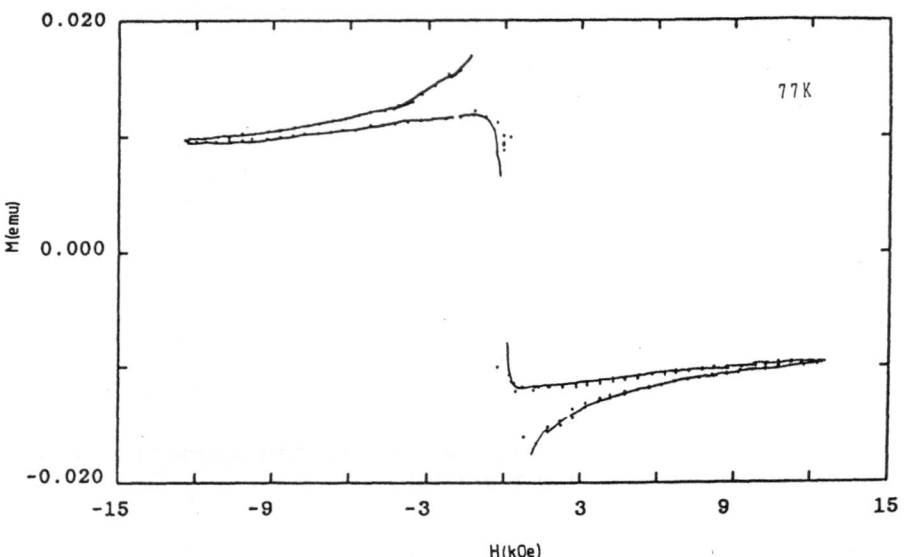

Fig. 5.   A magnetization versus magnetic field curve at 77K
for the BPSCCO pellet to which 1 wt% Al_2O_3 has been
added.

for these negative results at 77K. First, the added Ag particles
might be too coarse. According to the Kramer's theory[9] the flux
pinning force, f_0, between one fluxoid and one pinning center
becomes a maximum when the size of the pinning center is comparable to the coherence length ξ. In Fig. 4, the size of both
the Ag particles and the Bi-poor regions are of the order of
$1\mu m$. Their size must be ~100 times smaller in order to produce
the maximum f_0. Second, the flux melt may really occur below
77K, although our experimental data on U_o argue against flux
melting. Third, the weak link problem may still remain and
cause a degradation in U_o. Further investigation will be needed
on the problem of flux creep at 77K for the BPSCCO material.

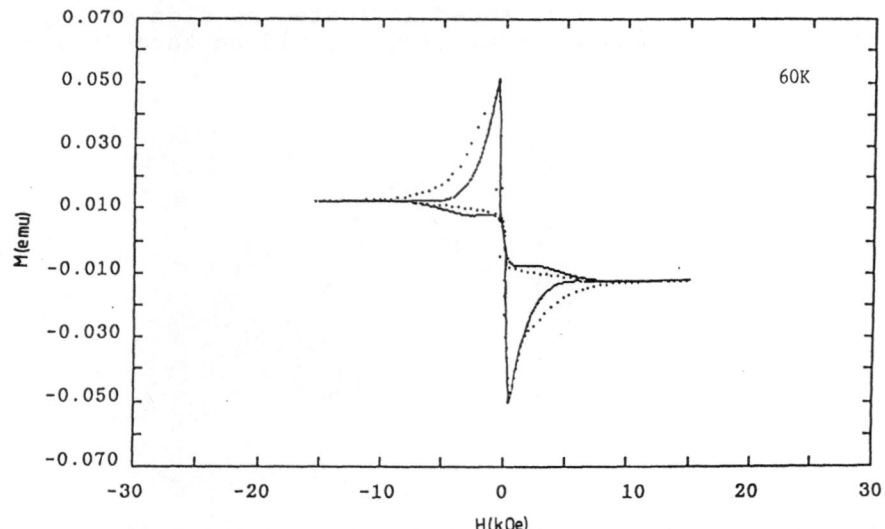

Fig. 6.   A magnetization versus magnetic field curve at 60K
for the 46-filament BPSCCO tape.

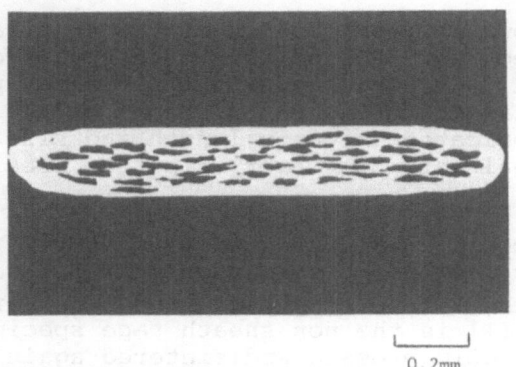

0.2mm

Fig. 7.   A cross section of the 46-filament BPSCCO tape which
has been cold worked, sintered, cold worked and then
sintered.

Figure 5 shows a magnetization versus magnetic field curve
at 77K for a BPSCCO pellet to which 1 wt%Al_2O_3 particles has
been added after the formation of the high T_c phase.   In
Fig. 5, a width to the hysteresis loop seems to have appeared.
However, Al_2O_3 tends to react with the BPSCCO matrix and to
degrade T_c and J_c.

Figure 6 shows a magnetization versus magnetic field curve
at 60K for the BPSCCO multifilamentary tape.   In Fig. 6, the
solid line and dotted line show the magnetization curves swept
at 2T/min and 0.2T/min, respectively. This result indicates
that flux creep occurs in the multifilamentary tape in the same
manner as in the pellets and monofilamentary tapes.

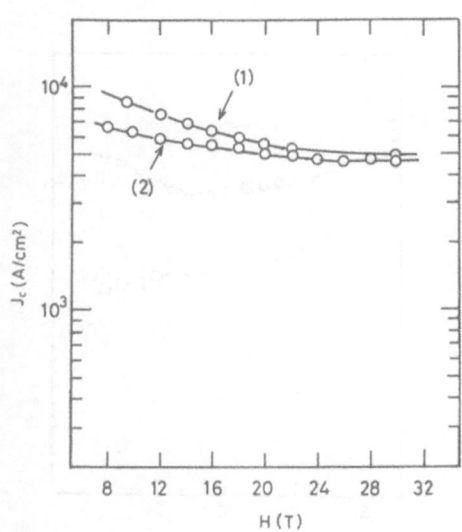

Fig. 8. J_c-H curves measured at 4.2K for the BPSCCO tape
specimens by a resistive method in magnetic fields
up to 30T. (1): the non-sheath tape and (2): the Ag-
sheath monofilamentary tape.

Figure 7 shows a cross section of the 46-filament BPSCCO tape which has been cold worked, sintered, cold worked again and then sintered. We have shown in our earlier paper[7] that the BPSCCO multifilamentary wires or tapes with a Ag sheath can be easily fabricated into any length and size by applying intermediate anneals at 150°C every time the areal reduction ratio becomes 10.

Figure 8 shows a J_c-H curve measured at 4.2 K for the BPSCCO tape specimens by a standard resistive method in magnetic fields up to 30T using the hybrid magnet of Tohoku University. In Fig. 8, specimen (1) is the non-sheath tape specimen sintered at 845°C for 36h, cold pressed and sintered again at 845°C for 36h. Specimen (2) is the monofilamentary tape (with Ag sheath) sintered at 840°C for 36h, cold worked with a flat roll and sintered again at 840°C for 36h. Specimen (1) was measured with a brass shunt for protection from heating. It is noteworthy that the J_c values of both specimens are almost constant in the range of magnetic field from 20T to 30T. This tendency in the J_c-H curve over 20T is quite a contrast to that of the conventional intermetallic superconductors. In general, J_c of the intermetallic compound superconductors decreases very rapidly above 20T. On the other hand, the upper critical field, H_{c2}, of the BPSCCO is ~100T at 4.2K when the magnetic field is applied parallel to the a-b plane, and so, this material is considered to have excellent J_c-H characteristics up to 80T-100T.

Figure 9 shows J_c-H curves measured at 4.2K for the BPSCCO mono- and multifilamentary tape specimens by a standard resistive method in magnetic fields up to 23T using the hybrid magnet at Tohoku University. In Fig. 9, specimen (a) is the non-sheath tape sintered at 845°C for 36h, cold pressed and sintered again at 845°C for 36h. Specimen (b) is the non-sheath tape for which cold pressing and sintering have been repeated twice after the

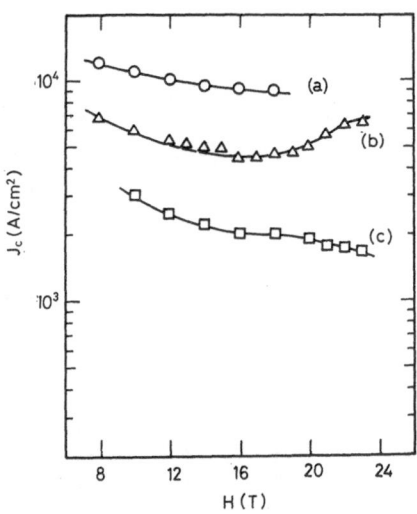

Fig. 9. J_c-H curves measured at 4.2K for the BPSCO mono- and multi-filamentary tape specimens by a resistive method in magnetic fields up to 23T. (a) and (b)are the non-sheath tape, and (c) the 46-filament tape.

first sintering, and specimen (c) is the multifilamentary tape sintered at 840°C for 36h, cold rolled and sintered again at 840°C for 36h. In specimens (a) and (b), the J_c values are as high as 10^4 A/cm^2 at 18T and 7×10^3 A/cm^2 at 23T, respectively. The J_c-H curve of specimen (b) shows a slight peak effect. Curve (c) shows that the J_c of multifilamentary tape is also independent of magnetic field at high fields.

Results of Fig. 8 and Fig. 9 indicate that the BPSCCO superconducting tapes have potential for practical use in extremely high magnetic fields at 4.2K. For example, they could be used as the innermost coil of a superconducting magnet of 25T or so in the near future. Although the J_c of 7000 A/cm^2 is not yet high enough, practical use will become possible if this value is doubled by optimizing such conditions as composition, fabrication process, sintering, grain size and so on.

CONCLUSIONS

Bi-(Pb)-Sr-Ca-Cu-O superconductor was fabricated into (a) tapes without a sheath and (b) mono- and multifilamentary tapes with a Ag sheath. In these specimens, the c axis tended to align well. Measurements of magnetization and J_c in magnetic fields up to 30T were made. The magnetization measurement revealed that, at 4.2K, these tape specimens had excellent J_c-H characteristics while at 77K flux pinning in the tape specimens was reduced to almost zero due to flux creep. In the resistive J_c measurement at 4.2K, the tape specimens without a sheath shows a J_c of 10^4 A/cm^2 at 18 T, and 7×10^4 A/cm^2 at 23T. The mono and multifilamentary tape specimens with a Ag sheath shows J_c's of 5×10^3 A/cm^2 at 30T and 2×10^3 A/cm^2 at 23T, respectively. In both the tapes without a sheath and the ones with a Ag sheath, the J_c values scarcely depended on the magnetic field from 20T to 30T. These results indicate that the BPSCCO tapes could be practically used in magnetic fields near 30T, or in higher fields, if the present J_c values are doubled by optimizing such conditions as composition, fabrication process, sintering, grain size and so on. The BPSCCO tapes could be used, for example, for the innermost coils of an extremely high-field superconducting magnet in the near future.

ACKNOWLEDGEMENTS

The authors would like to thank prof. Watanabe of Tohoku University for giving us a chance to measure J_c with the hybrid magnet.

REFERENCES

1. H. Maeda, Y. Tanaka, M. Fukutomi and T. Asano, A new high-T_c oxide superconductor without rare earth element, Jpn. J. Appl. Phys. 27:L209 (1988).
2. R. J. Cava, B. Batlogg, S. A. Sunshine, T. Siegrist, R. M. Fleming, K. Rabe, L. F.Schneemeyer, D. W. Murphy R. B. van Dover, P. K. Gallaghers, S. H. Glarun, S. Nakahara, R. C. Farrow, J. J. Krajewski, S. M. Zahurak, J. V. Waszczak, J. H. Marshall, P. Marsh, L. W. Rupp,Jr., W. F. Peck and E. A. Rietman, Studies of oxygen-deficient

Ba$_2$YCu$_3$O$_{7-\bar{\delta}}$ and superconductivity Bi(Pb)-Sr-Ca-Cu-O, Physica C 153-155:560 (1988).

3. S. Green, C. Jian, Yu Mei, H. L. Luo and C. Politis, Zero resistance at 107 K in the (Bi, Pb)-Ca-Sr-Cu oxide system, Phys. Rev. B 38:5016 (1988).

4. M. Takano, J. Takada, K. Oda, H. Kitaguchi, Y. Miura, Y. Ikeda, Y. Tomii and H. Mazaki, High-T$_c$ phase promoted and stabilized in the Bi,Pb-Sr-Ca-Cu-O system, Jpn. J. Appl. Phys. 27:L1041 (1988).

5. U. Endo, S. Koyama and T. Kawai, Preparation of the high-T$_c$ phase of Bi-Sr-Ca-Cu-O superconductor, Jpn. J. Appl. Phys. 27:L1476 (1988).

6. S. Koyama, U. Endo and T. Kawai, Preparation of single 110 K phase of Bi-Sr-Ca-Cu-O superconductor, Jpn. J. Appl. Phys. 27:L1861 (1988).

7. H. Sekine, K. Ogawa, K. Inoue, H. Maeda and K. Numata, Metallurgical studies and optimization of critical current density in Bi-(Pb)-Sr-Ca-Cu-O superconductors, Jpn. J. Appl. Phys. to be published in Aug. or Sept. in 1989.

8. P. L. Gammel, L. F. Schneemeyer, J. W. Waszczak and D. J. Bishop, Phys. Rev. Lett. 61:1666 (1988).

9. E. J. Kramer, C. L. Bauer, Phil. Mag. 15:1189 (1967).

SUPERCONDUCTING PROPERTIES OF Bi-Pb-Sr-Ca-Cu-O SYSTEM

M. Wakata, S. Miyashita, S. Kinouchi, T. Ogama,
K. Yoshizaki

Materials & Electronic Devices Lab.
Mitsubishi Electric Corp.
Sagamihara, Kanagawa 229, Japan

S. Yokoyama, K. Shimohata, and T. Yamada

Central Research Lab., Mitsubishi Electric Corp.
Amagasaki, Hyogo 661, Japan

ABSTRACT

The c-axis oriented Bi-Pb-Sr-Ca-Cu-O samples have been prepared by
sintering and several times of pressing and re-sintering procedure. The
repeat of pressing and sintering improved the critical current density as
a function of the field (B), $J_c(B)$. It is found that $J_c(B)$ is given by
$J_c(0)(B/B^*)^{-a}$ at low magnetic fields (a few mT < B < several 10mT) and by
$J_0 \exp(-B/B_0)$ at high fields (0.05T < B < several T) at 77K. By the
pressing and sintering procedure, the fitting parameters of B^*, B_0 and J_0
are increased and the power index of a is decreased. The field direction
dependence of those parameters is also shown. The V-I characteristics up
to a current density much higher than $10^2 \times J_c$ have been measured by using
a pulsed current. It is found that the intra-grain J_c is at least higher
than 2.3×10^4 A/cm^2 at 77K and zero external field and the resistance in
the weak link regions is decreased as decreasing the temperature. The
latter fact is consistent with our proposal that the weak link regions are
not any insulators but a superconductor.

INTRODUCTION

In Bi-Sr-Ca-Cu-O system,[1] there exist two superconducting phases with
the critical temperature (T_c) of 110K (2223 phase) and 80K (2212 phase)
and a semiconducting or superconducting phase with T_c of about 8K[2,3] (2201
phase). Lead doping is effective to synthesize 2223 phase and to obtain
$T_c(\rho=0)$ higher than 100K.[4] In our previous study,[5,6] another superconducting transition at 40 to 60K is found in the temperature dependence
of the susceptibility for Pb-free samples. We call it a-1 phase. The
structure can not be detected by X-ray diffraction analysis (XD) and the
T_c depends on the nominal composition and the sintering condition. By Pb
doping the transition due to a-1 phase disappears. However, another
transition at about 95K appears (a-2 phase). Similar to a-1 phase the
structure can not be detected by XD and the T_c depends on the nominal

Table 1. The preparation procedure of the samples

Series No.	Process No.	Process
1	0	Sintered at 845°C for 60hr
	1	Process 1-0 + P + Process 1-0
	2	Process 1-1 + P + Process 1-0
2	0	Sintered at 845°C for 30hr
	1	Process 2-0 + P + Process 2-0
	2	Process 2-1 + P + Process 2-0

P : Pressed at 2,000kgf/cm^2

composition and the sintering condition. In our recent study, it is found that the T_c is varied from 80 to 110K. The transition point (onset) and the width strongly depend on the magnetic field. Furthermore, such a transition can not be observed for the powder obtained by grinding sintered samples.

In oxide superconductors, it is well known that critical current densities (J_c) for the sintered materials are restricted by weak links. In Bi-Pb-Sr-Ca-Cu-O system we have proposed that a-2 phase should be responsible for the weak links. Recently, it has been reported that several times of pressing and resintering are effective to increase J_c and to improve the field dependence.[7] Such procedure enhances the crystal preferred orientation and then the density of the sample because each grain has a plate like shape and the surface is perpendicular to the c-axis. It is interesting to know how the weak links are affected by the preferred orientation and/or densification of the samples.

In this study, we have prepared oriented samples with various J_c's and measured the field dependence of the J_c in order to investigate how the behavior due to the weak links is changed by improving the J_c and how it depends on the field direction. The voltage-current (V-I) characteristics up to much higher current region have been also measured by using a pulsed current in order to estimate the intra-grain current density and the temperature and the field dependence of the resistance in weak link regions.

SAMPLE PREPARATION

The raw materials are Bi_2O_3, PbO, $SrCO_3$, $CaCO_3$ and CuO powders with a purity of 4N and a mean diameter of a few μm. The nominal compositions are $Bi_{1.6}Pb_{.4}Sr_2Ca_3Cu_4O_y$ (Sample A) and $BiPb_{.2}SrCaCu_2O_y$ (Sample B). These powders were mixed by ball milling, pressed at 500kgf/cm^2, calcined at 750°C for 12hr in air and ground by shaking milling. The obtained powders were pressed at 1,000kgf/cm^2, sintered at 845°C for 120hr in air and reground by shaking milling. The scanning electron microscopic observation and X-ray diffraction analysis revealed that the obtained powder particles had plate like shapes and the surface of the plates was parpendicular to c-axis. The powders are pressed at 1,000kgf/cm^2 into bar shaped pellets with 2mm in width, 0.5 to 1mm in thickness and 20mm in length. Subsequent two series of processes are shown in Tabel 1, where the sintering is performed in air and every pressing procedure is done by using the same dies. The sample name is indicated as the nominal composition (A or B)-series number (1 or 2)-process number (0, 1 or 2).

Table 2. The properties of the samples with series No. 1

Sample No.	A-1-0	A-1-1	A-1-2	B-1-0	B-1-1	B-1-2
$T_c(\rho=0)$; K	104.7	106.5	106.6	102.7	105.0	105.7
$\rho(T_c^{on})$; m$\Omega\cdot$cm	1.76	0.870	0.828	0.909	0.543	0.525
$T_c(X'')$; K	103.2	105.0	105.5	101.9	104.7	104.9
$J_c(0)$; A/cm^2	351	807	789	257	620	741

EXPERIMENTAL RESULTS

Samples with the series number 1 were used for the measurements of the temperature dependence of the resistivity and susceptibility, and the J_c in high field range up to 0.4T. The critical currents were determined by 1µV/cm criterion. Table 2 shows the summary of the data of the T_c ($\rho=0$), the resistivity at T_c^{on}, $\rho(T_c^{on})$, the temperature at which the imaginary part of the susceptibility is maximum at the field of 0.02Oe, $T_c(X'')$ and the J_c at 77K and zero externel field, $J_c(0)$. As shown in this table the repeat of pressing and sintering increases $T_c(\rho=0)$, $T_c(X'')$ and $J_c(0)$ and decreases $\rho(T_c^{on})$.

Figure 1 shows the field dependence of the J_c normalized by the $J_c(0)$, where P.S. means the pressed surface of the sample. As shown in this figure almost all the data in higher field than 0.05T are followed by the relation:

$$J_c(B) = J_0 \exp(-B/B_0) \tag{1}$$

The calculated parameters, B_0 and J_0 are summarized in Table 3. The

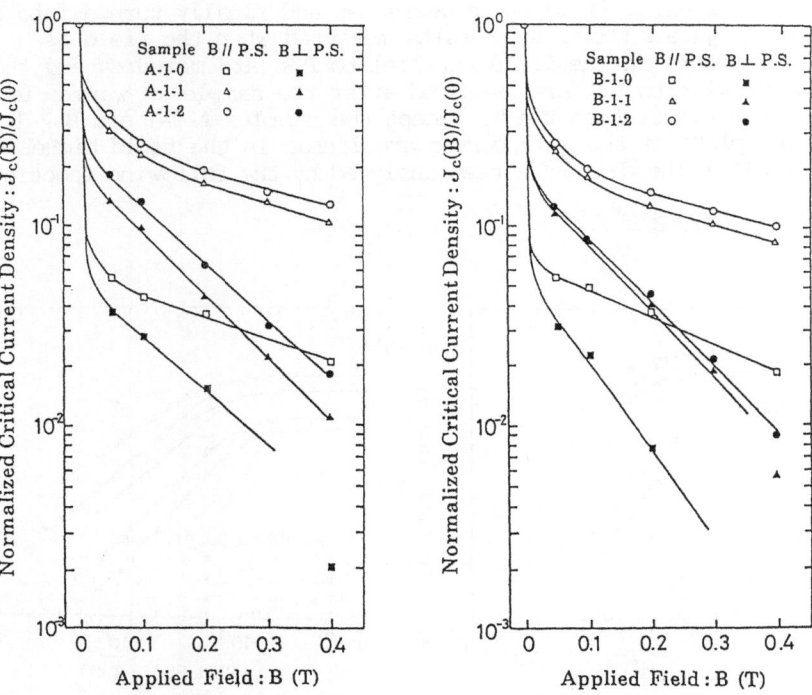

Fig. 1 The J_c-B properties at high fields for the samples with series No.1.

Table 3. The fitting parameters of the J_c-B properties for the samples with series No. 1. (0.05T < B < 0.4T)

Sample No.	B // P.S.		B ⊥ P.S.	
	J_0 (A/cm^2)	B_0 (T)	J_0 (A/cm^2)	B_0 (T)
A-1-0	22.3 ± 0.6	0.36 ± 0.02	22.8 ± 2.8	0.118 ± 0.007
A-1-1	181 ± 3	0.46 ± 0.01	164 ± 1	0.138 ± 0.001
A-1-2	210 ± 16	0.50 ± 0.06	212 ± 2	0.141 ± 0.007
B-1-0	18.0 ± 0.5	0.30 ± 0.01	14.8 ± 1.3	0.102 ± 0.006
B-1-1	125.7 ± 0.1	0.451 ± 0.001	117 ± 3	0.132 ± 0.002
B-1-2	131 ± 5	0.525 ± 0.004	135 ± 3	0.141 ± 0.002

characteristic fields of B_0 are 0.30 to 0.53T in the field parallel to the P.S., which are more than 3 times of those in the field perpendicular to the P.S.. By the repeat of pressing and sintering B_0 is increased by 40 (20) % for the sample A and by 75 (38) % for the sample B in the field parallel (perpendicular) to P.S., respectively. On the other hand, J_0 is almost independent of the direction of the field. However, the value is increased about 10 times by the repeat of pressing and sintering. In this analysis we used the data up to 0.4T. But it is confirmed that the relation (1) holds at least up to a few T.

As shown in Fig. 1, the improvement of the field dependence of J_c by the repeat of pressing and sintering is remarkable, especially in the field lower than 0.05T. Then the J_c's in lower field have been measured for samples with the series number 2. The J_c-B curves have a hysterysis, that is to say, the J_c's measured in the decreased field are higher than those in the increasing field. The value of the hysterysis is initially increased as decreasing field, then decreased and finally turned into a minus value. Figure 2 shows the results measured when the field is increased. The data in the field parallel to P.S. are measured and then those perpendicular to P.S. are measured after the sample is warmed up to the temperature higher than the T_c except the samples A-2-1 and B-2-1. These log-log plots of the J_c-B curves are linear in the field higher than a few mT, so that the data have been analyzed by the following function:

$$J_c(B) = J_c(0) \, (B/B^*)^{-a} \tag{2}$$

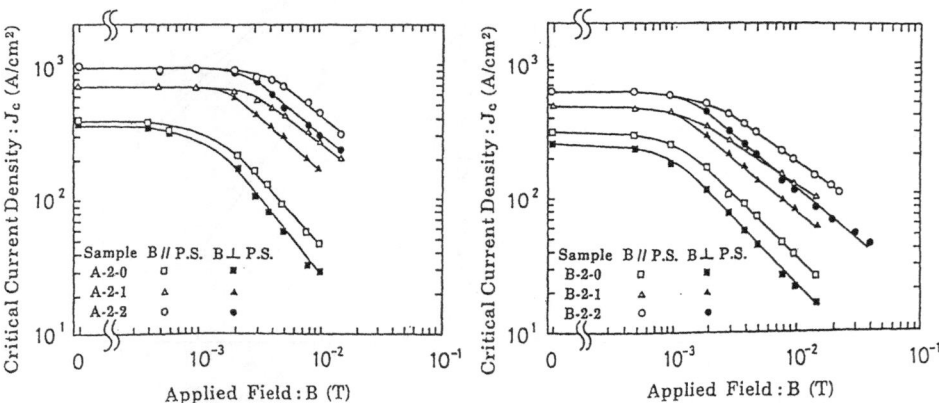

Fig. 2 The J_c-B properties at low fields for the samples with series No.2.

Table 4. The fitting parameters of the J_c-B properties of the samples with series No. 2. (1mT < B < 30mT)

| Sample No. | B // P.S. | | | B ⊥ P.S. | | |
	a	B^* (mT)	$J_c(0)$ (A/cm^2)	a	B^* (mT)	$J_c(0)$ (A/cm^2)
A-2-0	1.02 ± 0.04	1.22	383	1.26 ± 0.05#	1.12#	360#
A-2-1	0.68 ± 0.02	2.47	696	0.76 ± 0.02	1.57	694
A-2-2	0.74 ± 0.09	3.14	983	0.74 ± 0.04	2.07	967
B-2-0	0.92 ± 0.02	0.98	307	1.04 ± 0.02#	0.90#	251#
B-2-1	0.63 ± 0.03	1.09	485	0.80 ± 0.02	0.97	489
B-2-2	0.67 ± 0.01	1.57	614	0.84 ± 0.02	1.23	621

: include residual magnetization.

The calculated parameters, a and B^*, are shown in Table 4 with the data of $J_c(0)$. In this case the characteristic field, B^* has order of 1mT and depends on the field direction. However, such dependence is smaller than that of B_0. The value of B^* is increased by pressing and sintering. On the other hand the power index, a has a tendency to be decreased by such procedure. The dependence on the field direction is also small.

Ekin et al.[8] have reported that the V-I curves for sintered Y-Ba-Cu-O samples are linear in a current range higher than those critical currents, however, the resistivities are much lower than those normal state resistivities. We have also observed the similar phenomena, so measured the V-I characteristics up to much higher current range and found two-step transition as shown in Fig. 3.[9,10] In this figure we interpret that ρ_1 is the resistivity when only the weak link regions are in normal state and J_{c2} is the critical current density when the superconductivity in the high J_c regions begins to be broken.

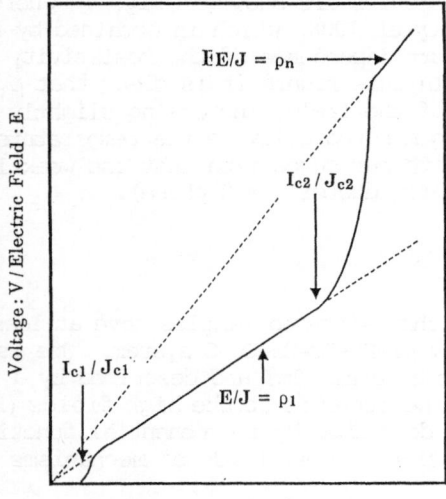

Fig. 3 Typical V-I characteristics for a poly-crystalline oxide superconductor.

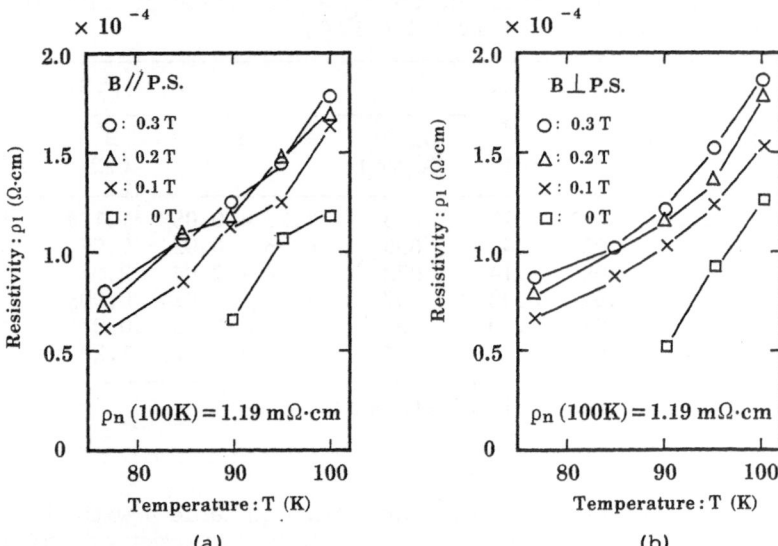

Fig. 4 The temperature and field dependence of ρ_1 for the sample A-1-1. The field is parallel (a) and perpendicular (b) to the P.S.

In order to estimate the J_{c2} in Bi-Pb-Sr-Ca-Cu-O system we have been trying measurements by using a pulsed current to avoid the heat generated mainly on current contacts. Unfortunately, we have obtained the data only for the non-oriented and porous samples with the density less than a half of the theoretical value. The J_{c2} at 77K and zero external field was more than 2.3 x 10^4 A/cm^2. Therefore, it can be estimated that the attainable J_c in Bi-Pb-Sr-Ca-Cu-O system is the same order as that in Y-Ba-Cu-O system at 77K and zero external field or much higher.

In this paper we report on the temperature and the field dependence of the ρ_1 for the sample A-1-1. The half width of a pulsed current was 0.1msec. The temperature was controlled by changing the pressure of liquid nitrogen. The results are shown in Fig. 4, where $\rho_n(100K)$ is the normal state resistivity at 100K, which is obtained by the linear extrapolation of the temperature dependence of the resistivity at higher temperature than T_c(onset). In this figure it is clear that ρ_1 is almost independent of the direction of the field, increasing slightly as the field increases, and decreasing monotonically as the temperature decreases. These facts are consistent with our suggestion that the weak link regions are not insulators but superconductors (a-2 phase).

DISCUSSION

It becomes clear that sintered samples have at least two kinds of the J_c-B characteristics in Bi-Pb-Sr-Ca-Cu-O system. One is at the low fields (L.F.) from a few mT to several 10mT and described by a power law as shown in the equation (2). The other is at the high fields (H.F.) from about 0.05T to several T and described by a exponential function as shown in the equation (1). The question is what kinds of mechanisms determine such J_c-B characteristics.

Peterson and Ekin[11] have reported that the J_c's for sintered Y(Ho)-Ba-Cu-O fit well with $B^{-3/2}$ at L.F. (<10mT). They assumed the weak links to be Josephson type junctions and analyzed the field dependence of the J_c

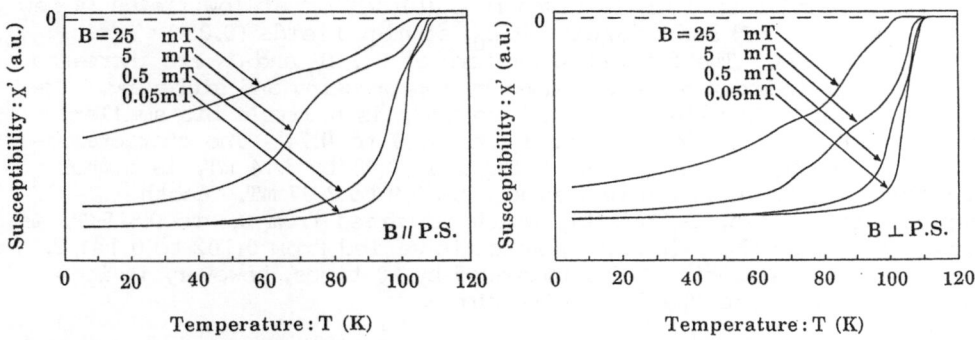

Fig. 5 The temperature and field dependence of the susceptibility for a typical oriented sample.

as follows: If the junction is a circular disk, it is given by the Airy diffraction pattern, varies as $B^{-3/2}$ apart from the oscillation. If the junction is rectangular, it is given by the generalized Fraunhofer pattern, varies as B^{-1} to B^{-2}, depending on the field direction. In our results at L.F. in Bi-Pb-Sr-Ca-Cu-O system as shown in Table 4, the power index, a, depends on the sample and the field direction, and varies from 1.26 to 0.63. Therefore, such field dependence can not be determined by any Joshephson junctions.

In order to discuss our results we have to consider the self field and the magnetic properties in weak link regions. In our samples the self field is about 1mT for the current density of 300A/cm^2. Therefore, the characteristic field, B^* is similar to the self field for the current density of $J_c(0)$. It is possible that the J_c's at lower field than 1mT do not fit with the relation (2) because the effect of the self field is strong only in that field range.

The ρ_1, which can be interpreted to be the resistivity when only weak link regions are in normal state, is decreased as decreasing the temperature. Therefore, it can be concluded that the weak link regions are not any insulators. It is consistent with our proposal that the weak link regions are a superconductor (a-2 phase). Then the field dependence of a-2 phase is important to discuss the field dependence of J_c. Figure 5 shows the temperature and field dependence of the susceptibility for a typical oriented sample. Two superconducting transitions are clearly seen, which are due to 2223 phase and a-2 phase. The ratio of the change of the susceptibility due to 2223 phase depends on the field direction, but the field at which the superconductivity of a-2 phase is broken does not and is about 25mT at 77K. This field is in an intermediate range between L.F. and H.F.. This fact suggests that the field dependence of the J_c in L.F. reflects the superconducting properties of a-2 phase and that in H.F. is determined by S-N-S junctions. However, further investigations are needed to make it clear, for example, the temperature dependence of the $J_c(B)$.

CONCLUSIONS

By using superconducting powders of which particles have plate-like shape, the c-axis oriented Bi-Pb-Sr-Ca-Cu-O samples have been prepared by sintering and 0, 1 and 2 times of pressing and re-sintering procedure. The repeat of pressing and sintering improved J_c and the field dependence.

It is found that $J_c(B)$ is given by $J_c(0)(B/B^*)^{-a}$ at low fields (a few mT < B < several mT) and by $J_0 exp(-B/B_0)$ at high fields (0.05T < B < several T) at 77K. The fitting parameters of B^*, B_0 and J_0 are increased and the power index of a is decreased by the pressing and sintering. The a(B//P.S.), which is varied from 1.02 to 0.63, is a little bit smaller than the a(B⊥P.S.), which is varied from 1.26 to 0.74. The characteristic field, B^*(B//P.S.), which is varied from 0.98 to 3.14 mT, is higher than the B^*(B⊥P.S.), which is varied from 0.9 to 2.07 mT. Another characteristic field, B_0(B//P.S.), which is varied from 0.3 to 0.525 T, is much higher than the B_0(B⊥P.S.), which is varied from 0.102 to 0.141 T. By pressing and sintering J_0 is increased by 10 times, however, it is almost independent of the field direction.

In this system J_c is limitted by the weak links which are not any insulators but may be a superconductor (a-2 phase). The T_c depends on the nominal composition and the sintering condition and varies from 80 to 110K. It is very sensitive to the external magnetic field. In this experiment the superconductivity is broken by a field of about 25mT at 77K. Therefore, the $J_c(B)$ at low fields should be reflected by the superconducting properties of a-2 phase. Furthermore, it is estimated that the $J_c(B)$ at high fields is limited by the weak links of S-N-S type. Further investigations are needed to make it clear, for example, the temperature dependence of the J_c-B characteristics.

In this study it is found that the crystal preferred orientation and densification is effective to suppress the effect of weak links and to improve the J_c-B characteristics in Bi-Pb-Sr-Ca-Cu-O system. If the weak links can be removed by further densification, the J_c much higher than 2.3 x 10^4 A/cm² should be achieved at 77K and zero external field.

REFERENCES

1. H. Maeda, Y. Tanaka, M. Fukutomi, and T. Asano, Jpn. J. Appl. Phys. Lett. 27 (1988) L209
2. J. Akimitsu, A. Yamazaki, H. Sawa, and H. Fujiki, Jpn. J. Appl. Phys. Lett. 26 (1987) L2080
3. C. Michel, M. Hervieu, M. M. Borel, A. Grandin, F. Deslandes, J. Provost, and B. Raveau, Z. Phys. B68 (1987) 421
4. M. Takano, J. Takada, K. Oda, H. Kitaguchi, Y. Miura, Y. Ikeda, Y. Tomii, and H. Mazaki, Jpn. J. Appl. Phys. Lett. 27 (1988) L1041
5. S. Miyashita, M. Wakata, A. Nozaki, K. Egawa, T. Ogama, and K. Yoshizaki, IEEE Trans. on Mag. MAG-25 (1989), to be published
6. M. Wakata, S. Miyashita, K. Egawa, H. Higuma, T. Ogama, K. Yoshizaki, S. Yokoyama, K. Shimohata, M. Morita, and T. Yamada, Program & Abstracts of ISTEC Workshop on Superconductivity (Oiso, 1989) p39
7. T. Asano, Y. Tanaka, M. Fukutomi, K. Jikihara, and H. Maeda, Jpn. J. Appl. Phys. Lett. 28 (1989) L595
8. J. W. Ekin, A. I. Braginski, A. J. Panson, M.A. Janocko, D. W. Capone II, N. Zaluzec, B. Flandermeyer, O. F. de Lima, M. Hong, J. Kwo, and S. H. Liou, J. Appl. Phys. 62 (1987) 4821
9. M. Wakata, H. Higuma, S. Matsuno, K. Egawa, and K. Yoshizaki, The Fifth Seminar on Frontier Technology, -Chemical Aspects of High-T_c Superconductors-, The Association for the Progress of New Chemistry, (Shuzenji, 1988)
10. K. Shimohata, S. Yokoyama, M. Morita, T. Yamada, and M. Wakata, Proc. of Int. Symp. on New Developments in Applied Superconductivity (Osaka, 1988) p238
11. R. L. Peterson and J. W. Ekin, Physica C (1988), to be published

INVESTIGATION OF NOBLE METAL SUBSTRATES AND BUFFER
LAYERS FOR BiSrCaCuO THIN FILMS

M.M. Matthiesen, L.M. Rubin, K.E. Williams,
D.A. Rudman

Department of Materials Science and Engineering
Massachusetts Institute of Technology
Cambridge, MA 02139

ABSTRACT

Noble metal buffer layers and substrates for $Bi_2Sr_2CaCu_2O_8$ (BSCCO) films were investigated using bulk ceramic processing and thin film techniques. Highly oriented, superconducting BSCCO films were fabricated on polycrystalline Ag substrates and on Ag/MgO and Ag/YSZ structures. Such films could not be produced on Au or Pt substrates under any annealing conditions. In addition, superconducting BSCCO films could not be produced on Ag/Al_2O_3, $Ag/SiO_2/Si$, or Ag/(Haynes 230 alloy) structures using high annealing temperatures (870°C). However, oriented, although poorly connected, superconducting BSCCO films were fabricated on Ag/Al_2O_3 structures by using lower annealing temperatures (820°C). Once lower processing temperatures are optimized, Ag may be usable as a buffer layer for BSCCO films.

INTRODUCTION

There are many potential applications for thin films of high temperature oxide superconductors. However, many of these applications require specific substrates. For example, interconnect applications require substrates such as silicon, gallium arsenide, or other materials, such as silicides, insulators, and metals, that are compatible with these semiconductor materials. Similarly, strip line resonator applications require substrates, such as sapphire (Al_2O_3), that exhibit low loss at microwave frequencies (10 GHz), and RF cavity applications require structural materials, such as steel or other alloys, that can be fabricated into complicated forms.

Currently, the choice of substrates for high temperature oxide superconductors, and therefore their applications, have been severely limited by the reactive nature of these materials. For example, while Bi-Sr-Ca-Cu-O (BSCCO) superconductor films have been successfully fabricated on $SrTiO_3$, MgO, and yttria-stabilized zirconia (YSZ) substrates.[1,2] BSCCO films have been

Advances in Cryogenic Engineering (Materials), Vol. 36
Edited by R. P. Reed and F. R. Fickett
Plenum Press, New York, 1990

611

found to react with Al_2O_3, Si, and SiO_2 substrates.[1,3] Clearly, buffer layers that are compatible with both BSCCO films and these technologically important substrates must be identified. To date, most of the work on buffer layers has focused on insulating materials such as ZrO_2, $SrTiO_3$, and MgO.[3,4,5] However, applications such as interconnects will require conducting buffer layers, and some work has begun on metallic materials.[6,7]

This investigation focused on the use of noble metals, such as silver, gold, and platinum, as substrates and buffer layers for BSCCO films. The chemical compatibility of the noble metals with BSCCO was determined using two methods. Bulk studies were conducted using noble metal and BSCCO powders that were pressed into pellets and then sintered.[8] Thin film studies were conducted by fabricating BSCCO films on noble metal substrates. In both studies, x-ray diffraction and magnetic susceptibility measurements were used to evaluate the samples. In addition, the effectiveness of noble metal buffer layers between BSCCO films and a variety of substrates was investigated by fabricating and annealing BSCCO/Ag/substrate structures. X-ray diffraction and resistive transport measurements were used to evaluate the BSCCO films in these multilayer structures.

These studies indicated that only Ag is promising as either a substrate or a buffer layer for BSCCO films. Preliminary results, which are in agreement with those of other researchers,[6] showed that both Au and Pt react with BSCCO. Consequently, only studies involving Ag will be presented in this paper. Highly oriented superconducting BSCCO films were fabricated on polycrystalline Ag substrates and on Ag/MgO and Ag/YSZ structures. However, it was not possible to fabricate such films on Ag/Al_2O_3, Ag/SiO_2, or Ag/(Haynes 230 alloy). In these cases, the Ag layer failed to protect the BSCCO films from the underlying substrate. The mechanism by which the Ag allowed the BSCCO film to react with the underlying substrate is currently under investigation.

$Bi_2Sr_2CaCu_2O_8$ FILMS

Amorphous BSCCO films approximately 0.75 μm thick were deposited by triode magnetron sputtering from three metal targets, in the presence of oxygen, onto room temperature substrates. A more detailed discussion of the sputtering system has been presented in an earlier publication.[1] All films had a composition near $Bi_2Sr_2CaCu_2O_x$. After deposition, the amorphous BSCCO films were annealed in a tube furnace with a flowing mixture of oxygen and argon. Because the processing window for forming the superconducting 2212 phase is quite large, several combinations of anneal temperature and anneal ambient were used to obtain this desired phase. In general, lower oxygen partial pressures require lower anneal temperatures, while higher oxygen partial pressures require higher anneal temperatures.[2]

In this study, two types of noble metal structures were used: 1) polycrystalline Ag foils (0.25 mm); and 2) polycrystalline Ag films deposited on a variety of substrates. Prior to deposition, some of the foils were annealed in 100% Ar at temperatures between 700°C and 900°C for 1 hour. The purpose of this pre-anneal was to remove any dissolved gases and any impurities that become volatile at high temperatures. The Ag

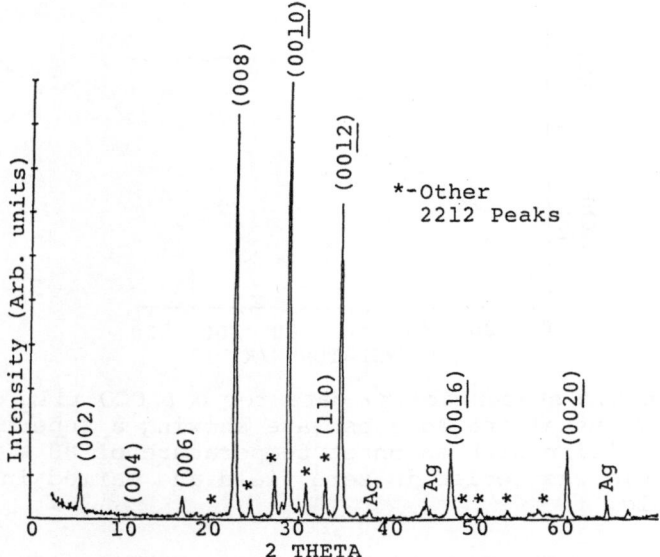

Figure 1. X-ray diffraction pattern for a BSCCO film on a
polycrystalline Ag substrate. The 2212 phase in the
BSCCO film is highly oriented, as indicated by the
(0,0,10):(1,1,0) ratio of 6.

buffer layers were deposited by electron beam evaporation onto
non-reactive substrates, including (100) MgO and (100) YSZ, and
reactive substrates, including Al_2O_3, SiO_2/(100)Si, and
polycrystalline Haynes 230, which is a nickel-based alloy.[9] Both
the BSCCO/substrate and BSCCO/Ag/substrate structures were then
annealed under a variety of conditions in order to investigate
the effects of temperature and oxygen partial pressure on the
phase formation and morphology of these multilayer structures.

Noble Metal Substrates: Highly oriented,
polycrystalline superconducting BSCCO films were fabricated on
polycrystalline Ag substrates. Figure 1, which presents the x-
ray diffraction pattern for a BSCCO/Ag sample that was annealed
at 870°C for 30 minutes in 20% O_2/80% Ar, shows that the BSCCO
film contains only the 2212 phase, most of which is oriented
with the c-axis normal to the plane of the film. The degree of
orientation was evaluated in terms of the ratio of the (0,0,10)
peak to the (1,1,0) peak of the 2212 phase. For a randomly
oriented 2212 sample, the (0,0,10):(1,1,0) ratio is 0.22.[10] For
the sample in figure 1, the ratio is 6, indicating a very high
degree of preferential orientation. Figure 2, which presents
the magnetic susceptibility data for the same sample, indicates
that the onset temperature of the superconducting transition is
80 K, but that the transition is somewhat broad. Thus,
polycrystalline Ag shows a great deal of promise as a substrate
for superconducting BSCCO films.

However, as shown in figure 3a, BSCCO films on Ag
substrates exhibit unusual morphologies. Although the amorphous
BSCCO films are smooth, the annealed BSCCO films develop hollow
bubbles, approximately 10 μm to 20 μm in diameter, which are
covered with platelet grains that are characteristic of the 2212
phase. These bubbles formed on all BSCCO/Ag samples that were
annealed at 870°C, regardless of the previous thermal histories

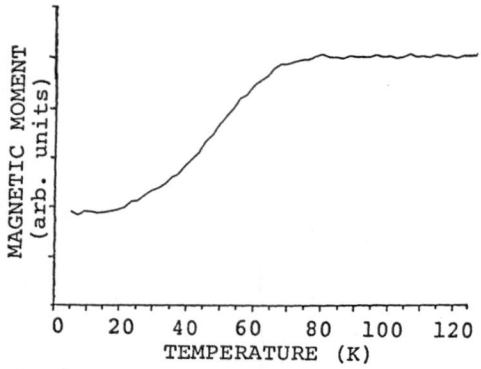

Figure 2. Magnetic susceptiblity data for a BSCCO film on a
polycrystalline Ag substrate showing a superconducting
transition with an onset temperature of 80 K. The
sample was cooled in zero field and warmed in a 50 G
field.

of those samples. The formation of these bubbles also appeared
to be independent of the oxygen partial pressure during the
anneal. In contrast, lower temperature anneals (820°C) produced
smooth BSCCO films. However, these films exhibited less
preferential orientation and degraded superconducting
properties. From these observations, it was concluded that the
bubbles were caused by the volatilization of impurities in the
Ag foils. Based on this conclusion, Ag foils were pre-annealed
in order to outgas any volatile impurities.

When a BSCCO film on a pre-annealed Ag substrate was
annealed at 870°C for 30 minutes in 20% O_2/80% Ar, the resulting
morphology was significantly different. Although the film was
not completely smooth, figure 3b shows that the film contained
only small, partially formed bubbles, instead of the large,
fully formed bubbles that are evident in figure 3a. In
addition, x-ray diffraction showed that the film in figure 3b is

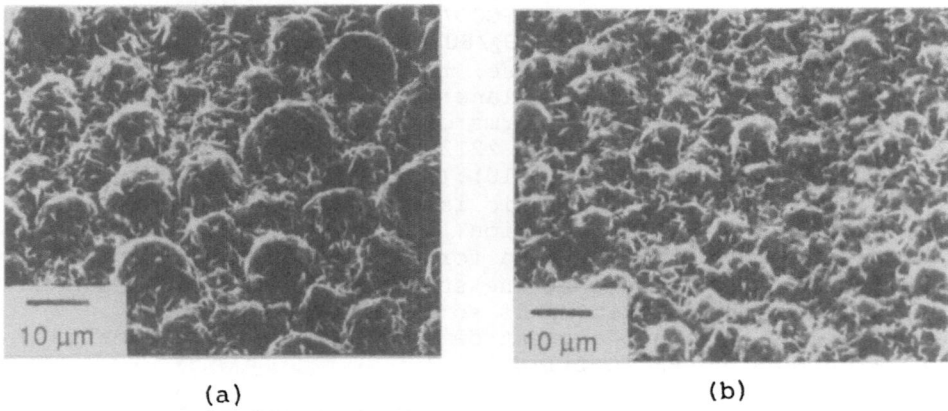

(a)   (b)

Figure 3. SEM micrographs of BSCCO films on Ag substrates that
were annealed at 870°C for 30 minutes in 20% O_2/80%
Ar. a) The BSCCO film on unannealed Ag contains
large, fully formed bubbles that are covered with
platelet grains of 2212. b) The BSCCO film on pre-
annealed Ag contains only small, partially formed
bubbles.

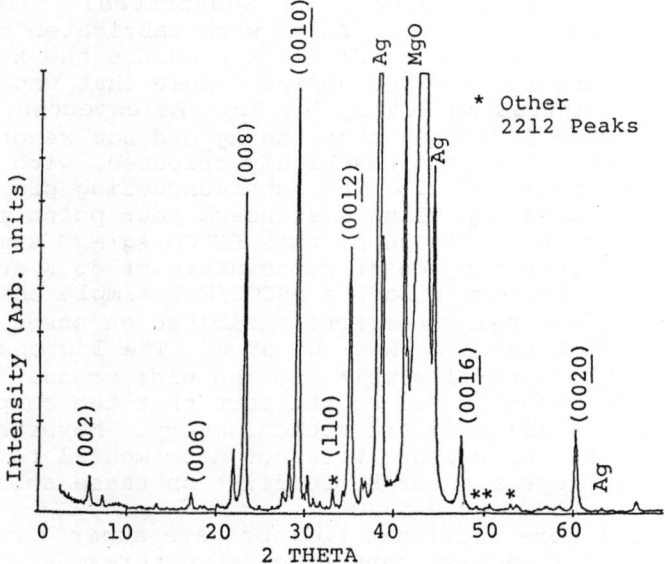

Figure 4. X-ray diffraction pattern for a BSCCO film on a Ag/MgO
structure. The 2212 phase in the BSCCO film is highly
oriented, as indicated by the (0,0,10):(1,1,0) ratio
of 21.

more highly oriented than the film in figure 3a. Specifically,
the film on the pre-annealed substrate exhibited a
(0,0,10):(1,1,0) ratio of 12, while the film on the unannealed
substrate exhibited a ratio of 6. This almost two-fold increase
in the degree of preferential orientation is most probably
caused by the difference in the morphology of the two films.
Magnetic susceptibility measurements showed that both samples
had essentially the same superconducting properties, with onset
temperatures of 80 K. If impurities are indeed responsible for
the formation of these bubbles, it is expected that the use of
purer Ag will allow the fabrication of smooth, highly oriented,
superconducting BSCCO films on polycrystalline metal substrates.

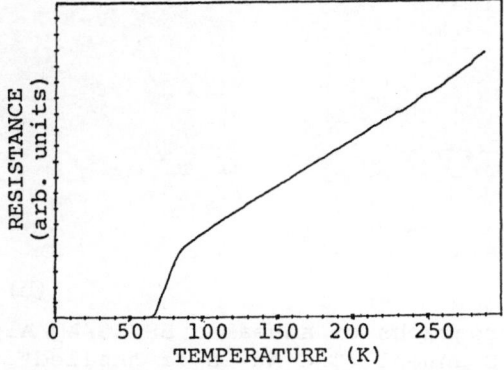

Figure 5. Resistance vs. temperature curve for a BSCCO film on
an Ag/MgO structure. The superconducting transition,
which is similar to that of a BSCCO film on MgO, has
an onset of 85 K and a T_c(R=0) of 65 K

Noble Metal Buffer Layers on Substrates: Highly oriented, superconducting BSCCO films were fabricated on both Ag/MgO and Ag/YSZ structures. Figure 4 presents the x-ray diffraction pattern for a BSCCO/Ag/MgO sample that was annealed at 870°C for 30 minutes in 20% O_2/80% Ar. As expected, the x-ray diffraction pattern indicates that the Ag did not react with the BSCCO film and that the film was highly oriented, with a $(0,0,10):(1,1,0)$ ratio of 21. The superconducting properties of this sample were measured using a standard four point probe technique. As shown in figure 5, this BSCCO/Ag/MgO sample exhibits a superconducting onset temperature of 85 K and a $T_C(R=0)$ of 64 K. In comparison, a BSCCO/MgO sample that was annealed under identical conditions exhibited an onset temperature of 72 K and a $T_C(R=0)$ of 65 K. The low onset temperature of the control sample and the wide transitions of both samples are probably due to the fact that the composition of these films was slightly off stoichiometry. Nevertheless, polycrystalline Ag underlayers were not detrimental to the superconducting properties of BSCCO films on these substrates.

It was much more difficult to fabricate superconducting BSCCO films on Ag/(reactive substrate) structures. Initially, several reactive substrates, including Al_2O_3, SiO_2/Si, and Haynes 230 alloy, were investigated. Because the results indicated that the Ag buffer layer was not acting effectively on any of these substrates, only one substrate, Al_2O_3, was investigated in depth. Figure 6 compares the morphologies of two BSCCO/Ag/Al_2O_3 structures that had 0.5 μm thick buffer layers. The first sample was annealed at 870°C for 30 minutes in 20% O_2/80% Ar. As shown in figure 6a, the Ag "balled up," allowing the BSCCO film to react with the exposed Al_2O_3 substrate. Energy dispersive x-ray analysis (EDX) was used to determine that the "balls" are pure Ag. This morphology was observed for all samples that were annealed at 870°C, regardless of the oxygen partial pressure during the anneal. Although thicker Ag layers (2.0 μm) did not

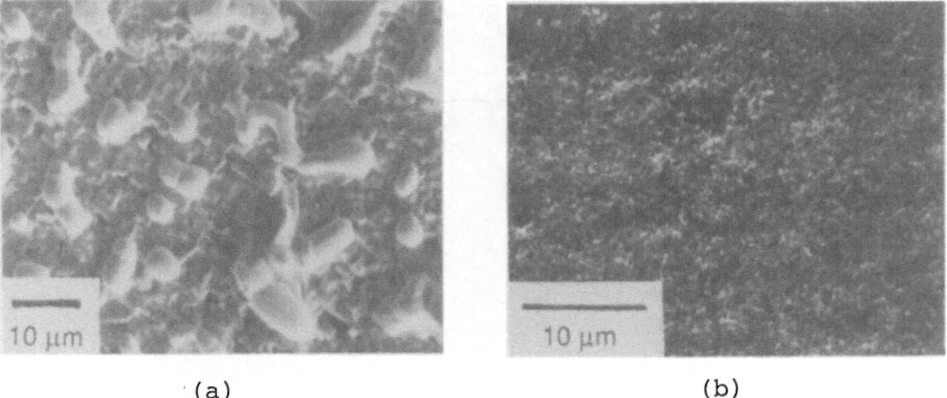

(a)   (b)

Figure 6. SEM micrographs of annealed BSCCO/Ag/Al_2O_3 structures.
a) 870°C anneal: The Ag layer "balled" up and the BSCCO film reacted with the underlying substrate.
b) 820°C anneal: The Ag layer protected the BSCCO from the substrate, allowing a superconducting film to form.

"ball up" during 870°C anneals, they did not prevent the BSCCO films from reacting with the underlying Al_2O_3 substrates. X-ray diffraction indicated that none of these samples contained the 2212 phase.

In contrast, lower anneal temperatures significantly improved the performance of the Ag buffer layer. Figure 6b shows that, when a $BSCCO/Ag/Al_2O_3$ sample was annealed at 820°C for 30 minutes in 2% O_2/98% Ar, the Ag layer did not "ball up" during this anneal. Moreover, the microstructure of the BSCCO film is characteristic of the early stages of the 2212 phase formation, before secondary grain growth occurs. X-ray diffraction showed that the 2212 phase is present in the film, and that it is slightly oriented with a $(0,0,10):(1,1,0)$ ratio of 1.5. However, the sample has a superconducting transition with an onset temperature of 65K and a $T_c(R=0)$ of 10 K. Samples that were annealed at 820°C in higher oxygen partial pressures (10% O_2) exhibited similar morphologies and properties. These results indicate that, for these annealing conditions, the Ag buffer layer protected the BSCCO film from the Al_2O_3 substrate. However, it did not allow a BSCCO film with well connected and highly oriented 2212 grains to form.

In summary, the performance of the Ag buffer layer in BSCCO/Ag/substrate structures is affected by the choice of substrate and the anneal temperature. Although the morphology of the Ag layer was affected by the Ag thickness, the overall performance was not. Because it is possible to fabricate highly oriented superconducting BSCCO films on Ag substrates, these results suggest that Ag/substrate interactions, rather than BSCCO/Ag interactions determine the effectiveness of the Ag buffer layer. The mechanism by which the Ag layer fails to protect the BSCCO film from the underlying substrate is not yet known. Possibilities include hillock formation and agglomeration in the Ag layer[11] and rapid diffusion between the BSCCO film and the substrate, probably along grain boundaries in the Ag layer.

CONCLUSIONS

The use of noble metal buffer layers and substrates for $Bi_2Sr_2CaCu_2O_8$ (BSCCO) films was investigated. Thin film techniques were used to evaluate Ag, Au, and Pt substrates for BSCCO films. Superconducting BSCCO films could not be fabricated on either Au or Pt substrates. In contrast, highly oriented superconducting BSCCO films were fabricated on polycrystalline Ag substrates. However, the exsolution of gaseous species or the volatilization of impurities from the Ag substrates resulted in the formation of bubbles in the BSCCO films. It is expected that superconducting BSCCO films that are both highly oriented and "bubble-free" can be fabricated using purer Ag substrates.

Highly oriented, superconducting BSCCO films were also produced on Ag/MgO and Ag/YSZ structures. However, it was not possible to fabricate superconducting BSCCO films on Ag/Al_2O_3, $Ag/SiO_2/Si$, or Ag/(Haynes 230 alloy) structures using the same high temperature annealing conditions. In each case, the Ag became discontinuous and the BSCCO film reacted with the underlying substrate. At lower annealing temperatures, the Ag was more effective as a buffer layer between BSCCO and Al_2O_3.

This suggests that, as *in situ* fabrication techniques are used to lower the processing temperatures required to form superconducting BSCCO films, Ag is likely to be an effective conducting buffer layer for many technologically useful substrates.

ACKNOWLEDGEMENTS

We gratefully acknowledge the assistance D.W. Face in developing our thin film fabrication techniques. We also acknowledge the help of R.M. Rose and several members of our group: J.T. Kucera, D. Steel, D. Curd, K. Rigby, G. Somer, R. Wilkinson, J.M. Graybeal, and T.P. Orlando. This research was supported by the U.S. Army Strategic Defense Command through Babcock and Wilcox.

REFERENCES

1. D.W. Face, M.J. Neal, M.M. Matthiesen, J.T. Kucera, J. Crain, J.M. Graybeal, T.P. Orlando, D.A. Rudman, Appl. Phys. Lett. **53** (3) 18 July 1988, 246.
2. D.W. Face, J.T. Kucera, J. Crain, M.M. Matthiesen, D. Steel, G. Somer, J. Lewis, J.M. Graybeal, T.P. Orlando, and D.A. Rudman, IEEE Transactions on Magnetics, MAG-25, 1989.
3. L.S. Hung, J.A. Agostinelli, G.R. Paz-Pujalt, J.M. Mir, Appl. Phys Lett. **53** (24), 12 December 1988, 2450.
4. H. Nasu, H. Myoren, Y. Ibara, S. Makida, Y. Nishiyama, T. Kato, T. Imura, Y. Osaka, Jap. J. Appl. Phys. **27** (4) April 1988, L634.
5. Y.M. Chiang, private communication.
6. S. Jin, R.C. Sherwood, T.H. Tiefel, G.W. Kammlott, R.A. Fastnacht, M.E. Davis, S.M. Zahurak, Appl. Phys. Lett. **52** (19), 9 May 1988, 1628.
7. R.C. Baker, W.M. Hurng, H. Steinfink, Appl. Phys. Lett. **54** (4), 23 January 1989, 371.
8. K.E. Williams, M.M. Matthiesen, L.M. Rubin, D.A. Rudman, unpublished.
9. Haynes International, Inc., Kokomo, IN 46902.
10. T. Kogure, private communication.
11. S.K. Sharma, J. Spitz, Thin Solid Films **65**, 1980, 339.

BREAK JUNCTION TUNNELING SPECTROSCOPY OF SINGLE-CRYSTAL

BISMUTH-BASED HIGH-TEMPERATURE SUPERCONDUCTORS[†]

John Moreland
National Institute of Standards and Technology
Electromagnetic Technology Division
Boulder, CO 80303

C. K. Chiang
National Institute for Standards and Technology
Polymer Division
Gaithersburg, MD 20899

L. J. Swartzendruber
National Institute for Standards and Technology
Metallurgy Division
Gaithersburg, MD 20899

ABSTRACT - We have measured the tunneling spectra of some high temperature superconducting crystal break junctions at 4 K. The samples were thin plates of $Bi_2SrCa_2Cu_2O_8$ compound. The tunneling spectra (conductance versus voltage) were not typical of BCS superconductor tunneling electrodes. The spectra of higher-resistance break-junction settings (R > 1 MΩ) show a tunneling gap on top of a linearly increasing conductance background signal. "Harmonic" dip features in the spectra of lower resistance break junction settings (R < 1 MΩ) indicated tunneling between multiple particles in the vicinity of the primary (highest resistance) contact of the junction. The dips occurred at about the same current but shifted in voltage when the resistance of the break junction was continuously adjusted to new settings.

Several laboratories have fabricated tunnel junctions having high-temperature superconducting electrodes. The collected data show some correlation among I-V characteristics,[1-3] but there are still inconsistencies between ostensibly similar tunnel junctions. Gap features in the tunneling current-voltage (I-V) characteristics and Josephson weak links have been observed for several types of junctions. The precise size of the tunneling gap and the I_cR products of weak links, however, depend not only on electrode materials, but on the method used to fabricate the junction. This paper is part of a systematic study to determine the causes of the inconsistencies in tunneling I-V curves of high temperature superconductor junctions. We think that some data variability stems from uncertain material quality. Tunneling measurements are very sensitive to

---

[†] Contribution of the National Institute of Standards and Technology, not subject to copyright.

Advances in Cryogenic Engineering (Materials), Vol. 36
Edited by R. P. Reed and F. R. Fickett
Plenum Press, New York, 1990

619

the superconducting properties within a superconducting coherence length of the surface of the tunnel junction electrodes. This fact combined with estimated coherence lengths of less than 2 nm for high temperature superconducting copper oxides,[4] implies that the crystallinity and stoichiometry of the compounds should be maintained very near the tunnel barrier interfaces. We propose, then, that tunnel junction structures consisting of two pristine single crystal electrodes separated by a "vacuum" barrier, or an epitaxial trilayer with two thin films separated by an inert barrier, will be required to obtain "good" tunneling data for high temperature superconducting materials. Freshly cleaved crystal surfaces used as tunneling electrodes in a single-crystal break junction, for example, may provide consistent data representative of the intrinsic bulk superconducting properties of the crystal.

The tunneling data presented in this paper are for some single-crystal $Bi_2SrCa_2Cu_2O_8$ break junctions. Previous results for single-crystal $HoBa_2Cu_3O_{7-\delta}$ and $Tl_2CaBa_2Cu_2O_7$ break junctions are reported in references 5 and 6, respectively. The tunneling data for the $HoBa_2Cu_3O_{7-\delta}$ crystal break junctions were inconclusive. Only weak tunneling gaps were observed in the I-V curves and low-resistance contacts were normal. The results for $Tl_2CaBa_2Cu_2O_7$ break junctions, on the other hand, showed clear tunneling gaps and low-resistance contacts were superconducting.

The crystals of $Bi_2SrCa_2Cu_2O_8$ were grown from a nonstoichiometric melt of $Bi_{2.5}SrCa_2Cu_2O_8$. The excessive Bi_2O_3 was intended as a crystal growth flux. Powders of $SrCO_3$, $CaCO_3$, and CuO were ground together to form the starting mixture. The mixture was then sintered in air at 800°C for 15 min and then at 950°C for 4.5 h. The reacted powder was mixed with Bi_2O_3 and ground again. This mixture was then melted in a heated alumina crucible at 900°C for 5 h. The melt was cooled slowly at 1°C/h to 200°C in air.

Fig. 1. SEM micrograph of crystal plates removed from the $Bi_2SrCa_2Cu_2O_8$ melt. These crystals are typical of the ones used for break junction tunneling measurements. They are thin mica-like sheets approximately 50 μm thick.

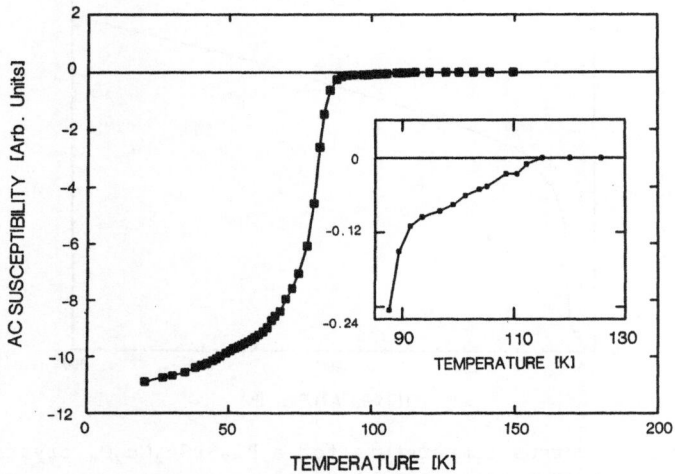

Fig. 2. The ac susceptibility of $Bi_2SrCa_2Cu_2O_8$ crystal mass as a function of temperature.

We used the break junction method[7] to generate a mechanically adjustable tunnel junction within the fracture of the crystals. Each crystal was mounted between two glass plates fastened to the surface of a glass beam. The crystal was subsequently fractured by bending the beam. Precise control of the bending force allowed us to establish a tunneling contact within the freshly exposed surfaces of the fracture. The junctions could be continuously adjusted from an open contact (R > 100 MΩ) to point contact. The crystals were broken in liquid helium at 4 K to minimize contamination of the fractured surfaces serving as the electrodes of the tunnel junction. Figure 1 is a scanning electron microscope (SEM) micrograph of the type of $Bi_2SrCa_2Cu_2O_8$ crystals used in our experiment. The larger flakes (approximately $1×1×0.01$ mm^3) were removed from the mica-like mass of the preparation melt with tweezers. The crystals were then fastened to the break junction substrate with epoxy. Silver contacts were painted to the corners of the crystals for electrical contact. Contact resistances were typically around 10 Ω.

The ac susceptibility measurement of a melt chunk indicated that the onset temperature for superconductivity was 110 K (see Fig. 2). The transition has a large diamagnetic signal starting at 90 K. The mid-point of the transition was at 80 K. The fraction of the 110 K phase estimated from the susceptibility data was 1%.

The ac resistance of the crystals was measured during the initial cooling of the break junction apparatus. The results are shown in figure 3. The mid-point resistive transitions occurred at 110 K with the onset of the transitions beginning above 120 K. At 100 K the resistance of the sample fell below the detection limit of the measurement of 10 $\mu\Omega$. The broad resistive transitions and the 110 K onset evident in the susceptibility data indicate that the crystals may be multiphased containing intergrowths with a distribution of transition temperatures. Multiple phase intergrowth has been observed previously in the Bi an Tl based superconducting CuO compounds.[8]

The I-V curves and the corresponding dynamic conductance as a function of voltage (G-V) were measured simultaneously using the circuit described

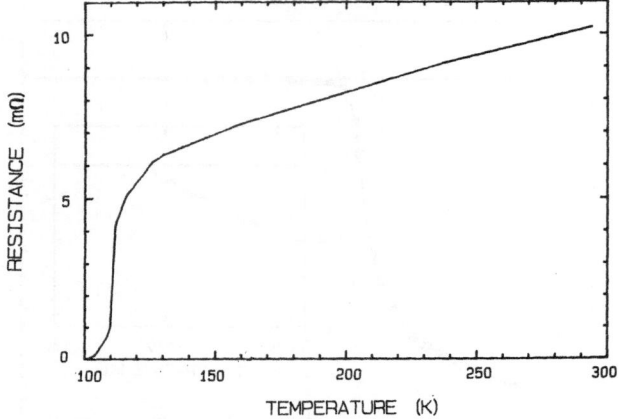

Fig. 3. Resistance versus temperature for a $Bi_2SrCa_2Cu_2O_8$ crystal in the break junction apparatus before fracture.

in Ref. 9. A 137 Hz, 1 mV modulation signal was used in the conductance measurement. Figure 4 shows the I-V and G-V curves of a $Bi_2SrCa_2Cu_2O_8$ break junction at 4 K. Conductance peaks are located at ± 63 mV. The difference between these numbers should be roughly equivalent to $4\Delta/e$, were Δ is defined to be the BCS energy gap of the electrode surfaces. Thus we conclude that $\Delta = 31.5 \pm 1$ meV for this junction setting. Other settings of this junction resulted in smaller tunneling gaps or gapless I-V curves. At lower-resistance settings of the junctions the G-V curves showed sharp dips in a linearly rising conductance with increasing voltage. Figure 5 shows the I-V and G-V curves for two lower-resistance settings of a $Bi_2SrCa_2Cu_2O_8$ break junction. The junction resistance increases as shown in figures 5a and 5b, when the break junction substrate is continuously flexed to open the tunneling contact. The G-V curves show a generally increasing conductance with increasing voltage similar to the higher-

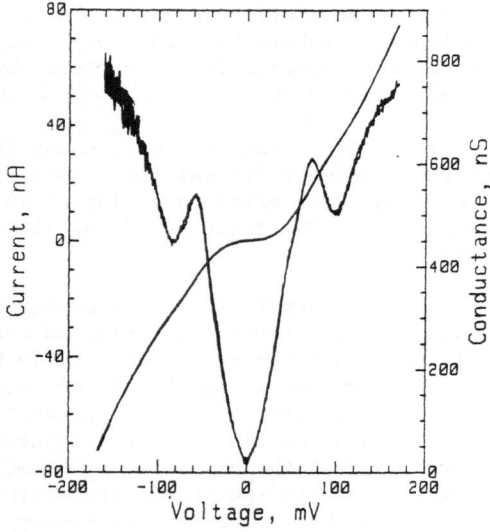

Fig. 4. The I-V and G-V curves for a $Bi_2SrCa_2Cu_2O_8$ break junction set at a relatively high resistance in liquid helium at 4 K.

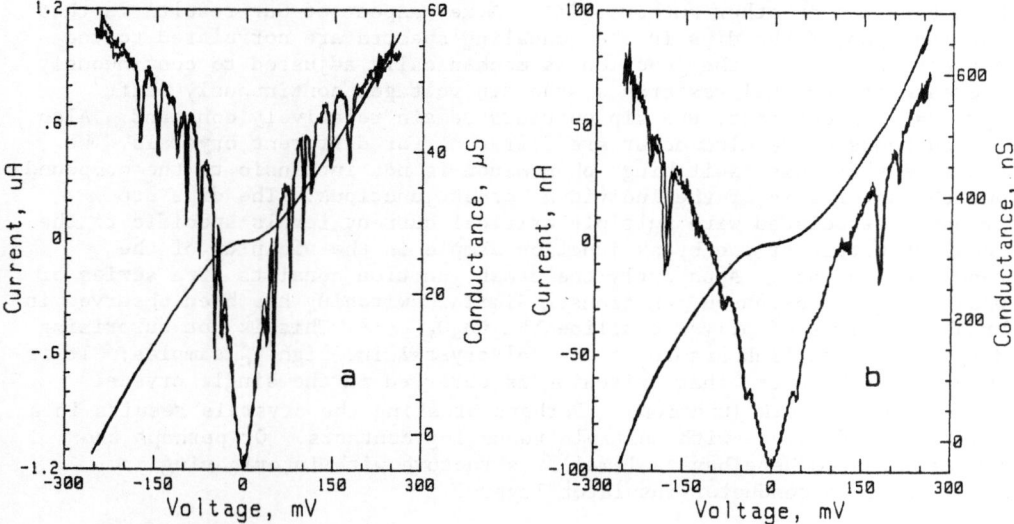

Fig. 5. Tunneling data for a $Bi_2SrCa_2Cu_2O_8$ break junction immersed in
liquid helium at 4 K. The I-V curves correspond to a progression of
mechanical settings of the junction. As the nominal resistance of the
junctions increases (5a to 5b), the position of the sharp dip features in
the corresponding G-V curves shift to higher biases.

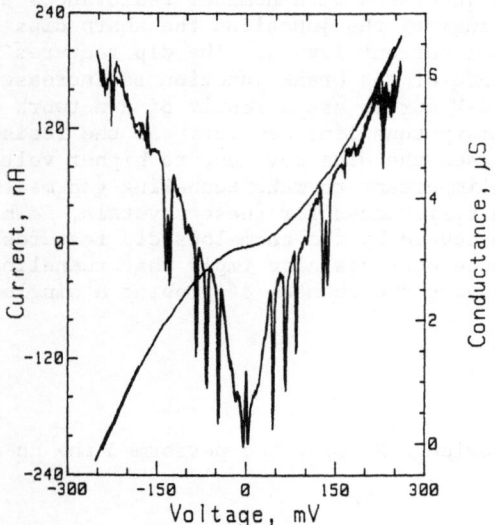

Fig. 6. Low-resistance setting for
the $Bi_2SrCa_2Cu_2O_8$ break junction
discussed in fig. 4. A zero bias
conductance peak is evident in the
G-V curve.

resistance settings of the junctions. Also, there are sharp dips in the G-V curves that sometimes occur at what appear to be harmonic voltage intervals. Similar tunneling spectra of Bi-based CuO superconductors have been reported by other authors.[10,11] A key aspect of our results is that the location of the dips in the tunneling spectra are correlated to the current levels. As the junction is mechanically adjusted to continuously increase its nominal resistance, the dip voltages continuously shift upwards. In contrast, the dip currents remain relatively constant. Also, the currents where dips occur are different for different crystals. We conclude that this "switching" phenomenon is not intrinsic to the compound but characteristic of the individual break junctions. The dips are probably associated with multiple critical current levels specific to the local structure of the break junction sample in the vicinity of the tunneling contact. Apparently the break junction consists of a series of weak links or Josephson junctions. Similar switching has been observed in break junctions of polycrystalline $YBa_2Cu_3O_{7-\delta}$.[12] This is not surprising due to the weak link nature of the polycrystalline high T_c samples. It is surprising, however, that switching is observed in the single crystal $Bi_2SrCa_2Cu_2O_8$ break junctions. Perhaps breaking the crystals results in a complicated fracture with multiple tunneling contacts. Or perhaps the crystals tested here have a lamellar structure with interleaving or syntactic superconductor-insulator layers.

A lower resistance setting of the $Bi_2SrCa_2Cu_2O_8$ break junction discussed above (see Fig. 5) resulted in the data shown in Fig. 6. A zero bias conductance peak revealing the presence of a weak supercurrent is evident in the G-V curve. This is further evidence that the material is superconducting in the vicinity of the tunneling contact of the break junction.

In conclusion, we have observed tunneling gaps in $Bi_2SrCa_2Cu_2O_8$ crystal break junctions. The tunneling gaps are clearly visible in the G-V tunneling spectra for junctions with nominal resistances above 1 MΩ. For lower resistance settings of the junctions the sharp dips in the G-V curves occur at given critical current levels. The dip features shift to higher biases as the resistance of the break junction is increased. We speculate that the dips in the G-V curves are a result of a network of weak links in the vicinity of a primary tunneling contact. As the resistance of the primary contact increases the dips move out to higher voltages. We think, therefore, that it is important to make tunneling gap measurements at higher break junction resistances for these crystals. This insures that the gap will not be obscured by the anomalous dip features in the G-V curves. High resistance contacts also imply that tunneling is limited to a small area, thus improving the chances of probing a single phase of the compound.

Acknowledgement

The authors acknowledge R. Drew who performed the ac susceptibility measurements.

References

1. J. Moreland, L. F. Goodrich, J. W. Ekin, T. E. Capobianco, and A. F. Clark, Anomalous behavior of tunneling contacts in superconducting perovskite structures, in: "Advances in Cryogenic Materials, vol. 34," A. F. Clark and R. P. Reed, eds., Plenum, New York, pp. 625-632 (1988).

2. K. E. Gray, Tunneling spectroscopy of novel superconductors, in: "Novel Superconductivity," S. A. Wolf and V. Z. Kresin, eds., Plenum, New York, pp. 611-625 (1988).

3. M. Lee, D. B Mitzi, A. Kapitulnik, and M. R. Beasley, Electron tunneling and the energy gap in $Bi_2Sr_2CaCu_2O_x$, Phys. Rev. B, 39:801 (1989).

4. T. P. Orlando, K. A. Delin, S. Foner, E. L. McNiff, Jr., J. M. Tarascon, L. H. Greene, W. R. McKinnon, and G. W. Hull, Upper critical fields of high-T_c superconducting $Y_{2-x}Ba_xCuO_{2-y}$, Phys. Rev. B, 35:7249 (1987).

5. J. Moreland, A. F. Clark, M. A. Damento, and K. A. Gschneidner, Jr., Single crystal $HoBa_2Cu_3O_x$ break junctions, Physica C, 153-155:1383 (1988).

6. J. Moreland, D. S Ginley, E. L. Venturini, and B. Morosin, Break junction measurement of the tunneling gap of a thallium based high temperature superconductor crystal, submitted to Appl. Phys. Lett. (1989).

7. J. Moreland and J. W. Ekin, Electron tunneling experiments using Nb-Sn "break" junctions, J. Appl. Phys., 58:3888 (1985).

8. D. S. Ginley, B. Morosin, R. J. Baughman, E. L. Venturini, J. E. Schriber, and J. F. Kwak, Growth of crystals and effects of oxygen annealing in the Bi-Ca-Sr-Cu-O and Tl-Ca-Ba-Cu-O superconductor systems, J. Crystal Growth, 91:456 (1988).

9. A. F. Hebard and P. W. Shumate, A new approach to high resolution measurements of structure in superconducting tunneling currents, Rev. Sci. Instrum., 45:529 (1974).

10. G. Briceno and A. Zettl, Tunneling spectroscopy in $Bi_2Sr_2CaCu_2O_8$: Is the energy gap anisotropic?, Solid State Comm. 70:1055 (1989).

11. R. Esudero, F. Morales, F. Estrada, and R. Barrio, Tunneling measurements in the B: superconductors, Mod. Phys. Lett. 3:73 (1989).

12. J. Moreland, L. F. Goodrich, J. W. Ekin, T. E. Capobianco, and A. F. Clark, Electron Tunneling Measurements of High T_c compounds using break junctions, Jap. J. Appl. Phys., 26-3:999 (1987).

GROWTH OF LARGE $(La_{1-x}Sr_x)_2CuO_4$ SINGLE CRYSTALS

AND REDUCTION OF OXYGEN DEFICIT

S. N. Barilo, A. P. Ges', S. A. Guretskii,
D. I. Zhigunov, A. A. Ignatenko, I. D. Lomako,
A. M. Luginets, and V. N. Shambalev

Institute of Solid State and Semiconductor Physics
Belorussian Academy of Sciences
Minsk, BSSR

INTRODUCTION

$(La_{1-x}Sr_x)_2CuO_4$ quasiternary compounds possess a number of unique physical properties related to structural, magnetic, metal-dielectric and superconducting phase transitions. The problem of single crystal growth of these cuprates is very important for their comprehensive study. Its solution is complicated by incongruent melting and by the fact that all the materials of the technological equipment are heavily corroded by the fluxes. Lanthanum-strontium cuprates are grown from the lithium borate or sodium borate fluxes or stoichiometric copper oxide[1-3]. The disadvantage of the borate solvents is the incorportion of impurity ions into the growing crystals, for instance, up to 8 at. % Li[1]. The use of superstoichiometric CuO as a solvent excludes the incorporation of the impurities into the single crystals, although the decompositon of cupric oxide at elevated temperatures hampers the formation of single crystals with stoichiometric oxygen. The best results have recently been reported on the technique of slow pulling of seeds from the surface of the superstoichiometric flux-melt[2].

EXPERIMENTS

In this work we employed the technique of growth of a limited number of seeds on a platinum crystal holder in the dynamic regime. Cupric oxide or $Li_4B_2O_5$ was used as a solvent. In both the systems of flux melts the lanthanum-strontium cuprates crystallize only at high concentrations of the crystal-forming oxides (more than 40 wt. %), which prevents their super-cooling. The width of the metastable range was 1-1.5°C. With a view to limiting the number of seeds, a special crystal holder was designed and the minimum rate of temperature reduction was 1-2°C/24 hrs. A small width of the metastable range specifies the following requirements on the crystal holder:

Advances in Cryogenic Engineering (Materials), Vol. 36
Edited by R. P. Reed and F. R. Fickett
Plenum Press, New York, 1990

627

1. To avoid the formation of parasite crystals the rod of the crystal holder should be overheated with respect to the flux melt.
2. The crystal-forming surface, on the contrary, should be overcooled and washed regularly and uniformly by the flux-melt stirring.
3. The crystal holder should stir the solution intensively for a long time, i.e. its construction should be sufficiently rigid.
4. The crystal-forming surface should be convenient enough for removing the as-grown crystals.

The supercooling of the seeding surface was achieved by placing the crystal holder into the coolest range of flux melt near the axis of the cylinder platinum crucible with a positive radial temperature gradient. The dynamical regime ensured intensive washing of the crystals by the flux melt during growth.

High-quality bulk $La_2CuO_{4-\delta}$ crystals (Fig. 1) were synthesized in the following way. The flux melt was homogenized at temperatures 15-30°C higher than the saturation temperature, and thereafter rapidly cooled to temperatures 1-2°C higher than that of spontaneous crystallization. Then the temperature was lowered at a rate of 1-2°C/24 hrs, with the flux melt mass being 500-1500 g. In this case the initially nucleated crystals on the crystal-forming platinum surface attracted the maximum of crystal-forming oxides, thus minimizing the possibility of formation of parasite centers of crystallization. The number of these centers and the final size of the as-grown crystals are determined by the accuracy of the temperature adjustment, the quality of the initial reagents and the nucleation surface, the temperature and concentration gradients in the flux melt, the rate of temperature decrease and, finally, by the total duration of the process.

While growing lanthanum cuprates substituted by Sr from both solvent systems, some additional difficulties arose due to the increasing viscosity of the flux melts and large distribution coefficients of the Sr^{2+} in them. Thus, for the $(La_{1-x}Sr_x)_2CuO_{4-\delta}$ -$Li_4B_2O_5$ system this coefficient varies from 10 with $x<0.05$ to 4 with $x>0.05$, while the formation of single crystals with $x>0.08$ is extremely difficult because of sharp deterioration of flux-melt properties and precipitation of the accompanying phase of lanthanum borate, $LaBO_3$. The attempted reduction of the flux-melt viscosity through the introduction of excess Li_2O resulted in the appearance of single crystals

Fig. 1. Single crystals of $La_2CuO_{4-\delta}$

628

Table 1. Flux-melt composition, technological parameters and distribution coefficient K for single crystal growth

Sample, No.	Flux composition, mol %					Cooling rate, °C/24hrs	K
	La_2O_3	SrO	CuO	Li_2O	B_2O_3		
1	17.7	4.6	19.4	38.9	19.4	0	4.8
2	14.1	6.3	16.2	42.9	21.1	40	5.0
3	17.5	5.6	19.2	38.5	19.2	1.2	4.6
4	17.5	3.4	19.7	39.3	19.7	0	4.7
5	14.1	6.3	16.2	42.9	21.1	8	5.0
6	13.2	11.3	18.9	37.7	18.9	0	2.6
7	13.0	–	87.00	–	–	0	–
8	13.5	1.5	85.0	–	–	1	5.5
9	13.4	2.9	83.7	–	–	1	5.4
10	12.6	4.2	83.2	–	–	1	4.6

of lanthanum-strontium cuprate-platinate with a body-centered cubic structure and elevated concentrations of lanthanum and platinum which had not been observed earlier. The study of chemical composition and kinetic properties of single crystals are underway.

For the stoichiometric CuO flux the Sr distribution coefficient was approximately 5. Some flux-melt compositions and growth parameters for various experimental runs are listed in Table 1. It is essential to note that the corrosion of platinum equipment increased with increasing Sr excess in the flux melt and wire increasing rate of stirring. Accordingly, the growth regime was chosen depending on the requirements imposed by size and quality of single crystals.

ANALYSIS OF CHEMICAL COMPOSITION AND STRUCTURE

Due to the very large coefficient of the Sr distribution and imperfection of the $(La_{1-x}Sr_x)_2CuO_{4-\delta}$ single crystals for oxygen qualitative estimation of their chemical composition as a function of the Sr excess, the growth temperature range, and quenching temperature of the samples was made by X-ray fluorescence analysis. A polycrystalline superconducting $La_{1.8}Sr_{0.2}CuO_4$ pellet was used as a reference. Since the Sr concentration in the samples is not large, the change of its concentration insignificantly influenced the mass coefficient of absorption μ. Therefore the Sr($K\alpha$) line intensity, normalized to the intensities of the peaks of copper and lanthanum, in the first approximation, is proportional to the Sr^{2+} concentration in the samples and in the reference.

The results are summarized in Table 2. Qualitative estimation of the chemical composition of the as-grown single crystals was performed by neutron activation analysis. The results are presented in Table 3, from which stability of the elemental composition of the samples irrespective of the quenching conditions of the samples is evident. In this case the La^{3+}

Table 2. Single crystals $La_{2-x}Sr_xCuO_{4-}$ X-ray fluorescent analysis data

Sample, No.	Ratio of peaks area		Sr^2 concentration, X	
	Sr/Cu	Sr/La, 10^{-3}	for SrCu	for Sr/La
1	0.79	4.3	0.057	0.066
2	0.76	3.3	0.055	0.051
3	1.17	5.3	0.084	0.081
4	0.46	2.6	0.033	0.040
5	0.72	3.4	0.052	0.052
6	1.94	12.0	0.140	0.180
8	0.30	3.2	0.025	0.032
9	0.60	6.1	0.050	0.060
10	0.87	6.6	0.070	0.067

concentration exceeds by 0.5-0.6 wt. % that estimated. The oxygen concentration in the samples was determined by subtraction of the lanthanum and copper mass from the total mass.

The oxygen concentration in most cases is lower than the stoichiometric one and varies greatly depending on which part (near the surface or bulk) of the crystal was analysed. The most homogenous crystals with respect to oxygen are those grown at minimum cooling rates (about 1°C/24 hrs) and quenched from 750°C. To improve the analysis conditions for Sr and Pt, the samples were irradiated in cadmium containers. The analysis data are summarized in Table 4. These samples are distinguished by a lower Sr distribution coefficient compared to those grown from the CuO solvent, however, they indicate the presence of platinum in their compositon. The source of the latter is corrosion of the platinum incorporates, as PtO_2 in copper vacancies of the crystal lattice. This is confirmed in particular by analysis of the crystals of the new cuprate-platinate phase, in which the platium concentration appreciably exceeds that of copper. The tentative composition of these crystals can be expressed by the general formula $La_{2-x}Sr_xPt_{1-y}Cu_yO_{7-z}$. In the single crystals of lanthanum strontium cuprate grown from the solution of superstoichiometric copper oxide no paltinum has been detected within the limits of the analysis. The crystal structure of the $(La_{1-x}Sr_x)_2CuO_{4-\delta}$ single crystals was examined by X-ray diffraction (Cu K_α-radiation). The pyramid-like $La_2CuO_{4-\delta}$ single crystals were faceted in the rhombic indication by the [111] and [001] planes. The incorporation of strontium into the crystal affected the morphology: the growth rate along the [001] direction was reduced and the samples were plate-like. The single crystals of the lanthanum-strontium cuprate-platinate phase exhibited well-developed faces (001), (110) and (101) peculiar to the body-centered cubic lattice (a=12.299 Å).

The dependence of the quantity and size of the $La_2CuO_{4-\delta}$ crystals on both the rate of temperature reduction during growth and on quenching temperture has been found. It has been revealed that an increase of the quenching temperature from 400 to 1150°C leads to the formation of small crystals characterized by dense structure and good quality. Large $La_2CuO_{4-\delta}$ crystals

Table 3. La_2CuO_{4-} single crystal neutron activation analysis data vs. quenching temperature, T_q, of sample 7

T_q, °C	Experimental concentration of the element, wt. %			Calculated concentration of the element (for stoichiometric O^{2-}), wt. %		
	La	Cu	O	La	Cu	O
1150	71.5	14.8	14.3	69.8	14.4	15.8
750	75.2	17.0	7.8	68.7	16.2	15.8
400	69.1	14.3	16.4	69.7	14.5	15.8

are formed at low quenching temperatures (~400°C), with deterioration in the quality of the grown crystals being connected both with nonuniform distribution of oxygen (or mixed-valency copper) in the crystal and changes of the rhombic lattice parameters over a wide range (a=5.345-5.335Å, b=5.396-5.380 Å, c=13.130Å). After thermal treatment of the samples in flowing oxygen at 750°C for 72 hrs, a noticeable decrease in the peak widths has been observed (3 fold) and a good separation ($K\alpha_1$, $K\alpha_2$) for the lines (333). Observation of crystals with a small amount of strontium has shown the parameters a and c of the rhombic lattice to increase slightly due to the small coefficient of the Sr distribution (a_o=5.353 Å, a_1=5.364 Å, c_o=123.145Å, c_1=13.174 Å).

THERMAL TREATMENT

The as-grown single crystals are characterized by oxygen deficiency that reduces structure-sensitive material properties, for instance the electrical resistance rises to 10^6 ohm. The filling of vacancies in the anion sublattice of the oxide material, in accordance with the thermodynamics of equilibrium processes, is possible at fixed temperature through changing

Table 4. Neutron activation analysis data for single crystals of $(La_{1-x}Sr_x)_2CuO_{4-\delta}$ grown from lithium borate and cupric oxide as solvent

Sample, No.	Solvent	Concentration of the element, wt. %				
		La	Cu	Sr	Pt	O
1	$Li_4B_2O_5$	62.5	15.2	2.0	0.7	19.6
4	$Li_4B_2O_5$	62.0	14.3	0.4	1.7	21.6
6	$Li_4B_2O_5$	60.7	13.9	1.7	2.2	21.8
7	CuO	71.4	12.3	–	0.09	16.3
8	CuO	73.7	12.0	0.6	0.09	13.7
9	CuO	68.7	13.7	0.9	0.09	16.7
10	CuO	72.5	13.6	0.8	0.09	13.1

the chemical potential of oxygen in the atmosphere. Since the oxygen saturation process of single crystal $(La_{1-x}Sr_x)_2CuO_{4-\delta}$ is primarily specified by the oxygen diffusion coefficient which is small (no more than 10^{-11} - 10^{-12} cm^2/ s) at temperatures of solid phase reactions, the intensification of the process is only possible when the partial oxygen pressure increases exponentially. One of the promising means of increasing the chemical potential of oxygen at the single crystal interface is the forced oxygen transport technique under constant electric field at the superionic conductor-oxygen containing material boundary. We employed the open system

$$O_2(Pt)\ [Bi_{1.5}Y_{0.5}O_3\ -(La_{1-x}Sr_x)_2CuO_{4-}\ -Bi_{1.5}Y_{0.5}O_3]\ (Pt)O_2$$

with inert porous platinum electrodes which were placed in a heating chamber with P_{O_2} = 2.110^4 Pa. The heating was carried out at temperatues from 100 to 800°C, the density of the current passing through the system varied over the range 0.01–12 mA/cm^2. After a certain period of relaxation the samples were quenched in air. Next, the samples were examined in an X-ray diffractometer and their electrical resistance was measured.

It has been established that the optimal temperature, at which the reaction is most intensive, is 550°C. That corresponds to the temperature of crystal structure transformation.

Increasing the current density caused a decrease of electrical resistance of the single crystals. In this case the reaction time was reduced as compared to that of the initial exposure (60 minutes) due to the oxygen supersaturation of the crystal surface resulting from the prolonged thermal treatment and the decay of the solid phase forming a porous structure. At a current density of 1.5–2 mA/cm^2 and an exposure time of 20 minutes, the material became a semiconductor with a resistance on the order of 10^2 ohm. Increasing the current density to 5 mA/ cm^2 brought about a decrease of electrical resistance by an order of magnitude. However, appreciable anisotropy of electrical resistance persists along different directions in the crystal irrespective of the arrangement of the solid electrolytes relative to the single crystal during the reaction.

ACKNOWLEDGMENTS

The authors are thankful to S. V. Shiryaev, V. V. Komar, and A. N. Igumentsev for help with the X-ray analyses.

REFERENCES

1. R. J. Birgeneau, C. Y. Chen, D. R. Gable, H. P. Jenssen, M. A. Kastner, C. J. Peters, P. J. Picone, T. Thio, T. R. Thurston, and H. L. Tuller, Soft-phonon behavior and transport in single crystal La_2CuO_4, Phys. Rev. Lett. 12:1329 (1987).
2. P. J. Picone, H. P. Jenssen and D. R. Gabbe, Phase diagram and single crystals growth of pure and Sr doped La_2CuO_4, J. Cryst. Growth 91:463 (1988).
3. Y. Hidaka, Y. Enomoto, M. Suzuki and T. Murakami, Single crystal growth and properties of high T_c oxide superconductors, Rev. Electr.Comm. Lab. 36:567 (1988).

INFLUENCE OF DOPING BY MICROAMOUNT TECHNETIUM ON SUPERCONDUCTIVITY

AND MECHANICAL PROPERTY FOR THALLIUM SYSTEM*

R.S. Wang, H.M. Lee and P. Nar

Department of Chemical Engineering, Tianjin University
Tianjin 300072, P.R. China

INTRODUCTION

On the base of the Tl-system high Tc superconductor[1-2], a new Tl-Tc-Ba-Ca-Cu-O system, which has good mechanical and electromagnetic properties and fine chemical stability, has been developed by doping with microamount of technetium (Tc) in the exploration of increasing superconductive transition temperature.

EXPERIMENTAL

A typical procedure for preparing the samples is the following. Appropriate amount of high purity (99.99%) Tl_2O_3, BaO_2, CaO, CuO and microamount of Tc_2O_7 were weighed stoichiometrically, ground and thoroughly mixed in an agate mortar. The well-mixed oxides were calcined at 880°C in air for 8h, then reground and pressed into pellets with a diameter of 12mm and a thickness of 2mm. The pellets wrapped with gold foils were put in a sealed quartz tube containing 1 atmosphere of O_2. It's much better to put the pellets on the inert samples prepared previously to avoid inducing impurities, referring to Fig.1. The tube was then put into a tube furnace, which had been heated to 880°C, and was sintered for 10 minutes. The samples were then cooled by furnace colling to room temperature.

The preparation of powder has also used the chemical method. The oxides powder prepared by citric acid complexation evaporating method has homogeneous composition and granularity, good reaction activity, high purity. This can avoid inducing impurity element Si, which has an adverse effect on the superconductive property of the material, in grinding process.

---

* Project sponsored by the National Natural Science Foundation of China.

Advances in Cryogenic Engineering (Materials), Vol. 36
Edited by R. P. Reed and F. R. Fickett
Plenum Press, New York, 1990

633

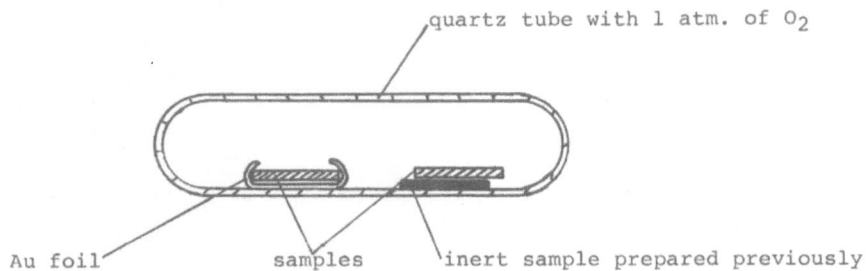

quartz tube with 1 atm. of O_2

Au foil   samples   inert sample prepared previously

Fig.1   Experimental arrangment for reaction of Tl-Tc-Ba-Ca-Cu-O system

RESULTS AND DICUSSIONS

    The result of resistance versus temperature curve is shown in Fig.2. The resistance of the sample prepared by the chemical method has positive relation above 144K, namely $d\rho/dT>0$, and a great descent at 138K, but it trials a tail to reach zero resistance until 121K. This forebodes a possible extraordinary phenomenon of superconductive behaviour at about 138K.

    Superconductivity persisted in some samples after having been placed in air for five months, and superconductivity at 96K still retained in a sample boiled in boiling water for two hours and then exposed in air for two weeks. This showed a fine antihydrolysis behaviour.

    By measurement, we note that doping with microamount of Tc makes the superconductor improve its mechanical properties obviously. When molar ratio of Tc to Tl is 0.002, its microhardness is increased by 6.2 times, while doping has only a little influence on the zero resistance. We think, therefore, that doping of Tc plays an important role in improving the mechanical properties and the chemical stability. By doping experiments of some microamount elements, we can derive some information about the micromechanism of the superconductivity[3].

    Some peculiar electromagnetic properties due to doping with Tc have been observed in this material.

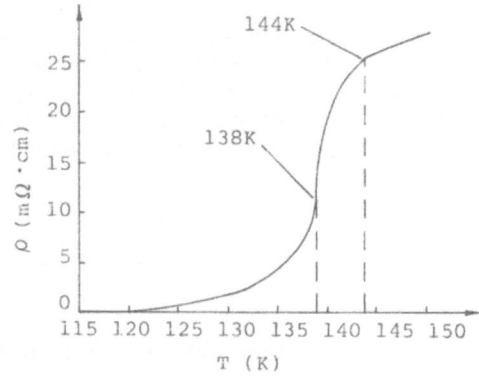

Fig.2   Resistance versus temperature curve of the sample

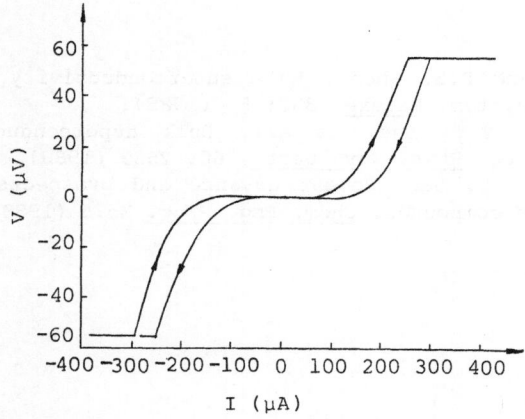

Fig.3   I-V characteristic curve of a sample at 77K under zero field

The I-V characteristic curve shows a clear hysteresis, and the critical current and resistivity are obviously related to the variation process of the applied magnetic field and the measuring current, referring to Fig.3. Some authors explained the hysteresis by the capture of weak links among grain superconductors to the magnetic flux. This explanation implies that the hysteresis is connected with the superconductivity of grains. We have measured the I-V characteristic curve at room temperature (303K) under zero field, noting the existence of the hysteresis, referring to Fig.4. It shows that maybe the explanation above isn't all-round enough, or there are probably some superconductive evidences at a higher temperature.

In a word, Tl-system samples prepared by doping with microamount of technetium have such advantages as high Tc, good mechanical and electromagnetic properties and fine chemical stability. We think that doping with technetium makes the sample improve its mechanical properties, and doping with Tc is probably the essential reason of the peculiar electromagnetic properties. Futher study to this new material is still in progress.

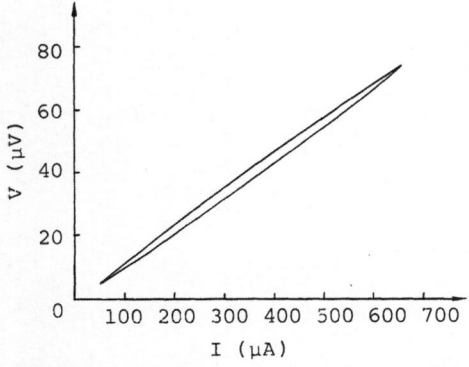

Fig.4   I-V characteristic curve of a sample at room
temperature (303K) under zero field

REFERENCES

1. A.M. Hermann and Z.Z. Sheng, Bulk superconductivity at 120K in the Tl-Ca-Ba-Cu-O system, <u>Nature</u>, 332: 55 (1988).
2. S.S.P. Parkin, V.Y. Lee, et al., Bulk superconductivity at 125K in $Tl_2Ca_2Ba_2Cu_3O_x$, <u>Phys. Rev. Lett.</u>, 60: 2539 (1988).
3. R.S. Wang and H.M. Lee, Recent advance and prospects in the high-Tc superconductive compounds, <u>Chem. Eng. Prog.</u> No.5 (1989).

$Pb_2Y_{0.5}Ca_{0.5}Sr_2Cu_3O_{8+x}$ SUPERCONDUCTING CERAMICS FROM GLASS CRYSTALLIZATION

Haixing Zheng and J.D.Mackenzie

Department of Materials Science and Engineering
University of California–Los Angeles
Los Angeles, CA 90024, USA

ABSTRACT

$Pb_2Y_{0.5}Ca_{0.5}Sr_2Cu_3O_y$ forms a glass. The glass transition temperature and the first crystallization temperature of the glass are $344^\circ C$ and $387^\circ C$, respectively. The density of the glass is 6.3146 g/cm^3. Heat treatments of the glass in air do not produce $Pb_2Ca_{0.5}Y_{0.5}Sr_2Cu_3O_y$ superconducting phase at all temperatures, but heat treatments of the glass in argon yield the $Pb_2Ca_{0.5}Y_{0.5}Sr_2Cu_3O_y$ phase starting at $430^\circ C$. After being heat treated at $660^\circ C$ for 8 hours in argon, the sample is superconducting with an onset T_c of 80 K.

INTRODUCTION

Since high-transition-temperature superconducting La–Ba–Cu–O ceramics were reported[1], a series of new high-Tc superconducting ceramics were discovered, which included: $RBa_2Cu_3O_{7-x}$ (R = Y, Rare Earth)[2], $Y_2Ba_4Cu_8O_y$[3], Bi–Ca–Sr–Cu–O ($Bi_2Sr_2CuO_6$, $Bi_2CaSr_2Cu_2O_8$ and $Bi_2Ca_2Sr_2Cu_3O_{10}$)[4,5], Tl–Ca–Sr–Cu–O ($Tl_2Ba_2CuO_6$, $Tl_2CaBa_2Cu_2O_8$, $Tl_2Ca_2Ba_2Cu_3O_{10}$)[6,7] and Bi–K–Ba–O[8] systems. At the end of last year, superconductivity in the Pb–Sr–R–Cu–O system was reported for $Pb_2Sr_2RCu_3O_{8+x}$ (R is a lanthanide or a mixture of Ln+Sr or Ca) with an onset Tc around 77 K[9,10]. The preparation conditions for this new family of compounds were more strict: direct synthesis by the reaction of the component metal oxides or carbonates in air or oxygen at < $900^\circ C$ is not possible. Successful synthesis was achieved by the reaction of PbO with prereacted (Sr,Ca,R) oxide precursors in N_2 (1% O_2)[9]. Also the superconducting transition of the polycrystalline samples was broad:

Advances in Cryogenic Engineering (Materials), Vol. 36
Edited by R. P. Reed and F. R. Fickett
Plenum Press, New York, 1990

637

$T_{c(onset)}$ = 79 K and $T_{c(zero)}$ = 32 K of Cava et al's samples[9], and
$T_{c(onset)}$ = 76 K and $T_{c(zero)}$ = 55 K of Subramanian et al.'s samples[10].
Subramanian et al. considered that the narrower transition of their
samples might be attributed to the different synthetic method. Since the
pure single phase of superconducting materials should have a much
narrower transition width (ΔT = 1-2 K), alternate techniques to prepare
this new family of high Tc superconductors should be investigated.

The Bi-Ca-Sr-Cu-O system of high Tc superconductors has been
prepared by glass crystallization[11-14]. The experimental results show
several advantages of this technique: the practical pore-free
superconducting ceramics[14], the control of the microstructure of
ceramics[15] and the ease of fabrication of different shapes of
ceramics[16,17]. PbO is a conditional glass former like Bi_2O_3, therefore
the Pb-Y-Ca-Sr-Cu-O system should have a glass-forming region. In this
work we report glass formation in the $Pb_2Y_{0.5}Ca_{0.5}Sr_2Cu_3O_y$ composition.
The crystallization process of $Pb_2Y_{0.5}Ca_{0.5}Sr_2Cu_3O_y$ glasses (P2YCSC
glass) in air and Ar atmosphere have been investigated. The P2YCSC glass
heat treated in Ar atmosphere yields $Pb_2Y_{0.5}Ca_{0.5}Sr_2Cu_3O_y$ phase and
shows superconductivity.

EXPERIMENT

Reagent grade $Pb(NO_3)_2$, Y_2O_3, $CaCO_3$, $SrCO_3$ and CuO were mixed with
the ratio Pb:Y:Ca:Sr:Cu = 2:0.5:0.5:2:3. 5 g batches were melted in a
platinum crucible at 1200°C for 15 minutes. The melt was quenched by a
brass twin-roller, which gives a cooling rate of ~10^7 K/sec. The
thickness of the resulting glasses was about 0.1 mm.

Densities of glasses were measured by the Archimedean method using
toluene as a medium. The glass transition temperature (T_G) and the onset

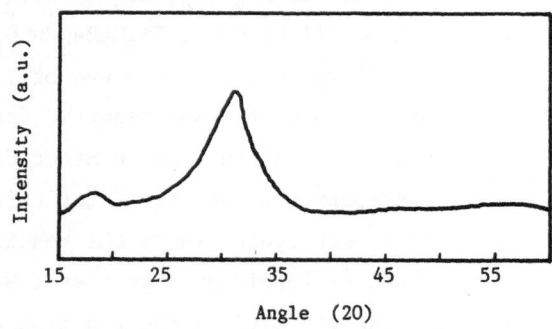

Fig.1 X-ray diffraction pattern of the quenched
$Pb_2Y_{0.5}Ca_{0.5}Sr_2Cu_3O_y$ materials

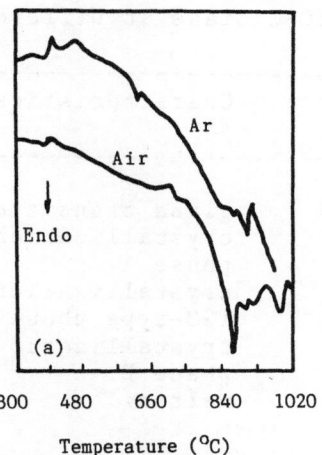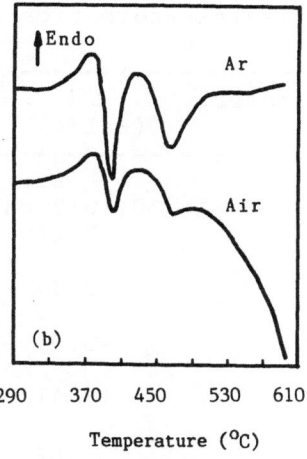

Fig.2   DTA (a) and DSC (b) of P2YCSC glass in air and Ar

crystallization temperature (T_x) were analyzed using the thermal
analysis system (Perkin-Elmer DTA 1700 and DSC) at a heating rate of
$10^{\circ}C$/min. X-ray diffraction was carried on the powder samples using Cu Kα
radiation. Electrical resistivities of the samples were measured by the
standard four-probe technique (the current density for the measurement
was 50 mA/cm^2, a.c. 10 Hz). The temperature was monitored using a diode
sensor with a Model DRC 80 Temperature Controller (Lake Shore
Cryotronics Inc.).

EXPERIMENTAL RESULTS AND DISCUSSION

A. $Pb_2Y_{0.5}Ca_{0.5}Sr_2Cu_3O_y$ Glass

The x-ray diffraction pattern of the quenched material is shown in
Fig.1, which indicates the material is non-crystalline. The thermal
analyses (DTA and DSC) of the quenched material shown in Fig.2 further
support the glass formation by showing the glass transitions. The glass
transition temperatures (T_G) and the crystallization temperature (T_x)
are $344^{\circ}C$ and $387^{\circ}C$, respectively. The density of the glass is 6.3146
g/cm^3.

The difference between T_x and T_G is one of the indicators for
glass stability. The ($T_x - T_G$) of $Pb_2Y_{0.5}Ca_{0.5}Sr_2Cu_3O_y$ glass (P2YCSC
glass) is $43^{\circ}C$. Since normal stable glasses have (T_x-T_G) > $50^{\circ}C$, the
($T_x- T_G$) of $43^{\circ}C$ implies that the P2YCSC glass is unstable. As it is
known, PbO can not form glass by itself. However Pb^{2+} is a highly
polarisable cation. In the presence of strong polarization cations, Pb^{2+}
cations may change the coordination number and enter the glass network.
This may be why P2YCSC can form a glass, but the glass is not stable.

Table 1 Phase Transformation of P2YCSC Glass at Different
         Atmosphere

Onset Temperature ($^\circ$C)	ΔH (cal/g)	Characteristics
	Air	
344.0	0.1012 cal/g.$^\circ$K	glass transition
389.7	-3.21	crystallization of phase A
457.5	-2.34	crystallization of 123-type phase
688.7	-3.01	crystallization of phase B
886.2	48.41	melting
	Ar	
344.1	0.2077 cal/g.$^\circ$K	glass transition
387.1	-7.55	crystallization of $Pb_2Ca_{0.5}Y_{0.5}Sr_2Cu_3O_y$-like phase
444.5	-8.25	phase transformation
625.5	1.15	partial melting
900.2	2.82	melting

B. Crystallization of $Pb_2Y_{0.5}Ca_{0.5}Sr_2Cu_3O_y$ Glass

As shown in Fig.2, P2YCSC glass has several endothermic or
exothermic peaks above the glass transition temperature (T_G). Table 1
lists temperatures of these peaks and the heat of transformations. In
order to understand the charateristics of these peaks, P2YCSC glasses
are heat treated at different temperature (just above each peak
temperature) for four hours, in air or in argon. The x-ray diffraction
patterns of these samples are shown in Fig.3 and 4.

In air, the sample still showed the amorphous x-ray diffraction
pattern after being heat treated at 430°C. Since the releasing energy of
the peak is small (-3.21 cal/g), the change in the structure should be
small. When heat treated at 520°C, the sample consists of 123-type
$(Y,Ca)Sr_2(Pb,Cu)_3O_x$, CuO, and amorphous phase. The amount of 123-type
$(Y,Ca)Sr_2(Pb,Cu)_3O_x$ and CuO increases, while the amorphous phase
vanishes when heat treated at 680°C. Therefore the exothermic peak at
470°C is the crystallization peak of 123-type $(Y,Ca)Sr_2(Pb,Cu)_3O_x$ phase.
Comparision of the patterns of the glasses heat treated at 820°C and
680°C indicate the formation of a new phase, which corresponds to the
exothermic peak at 720°C. Since we have observed high intensity of x-ray
diffraction peaks of CuO in a more PbO-contained sample:
$Pb_3Y_{0.5}Ca_{0.5}Sr_2Cu_3O_y$[18], lead cations substituting copper cations in 123-

640

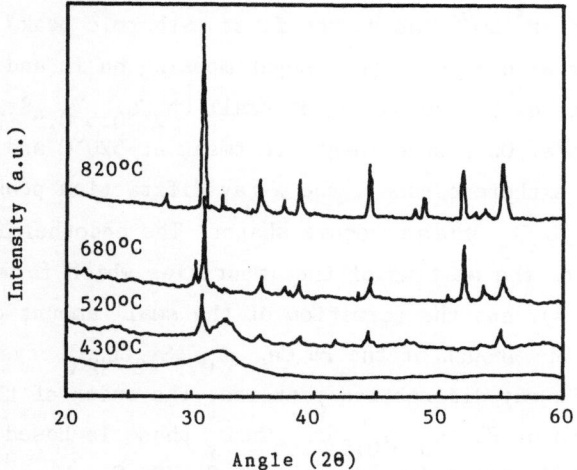

Fig.3 X-ray diffraction patterns of P2YCSC glasses
heat treated at different temperature for
four hours in air

type $(Y,Ca)Sr_2(Pb,Cu)_3O_x$ phase is expected. There is no formation of
$Pb_2Y_{0.5}Ca_{0.5}Sr_2Cu_3O_y$ superconducting phase from the P2YCSC glass at
temperature in air.

The crystallization process of $Pb_2Ca_{0.5}Y_{0.5}Sr_2Cu_3O_y$ glass in argon
is completely different (Fig.4) although the first two exothermic peaks
are located at almost the same temperatures. The heats of transformation
of the first two exothermic peaks in Ar, however, are larger than those
of the peaks in air (Table 1). The x-ray diffraction pattern of P2YCSC

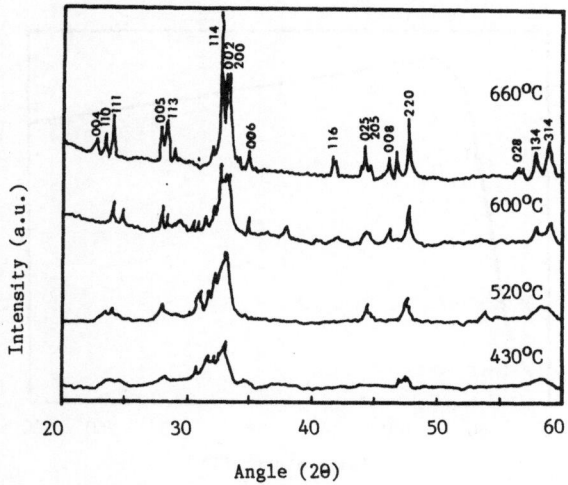

Fig.4 X-ray diffraction patterns of P2YCSC glasses
heat treated at different temperature for
four hours in Ar

glass heat treated at $430^{\circ}C$ (above the first exthermic peak) in argon shows a broad peak with a sharp peak superimposing on it and a peak at $-47^{\circ}C$, which indicates the formation of small $Pb_2Ca_{0.5}Y_{0.5}Sr_2Cu_3O_y$ crystal in the glass. On further heat treatment at $520^{\circ}C$ and $600^{\circ}C$ (above the second exthermic peak), the x-ray diffraction peaks of the $Pb_2Ca_{0.5}Y_{0.5}Sr_2Cu_3O_y$ phase become shaper. The endothermic peak at $625^{\circ}C$ may be due to the melting of the impurities which formed at low temperatures (Fig.4), and the formation of the small amount of the melt at $625^{\circ}C$ assists the growth of the $Pb_2Ca_{0.5}Y_{0.5}Sr_2Cu_3O_y$ crystals which gives the perfect x-ray diffraction patterns. The index of this diffraction pattern of $Pb_2Ca_{0.5}Y_{0.5}Sr_{0.5}Cu_3O_y$ phase is based on the published x-ray diffraction data for the $Pb_2Sr_2YCu_3O_8$ phase[9]. The sample is completely melted when heat treated at $860^{\circ}C$. It is clear that crystallization of the $Pb_2Ca_{0.5}Y_{0.5}Sr_2Cu_3O_y$ phase from the P2YCSC glass is a gradual transformation, since at all temperatures the x-ray diffraction patterns show $Pb_2Ca_{0.5}Y_{0.5}Sr_2Cu_3O_y$ phase diffraction peaks with minor differences in the separation of the x-ray diffraction peaks.

C. Superconductivity

Fig.5 shows the relative resistivity (R/R_{300K}) of the P2YCSC glass (heat treated at $660^{\circ}C$ for eight hours in Ar) as a function of temperature. The resistance of the sample is about 5 ohms, which is much higher than other high T_c superconducting oxides. As shown in Fig.5, the

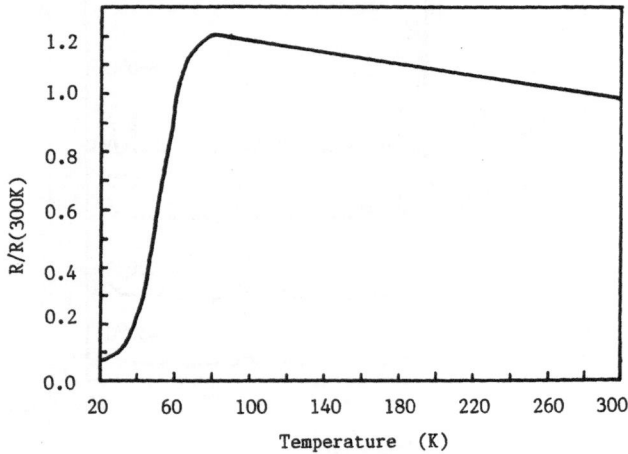

Fig.5 The dependence of the relative resistivity of the
P2YCSC glass ceramics (heat treated at $660^{\circ}C$ for
eight hours in Ar) on temperature

resisvity of the sample increases with decreasing temperature, behaving like a semiconductor. At 80 K, the resistivity starts decreasing, then shows a large drop. However the decrease of resisvity becomes smooth at 50 K. The sample does not show zero resistance even at 20 K. The reason still needs to be investigated.

D. Summary and Conclusion

There is a glass formation in $Pb_2Y_{0.5}Ca_{0.5}Sr_2Cu_3O_y$. The glass transition temperature and the first crystallization temperature of the glass are $344^\circ C$ and $387^\circ C$, respectively. The density of the glass is 6.3146 g/cm^3. Heat treatments of the glass in air do not produce the $Pb_2Ca_{0.5}Y_{0.5}Sr_2Cu_3O_y$ superconducting phase at all temperatures, while the heat treatments of the glass in argon yield the $Pb_2Ca_2Y_{0.5}Sr_2Cu_3O_y$ phase starting at $430^\circ C$. After being heat treated at $660^\circ C$ for 8 hours in argon, the sample becomes superconducting with onset T_c of $80^\circ K$.

Acknowledgements: The financial supports of the Strategic Defense Initiative Organization and the Air Force Office of Scientific Research, Directorate of Chemical and Atmosphere Science are grately appreciated.

REFERENCES

1. J.G.Bednorz and K.A.Muller, Z.Phys.B64, 189(1986)
2. M.K.Wu, J.R.Ashburn, C.J.Torng, P.H.Hor, R.J.Meng, L.Gao, Z.J.Huang, Y.Q.Wang and C.W.Chu, Phys.Rev.Lett., 58, 908(1987)
3. A.F.Marshall et al., Phys.Rev.B37, 9353(1988)
4. H.Maeda, Y.Tanaka, M.Fukutomi and T.Asano, Jpn.J.Appl.Phys., 27, L209(19888)
5. C.W.Chu, J.Bechtold, L.Gao, et al., Phys.Rev.Lett., 60, 941(1988)
6. Z.Z.Sheng and A.M.Hermann, Nature 332, 138(1988)
7. R.M.Hazen, et al., Phys.Rev.Lett., 60, 1657(1988)
8. L.F.Mattheiss, E.M.Gyorgy and D.W.Johnson, Jr., Phys.Rev.B 37, 3745(1988)
9. R.J.Cava, B.Batlogg, J.J.Krajewski, et al., Nature, 336, 211(1988)
10. M.A.Subramanian, J.Gopalakrishnan, C.C.Torardi, et al., Physica C, 157, 124(1989)
11. D.G.Hinks, L.Soderholm, D.W.Capone II, et al., Appl.Phys.Lett., 53, 423(1988)
12. T.Komatsu, R.Sato, K.Imai, et al., Jpn.J.Appl.Phys., 27, L550(1988)
13. T.Minami, Y.Akamatsy, M.Tatsumisago, et al., Jpn.J.Appl.Phys., 27, L777(1988)

14. Haixing Zheng and J.D.Mackenzie, Phys.Rev., B38, 7166(1988)

15. Ren Xu, Haixing Zheng and J.D.Mackenzie, to be published in J.Non-Cryst. Solids

16. Y.Abe, H.Hosono, M.Hosoe, J.Iwase and Y.Kubo, Appl.Phys.Lett., 53, 1431(1988)

17. Haixing Zheng and J.D.Mackenzie, to be published in the Proceeding of the XV International Congress on Glasses, Leiningrad, USSR, July, 1989

18. Haixing Zheng and J.D.Mackenzie, to be published

THE ORIGIN AND FUTURE OF COMPOSITE ALUMINUM CONDUCTORS

C. E. Oberly[1] and J. C. Ho[2]

[1]Wright R&D Center, Wright-Patterson AFB, OH 45433

[2]Wichita State University, Wichita, KS 67208

ABSTRACT

A high purity aluminum composite conductor with an aluminum alloy matrix has been developed which retains both a very high resistance ratio and yield strength when cooled by liquid hydrogen. In typical cryogenic magnet applications, the composite aluminum approach is lighter in weight and more versatile than superconductors at any temperature as long as large flows of cryogenic hydrogen are permissible.

INTRODUCTION

Potential applications of pure metal conductors with extremely low electrical resistance at cryogenic temperatures were considered in the 1960's.[1-5] The promise of high purity aluminum for achieving high operating electrical resistance ratios (RR=resistance at room temperature/resistance at operating conditions) with high residual resistance ratio pure aluminum (RRR=resistance at room temperature/resistance at 4.2K) was not immediately delivered.[6-9] Retention of manufactured properties in the ultra-pure aluminum (5-9's to 6-9's pure) was difficult in practical environments because of extremely low strength[10], large grain effects[8], magnetoresistance[6,9], work hardening during magnet winding, uncontrolled annealing (even at room temperature and below) and, cyclic strain induced increases in resistance [11-13].

In 1961, the famous International Conference on High Magnetic Fields was held at M.I.T. in Cambridge MA where both Type II superconducting magnet demonstrations[14] and large liquid cryogen cooled magnet demonstrations [1,2,15] were announced. Water cooled magnets had reached practical cooling and power limits and were difficult to extend to higher field and larger volumes[16]. The trade off between cryogenic resistive magnets and superconducting magnets became more and more favorable for superconducting magnets cooled by liquid helium. The best cryogenic coolants to provide maximum benefit for resistive magnets were liquid hydrogen at 21K[1,4,5,15] and liquid neon at 27K[2,3]. The impact of liquid hydrogen hazards and liquid neon

Advances in Cryogenic Engineering (Materials), Vol. 36
Edited by R. P. Reed and F. R. Fickett
Plenum Press, New York, 1990

645

costs ($250 1/liter today) on cryogenic resistive magnet operation combined with successful demonstration of stabilized superconductive conductors and magnets after 1968 resulted in superconductors dominating the laboratory and commerical market because thay were cheaper and easier to build and operate.

Depending on the application and the superconducting material, an operating temperature of 2 to 10 K is mandatory to provide thermal stability. The helium coolant and refrigeration cycles required to maintain low temperature impose tremendous penalties for special applications such as space power systems, associated with maintainability, reliability, electrical power consumption, heat rejection radiators, as well as the extra weight of peripheral equipment. For these reasons, utilization of liquid hydrogen, often available in space vehicles [17], makes pure metal conductors again attractive. It has been long recognized [4,5] that aluminum is the best material for this purpose with low weight and magnetoresistance among the main benefits. For example, a pure aluminum conductor with a residual resistance ratio of less than 1000 can perform the same job as a superconductor. Moreover, aluminum is only 40% of the mass of typical superconductors or other pure metals and could carry much higher current densities than the practical limit of 2×10^8 A/m^2 expected of lightweight superconductors. However, application of aluminum was stymied by unreliable mechanical and electromagnetic properties.

A broad range of data in terms of impurity, magnetic field, size, strain, and fatigue effects on low temperature resistivity of high purity aluminum is now available[8-13]. The major disadvantage of an aluminum conductor is its very low strength. This fact is even more serious for such applications as pulsed power devices where, in order to allow full current penetration during a short pulse, the conductors must be in the form of fine filaments. Since a braid of the filaments cannot be structurally supported, a better approach would be to embed aluminum filaments in a strong alloy matrix. In manufacturing such composite materials, enormous residual strain and deformation of the pure aluminum induces work hardening, causing a large increase in electrical resistance. Annealing of the heavily worked composite can remove most of the residual strain, but this causes diffusion of impurities or alloying elements from the matrix into the pure aluminum filaments and lowers the RRR value. Incorporation of diffusion barriers such as tantalum or niobium could alleviate these conditions, but these very hard materials cannot be processed in a practical manner with the very soft, ductile, pure aluminum in the composite. Catastrophic sausaging of aluminum filaments or fracture of the entire work piece usually prevails. Male and Iyer at Westinghouse[18] pioneered multifilament aluminum using a structural aluminum alloy matrix with some success, but impurity diffusion remained a serious problem.

The breakthrough in producing composite aluminum came about when a new Al-Fe-Ce alloy [19,20] was tested as the matrix at the Wright Research and Development Center [21-24] following the work at Westinghouse[18]. This lightweight, high strength material with favorable thermal and electrical properties was initially developed at ALCOA for high temperature applications[19-20]. With 8.4 and 3.6 wt.% Fe and Ce, respectively, it derives its strength from densely dispersed fine intermetallic particles, particularly in the vacuum hot-pressed (VHP) condition following cold-compaction of powder snythesized products. Some of the relevant properties were reviewed earlier [21]. The utility of the thermal stability of the dispersed Al-Fe-Ce intermetallic particles is also very valuable in reducing diffusion to pure aluminum filaments during annealing of a composite.

EXPERIMENTAL PROCEDURES

Feasibility studies of the composite conductor approach with an Al-Fe-Ce matrix began in 1984 with extrusion of composite billets using streamlined dies. Processing parameters (temperature and strain rate) were selected based on dynamic materials modeling such that dynamic recrystallization of the original powder-metallurgy matrix would occur[22]. In comparison with the VHP condition, dynamic recrystallization resulted in a product having lower room-temperature yield strength but much improved microstructure in terms of elimination of prior powder particle boundaries and a more homogeneous intermetallic particle distribution, which then allowed subsequent extrusion to be carried out at lower temperatures. Two sets of composite aluminum experiments, which had previously been reported, [23-24] can be summarized in the following two paragraphs.

Rods of 68.9 cm (19 1/4")diameter of commercially-pure (99.8%) Al were first inserted into drilled-through holes in a 5.1 cm (2")-diameter x 10.2 cm (4")-long Al-Fe-Ce alloy billet. The composite billet was hot-extruded to an area reduction of 12:1. Seven sections of the extrusion product were re-stacked in an Al matrix and re-extruded at 12:1. Finally, part of the 133-filament composite was further extruded in multisteps at room temperature to a final diameter of 7.9 mm (0.31") at Westinghouse [18]. This represented an overall area reduction of more that 100,000 times for each filament. Micrographs showed well defined boundaries between the matrix and filaments without observable cracks, indicating that the alloy and pure Al can be co-processed successfully.

A second billet was prepared in the same way except that the nineteen commercially-pure Al rods were replaced by seven high purity (Puratronic grade, 99.98%) Al rods from Johnson Matthey Chemical, Ltd. so that Fe and Ce diffusion evaluations could be made. Following a 16:1 area reduction, no detectable Fe and Ce was found in the Al filaments by electron microprobe analysis, setting an upper limit of their concentration at 100 ppm. Additional verification of this observation was based on residual resistivity ratio determinations at room temperature and 4.2 K. An electrical resitivity sample was prepared by machining a 22.9cm (9")-long section of the extrusion product to a 0.8cm (5/16")-diameter; the seven Al filaments comprised 1/4 of the total cross section with the remaining 3/4 being the alloy matrix. Following a 2-hour anneal at 200C, the RRR value was found to be 400. By coupling this number with that of the matrix (RRR=17), it was concluded that the RRR value for Al filaments alone would be close to 900. This RRR is comparable to that of stress-free Al with a 99.99% purity.

Described here are results of an unreported third set of experiments which further demonstrate the feasibility of this technology. Figure 1a shows a 7.6 cm (3")-diameter x 7.6cm (3")-long alloy ingot containing an array of sixty-one 0.635cm

(1/4")-diameter high purity Al rods from ALCOA which represents 50% of the cross sectional area. The cross section of a 20:1 extrusion product is shown in Figure 1b, again without appreciable distortion of the geometry. Following successful re-stacking and similar 20:1 extrusion procedures, Figure 1c and Figure 1d show the products containing 427 and 2,989 filaments, respectively. The latter product would contain filaments so small that resistance size effects would contribute.

The feasibility of reducing the composites to flat conductors by rolling was examined. Figure 2 shows cross sections of the above-mentioned extrusion products with a sequence from round to a high aspect ratio flat strip which still contains seven separated high-purity filaments. Initial attempts at rolling with an external shell of Al-Fe-Ce resulted in fish scale fractures because of the high hardness and low deformation character of the material as shown in Figure 3. Subsequent cladding of the extrusion shell by a layer of softer aluminum alloy such as 1100 resulted in success. The bottom picture in Figure 2 shows the external cladding before it is etched from the surface. Figure 3 illustrates the smooth surface after cladding beside the failure with an Al-Fe-Ce surface.

DISCUSSION

The availability of the composite aluminum has led to a number of development efforts conducted by others in the areas of energy storage [25-26] and power generation [27-29]. These proceedings also include a number of papers describing properties of the composite aluminum[30-37]. Figure 4 illustrates various reduction products developed by Innovare Inc [36]. In

(a) (b)

(c) (d)

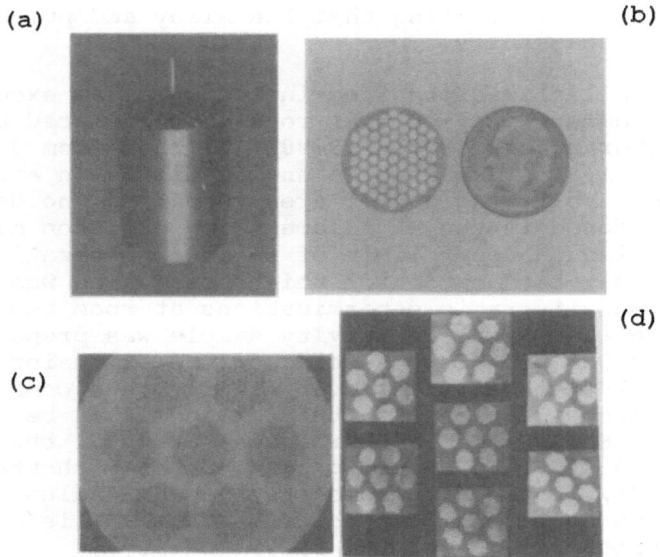

Fig. 1   Successive reductions and restacking of (a) pure
         aluminum rods in an Al-Fe-Ce billet with 61 filaments,
         (b) after reduction, (c) restacking and reduction with
         427 filaments and (d) with 2989 filaments.

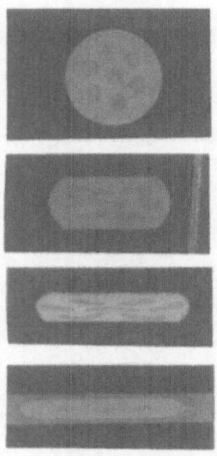

Fig. 2 Successive reduction of
a 7-filament composite
by Rolling.

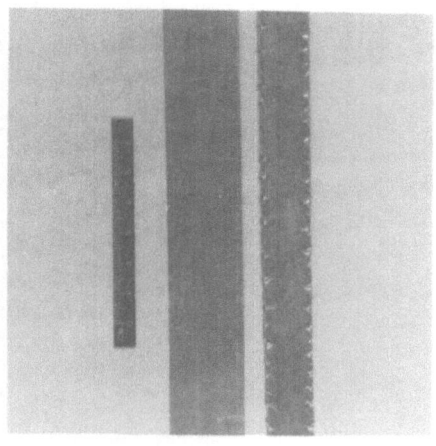

Fig. 3 Rolled 7-filament composite with cladding on
right and without
cladding on left.

addition the heat transfer to hydrogen has been re-examined for
the low temperature difference required to maintain low RR,
typically in the range of 2 to 5K [38-40].

The development of composite aluminum has not been without
problems, and a complete understanding of anomalous magnetoresistance induced in the composite is still not in hand [30,31,35].
In addition to the amonalous magnetoresistance problem, a clear
design procedure for establishing an operational resistance at a
given temperature, magnetic field and strain history is not
completely developed for the composite.

The aluminum composite is also subject to coupling losses
in transient magnetic field conditions similar to that
experienced by superconductors. Carr[41] and others [30] at
Westinghouse have analyzed losses resulting from aluminum
filaments of very high purity embedded in the structural Al-Fe-
Ce matrix which is still a very good conductor [21,33]. A simpler
method for significantly reducing losses is to incorporate
resistive barriers around the filaments [25,41]. The successful
incorporation of the resistive barriers is a daunting task that
remains for future research to resolve.

CONCLUSIONS

Although aluminum is much lighter weight than
superconductors, either metallic or ceramic, it will compete
well against either type of superconductor only in environments
where liquid hydrogen is readily available such as aboard the
Space Shuttle. Coolant requirements for large generators or
inductors will be many times less than that required for the
fuel flow to the Space Shuttle main engine, and the conductor
coolant will be rejected at 2 to 5 Kelvin above the storage tank
supply temperature. Large power and magnet devices will still
require 100's of grams to kilograms of liquid hydrogen flow per
second which is a severe penalty for most applications.
Consequently, ceramic superconductors will generally be more

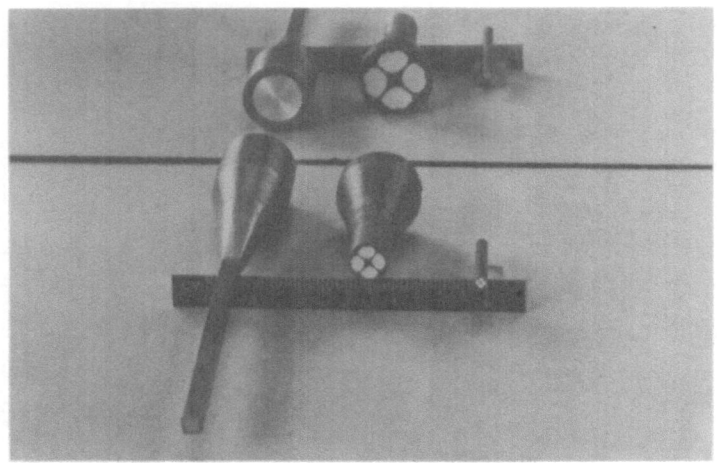

Fig. 4 Extruded aluminum composites produced by Innovare Inc.[36]

practical to cool in similar applications when they achieve the necessary development status.

The composite aluminum will always be the lightest weight approach for magnet and power applications as one of the authors stated long ago [4]. The new aluminum composite conductor reported here has the potential of providing a readily, if not easily, fabricated product in a wide variety of forms that can provide unique properties at 21K in a conductor at a yield strength of 0.35GPa (50Ksi) and an operating resistance ratio of 400 to 500 in magnetic fields up to 10 Tesla at a limiting mechanical strain of 0.1 to 0.2%.

ACKNOWLEDGMENTS

H.L. Gegel (now at Universal Energy Systems Inc., in Dayton, OH) suggested use of an Al-Fe-Ce matrix for low impurity diffusion and high strength in the matrix. Don Linder of the University of Dayton Research Institute, Dayton, OH, cheerfully and without mistake, fabricated and assembled the composite billets. A number of people in the High Temperature Materials Branch of the Materials Laboratory at the Wright Research and Development Center provided helpful advice and operated the extrusion press. We also thank A.T. Male and N. Iyer for their pioneering work that showed the way and for the initial drawdown of the composite to the fine wire.

REFERENCES

1. J.R. Purcell, "An Aluminum Magnet Cooled with Liquid Hydrogen," Proceedings of the International Conference on High Magnetic Fields at Cambridge, MA on 1-4 November 1961, publ. as High Magnetic Fields, ed. by H. Kolm, B. Lax, F. Bitter and R. Mills, Wiley & Sons, Inc., New York, (1962) 166-169.
2. J.C. Laurence and G.V. Brown, "A Large Liquid-Neon-Cooled Electromagnet," ibid, 170-179.

3. J.C. Laurence and W.D. Coles, "Design Construction and Performance of Cryogenically Cooled and Superconducting Electromagnets, Proceedings of the International Symposium of Magnet Technology, ed. by H. Brechna and H.S. Gordon, 8-10 September 1965 at Stanford, CA, Published by the National Bureau of Standards, Springfield, VA, Report No. CONF-650922, (1965) 574-587.

4. R.R. Barthelemy and C.E. Oberly, "Optimum Magnets for MHD Generators," International Conference on MHD, Salzburg, Austria in 1966, as published in "Electricity Form MHD," Vol. III, International Atomic Energy Agency, Vienna, (1966).

5. V.D. Arp, M.B. Kasen and R.P. Reed, "Magnetic Energy Storage and Cryogenic Aluminum Magnets," Report AFAPL-TR-68-87, Air Force Aero Propulsion Laboratory, Wright-Patterson AFB (1969).

6. J.R. Purcell, Personal Communications.

7. G.V. Brown and W.D. Coles, Personal Communication, 1989.

8. F.R. Fickett, "Aluminum 1: A Review of Resistive Mechanisms in Aluminum," Cryogenics 11 (1971) 349-367. (194 References).

9. F.R. Fickett, "Magnetoresistance of Very Pure Polycrystalline Aluminum," Phys. Rev. B3 (1971) 1941 (38 References).

10. R.P. Reed, "Aluminum 2: A Review of Deformation Properties of high Purity Aluminum and Dilute Aluminum Alloys," Cryogenics 12 (1972) 259-291. (471 References)

11. K.T. Hartwig, G.S. Yuan and P. Lehman, "Strain Resistivity at 4.2K in Pure Aluminum," Advances in Cryogenic Materials, Vol. 32, edited by R.P. Reed and A.F. Clark, Plenum Publ. Co., (1986), 405-412.

12. G.S. Yuan, P. Lehman and K.T. Hartwig, "The Effects of Prestrain on Low Temperature Fatigue Induced Resistivity in Pure Aluminum," Advances in Cryogenic Materials, Vol. 32, edited by R.P. Reed and A.F. Clark, Plenum Publ. Co., New York, NY (1986) 413-419.

13. K.T. Hartwig and G.S. Yuan, "Strength and Resistivity Changes Caused by Cyclic Strain at 4.2K in Pure Aluminum," Proceedings of the International Cryogenic Conference as published in Cryogenic Materials '88, Vol. II, "Structural Materials," (1988), pg. 677.

14. J.E. Kunzler, "Superconductivity in High Magnetic Fields at High Current Densities", Proceedings of the International Symposium on Magnetic Technology, publ. as, High Magnetic Fields, ed. by H. Kolm, B. Lax, F. Bitter and R. Mills, Wiley & Sons, Inc., New York (1962) 574-579.

15. H.L. Laquer, "The Cryogenic Magnet Program at Los Alamos," ibid., 156-165.

16. F. Bitter, "Water-Cooled Magnets," ibid., 85-100.

17. NASA Design Monogram (NASA SP-8048), "Liquid Rocket Engine Turbopump Bearings," March 1971.

18. A.T. Male and N. Iyer , Westinghouse Research and Development Center, Pittsburgh PA, personal communication.

19. W.M. Griffith, R.E. Sanders, Jr., and G.J. Hildeman, "Elevated Temperature Aluminum Alloys of Aerospace Applications," in: "High Strength Powder Metallurgy Aluminum Alloys." M.J. Koczak and G.J. Hildeman , eds., The Metallurgical Society of AIME, Warrendale, PA, (1982) 209-224.

20. S.D. Kirchoff, R.H. Young, W.M. Griffith, and Y. Kim, "Microstructure/Strength/Fatigue Crack Growth Relations in High Temperature P/M Aluminum Alloys," ibid., 237-248.

21. J.C. Ho, C. E. Oberly, H.L. Gegel, W.M. Griffith, J.T. Morgan, W.T. O'Hara and Y.V.R.K. Prasad, "A New Aluminum-Base Alloy with Potential Cryogenics Applications, <u>Advances in Cryogenic Engineering- Materials</u>, edited by R.P. Reed and A.F. Clark, Plenum Publ. Co., <u>32</u> (1986) 437-442.

22. Y.V.R.K. Prasad, H.L. Gegel, S.M. Doraivelu, J.T. Morgan, K.A. Lark, and D.R. Barker, "Modeling of Dynamic Material Behavior in Hot Deformation: Forging of Ti-6242," <u>Metal. Trans.</u>, <u>15A</u> (1984) 1883-1892.

23. J.C. Ho, C.E. Oberly, H.L. Gegel, W.T. O'Hara, J.T. Morgan, Y.V.R.K. Prasad, and W.M. Griffith, "Composite Aluminum Conductors for Pulsed Power Applications at Hydrogen Temperatures," Proceedings for the 5th IEEE Pulsed Power Conference, Arlington, VA (1985) 627-629.

24. C. E. Oberly, J.C. Ho, and H.L. Gegel, "Composite Aluminum Conductor for Pulsed Power Applications at Liquid Hydrogen Temperature," U.S. Patent No. 4,711,825, Awarded 8 Dec 1987.

25. S. K. Singh, W.J. Carr, Jr., T.J. Fagan, Jr., T.D. Hordubay, and H.L. Chuboy," Compact Cryogenic Inductors," Proceedings of the 1989 Cryogenic Engineering Conference, to be published in <u>Advances of Cryogenic Engineering</u>, Vo. 35, Plenum Press Publ. Co., New York, NY., 1990.

26. D.E. Johnson and E. Ruckstadter, "A Hyperconducting 5 Mj, 1MA Pulse Transformer," ibid.

27. P.W. Eckels, T.J. Fagan, Jr., J.H. Parker, Jr., A. Patterson and L.J. Long, "Cryogenic Generator Cooling," ibid.

28. C.E. Oberly and R. L. Schlicher, "Cryogenic Aluminum Wound Rotor Concept," IEEE Proceedings of the 5th IEEE Pulsed Power Conference, Arlington, VA (1985) 752-755.

29. C.E. Oberly and R.L. Schlicher, "Cryogenic Aluminum Wound Rotor for Lightweight, High Voltage Generators," U.S. Patent No. 4,739,200, Awarded 19 April 1988.

30. P.W. Eckels, N.C. Iyer, A. Patterson, A.T. Male, J.M. Coltman, and J.H. Parker, Jr., "Magnetoresistance: The Hall Effect in Composite Aluminum Cyroconductors," Cyrogenics <u>29</u> (1989) 748.

31. F.R. Fickett and C.A. Thompson,"Anomalous Magnetoresitance in Al/Al Alloy Composites,", Proceedings of the 1989 International Cryogenic Materials Conference, to be published in <u>Advances of Cryogenic Engineering-Materials,</u> Vo. 36, Plenum Press Publ. Co., New York, NY, 1990.

32. C.A. Thompson, A.J. Zink and F.R. Fickett, "Magnetoresitance Studies of Metal/Metal Composite Conductors," ibid.

33. K.T. Hartwig and R.J. DeFrese,"Cyrogenic Mechanical and Electrical Testing of Aluminum Cyroconductors", ibid.

34. M.K. Premkumar, F.R. Billamn, D.J. Chakrabarti, R.K. Dawless and A.R. Austen "Composite Aluminum Conductor for High Current Density Applications at Cyrogenic Temperatures", ibid.

35. P.W. Eckels, N.C. Iyer, A. Patterson, A.T. Male, J.H. Parker, Jr., and J.W. Coltman," Magnetoresistance: The Hall Effect in composite Aluminum Cyroconductors," ibid.

36. A.R. Austen and W.L. Hutchinson, "Semi-Continuous Hydrostatic Extrusion of Composite Conductors," ibid.

37. W.N. Lawless, "Thermal Properties of Composite Aluminum Conductor in the 13-30K Range," ibid.

38. B. Louie, "Onset of Nucleate and Film Boiling Resulting in Transient Heat Transfer to Liquid Hydrogen," Proceedings of the 1989 Engineering Conference, to be published in

Advances of Cryogenic Engineering, Vo. 35, Plenum Press
Publ. Co., New York, NY , 1990.

39. M.W. Dew, J.M. Strohmayer and E.R. Johnson, "Heat Transfer
    in Pressurized Liquid Hydrogen at Low Values of Film Delta-
    T," ibid.

40. S.S. Kang and C.J. Crowley, "Effect of Current Pulses on the
    Resistivity of a Cyrogenic High-Purity Aluminum Conductor,"
    ibid.

41. W.J. Carr, Jr., "AC Loss in a Composite Hyperconductor,"
    ibid.

MAGNETORESISTANCE IN COMPOSITE CONDUCTORS

P. W. Eckels and J. H. Parker, Jr.

Westinghouse Electric Corporation
Pittsburgh, PA 15235-5098

ABSTRACT

A recent paper recorded[1] the authors' participation in a study which hypothesized the existence of a Hall generation contribution to the apparent magnetoresistivity of an aluminum composite conductor. The present effort developes a more exact mathematical analysis of the Hall generation and compares the model's predictions with experimental data[2] unavailable at the time of earlier publication. The new model, being more rigorous, supports the hypothesis that Hall generation losses will contribute to the apparent magnetoresistivity whenever field and current interact to produce a nonuniform Hall voltage. Hall generation losses, then, could be expected in conductors in a variety of circumstances: in composites due to current maldistribution associated with dissimilar resistivities of the components, in composites due to dissimilar Hall coefficients of different materials such as copper and aluminum, and in homogeneous conductors due to current maldistribution during the transient diffusion of current. The present analysis shows improved comparison with the experimental data at 4.2K.

INTRODUCTION

Figure 1 defines the terms and coordinate system for the Hall voltage which is written as

$$\bar{V}_H/t = \bar{J} \times \bar{B} \tag{1}$$

Neglecting the tensorial nature of the Hall phenomena, a single dimension coefficient, R_H called the ordinary Hall coefficient, is defined as

$$R_H = (V_H/t)/(\bar{J} \times \bar{B}) \tag{2}$$

Extending the analogy to Ohms law, R_H is frequently identified as the Hall resistance; with suitable choices for area and length,

Advances in Cryogenic Engineering (Materials), Vol. 36
Edited by R. P. Reed and F. R. Fickett
Plenum Press. New York. 1990

655

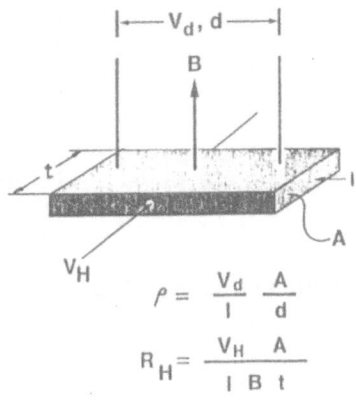

$$\rho = \frac{V_d}{I}\frac{A}{d}$$

$$R_H = \frac{V_H}{I}\frac{A}{B\,t}$$

Fig. 1. Definition of the Hall coefficient.

the Hall resistivity from

$$\rho_H = R_H/(L/A) \qquad\qquad 3$$

and the inverse of resistivity, the Hall conductivity, σ_H. Typical values for the Hall resistivity of bulk polycrystalline Al and Cu at 4K, two materials of interest as stabilizers for super-conductors, are 10^{-10} and -10^{-10} Ohm-m/Tesla respectively.

Now if the conductor of Figure 1 is considered a small differential element of a larger conductor, it is evident that the Hall voltage is a volumetric voltage generated by local field and local current density, which may, vary locally and if a current path is provided, produce current flow.

Our interest in Hall generation within a conductor was aroused by the composite conductor shown in Figure 2, which was first described by Eckels et al.[1] That composite conductor or a

Fig. 2. Composite conductor with a core of pure Al and a protective sheath of Al-Fe-Ce.

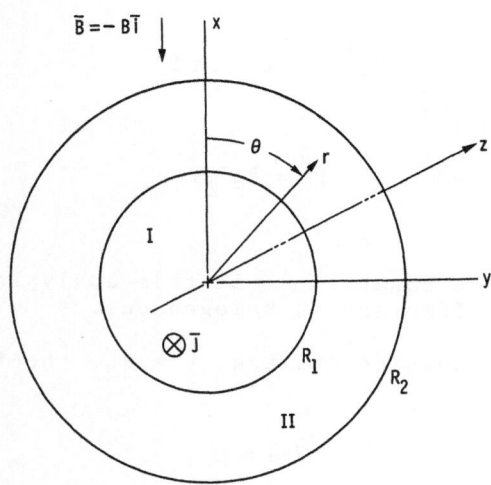

Fig. 3. Geometry and coordinate system
for the composite conductor.

variant thereof may satisfy some of the requirements of the
conductor identified by Schlicter and Oberly.[3] The Hall generation analysis and high purity Al magnetoresistivity data
presented by Eckels et al. were sufficient to explain some of the
anomalous behavior of the magnetoresistivity of the composite
conductor. This paper is the second on that subject. Here the
Hall generation is mathematically treated and more recent higher
field data[3] is used as the basis for comparison of theory and
experiment.

THEORY

For simplicity, two concentric conducting right circular
cylinders of different conductivity but without interface resistance are considered as shown in Figure 3. In this case the pair
carry a transport current which is principally confined to the
inner cylinder due to its much higher conductivity. Figure 3
illustrates the geometry, including the externally imposed magnetic field, $-B$ i, and the terminology used in the following
development. At the interface, continuity of current flow normal
to the boundary and continuity of electric field along the boundary are imposed as conditions. With these boundary conditions,
the solutions to the differential equations for the inner and
outer domains, $r < R_1$ and $R_1 < r < R_2$, with their differing Hall
voltages are matched. The Hall generation loss is then the integral of E·J over the domain. As solutions to $\nabla^2 \psi = 0$, we have

$$\psi^I = a_1 r \sin \theta \qquad\qquad 4$$

$$\psi^{II} = (a_2 r + b/r) \sin \theta \qquad\qquad 5$$

The electric field is $- \nabla \psi$ and the Hall electric field is

$$\bar{E}^H = - E^H j \qquad\qquad 6$$

or

$$E_r^H = -E^H \sin \theta \qquad 7$$

where

$$E^H = R_H J_Z B \qquad 8$$

in accordance with Equation 2. In this analysis we shall assume that the Hall coefficient is homogeneous.

Now, at the outside surface, $r = R_2$, the following relation exists:

$$\nabla \cdot \bar{J} = 0 \qquad 9$$

or

$$\sigma E_r^I = 0 \qquad 10$$

From Equation 5 we develop a relation between two of the three arbitrary integration constants.

$$a_2 - b/R_2^2 = 0 \qquad 11$$

Conditions at the interface are now used to further specify the constants. Continuity of the radial current at the interface expresses as ▾ · J = 0 and equality of the azimuthal electric field expressed as ▾ X Ē = 0 yields the following set of relationships.

$$\sigma^I(E_r^I + E_r^H) = \sigma^{II} E_r^{II} \qquad 12$$

or

$$\sigma^I(a_1 + E^H) = \sigma^{II}(a_2 - b/R_1^2) \qquad 13$$

$$E_\theta^I = E_\theta^{II} \qquad 14$$

or

$$a_1 = a_2 + b/R_1^2 \qquad 15$$

Letting $\beta = R_2/R_1$ $\qquad 16$

and evaluating the integration constants using equations 11, 13 and 15 we have:

$$a = \frac{-E^H}{(1 + \beta^2) + (\sigma^{II}/\sigma^I)(\beta^2 - 1)} \qquad 17$$

$$b = a_2 R_2^2 \qquad\qquad 18$$

$$a_1 = a_2[1 + \beta^2] \qquad\qquad 19$$

The scalar potentials are then written as

$$\psi^I = -\alpha E_H r \sin \theta \qquad\qquad 20$$

where

$$\alpha = [1 + (\sigma^{II}/\sigma^I)(\beta^2 - 1)/(\beta^2 + 1)]^{-1} \qquad\qquad 21$$

and

$$\psi^{II} = \frac{\alpha E_H}{1 + \beta^2}\left[r + \frac{R_2^2}{r}\right]\sin \theta \qquad\qquad 22$$

Note that if $\sigma^I \gg \sigma^{II}$, $\alpha \sim 1$. \qquad\qquad 23

The electric field is written as

$$E_r = -\frac{\partial \psi}{\partial r} \qquad\qquad 24$$

$$E_\theta = -\frac{1}{r}\frac{\partial \psi}{\partial \theta} \qquad\qquad 25$$

If we consider the case of $\sigma^I \gg \sigma^{II}$, , or $\alpha \rightarrow 1$, most of the Hall generation losses are confined to region II and the Hall power loss is given as

$$P = \int_o^{2\pi} d\theta \int_{R_1}^{R_2} \sigma^{II}\left(E^{II}\right)^2 r dr \qquad\qquad 26$$

Evaluating the integral yields

$$P = \pi\sigma^{II}E_H^2 R_1^2(\beta^2 - 1)/(\beta^2 + 1)\,[W/m] \qquad\qquad 27$$

Equation 27 is the relation sought describing the Hall power generation, or loss, if the conductivity of the inner core is much greater than that of the outer, i.e., high conductivity Al within a structural sheath. Similar analysis can describe materials of like conductivity with different Hall coefficients such as Cu and Al. Fourier analysis can be used to solve the similar problem where the conductor is rectangular.

Equation 27 compares favorably with the simple elemental circuit model equation presented earlier. The geometric term $\pi R_1^2(\beta^2 - 1)/(\beta^2 + 1)$ and the earlier geometric term, $(2t\delta)$, are in the ratio of 0.6. Due to the large sample corner radii shown in Figure 2, the circular geometry analysis is used in this comparison of the theoretical and experimental results.

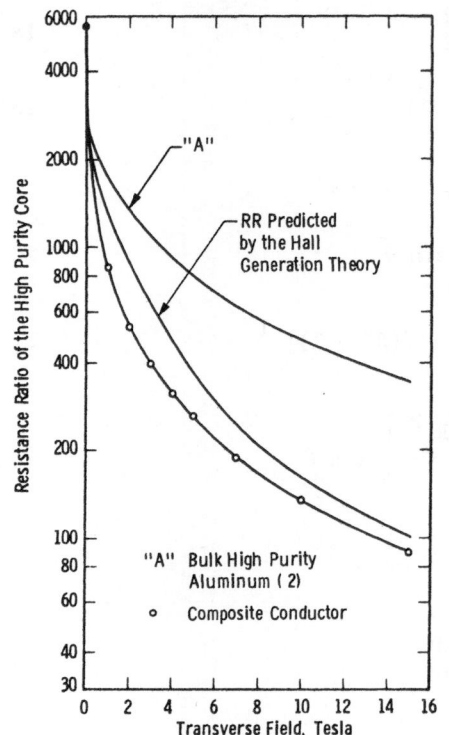

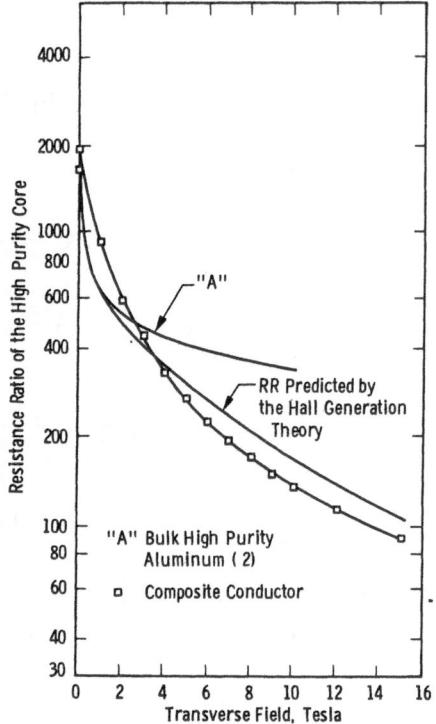

Fig. 4. Comparison of the Hall generation theory with the experimental data at 4.2K.

Fig. 5. Comparison of the Hall generation theory with the experimental data at 20.3K.

EXPERIMENTAL DATA

The experimental magnetoresistance data for the composite conductor is in publishing and in that paper is compared to the Hall generation theory using bulk polycrystalline high purity aluminum experimental data reported by Fickett.[4] Fickett acquired his data using the transport current method. Hartwig used the eddy current decay method in the majority of his experiments. He developed a Kohler type relation by modifying an expression developed by Corruccini which represented his experimental data well in the range of our interest. We have used Hartwig's expression,

$$\rho_H = \left\{ \frac{2H_*^2 (1 + 0.006 \, H_*)}{(4 + 3H_* + H_*^2)} + 1 \right\} \rho \qquad 28$$

to provide the resistivity of bulk pure Al at the applied field and at the sample temperature. In our comparison, the bulk resistivity at field and temperature is summed with the apparent resistivity due to Hall generation computed from Equation 29.

$$\rho = \frac{P}{I^2} \frac{A}{L} \qquad 29$$

From this sum, the resistance ratio intended for comparison to the experimental values determined for the composite core[1] is determined as

$$RR = \frac{\rho_{300K}}{(\rho + \rho_H)}$$

The formalism for determining the resistance ratio of the composite core of pure Al from the resistance ratio of the composite conductor appears in Reference 1.

DISCUSSION

Figure 4 shows the Hall generation theory compared to experimental data with the sample at 4.2K. Figure 5 is a similar comparison for 20.3K test conditions. The upper curve in Figure 3 is the resistance ratio at field and temperature of bulk pure Al as given by Hartwig. The lower curve in the same figure is the experimentally determined resistance ratio of the core of the composite.[1] Certainly the composite core is displaying much larger magnetoresistivity effects than expected. In each figure, the resistance ratio computed by the Hall generation theory is also shown. It is evident that internal Hall generation can account for most of the anomalous magnetoresistivity of the composite's core. The agreement of the theoretically extrapolated data with the experimental composite data is sufficient to explain the very large magnetoresistivity of the composite conductor.

Using eddy current decay data to provide the basis magnetoresistivity for bulk Al from which the composite magnetoresistivity is predicted by the Hall generation theory produces better agreement between theory and experiment. This probably results from both the eddy current data and the Hall current generated being peripheral or transverse current. If so, then there is a systematic variation between transverse and longitudinal components in the magnetoresistivity tensor.

CONCLUSION

The anomalously large magnetoresistance of the Al composite conductor is due to internal Hall current generation. Theory is presented which enables the magnetoresistance of composites to be predicted. Preliminary indications are that the longitudinal and transverse components of magnetoresistivity as measured by the eddy current decay and transport current techniques are systematically different in drawn or extruded conductors.

NOMENCLATURE

A	Conductor area, m^2
a_1	Integration constant
a_2	Integration constant
B	External field, Tesla
b	Integration constant
E_H	Electric field, V
E_I	Hall electric field, V
E^I	Electric field in region I, V
H_*	$B\rho_{300K}/(100\rho)$, 100 x Tesla
J	Current density, A/m^2
L	Conductor length, m
$R_{1\&2}$	Interface and outer radii
r	Radius, m
α	See Equation 21

β	R_2 / R_1
ρ	Resistivity
ρ_H	Resistivity at field
α	Conductivity
θ	Angular displacement
ψ	Scalar potential

REFERENCES

1. Eckels, P. W., Iyer, N. C., Patterson, A., Male, A. T., Parker, J. H., Jr, and Coltman, J. W., "Magnetoresistance: The Hall Effect in Composite Aluminum Cryoconductors," Cryogenics, Vol. 29, No. 7, July 1989, Pg. 748.

2. Hartwig, K. T., Personal Communication of the literature review on properties of pure aluminum by Hartwig and Yuan, 1988.

3. R. L. Schlicher and C. E. Oberly, Cryogenic Aluminum Wound Generator Rotor Concept, Fifth IEEE Pulsed Power Conf., Paper PTTT-12, June 1985.

4. Fickett, F. R., "Magnetoresistance of Very Pure Polycrystalline Aluminum," Physical Review B, Vol.3, No. 6, Mar. 15, 1971, Pg. 471.

MAGNETORESISTANCE OF MULTIFILAMENT Al/Al-ALLOY CONDUCTORS*

C. A. Thompson and F. R. Fickett

Electromagnetic Technology Division
National Institute of Standards and Technology
Boulder, Colorado

ABSTRACT

Previously we have shown that composite monofilament conductors consisting of very pure aluminum confined in an Al-Fe-Ce alloy sheath show an anomalously high magnetoresistance compared to pure aluminum. Some monofilament conductors showed values of $\Delta R/R$ in excess of 50 at 4 K in fields of 10 T, whereas pure aluminum values are usually an order of magnitude smaller. Concerns that similar anomalous behavior might occur in multifilament wires of the same materials prompted this study. Multifilamentary conductors with pure aluminum filaments contained in an Al-Fe-Ce matrix have been investigated.

INTRODUCTION

Composite conductors consisting of multiple filaments of high purity metal extruded within an alloy matrix are the subject of this research program. These conductors, sometimes called hyperconductors, are designed to achieve high mechanical strength and high electrical conductivity at cryogenic temperatures. The first use of Al-Fe-Ce alloy in wires, similar to those investigated here, was by Air Force Wright Research and Development Center.[1] The filaments are pure aluminum with a typical residual resistance ratio (RRR = resistance at room temperature/resistance at 4 K) of 1500 before extrusion.

Square monofilament conductors consisting of very pure aluminum confined in an Al-Fe-Ce alloy sheath[2] typically show a transverse magneto-resistive behavior quite unlike that previously observed for pure aluminum.[3] Concerns that similar behavior in multifilament conductors could significantly reduce in-field performance motivated this research. Magnetoresistance measurements were made on a variety of Al/Al-Fe-Ce multifilament wires prepared by an outside vendor. Data on these wires span the temperature range from 4 to 30 K with a magnetic field variation of 0 to 9 T.

---

*Research funded by AFWAL under MIPR# FY1455-89-N0606.

Advances in Cryogenic Engineering (Materials), Vol. 36
Edited by R. P. Reed and F. R. Fickett
Plenum Press, New York, 1990

663

Table 1. Composite Multifilament Wires

Composition Core/Sheath	Diameter mm (mils)	Number of Filaments	Anneal Temp,Time °C, min	RRR	ΔR/R 4 K, 6 T
Al/Al8Fe4Ce	0.8 (30)	19	As extruded	428	2.3
Al/Al8Fe4Ce	0.8 (30)	19	400, 20	544	3.4
Al/Al8Fe4Ce	0.8 (30)	4	As extruded	364	2.5
Al/Al8Fe4Ce	0.8 (30)	4	400, 10	764	3.1
Al/Al4Fe2Ce	0.8 (30)	19	As extruded	413	2.6
Al/Al4Fe2Ce	0.8 (30)	19	400, 8	558	3.1
Al/Al8Fe4Ce	1.3 (50)	19	As extruded	398	2.7
Al/Al8Fe4Ce	1.3 (50)	19	400, 20	683	3.7

Although the magnetoresistance of the multifilament wires is anomalous, it is not of the magnitude previously observed in the square cross-section monofilament composite conductors and it is somewhat predictable if the magnetoresistive properties of the individual component materials are known.

EXPERIMENT

Samples

Eight multifilament wires in various configurations were measured (see Table 1). Sheath material for three pairs of samples is Al8Fe4Ce with Al4Fe2Ce used for the fourth pair. Prior to wire fabrication, the pure aluminum filament material's RRR was reported to be 1500. No sample processing was done on these conductors at NIST. Wire diameters are either 0.8 mm or 1.3 mm and the number of filaments either 4 or 19. In all samples, the filament area is designed to be 50% of the total conductor cross section. Filament shape was not thoroughly analyzed, but some distortion is apparent. However, all filaments appear to be continuous along the entire length of the test samples.

Variable Temperature Apparatus

The apparatus was designed for variable temperature transverse magnetoresistance measurements on two wire samples using a four wire method. The magnetic field was applied using a 6.4 cm bore, 10 T radial access magnet with a uniform field (to 1%) along the central 5 cm of the samples. The sample cryostat consists of a single-wall tube containing low pressure helium gas and a heated sample block. A calibrated carbon glass resistor and a capacitance thermometer sensor were centrally mounted in the sample block for temperature measurement and control. Straight samples were mounted on the block with mechanical current contacts and spot-welded voltage taps having a separation ~4.5 cm. The current lead design prevents straining of the sample due to thermal contraction while proper lateral support holds the samples securely against the Lorentz force.

Sample block temperature was measured using the carbon glass sensor at zero magnetic field. The capacitance sensor was used for control of the block temperature while the magnetic field was applied. The temperature was rechecked at zero field at the end of each run. Data obtained at 4 K were taken with the samples in liquid helium.

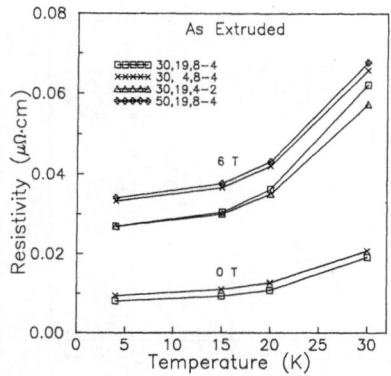

Fig. 1. Resistivity versus temperature at 0 and
6 T for as extruded wires.

The sample current was provided by a constant current source with a current reversal switch. Sample current levels were minimized to reduce Lorentz force and stay well below the range of sample self heating, but were adequate to maintain a sufficient voltage signal. Sample voltages were switched through a low thermal voltage scanner to a nanovoltmeter. To minimize thermal voltages, the sample voltage leads were run continuously out of the cryostat to a stable room temperature connection. The scanner provided the computer interface for the sample current reversing switch. Sample current was calculated from the voltage obtained from a calibrated resistor in series with the samples. All instruments except the constant current source were computer interfaced.

Data Acquisition

Data were obtained through a computerized data acquisition system. The system allowed for rapid accumulation of large amounts of data while maintaining tight control over system parameters. The data acquisition system was thoroughly tested before each run to be certain the numbers which poured forth from the computer were reasonable.

At a given field and temperature, the measurement was repeated once for each sample and the results averaged. Each measurement consisted of three data sets: current positive, current negative, and current positive.

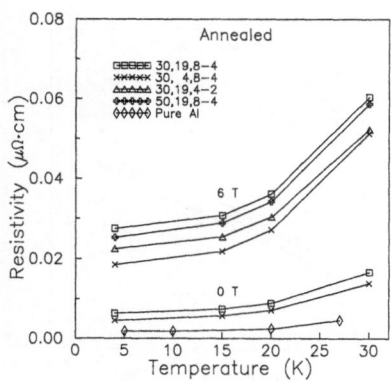

Fig. 2. Resistivity versus temperature at 0 and
6 T for annealed wires with 1500 RRR
aluminum shown for comparison.

665

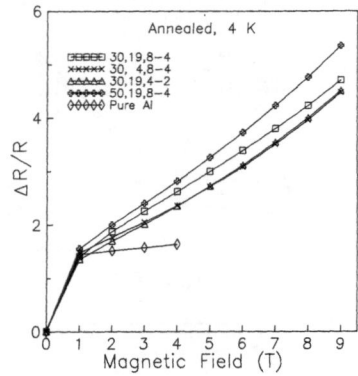

Fig. 3. Magnetoresistance for annealed wires at 4K
with 2000 RRR aluminum shown for comparison.

The temperature was monitored before, during, and after each data set. A
data set consisted of thirty sample voltage and current measurements. The
current was reversed once to allow the averaging of thermal voltages and a
second time as a check of the stability of the thermal voltages and the
instrumentation. Any significant instabilities in temperature, sample
current, or sample voltage during the data acquisition routine caused a
rejection of the data set and a repeat of the measurement.

RESULTS

Resistivity-versus-temperature data for the composite wires in the as-
extruded condition are shown in Figure 1. These data are not the adjusted
filament resistivity but rather the bulk wire resistivity. The top four
curves are 6 T data and the bottom two are zero-field values. Figure 2
shows the same wires in an annealed condition including resistivity values
for 1500 RRR aluminum calculated from the residual resistivity and the known
variation of the intrinsic resistivity with temperature.[4] A typical legend
on these figures gives the following information: wire diameter in mils,
number of filaments, and sheath composition.

Transverse magnetoresistance curves for these wires are shown in
Figure 3 and Figure 4 for the annealed wires at 4 K and 20 K. Aluminum with
a RRR of 2000 is shown here for comparative purposes.[3]

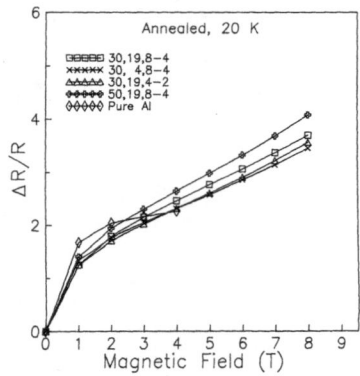

Fig. 4. Magnetoresistance for annealed wires at 20K
with 2000 RRR aluminum shown for comparison.

The filament sizes for these wires cover a fairly large range. Assuming that the total filament area is 50% of the total cross section, the filament diameters are calculated to be 124 μm for the 30 mil diameter 19 filament wire, 206 μm for the 50 mil diameter 19 filament wire, and 269 μm for the 30 mil diameter 4 filament wire.

A piece of data essential to the interpretation of the results is the RRR of the sheath material. A lack of available material for testing, especially in a condition resembling that expected of the wire sheath material, has created a problem in that regard. We have measured two samples of the Al8Fe4Ce material in bulk form (with quite different processing parameters) and found RRR values of 17 and 18 with room temperature resistivities of 2.98 $\mu\Omega\cdot$cm and 4.15 $\mu\Omega\cdot$cm respectively. A magnetoresistance measurement on the higher resistance material was made at 4 K and, as expected, ΔR/R was very low, reaching only 0.5 at 10 T. No sample of the 4-2 material has been available.

DISCUSSION

The temperature dependence of the resistivity of the wires is relatively normal, even in high magnetic fields. In the as-extruded state, the wires with the largest filament diameter have the highest resistivity while the opposite is true after annealing. In fact, annealing seems to have essentially no effect on the small filament wire (30,19,8-4), although the same wire with the Al4Fe2Ce matrix does show an effect. This reversal is most probably due to a combination of size effect, grain boundaries, and perhaps contamination of the pure aluminum with impurities from the matrix. Some of the variations observed with the annealed samples may result from the fact that they were annealed for different times. There is some evidence that RRR is very sensitive to anneal time and anneal temperature.

The magnetoresistance effect is large for the materials involved (compare the pure aluminum data to that for the composites), but does not appear to be large enough to be of concern for most applications. Furthermore, the magnetoresistance effect decreases with increasing temperature. The rapid rise of ΔR/R with field is not typical behavior for pure aluminum, but is always observed in these composite conductors. This effect is discussed in detail in another paper.[2] It appears to be due to the transit of electrons from the low resistance core into the high resistance sheath under the influence of the applied field. Magneto-resistance data for the as-extruded conductors is not shown, but has the same general behavior with somewhat lower values at high field in all cases.

The dc size effect is an important parameter for these conductors. At 4 K, the mean free path of the electrons in the filaments (assuming a filament RRR = 1500) is ~40 μm. Thus, in the smallest filaments, nearly 90% of the electrons are within one mean free path of the interface. The exact interaction at the interface is unknown, but certainly it is not complete reflection.

One way to investigate the importance of the dc size effect is to calculate the RRR of the filaments under the various anneal conditions. If we assume that the filament material actually returns to its original RRR of 1500 upon annealing, then any deviation of the calculated RRR from this value should indicate the presence of other effects. Calculation of the filament RRR was done using a parallel resistor model, which should be appropriate in zero field but more questionable otherwise. For the situation where the pure aluminum and the matrix alloy have equal areas, it can be shown that the expression for the filament RRR is

Table 2. Measured wire RRR, calculated filament RRR,
and filament diameters

Sample	RRR	RRR_f	D, μm
30,19,8-4 Annealed	544	883	124
30, 4,8-4 "	764	1245	269
30,19,4-2 "	558	906	124
50,19,8-4 "	683	1112	206
30,19,8-4 As-extruded	428	692	124
30, 4,8-4 "	364	587	269
30,19,4-2 "	413	668	124
50,19,8-4 "	398	643	206

$$RRR_f = [\{(\rho_m/\rho_f) + 1\}\cdot RRR + RRR_m]/(\rho_m/\rho_f), \tag{1}$$

where RRR_f is the filament value, ρ_m is the measured matrix resistivity at 295 K, ρ_f is the filament resistivity at 295 K, RRR is the measured value for the conductor, and RRR_m is the measured matrix value.

Table 2 is generated by substituting a matrix resistivity of 4.15 $\mu\Omega\cdot cm$, a filament resistivity of 2.67 $\mu\Omega\cdot cm$, and a matrix RRR of 18.4 into equation (1) (where D is the calculated filament diameter). Note that there is no size effect correction here.

None of the annealed samples have a filament RRR of 1500, although the large filaments are close. If we make a maximum size effect correction[4] using the mean free path appropriate for aluminum with RRR = 1500, the two highest values rise to 1425 and 1322, still not high enough. This indicates that other mechanisms are operating to reduce the RRR in these wires and they are more important for the smaller filament materials. The best guess for now is that grain boundary effects may be more important than a conventional calculation would indicate. There is some justification for this from early work done on aluminum grain boundary resistivity, and that is discussed in some detail in Ref. 4.

CONCLUSIONS

The multifilament composite conductors are relatively well behaved in their low temperature resistive behavior. The effect of annealing is uncertain, and the apparent final RRR of the filaments is not predictable except that larger filaments anneal closer to the maximum RRR expected. The magnetoresistance values are not wildly anomalous, but are not typical of pure aluminum. Filament size does not change the high field behavior to a great degree. It appears that, for all but ac applications, fewer, larger filaments are better.

ACKNOWLEDGMENTS

The authors thank Dhruba Chakrabarti of Alcoa for providing the wire samples used in this research.

REFERENCES

1.   J. C. Ho, C. E. Oberly, H. L. Gegel, W. M. Griffith, J. T. Morgan, W. T. O'Hara and Y. V. R. K. Prasad, "A New Aluminum-Base Alloy with

Potential Cryogenic Applications," in: <u>Advances in Cryogenic Engineering (Materials)</u>, Vol. 32, edited by R. P. Reed and A. F. Clark, Plenum, New York, (1986), pp 437-442.

2.   F. R. Fickett and C. A. Thompson, "Anomalous Magnetoresistance in Al/Al-Alloy Composite Conductors," Paper DY-01, this conference.

3.   F. R. Fickett, Magnetoresistance of very pure polycrystalline aluminum, <u>Phys. Rev. B 3</u>:1941 (1971).

4.   F. R. Fickett, Aluminum 1. A review of resistive mechanisms in aluminum, <u>Cryogenics 11</u>:349 (1971).

ANOMALOUS MAGNETORESISTANCE IN Al/Al-ALLOY COMPOSITE CONDUCTORS[*]

F. R. Fickett and C. A. Thompson

National Institute of Standards and Technology
Boulder, Colorado 80303

ABSTRACT

The transverse magnetoresistance of several composite conductors
containing a large single filament of pure aluminum in a matrix of Al-Fe-Ce
has been measured at 4 K to fields of 10 T. The magnetoresistance ($\Delta R/R$)
of the composite is very large, rising to 55 in the "worst" case. Previous
measurements on pure polycrystalline aluminum have always shown a rapidly
saturating behavior with a very small linear component; $\Delta R/R$ rarely exceeds
a value of 5-6. In addition, the magnetoresistance of the composite
samples shows a structure as the field is rotatated around the current
axis.

INTRODUCTION

Conductors which attempt to achieve both high strength and high
electrical conductivity at low temperatures by using a pure metal core
surrounded by a high strength alloy sheath have come to be known as
hyperconductors. Such materials, based on aluminum, are the subject of
this research. The sheath material is one of two alloys of Al-Fe-Ce
developed at the Air Force Wright Research and Development Center.[1] The
core material is pure aluminum. Aluminum is available in a wide range of
purity with residual resistance ratios (RRR = room temperature
resistance/resistance at 4 K) to 40,000. Further details of the structure
of these conductors are given elsewhere in this volume.[2,3]

The magnetoresistive behavior of pure polycrystalline aluminum at low
temperatures is well known[4,5] and, since the sheath material is a
relatively high resistance alloy, with an accompanying low magnetoresistance, we expect the magnetoresistive behavior of the composite to be
much like that of pure aluminum. The magnetoresistance is expressed as
$\Delta R/R$, the change in resistance due to the field divided by the zero field
value. It first rises rapidly with field and then levels off, with a very
slight linear rise, to the highest fields measured. The high field value
of $\Delta R/R$, even for very pure aluminum, seldom exceeds 5-6. Our measurements
of a number of hyperconductor samples with square cross section indicate an

---

[*] Research funded by AFWAL under MIPR # FY1455-89-N0606

Advances in Cryogenic Engineering (Materials), Vol. 36
Edited by R. P. Reed and F. R. Fickett
Plenum Press, New York, 1990

671

anomalous behavior with a value as high as 55 at 10 T in one instance. Furthermore, a rotational structure is seen in the magnetoresistance. Here we present our observations and describe a number of experiments performed to investigate the behavior of this anomalous magnetoresistance.

The theoretical explanation for the effect is elusive. Here we consider a possible electron free path scenario that may explain the data, and another paper in this volume looks at the problem from a different angle.[6] A comprehensive experiment involving a large number of samples in which parameters can be controlled is not possible at present, so that much of the conjecture presented here is just that. From a practical standpoint, it is important to realize that a high magnetoresistance value is not necessarily detrimental to the operation of a device. Because the highest $\Delta R/R$ values are usually found in the materials with the lowest zero-field resistance, the actual resistance of the sample in the magnetic field may well be lowest for the high $\Delta R/R$ samples.

EXPERIMENT

Samples

Two hyperconductor samples (#2 and #4) were measured. In addition, a sample of the sheath material and a sample of the core material in the form of a composite sample from which the sheath had been chemically removed were measured. The composites were 3.8 mm square with a 2.5 mm square core. The starting core material was said to have a RRR of 30 000[*]. In the as-processed (unannealed) state, the core RRR for sample #2 was calculated to be near 9000 (the as-measured RRR = 4370). For sample #4, the core RRR was 3600 (as-measured RRR = 1980). This calculation considers the core and sheath to form a set of parallel resistors.[7] None of the composite samples was annealed after receipt. The core material sample had a very pitted and rough surface as a result of the chemical etch. Its RRR was 7830 as received and rose to 10,300 after a 30 minute anneal at 300°C (no size effect correction).

Two special sample modifications were made in the course of the investigation. In the first, the sheath of sample #2 was slotted down to the core surface so that voltage taps could be attached to the core material itself. This was an attempt to see whether the electrical behavior of the core was significantly different from that of the composite. The second modification involved removing the entire sheath from one of the four sides of sample #4 in an attempt to determine the mechanism for the rotational variation in $\Delta R/R$.

Apparatus

The experiments described here are classical magnetoresistance measurements in which the voltage is measured in a four-probe configuration.[8] The magnet is a transverse-access superconducting magnet. The sample holder is designed to hold the sample securely against the Lorentz force. Complete rotation of the sample in the field is possible with the field remaining normal to the sample. Voltage taps are attached to the center of the sample with a spacing of ≈4 cm by spot welding. Temperature control is provided by a combination of a calibrated carbon glass

---

[*] A comment needs to be made about the RRR in pure aluminum. Much above 5000, the resistance of aluminum is not residual (temperature independent) at 4 K. The temperature dependent part can make up to 35% of the resistivity of very pure (30 000) material.

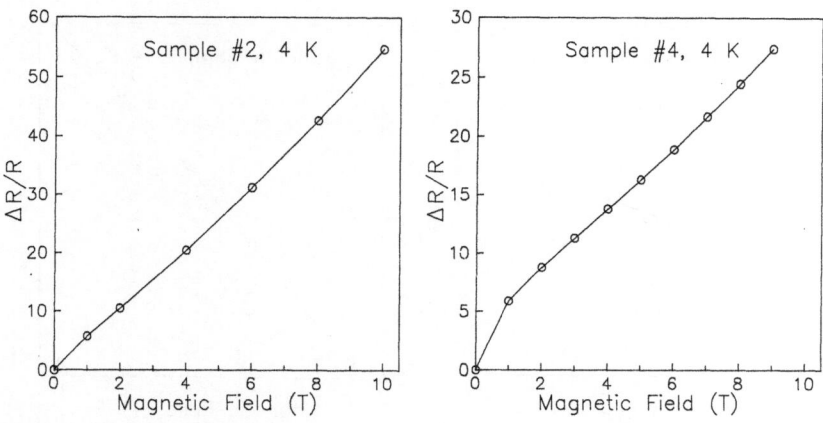

Fig. 1. Sample #2 magnetoresistance. Fig. 2. Sample #4 magnetoresistance.

thermometer for zero-field temperature determination and a capacitance thermometer for maintaining control in the applied field. The sample space is isolated from the helium bath by a single-wall vacuum space. The details of the data acquisition and analysis system are provided elsewhere.[7]

DATA

Magnetoresistance of unmodified samples

Figure 1 shows the magnetoresistance of sample #2. Note that the 10 T value is 55. Furthermore, the curve shows a slight upward slope at the higher fields -- not at all like the behavior of pure aluminum. Figure 2 shows the behavior of sample #4 with a lower RRR in the core material. It still shows a large ΔR/R at high fields and the slight upward curvature that has turned out to be characteristic of all composite conductors.

Magnetoresistance measured on the unannealed core sample is shown in Fig. 3. Its behavior is more what we expect for pure aluminum <u>except</u> that the value of ΔR/R is extemely large and the approach to linear behavior takes place at much higher fields than would be expected. This sample also shows a strong rotational variation in the magnetoresistance as described below.

The magnetoresistance measured on a round rod of sheath material is shown in Fig. 4. There is no reason to believe that this sample actually is in the same state as the sheath of our composite conductors, since the processing history was not available. Regardless, the general behavior is probably the same, namely a very small magnetoresistance that is probably negligible for our purposes except that the high resistance of the Al-Fe-Ce makes even a small ΔR in the sheath larger than the largest ΔR in the core and this fact may well have implications for the explanation of the large magnetoresistance of the composite.

Magnetoresistance of cut samples

In an attempt to see whether the core was behaving differently from the whole composite, slots about 2 mm in length were cut to the core of sample #2 at the location of the voltage taps and the taps were reattached directly to the core. The resulting magnetoresistance curve did not change

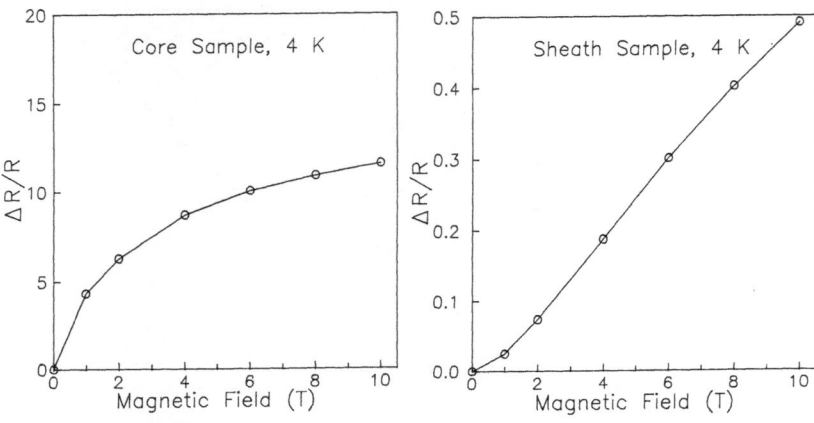

Fig. 3. Core magnetoresistance.   Fig. 4. Sheath magnetoresistance.

from that shown in Fig. 1, and no change was observed in the rotational behavior (see below).

Sample #4 was cut to completely remove the sheath from one surface by careful milling in a specially prepared jig. The cut actually removed a bit of the core material, so the new dimensions were 3.8 x 2.9 mm. Voltage taps were attached to both the exposed core and to the sheath at 90° to the cut surface. Figure 5 shows the magnetoresistance measured after this operation. The field orientation with respect to the tap sets is shown in the inset. The behavior is similar to that of the uncut sample (Fig. 2), but the values are lower. In addition, the core taps show a smaller $\Delta R/R$ than the sheath taps. If the sample is rotated 90° CCW, two interesting effects are seen: first, both the core and sheath curves become the same. Second, the curve is essentially that for the uncut sample (Fig. 2); that is, it is significantly higher than the curves seen in the 0° orientation.

Rotational behavior of the magnetoresistance

All the square samples measured showed a variation of magnetoresistance with angle. Typical behavior is shown by the data for sample #4 in Fig. 6. All composite samples have a total variation in $\Delta R/R$

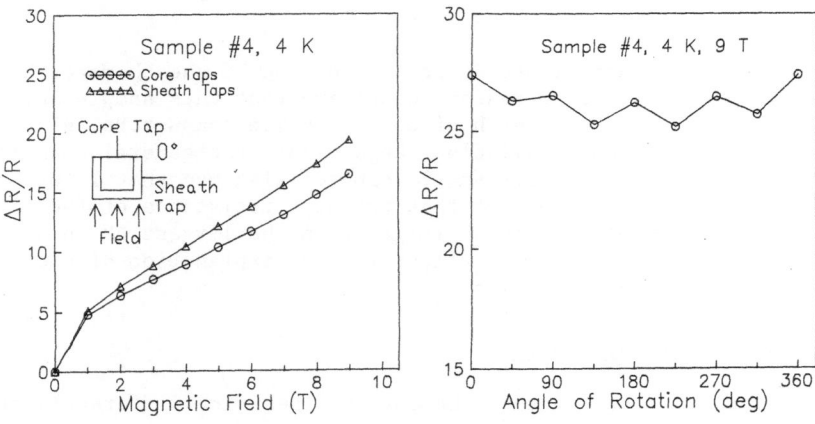

Fig. 5. Magnetoresistance of sample   Fig. 6. Effect of rotation on the
        #4 with one surface removed.   magnetoresistance.

of about 10%. The core sample variation was much larger, 58% (at 10 T). After the sheath was removed from one face of #4, the variation increased significantly as shown in Fig. 7 for both the core and sheath taps.

Temperature dependence of the magnetoresistance

The magnetoresistance of sample #4 was measured up to 20 K. The zero-field resistance rises significantly over this range. The results are shown in Fig. 8. The behavior as a function of temperature is about as would be expected for an alloy, the magnetoresistance decreases with increasing temperature and becomes more nearly linear. However, once again, this is opposite to the behavior observed previously for pure aluminum in this temperature range.

DISCUSSION

A large number of parameters may affect the electrical behavior of pure metals. These have been reviewed earlier.[9] In most instances, the effects are related in one way or another to the long mean free path, ℓ_b, in the bulk material. For relatively pure aluminum:

$$\rho(295 \text{ K}) = 2.666 \ \mu\Omega \cdot \text{cm}$$

$$\rho(4 \text{ K}) = \rho_0 \quad \text{(the residual resistivity)}$$

For aluminum with RRR = 10 000, $\rho_0 = 2.67 \times 10^{-4} \ \mu\Omega \cdot \text{cm}$

The quantity $\rho_b \ell_b$, where ρ_b is the bulk resistivity, is assumed to be a constant independent of temperature or purity. The assumption is not exactly true, but it provides an approximation that allows us to calculate the electron mean free path at various temperatures. This is the average distance that an electron travels before suffering a scattering interaction with the lattice, with an impurity or defect, or with the surface. These interactions are the source of electrical resistance.

For aluminum we choose: $\rho_b \ell_b = 0.7 \times 10^{-11} \ \Omega \cdot \text{cm}^2$. This value is very uncertain and values 40% on either side of this have been reported in the literature. It is a reasonable compromise.

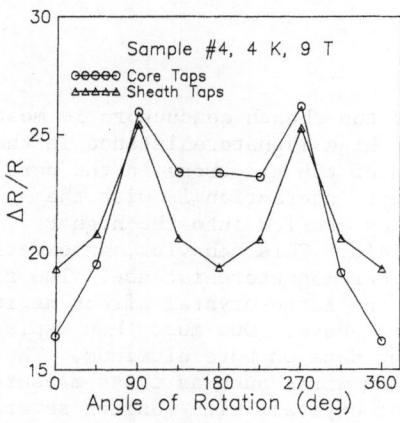

Fig. 7. Rotational behavior of the magnetoresistance.

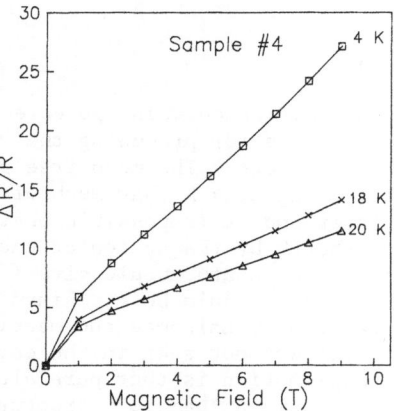

Fig. 8. Temperature dependence of magnetoresistance.

Thus, the mean free path value in bulk aluminum (no size effect) with RRR = 10 000 is: $\ell_b(4 \text{ K}) = 2.6 \times 10^{-2}$ cm = 0.26 mm.

This says that, in the core of the composite conductor, the electrons in 37% of the cross section are within one mean free path of the interface. A similar calculation for a core RRR of 5000 gives a value of 20%.

Source of the large magnetoresistance

The source of the large magnetoresistance almost certainly lies in the behavior of electrons near the interface between the core and the sheath. A calculation of the fraction of electrons that need to enter the sheath in order to create the voltage associated with the observed ΔR/R shows that it is well within the range of those available from the core boundary region. The mechanism by which this transfer occurs and is maintained is still being determined. Small angle scattering is common in aluminum[10] and may well play a significant role. We do not, however, think that magnetic breakdown, which occurs only for very limited directions within the crystal,[11] can explain the observations.

The shape of the magnetoresistance curves (slight upward curvature) at temperatures generally below 20 K is consistent with an early development by Balcombe,[12] who averaged the magnetoelectric tensor over a variety of crystal directions. This argues for an effect due to large, well defined crystallites, certainly a possibility given the large crystal dimensions observed in the core of these composites.

In a few of the earlier works on the magnetoresistance of aluminum, it was suggested that the surface condition of the conductor might significantly affect the behavior of the magnetoresistance. Almost no data were presented to back up the observation, but this possibility needs to be revisited especially in light of the behavior of the core material sample.

Source of the rotational behavior

The rotation plots further support the mean free path argument in that the anisotropy is greatly increased when one surface is removed. The lower value of ΔR/R seen at 0° represents a significant decrease in the effect of the field when it is normal to only one sheath face. The lack of a decrease at 180° on the core taps to the 0° value is not understood at present. Earlier work on the longitudinal magnetoresistance of copper samples[13] showed that very large effects can result from very small misalignments of voltage taps and perhaps that is the situation here.

CONCLUSIONS

The interaction between the core and the sheath conductors is most likely responsible for producing the very high magnetoresistance in these composite conductors. The mean free path of the electrons in the core is sufficiently long at 4 K that much of their interaction is with the boundary layer and it is possible that they are led into the higher resistance sheath by the action of the field. This behavior, an excess of electrons in the sheath, would give a higher magnetoresistance. The slight upturn in the data could be explained by the large-crystal effect as in the theory put forth by Balcombe and mentioned above. One must then explain why the effect was not seen in the earlier data on pure aluminum. The most plausible explanation is that pure aluminum wires such as those measured in [4] tend to assume a "bamboo" structure of crystals with lengths several times the diameter of the wire and, perhaps, with a preferred orientation. In the case of the composite conductors, there is a more random distribution of grain sizes in the core. Also, the earlier data extended only to fields of 4 T and the effect is obvious only well above that.

We have proposed that the anomalous magnetoresistance is associated with electron transfer across the core-sheath boundary. The logical way of changing this situation is to create a nonconducting boundary. The problems with this approach are several: we still require good thermal conductivity between the core and the sheath, and few materials are good thermal conductors and poor electrical conductors, although a very thin film of insulating material might work. However, a good mechanical bond between the core and sheath is critical to many applications and an insulating layer might well degrade that.

ACKNOWLEDGEMENTS

The samples investigated here were provided by P. Eckels and colleagues at the Westinghouse Research and Development Center. We are grateful for their assistance and for numerous spirited discussions on the physics of electrical conduction in metals.

REFERENCES

1. J. C. Ho, C. E. Oberly, H. L. Gegel, W. M. Griffith, J. T. Morgan, W. T. O'Hara and Y. V. R. K. Prasad, A new aluminum-base alloy with potential cryogenic applications, Adv. in Cryo. Eng., Materials, 32:437 (1986).

2. C. E. Oberly, The origin and future of composite aluminum conductors, paper CY-01, this conference.

3. N. C. Iyer, A. T. Male and P. W. Eckels, Composite aluminum conductors for power applications, paper CY-03, this conference.

4. F. R. Fickett, Magnetoresistance of very pure polycrystalline aluminum, Phys. Rev. B 3:1941 (1971).

5. R. J. Corruccini, The electrical properties of aluminum for cryogenic electromagnets, National Bureau of Standards Technical Note No. 218 (1964).

6. P. W. Eckels, N. C. Iyer, A. Patterson, and A. T. Male, Magneto-resistance: the hall effect in composite aluminum cryoconductors, paper DY-02, this conference.

7. C. A. Thompson and F. R. Fickett, Magnetoresistance of multifilament Al/Al-alloy conductors, paper DY-03, this conference.

8. F. R. Fickett, Electrical properties, chapter 5 of "Materials at Low Temperatures," R. P. Reed and A. F. Clark, eds., American Society for Metals, Metals Park, Ohio (1983).

9. F. R. Fickett, Aluminum 1. A review of resistive mechanisms in aluminum, Cryogenics 11:349 (1971).

10. A. B. Pippard, "Magnetoresistance in Metals," Cambridge University Press, (1989).

11. W. C. Mitchel and R. S. Newrock, Magnetic breakdown in very dilute aluminum-silver alloys, Phys. Lett. 73A:336 (1979).

12. R. J. Balcombe, The magneto-resistance of aluminum, _Proc. Roy. Soc. (London)_ A275:113 (1963).

13. F. R. Fickett and A. F. Clark, Longitudinal magnetoresistance anomalies, _J. Appl. Phys._ 42:217 (1971).

MEASUREMENT OF THE ELECTRICAL RESISTIVITY AND THERMAL

CONDUCTIVITY OF HIGH PURITY ALUMINUM IN MAGNETIC FIELDS

J. P. Egan and R. W. Boom

Applied Superconductivity Center, University of Wisconsin

Madison, WI 53706

ABSTRACT

Experimental measurements on three samples of high-purity aluminum (RRR = 172, 1530, and 10,000) provide tables and graphs of $\rho(B,T)$ and $\lambda(B,T)$ vs. RRR for determining the electrical resistivity and thermal conductivity as a function of purity in transverse magnetic fields to 4 T for temperatures from about 7.5 K to 50 K.

INTRODUCTION

The electrical resistivity (ρ) and thermal conductivity (λ) of bulk-polycrystalline, high-purity-aluminum (H.P.Al) wire samples (dia. 3 mm, length 15 cm) with residual-resistance ratios (RRR = R(273 K)/R(4.2 K)) of 172, 1530, and 10,000 were measured in transverse magnetic fields (B) to 4 T, for temperatures (T) from about 4.2 K to 90 K. Each sample remained fixed for undisturbed measurements of both $\rho(B,T)$ and $\lambda(B,T)$. Transverse magnetic fields were used since they influence the transport properties by roughly twice that of longitudinal magnetic fields and because most practical applications utilize transverse fields. The transverse fields were produced by a dipole magnet with a field uniformity of 2% along the 9 cm sample test length.

RESULTS

Table I lists measured values of $\rho(B,T)$ for temperatures from 7.5 K to 50 K in transverse magnetic fields up to 4 T. Measurements show that for any field strength, $\rho(B,T)$ diverges from its zero-field value and then converges to $\rho(0,T)$ as the temperature is raised. Above about 90 K, all H.P. Al samples have equivalent ρ values. $\rho(4\ T, 90\ K)$ is only 4% above its zero-field equivalent of 350 nΩ cm.

Table II lists measured values of $\lambda(B,T)$ for the three samples. $\lambda(0,70\ K) \approx 5.0$ W/cm K for all H.P. Al and this is reduced less than 7% by a 4 T transverse magnetic field. The RRR 10,000 sample had zero-field size-effect corrections estimated to be about 10% for ρ and 5% for λ at the lowest temperatures measured.[1,2] Since the size effects are small, no surface-scattering corrections are made to the data.

Advances in Cryogenic Engineering (Materials), Vol. 36
Edited by R. P. Reed and F. R. Fickett
Plenum Press, New York, 1990

TABLE 1. MAGNETO-ELECTRICAL RESISTIVITY OF ALUMINUM

RRR	T=7.5K	10K	15K	20K	25K	30K	35K	40K	50K
				ρ (nΩ cm)					
				B = 0T					
172	14.2	14.4	14.8	15.7	17.6	20.9	26.5	35.1	64.9
1530	1.60	1.72	2.06	2.72	4.02	6.66	11.6	19.6	48.3
10000	.269*	.318	.486	.895	2.02	4.65	9.68	18.0	47.5
				B = .03T					
172	14.2								
1530	1.73	1.82	2.06	2.76	4.02				
10000	.446	.460	.666	1.06	2.10				
				B = .25T					
172	15.1	15.2	15.5	16.5	18.3	21.5			
1530	3.21	3.28	3.63	4.28	5.60	8.09		20.5	
10000	.724	.928	1.58	2.61	4.09	6.55	11.1	18.8	47.9
				B = 0.5T					
172	17.0	17.1	17.4	18.2	19.8	22.8	28.2	36.6	65.5
1530	3.78	3.98	4.57	5.62	7.40	10.2		22.1	49.4
10000	.770	.996	1.88	3.44	5.65	8.65	13.3	20.7	48.7
				B = 1.0T					
172	21.0	21.1	21.3	22.1	23.6	26.8	31.9	39.9	68.2
1530	4.14	4.42	5.36	7.11	9.70	13.6		26.3	52.5
10000	.824	1.04	2.09	4.16	7.39	11.7	17.3	25.1	51.6
				B = 1.5T					
172	24.3	24.3	24.4	25.5	27.2	30.6	35.6	43.9	71.6
1530	4.29			7.83	11.1	16.0		30.5	
10000			2.20	4.45	8.30	13.7	20.7	29.2	56.1
				B = 2.0T					
172	26.8	26.8	27.1	28.3	30.4	34.1	39.5	47.9	75.3
1530	4.35	4.73	5.91	8.24	12.0	17.8		33.9	60.9
10000	.843	1.072	2.29	4.69	8.82	15.0	22.9	32.7	60.6
				B = 2.5T					
172	28.4	28.4	29.1			36.8	42.7	51.7	78.6
1530	4.41			8.43	12.5				
10000			2.35	4.90		15.8	24.5		64.5
				B = 3.0T					
172	29.8	29.8	30.7	32.5	34.8	39.1	45.8	55.4	82.1
1530	4.46	4.28	6.06	8.61	13.0	20.1		39.8	69.4
10000	.845	1.078	2.40	5.11	9.83	16.6	25.9	37.3	68.5
				B = 3.5T					
172	31.1	31.2	32.4	34.2	36.6	41.2	48.9	59.0	85.5
1530									
10000				5.32	10.3		27.3	39.8	72.4
				B = 4.0T					
172	32.4	32.5	33.5	35.5	38.5	43.2	52.1	62.4	89.0
1530	4.52	4.90	6.18	8.96	13.6	21.5		44.3	76.2
10000**	.847	1.084	2.50	5.53	10.8	18.2	28.7	42.0	76.2

* ρ(4.2K) = 0.245 nΩ cm

** Linearly extrapolated values at 7.5K, 10K, 20K, and 25K.

TABLE 2. MAGNETO-THERMAL CONDUCTIVITY OF ALUMINUM

RRR	T=7.5K*	10K	15K	20K	25K	30K	35K	40K	50K
				λ (W/cm K)					
				B = 0T					
172	12.9	16.7	22.5	25.1	24.6	22.1	18.1	14.9	9.56
1530	91.9	105.	103.	84.0	61.7	42.0	28.3	20.1	11.0
10000	331.	319.	219.	134.	78.8	47.3	30.6	20.3	11.2
				B = .03T					
172	12.9								
1530	87.7	102.	102.	83.8	61.7				
10000	253.	256.	202.	131.	78.8				
				B = .25T					
172	12.1	15.7	21.3	24.1	23.9	21.5			
1530	49.4	59.4	66.7	63.8	52.8	39.3	27.6		
10000	115.	108.	93.7	80.8	62.1	42.5	29.2	20.0	
				B = 0.5T					
172	10.8	14.2	19.5	22.4	22.6	20.7	17.5	14.4	
1530	40.1	46.3	49.6	48.5	43.6	34.7	25.8	19.0	
10000	101.	89.4	67.4	58.3	48.4	36.4	27.0	19.4	
				B = 1.0T					
172	8.78	11.5	16.0	18.9	19.6	18.5	16.1	13.5	9.25
1530	35.2	38.7	37.5	35.1	32.3	27.6	22.2	17.2	10.5
10000	92.2	78.8	52.0	41.3	35.1	28.8	23.1	17.6	10.6
				B = 1.5T					
172	7.58	9.91	13.8	16.3	17.0	16.8	14.8	12.7	
1530	33.6	36.1	33.0	29.3	26.3	23.3	19.6		
10000	88.8	74.7	46.2	34.6	28.9	24.3	20.2	16.2	
				B = 2.0T					
172	6.83	8.91	12.4	14.6	15.3	15.1	13.7	11.9	8.68
1530	32.6	34.7	30.5	26.1	23.3	20.5	17.7	14.6	9.70
10000	86.7	72.2	43.1	31.0	25.1	21.2	18.0	14.8	9.85
				B = 2.5T					
172	6.36	8.27	11.4	13.4	14.0	13.9	12.7	11.2	
1530									
10000	85.0	70.4	41.0	28.8	22.8	19.1	16.3	13.6	
				B = 3.0T					
172	6.04	7.81	10.7	12.4	13.1	12.9	11.9	10.7	8.19
1530	31.3	33.0	27.9	22.7	19.3	16.9	14.6	12.6	9.03
10000	83.5	69.0	39.5	27.2	21.2	17.6	15.1	12.7	9.07
				B = 3.5T					
172	5.78	7.48	10.2	11.7	12.3	12.0	11.3	10.1	
1530									
10000	82.4	67.7	38.4	26.0	19.9	16.3	14.0	12.0	8.69
				B = 4.0T					
172	5.61	7.24	9.78	11.2	11.5	11.4	10.6	9.73	7.61
1530	30.4	32.0	26.4	21.0	17.2	14.8	12.7	11.2	8.41
10000	80.8	66.7	37.4	25.0	19.0	15.5	13.2	11.3	8.36

* λ of RRR 10,000 sample measured at 8.5K not 7.5K.

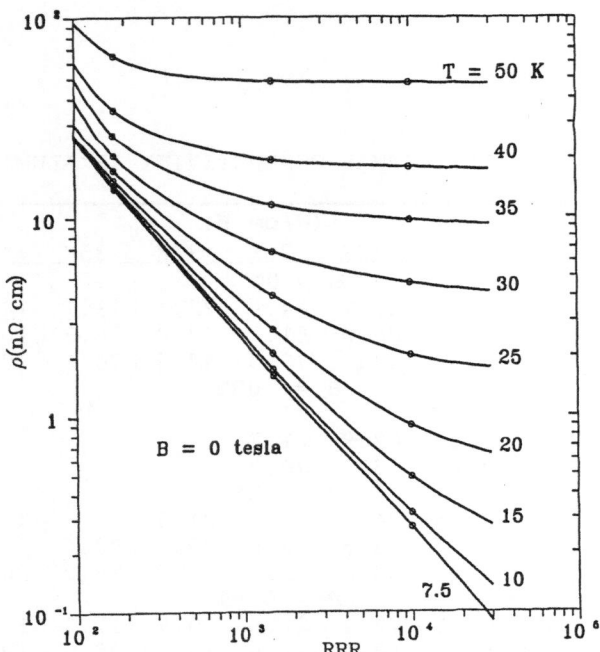

Fig. 1a. Electrical resistivity vs. RRR in zero field for temperatures from 7.5 K to 50 K.

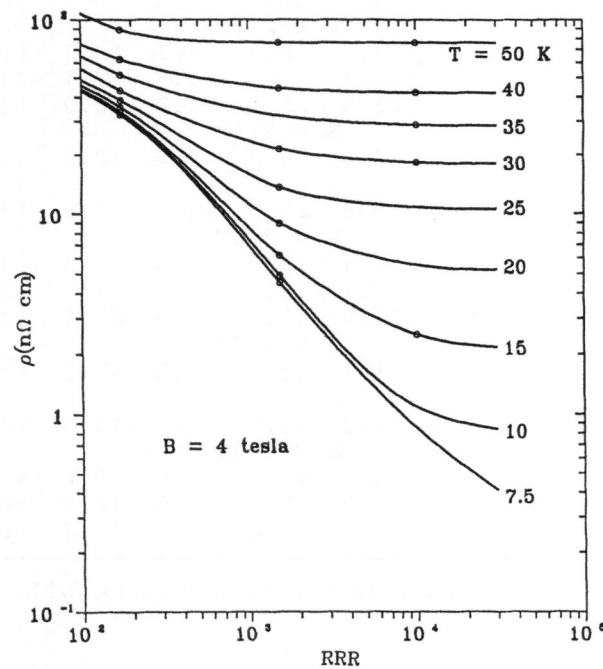

Fig. 1b. Electrical resistivity vs. RRR from 7.5 K to 50 K in a 4 T transverse magnetic field.

The data contained in Tables 1 and 2 are generalized to different resistance ratios by plotting ρ and λ vs. RRR, Figures 1 and 2. Curves fitted through the data points at RRR's of 172, 1530, and 10,000 span purities of aluminum from $100 < RRR < 30,000$. The curves of $\rho(B,T)$ vs. RRR agree to within 10% of most published values.[3-6] There are no reliable bulk H.P.Al magneto-thermal-conductivity data with which to make a comparison. Zero-field values of ρ and λ have been taken from the literature to more accurately plot Figures 1 and 2.[7,8]

Figures 1a and b are graphs of ρ vs. RRR for temperatures from 7.5 K to 50 K in zero field and in a 4-T transverse magnetic field, respectively. The lowest temperature magnetoresistivity data was measured at 7.5 K for the RRR 172 and 10,000 samples. The RRR 1530 sample gave equivalent $\rho(B,T)$ results for measurements taken from 4.2 K to 7 K. This is due to the saturating nature of ρ at low T, called the residual resistivity, where electron-phonon scattering is negligible and impurity/defect collisions become the predominant mechanism for electron scattering. The lower the resistance ratio, the higher the temperature at which ρ saturates and becomes a constant with falling temperature. It is estimated that $\rho(B, 4.2 K)$ for the RRR 10,000 sample will be about 10% below the measured $\rho(B, 7.5 K)$ values.

As a rule of thumb, the longitudinal magnetoresistivity component of the total resistivity is roughly half its transverse counterpart, $[\rho(B,T) - \rho(0,T)]_{Trans} = 2[\rho(B,T) - \rho(0,T)]_{Long}$. This was confirmed through ρ and λ measurements on the RRR 10,000 sample in longitudinal magnetic fields to 5 T.

Figures 2a-d are graphs of λ vs. RRR for transverse magnetic fields to 4 T at temperatures of 7.5, 15, 20, and 30 K. The thermal conductivity graphs are plotted at constant T, different from the constant B electrical resistivity graphs, to avoid crossing of the curves due to the peak in curves of λ vs. T. The lowest-temperature magnet-field measurements of λ for the RRR 10,000 sample were taken at 8.5 K, not 7.5 K, as shown in Table 2. The peak occurs at $\lambda(0,7.8 K) = 334$ W/cm K for the RRR 10,000 sample and, $\lambda(B,7.5 K)$ is within 1% of the $\lambda(B,8.5 K)$ values contained in Table 2. As shown in Figures 3b and 4a-d, there is a diminishing return in purifying aluminum beyond 10,000 RRR.

DISCUSSION

The electrical and thermal $(W=1/\lambda)$ magnetoresistivities can be extended to higher fields than 4 T, at low enough temperatures, by simply determining the slope in the high-field linear region and extrapolating. This is possible because the slope of the magnetoresistivity for fixed T has never been shown to deviate from linearity in this regime. These slopes can be calculated with the aid of Tables 1 and 2 for the RRR 1530 and 10,000 samples for temperatures below 20 K and 25 K for ρ and below 15 K and 20 K for W, respectively, for the two samples. This was done to obtain the values of $\rho(4 T)$ for the RRR 10,000 sample at temperatures of 7.5, 10, 20, and 25 K, see Table 1. Additional plots of ρ and λ vs. RRR can be generated with Tables 1 and 2. It is helpful to use graphs of $\rho(B)$ and $\lambda(B)$ vs. T in conjunction with Figures 1 and 2 to obtain a more accurate interpolation for the desired value of $\rho(B,T)$ or $\lambda(B,T)$.

In this paper, the RRR values refer to resistance, not resistivity, which due to thermal contraction are a mere 0.4% smaller than the more common residual-resistivity

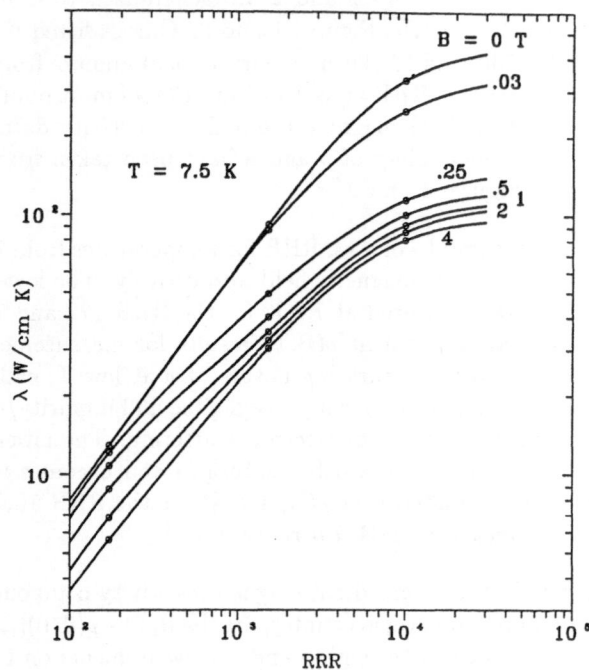

Fig. 2a. Thermal conductivity vs. RRR at 7.5 K
for transverse magnetic fields to 4 T.

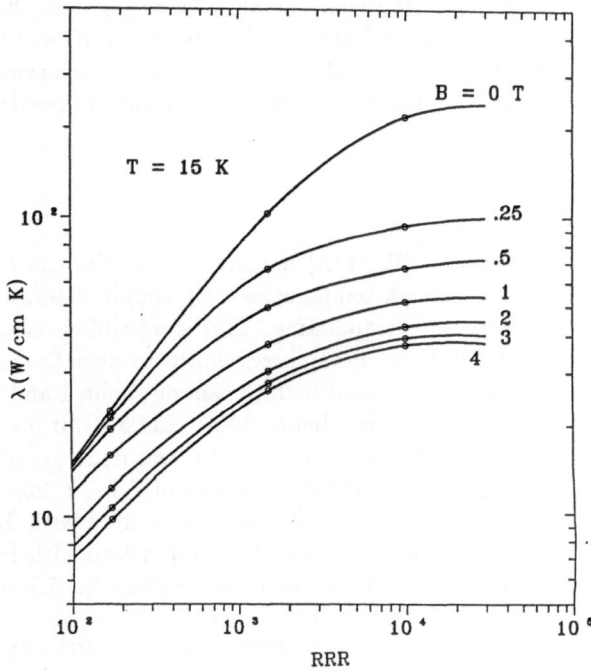

Fig. 2b. Thermal conductivity vs. RRR at 15 K
for transverse magnetic fields to 4 T.

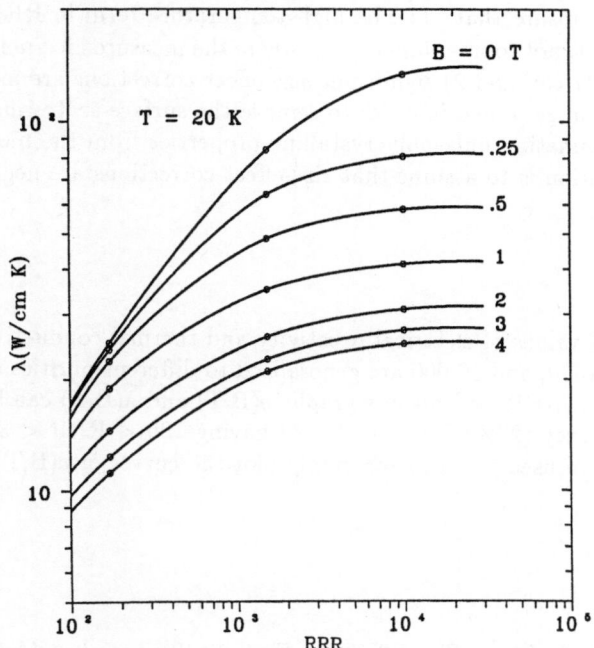

Fig. 2c. Thermal conductivity vs. RRR at 20 K
for transverse magnetic fields to 4 T.

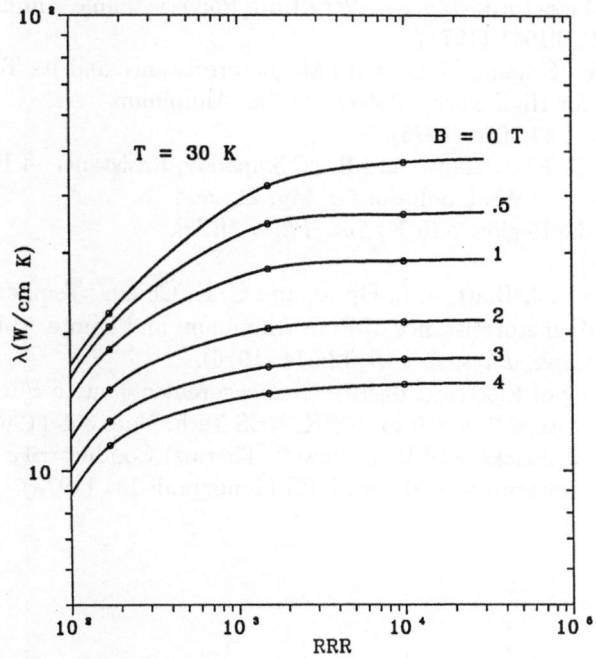

Fig. 2d. Thermal conductivity vs. RRR at 30 K
for transverse magnetic fields to 4 T.

ratio (RRR $= \rho(273\ K)/\rho(4.2\ K)$). In making comparisons with published data care must be exercised to insure that: 1) The high temperature term in RRR was determined for the ice point, not room temperature, where the measured ice-point resistivity is $\rho(273\ K) = 2.44\ \mu\Omega$ cm, and 2) significant size-effect corrections are included when calculating R(4.2 K) in zero field in order to remove the surface-scattering term in resistance, thus approximating bulk-polycrystalline properties from the measured data. The standard convention is to assume that size-effect corrections are negligible in applied magnetic fields.

SUMMARY

The measured values of electrical resistivity and thermal conductivity for samples with RRR 172, 1530, and 10,000 are generalized to different purities of aluminum by plotting ρ and λ vs. RRR. With these graphs $\rho(B,T)$ and $\lambda(B,T)$ can be estimated with reasonable accuracy (25%) for bulk H.P.Al having $100 < RRR < 30,000$. Published values have been used to more accurately plot the curves of $\rho(B,T)$ and $\lambda(0,T)$ vs. RRR.

REFERENCES

1. G. Brandli and J. L. Olsen, Size Effects in Electron Transport in Metals, *Mater. Sci. Eng.*, 4:61 (1969).
2. C. L. Tien, B. F. Armaly, and P. S. Jagannathan, Thermal Conductivity of Thin Metallic Films and Wires at Cryogenic Temperatures, "Thermal Conductivity, Proc. of the 8th Intern. Thermal Cond. Conf.," Plenum Press, New York (1968), p. 13.
3. F. R. Fickett, Magnetoresistance of Very Pure Polycrystalline Aluminum *Phys. Rev. B*, 3:1941 (1971).
4. B. Krevet and W. Schauer, Transverse Magnetoresistance and Its Temperature Dependence for High-Purity Polycrystalline Aluminum *J. Appl. Phys.*, 47:3656 (1976).
5. M. N. Klopkin, G. Kh. Panova, and B. N. Samoilov, Resistance of Pure Aluminum and of Weak Solutions of Mg, Zn, and Ga in Al in the Region 2-40°K, *Sov. Phys. JETP*, 45:287 (1977).
6. M. L. Snodgrass F. J. Blatt, J. L. Opsal, and C. K. Chiang, Temperature-Dependent Magnetoresistance of Pure Aluminum and Dilute Al-Ga and Al-Mg Alloys, *Phys. Rev. B*, 13:574 (1976).
7. L. A. Hall, Survey of Electrical Resistivity Measurements on 16 Pure Metals in the Temperature Range 0 to 273°K, NBS Tech. Note 365 (1968).
8. G. E. Childs, L. J. Ericks, and R. L. Powell, Thermal Conductivity of Solids at Room Temperature and Below, NBS Monograph 131 (1973).

THERMAL PROPERTIES OF COMPOSITE ALUMINUM CONDUCTORS IN THE

TEMPERATURE RANGE 13 - 30 K

W. N. Lawless

CeramPhysics, Inc.
921 Eastwind Drive, Suite 110
Westerville, Ohio 43081

ABSTRACT
     The specific heats and thermal conductivities of 4- and 19-filament composite conductors have been measured, 13 - 30 K. The filaments are high-purity aluminum, and the cladding is an 88Al-8Fe-4Ce alloy; the thermal properties of pure aluminum and the alloy have also been measured. Simple thermal models can be used to explain satisfactorily the thermal properties of the composite conductors based on the thermal properties of the pure aluminum and the alloy, suggesting that interactions between the filaments and the alloy are minor. Derivative data for the thermal diffusivity and for an intermediate thermal group parameter are also given for the composites.

INTRODUCTION

     High-purity and stress-free aluminum has an extremely low electrical resistivity in the hydrogen-temperature range, even under strong magnetic fields.[1] To allow fast current penetration in pulsed-power applications, fine aluminum filaments are required, and this in turn requires embedding the filaments in a high strength matrix alloy. Recent progress has been made in embedding aluminum filaments in an Al-Fe-Ce alloy,[2] and the advantage of this alloy is that there is minimal diffusion alloying with the high-purity aluminum filaments during processing. In addition, this alloy has a workability compatible with the high purity aluminum and maintains the lightweight advantage of aluminum.

     In any pulsed-power application of these composite conductors in the hydrogen-temperature range, the thermal management of the conductor is very important. Therefore, we have undertaken the measurement of the thermal properties of 4- and 19-filament composite conductors in the range 13 - 30 K. In addition, the thermal properties of high-purity aluminum and of the cladding alloy have also been measured for the purpose of modeling the data on the composite conductors. The cladding alloy involved in these measurements is 88Al-8Fe-4Ce.

Advances in Cryogenic Engineering (Materials), Vol. 36
Edited by R. P. Reed and F. R. Fickett
Plenum Press, New York, 1990

687

Thermal Conductivity Measurements

The thermal conductivity measurements were performed in a closed-cycle refrigerator (Cryomech, Model GBO4) using the linear-heat-flow method. All thermal-conductivity samples were in the form of wires, 0.076-cm diameter, except for the alloy sample which was in the form of a long bar (0.32 x 0.16 x 7 cm³). In order to gain a long length of wire, the wire samples were threaded through thin-walled capillary tubes in a serpentine fashion for mechanical support, and an annealing step was performed with the wires in the tubes (400 °C for 10 minutes). In the case of the alloy sample, the thermal conductivity was expected to be low enough that a 4-cm long bar was used. The high-purity aluminum wire sample was 64 cm long, and the composite-conductor samples were 10 cm long. All samples for these measurements were prepared by the Alcoa Technical Center. The composite conductor samples were fabricated by extrusion methods, and the diameter of the aluminum filaments in the 4-filament conductor is 0.0262 cm, and in the 19-filament conductor, 0.0121 cm.

Thermal conductivity data for the high purity aluminum and for the Al-Fe-Ce alloy are shown in Fig. 1. The alloy has a thermal conductivity about two orders of magnitude smaller than that of the pure aluminum and the opposite temperature dependence. The data for pure aluminum agree well with literature data.[3] Thermal conductivity data for the 4- and 19-filament composite conductors is shown in Fig. 2. The 4-filament conductor has the larger thermal conductivity because of the larger content of pure aluminum in the filaments. On comparing the data in Figs. 1 and 2 it is seen that the filaments of pure aluminum dominate the thermal conductivity of the composite conductors.

Specific Heat Measurements

Specific heat measurements were performed in an adiabatic calorimeter using the pulse methods described elsewhere.[4] Samples

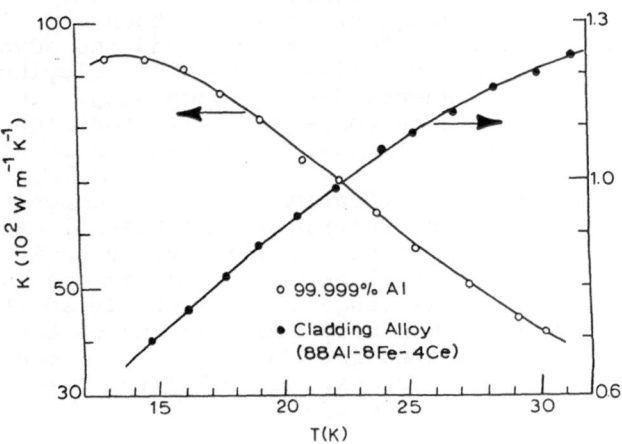

Fig. 1. Thermal conductivity data measured on pure aluminum and on the cladding alloy.

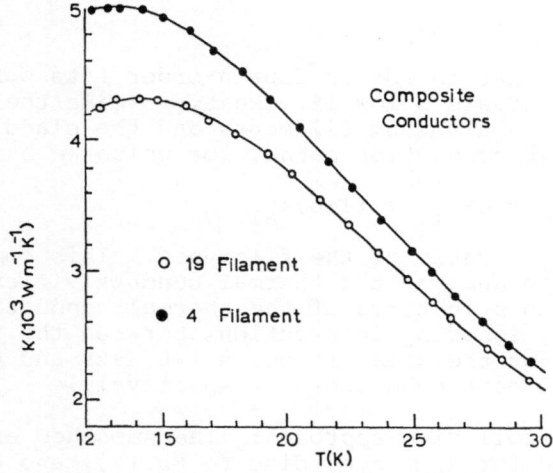

Fig. 2. Thermal conductivity data measured on the 4- and 19-filament composite-conductor wires.

in the form of pellets (approx. 1-cm diam x 1-cm height) were measured. Specific heat data for the pure aluminum sample and for the cladding-alloy sample are shown in Fig. 3, and data for the 4- and 19-filament conductor samples are shown in Fig. 4. The specific heat data for the samples in Figs. 3 and 4 are very similar, due to the large aluminum contents in all samples. The data in Fig. 3 for pure aluminum agree very well with literature data.[5]

DATA ANALYSES

Thermal Conductivity

The first step in analyzing the data in Figs. 1 and 2 is to expand the experimental data in the power series

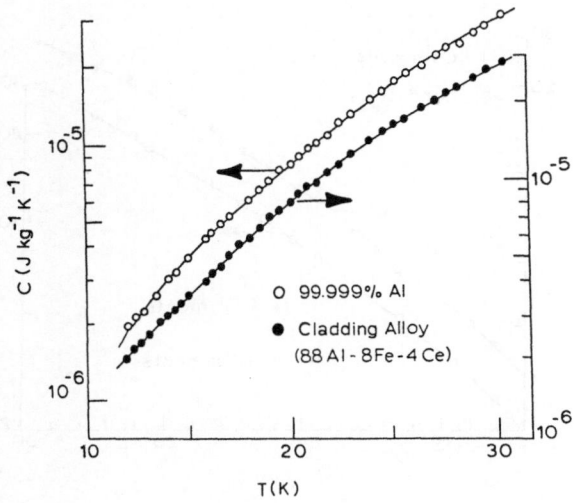

Fig. 3. Specific heat data measured on pure aluminum and on the cladding alloy.

$$K = \Sigma a_n T^n \qquad (1)$$

and it was found that third- or fourth-order fits were sufficient to reduce the residuals below 1%. Next, we make the simple assumption that the aluminum filaments and the cladding alloy form parallel heat conduction paths, for which we have

$$K_{comp} = eK_{fil} + (1-e)K_{alloy} \qquad (2)$$

where e is the area ratio of the filaments. This composite relation allows us to analyze the thermal conductivities of the composite conductors in terms of the thermal conductivities of the constituents, ignoring interactions between the filaments and the alloy. For the area ratios, e = 0.4929 and 0.5065 for the 4- and 19-filament conductors, respectively.

We take the following approach: The smoothed experimental data are analyzed for K_{fil} according to Eq.(2), and K_{fil} is examined relative to K_{Al}, the smoothed thermal conductivity data for pure aluminum. Data for the ratio K_{fil}/K_{Al} are given in Table 1. An interesting correlation is seen in Table 1; namely, the K_{fil}/K_{Al} ratio is uniformly greater than unity for the 4-filament conductor and less than unity for the 19-filament conductor. In all cases, however, the variation from unity is < 12%, which indicates that interactions between the filaments and the alloy are relatively minor from the standpoint of thermal conductivity. The $(1-e)K_{alloy}$ term in Eq.(2) represents a small contribution -- the K_{fil}/K_{Al} ratios change by only 1% if this term is ignored altogether. Therefore, the results in Table 1 are not dependent on the parallel-path assumption in Eq.(1).

Specific Heat

The first step in analyzing the specific heat data in Figs. 3 and 4 was to expand the experimental data in a power series in temperature, and it was found that third-order fits to the logarithmic form

$$\log C = \Sigma b_n T^n \qquad (3)$$

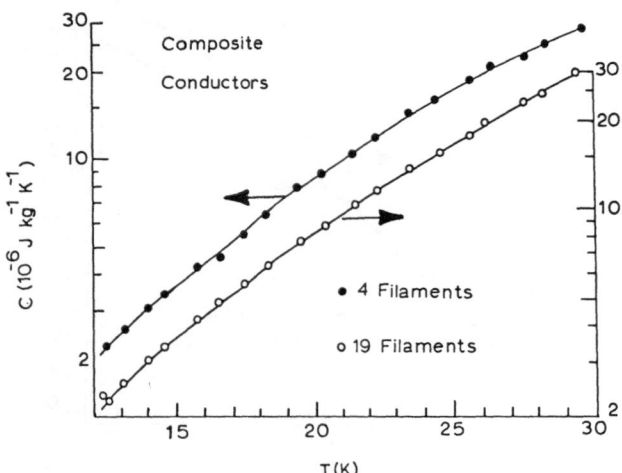

Fig. 4. Specific heat data measured on the 4- and 19-filament composite conductors.

Table 1. Thermal Conductivity Ratios

T(K)	K_{fil}/K_{Al} 4-Filament	19-Filament
12	1.083	0.882
14	1.071	0.902
16	1.059	0.914
18	1.048	0.923
20	1.037	0.931
22	1.027	0.939
24	1.017	0.949
26	1.010	0.958
28	1.007	0.964
30	1.014	0.959

described the experimental data very well (residuals < 0.5%). The smoothed experimental data for pure aluminum and for the cladding alloy were analyzed for the effective Debye temperatures using a published table of the Debye function,[6] and these Debye temperatures are given in Table 2. Literature data for the Debye temperature of pure aluminum are in the range 400 - 426 K.[6] The Debye temperatures in Table 2 increase slightly with increasing temperature, and the reason for this is that the electronic contribution to the specific heat becomes negligibly small above about 18 K. The ratio of the Debye temperatures in Table 2 between pure aluminum and the cladding alloy is about 1.08. From the Lindemann relation[6] one expects this ratio to vary inversely with the square root of the ratio of the atomic weights, from which we find 1.12, in reasonably good agreement with the experimental data.

Because there is minimal diffusion-alloying of the cladding alloy with the aluminum filaments, we expect the specific heat of the composite conductors to follow a simple additivity relation,

$$C_{pred} = \varepsilon C_{Al} + (1-\varepsilon)C_{alloy} \qquad (4)$$

where ε is the weight % of aluminum in the conductor. For the 4- and 19-filament conductors involved here, ε = 47.29 and 48.65%, respectively, based on the starting billets used in the extru-

Table 2. Effective Debye Temperatures

T(K)	Pure Aluminum	Cladding Alloy
12	397 K	366 K
14	405	375
16	410	381
18	412	383
20	412	383
22	412	383
24	411	382
26	410	382
28	408	382
30	407	382

Table 3. Comparison of Predicted and Measured Specific Heat Data

T(K)	C_{pred}/C_{actual} 4-Filament	19-Filament
12	0.9579	0.9692
14	0.9810	0.9732
16	0.9910	0.9789
18	0.9909	0.9844
20	0.9842	0.9880
22	0.9749	0.9882
24	0.9659	0.9820
26	0.9616	0.9695
28	0.9654	0.9482
30	0.9811	0.9177
Mean	0.9754	0.9699

sions. The smoothed specific heat data were analyzed according to Eq.(4) for the ratio C_{pred}/C_{actual}, and these ratio data are given in Table 3. As seen in Table 3, the additivity relation Eq.(4) describes the experimental data very well, although the clear indication is that the measured specific heat data are 2 - 3% larger than the predicted data. This discrepancy is not due to a systematic error in the measurements because all samples were measured under identical conditions. Nor is this finding associated with the electronic contribution to the specific heat because there is no clear temperature dependence to the data in Table 3, in contrast to the data in Tables 1 and 2. We conclude this small effect is probably due to the minimal diffusion-alloying between the alloy and the aluminum filaments during processing.

Thermal Parameters

Thermal management involves four thermal parameters, and which parameter is most important depends on the time scale of the thermal disturbance under consideration. The thermal parameters are: (1) Specific heat; (2) Thermal conductivity; (3) Thermal diffusivity; and (4) An intermediate thermal group parameter. We have a database from the above measurements that allows the evaluation of these latter two parameters. The thermal diffusivity is defined by

$$k = K/\rho C \tag{5}$$

where ρ is density. The densities for the 4- and 19-filament conductors are 2816 and 2813 kg/m, respectively. Using the smoothed K- and C-data for the composite conductors in Eq.(5), the estimated thermal diffusivity data for the composite conductors are shown plotted in Fig. 5. The thermal diffusivities decrease rapidly with increasing temperature because both K and 1/C similarly decrease (Figs. 2 and 4). For example, the thermal relaxation time at 30 K is about 30 times smaller than that at 12 K. The 4-filament conductor has the larger thermal diffusivity in Fig. 5 due to the larger thermal conductivity, Fig. 2.

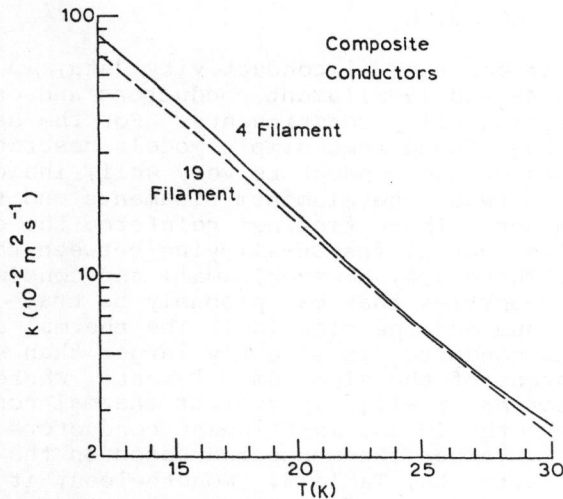

Fig. 5. Thermal diffusivity data for the
4- and 19-filament composite conductors derived from the (smoothed) thermal data in Figs. 2 and
4 according to Eq. (5).

The intermediate thermal group parameter is defined by

$$\eta = (K\rho C)^{1/2}, \tag{6}$$

and proceeding as above the estimated data for this parameter are
shown plotted in Fig. 6. The η-data in Fig. 6 increase uniformly
with increasing temperature because the strong positive
temperature dependence of C overwhelms the negative temperature
dependence of K for both conductors. Here also, the 4-filament
conductor has the larger thermal group parameter.

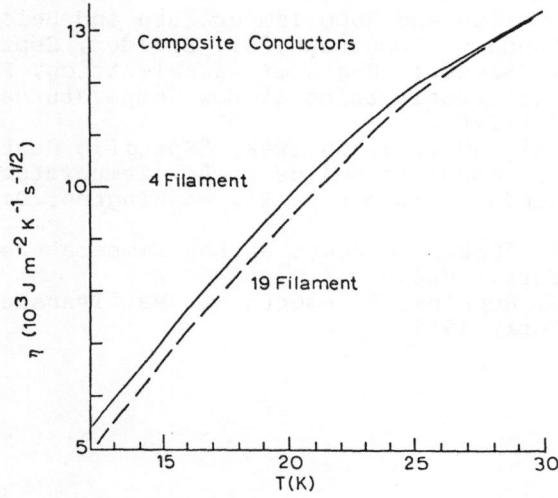

Fig. 6. Thermal group parameter data for
the 4- and 19-filament composite
conductors derived from the
(smoothed) thermal data in Figs.
2 and 4 according to Eq. (6).

CONCLUSIONS AND DISCUSSION

Specific heat and thermal conductivity data, 13 - 30 K, have been measured on 4- and 19-filament conductors and on the pure aluminum and cladding alloy constituents. For the samples studied, it has been found that simple models describe the thermal properties of the conductors very well, indicating that the interactions between the aluminum filaments and the cladding alloy are very minor. These findings reinforce the original metallurgical results that diffusion-alloying between these two materials is small.[2] There are, however, small and consistent effects in the thermal properties that can probably be traced to alloying between the aluminum and the cladding: The thermal conductivity of the 4-filament conductor is slightly larger than expected based on the content of the aluminum filaments, whereas the 19-filament conductor has a slightly smaller thermal conductivity, Table 1. And both the 4- and 19-filament conductors have specific heats somewhat larger than expected based on the specific heats of the constituents, Table 3. Nonetheless, it is expected that the thermal properties of any composite conductor in this system can be reliably estimated from the data presented here for pure aluminum and the cladding alloy using Eqs.(2) and (4). Finally, using smoothed experimental data for the specific heat and thermal conductivity of the composite conductors, the thermal diffusivity and the intermediate thermal group parameter have been estimmated for the 4- and 19-filament conductors, Figs. 5 and 6.

References

1. F. R. Fickett, "Aluminum 1: A Review of Resistive Mechanisms in Aluminum," Cryogenics 11:349 (1971).
2. J. C. Ho, C. E. Oberly, H. L. Gegel, W. M. Griffith, J. T. Morgan, W. T. O'Hara, and Y. V. R. K. Prasad, "A New Aluminum-Base Alloy with Potential Cryogenic Applications," Adv. Cryogenic Eng. Mat. 32:437 (ed., R. P. Reed and A. F. Clark, Plenum Publ. Co., 1986).
3. G. E. Childs, L. J. Ericks, and R. L. Powell, "Thermal Conductivity of Solids and Room Temperature and Below," Nat'l. Bureau of Standards Monograph 131, Boulder, Sept. 1973.
4. W. N. Lawless, "Specific Heats of Paraelectrics, Ferroelectrics, and Antiferroelectrics at Low Temperatures," Phys. Rev. B14:134 (1976).
5. R. J. Corruccini and J. J. Gniewek, "Specific Heats and Enthalpies of Technical Solids at Low Temperatures," Nat'l. Bureau of Standards Monograph 21, Washington, D. C., Oct. 1960.
6. E. S. R. Gopal, "Specific Heats at Low Temperatures," Plenum Press, New York, 1966.
7. M. Jakob and G. Hopkins, "Elements of Heat Transfer", Wiley Publ., New York, 1957.

AC LOSS IN A COMPOSITE HYPERCONDUCTOR*

W. J. Carr, Jr.

Westinghouse Electric Corporation
Research and Development Center
1310 Beulah Road
Pittsburgh, PA 15235

ABSTRACT

For high magnetic field operation at cryogenic temperatures near 20°
K the best lightweight conductor presently available is high-purity
aluminum, sometimes referred to as a hyperconductor. However, because of
its high purity the material is mechanically weak, and because of its high
conductivity the conductor must be fabricated as fine twisted strands to
reduce ac eddy current losses. Thus a construction whereby fine filaments
are embedded in a high-strength, high-resistivity matrix has been
suggested. The ac loss in such a conductor is computed and shown to be
similar in some respects to that for a multifilamentary superconductor.

INTRODUCTION

Applications involving magnetic energy storage exist for electrical
conductors operating in very high magnetic fields at cryogenic
temperatures. For operation at liquid hydrogen temperature in pulsed
magnetic fields of 15 to 20 T the best conductor presently available seems
to be high-purity aluminum. Highly pure metals of this type with
relatively large Debye temperatures are sometimes referred to as
hyperconductors. Because of the high purity the conductors are soft and
must be supported mechanically; and this fact along with the desirability
of breaking up large conductors into fine filaments to reduce the eddy
current loss has led to a construction similar to that used for
superconductors. Figure 1 which shows a conductor strand made up of
aluminum filaments embedded in an aluminum alloy matrix illustrates the
principle. The size of the filament is fixed by the diameter needed to
keep the eddy current loss within the filament to a negligible value. The
purpose of the matrix is two-fold: to provide mechanical strength and to
partially insulate the filaments. Thus a high-strength, high-resistivity
material at cryogenic temperatures is called for. One possibility for the
matrix is an Al-Ce-Fe material described by Ho et al[1] designed as a
lightweight matrix which will not contaminate the aluminum filaments. The
resistivity of the Al-Ce-Fe alloy is the order of 10^{-5} ohm-cm as compared

---

*Work supported by the Wright-Patterson Air Force Base, Ohio 45433, under
Contract No. F33615-86-C-2681.

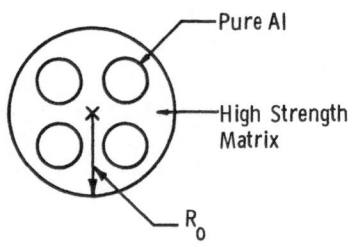

Figure 1 — Multifilamentary
hyperconductor strand.

with the order of 10^{-8} for the aluminum filaments at 20°K in large
magnetic fields. This matrix appears to be satisfactory for some
applications but in general a higher resistivity material would be
desirable inasmuch as the diameter of the strand (and therefore the amount
of current it can carry) for a given allowed eddy cuirrent loss depends
upon the matrix resistivity. The conductor itself is made by insulating
the strand and forming a cable of strands; thus relatively large strands
reduce the problem of cable fabrication. One method for obtaining higher
resistivity is to place a barrier around the filaments as in Figure 2. In
principle the barrier can be used in two ways: (a) to prevent diffusion,
which allows a wider range of alloys to be considered for the matrix and
(b) as an insulating layer which increases the effective resistivity of
the matrix by changing the geometry of eddy current flow.

In order for the subdivision of the strand into filaments to be
effective in reducing the eddy current loss for a changing transverse
magnetic field, the strand must be twisted. The reason is illustrated in
Figure 3. The changing transverse magnetic field tends to cause axial
current to be induced whose pattern of flow is broken up by the twist.
Induced current at the top of Figure 3 tends to flow to the right, while
that at the bottom tends to flow to the left. If the current followed the
low resistance path of the filaments a divergence would occur at the
middle, and it follows that the current must leave the filaments and flow
across the matrix to filaments on the other side. Thus for a twisted
strand the eddy current flow due to a changing transverse magnetic field
is largely across the diameter of the strand in analogy with the flow in a
superconductor. It is this fact which allows an insulating barrier around
the filaments to greatly increase the effective resistivity of the matrix
as shown in Figure 4.

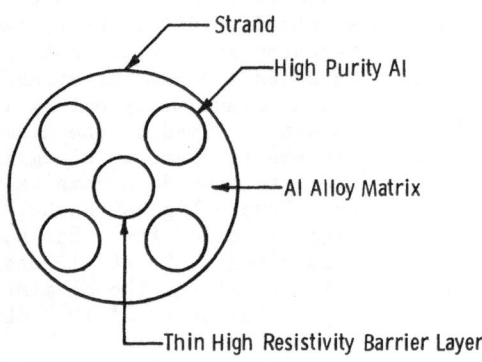

Figure 2 — Hyperconductor strand
with diffusion barrier.

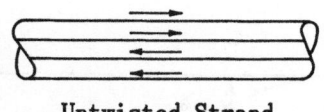

Untwisted Strand

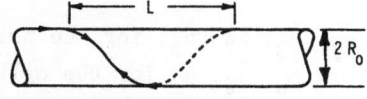

Twisted Strand

Figure 3 — Current in Filament for
Twisted and Untwisted Strands.

EDDY CURRENT LOSS IN THE STRAND

Let it be assumed that the pulsed current through the conductor can
be approximated by a dc current plus an alternating current of angular
frequency ω, where the amplitude of the alternating current is small
relative to the dc. Then the loss in a conductor which is used to
construct a coil is given approximately by the loss from the dc transport
current plus the eddy current loss from the alternating transverse
magnetic field of the coil. The latter has been calculated from Maxwell's
equations using an anisotropic continuum model described previously by the
author.[2] The model imagines the filaments of Figure 1 to be uniformly
smeared out over the strand giving a continuum with high conductivity $\sigma_{\parallel}$
along the direction of the filament axis, and low conductivity $\sigma_{\perp}$
transverse to the filament axis. The difficult part of the problem comes
from solving the Maxwell equations subject to the spiral symmetry of the
twisted strand, since this symmetry completely changes the nature of the
solution.

The equation to be solved is

$$\text{Curl}^2 \, \vec{E} = -\frac{4\pi}{c^2} \frac{\partial \vec{j}}{\partial t} \tag{1}$$

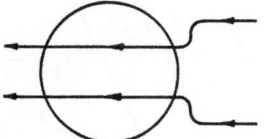

Filament Without Barrier

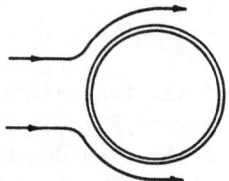

Filament With Barrier

Figure 4 — Induced Current Flow
Pattern in a Strand.

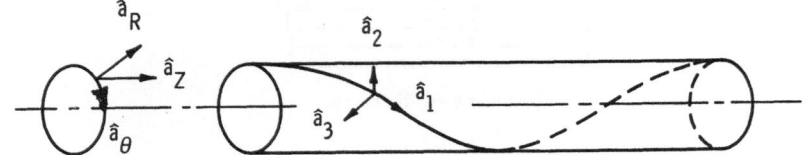

Figure 5 – Unit vectors $\hat{a}_1$, $\hat{a}_2$, and $\hat{a}_3$, for the twisted filaments and the unit vectors $\hat{a}_R$, $\hat{a}_\theta$, $\hat{a}_z$ for the cylindrical conductor.

coming from combining Maxwell's equations with obvious nomenclature. The equation is solved for a long strand where $\frac{\partial}{\partial z} = 0$ with z the axis of the strand, and subject to an anisotropic ohms law

$$\vec{j} = \sigma_{||}\vec{E}_{||} + \sigma_\perp \vec{E}_\perp \qquad (2)$$

where the parallel and transverse directions refer to the twisted filaments in the strand. The power loss at any point is

$$\vec{E} \cdot \vec{j} = \sigma_{||}E_1{}^2 + \sigma_\perp E_2{}^2 + \sigma_\perp E_3{}^2 \qquad (3)$$

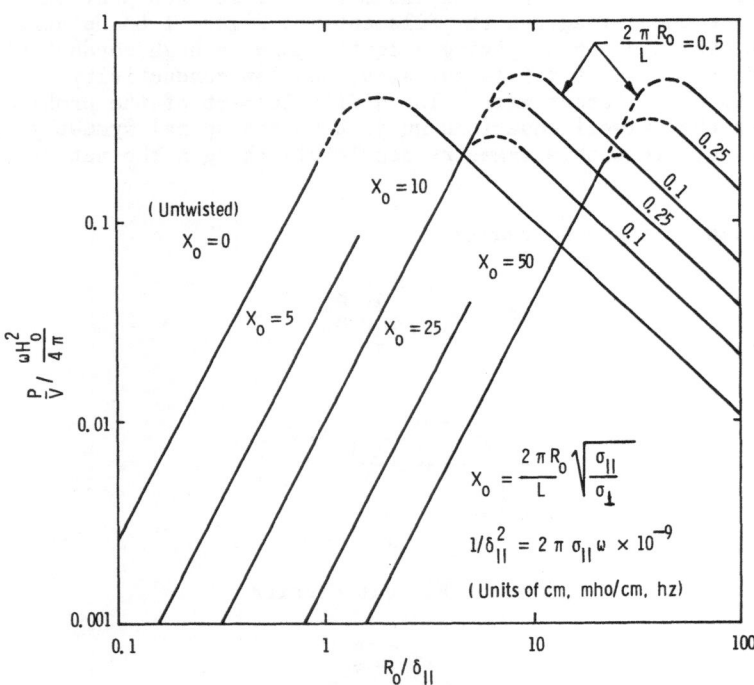

Figure 6 — Dimensionless plot of the transverse field eddy current loss per unit volume P/V versus $R_o/\delta_{||}$, where R_o is the radius of the strand and $\delta_{||}$ is the skin depth due to the conductivity $\sigma_{||}$. The loss given by the ordinate is reduced by the factor factor $\omega H_o{}^2/4\pi$ where ω is 2π times the frequency, and H_o is the peak amplitude of the ac transverse magnetic field. R_o/L is the ratio of strand radius to twist pitch.

where the 1, 2, 3 axes are indicated in Figure 5. Two skin depths $\delta_{||}$ and $\delta_{\perp}$ may be defined by

$$\frac{1}{\delta_{||}^2} = 2 \pi \delta_{||} \omega \times 10^{-9} \tag{4}$$

$$\frac{1}{\delta_{\perp}^2} = 2\pi\sigma_{\perp}\omega \times 10^{-9} \tag{5}$$

in units of cm, mhos-cm and hertz.

In expressing the results it is useful to define a parameter X_o by

$$X_o = \frac{2\pi R_o}{L} \sqrt{\frac{\sigma_{||}}{\sigma_{\perp}}}$$

where R_o is the radius of the strand and L is the twist pitch. Then if H_o is the amplitude of the ac magnetic field the principal results in watts/cm^3 are

(a) <u>Low Frequency (No Skin Effect)</u>
   The power loss per unit volume of composite conductor

assuming that $\dfrac{\sigma_{\perp}}{\sigma_{||}} \ll \left(\dfrac{2\pi R_o}{L}\right)^2 \ll 1$, (i.e., $X_o^2 \gg 1$) is

$$\frac{P}{V} \approx \frac{\omega H_o^2}{4\pi} \left[\frac{L}{2\pi\delta_{\perp}}\right]^2 \times 10^{-7}$$

$$= \frac{\omega H_o^2}{4\pi} \frac{1}{X_o^2} \left[\frac{R_o}{\delta_{||}}\right]^2 \times 10^{-7} \tag{6}$$

(b) <u>High Frequency Case (Skin Effect)</u>

$$\frac{P}{V} \approx \frac{\omega H_o^2}{4\pi} \left[\frac{\delta_{||}}{R_o} + \left(\frac{2\pi R_o}{L}\right)^2 \frac{\delta_{\perp}}{R_o}\right] + 10^{-7} \tag{7}$$

The transition between low frequency and high frequency behavior is given roughly (within about a factor of two) by the point where $R_o/\delta_{||} = (X_o^2/2 + 1)^{1/2}$. The results are shown on a log-log plot in Figure 6 where the lines of positive slope are the low frequency curves and those with negative slope are the high frequency curves. The changeover occurs when skin effect begins to occur. The right-hand side of the plot is of interest only for strands of very large radius or for quite large power frequencies, and only the left-hand side will be considered here.

TYPICAL PARAMETERS

According to (6) the transverse field loss decreases as L^2 and therefore a well-twisted strand is needed for low loss. In multifilament superconductors it is found that the twist pitch cannot be made much smaller than 10 to 20 times the radius R_o, for otherwise the filaments begin to break. While it might be possible to twist the soft aluminum more than this, it is not desirable for other reasons. The twist increases the path length of the dc transport current and causes the effective resistance to increase. The ratio of the resistance for a given L to that for an untwisted strand is easily computed to be

$$\frac{r(L)}{r(\infty)} = \frac{1}{\left(\frac{L}{2\pi R_o}\right)^2 \ln\left[1 + \left(\frac{2\pi R_o}{L}\right)^2\right]} \tag{8}$$

which for $L/R_o = 10$ is already 1.186 indicated a nearly 20% increase in dc loss.

The loss in an isolated filament per unit volume is easily computed to be

$$\left(\frac{P}{V}\right)_f = \frac{\omega H_o^2}{4\pi} \frac{1}{4} \left(\frac{d}{2\delta_f}\right)^2 \times 10^{-7} \tag{9}$$

where δ_f is the skin depth for the aluminum and d is the filament diameter. If a frequency of 10 hz is assumed $\delta_f \approx 0.13$ cm, and if H_o is assumed to be 1.5 T the loss within the filament is about 0.17 W/cm^3 for d = 0.02 cm. Since 10 to 20 times this much heat can easily be carried away, no gain is made in making the filament diameter smaller than about 0.02 in this example, and furthermore filaments appreciably smaller may exhibit a size effect in their resistance.

The conductivities $\sigma_{||}$ and $\sigma_\perp$ may be estimated from the expressions

$$\sigma_{||} = \lambda \sigma_f + (1 - \lambda) \sigma_m \tag{10}$$

and

$$\sigma_\perp \approx \frac{1 + \lambda}{1 - \lambda} \sigma_m \tag{11}$$

where λ is the fraction of pure Al in the strand, and σ_f and σ_m are the conductivities of the filament and matrix. A typical value for λ is assumed to be about 0.6. In the case of a barrier around the filaments

$$\sigma_\perp \approx \frac{1 - \lambda}{1 + \lambda} \sigma_m \tag{12}$$

The strand diameter is chosen to keep the ac loss within reasonable bounds, which is considered to be in the neighborhood of 0.5 W/cm^3.

For the case of σ_m the order of 10^{-6} Ω-cm with no barrier a typical strand would have only about four filaments and carry about 50 A. For a decrease in matrix conductivity by a factor of ten obtained with a barrier or with a different matrix material, one can increase the strand diameter so that the number of filaments and the current carrying capacity of the strand is increased by a factor of ten.

REFERENCES

1. J. C. Ho, C. E. Oberly, H. L. Gegel, W. M. Griffith, J. T. Morgan, W. T. O'Hara, and Y. V. R. K. Prasad, Ad. Cry. Eng 32, 437 (1986).

2. W. J. Carr, Jr., J. Appl. Phys. 45, 929 (1974).

CRYOCONDUCTOR MATERIALS TESTING SYSTEM

A. Lumbis, J. Roelli, D. Frutschi,
J.T. Gehan and K.T. Hartwig

Texas A&M University
Department of Mechanical Engineering
College Station, TX 77843-3123

ABSTRACT

A materials testing system was designed and constructed to perform cyclic-strain resistivity testing in magnetic fields to 8 Tesla and at temperatures from 4.2 K to 77 K. The system has a load capacity of ±8.9 kN (2000 lb). Helium gas cooling is used to maintain a constant temperature in the 15 K to 50 K region. Data acquisition, including stress-strain information and resistivity measurements, is controlled by a personal computer. An overall helium consumption rate of 8 l/hr is achieved for a 20 K test in magnetic fields to 8 Tesla.

INTRODUCTION

Liquid hydrogen can be a coolant for the operation of high powered cryogenic magnet systems.[1] Such systems might operate at a temperature between 14 and 30 K. Hydrogen freezes at 13.8 K and boils under one atmosphere of pressure at 20.3 K. The upper bound on the operating temperature is a practical limit established by the intrinsic resistivity of typical pure metal cryoconductors such as copper and aluminum.

A materials test facility has been developed at Texas A&M University in support of the Air Force Cryogenic Inductor Project.[1] Design of the mechanical system is similar in some respects to other systems developed elsewhere.[2] The purpose of the present facility is to be able to subject pure aluminum and composite aluminum cryoconductors to conditions similar to those expected in space based devices where liquid hydrogen would be the coolant. Such conditions may include monotonic and cyclic stress, transverse magnetic fields to 5 or 6 Tesla and an operating temperature of near 20 K. The test system capability includes the simultaneous application of these conditions. In order to test cryoconductor samples under the range of conditions possible, the test facility was developed with five key working elements: mechanical fixtures for applying stress, a variable temperature helium gas supply system, an 8 Tesla solenoidal superconducting magnet, dewar systems for containment, and a supporting computer system for experiment control and data acquisition. Each of these systems is described in detail below with key procedures and performance characteristics of the combined system.

Advances in Cryogenic Engineering (Materials), Vol. 36
Edited by R. P. Reed and F. R. Fickett
Plenum Press, New York, 1990

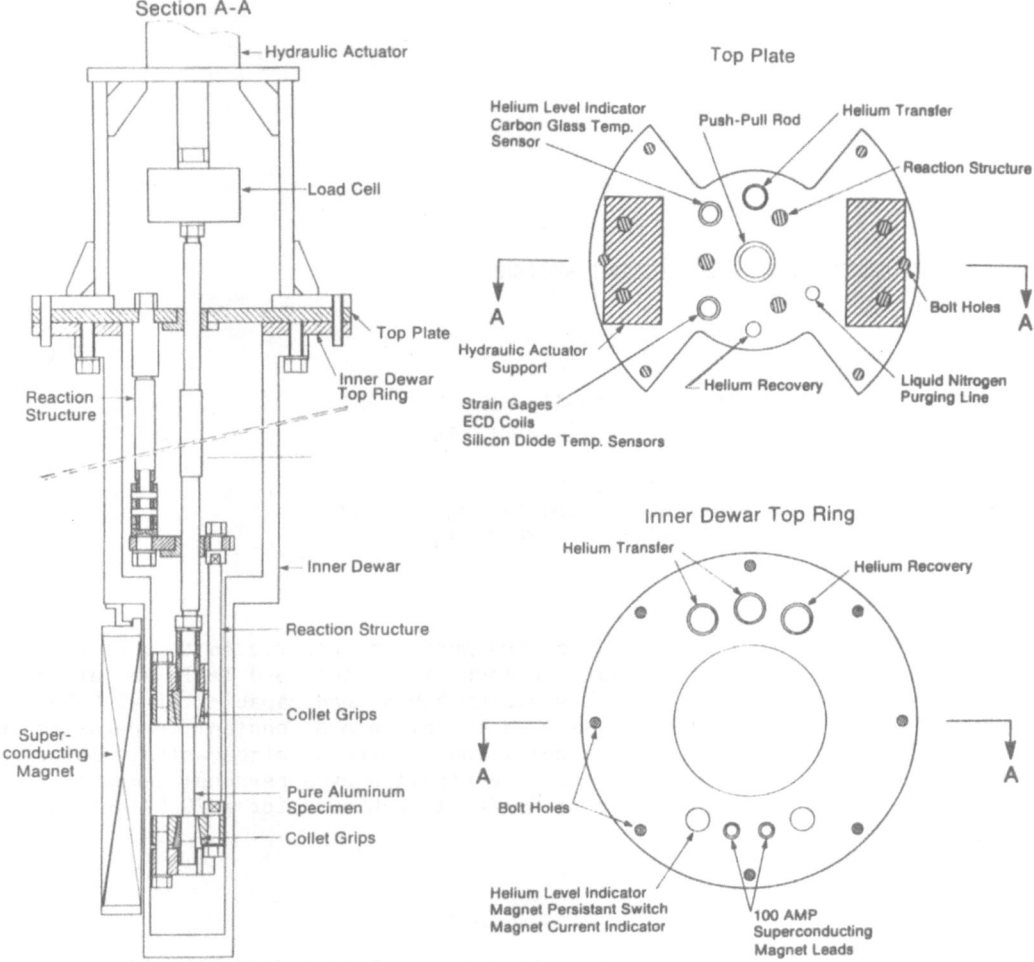

Fig. 1.   Mechanical Load Fixture.

SYSTEM FEATURES

Load Frame and Fixtures

The mechanical load fixture shown in Figure 1 is used to apply sufficient force to induce plastic strains up to 0.3 percent in one centimetre
diameter specimens of pure aluminum. The MTS model 244.12 hydraulic actuator can apply a maximum load of ±24.5 kN (±5500 lb) however the frame that
it is attached to will only support a 8.9 kN (2000 lb) load. These loads
are measured using a 8.9 kN (2000 lb) Lebow Model 3169 load cell. The load
cell is mounted between the actuator and the push-pull rod. The reaction
structure is mounted to the top plate. The components of the push-pull rod
and the reaction structure are made of either Nitronic-60 or 304 stainless
steel. The top plate is used to fasten the hydraulic actuator to the dewars
which are used to insulate the test structure. Electrical connection points
for the strain, resistivity, magnet current and control, and helium level
signals are provided for in the top plate. Sealed fittings for the transfer
of cryogenic fluid are also installed in the top plate. The cost of liquid
helium justified venting the test structure to a helium recovery system.

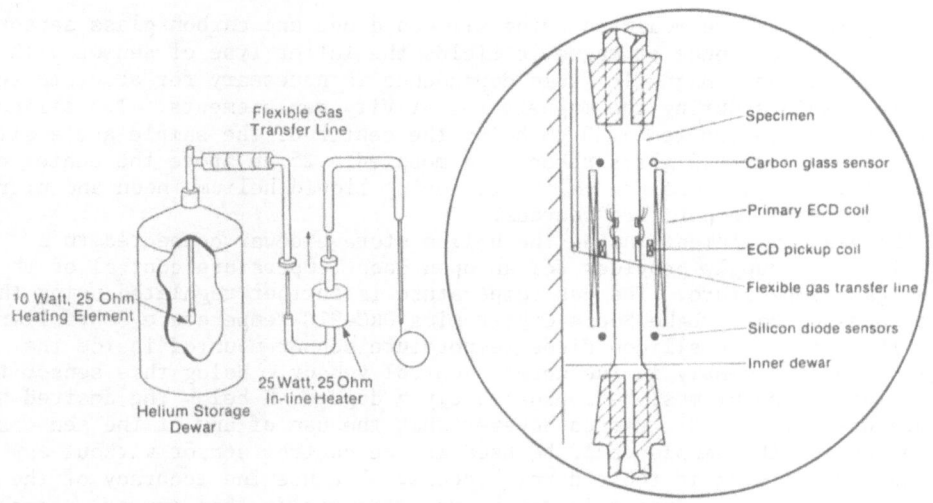

Fig. 2. Variable Temperature Gas System.

Because of the large temperature gradient that exists across this top plate the cross-sectional area of the push-pull rod and the reaction structure that connects to the top plate where kept to a minimum. For the reaction structure this was accomplished by using tubular instead of solid cross-sections. The cross-sectional area of the push-pull rod and the reactions structure is 6.93 cm^2. The test sample is connected to the push-pull rod and the reaction structure using collet grips made of Nitronic-60 stainless steel. The grips are modified versions of the design described by Lehmann.[3]

Load Control

The MTS Micro-controller Model 458.20 connected to the hydraulic actuator allows for displacement, load or strain control. Micro Measurements S-T-C (self temperature compensated WK-13-125BT 350) 350 ohm strain gages mounted diametrically on the sample allow for the strain levels induced in the sample to be measured. Continuous measurements of load and strain by the controller using the load cell and strain gages, respectively, are accurate to within ±1 percent of full scale. Since the experiment involves cycling the sample between a set strain limit the actuator movement is controlled according to the strain gage signals using a feedback loop.

Variable Temperature Gas System

The variable temperature gas system incorporated into the test structure allows for testing at temperatures from about 10 K to 77 K (Figure 2). The 10 watt, 25 ohm resistive heating element mounted in the helium storage dewar produces helium gas which is transferred into the inner dewar using the flexible gas transfer line. Enroute the gas is further heated using an 25 watt, 25 ohm in-line heater. The heated helium gas is discharged into the test dewar on one side of the test sample. The alignment of the discharge nozzle is tangential to the test sample creating a cyclonic air movement. This setup proved to be better than a radial alignment because helium gas exiting directly toward the sample was obstructed by the eddy current decay coils resulting in higher temperature gradients in the test specimen area.

Temperatures are measured using silicon diode and carbon glass sensors. Because of the presence of magnetic fields the latter type of sensor with its relatively weak magnetic field dependence is necessary for accurate temperature readings during the magneto-resistivity measurements. Two silicon diode sensors are mounted 4.25 cm below the center of the sample and a silicon diode and a carbon glass sensor are mounted 4.25 cm above the center of the sample. The sensors are calibrated using liquid helium, neon and nitrogen as reference temperature sources.

The heating element inside the helium storage dewar connected to a variable power supply provides for an open-loop temperature control of the helium gas temperature. The gas temperature is further regulated using the microprocessor based Lake Shore Cryotronics DRC-91C Temperature Controller with PID control. A silicon diode temperature sensor mounted inside the in-line heater assembly is the normal control sensor. Using this sensor the controller set point must be approximately 5 degrees K below the desired gas temperature. It was discovered however that the use of any of the sensors mounted beside the sample could be used as the control sensor without any difficulties arising in the control process. Because the accuracy of the carbon glass sensor is not affected by magnetic fields this sensor is generally chosen for the control sensor. The use of this sensor ensures that the control process is not disrupted by any magnetic fields that may be present during testing.

Superconducting Magnet

A 3.5 inch inner diameter, 8 Tesla, solenoidal superconducting magnet with vapour cooled leads is mounted in line with the test sample on the outside of the inner dewar as shown in Figure 3. This allows magneto-resistivity measurements of the test sample to be made.

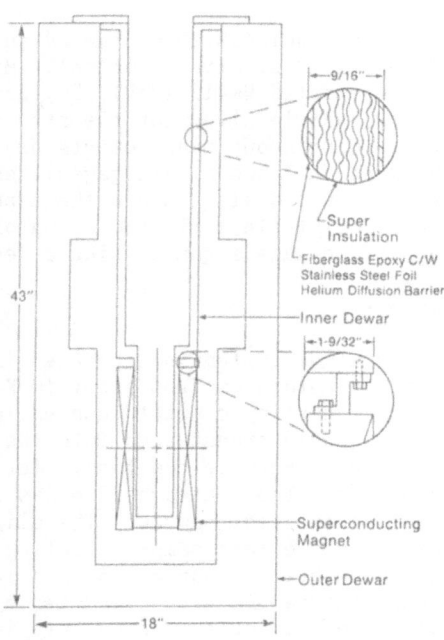

Fig. 3. Containment Dewars.

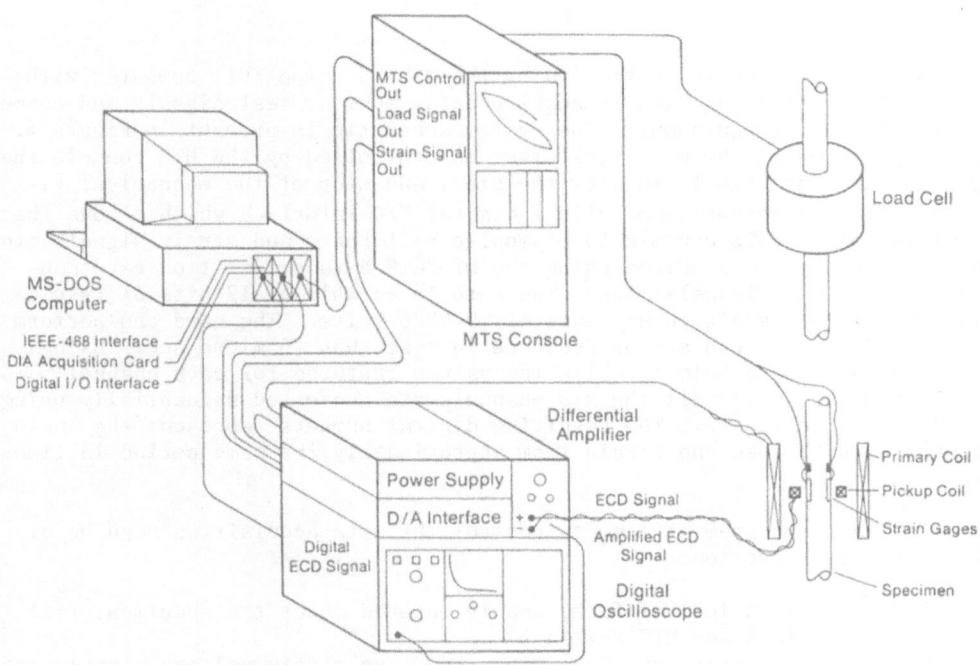

Fig. 4. Schematic of the Data Collection System.

Containment Dewars

The mechanical fixture is inserted into the two concentric superinsu-
lated dewars shown in Figure 3. The outer dewar constructed of
composite-aluminum materials is 40 inches deep and has a 9 inch diameter
neck. To increase the storage capacity, the outer dewar has a 13 inch diameter belly. To isolate the test chamber and to provide a support apparatus
for the superconducting magnet an inner dewar was used. The walls of the
neck and outer vessel of the inner dewar are constructed of fiberglass epoxy
with a stainless steel foil helium diffusion barrier. Super-insulation
fills the space between the neck and the outer vessel. An aluminum alloy is
used for the reduced end of the inner dewar and the magnet support structure. Liquid helium sensors mounted in both dewars monitor the level of
liquid helium.

Electrical Resistivity

Electrical resistivity measurements[4] are made with an eddy current
decay (ECD) system which is composed of two concentric solenoid coils that
surround the specimen. A DC magnetic field is generated in the outside coil
and the resulting eddy current signal is picked up by the inner coil. The
coil system should be mounted securely to the reaction structure to prevent
movement or vibration. Probe vibration in a strong magnetic field will
result in signal distortion and inaccurate measurements due to the movement
of the coils normal to the magnetic field lines.[5] After the eddy current
signal is amplified by a Tektronix AM502 differential amplifier it is captured on a Nicolet 4094B digital oscilloscope and then is transferred to a
MS-DOS compatible computer using a IEEE-488 interface. Using the least
squares fit method the time constant of the eddy current signal is calculated and the resistivity of the specimen determined.

Data Collection

Data collection is controlled by the MS-DOS compatible computer with three additional cards used to acquire data, monitor test signals and communicate with other equipment. The system schematic is present in Figure 4. Since operation of the mechanical test is controlled by the MTS console the computer must be able to monitor the start and stop of the mechanical cycling. This is accomplished with a digital I/O interface which allows the condition of the MTS console to be monitored. Force and strain signals from the MTS console are acquired using the DT-2818 data acquisition card manufactured by Data Translations. The card is capable of 12 bits of resolution with a full scale input range of ± 9.9976 volts. The card can perform a complete 2 channel conversion sequence in less than 75 micro-sec. Using separate sample and hold circuits the values captured for each channel remain constant while all the A/D channels are converted sequentially using a single A/D converter. The resulting digital numbers represent the analog signals of both load and strain from approximately the same period in time (<5 ns).

The BASIC program written to control the data acquisition[6] can be divided into four sections.

1. Initialization - Information is entered about the specimen, test conditions and MTS settings.
2. Data Collection and Analysis - Load and strain voltage signals are sampled and converted to the appropriate units. Plastic strain per cycle, stress levels and total plastic strain are stored for later analysis.
3. Resistivity Measurements - Eddy current decay curves are transferred from the digital oscilloscope to the PC. From these curves the resistivity is calculated and stored for later analysis.
4. Post Processing of Data - After completion of the test data is sorted for later analysis.

TEST PROCEDURES

Liquid Helium Testing

Once the specimen is mounted in the test structure, the controller is put in load control to ensure that any thermal strains that are induced in the specimen during the cooldown process do not inadvertantly damage it. The test structure is lowered into the inner dewar and pre-cooled with liquid nitrogen. To prevent the test structure from warming up during the subsequent helium transfer process the initially warm gas is allowed to flow through the transfer line directly to the recovery system. The helium discharging from the transfer line should be below 77 K before being transferred into the test structure region. Once an adequate volume of helium has been transferred the MTS controller is switched to strain control and zeroed.

The specimen is loaded through a set strain range for 3000 cycles. Load and strain voltages are sampled and converted into the appropriate units by the computer. Because of the large changes in resistivity of the test sample, the majority of the resistivity measurements are taken within the first 100 cycles. The calculated resistivity is recorded by the computer. The cycling speed at the beginning of testing is 0.02 Hz with a corresponding sampling rate of 4 Hz for the computer. This is increased to 0.2 and 10 Hz, respectively, by the 200th cycle. The corrsponding data collection rates are 3 and 18 measurements per second.

Once the testing is complete the MTS controller is put back in load control and any remaining liquid helium is boiled off. The test structure is then removed from the dewar and allowed to warm to room temperature before the test specimen is removed from the structure.

Helium Gas Testing

Preparations for a helium gas test are identical to those for a liquid helium test. Connecting the flexible gas transfer line to the recovery system the helium gas is cooled to below 77 K before introducing the gas into the test structure. Once introduced the gas is further cooled to a temperature usually 5 to 10 K below the desired test temperature. This will remove any latent heat still present in the system. The gas is then warmed to the desired test temperature. Stabilization of the temperature in the test dewar takes between 1 and 2 hours. After stabilization the gas temperature remains constant to within ± 0.1 K and thermal gradients in the gas at the region of the test specimen are below 0.012 K/cm.

Magneto-Resistivity Testing

Preparation of the magnet for magneto-resistivity tests involves pre-cooling the outer dewar with liquid nitrogen and then transferring liquid helium into the outer dewar. Magneto-resistivity measurements are made at several magnetic fields up to 7 Tesla. The superconducting magnet is put in its persistent mode when the actual measurements are taken. The hydraulic actuator is depressurized during these tests because the high magnetic fields can cause adverse behaviour in the system's hydraulic components.

PERFORMANCE CHARACTERISTICS

The materials test facility described in this paper was developed to subject pure aluminum to conditions of montonic and cyclic stress, and transverse magnetic fields at temperatures near 20 K. The load capacity of the structure is ± 8.9 kN (± 2000 lb). Magnetic fields up to 8 Tesla can be created. Maximum cycling rate of the structure is 0.2 Hz with data acquisition. The maximum crosshead speed of the system is .88 in/sec. Temperatures from 4.2 K to 50 K have been achieved with a stability of ± 0.1 K and a test region gas temperature gradient of .012 K/cm. Target temperature stabilization takes 1 to 2 hours. Helium consumption rates are 4 l/hr for liquid helium tests, 6 l/hr for 20 K helium gas tests and 2 l/hr are required to operate the superconducting magnet.

ACKNOWLEDGEMENTS

Support for this work was provided by the U.S. Air Force under Contract F33615-86-C-2683.

REFERENCES

1. C. E. Oberly, The Origin and Future of Composite Aluminum Conductors, to be published in Adv. Cryo. Eng., Vol 36.
2. R. P. Reed, A Cryostat for Tensile Tests in the Temperature Range 300 to 4 K, Adv. Cryo. Eng., 7:449 (1962).
3. P. Lehmann, G. S. Yuan, and K. T. Hartwig, Grip for Fatigue Testing Pure Aluminum, Cryogenics, 25:164 (1985).
4. A. F. Clark et al, Standard Reference Materials: The Eddy Current Method for Resistivity Characterization of High Purity Metals, NBS STP 260-39, May 1972.

5. C. Y. Hua and K. T. Hartwig, Resistivity Measurements in High Magnetic Fields by the Eddy Current Decay Method, accepted for publication in Cryogenics.

6. J. T. Gehan, Total Plastic Strain and Electrical Resistivity in High Purity Aluminum Cyclically Strained at 4.2 K, M.S. Thesis, Texas A&M Univ., Dec. 1988.

MECHANICAL AND ELECTRICAL TESTING

OF COMPOSITE ALUMINUM CRYOCONDUCTORS

K.T. Hartwig and R.J. DeFrese

Texas A&M University
Department of Mechanical Engineering
College Station, TX 77843-3123

ABSTRACT

Results are reported of monotonic and cyclic-strain resistivity experiments at 4.2 and 20 K on Alcoa fabricated composite aluminum cryoconductors. The conductors contain filaments of nominal 99.999% pure aluminum in an Al-Fe-Ce matrix. Initial conductor resistance ratios (RR = R(273 K)/R(T)) range from 340 to 800 at 4.2 K and from 259 to 500 at 20 K. Sample resistivity increases nearly linearly with monotonic plastic strain at a rate of 7 ± 2 nΩcm / % strain. Force controlled cyclic loadings cause the resistivity to increase rapidly to a saturation value reached after about 500 cycles.

INTRODUCTION

This work supports an Alcoa project funded by the U.S. Air Force to develop light weight, high current conductors for space-based applications. Earlier encouraging research on multifilamentary aluminum composite conductors[1] was the effort from which this project began.

The objective of the conductor development program is to fabricate long lengths of cryoconductor wire for use in inductor devices operating at near 20 K. To help establish which conductor design and fabrication steps give satisfactory overall performance, conductors containing different numbers of filaments, and different matrix compositions were prepared by Alcoa. Our part of the project is to provide 4.2 and 20 K mechanical and electrical testing support and to help identify promising conductor design features and fabrication methods.

Of major concern to the operation of a cryogenic inductor is strain in the conductor windings during use. In particular, plastic strain at 4.2 K in pure aluminum causes a dramatic increase in resistivity.[2,3] Because the main conducting component in the Alcoa cryoconductors is pure aluminum, there is a need to know how the composite conductor resistivity will be affected by strain at 20 K. The major thrust of our activities has been to gather cryogenic-strain resistivity information on various cryoconductor samples. Key results from monotonic-strain and cyclic-strain resistivity tests are reported.

Advances in Cryogenic Engineering (Materials), Vol. 36
Edited by R. P. Reed and F. R. Fickett
Plenum Press, New York, 1990

709

MATERIALS AND METHODS

The cryoconductor samples are filamentary composites fabricated by Alcoa.[4] They consist of an Al-Fe-Ce matrix with nominal 99.999% pure aluminum fibers (residual resistivity ratio (RRR) ~ 1500). Two types of alloy matrix material are used: Al-8wt.%Fe-4wt.%Ce (Al-8Fe-4Ce) and Al-4Fe-2Ce. The number of filaments in the cryoconductor is either four or nineteen, the volume fraction of each being about one-half. Samples are prepared in both the annealed (400°C/20 min/air) and cold extruded conditions. Wire diameters are either 0.74 mm (0.029 in) or 1.22 mm (0.048 in).

Tests were conducted by first attaching two small drill chucks on either end of a four inch long wire sample. One end was fastened to the bottom of a reaction tube by a spherical nut while the other end was attached to a pull rod connected to a hydraulic actuator through a load cell. Actuator movement was controlled by an MTS 458-20 micro console. Tests were carried out under both load and displacement control.

A helium gas system was used to create a 20 K environment around the sample. Cold gas was generated by warming liquid helium inside a storage dewar. The resulting gas was passed through a transfer line with an in-line heater. The temperature was regulated to within ± 0.2 K at the sample by a Lake Shore Cryotronics DRC-91C temperature controller.

The four point potential difference method was used to determine sample resistivity. A current, usually one or two amperes, was supplied by a Hewlett Packard 6284 A power supply and passed through the sample. Voltage across the sample was measured using a Keithly 181 nanovoltmeter. To accommodate thermal EMF effects, a current reversing switch was used and an average voltage drop was taken between the two current directions. To determine sample strain, an extensometer with a gage length of one inch, was affixed to the sample. A silicon diode temperature sensor was placed within one inch of the sample and was used to measure the temperature in the sample environment. A schematic of the sample fixturing and instrumentation is shown in Fig. 1.

The cryoconductor samples were subjected to two different kinds of tests: monotonic-strain resistivity and cyclic-strain resistivity. In the

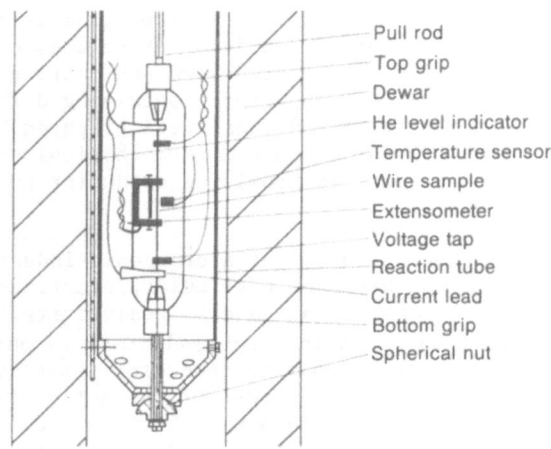

Pull rod
Top grip
Dewar
He level indicator
Temperature sensor
Wire sample
Extensometer
Voltage tap
Reaction tube
Current lead
Bottom grip
Spherical nut

Fig. 1. Schematic of sample connections and sensor locations for strain-resistivity tests.

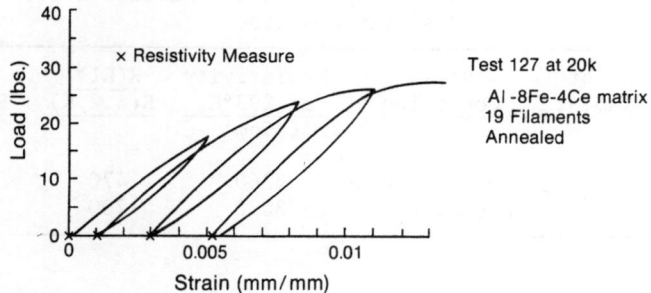

Fig. 2. Loading description for monotonic-strain resistivity measurements

monotonic tests the sample was strained repeatedly at a rate between 5×10^{-4} and 3×10^{-3} in/in/sec through successively higher amounts of total plastic strain. Fig. 2 shows a typical set of load–unload cycles. After each cycle, the sample resistivity was measured. This procedure was continued until sample fracture occurred. For the cyclic-strain resistivity experiments, the wire samples were subjected to cyclically varying stresses between ten and one-hundred percent of their yield strength. The strain rate here did not exceed about 10^{-3} in/in/sec. The resistivity was measured after selected cycles, frequently at the beginning and less frequently toward the end of the test. Fig. 3 shows a stress vs. strain and a stress vs. time curve for a typical cyclic test.

The specimen resistivity at 4.2 or 20 K is determined by multiplying the resistivity measured at room temperature ($\rho(RT)$) by the voltage ratio $V(T)/V(RT)$ where $V(T)$ and $V(RT)$ are the voltage drops measured at the test temperature and room temperature respectively. The monotonic-strain resistivity values are corrected as shown in Eq. 1 due to an increase in sample length and a decrease in sample diameter as plastic (engineering) strain (ε_E) accumulates:

$$\rho(T)^c = \left[\frac{V(T)}{V(RT)} \left(\frac{1}{(1+\varepsilon_E)^2} \right) \right] \rho(RT) \qquad (1)$$

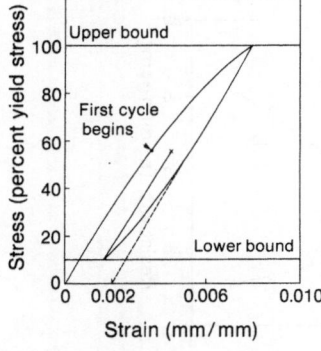

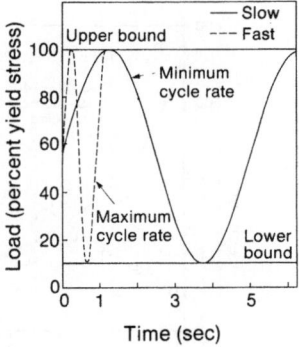

Fig. 3 a) Description of initial loading for cyclic-strain resistivity tests.

Fig. 3 b) Stress-time history for cyclic tests.

Table 1. Representative Cryoconductor Resistivities
and Resistance Ratios

Matrix Alloy	No. of Filaments	Wire Condition	Resistivity at 293°K (nΩcm)	R(RT) / R(4.2 K)	R(RT) / R(20 K)
Al-8Fe-4Ce	19	Cold Extruded	3360	470	320
Al-4Fe-2Ce	4	Annealed	3020	700	465

RESULTS AND DISCUSSION

Tensile Tests and Resistance Ratio Measurements

Initial experiments were carried out to determine the basic cryogenic mechanical and electrical characteristics of the composite wire. All wires behaved similarly at 20 K from a mechanical standpoint: the 0.2 percent offset yield strength (YS) exceeds 165 MPa (24 ksi), the ultimate tensile strength (UTS) exceeds 276 MPa (40 ksi), and elongation at fracture varied from 1.5 to 8 percent. The variation in elongation may be indicative of a sensitivity to surface flaws in the matrix component of the wire.

Resistivity and resistance ratio measurements on two conductor types are listed in Table 1. In general, conductors having 4 filaments or in the annealed condition have lower resistivities than the 19 filament or cold extruded wires, respectively. Conductors having the Al-8Fe-4Ce matrix alloy are generally stronger but do not necessarily have lower resistivities.

Monotonic-Strain Resistivity

Strain-resistivity tests were run on wire samples having all combinations of matrix type, filament number and wire condition. The typical response shown in Fig. 4 is for resistivity ratio to fall as total plastic strain increases. The associated resistivity increase with strain is found to be approximately linear. See Fig. 4. This linear nature is also observed by plotting log ($\Delta\rho$) vs. log (ε_t) and noting that the exponential term is close to one.

The strain-resistivity data from each test were fit to two equations:

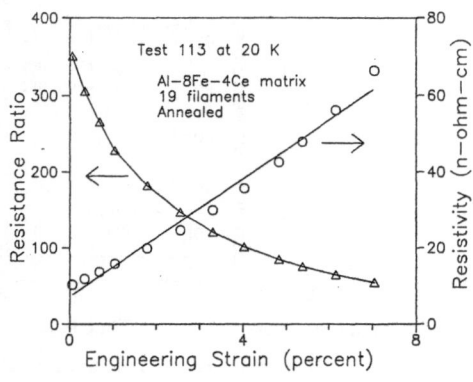

Fig. 4. Resistance ratio and resistivity at 20 K.

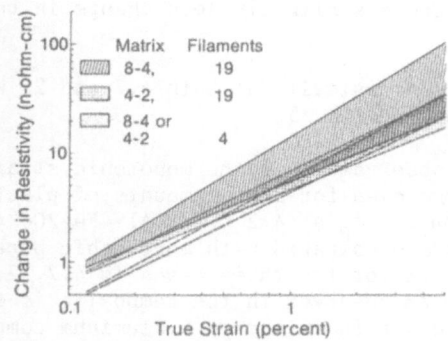

Fig. 5. Resistivity increase trends at 4.2 and 20 K
for different conductor types. Categories
include annealed and cold extruded conductor
conditions.

$$\rho = m(\varepsilon_E) + B \tag{2}$$

$$\Delta\rho = C(\varepsilon_t)^D \tag{3}$$

where ε_E and ε_t are engineering and true strain respectively and m, B, C and
D are constants. Average values of m and D for all tests are near 7 nΩcm /
%ε_E and 1.0, respectively. There are, however, noticeable differences in
the strain-resistivity behavior between the various specimen types:

- The Al-8Fe-4Ce matrix, cold extruded and 19 filament conductors degrade more with total plastic strain than their 4Fe-2Ce, annealed and
  4 filament counterparts.

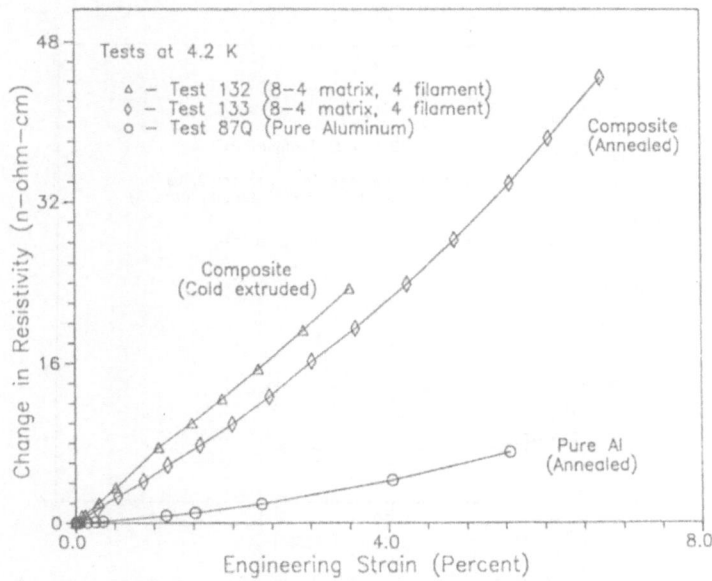

Fig. 6. Change in resistivity versus engineering strain at
4.2 K for pure aluminum (Test 87Q) and aluminum
composite conductors (Tests 132 and 133). Composites
have 8Fe-4Ce matrix and 4 filaments.

- Strain at 20 K produces slightly less change in resistivity than at 4.2 K.

General trends of strain resistivity at both 4.2 and 20 K for several hyper-conductor types are shown in Fig. 5.

A further general observation of the monotonic-strain resistivity results is that $\Delta\rho$ is large even for small amounts of plastic strain. This is shown in Fig. 6 where $\Delta\rho$ vs. ε_E at 4.2 K for Al-4Fe-2Ce/4 filament annealed and cold extruded wires are compared with monolithic pure aluminum. For one percent plastic strain, $\Delta\rho$ for the three cases is 4.7, 7.0 and 0.5 nΩcm, respectively. The larger $\Delta\rho$ observed in the composite is due in part to a larger amount of plastic strain in the pure aluminum component. The reason for this is because a much larger total strain is required in the composite as compared to pure aluminum to produce a given amount of permanent strain.

Cyclic-Strain Resistivity

Fig. 7 shows that the RR drops rapidly at first, in the composite, then levels off after about 500 strain cycles. This trend is common to all conductor types and is quite different from the results observed in constant strain-range experiments on monolithic pure aluminum where equilibrium is not reached until after 1000 cycles. See Fig. 7. This difference stems from the effect of strain environment. The composite material is cycled under stress control which subjects the pure aluminum component to rapidly decreasing amounts of cyclic plastic strain.

Additional general trends are noticed in the cyclic-strain results: 1) annealed wire shows about twice the increase in resistivity as seen in the cold extruded wire, and 2) resistivity increases appear to be slightly higher at 20 K than at 4.2 K for annealed wires. Point 1 is consistent with earlier work.[5] Point 2 is under further investigation.

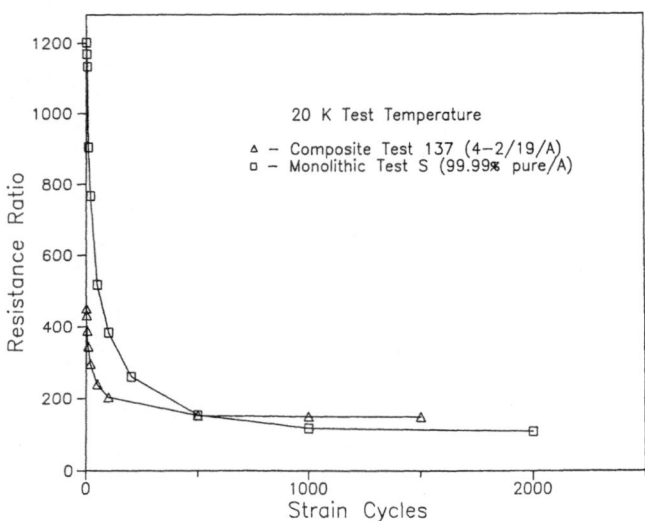

Fig. 7. Resistance ratio versus number of strain cycles at 20 K. Composite curve is from Test 137 on 4Fe-2Ce/19 filament/annealed material. Monolithic results are from Test S on 2430 RRR annealed 99.99% Al for a strain range of 0.3%.

CONCLUSIONS

Phase I experiments lead to the following main findings:

1. The YS, UTS and elongation values at 20 K for all conductors tested exceed 165 MPa (24 ksi), 276 MPa (40 ksi) and 1.5%, respectively.

2. The monotonic strain-resistivity behavior of aluminum composite conductors obeys either of two relationships reasonably well:

$$\rho = m(\varepsilon_E) + B$$

$$\Delta\rho = C(\varepsilon_t)^D$$

where ε_E and ε_t are engineering and true strain respectively. Sample resistivity increases approximately linearly for all samples at a rate of 7 ± 2 nΩcm per percent engineering strain.

3. A larger $\Delta\rho$ per unit plastic strain is seen in composite wires than in pure aluminum. The effect is due in part to the larger total strain required in the composite to produce a given amount of observed permanent strain.

4. The resistivity increase in composite conductors caused by force-controlled cyclic strain reaches a maximum after about 500 cycles. This is in contrast to results from strain-controlled cyclic testing on monolithic pure aluminum samples where equilibrium is not reached until after 1000 cycles.

ACKNOWLEDGMENT

Financial support through Alcoa Technical Center from the U.S. Air Force under contract F33615-86-C-2682 is gratefully acknowledged.

REFERENCES

1. J.C. Ho and C.E. Oberly, "Composite Aluminum Conductors for Pulsed Power Applications at Hydrogen Temperatures," Proceedings of the Fifth IEEE Pulsed Power Conference, Arlington, Va., 627 (1985).

2. K.T. Hartwig, G.S. Yuan and P. Lehmann, "Strain Resistivity at 4.2 K in Pure Aluminum," Adv. Cryo. Engr., Plenum Press, New York, 32:405 (1986).

3. K.T. Hartwig and G.S. Yuan, "Strength and Resistivity Changes Caused by Cyclic Strain at 4.2 K in Pure Aluminum," Cryogenic Materials '88 Vol. 2. Structural Materials, 1988 International Cryogenic Materials Conference, Boulder, CO, 2:677 (1988).

4. M.K. Premkumar, F.R. Billman, D.J. Chakrabarti and R.K. Dawless, "Composite Aluminum Conductor for High Current Density Applications at Cryogenic Temperatures," to be published in Adv. Cryo. Engr., Vol. 36.

5. G.S. Yuan, P. Lehmann and K.T. Hartwig, "The Effect of Prestrain on Low Temperature Fatigue Induced Resistivity in Pure Aluminum," Adv. Cryo. Engr., Plenum Press, NY, 32:413 (1986).

TOTAL PLASTIC STRAIN AND ELECTRICAL RESISTIVITY IN

HIGH PURITY ALUMINUM CYCLICALLY STRAINED AT 4.2 K

J. T. Gehan[*] and K. T. Hartwig

Texas A&M University
Department of Mechanical Engineering
College Station, TX 77843-3123

ABSTRACT

The objective of this research is to correlate the resistivity
increases in pure aluminum with the total accumulated plastic strain in
cyclic strain experiments performed at 4.2 K. Monolithic tensile specimens
were cycled to 0.1%, 0.2% and 0.3% strain for 3000 cycles. Resistivity
measurements were made periodically during the experiments using the
Eddy-Current Decay method. Resistivity increases are found to be linearly
proportional to total plastic strain over much of the region for which
plastic strain accumulates.

INTRODUCTION & BACKGROUND

While there are many parameters that influence the selection of
stabilizer materials for superconducting magnet systems, a primary
consideration is the degradation of electrical properties of the material
due to strain effects. Mechanical stresses in the composite conductor
result from the magnetic fields produced during magnet operation. These
stresses cause the wire to undergo elastic as well as plastic deformation.
Depending upon the particular application, a supermagnet may be electrically
loaded and unloaded many times during its life. Consequently, the
mechanical stresses in the conductor are considered to be cyclic.
Investigations on the tensile and cyclic effects of plastic deformation on
stabilizer materials (aluminum) have shown that the electrical resistivity
increases substantially with the addition of plastic deformation.[1]

A study of the cyclic strain effects on the resistivity of pure
aluminum was performed by Hartwig et al.[1,2,3,4] The results show that the
RRR(rho 273 K/rho 4.2 K) decreased substantially during strain cycling. The
resistivity and maximum stress per cycle were measured periodically during
cycling to record the rate of increase with the number of cycles. In 0.1,
0.2, and 0.3% strain range tests of 99.997 and 99.992% aluminum (nominal
1000 RRR and 300 RRR respectively) the resistivity increased significantly
during the first 1000 cycles and reached saturation after 1500 cycles. The
maximum stress per cycle increased similarly and saturated at about the same
point. From further analysis of one set of cyclic data, Hartwig computed

---

[*]Now with McDonnell Douglas Space Systems, Houston, Tx.

Advances in Cryogenic Engineering (Materials), Vol. 36
Edited by R. P. Reed and F. R. Fickett
Plenum Press, New York, 1990

the total plastic strain for ten strain cycles and related it to the resistivity increase. The results showed that resistivity increased proportionally to the total plastic strain.[5]

Strain controlled cyclic tests can be used to examine several work hardening and fatigue properties of metals. A test of this type is used to mechanically load a specimen between two fixed strain limits repeatedly. The distance between the two strain limits is known as the strain range. Each cycle is characterized by a hysteresis loop formed from the stress induced on a sample over a particular strain range. The size and shape of the hysteresis loop illustrate several properties of the material being tested (i.e. yield strength, ductility, modulus of elasticity). When the metal is cycled repeatedly, the changes in these properties can be observed as the number and distribution of defects within the metal increase.

A concern for cryoconductors is the effect of deformation on electrical resistivity. While this effect has been investigated by examining the resistivity increase with strain cycles, there is not a one to one correlation between these two parameters. The portion of deformation per cycle which is plastic and hence the resistivity increase per cycle decreases rapidly until the saturation of work hardening occurs at about 1500 cycles.

The objective of the work described here is to relate the accumulation of plastic strain in high purity aluminum with changes in its electrical resistivity. Further details are available elsewhere.[6]

EXPERIMENTAL SETUP & METHODS

The aluminum samples were gripped by a mechanical fixture fabricated for this research. Force was applied to the specimens through an MTS 244.12 hydraulic actuator attached to the mechanical structure. The actuator displacement and resulting strain on the sample were controlled by an MTS 458.20 microconsole controller. To maintain a test temperature of 4.2K, the mechanical fixture was immersed in a liquid helium filled dewar. A more complete description of this mechanical system is available in the literature.[7]

Electrical resistivity of the samples was measured using the Eddy Current Decay method.[8] The sample geometry is similar in size and shape to the specification used for ASTM tensile testing of metals.

Measurements of the electrical resistivity and mechanical load & strain were collected and analyzed by a personal computer. A software program was developed to record the load–strain data, compute specimen resistivity from data collected and calculate the plastic strain per cycle and the total plastic strain for each experiment. This program was developed and run on an IBM AT class personal computer in the QuickBASIC language. Special peripheral cards were added to the computer to convert the analog load & strain signals to digital format and collect eddy current decay curve data for resistivity measurements from a digital oscilloscope.

Each sample was cycled between fixed strain limits 3000 times and measurements of its electrical resistivity were taken periodically during the experiment. The load and strain signals form hysteresis loops which illustrate changes in the mechanical properties with increased cycling. By measuring the width of the hysteresis loop along the zero load point, an indication of the amount of plastic deformation can be quantified. The plastic strain per cycle (ep/cycle) represents a computed value that is based upon the points at which the hysteresis loop intersects the strain

Table 1. Test Variables of Experiments

Table 1. Test Variables of Experiments

Test	Strain Range [%]	Cycles	Diameter [inch]	Initial RRR [-]
D	0.2	3000	0.392	843
E*	0.1,0.3	4500	0.388	876
F	0.3	3000	0.391	887

*Cycles 1–3000 at 0.1% Er; cycles 3001–4500 at 0.3% Er.

axis. These intercept points are determined from the load–strain data collected. Each intercept is found by taking the two closest data points to the strain axis for that portion of the loop (one point above and one below the axis) and, using a linear interpolation equation, computing the value of strain at the point of zero stress. The plastic strain per cycle is calculated from the intersection points of the current cycle as well as the last intersection from the previous cycle.

RESULTS & DISCUSSION

The variables of each experiment are listed in **Table 1**. Test E was run at 0.1% strain range for 3000 cycles followed by 1500 cycles at 0.3% strain

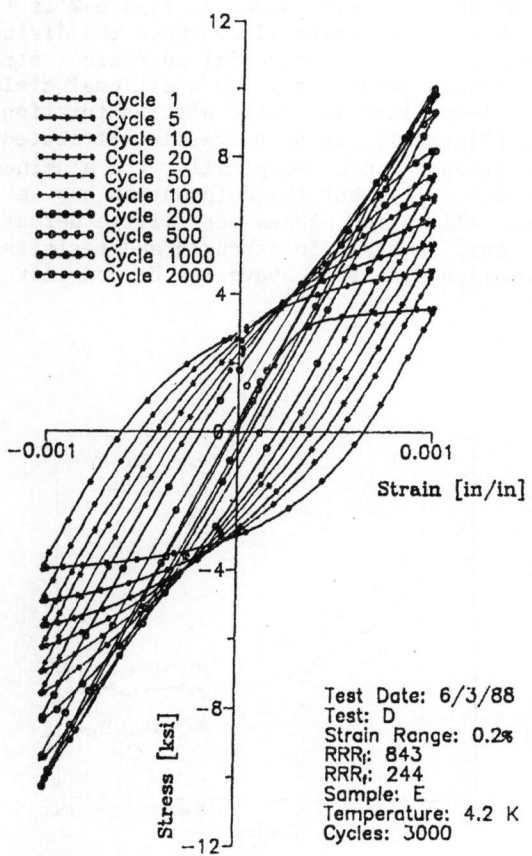

Cycle 1
Cycle 5
Cycle 10
Cycle 20
Cycle 50
Cycle 100
Cycle 200
Cycle 500
Cycle 1000
Cycle 2000

−0.001

0.001

Strain [in/in]

Test Date: 6/3/88
Test: D
Strain Range: 0.2%
RRR$_i$: 843
RRR$_f$: 244
Sample: E
Temperature: 4.2 K
Cycles: 3000

Stress [ksi]

Fig. 1. Test D Digital Stress–Strain Data

range to illustrate the dependence of strain range on the rate of resistivity increase. Only the first 3000 cycles of Test E will be used in comparison with Tests D and F.

Figure 1 illustrates the hysteresis loop data from Test D collected by computer. The first several cycles of Test D are characterized by large amounts of deformation on the sample. The loop width is relatively wide at the intersection points with the strain axis during the first several cycles. At about 1000 cycles, the sample exhibited very little plastic deformation and a finite loop width remained almost constant in size from cycle 1500 to cycle 3000. These trends were also observed in Tests E and F.

The plastic strain per cycle is plotted as a function of cycles for these tests in **Figure 2**. A logarithmic scale is used to illustrate the differences in the final plastic strain per cycle values. While the initial plastic strain per cycle values were relatively large, they decreased rapidly until a saturation of work hardening occurred at about 1500 cycles.

Studies on aluminum have shown that the primary lattice defects responsible for strengthening observed during work hardening are line defects (dislocations).[9] During the plastic deformation process, many dislocations are formed and begin to collide with one another and become tangled. With sufficient deformation, the tangles form areas of high dislocation density and develop into cells. The cell size has been found to decrease with large amounts of plastic deformation and an increased number of fatigue cycles.[9]

The saturation of the ep/cycle values in **Figure 2** is the result of cell substructures reaching a steady state size. Once the dislocations have developed into cells which give the material sufficient strength to behave elastically for that strain range, very few additional dislocations are produced. The finite loop width exhibited after saturation in the hysteresis loop data (**Figure 1**) can be primarily attributed to cell motion (sliding) within the grains.[10] Macroscopically, the aluminum appears to continue to deform plastically, but the deformation has very little affect on its strength. The cell motion causes very little defect production and therefore little, if any, increase in strength and resistivity. The cell motion type of deformation described above may be considered to be inelastic deformation.

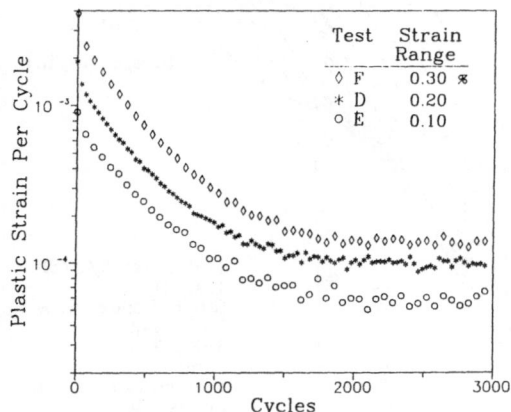

Fig. 2. Plastic Strain per Cycle as a Function of Cycles for 99.99% Aluminum at 4.2 K

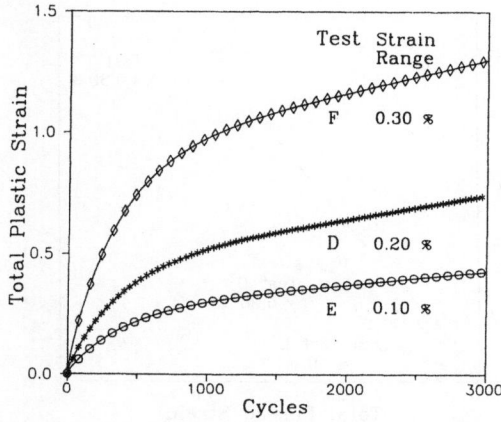

Fig. 3. Total Plastic Strain as a Function of
Cycles for 99.99% Aluminum at 4.2 K

The total plastic strain (TPS) as a function of cycles is presented in
Figure 3. As in **Figure 2**, the amount of TPS increases rapidly until the
point of mechanical saturation after which there is an almost constant value
of ep/cycle added to the total for the remainder of the experiment. Above
1500 cycles, the total plastic strain values increase at a linear rate.

The electrical resistivity increase (Dp) due to strain cycling is
presented in **Figure 4**. The Dp value is obtained by subtracting the initial
resistivity value from all subsequent values. The Dp values vary according
to a pattern similar to the ep/cycle values with steady state values reached
by 1500 cycles. The Dp versus TPS relationship is presented in **Figure 5**.
The numbers beside the data points correspond to the appropriate cycle
number. For each strain range, the slope of the linear portion of the curve
is different. For Tests D (0.2% strain range), E (0.1% strain range) and F
(0.3% strain range), the rates of resistivity increase are 13.0, 6.86 and
19.8 [(nOhm-cm)/(in/in)] over the linear region of the Dp-TPS curves. The
resistivity increase over the linear region of the Dp-TPS for Tests D, E and
F can be expressed in a general equation as follows,

$$Dp = TPS(65 \ Er + 0.3)$$

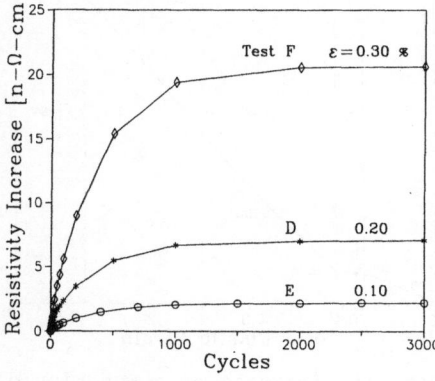

Fig. 4. Resistivity Increase as a Function of
Cycles for 99.99% Aluminum at 4.2 K

721

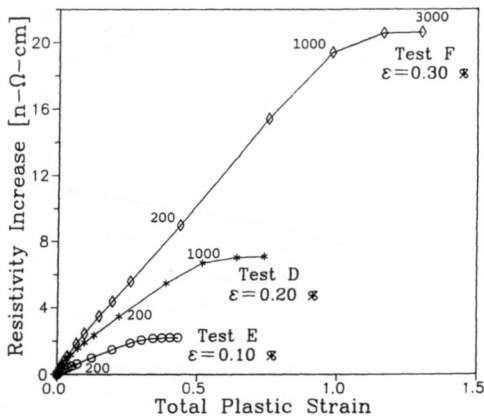

Fig. 5. Resistivity Increase Versus Total Plastic Strain
after 3000 Cycles for 99.99% Aluminum at 42 K

where Er is the strain range in percent, TPS is the total plastic strain and
Dp is in nOhm-cm.

To illustrate the Dp-TPS slope dependence on the strain range, the
sample used in Test E was cycled 1500 times at 0.3% strain range after the
initial 3000 cycles at 0.1% strain range. The data from the additional 1500
cycles was added to the initial 3000 cycles and the cumulative Dp-TPS curve
from cycles 1-4500 is presented in **Figure 6**. The resistivity increases at a
linear rate during Stage I (cycles 1-1000) followed by almost no increase
during cycles 1000-3000 (Stage II). The knee at data point 3000 is the
starting point of the 0.3% strain range cycling. In Stage III, the Dp is
again linear, but increases at a different rate than Stage I. Saturation of
the Dp values begin to appear again at 4000 cycles (Stage IV). It should be
noted that the slope of Stage III of this curve is similar to the slope of
the linear region of Test F where the slopes are 17.7 [nOhm-cm/(in/in)] and
19.8 [nOhm-cm/(in/in)] respectively. The differences in slope may be
attributed to the strain history of the first sample.

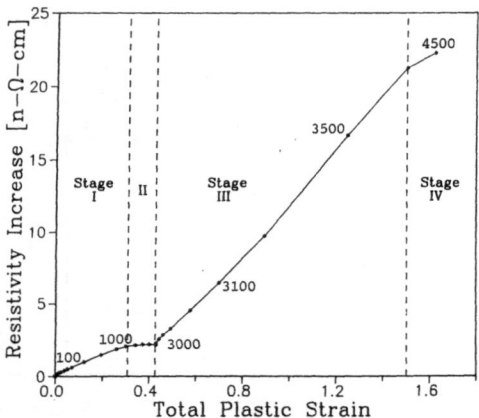

Fig. 6. Resistivity Increase as a Function of Total
Plastic Strain for Cycles 1-4500 of Test E
(Stages I and II at 0.1%, III and IV at 0.5%)

CONCLUSIONS

1. The resistivity increase in high purity aluminum due to cyclic strain is linearly related to the total plastic strain for large amounts of plastic deformation. This relationship is dependent upon the strain range experienced by the sample.

2. For nominal 870 RRR aluminum cyclically strained between 0.1 and 0.3%, the resistivity increase - total plastic strain relationship can be approximated by the following equation,

$$Dp = TPS(65 \ Er + 0.3)$$

where Er is the strain range, TPS is the total amount of plastic strain and Dp is in nOhm-cm. This relationship is valid for the first 1000 strain cycles. After this point, the mechanical and electrical properties of aluminum begin to stabilize and only small increases in electrical resistivity are observed.

ACKNOWLEDGEMENT

Support for this work was provided by the U.S. Air Force under Contract F33615-86-C-2683.

REFERENCES

1. K. T. Hartwig, G. S. Yuan and P. Lehmann, The Effects of Low Temperature Fatigue on the RRR and Strength of Pure Aluminum, IEEE Trans. on Magnetics, MAG-1(2):161 (1985).
2. G. S. Yuan, P. Lehmann and K. T. Hartwig, The Effect of Prestrain on Low Temperature Fatigue Induced Resistivity in Pure Aluminum, Advances in Cryogenic Eng., 32:413 (1985).
3. K. T. Hartwig, G. S. Yuan and P. Lehmann, Strain Resistivity at 4.2 K in Pure Aluminum, Advances in Cryogenic Eng., 32:405 (1985).
4. K. T. Hartwig and G. S. Yuan, Strength and Resistivity Changes Caused by Cyclic Strain at 4.2 K in Pure Aluminum, Cryogenic Materials '88, Volume 2. Structural Materials, Proceedings of the International Cryogenic Materials Conference, June 7-10, 1988, Shenyang, China, 677 (1988).
5. K. T. Hartwig, unpublished data, Mechanical Engineering Dept., Texas A&M University.
6. G. T. Gehan, "Total Plastic Strain and Electrical Resistivity in High Purity Aluminum Cyclically Strained at 4.2 K," M.S. Thesis, Texas A&M University (1988).
7. A. Lumbis, J. Roelli, D. Frutschi, J.T. Gehan and K. T. Hartwig, Cryoconductor Materials Testing System, to be submitted to Adv. Cryo. Engr., Vol. 36.
8. A. F. Clark, "Standard Reference Materials: The Eddy Current Decay Method for Resistivity Characterization of High Purity Metals," NBS Special Publication 260-39, U.S. Department of Commerce, National Bureau of Standards, Boulder, CO, May 1972.
9. R. P. Reed, Aluminum 2. A Review of Deformation Properties of High Purity Aluminum and Dilute Aluminum Alloys, Cryogenics, 12(4):259 (1972).
10. Conversations with Dr. T. Pollack, Aerospace Engineering Dept., Texas A&M University, July-August 1988.

CYCLIC-STRAIN RESISTIVITY

IN PURE ALUMINUM AT 20 K

K.T. Hartwig, A. Lumbis and J. Roelli

Texas A&M University
Department of Mechanical Engineering
College Station, TX 77843-3123

ABSTRACT

Strain induced resistivity increases are examined in nominal 99.99 and 99.999% pure aluminum under conditions of cyclic strain at 20 K. Samples having residual resistivity ratios (RRR = ρ(273 K)/ρ(4.2 K)) of 1000, 2420 and 9440 were prepared by cold working and annealing bars originally 25.4 mm in diameter. Constant strain range (ε_r) tests were performed for 0.05%<ε_r<0.3% through 3000 strain cycles. Discussions of the effects of strain range, strain temperature (20 K vs. 4.2 K) work hardening and post cyclic strain annealing on the cyclic-strain induced resistivity increase are presented.

INTRODUCTION

The problem under investigation is cyclic-strain resistivity in pure aluminum at 20 K. This temperature is of interest because of proposed inductor devices operating in a liquid hydrogen environment.[1] These devices could be wound from composite cryoconductors employing highly pure aluminum as the main conducting element and might be attractive for space based applications where low mass is important. Such space based magnets could be designed with an amount of structure dependent to a large extent on conductor strain limits.[2] The strain limits will of course depend on the specific conductor chosen. For a composite aluminum cryoconductor, the strain limit may very well be set by the amount of cyclic-strain resistivity increase that is acceptable to maintain a desired level of conductor and device performance. This is related to the fact that a resistivity increase will decrease the allowable transport current due to ohmic heating and heat transfer limitations. It is because of this reasoning that it is important to understand how the resistivity of pure aluminum is affected by cyclic strain at 20 K.

The phenomenon of cryogenic cyclic-strain resistivity stems from repeated plastic deformations of the metal.[3,4] As the level (or range) of plastic strain decreases, so does the associated resistivity increase.[5] The magnitude of the effect depends, at least at 4.2 K, primarily on two factors: the initial strain level (range) and the degree of work hardening prior to cyclic strain.[6] The effect of metal purity is secondary for purities corresponding to the RRR (2428 $n\Omega cm/\rho$(4.2 K)) range of 150 to 4500.[4]

Advances in Cryogenic Engineering (Materials), Vol. 36
Edited by R. P. Reed and F. R. Fickett
Plenum Press, New York, 1990

725

Table 1. Aluminum Specifications

Grade[a]	Lot No.	Heat Treatment (hr) (C)		Nominal RRR[b]	Measured Impurities[c]
99.99	B-416	1	225	1020	Si(6.6),Fe(<0.7),Cu(4)
		1	500	2420	Mg(0.6),Ca(0.7),Ba(0.5)
Kryal-S	F-647	1	500	9440	Si(1),Fe(<0.7),Cu(0.5), Mg(0.5),Ca(0.6)

[a] purities are estimated to be 99.9984 and 99.9996.
[b] taking $\rho(273 \text{ K}) = 2.428 \text{ } \mu\Omega\text{cm}$.
[c] values in ( ) are ppm by weight and were measured by emission spectroscopy.

Nonetheless, after high levels of cyclic strain, a lower purity aluminum may have a lower total resistivity than a high purity metal given the same strain environment.[4]

The temperature at which strain is applied may also affect the associated resistivity increase. This statement is based on the isochronal annealing behavior of 99.995% aluminum deformed at 78 K.[7] A series of recovery stages are observed beginning at about 80 K and ending at near room temperature. Each stage of recovery is associated with the movement of different types of defects. Such behavior leads one to believe that the response to strain at different cryogenic temperatures will depend on the temperature at which the strain is applied. If partial recovery occurs at temperatures as low as 20 K, one would expect a different strain resistivity than is found at 4.2 K where large amounts of strain-resistivity data already exist. Whatever the effect of temperature, it is likely magnified at high levels of plastic strain.

The objective of this study is to determine how cyclic strain at 20 K will influence the resistivity of pure aluminum. The affects of strain level, strain rate, strain temperature, metal purity, and annealing are being evaluated. The method of the investigation follows that of an earlier similar study on pure aluminum done at 4.2 K.[4]

MATERIALS AND METHODS

The aluminum tested is supplied by Vereinigte Aluminium-Werke (VAW) of West Germany. Material chemistries and heat treatments are listed in **Table 1.** Polycrystalline cylindrical samples are fabricated by swaging 25.4 mm diameter bar stock to 14.8 mm, machining specimens with a 10 mm diameter by 80 mm long gage length, heat treating and cleaning appropriately. The nominal 99.99 specimens contain grains with average diameters of 0.3 and 0.7 mm for the 225 and 500°C heat treatments respectively.

Sample strain is measured by two strain gages mounted 180° apart at the sample midsection. Two dummy gages attached to a stress free length of pure aluminum and located near the sample complete a Wheatstone bridge circuit. Sample resistivity is also measured at the sample midsection. This is accomplished by eddy current decay coils located around the sample and aligned over the strain gages. See Ref. 8 for further details.

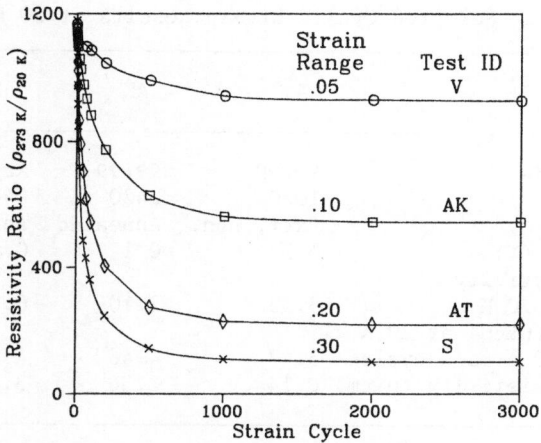

Fig. 1.   Resistivity ratio versus strain cycles at 20 K for
          annealed 99.99 aluminum.

The constant temperature environments of 4.2 K and 20 K are maintained by containing boiling liquid helium or temperature regulated helium gas respectively in a super-insulated composite and aluminum dewar. For gas tests, the gas temperature is maintained to ± 0.1 K in the test region and is measured by either silicon diode or carbon glass sensors located in the gas within one centimeter of the specimen. A thorough description of the cold helium gas supply system is presented elsewhere.[8]

Specimens are securely held in the loading fixture by collet grips. Fully reversed cyclic stress is applied by a hydraulic actuator-feedback control system operated in the strain-control mode. The force applied is measured by a 2000 lb. load cell located outside the test chamber. Stress and strain signals are recorded in real time by an X-Y plotter and by a computer data acquisition system for immediate display and analysis at a later time.

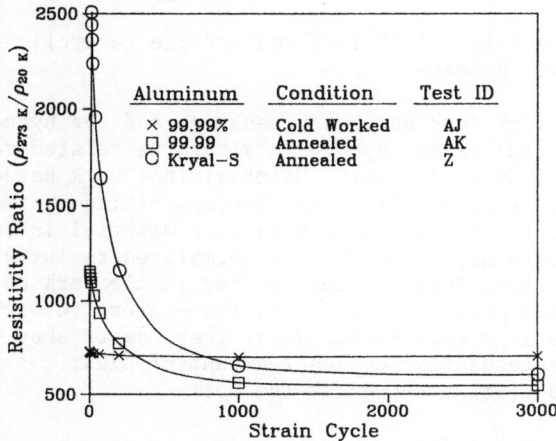

Fig. 2.   Resistivity ratio versus strain cycles at 20 K for
          a strain range of 0.1 percent.

Table 2.   Selected Cyclic Strain Results at 20 K

TEST ID	AJ	AK	Z
Aluminum Grade	99.99	99.99	Kryal-S
Nominal RRR	1020	2420	9440
Condition	Part. Ann.	Annealed	Annealed
Strain Range (%)	0.1	0.1	0.1
Initial Resistivity (nΩcm) at 20 K	3.33	2.10	0.97
Resistivity (nΩcm) at 20 K after 3000 strain cycles	3.47	4.46	4.04
Change in resistivity (nΩcm)	0.14	2.36	3.07

RESULTS AND DISCUSSION

Cyclic-Strain Resistivity at 20 K

Resistivity increases rapidly at first then reaches a saturation value during a cyclic strain test at 20 K. As shown in **Fig. 1** for 99.99% Al, when the resistivity ratio (RR = 2428 nΩcm/ρ(20 K) is plotted versus the number of strain cycles (N), the change in resistivity is lower for lower strain ranges.

Fig. 2 presents results of RR vs. N at a strain range of 0.1 percent for fully annealed 99.99 and Kryal-S aluminum and for the 99.99 material in a partially annealed condition (one hour at 225 C). Note that the resistivity of the partially annealed sample is least affected by cyclic strain. Although the final resistivity of the annealed 99.99 material is the highest, it is the annealed Kryal-S material that shows the largest increase in resistivity after 3000 strain cycles ($\Delta\rho$(3000 Cy)). **Table 2** presents numerical comparisons.

The results shown in Table 2 are consistent with behavior seen at 4.2 K in pure aluminum and support two earlier findings:

1.   Annealed material with a higher initial RRR shows higher levels of cyclic-strain resistivity degradation.

2.   Cold worked material is less susceptible to cyclic-strain resistivity increases.

Both findings stem from work hardening behavior and the hypothesis that the resistivity change ($\Delta\rho$) caused by cyclic strain is related nearly linearly to the accumulated plastic strain.[9] Material that work hardens rapidly during cyclic strain will exhibit a smaller associated resistivity increase because the accumulated plastic strain in this material is less. A plot of the results from tests AJ, AK and Z on a normalized resistivity axis (i.e. $\Delta\rho$/$\Delta\rho$(3000 Cy)) confirms that the cold worked sample work hardens quickest and that the annealed Kryal-S material is the slowest to work harden. Plots of accumulated plastic strain versus N for these cases show that the work hardened material accumulates the least amount of plastic strain while the annealed Kryal-S material accumulates the most.

Comparison with 4.2 K Results

Fig. 3 shows the resistivity increase associated with cyclic strain at both 4.2 and 20 K for 99.99 Al. The strain range is 0.1 percent. Notice

Table 3. Values of the Constants A and B
for Cyclic-Strain Resistivity Increases
$(\Delta\rho(3000 \text{ Cy}) = A \ \varepsilon_r^{B})$

Results	Temp. (K)	A	B
UW-Madison(4)	4.2	2.3×10^6	2.0
TAMU	4.2	7.4	2.2
TAMU	20	4.3	2.1

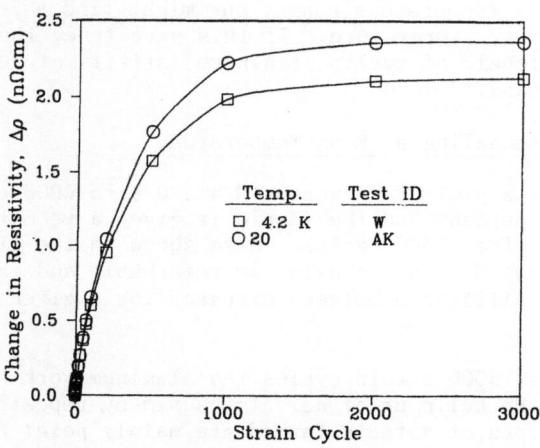

Fig. 3. Change in resistivity versus strain cycles for
99.99 Al at 4.2 and 20 K.

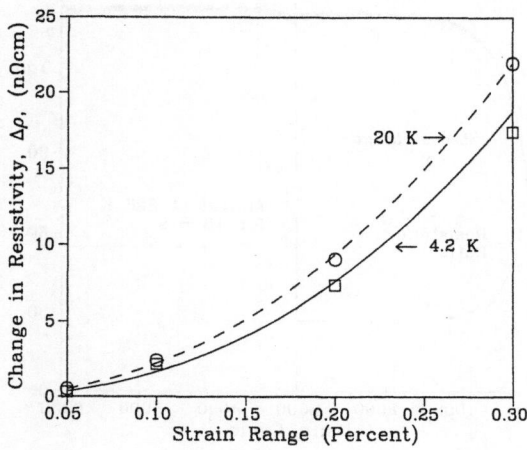

Fig. 4. Change in resistivity after 3000 strain cycles
versus strain range at 4.2 and 20 K.

that the resistivity increase is nearly the same until about 100 cycles after which $\Delta\rho(20\ K)$ is significantly higher. Similar but more pronounced trends are seen for higher strain ranges. Shown in **Fig. 4** are plots of $\Delta\rho(3000\ Cy)$ versus strain range for the 4.2 and 20 K cases. It is clear that cyclic strain is worse at 20 K and for higher strain ranges. Both curves are fit reasonably by the expression

$$\Delta\rho(3000\ Cy) = A\ (\varepsilon_r)^B \tag{1}$$

where A and B are constants, ε_r is taken as the dimensionless fraction, and $\Delta\rho(3000\ Cy)$ has units of nΩcm. Values for A and B are listed in **Table 3**.

The fact that cyclic strain at 20 K produces slightly higher resistivity increases is probably due to a higher rate of work hardening at the lower temperature. This statement is supported by the normalized resistivity increase and the accumulated plastic strain curves. It also makes sense if dynamic recovery processes are considered. Deformation will produce interstitials, vacancies, and dislocations, and since some recovery may occur at the 20 K temperature range, one might find a lower rate of work hardening at the higher temperature. If this were true, then one would expect even higher levels of cyclic-strain resistivity at temperatures above 20 K, which is the case.[10]

Post Cyclic-Strain Annealing at Room Temperature

Fig. 5 presents a plot of RR versus N at 20 K to 6000 cycles. The strain range is 0.2 percent and the sample is given a room temperature anneal for 18 hours after 3000 cycles. Also shown in the figure is a plot of stress range versus N. The behavior is remarkable and easily understood. Stress range is the difference between stresses for a given cycle at the strain range limits.

During the first 3000 strain cycles the aluminum work hardens and gains resistivity due to the build up of defects caused by repeated plastic deformation. The types of defects formed are mainly point defects and dislocations. The point defects contribute more to the resistivity increase than do the line defects. This is true because about 70% recovery occurs within minutes after a room temperature anneal and point defects are quite

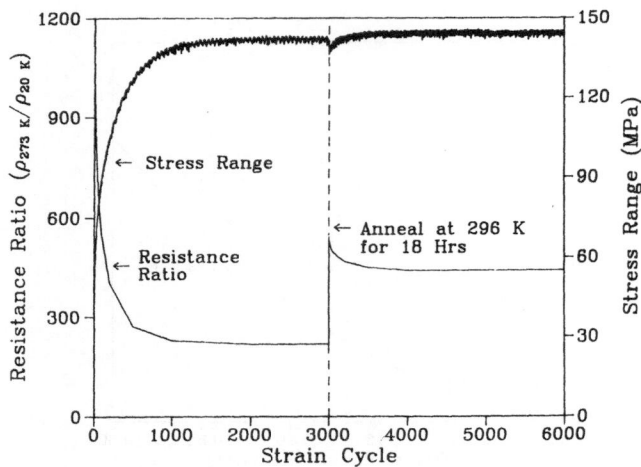

Fig. 5. Resistivity ratio and stress range versus strain cycles at 20 K for 99.99 aluminum.

mobile at 300 K.[7] The dislocations on the other hand do not anneal so rapidly. They also are more effective in strengthening the metal. After the relatively short room temperature anneal, the number of dislocations is only slightly different from that immediately after the 3000 strain cycles at 20 K. Some reorganization in the dislocation substructure probably takes place as point defects migrate to sinks and as line defects move with the added thermal energy available.

The results shown in **Fig. 5** could have important consequences to the operation of devices containing pure aluminum, operating at cryogenic temperatures and subjected to cyclic loading. Warming to room temperature after a period of operation will result in a pure aluminum component that: a) is quite resistant to further cyclic-strain resistivity degradation because it has effectively been work hardened and b) has a RR significantly above that for unwarmed material.

CONCLUSIONS

1. Cyclic plastic strain at 20 K in pure aluminum causes resistivity changes similar in character but slightly larger in magnitude than those observed at 4.2 K.
2. Recovery processes probably influence strain-resistivity effects at 20 K in pure aluminum.
3. Warming pure aluminum to room temperature after cryogenic cyclic plastic strain causes substantial recovery without a significant drop in strength.

ACKNOWLEDGMENT

This research was supported by the Department of the Air Force, Air Force Systems Command, Aeronautical Systems DIV/PMRSA Wright-Patterson AFB, OH, under contract F33615-86-C-2683 with Texas A&M University.

REFERENCES

1. J.C. Ho and C.E. Oberly, "Composite Aluminum Conductors for Pulsed Power Applications at Hydrogen Temperatures," Proc. Fifth IEEE Pulsed Pow. Conf., Arlington, VA, 627 (1985).
2. Y. Eyssa, R.W. Boom, G.E. McIntosh, and Q.F. Li, "A 100 kWh Hour Energy Storage Coil for Space Application," IEEE Trans. Mag., 19:1081 (1983).
3. S. Ceresara, H. Elkholy, and T. Federighi, "Resistivity Increase in Polycrystalline Al Heavily Cold-worked at 78 K," Phys. Stat. Sol., 8:509 (1965).
4. K.T. Hartwig and G.S. Yuan, "Strength and Resistivity Changes Caused by Cyclic Strain at 4.2 K in Pure Aluminum," Cryo. Mat. 88 Vol. 2 Structural Materials, 677 (1988).
5. K.T. Hartwig and J.T. Gehan, "A Graphical Model of Cyclic Plastic Strain in Pure Aluminum," Cryogenic Materials '88 Vol. 2 Structural Materials, 925 (1988).
6. G.S. Yuan, P. Lehmann, and K.T. Hartwig, "The Effects of Prestrain on Low Temperature Fatigue Induced Resistivity in Pure Aluminum," Adv. Cryo. Engr., 32:413 (1986).
7. S. Ceresara, H. Elkholy, and T. Federighi, "Influence of the Amount of Strain at 78 K on the Recovery Process in Al 99.995%," Phil. Mag., 12:1105 (1965).
8. A. Lumbis, J. Roelli, D. Frutschi, J.T. Gehan, and K.T. Hartwig, "Cryoconductor Materials Testing System," to be submitted to Adv. Cryo. Engr., Vol. 36.

9. J.T. Gehan and K.T. Hartwig, "Total Plastic Strain and Electrical
    Resistivity in High Purity Aluminum Cyclically Strained at 4.2 K," to
    be submitted to Adv. Cryo. Engr., Vol. 36.
10. K.T. Hartwig, unpublished results.

COMPOSITE ALUMINUM CONDUCTOR FOR HIGH CURRENT DENSITY

APPLICATIONS AT CRYOGENIC TEMPERATURES

M. K. Premkumar, F. R. Billman, D. J. Chakrabarti, R. K. Dawless,
and A. R. Austen*

Alcoa Laboratories
Alcoa Center
Pennsylvania 15069

*Innovare Inc.
Bath, Pennsylvania 18014

ABSTRACT

Electrical resistivity of high purity aluminum decreases by several orders of magnitude below a temperature of approximately 25K and consequently it can be used for high current density applications at these low temperatures. In addition, the low density of aluminum makes it competitive with conventional superconductors in weight critical applications at cryogenic temperatures. However, high purity aluminum does not have the requisite strength to withstand magnetic forces produced by the high currents and hence needs to be structurally supported. A composite conductor consisting of high purity aluminum filaments supported in a high strength powder metallurgy processed Al-Fe-Ce alloy matrix has been successfully fabricated to meet these requirements by a combination of hot extrusion through streamline dies followed by cold drawing to wire gauges. Two different arrangements of filaments and two matrix alloy compositions have been examined. Effects of hot extrusion and cold drawing as well as the influence of superimposed hydrostatic stresses during cold working on the filament shape are considered in terms of deformation principles. Results of microstructural observations and theoretical analyses of diffusional contamination of the high purity aluminum filaments during processing are presented.

INTRODUCTION

Electrical resistance of pure metals at cryogenic temperatures is extremely low and, consequently, this makes them attractive as cryogenic conductors. Of all the pure metals, aluminum is particularly promising for a number of reasons. Primary among them is its low density which makes it an optimum choice for weight critical applications such as space based systems. Secondly, aluminum has very high electrical and thermal conductivity at cryogenic temperatures and also can be more economically produced in a high purity form than the other conductor material most commonly used, copper. The third most important advantage aluminum possesses over copper is its behavior in a strong magnetic field (>2 Tesla). In the presence of a strong magnetic field, the electrical resistance of most pure metals increases. However, in the case of aluminum, the resistance approaches a saturation value at high magnetic fields or in essence has a very small linear increase with field.[1] In comparison, copper shows a strong increase in resistance with field.

High purity aluminum conductors have very favorable electrical properties at the boiling point of hydrogen, 20K, and hence they are excellent candidate materials for use at this temperature. The resistance of high purity aluminum at this temperature is 1/500th its resistance at room temperature

Advances in Cryogenic Engineering (Materials), Vol. 36
Edited by R. P. Reed and F. R. Fickett
Plenum Press, New York, 1990

and, in addition, it has been shown[2] that the advantage of reducing resistivity of a cooled high purity aluminum conductor exceeds the energy spent to obtain the low temperatures. The maximum advantage for 99.999 percent pure aluminum occurs at 20K. Such conductors would be ideally suited for space based applications where liquid H_2 is available as a fuel source and can be readily used as a cryogen. Additionally, aluminum conductors would be preferred over conventional superconductors in certain applications. Conventional superconductors require liquid He (4.2K) for their operation and the equipment for liquefaction and handling of liquid He adds complication and weight to the overall system.

One major disadvantage of high purity aluminum is its extremely low strength. In windings made of cryogenic conductors, mechanical stresses come from two sources:[2] the interaction of the magnetic field with the flowing currents and residual stresses during cooling. The stresses due to magnetic forces can be very significant and the high plasticity of aluminum can result in severe permanent deformation. Besides causing physical damage to the windings, the plastic strain also increases the resistance of the conductor significantly. Hence, in order to overcome the negative effects of low strength, the aluminum conductors have to be structurally reinforced by some means.

Besides braiding the high purity Al conductor with some high strength material to provide mechanical support, a more preferred approach with economic and technical benefits is to embed the Al in a high strength matrix and cofabricate them. The matrix material supporting the conductor in the composite assembly must have high strength, good thermal conductivity to remove heat generated in the conductor due to passage of current, reasonably high electrical resistivity to minimize eddy current losses in the matrix and workability compatible with high purity aluminum. Probably the most important requirement is that the alloying elements in the matrix must have very low diffusion rates in aluminum to prevent contamination of the high purity Al conductor during processing. This precludes the use of most commercial aluminum alloys.

Powder metallurgy (P/M) processed Al-Fe-Ce alloys[3] satisfy the above requirements for a matrix material. These high strength dispersion strengthened alloys were designed for elevated temperature applications and consist of thermodynamically stable aluminides and in addition, Fe and Ce are two of the slowest diffusing species in aluminum. Research efforts at the Aero Propulsion Laboratories, Wright Patterson Air Force Base, Dayton, Ohio have demonstrated the feasibility of coextruding a multifilament composite conductor, consisting of Al filaments in an Al-Fe-Ce matrix.[4,5] This laboratory scale success led to the current research activity at Alcoa aimed at commercial scale fabrication of a composite conductor wire. The present paper reports the results of the first phase of this research program.

EXPERIMENTAL PROCEDURE

Two composite geometries as dictated by potential end applications were chosen for this study, a 19 Al filament and a 4 Al filament conductor. Schematics of the composite conductor design for the two cases are shown in Figure 1. In both cases, the area fraction occupied by the current carrying high purity Al filaments is ≈50 percent. In order to evaluate the influence of matrix flow stress on the fabricability of the composite, two matrix alloy compositions were selected, namely Al-8% Fe-4% Ce and Al-4% Fe-2% Ce.

The matrix alloys were processed from atomized powders by Alcoa's proprietary practice. Powders were cold consolidated and vacuum hot pressed to full density to yield billets which were 15.5 cm diameter x 61.0 cm long. These billets were intentionally processed at a much higher temperature (500°C) to reduce their flow stress and minimize the flow stress mismatch between the matrix and filaments. Normally these alloys are processed in the 370-425°C temperature range for high strength applications. Hot pressed billets were then precision gun drilled to the required geometries. High purity aluminum (99.999 percent) with Residual Resistivity Ratio

(RRR = $\dfrac{\rho_{300K}}{\rho_{4.2K}}$) of approximately 1500 was used for the filaments. Cast and hot extruded 12.7 cm diameter ingots were produced by a proprietary refining process and then further hot extruded to either 2.5 cm diameter or 5.5 cm diameter These were the starting Al rods for the 19 filament and 4 filament composite geometries, respectively.

The composite billet consisting of the high purity Al filaments in the drilled Al-Fe-Ce alloy matrix was extruded at 482-510°C from a 16.2 cm diameter cylinder to 1.9 cm diameter rod at an

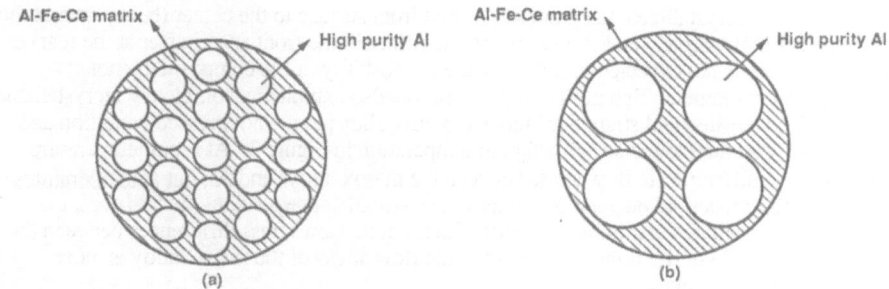

Fig. 1. Schematics illustrating the composite conductor design.
(a) 19 filament configuration; (b) 4 filament configuration.

extrusion reduction ratio of 72:1. In order to minimize longitudinal shearing between matrix and filaments and to maximize overall strain uniformity, the composite was extruded through a streamline die. The die profile used was based upon the theory developed by Richmond et al.[6,7] The 1.9 cm diameter redraw rod was then cold drawn to final wire gauges, i.e., 0.25-0.08 cm diameter. Poor formability of the matrix alloy necessitated numerous intermediate annealing treatments. Annealing temperature and time were optimized as excessive thermal exposure of the composite wire can result in contamination of the high purity Al and hence a decrease in the RRR of the wire. The redraw rod was also fabricated into wire by an alternative processing technology, cold hydrostatic extrusion. The large compressive hydrostatic stresses generated during hydrostatic extrusion increase the formability of the matrix alloy and allow fabrication without the need for any annealing treatments. The influence of hydrostatic stresses on distortion of Al filaments in the final wire was also examined.

RESULTS AND DISCUSSIONS

Optical micrographs representative of the initial matrix billet and extruded aluminum rod are shown in Figure 2. The microstructure of the hot pressed Al-8% Fe-4% Ce matrix, Figure 2a, is fairly uniform. The Zone A (fine optically featureless areas) and Zone B (coarser areas) type regions[8] typically observed in these alloys processed at lower temperatures are not discernible. The high temperature processing has coarsened and homogenized the structure considerably. The lower solute alloy was observed to be similar microstructurally. Figure 2b is a longitudinal section of the 2.5 cm diameter extruded aluminum rod and illustrates a fine grained recrystallized structure.

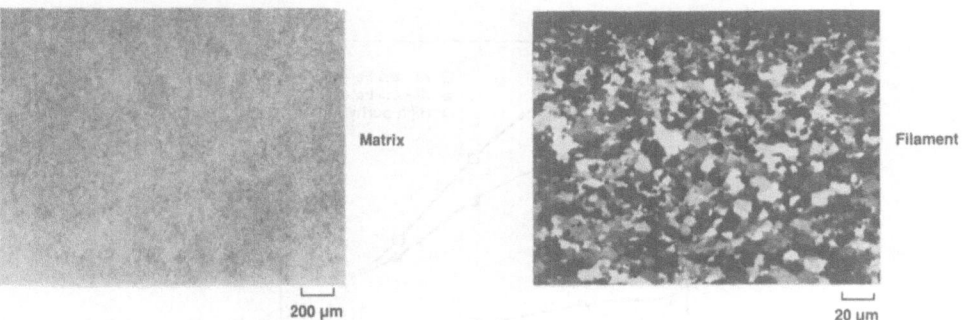

Fig. 2. Optical micrographs showing the microstructure of
the starting components of the composite conductor.
(a) Hot pressed Al-8% Fe-4% Ce matrix billet;
(b) Hot extruded high purity Al filament, longitudinal section.

Although completely recrystallized, the grain size varied from surface to the center (being finer at the surface and coarser in the center) and front to rear (being finer in the front and coarser at the rear) of the extrusions, but this is not expected to influence the fabricability of the composite by hot extrusion. The 5.5 cm diameter high purity Al filament rod also exhibited a completely recrystallized microstructure. The tensile yield strength of the two matrix alloys in the hot pressed condition and extruded high purity Al are plotted as a function of temperature in Figure 3. At room temperature there is considerable difference in flow stress between the matrix alloys and Al, but at temperatures above 425°C the differences are decreased. Hence at the extrusion temperature (482-510°C), the absolute flow stress mismatch is minimal. Besides the absolute flow stress differences between the two components, the ratio of the flow stress of Al to the flow stress of the matrix alloy is more important for uniform deformation.

Figure 4 shows representative cross-sectional views of extruded 1.9 cm diameter composite rod with 19 and 4 filaments. The originally round individual filaments are distorted although the overall arrangement of filaments is maintained. This distortion indicates that the deformation in the streamline die was nonuniform across the cross-section due to the flow stress difference between the two materials. Although the Alcoa streamline die has been designed for minimum shear and uniform strain, in reality, friction at the die surface causes nonuniform deformation through the section. This is further aggravated by the severe extrusion reduction ratio, 72:1. Minimizing this distortion would require improved lubrication and die design to further minimize shear as well as changes in shape and spatial distribution of the filaments.

The interface between the matrix and the high purity Al is very important from the application point of view. A good interfacial bond is necessary in the final conductor for effective heat transfer from the Al filaments (which are the prime current carriers) to the matrix so as to maintain the overall temperature of the conductor at liquid hydrogen temperature for maximum efficiency. Additionally, a good bond is necessary for transferring mechanical stresses (due to magnetic fields) to the matrix. One drawback of a good metallurgical bond is its tendency to increase diffusional contamination of the Al filaments during annealing treatments. After hot extrusion the matrix-filament interface is very sound and appears to be metallurgically bonded, Figure 5. No porosity or discontinuity of any kind were observed.

Cold drawing of the extruded rod by conventional wire drawing techniques is more complicated due to the much larger difference in flow stress between the two materials at room temperature. Additionally, during drawing the matrix work hardens significantly while the high purity aluminum dynamically recovers and recrystallizes thus increasing the flow stress mismatch. Formability of the Al-Fe-Ce alloy is also inherently poor and hence numerous intermediate annealing treatments are required for successful codrawing of the composite. Intermediate annealing was carried out at 400-450°C. Higher annealing temperatures are preferred to produce measurable softening of the matrix alloy but caution has to be exercised as unduly high temperatures can contaminate the Al filaments significantly.

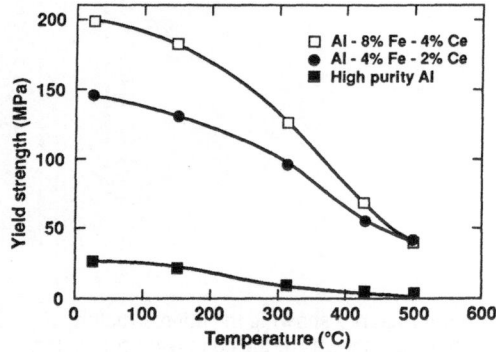

Fig. 3. Elevated temperature yield strength response of Al-Fe-Ce alloys and high purity Al.

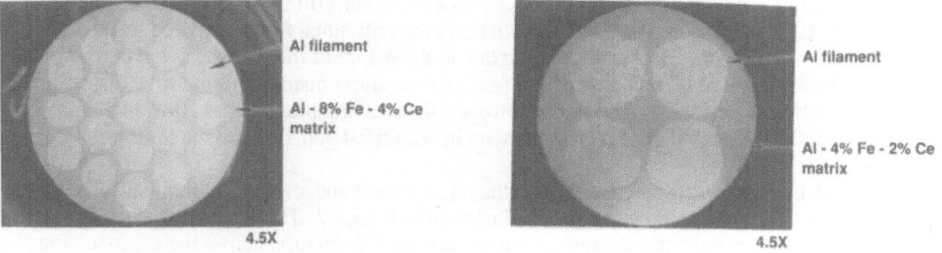

Fig. 4. Optical Micrographs showing the cross-sections of
1.9 cm dia. streamline die extruded composite rods.
(a) 19 filament configuration; (b) 4 filament configuration.

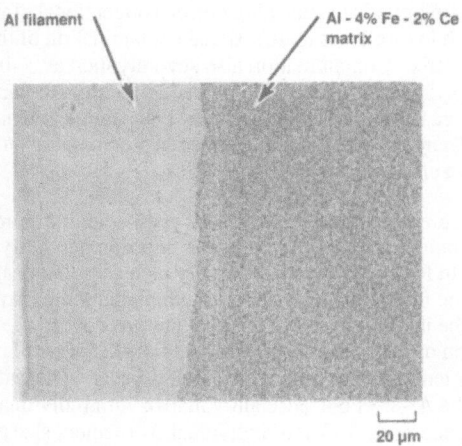

Fig. 5. Optical micrograph of the matrix-filament interface
in the extruded composite conductor.

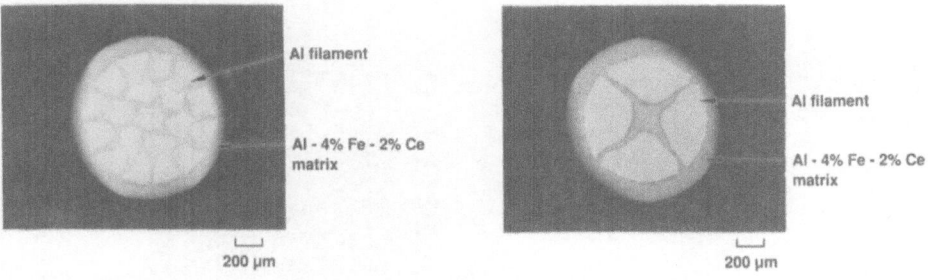

Fig. 6. Optical micrographs showing cross-sections of cold drawn 0.13 cm dia. composite wires.
(a) 19 filament configuration; (b) 4 filament configuration.

737

Cross-sections of typical drawn composite wire of 0.13 cm diameter are presented in Figure 6. Further extensive distortion of the individual filaments is observed as compared to the extruded structures. The 4 filament composite shows a more uniform arrangement of filaments as opposed to the 19 filament composite and in both cases the individual filaments were seen to be separated from one another. Extensive shearing due to the nature of the stresses in the deformation zone during drawing, aided by the large differences in flow stress result in progressive distortion of the filaments during the drawing operation. However, the drawing practice has allowed the production of long lengths (>90 meters) of composite wire in both the 4 and 19 filament configurations.

The influence of hydrostatic stresses on the arrangement and shape of individual filaments during fabrication by hydrostatic extrusion is illustrated in Figure 7. Hydrostatic state of stress minimizes the shear component and provides more uniform deformation across the cross-section thereby retaining to a large extent, the original extruded filament arrangement and shapes. In addition, due to hydrodynamic lubrication conditions in the hydrostatic extrusion process, friction is minimal while it plays a big part in conventional wire drawing. One current restriction on the hydrostatic extrusion process is its inability to produce long commercial lengths of conductor. This, however, would change with the development of a semicontinuous hydrostatic extrusion process which is presently being pursued.

The cold work imparted to the wires during drawing has to be annealed out prior to the end use of the wire as any plastic strain drastically increases the resistivity of high purity aluminum. This final anneal as well as intermediate annealing during conventional drawing requires high temperature exposure which in turn leads to diffusional contamination of the high purity Al filaments by Fe, Ce, etc., from the matrix. Contamination also severely increases the resistivity. Hence during annealing two competing effects are active in the high purity Al; annealing eliminates defects due to cold work which decreases the resistivity of the Al; annealing contaminates Al thereby increasing its resistivity. Therefore, the choice of annealing temperatures/times must be dictated by a trade-off between these two effects.

The above concepts are illustrated by the results presented in Figures 8 and 9. They show the effects of annealing temperature, annealing times and composite geometry on the RRR of 0.08 cm diameter composite wire. In Figure 8, the RRR initially increases to a peak and then starts to decrease with annealing time for all 3 temperatures. With higher temperatures, the peak is shifted to shorter annealing times. The initial increase in RRR is due to recovery of Al filaments while subsequently, contamination of the filament contributes to the decrease of the RRR and this effect is more pronounced at higher temperatures where diffusion is faster. These data represent the 19 filament composite in the Al-8% Fe-4% Ce alloy matrix with individual filaments of 0.01 cm diameter in the final composite wire. Effect of individual Al filament size on annealing behavior is seen in Figure 9 where the 4 and 19 filament composite wires are compared. In the 4 filament composite where the individual filaments are 0.03 cm diameter, the influence of contamination is less severe as the relative percentage of the cross-section contaminated is smaller than in the 19 filament case for a given contamination depth. Consequently, the 4 filament composite wire recovers to a much higher RRR before it decreases. The initial differences in the RRR before annealing are due to processing variations.

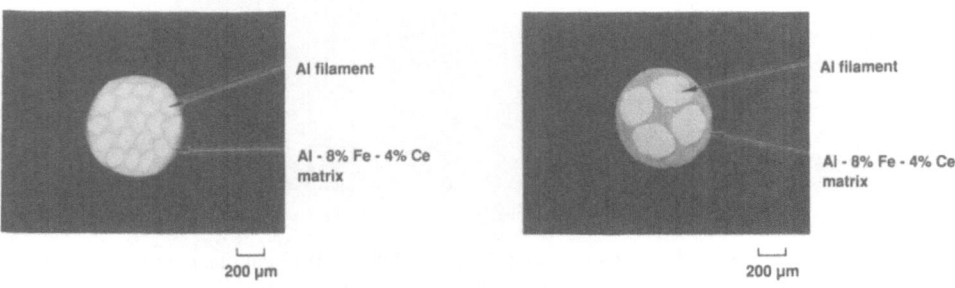

Fig. 7. Optical micrographs showing the cross-sections of 0.08 cm dia.
composite wire produced by hydrostatic extrusion.
(a) 19 filament configuration; (b) 4 filament configuration.

Table 1. Room temperature tensile properties of
0.13 cm diameter composite wire

Matrix Alloy	# of Filaments	Yield Strength (ksi)	Ultimate Tensile Strength (ksi)	% Elongation
Al-8% Fe-4% Ce	19	25.0	29.2	10.0
Al-4% Fe-2% Ce	19	18.5	23.4	11.5
Al-8% Fe-4% Ce	4	23.2	28.0	11.5
Al-4% Fe-2% Ce	4	20.1	23.8	8.8

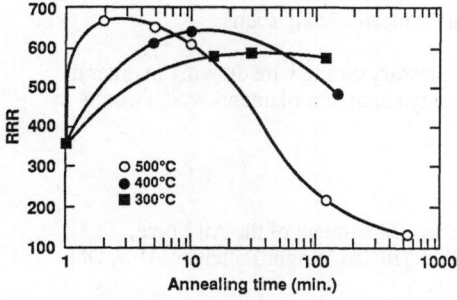

Fig. 8. Effect of annealing temperature and time on the RRR of 0.08 cm dia. 19 filament/Al-8% Fe-4% Ce matrix composite wire.

Fig. 9. Effect of annealing time at 400°C on the RRR of 19 and 4 filament/Al-8% Fe-4% Ce matrix 0.08 cm dia. composite wire.

Theoretically, the depth of penetration by diffusion can be calculated based on bulk diffusivity and hence the percentage of filament cross-section contaminated estimated. This is illustrated for the diffusion of Fe into Al in Figure 10a. It is clear that the effects are more severe for the 19 filament composite. However, in Figure 9, the peak times before the RRR decreases are of the order of 10 minutes and cannot be explained by Fe diffusion since for these short times, less than 2 percent of the individual filament cross-section is contaminated. This implies that there may be a faster diffusing species which contributes to the contamination. Silicon is usually present in commercial aluminum alloys and also has a much higher diffusivity than Fe and is a likely source of contamination. As Figure 10b shows, almost 20 percent of the cross-section can be contaminated by Si in the short time frames and can account for the observed decrease in RRR. This illustrates the importance of the purity of the base Al in the matrix alloy and the need to keep it free of fast diffusing species.

Representative room temperature tensile properties of 0.13 cm diameter composite wires in both the 19 and 4 filament configurations are given in Table 1. The strength of the matrix alloy is reflected in the composite; the lower solute matrix having a lower yield and tensile strength. The 19 and 4 filament wires do not show any differences in strength as expected, since the strength of the composite wires is essentially controlled by the matrix strength. The tensile elongations were in the range of 8.0-12%.

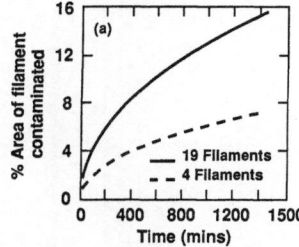
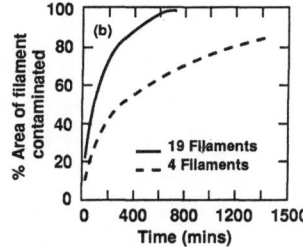

Fig. 10. Area of individual Al filaments contaminated by diffusion as a function of time at 400°C for the 19 and 4 filament configurations. (a) By diffusion of Fe; (b) by diffusion of Si.

739

CONCLUSIONS

- A composite conductor consisting of high purity aluminum filaments in an Al-Fe-Ce high strength matrix has been successfully fabricated on a commercial scale by a combination of hot extrusion in streamline dies and cold wire drawing.

- Composite conductors of two matrix alloys with different strength levels and two geometries consisting of 19 and 4 high purity aluminum filaments were fabricated in 0.13 cm diameter and 0.08 cm diameter sizes.

- Distortion of the filament shape in the final wire cross-section was less with hydrostatic extrusion than with conventional cold drawing.

- The interface between the matrix and the filaments is mechanically sound.

- Optimum annealing temperatures and times are necessary during wire drawing in order to minimize diffusional contamination of the high purity aluminum filaments which would reduce the RRR of the composite wire.

ACKNOWLEDGEMENT

The authors acknowledge the program support of the Department of the Air Force, Air Force Systems Command, Aeronautical Systems DIV/PMRSA, Wright-Patterson AFB, Ohio 45433-6503, under the direction of Dr. Charles E. Oberly.

REFERENCES

1. H. Nomura, M. Obata and S. Shimanoto, Cryogenics, Vol. II, No. 5, p. 396, 1971.

2. V. I. Gostishehev, Cryogenic Conductor Made of High Purity Aluminum, Fiz. Met. Metalloved, Vol. 62, No. 2, p. 303, 1986.

3. Alcoa Tech Brief, "Alloys CU78 and CZ42 for Elevated Temperature Applications", Aluminum Company of America, Pittsburgh, PA, 1986.

4. C. E. Oberly et al., U. S. Patent 4,711,825, 1987 December 08.

5. J. C. Ho, C. E. Oberly, H. L. Gegel, W. T. O'Hara, Y. U. R. K. Prasad and W. M. Griffith, "Composite Aluminum Conductors for Pulsed Power Applications at Liquid Hydrogen Temperatures," Fifth IEEE Pulsed Power Conference, Arlington, Virginia, 1985 June 11.

6. O. Richmond and M. L. Devenpeck, Proc. 4th U.S. Natl. Congr. Appl. Mechs., p. 1053-1057, 1962.

7. O. Richmond and H. L. Morrison, J. Mechs. Phys. Solids, Vol. 15, p. 195, 1967.

8. H. Jones, Mater. Sci. Eng., Vol. 5, p. 1, 1969.

SEMI-CONTINUOUS HYDROSTATIC EXTRUSION

OF COMPOSITE CONDUCTORS

A. R. Austen and W. L. Hutchinson

Innovare, Inc.
7277 Park Drive
Bath, PA 18014

Many high current density conductors are made of composites
which are difficult to fabricate into wire due to their limited
ductility. For decades, hydrostatic extrusion has been used for
processing of materials where the high compression deformation
environment induces extended ductility, but the practical
application of the technique has been limited. In conventional
hydrostatic extrusion the entire volume of material being
processed must be loaded into the pressure chamber at one time
limiting the starting material to chamber length billets or to
small diameter wires which can be coiled and loaded into the
chamber on spools. This paper describes experiments with semi-
continuous hydrostatic extrusion tooling in which a starting bar
of up to 20mm (.8 in.) diameter and any length can be
hydrostatically extruded at fluid pressures up to 1450 MPa (210
ksi). A portion of the bar is advanced into the unpressurized
chamber, gripped, the chamber pressurized, a short length of
material is extruded, the pressure is relieved and the grip
released to begin another cycle. Avoiding the potential for
defect formation associated with the stop-start extrusion
processing cycle is discussed. Also presented are examples of
single and multi-filament composite conductors which were
produced during the development of this apparatus.

BACKGROUND

Hyperconductor and superconductor wires are usually
composites for their electrical characteristics and/or mechanical
support of the conducting filaments. Often the composite
components' flow strengths are significantly different from each
other. This difference limits their formability which makes
producing wires more difficult. Even if the components are
individually ductile, deformation defects can occur when they are
combined into a composite. Usually these defects can be avoided
by controlling the deformation processing parameters within a
safe window of values (1,2). Also, the individual components may
have limited formability. Examples range from low ductility Al-
8Fe-4Ce alloy used as a stabilizer and strength reinforcing
constituent to brittle Nb_3Sn superconductor filament material
(3,4).

Advances in Cryogenic Engineering (Materials), Vol. 36
Edited by R. P. Reed and F. R. Fickett
Plenum Press, New York, 1990

741

Converting these composites into conductor wires requires extensive metalworking which is usually extrusion or drawing. Considering the previously described limited ductilities of these composites, extrusion has an advantage over drawing. Simply, drawing is a tensile process and extrusion is a compressive process. The superposition of a hydrostatic stress state during deformation extends the formability of the individual constituents (5,6) and of composites (7). Also the use of elevated temperatures to extend ductility is much easier to implement as hot extrusion. Hot extrusion is widely practiced while hot drawing has only limited applications in wire production of brittle metals such as tungsten and beryllium. Furthermore, extrusion, hot or cold, allows a much wider range of processing parameter choices to enhance formability. For example, having the freedom to specify die cone angles and reduction combinations is often critical in successful composite processing (8).

Typically, the hot or cold extrusion of a relatively large diameter short billet is the initial deformation processing step for producing conductor composites. This technique is practical for processing a volume of material sufficient to make the desired product in a continuous length. However, the result of this initial diameter breakdown is a bar or rod product which is generally too large in diameter for the conductor application. Thus, additional reduction to the finished product is required and this means reduction by drawing through a series of dies.

Multi-die drawing is a very good and widely practiced process, but it does impose limits on the composite product. The process demands sufficient formability to withstand the tensile stress state during deformation. No composite defects can form within the limited range of die cone angle and reductions available. Also, if annealing is required, the resistance to degradation of the composite with repeated intermediate annealing steps may be a factor. Given these drawbacks, it is natural to look at the possibility of using extrusion instead of drawing.

A serious limitation to using conventional extrusion as the second reduction step arises because the entire volume of material being processed must be loaded into the extrusion chamber at one time. This feature restricts the starting material volumes to small diameter, chamber length billets. To gain the benefits of extrusion and avoid the above drawback of conventional ram or hydrostatic extrusion, machines for small diameter long bar extrusion have been evaluated in various forms and levels of development over the last twenty five years. In the 1960's, intermittent or semi-continuous extrusion was introduced by Green (9), Pugh (10) and Alexander (11). Continuous hydrostatic extrusion by viscous drag was introduced by Fuchs (12) in the late 1960's and refined by him into the continuous wire extrusion process using gear driven chamber segments in the early 1970's (13). In the mid 1970's, Kobe Steel Company announced continuous hydrostatic-extrusion (14) using a wire pushing machine which requires the billet wrap around a rotating drum inside a pressure box which is limited to a pressure below the yield strength of the billet. Also in the mid 1970's Austen patented multi-stage extrusion (15). This CON-DIE[TM] process was envisioned as a way to go from the initial, large diameter billet to the final wire product in a

FIGURE 1   Innovare Model-12, 1780 kN (200 Ton) Force Press
with Semi-Continuous Hydrostatic Extrusion Tooling Package

series of tandem extrusion steps but it never received
sufficient support to warrant commercial development. In light
of this background, Innovare's work on semi-continuous
hydrostatic extrusion is a continuation of the 25 year effort to
provide a commercially attractive apparatus for extrusion
processing of long, small diameter composite bars into wires.

SEMI-CONTINUOUS EXTRUSION AT INNOVARE

    Specifically, Innovare has been working to develop a semi-
continuous extrusion process for 19mm (0.75") diameter aluminum
composite conductor bars based on high purity aluminum filaments
in an Al-8Fe-4Ce alloy (Alcoa CU 78) matrix. The approach was
to build on the existing semi-continuous extrusion concept and
overcome the pressure limitations we saw in the earlier designs.
This paper describes an experiment in which the goal is to take
a bar of up to 20mm (.8 in.) diameter and hydrostatically
extrude it using fluid pressures up to 1450 MPa (210 ksi). A
portion of the bar is advanced into the unpressurized chamber,
gripped, the chamber pressurized, a short length of material is
extruded, the pressure is relieved and grip released to begin
another cycle. This work is being carried out with the use of
an Innovare, Model 12 Advanced Metalworking System as shown in
Figure 1.

    The areas of interest are improvements to the grip and
pressurization mechanics to allow us to use extrusion pressures
well above the 1035 MPa (150 ksi) limit imposed by the earlier
designs. The first change was to abandon the billet bar grip
being actuated by the extrusion fluid pressure compressing the
grip sleeve around the billet inside the pressure chamber.
Stress analysis shows this type of grip mechanism to have a
maximum operating limit of 1035 MPa (150 ksi). The Model 12
uses a grip sleeve which is external to the pressure chamber and
is independently operated. Figure 2 shows schematic drawings
illustrating the two types of grip designs. Our initial grip
design performance target is set at 1450 MPa (210 ksi) operating

extrusion pressure with higher pressure limits for future designs.

Another difference from earlier machines is our method for controlling the extrusion fluid pressure during extrusion. Earlier designs required ports into the high pressure extrusion chamber to control the extrusion pressure. These ports were passages through the grip tooling and/or the high pressure chamber wall which again limited the operating pressure to the 1035 MPa (150 ksi) range. The Model-12 avoids such fluid access ports by controlling the chamber pressure with an independently movable annular ram through which the die support stem passes.

These experiments may lead to providing the added extrusion pressure capabilities required for the necessary reduction ratios for the aluminum composite conductor products. However, the external grip approach increases the cycle time by up to 20% because it requires more sequential operating steps than an internal grip. However, this trade off in slower production seems justified by the bigger extrusion reduction which in many cases allows production of the final product directly from the extruder. In other cases it means much less final drawing reduction to get the finished product.

AUTOMATIC CONTROL AND DATA ACQUISITION

Our development of a practical semi-continuous hydrostatic extrusion apparatus would have been a much more difficult and expensive undertaking were it not for Innovare's existing Advanced Metalworking System (AMS) and its programmable control and data acquisition features. Each subsystem and process step was developed and test operated on the AMS. It allowed the mechanical components, control procedures, and process monitoring data collection routines to be simultaneously developed as building blocks of the current Fully Automated Semi-continuous Tooling (F.A.S.T.) equipment.

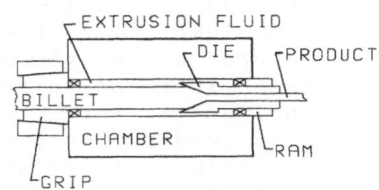

(a) Internal Billet Bar Grip (9)

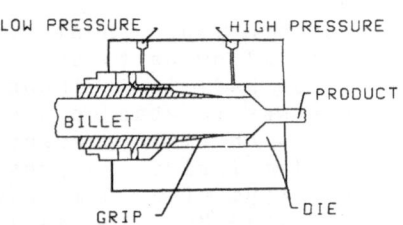

(b) External Billet Bar Grip

FIGURE 2 - Comparison of Internal and
External Grip Sleeve Designs.

The Control System is built around a BASIC language programmable control processor. That processor and its Analog and Digital Interfaces are packaged in an industrial operator's console along with enough DC power supplies and amplifiers to operate all the solenoids, proportional hydraulic valves, and instruments required by a piece of equipment as complex as the semi-continuous extruder. The console was not developed for the semi-continuous extrusion application. It is generic to the AMS.

This generic AMS control console consists of an array of programmably scanned and illuminated push buttons and program controlled digital read-outs. These interface the operator with the system. All the signal input and signal/power output capability passes through connectors on the side of the console. The functions of these connectors are programmable, so that instruments or actuators can readily be added or removed as the equipment requirements change. This same feature allows the console to easily be reconfigured for different applications allowing the AMS unit to operate all its optional tooling packages with the same sophistication of control and the same comprehensive data collection and display capability.

Software generic to the AMS control system consists of over 1400 lines of BASIC code which define the system I/O and all its interfacing functions, communication routines, etc. A skeleton main control program provides the framework for applications development, so much of the tedious work which would be required for such a sophisticated application was done prior to the actual semi-continuous extrusion F.A.S.T. program.

The data collection function is built into the AMS control software and works in conjunction with the companion software running on an IBM^{TM} compatible Personal Computer (PC). The PC is connected to the control console processor and a constant two way communication is maintained. The control system responds to requests for data from the PC and initiates messages to the operator keeping him or her advised of significant events or problems occurring during execution of its control program. All this communication is done without interfering with the control system's operation of the process. Indeed, the F.A.S.T. equipment can be operated without the PC with no effect other than the loss of data collection and detailed information passed to the operator.

The software on the PC utilizes the high resolution color graphics (EGA) screen to display up to 19 critical process variables simultaneously. This data is updated at rates of ~20 variables per second, so if you monitor all 19 possible variables you are looking at readings refreshed every second. Numerical and graphical displays are menu selectable. All data is passed on to disk files which can be recalled, edited, plotted or printed in any form or format. All this data acquisition is virtually automatic.

AVOIDING FLOW PATTERN DEFECT

A further development in this program was the study and avoidance of a composite flow pattern defect which could arise from the stop-start nature of semi-continuous extrusion.

It is well known experimentally and understood analytically that the grid distortion pattern on the product from an extruded split billet changes with die-to-billet friction changes. During semi-continuous hydrostatic extrusion, each time the extrusion is restarted, die friction decreases. This decrease in friction can be dramatic due to the transition from metal-to-metal contact to hydrodynamic film lubrication. This abrupt transition causes a major change in the slope of the transverse grid lines. To accommodate this change, metal must flow radially which will at least alter the composite geometry of the product in this transition zone. In the extreme, it can cause a surface rippling in the product.

This friction change induced surface rippling has been demonstrated in Innovare's stop-start experiments as shown in Figure 3. Also, using analytical modeling, the associated radial metal flow was predicted for a given friction change and set of process geometry parameters. From the experimental work, the analytically predicted large values of the radial distortion strain correlated with the tendency for the occurrence of the "stop-start" surface ripple defect shown in Figure 3.

Based on the knowledge gained in this study, we determined that the proper choices of tooling geometry and processing parameters could avoid this defect. When extruding composite samples using the conditions to prevent defects, we did not observe any surface ripple defects. The many of stop-start zones in the extruded composite products tested have not, to our knowledge, produced any evidence of such defects. Almost a dozen of the papers presented in this conference are giving data on Innovare extruded aluminum composite conductor samples. Also, none of the metallographic examinations turned up obvious filament contraction or dilation patterns.

ALUMINUM COMPOSITE CONDUCTORS

During our studies of the semi-continuous extrusion process and equipment Innovare produced a variety of aluminum composite conductor samples in lengths up to 7.6m (300 in.) using conventional hydrostatic extrusion tooling in it's 1.780 MN (200 ton), Model 12 Advanced Metalworking System. These were made for Alcoa from 19mm (.75 in.) diameter hot extruded bars of Al-8Fe-4Ce (CU 78 Alloy) with high purity filaments which Alcoa

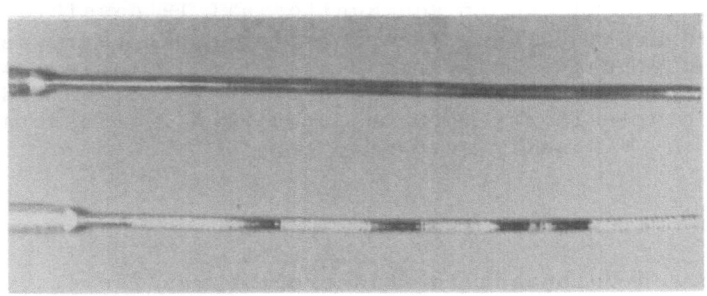

FIGURE 3 - "Stop-Start" Surface Ripple Defect

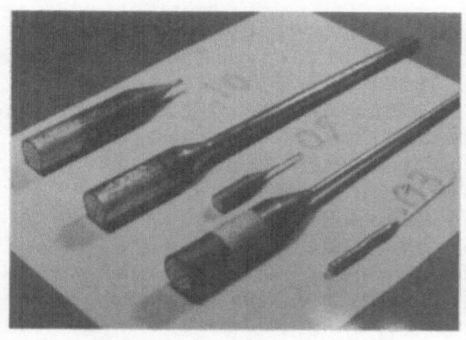 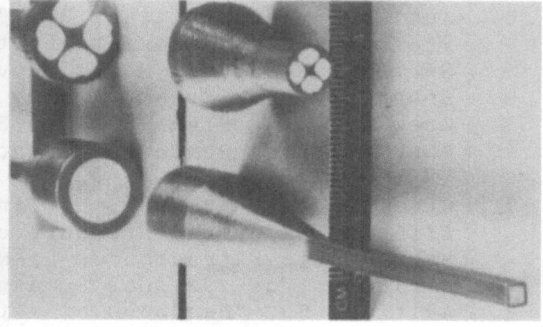

(a) 19 Filament Conductors   (b) Single & 4 Filament Conductors
FIGURE 4 - Extruded Aluminum Composite Conductors

supplied. Examples of the bars and products are shown in Figure 4. These samples were made in a single reduction for product diameters down to 2.5mm (0.10 in.) diameter. Two step reductions were required for smaller products.

DISCUSSION

This paper is a progress report on the experiments with semi-continuous hydrostatic extrusion because it is an ongoing effort with much still to be demonstrated and learned. After the demonstration of long length wire production, we need to assess its commercial value in the light of production costs. Beyond that evaluation, questions on the maximum reduction limits need to be addressed. Finally, we need to explore how much this process approach offers to allowing the use of higher strength, less ductile matrix alloys and greater proportions of high purity filament volume in the composites.

ACKNOWLEDGEMENT

Research supported by Department of the Air Force, Air Force Systems Command, Wright Research and Development Center, OH under contract F33615-86-C-2682 with The Aluminum Company of America.

REFERENCES

1. Zoerner, W., Austen, A., and Avitzur, B., "Hydrostatic Extrusion of Hard Core Clad Rod," Trans. ASME, Series D, Vol. 94, No. 1, March 1972, pp. 78-80.
2. Avitzur, B., Wu, R., Talbert, S., and Chou, T. T., "Criterion for the Prevention of Core Fracture During Extrusion of Bi-Metal Rod," Trans. ASME, Series B, Vol 104, No.3, 1982, pp 293-304.
3. Vereshchagin, L. F., Konyaev, Yu. S., Verzon, E. M., and Veller, M. V., "Effect of Hydrostatic Extrusion on the Composition and Properties of Nb_3Sn Compounds", Institute of Physics of High Pressure, 1972, pp. 447.
4. Avitzur, B., "Alternate Manufacturing Technologies for the Production of Multi-filamentary Superconducting Wire by the External Bronze Technique," Proceedings of the North American Manufacturing Research Conf., NAMRC X, pp. 218-224. Published by the SME, 1982.

5. Austen, A. and Avitzur, B., "Influence of Hydrostatic Pressure on Void Formation at Hard Particles," Trans. ASME, Series B. Vol. 96, No. 4, Nov. 1974, pp. 1192-1196.
6. Bobrowsky, A., Stack, E. A., and Austen, A., "Extrusion and Drawing Using High Pressure Hydraulics", Creative Manufacturing Seminar, Advances in Metal Forming, ASTME, Dearborn, Michigan, November 4, 1964.
7. Story, J. M., Avitzur, B., and Hahn, W. C., Jr., "The Effect of Receiver Pressure on the Observed Flow Pattern in Hydrostatic Extrusion of Bi-Metal Rods", Trans. ASME, Series B. Vol. 98, No. 3, Aug. 1976, pp. 909-913.
8. Avitzur, B., "Handbook of Metal Forming Processes", John Wiley, New York, 1983.
9. Slater, H. K. and Green D. Augmented hydrostatic extrusion of continuous bar. In High pressure engineering (London: The Institution of Mechanical Engineers, 1968), 217-22, (proc. Instm mech. Engtrs, 182, pt 3 C)
10. Pugh, H. LI. D., Recent Developments in Cold Forming, Bulleid Memorial Lecture, 1965, Vols IIIA, IIIb, Univ. of Nottingham Press, Nottingham, England, 1965.
11. Alexander, J. M. and Lengyel, B., Semi-continuous hydrostatic extrusion of wire, In Applied mechanics convention (London: The Institution of Mechanical Engineers, 1966), 317-27, (proc. Instn mech. Engrs, 180, pt 3 I).
12. Fuchs, F. J., Hydrostatic Wire Extrusion, Wire J., October 1970, pp 105-113.
13. Fuchs, F. J. and Schmehl, G. G., Continuous Hydrostatic Extrusion, in: NEL/AIRAPT Internal Conference on Hydrostatic Extrusion, University of Sterling, Scotland, June 13-15, 1973, pp. 334-338.
14. Matsushita, Tomiharu; Yamaguchi, Yoshihiro; and Takatsuka, Kora, A New Concept of Continuous hydrostatic Extrusion-Drawing, (in Japanese), in: Japan Spring Congress for Plastic Working, May 14-16, 1974, pp. 363-366.
15. Austen, A. R., "Method and Apparatus for Extrusion", U.S. Patent No. 3,999,415, December 28, 1976.

RESIDUAL RESISTANCE OF ULTRA HIGH PURITY COPPER BY SULFATE-ELECTROREFINING AND ZONE-REFINING

A. Kurosaka, H. Tominaga, T. Takayama, and H. Osanai

Material Research Laboratory, Fujikura Ltd.
1-5-1 Kiba, Koto-Ku, Tokyo 135, Japan

ABSTRACT

Ultra-high-purity (99.9999% or higher) copper rod having a RRR value of 5000 was prepared by a two-stage refining process, i.e., a combination of two processes, sulfate-electrorefining and zone-refining. In the first process, the potential of the cathode electrode was maintained at a level of 0.2V (vs NHE) at 25°C to prevent the deposition of more active metals than copper. In the second process, the width of the molten zone was made smaller than the diameter (16 mm) of the copper rod to obtain a higher refining efficiency. A wire of 0.5 mm in diameter was drawn from the refined rod. The RRR value of the ultra-high-purity copper wire annealed in a N_2 gas atmosphere at 500°C for two hours is as high as 4100, while that of conventional oxygen-free copper wire is usually 200. Ultra-high-purity copper will be suitable for applications in the superconducting field, electronic devices and so on.

INTRODUCTION

In general, copper is required to have a higher RRR value than usual when it is used as a stabilizing element for low-temperature superconductors or as a material for high-field magnet conductors. The cryogenic properties of oxygen-free copper have already been discussed in some papers[1-4], but there is virtually no report on the cryogenic properties of ultra-high-purity copper whose RRR value exceeds 4000. In the present investigation, we have measured some cryogenic properties of ultra-high-purity copper, high-purity (99.999%) copper and oxygen-free copper. The measured cryogenic properties are the electrical resistivity in liquid nitrogen and in liquid helium, the effect of grain size on the RRR value, and the tensile strength in liquid nitrogen. This paper describes the cryogenic properties of ultra-high-purity copper in comparison with those of high-purity copper and oxygen-free copper.

PREPARATION OF MATERIALS

The materials used for the present investigation were prepared in the following ways:

Ultra High Purity Copper (UHPC)

This material was prepared from 99.95% pure copper by a two-stage refining process, i.e., a combination of two processes, sulfate-electrorefining and zone-refining. In the first process, the potential of the cathode electrode was maintained at a level of 0.2V (vs NHE) at 25°C to prevent the deposition of more active metals than copper. In the second process, the width of the molten zone was made smaller than the diameter (16

Advances in Cryogenic Engineering (Materials), Vol. 36
Edited by R. P. Reed and F. R. Fickett
Plenum Press, New York, 1990

749

TABLE 1. GDMS analysis of UHPC, HPC, and OFC

Elements	B	Na	Mg	Al	Si	P	S	K	Ca	Cr
UHPC	ND	ND	ND	ND	0.03	ND	ND	ND	ND	ND
HPC	ND	0.01	ND	0.06	0.2	0.01	0.2	ND	ND	0.03
OFC	ND	0.04	ND	0.06	0.3	0.01	2.8	0.05	0.1	0.1

Elements	Fe	Ni	As	Se	Te	Ag	Cd	Sb	Pb	Bi
UHPC	0.02	ND	ND	ND	ND	0.03	ND	ND	ND	ND
HPC	1.0	0.05	ND	ND	ND	0.45	ND	ND	ND	ND
OFC	0.9	0.1	ND	ND	ND	13.5	ND	0.3	0.3	ND

ND; not detectable

mm) of the copper rod to obtain a higher refining efficiency. The material was a 16 mm diameter rod having a RRR ($\rho_{298K}/\rho_{4.2K}$) value of 5000.

High Purity Copper (HPC)

This material was prepared by vacuum casting after being refined by sulfate-electrorefining under the same conditions as the preparation of UHPC. It was in the shape of 20 mm diameter rod.

Oxygen Free Copper (OFC)

This material was one currently produced by the Dip Forming Process at Fujikura Ltd. It was in the shape of a 17 mm diameter rod.

Table I shows the results of the Glow Discharge Mass Spectrometry (GDMS) analysis of UHPC, HPC and OFC.

EXPERIMENTAL PROCEDURE

Wire specimens were prepared from these materials following the steps of:
1. Swaging to 5.0 mm diameter.
2. Etch-cleaning with nitric acid.
3. Drawing to 2.5 mm diameter.
4. Etch-cleaning with nitric acid.
5. Drawing to 0.5 mm diameter (special diamond dies were used for drawing UHPC specimen).
6. Aging at room temperature for more than 30 days.
   Etch-cleaning and drawing through special diamond dies were performed for the prevention of contamination.

For specimens annealed in a N_2 gas atmosphere at 300, 400, 500 and 600°C for two hours and specimens aged at room temperature, resistivity measurements were performed by the conventional four-terminal method at room temperature (298 K), in liquid nitrogen (76 K) and in liquid helium (4.2 K). Before annealing, these specimens were wound around a silica glass tube to prevent them from being strained during the measurement. Voltage taps were arranged with a spacing of 1 m. A current of 1.0 A was fed to the specimens at all temperature, and millivolt voltages were obtained at 298 K and 76 K and microvolt voltages at 4.2 K. The temperature dependence of the resistivity of each specimen annealed at 500°C was measured by a germanium thermometer calibrated for temperatures up to 100 K.

The specimens annealed at 300, 400, 500 and 600°C were observed for microstructure and then measured for grain size by the intercept method.

At room temperature and in liquid nitrogen, tensile tests were made on specimens annealed at 100 to 260°C for 30 minutes and specimens aged at room

TABLE II. RRR($\rho_{298K}/\rho_{4.2K}$) values of specimens aged at room temperature and annealed at 500°C

Specimens	UHPC	HPC	OFC
As aged at room termperature for more than 30 days	550	58	44
As annealed in a N₂ gas atmosphere at 500°C for 2 hours	4100 (3200)*	824	207

* no etch-cleaning

temperature for more than 30 days. The specimen length was 50 mm and the strain rate was 1.7 x 10⁻³s⁻¹.

RESULTS AND DISCUSSION

Electrical Resistivity

Table II shows the RRR ($\rho_{298K}/\rho_{4.2K}$) values of the specimens aged at room temperature and of the specimens annealed at 500°C. In spite of etch-cleaning and use of the special diamond dies, the UHPC specimens exhibited a lower RRR value than the starting material due to contamination in the drawing process. However, the RRR value of the UHPC specimen annealed at 500°C without etch-cleaning had a still lower RRR value of 3200.

Of the specimens aged at room temperature, UHPC exhibited a much higher RRR value than HPC and OFC. This distinct difference seems attributable to two reasons. Firstly, the quantity of impurities (Si, S, Fe, Ag, etc.), which increases the resistivity of copper, decreases extremely as shown in Table I. Secondly, the quantity of crystal defects (grain boundaries), which reduce the conductivity of copper, decreases because UHPC recrystallizes even at room temperature. The RRR value of UHPC aged at room temperature was higher than that of OFC annealed at 500°C. When used as a stabilizing element for low-temperature superconductors, UHPC, even unannealed, would give a higher conductivity than OFC.

Figure 1 shows the temperature dependence of the resistivity of the specimens annealed at 500°C. In the temperature region above 20 K, the

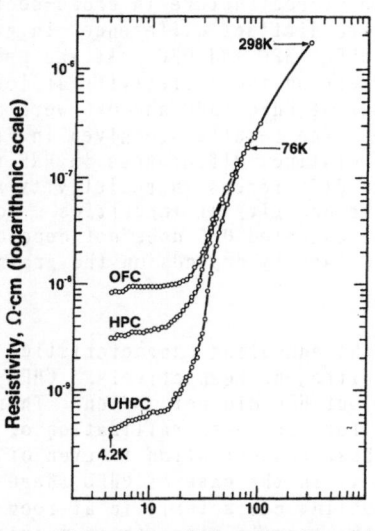

FIG. 1. Temperature versus resistivity of UHPC, HPC, and OFC annealed at 500°C.

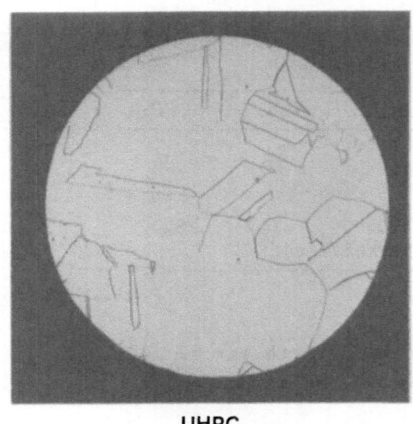

UHPC

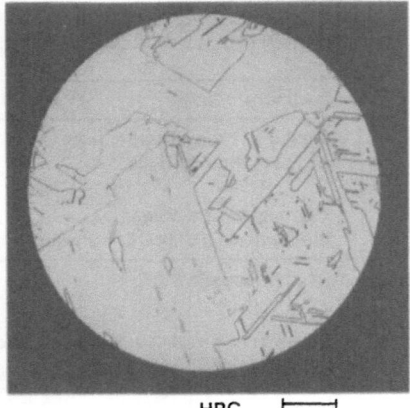

HPC

0.1 mm

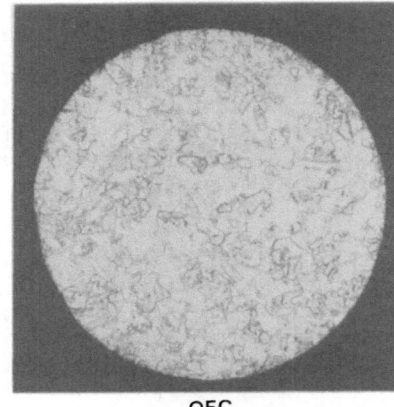

OFC

PHOTO. 1. Microstructure in cross-section of specimens annealed at 500°C.

differences in resistivity between the three specimens became smaller with increasing temperature and was virtually negligible in the vicinity of 76K. This result suggests that UHPC should be used at temperatures below 20 K in order to ensure the superiority in conductivity of this material at all times.

Photograph 1 shows the microstructure in cross-section of the specimens annealed at 500°C. There are distinct differences in grain size (area of grain boundaries) between UHPC, HPC and OFC. It was presumed from this fact that grain size exerted effect on the resistivity at low temperatures. In view of this, the RRR values of UHPC, HPC and OFC were determined as a function of the grain size. The results are given in Figure 2. Even with the same grain size, there are distinct differences in RRR value between the three specimens, therefore, these differences in resistivity at low temperatures seem to depend mainly on the quantity of impurities. Moreover, it has been found that the RRR value of annealed OFC does not depend on the grain size, while that of annealed UHPC largely depends on the grain size.

Tensile Strength

Figures 3 and 4 show the annealing characteristics measured at room temperature and in liquid nitrogen, respectively. UHPC softened simply by aging at room temperature, but HPC did not soften. Therefore, it can be said that the activation energy for the recrystallization of UHPC is much influenced by impurities whose concentration is even of the order of 0.1 ppm. As seen from Figures 3 and 4, in the case of UHPC there is virtually no difference between the annealing characteristic at room temperature and that in liquid nitrogen, while the tensile strength in liquid nitrogen is about 30 to 40% higher than that at room temperature. The same phenomena can be

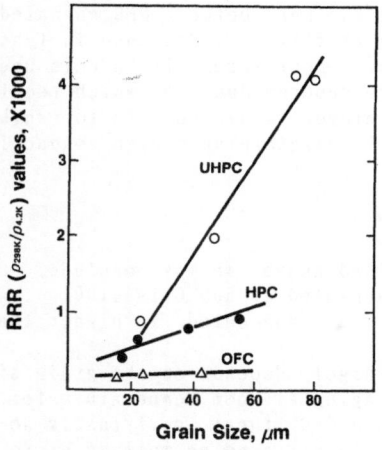

FIG. 2. RRR($\rho_{298K}/\rho_{4.2K}$) values of UHPC, HPC, and OFC as a function of grain size.

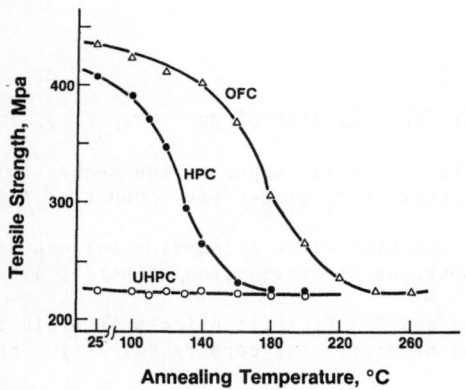

FIG. 3. Annealing characteristics measured at room temperature.

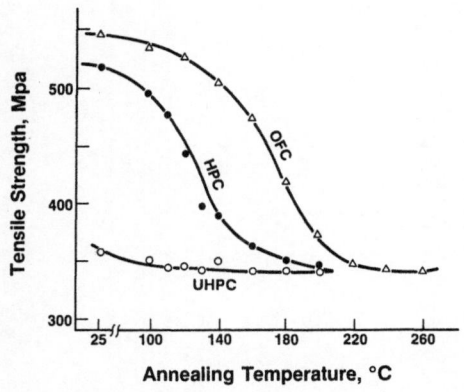

FIG. 4. Annealing characteristics measured in liquid nitrogen.

observed for OFC and HPC. Further, between OFC annealed at temperatures above 240°C and UHPC, there is virtually no difference in tensile strength in liquid nitrogen as well as at room temperature. If UHPC is used as a stabilizing element for low-temperature superconductors, which requires annealing after drawing, this stabilizing element would be able to exhibit a tensile strength similar to that of OFC while maintaining a high conductivity.

CONCLUSIONS

From the facts described above, we may conclude:
1. The RRR value of UHPC annealed at 500°C is 4100.
2. The RRR value of UHPC, even unannealed, is higher than that of OFC annealed at 500°C.
3. The RRR value of UHPC largely depends on the grain size.
4. UHPC softens simply by aging at room temperature for more than 30 days.
5. Between UHPC and softened OFC, there is virtually no difference in tensile strength in liquid nitrogen as well as at room temperature.

If these characteristics are given due consideration, UHPC would be much more suitable for use in the superconducting field. We are aiming to prepared an optimum grade of copper for applications in low-temperature use and think that this can be attained by making more studies on cryogenic properties, refining method and processing technique.

REFERENCES

1. T. Nara and Y. Yamada; Zone refining of pure copper, J. Japan Inst. Metals 24, 460 (1960).
2. F. R. Fickett; The effect of mill temper on the mechanical and magnetoresistive properties of oxygen-free copper at 4 K, Adv. Cryo. Eng. 30, 453 (1985).
3. F. R. Fickett and T. E. Capobianco; Relationships between mechanical and magnetoelectric properties of oxygen-free copper at 4 K, Adv. Cryo. Eng. 32, 421 (1986).
4. R. P. Reed, R. P. Walsh, and F. R. Fickett; Effects of grain size and cold rolling on cryogenic properties of copper, Adv. Cryo. Eng. 34, 299 (1987).

MAGNETIC SUSCEPTIBILITY OF INCONEL ALLOYS

718, 625, AND 600 AT CRYOGENIC TEMPERATURES

Ira B. Goldberg, Michael R. Mitchell, Allan R. Murphy

Rockwell International Science Center
Thousand Oaks, California 91360

Ronald B. Goldfarb, Robert J. Loughran

National Institute of Standards and Technology
Boulder, Colorado 80303

ABSTRACT

In June 1988, the Discovery Space Shuttle mission was delayed because of a malfunctioning hydrogen fuel bleed valve system. The problem was traced to the linear variable differential transformer (LVDT) which produced erroneous readings for the valve position. Near liquid hydrogen temperatures, Inconel 718 used in the armature of the LVDT became strongly magnetic. The AC magnetic susceptibility of three samples of Inconel 718 of slightly different compositions, one sample of Inconel 625, and one sample of Inconel 600 were measured as a function of temperature. Inconel 718 behaves as a spin glass. Its susceptibility reaches a maximum between 15 and 19 K, near the liquid hydrogen boiling point, 20 K. The susceptibility increases by an order of magnitude as the iron content increases by 1.2% and the nickel content decreases by 1.5%. The nominal composition is 12-20% iron and 50-55% nickel. Inconel 625, which contains about 4% iron, was paramagnetic. Inconel 600 exhibited spin glass properties below 6 K, short-range ferromagnetism between 6 and 92 K, and paramagnetism above 92 K.

INTRODUCTION

The launch of the Discovery Space Shuttle, the first flight after resumption of the space program in the summer of 1988, was delayed because of an apparently faulty hydrogen fuel bleed valve. This valve

Advances in Cryogenic Engineering (Materials), Vol. 36
Edited by R. P. Reed and F. R. Fickett
Plenum Press, New York, 1990

755

Table 1. Chemical compositions (mass percent) of Inconel alloys 718, 625, and 600

Element	Inconel 718-1153	Inconel 718-1094	Inconel 718-1	Inconel 625	Inconel 600
Nickel	52.08	52.80	53.59	60.61	72.0 min
Iron	19.23	18.46	18.02	4.39	6.0-10.0
Chromium	17.42	18.22	18.07	21.90	14.0-17.0
Molybdenum	2.93	2.99	2.99	8.56	
Niobium	5.14	5.05	5.19	3.71	
Manganese			0.14	0.15	1.00 max
Titanium	1.05	0.96	0.91	0.20	

appeared not to close completely. Inspection of both the valve and linear variable differential transformer (LVDT) position sensor in the laboratory showed that the valve did in fact close completely, but that the valve position was indicated incorrectly by the LVDT. This, in turn, led to close inspection of the LVDT and its components.

The LVDT consisted of three coils as shown in Fig. 1. The inner coil was the drive or primary, and the two outer coils were the differential pick-up or secondaries. A high permeability armature was connected to the valve by an Inconel 718 armature extender. With the valve closed, the Inconel armature extender was within the right secondary when the armature was inside the left solenoid (Fig. 1). However, with the valve open, the armature extender was outside the coil assembly. The fact that the LVDT performed well at temperatures greater than 35 K (-397°F), but not near its operating point of about 20 K (-424°F), led to this investigation of the magnetic properties of Inconel alloys.

Limited data exist on the magnetic properties of Inconel or nickel-chromium-iron alloys. The most current data for Inconel 718 is given in the Handbook on Materials for Superconducting Machinery.[1] Neither the geometry of the samples nor the composition is given. Magnetization plots for temperatures between 4.2 and 50 K, presumably corrected for demagnetizing field, show that the material does not saturate at fields up to

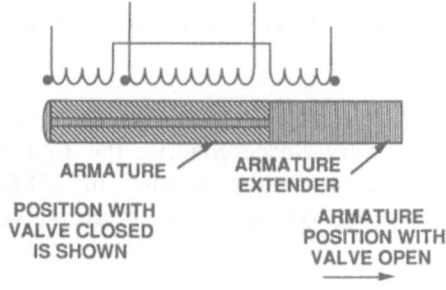

Fig. 1. Schematic diagram of a Linear Variable Differential Transformer (LVDT).

5.6 MA/m (70 kOe). One other report[2] provides useful information on Inconel 750, but it is not directly applicable to Inconel 718. No literature data were found for Inconel 625. Relevant studies have been carried out on Ni-Cr, Ni-Mo[3,4] and Ni-Fe-Cr[5,6] alloys. While the nature of the measurements is quite different from ours, they provide some useful guidelines for predicting trends in the magnetic properties of such materials.

SAMPLES AND EXPERIMENTAL MEASUREMENTS

Inconel 718 was chosen for the armature extender because of its high strength at low temperatures.[7,8,9] Thus, in addition to samples of Inconel from the armature extenders, other Inconel alloys were investigated. Three samples of Inconel 718 were examined. Two samples from the LVDT were part of the threaded portion that connects the armature to the valve assembly. One sample, designated 718-1153, was from a malfunctioning LVDT. A second, designated 718-1094, was cut from a useable LVDT. A third sample, designated 718-1, was a cylinder received from the manufacturer. Samples of Inconel alloys 600 and 625 were in the form of small rods, approximately 6 mm diameter by 12 mm. The Inconel 625 sample was received from the manufacturer, and the Inconel 600 was received from Argonne National Laboratory. The nominal compositions of the alloys can be found in Ref. 1. The analyses of the Inconel 718 and 625 samples used here are given in Table 1. The chemical analysis of Inconel 600 given in Table 1 is nominal.

Demagnetization factors were assumed to be average or magnetometric values. The Inconel 718 materials from the LVDTs were threaded rods, but were treated as cylinders. The average diameter, d, was calculated according to the equation

$$d = (4 \, m \, / \, \pi \, \rho \, L)^{1/2} \qquad (1)$$

where m is the mass, ρ is the density, and L is the overall length. The average demagnetization factor was determined from the ratio of the length to the diameter.[10]

Measurements on all samples were carried out using an AC susceptometer[11] and on several samples using a vibrating sample magnetometer (VSM). AC measurements were carried out at applied alternating magnetic fields of 8, 80, and 800 A/m (0.1, 1, and 10 Oe) rms, and frequencies of 10, 100, and 1000 Hz. No DC field was applied to the sample for AC susceptibility. Results were slightly dependent on field and frequency.

VIBRATING SAMPLE MAGNETOMETRY

The results obtained using the VSM were similar to those previously reported in that the samples did not saturate even at fields as high as 5.5 MA/m (70 kOe). Using a reference temperature of 4.4 K, at the maximum applied field, the data reported[1] for generic Inconel 718 indicates a magnetization of 1.13 x 10^5 A/m (113 emu/cm^3) while that of sample 718-1094 is 1.18 x 10^5 A/m (118 emu/cm^3) and of sample 718-1153 is 1.33 x 10^5 A/m (133 emu/cm^3). These results are in fairly good agreement with each other.

Table 2. Temperatures and values of the maximum susceptibility of Inconel 718 and 625 samples

	718-1153	718-1094		718-1	625
Applied field (A/m)	800	800	8	800	800
(Oe)	10.0	10.0	0.1	10.0	10.0
Temperature (K) at maximum susceptibility	19.0	16.1	15.7	15.2	< 5.3 K
χ at 10 Hz	13.3	3.26	3.94		0.0030 at 16K
χ at 100 Hz	12.8	3.19	3.70		0.0031 at 16K
χ at 1000 Hz	12.7	3.13		1.51	0.0034 at 16K

AC SUSCEPTIBILITY

Inconel 718 Alloys

AC susceptibility measurements were carried out, primarily because this mode of measurement most closely corresponds to that of the Inconel armature as it is used in the LVDT.

The three samples of Inconel 718 exhibited similarly shaped plots of susceptibility vs. temperature. However the magnitudes of the susceptibilities were significantly different as shown in Fig. 2, even though the elemental composition of the alloys varied by only 1.5% in nickel, 1.2% in iron, and 0.65% in chromium. It is noteworthy that the peak magnetic

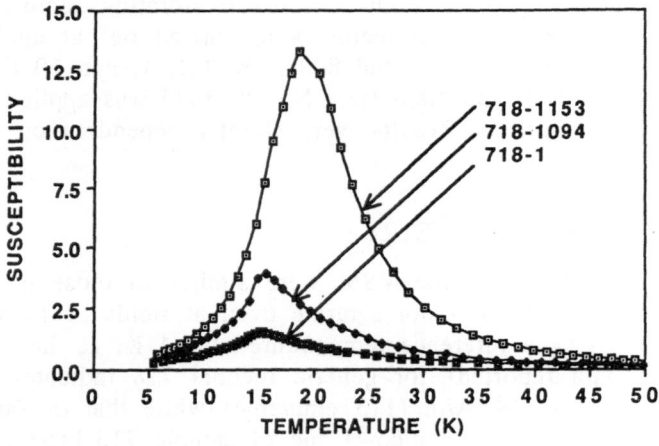

Fig 2. Susceptibility of three Inconel 718 alloys as a function of temperature. Alloy compositions given in Table 1.

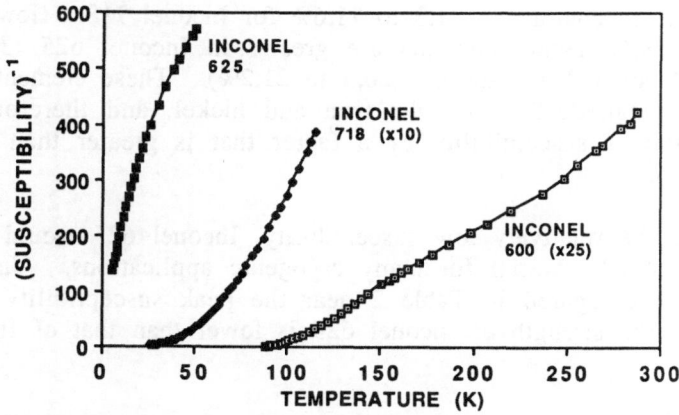

Fig. 3. Plots of the inverse AC susceptibility of three Inconel alloys. Data for Inconel 718-1153 are shown.

susceptibility varies by a factor of eight over this concentration range. The temperature at the maximum susceptibility also increases with higher iron content as shown in Table 2.

The Inconel 718 alloys appear to be spin glasses at temperatures below about 16 K and are paramagnetic above that temperature. Evidence that points to the spin glass state is derived from the VSM and susceptibility measurements. Key factors are that the magnetization of the sample does not saturate, even at fields of 5.5 MA/m (70 kOe), the AC susceptibility exhibits a maximum, and this maximum shifts slightly to lower temperature with increasing AC field.[12] The paramagnetic phase is demonstrated by nearly linear plots of the inverse susceptibility vs. temperature for temperatures significantly greater than that of the peak susceptibility (Fig. 3). This is consistent with Curie-Weiss behavior.

Whether or not Inconel 718 can be used as a nonmagnetic material will depend on the temperature, the exact composition, and the limit of magnetic susceptibility that can be tolerated for a given application. Ironically, the peak susceptibility is close to the boiling point of hydrogen at atmospheric pressure. The AC data suggest that iron exhibits the largest effect on susceptibility. In contrast, the VSM data show that the high-field magnetization is relatively unaffected by composition. Thus, Inconel 718, when used at liquid hydrogen temperatures, may interfere with certain measurements. The effect would be much larger if inductance (such as in the LVDT) rather than bulk magnetization is measured.

Inconel 625

Inconel 625 exhibits a decreasing susceptibility with temperature above 5.7 K. Measurements were not carried out at lower temperatures. The paramagnetic behavior is shown by the linear relationship between the inverse susceptibility and temperature in Fig. 3. In this case, the iron content is only 4.4% even though the nickel content is greater than in the samples of Inconel 718. The combined iron and nickel in the Inconel 625 sample is 64.6%, in contrast to 71.3 to 71.6% for Inconel 718. However, the chromium and molybdenum contents are greater in Inconel 625 (30.5%) than in any of the Inconel 718 materials (20.3 to 21.2%). These elements can couple antiferromagnetically with the iron and nickel, and therefore reduce the magnetization or susceptibility by a factor that is greater than they would as inert diluents.

Because of the relatively low susceptibility, Inconel 625 would be a useful "nonmagnetic" material for many cryogenic applications. Values of the susceptibility are compared in Table 2 near the peak susceptibility of Inconel 718. However, the strength of Inconel 625 is lower than that of Inconel 718.[1,7]

Inconel 600

Inconel 600 has greater nickel composition than either of the other alloys, but is intermediate in iron content as shown in Table 1. Both the real and the imaginary components of the susceptibility (χ' and χ'', respectively) are shown in Fig. 4. Three magnetic phases are evident. The first phase appears to be a spin glass below about 6 K. The second phase, which exists between 6 K and 92 K, appears to be a ferroglass.[13] Such phases are characterized by ferromagnetic correlation over a finite length, in general longer than that of a spin glass. At temperatures above 92 K, the material is paramagnetic. A plot of the inverse susceptibility as a function of temperature is shown in Fig. 3.

CONCLUSIONS

The most significant conclusion is that Inconel 718 alloys apparently exhibit a spin glass state below 16 K. Experimentally, this is concluded from data obtained using VSM and AC susceptibility. AC susceptibility shows that Inconel 600 exhibits three different magnetic phases. The lowest temperature phase (below 6 K) appears to be a spin glass somewhat similar

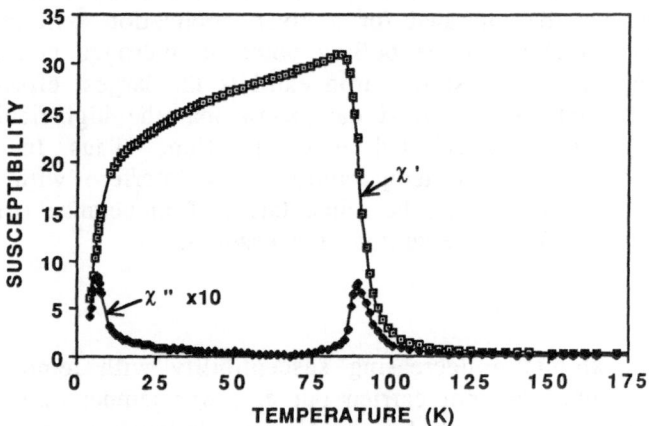

Fig. 4. Susceptibility of a sample of Inconel 600, corrected for the geometric demagnetization factor. Measurements were conducted with an applied field of 80 A/m (1 Oe) at 10 Hz. χ'' is multiplied by 10.

to that of Inconel 718. The intermediate temperature phase (6 to 92 K) may be ferroglass. At elevated temperatures, all of the Inconel alloys are paramagnetic.

The magnetic states of the Inconel alloys and the magnitude of the susceptibilities appear to be strongly dependent on the exact composition of the alloy. Not investigated in this study were the effects of cold working and of heat treatment or annealing, which may be significant.

ACKNOWLEDGMENTS

The authors thank R. W. Cross for carrying out AC susceptibility measurements on the Inconel 600. Research sponsored by Rocketdyne Division, Rockwell International Corporation, NASA Contract No. NAS8-400000, and Center for Electronics and Electrical Engineering, National Institute of Standards and Technology.

REFERENCES

1. "Handbook on Materials for Superconducting Machinery," Metals and Ceramics Information Center, Battelle Laboratories, Columbus, Ohio (1977).

2. E. W. Collings, F. J. Jelinek, J. C. Ho, and M. P. Mathur, Magnetic properties of commercial soft magnetic alloys at cryogenic temperatures, Adv. Cryogenic Engr., 22: 159 (1977).

3. V. Marian, Ferromagnetic Curie points and the absolute saturation of some nickel alloys, Ann. Phys. [11], 7:459 (1937).

4. S. Sadron, Ferromagnetic moments of the elements and the periodic system, Ann. Phys. [10], 17: 371 (1932).

5. L. R. Jackson and H. W. Russell, Temperature sensitive magnetic alloys and their uses, Instruments, 11: 280 (1938).

6. P. Chevenard, Alloys having high nickel and chromium contents, Rev. de Met., 25: 14 (1928).

7. Inco Alloys International, brochures "Inconel Alloy 718," "Inconel Alloy 625," and "Inconel Alloy 600," Huntington, West Virginia (1985).

8. W. F. Weston, H. M. Ledbetter, and E. R. Naimon, Dynamic low-temperature elastic properties of two austenitic nickel-chromium-iron alloys, Mater. Sci. Engr., 20: 184 (1975).

9. W. F. Weston and H. M. Ledbetter, Low-temperature elastic properties of a nickel-chromium-iron-molybdenum alloy, Mater. Sci. Engr., 20: 287 (1975).

10. G. W. Crabtree, Demagnetizing effects in the de Haas-van Alphen effect, Phys Rev., 16:1117 (1977).

11. R. B. Goldfarb and J. V. Minervini, Calibration of AC susceptometer for cylindrical specimens, Rev. Sci. Instrum., 55; 761 (1984).

12. K. Moorjani and J. M. D. Coey, "Magnetic Glasses," Elsevier, Amsterdam, (1984), Chapter 1.

13. R. B. Goldfarb, K. V. Rao, and H. S. Chen, Differences between spin glasses and ferroglasses, Solid State Commun., 54: 799 (1985).

THE MAGNETOCALORIC EFFECT IN NEODYMIUM

C.B.Zimm, P.M.Ratzmann, J.A. Barclay

Astronautics Corporation of America
Astronautics Technology Center
Madison, WI 53716

G.F. Green, J.N. Chafe

D.W. Taylor Naval Ship R&D Center
Bethesda, MD 21402

ABSTRACT

The adiabatic temperature change upon magnetization, ΔT_s, and the field dependent heat capacity of Nd have been measured between 5 K and 40 K in applied magnetic fils up to 7 T in order to assess its potential for use as a magnetic refrigerant in an active magnetic regenerator (AMR). The ΔT_s of Nd is proportional to temperature between 4 K and 10 K, with a maximal value of 2.5 K at 7 T and 10 K. Nd thus shows some applicability to AMR use.

INTRODUCTION

Neodymium is a rare earth metal which magnetically orders below 20 K. It has been suggested[1,2] for use in passive magnetic regenerators[3] in gas-cycle refrigerators due to its exceptionally large zero field heat capacity at temperatures below 20 K. Nd is malleable and machinable, and is thereby usable in geometries such as thin sheets which are difficult to fabricate for the more brittle intermetallic compounds. Because the large heat capacity is of magnetic origin, it is reasonable to consider the use of Nd in an active low temperature magnetic refrigerator stage. Such an active magnetic refrigerator[4] (MR) uses a changing externally applied magnetic field to cause a magnetic material to undergo a thermodynamic cycle. To date, only the zero or low field heat capacity of Nd has been measured.[5,6] In order to determine the applicability of Nd to magnetic refrigeration, the thermal response to an applied magnetic field must also be known as well as the zero field heat capacity. Therefore, we have measured the field dependent heat capacity, C_B, and the adiabatic temperature change upon application of an external field, ΔT_s, for Nd in applied fields up to 7 T. These quantities allow the thermodynamic cycle and performance of a MR to be estimated.[7]

EXPERIMENTAL METHOD

The sample of neodymium used in this study was machined from a polycrystalline ingot of 99.9% atomic purity with respect to other rare earths. The ingot was produced by Research Chemicals of Phoenix, AZ.

The heat capacity and temperature change of the sample upon change in applied field were measured by an adiabatic method. The sample was brought to internal thermal equilibrium at a temperature T_i, before either a known heat pulse, for C_B, or a known change in external magnetic field for ΔT_s, was applied, and then the final temperature T_f was observed. Correction for heat lost to the surroundings was done by a simple linear

Advances in Cryogenic Engineering (Materials), Vol. 36
Edited by R. P. Reed and F. R. Fickett
Plenum Press, New York, 1990

763

extrapolation to the middle of the heat pulse or field change. This correction was quite small because the time constant for internal thermal equilibrium of the sample was typically several orders of magnitude shorter than the time constant for equilibration with the surroundings. Excellent thermal isolation was ensured by suspending the sample via linen threads, using high resistance leads for the heater and thermometer, winding the heater directly on the sample, and placing the thermometer in a snug hole at the center of the sample. The sample was surrounded by a can maintained at a temperature intermediate between the initial and final temperatures of the sample. This can was surrounded in turn by two more isothermal cans sharing a common vacuum. The temperatures of the sample and the surrounding cans were measured by carbon-glass thermometers whose field dependence was less than 0.1 K. The sample thermometer was calibrated by the manufacturer to 40 mK and was checked at 4.2 K at the start of each run.

The corrections for the heat capacity of the thermometer, heater wire, threads and varnish attached to the 0.022 kg sample were typically less than 5%. For ΔT_s, the measured values were corrected for addenda by multiplying by the ratio of the total heat capacity of the neodymium and addenda to that of the neodymium alone. The heat capacity values used for the corrections were the average of the values at the initial and final applied fields. The approximately ellipsoidal sample had a length to diameter ratio of 3.0 and was oriented with its long axis parallel to the applied magnetic field. Demagnetization corrections to the applied field have not been made but magnetization data[8] indicate these corrections are small.

Because the time constant for conduction cooling the sample and surrounding cans is extremely long, all the data were taken under computer control in a stepwise warming mode, with the temperatures of the cans and the sample stabilized before each heat pulse or field change. Before each run, the sample was cooled from greater than 100 K to 4.2 K in zero applied field. For heat capacity measurements, the length of the heat pulse was about 1/4 of the thermal time constant of the sample and the power of the heat pulse was chosen to warm the sample by 0.2 - 1 K.

RESULTS

Figure 1 shows the measured field-dependent heat capacity of Nd. The two conspicuous magnetic transitions are due to ordering on the hexagonal sites (19 K) and on the cubic sites (8 K). The low temperature transition is not quite as sharp as seen previously,[5,6] presumably due to the nominal 0.1% impurity content of the sample.

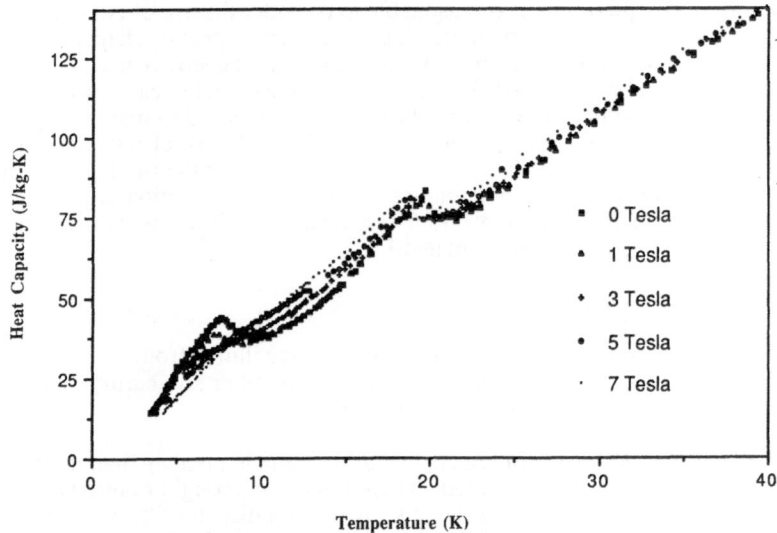

Fig. 1. The measured field-dependent heat capacity of Nd.

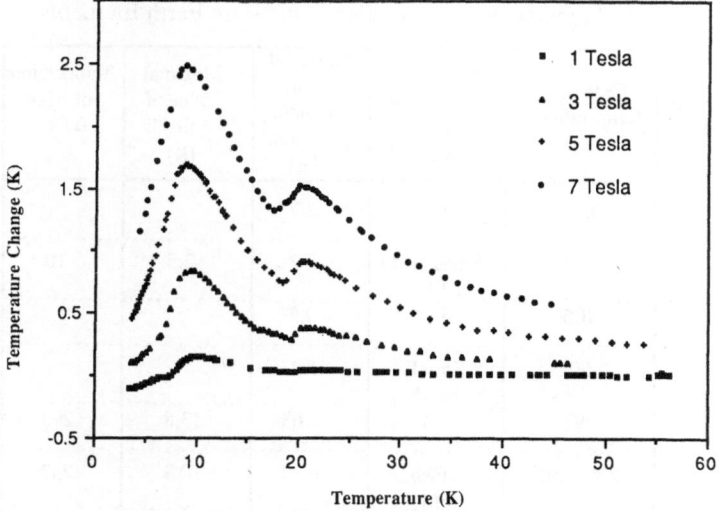

Fig. 2. The measured ΔT_s for Nd as the magnetic field is increased from zero up to the stated value.

Figure 2 shows the measured ΔT_s for Nd as the magnetic field is increased from zero up to the stated value. The two magnetic transitions correspond to the two peaks seen. Clearly, the 8 K transition on the cubic sites couples more strongly to the external field than the 19 K hexagonal site transition, a difference that may be partly explained in the polycrystalline specimen by the much stronger anisotropy observed in the hexagonal plane[8]. The large anisotropy ensures that only the component of the applied field in the plane of easy magnetization in a given crystallite contributes to the ordering for the 19 K transition, reducing the observed ΔT_s. Note also that below 8 K, Nd cools when a 1 T magnetic field is applied. This is typical behavior for an antiferromagnet,[9] for which an applied field hinders the ordering.

Figure 3 shows the entropy of Nd as a function of temperature for five applied magnetic fields. The curves for each field were calculated by numerically integrating the measured heat capacity at that field divided by temperature. The additive constants to the entropy due to the unknown values of the heat capacity at temperatures below the lowest temperature attainable by the apparatus were determined by requiring the horizontal spacing between the entropy curves for the various fields to be consistent with the measured adiabatic temperature changes. This was done by a least squares fitting procedure.

DISCUSSION

Neodymium has its maximal ΔT_s in the 5 K to 20 K temperature region, as expected from the presence of ordering transitions at approximately 8 K and 19 K. The maximal value of ΔT_s is 2.5 K, which is relatively small. Several mechanisms may explain the reduction in ΔT_s compared to a material such as Gd which has ΔT_s of 14 K for a 7T field change.

● Nd has the double-hexagonal close-packed structure. This means that there are two crystallographically inequivalent sites in the crystal. The Nd moments at these two sites order at distinct temperatures (8 K and 19 K), halving the amount of field-dependent entropy available at each ordering temperature relative to the total entropy of the crystal.

● Antiferromagnetic ordering geometry in Nd (also in Er, Tm, Ho) prevents a strong enhancement of the external field by the exchange field near the ordering temperature, i.e., the magnetic field fights the ordering rather than enhancing it; and

Table 1: The Magnetocaloric Properties of the Rare Earth Elements

Magnetic RE Element	Ordering Temperatures[8] (K)	Highest Temperature Ordering Geometry [8]	Observed Saturation Moment per Atom[8] (μ_B)	Maximal Value of ΔT_S at 7 T (K)	Temperature of Max ΔT_S (K)
Pr	0.03	AFM [a]	2.7		
Nd	8, 19	Modulated AFM	2.2	2.5	10
Sm	106	AFM	0.5		
Eu	90	Spiral AFM	5.1		
Gd	293	FM	7.63	13.8	293
Tb	220, 230	AFM/ FM	9.33	10.3	237
Dy	86, 178	Spiral AFM	10.3	11.5	180
Ho	19, 133	Spiral AFM	10.3	6.1	136
Er	19, 52, 84	Modulated AFM	9.0	4.8	40
Tm	32, 56	Modulated AFM	7.12	3	58

a antiferromagnet

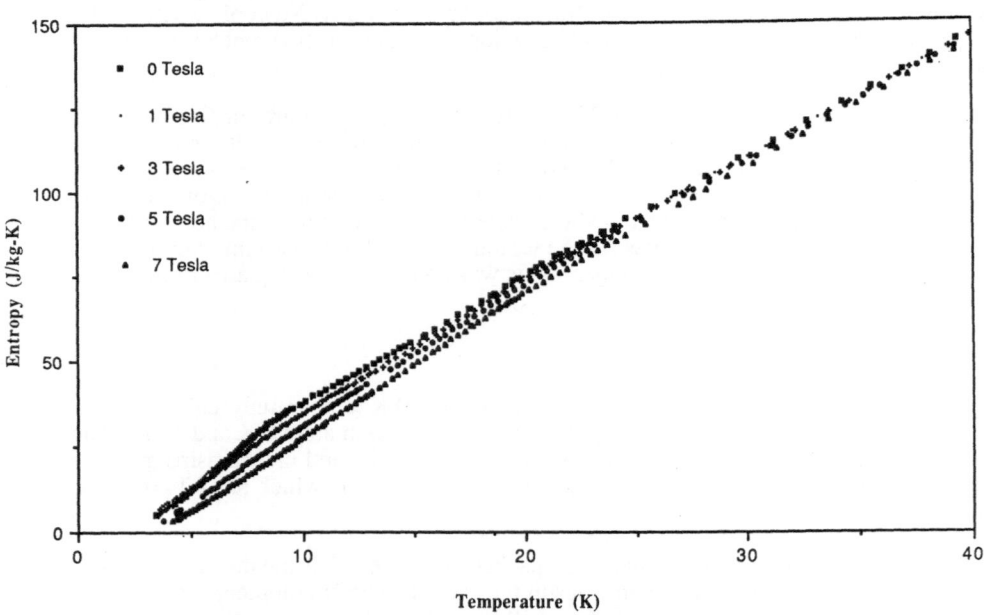

Figure 3. The entropy of Nd as a function of temperature for five applied magnetic fields. The curves were calculated by integrating the measured heat capacity divided by temperature.

• The crystal field interaction in the light rare earths is greater than the exchange interaction, opposite to the situation in the heavy rare earths. The weak exchange causes the ordering transition in Nd to occur at very low temperature, but apparently much of the entropy available in the magnetic system has already been removed by the crystal field above 20 K. The ground state of Nd in the crystal field of the pure metal is a doublet.[10] Because the total entropy of magnetic ordering is Rln(2J+1), the doublet ground state restricts the total entropy available at temperatures low compared to the crystal field splitting to Rln 2, instead of Rln 10 in the absence of a crystal field splitting. When a strong magnetic field is applied, the upper magnetic states may get mixed into the ground state, partly canceling the desired entropy reduction from the alignment of magnetic moments.

The third factor suggests that the other light rare earths (Ce,Pr,Sm) will probably not be effective magnetic refrigerants. Previous thermomagnetic property measurements have been done on Dy,Er,Tb,Ho,Tm, and Gd.[9,11,12,13,14] Europium is the only magnetic rare earth element left to be investigated. Europium undergoes an unusual first order antiferromagnetic transition at 94 K. However, magnetization experiments on europium show that only 15% of the full $7\mu_B$ saturation magnetization is obtained in fields of 7 T, both above and below the Neel point.[8] The small moment suggests that Eu will not be a good magnetic refrigerant at accessible magnetic fields.

A summary of the observed ΔT_s for the pure rare earths is given in Table 1. One can see that materials with significant ΔT_s are available for almost the whole temperature region from room temperature to about 10 K, but choosing a set of pure rare earths to match a specific ΔT_s versus T curve desired may be difficult.

ACKNOWLEDGMENTS

Technical support from Craig Chamberlain is gratefully acknowledged.

REFERENCES

1.   G.F. Green, W.G. Patton and J. Steven, Low Temperature Ribbon Regenerator, in Proc. Second Interagency Meeting on Cryocoolers, David Taylor Ship Research and Development Center, Annapolis, p191 (1986).

2.   K.H.J. Buschow, J.F. Olijhoek and A.R. Miedema, Extremely large heat capacities between 4 and 10 K, Cryogenics, 15:261 (1975).

3.   R.Li and T.Hashimoto, A new regenerator material $Er(Ni_{1-x}Co_x)_2$ with high specific heat in the range from 4.2 K to 20 K, in: Proc. Twelfth Intl. Cryo. Engr. Conf., Butterworth, Guildford, UK, p423 (1988).

4.   J.A. Barclay and W.A. Steyert, Active Magnetic Regenerator, US Patent 4332135, June 1, (1982).

5.   O.V. Lounasmaa and L.J. Sundstrom, Specific heat of La,Pr,Nd and Sm metals between 3 and 25 K, Phys. Rev., 158:591 (1967).

6.   E.M. Forgan, et.al., Measurements of the heat capacity of Nd in the range 2-10 K and zero magnetic field, J. Phys. F, 9:651 (1979).

7.   C.R. Cross et.al., Optimal temperature-entropy curves for magnetic refrigeration, in: "Advances in Cryogenic Engineering", 33, Plenum Press, New York, p767 (1988).

8.   K.A. McEwen, Magnetic and Transport Properties of the Rare Earths, in: "Handbook on the Physics and Chemistry of Rare Earths, I:Metals," North Holland, Amsterdam, p.411 (1978).

9.   S.M.Benford, "The magnetocaloric effect in dysprosium, <u>J. Appl. Phys.</u>, 50:1868 (1979).

10.   L. Kowalewski and A. Lehmann-Szweykowska, "Magnetic phase transitions in neodymium, in a phenomenological approximation, Part I," <u>Acta Physica Polonica</u>, A44:281 (1973).

11.   C.B. Zimm et.al., "The magnetocaloric effect in erbium," Proc. 5th Intl. Cryocooler Conf., Wright Research and Development Center, Wright-Patterson Air Force Base, Ohio, p49 (1988).

12.   G.F. Green, W.G. Patton and S. Stevens, The magnetocaloric effect of some rare earth metals, in: "Advances in Cryogenic Engineering," 33, Plenum Press, New York, p777 (1988).

13.   C.B. Zimm et.al., "The magnetocaloric effect in thulium," <u>Cryogenics</u> (to be published, August 1989).

14.   S.M. Benford and G.V. Brown, T-S diagram for gadolinium near the Curie temperature, <u>J. Appl. Phys.</u>, 52:2110 (1981).